Signal Processing and Analysis of Electrical Circuit

Signal Processing and Analysis of Electrical Circuit

Special Issue Editors

Adam Glowacz

Grzegorz Królczyk

Jose Alfonso Antonino Daviu

MDPI • Basel • Beijing • Wuhan • Barcelona • Belgrade • Manchester • Tokyo • Cluj • Tianjin

Special Issue Editors

Adam Glowacz
AGH University of Science and Technology
Poland

Grzegorz Królczyk
Opole University of Technology
Poland

Jose Alfonso Antonino Daviu
Universitat de València
Spain

Editorial Office
MDPI
St. Alban-Anlage 66
4052 Basel, Switzerland

This is a reprint of articles from the Special Issue published online in the open access journal *Electronics* (ISSN 2079-9292) (available at: https://www.mdpi.com/journal/electronics/special_issues/signal_circuit).

For citation purposes, cite each article independently as indicated on the article page online and as indicated below:

LastName, A.A.; LastName, B.B.; LastName, C.C. Article Title. *Journal Name* **Year**, *Article Number*, Page Range.

ISBN 978-3-03928-294-4 (Pbk)
ISBN 978-3-03928-295-1 (PDF)

Contents

About the Special Issue Editors

Adam Glowacz received his Ph.D. in Computer Science from the AGH University of Science and Technology, Cracow, Poland, in 2013. Adam Glowacz is the author/coauthor of 106 scientific papers (58 papers indexed by Web of Science) that correspond to a h-index of 21 and 1026 citations in Web of Science and a h-index of 23 and 1407 citations in Google Scholar. He has supervised 30 B.Sc. and 12 M.Sc. theses. Adam Glowacz is an Associate Editor of Symmetry, Electronics, Measurement, and Advances in Mechanical Engineering and has also authored 300 scientific reviews.

Grzegorz Krolczyk is Professor and Vice-Rector for Research and Development at Opole University of Technology and uuthor and coauthor of 180 scientific publications (100 JCR papers), as well as around 30 studies and industrial applications. His main scientific activities are in the analysis and improvement of manufacturing processes, surface metrology, and surface engineering. His research focuses on sustainable manufacturing as a tool for the practical implementation of the concept of social responsibility in the area of machining. Grzegorz Krolczykis is a member of several scientific organizations, including an expert in the Section of Technology of the Committee on Machine Building of the Polish Academy of Sciences. In addition, he is a member of several editorial committees of scientific journals. He has participated in advisory and opinion-forming bodies, including the advisory team of the Minister of Science and Higher Education. The coauthor of two patent applications, Grzegorz Krolczyk has been awarded on numerous occasions for his scientific activities in Poland and around the world.

Jose A. Antonino-Daviu received his M.Sc. and Ph.D. degrees in Electrical Engineering, both from the Universitat Politècnica de València, Valencia, Spain, in 2000 and 2006, respectively. He has worked for IBM, where he was involved in several international projects. He is currently Full Professor in the Department of Electrical Engineering, Universitat Politècnica de València. He was an Invited Professor at Helsinki University of Technology, Finland, in 2005 and 2007; Michigan State University, USA, in 2010; Korea University, South Korea, in 2014; Université Claude Bernard Lyon 1, France; and Coventry University, U.K., in 2016. He is a coauthor of more than 200 papers published in technical journals and conference proceedings and one international patent. Dr. Antonino-Daviu is Associate Editor of *IEEE Transactions on Industrial Informatics*, *IEEE Industrial Electronics Magazine*, and *IEEE Journal of Emerging and Selected Topics in Industrial Electronics.* He received the IEEE Second Prize Paper Award from the Electric Machines Committee of the IEEE Industry Applications Society (2013). He also received the Best Paper Award in the conferences IEEE ICEM 2012, IEEE SDEMPED 2011, and IEEE SDEMPED 2019 and "Highly Commended Recognition" of the IET Innovation Awards in 2014 and in 2016. He was the General Co-Chair of SDEMPED 2013 and is a member of the Steering Committee of IEEE SDEMPED. In 2016, he received the Medal of the Spanish Royal Academy of Engineering (Madrid, Spain) for his contributions in new techniques for predictive maintenance of electric motors. In 2018, he was awarded the prestigious 'Nagamori Award' from the Nagamori Foundation (Kyoto, Japan). In 2019, he received the SDEMPED Diagnostic Achievement Award (Toulouse, France) for his contributions to advanced diagnosis of electric motors.

electronics

Editorial

Signal Processing and Analysis of Electrical Circuit

Adam Glowacz [1,*] **and Jose Alfonso Antonino Daviu** [2]

1 Department of Automatic Control and Robotics, Faculty of Electrical Engineering, Automatics, Computer Science and Biomedical Engineering, AGH University of Science and Technology, al. A. Mickiewicza 30, 30-059 Kraków, Poland
2 Instituto Tecnológico de la Energía, Universitat Politècnica de València (UPV), Camino de Vera s/n, 46022 Valencia, Spain; joanda@die.upv.es
* Correspondence: adglow@agh.edu.pl

Received: 16 December 2019; Accepted: 20 December 2019; Published: 23 December 2019

1. Introduction

The analysis of electrical circuits is an essential task in the evaluation of electrical systems. Electrical circuits are made up of interconnections of various elements, such as resistors, inductors, transformers, capacitors, semiconductor diodes, transistors and operational amplifiers. Electrical signals, acoustic and vibrations carry useful information. They are known as diagnostic signals. Electrical circuits are used for equipment, circuit protection, circuit control, computers, electronics, electrical engineering, cars, planes and trains.

The analysis of signals is also essential. It is used for electrical engineering, sound recognition, speaker recognition, fault diagnosis, image processing, fast Fourier transform (FFT), wireless communication, control systems, process control, genomics, economy, seismology, feature extraction and digital filtering.

2. The Present Special Issue

This special issue with 34 published articles shows the significance of the topic "Signal Processing and Analysis of Electrical Circuit". The topic gained noticeable attention in recent time. The accepted articles are categorized into four different areas:

Signal processing and analysis methods of electrical circuits;
Electrical measurement technology;
Applications of signal processing of electrical equipment;
Fault diagnosis of electrical circuits;

The paper [1] describes the fault diagnosis of a commutator motor using signal processing methods and acoustic signals. Five commutator motors were analyzed: a healthy commutator motor, a commutator motor with a broken rotor coil, a commutator motor with shorted stator coils, a commutator motor with a broken tooth on sprocket and a commutator motor with a damaged gear train. Feature extraction method MSAF-15-MULTIEXPANDED-8-GROUPS (Method of Selection of Amplitudes of Frequency Multiexpanded 8 Groups) was introduced [1]. Processing and feature extraction of an underwater acoustic signal was shown in the paper [2]. The authors proposed a feature extraction method for an underwater acoustic signal. It was based on VMD (variational mode decomposition), DCO (duffing chaotic oscillator) and KPE (kind of permutation entropy) [2]. The next paper [3] presented two models (HOCTVL1 model and SAHOCTVL1 model) for solving the problem of image deblurring under impulse noise. The proposed models are good for recovering the corrupted images [3].

A multispectral backscattered light recorder of insects' wingbeats was presented in the paper [4]. The proposed device extracted a signal of the wingbeat event and color characterization of the insect. The authors of the paper analyzed the following insects: the bee (*Apis mellifera*) and the wasp (*Polistes* *gallicus*) [4]. A 13-bit 3 MS/s asynchronous SAR ADC with a passive resistor was described [5]. Passive resistors were adopted by the described delay cell. A delay error was less than 5 percent [5]. A miniaturized frequency standard comparator based on FPGA was presented. The noise floor of the analyzed comparator was better than $7.50 * 10^{-12}$ (1/s) [6]. A low-ripple switched-capacitor DC–DC Converter with parallel low-dropout regulator was proposed. The converter used a four-bit DCpM control and parallel low-dropout regulator [7]. A fuzzy logic system was proposed for the assessment of stator winding short-circuit faults in induction motors. The proposed approach achieved a positive classification rate of 98% [8]. A capacitance-to-time converter-based electronic interface was designed. The proposed interface is suitable for on-chip integration with sensors of force, humidity, position etc. [9]. The self-calibrating dynamic comparator was developed. The presented approach reduced the input offset by 10× [10]. There are also other interesting articles in the presented special issue. The proposed approaches and devices can be improved and used for the electrical systems in the future.

The proposed topics are essential for industry. Signal processing and analysis of diagnostic signals are used for fault diagnosis and monitoring systems [11–26]. Signal processing and image processing methods are used for many applications, for example medical applications [27–36]. Switched-Capacitor DC–DC converters are also an interesting topic of research [37–41].

3. Concluding Remarks

Acceleration of the development of electrical systems, signal processing methods and circuits is a fact. Electronics applications related to electrical circuits and signal processing methods have gained noticeable attention in recent time. The methods of signal processing and electrical circuits are widely used by engineers and scientists all over the world.

The presented papers have made a contribution to electronics. The presented applications can be used in the industry. The presented approaches require further improvements for industry and other applications.

Author Contributions: A.G. wrote original draft preparation. He was responsible for editing. J.A.A.D. was also responsible for editing. He also supervised the paper. All authors have read and agreed to the published version of the manuscript.

Acknowledgments: The Guest Editors would like to thank all authors, reviewers and editorial board of MDPI Electronics journal for their valuable contributions to this special issue.

Conflicts of Interest: The authors declare no conflict of interest.

References

1. Głowacz, A. Acoustic-based fault diagnosis of commutator motor. *Electronics* **2018**, *7*, 299. [CrossRef]
2. Li, Y.; Chen, X.; Yu, J.; Yang, X. A Fusion Frequency Feature Extraction Method for Underwater Acoustic Signal Based on Variational Mode Decomposition. Duffing Chaotic Oscillator and a Kind of Permutation Entropy. *Electronics* **2019**, *8*, 61. [CrossRef]
3. Xiang, J.; Ye, P.; Wang, L.; He, M. A Novel Image-Restoration Method Based on High-Order Total Variation Regularization Term. *Electronics* **2019**, *8*, 867. [CrossRef]
4. Rigakis, I.; Potamitis, I.; Tatlas, N.A.; Livadaras, I.; Ntalampiras, S. A Multispectral Backscattered Light Recorder of Insects' Wingbeats. *Electronics* **2019**, *8*, 277. [CrossRef]
5. Ju, H.; Lee, M. A 13-bit 3-MS/s Asynchronous SAR ADC with a Passive Resistor Based Loop Delay Circuit. *Electronics* **2019**, *8*, 262. [CrossRef]
6. Tang, S.; Ke, J.; Wang, T.; Deng, Z. Development of a Miniaturized Frequency Standard Comparator Based on FPGA. *Electronics* **2019**, *8*, 123. [CrossRef]
7. Lee, J.Y.; Kim, G.S.; Oh, K.I.; Baek, D. Fully Integrated Low-Ripple Switched-Capacitor DC–DC Converter with Parallel Low-Dropout Regulator. *Electronics* **2019**, *8*, 98. [CrossRef]
8. Mejia-Barron, A.; de Santiago-Perez, J.J.; Granados-Lieberman, D.; Amezquita-Sanchez, J.P.; Valtierra-Rodriguez, M. Shannon Entropy Index and a Fuzzy Logic System for the Assessment of Stator Winding Short-Circuit Faults in Induction Motors. *Electronics* **2019**, *8*, 90. [CrossRef]

9. De Marcellis, A.; Reig, C.; Cubells-Beltran, M.D. A Capacitance-to-Time Converter-Based Electronic Interface for Differential Capacitive Sensors. *Electronics* **2019**, *8*, 80. [CrossRef]
10. Ramkaj, A.; Strackx, M.; Steyaert, M.; Tavernier, F. An 11 GHz Dual-Sided Self-Calibrating Dynamic Comparator in 28 nm CMOS. *Electronics* **2019**, *8*, 13. [CrossRef]
11. Yan, X.P.; Xu, X.J.; Sheng, C.X.; Yuan, C.Q.; Li, Z.X. Intelligent wear mode identification system for marine diesel engines based on multi-level belief rule base methodology. *Meas. Sci. Technol.* **2018**, *29*. [CrossRef]
12. Stief, A.; Ottewill, J.R.; Orkisz, M.; Baranowski, J. Two Stage Data Fusion of Acoustic. Electric and Vibration Signals for Diagnosing Faults in Induction Motors. *Elektron. Elektrotechnika* **2017**, *23*, 19–24. [CrossRef]
13. Singh, G.; Naikan, V.N.A. Detection of half broken rotor bar fault in VFD driven induction motor drive using motor square current MUSIC analysis. *Mech. Syst. Signal Process.* **2018**, *110*, 333–348. [CrossRef]
14. Zhang, C.; Peng, Z.X.; Chen, S.; Li, Z.X.; Wang, J.G. A gearbox fault diagnosis method based on frequency-modulated empirical mode decomposition and support vector machine. *Proc. Inst. Mech. Eng. Part C* **2018**, *232*, 369–380. [CrossRef]
15. Michalak, M.; Sikora, B.; Sobczyk, J. Diagnostic Model for Longwall Conveyor Engines. In *Man-Machine Interactions 4, ICMMI 2015, Book Series: Advances in Intelligent Systems and Computing, Proceedings of the Man-Machine Interactions 4—4th International Conference on Man-Machine Interactions, ICMMI 2015, Kocierz Pass, Poland, 6–9 October 2015*; Springer: Berlin/Heidelberg, Germany, 2016; Volume 391, pp. 437–448. [CrossRef]
16. Glowacz, A.; Glowacz, W. Vibration-Based Fault Diagnosis of Commutator Motor. *Shock Vib.* **2018**, 7460419. [CrossRef]
17. Glowacz, A.; Glowacz, Z. Recognition of rotor damages in a DC motor using acoustic signals. *Bull. Pol. Acad. Sci. Tech. Sci.* **2017**, *65*, 187–194. [CrossRef]
18. Glowacz, A. Recognition of acoustic signals of induction motor using FFT. SMOFS-10 and LSVM. *Eksploat. Niezawodn.* **2015**, *17*, 569–574. [CrossRef]
19. Legutko, S. Development Trends in Machines Operation Maintenance. *Eksploat. Niezawodn.* **2009**, *2*, 8–16.
20. Hreha, P.; Radvanska, A.; Knapcikova, L.; Krolczyk, G.M.; Legutko, S.; Krolczyk, J.B.; Hloch, S.; Monka, P. Roughness Parameters Calculation by Means of On-Line Vibration Monitoring Emerging from AWJ Interaction With Material. *Metrol. Meas. Syst.* **2015**, *22*, 315–326. [CrossRef]
21. Liu, M.K.; Weng, P.Y. Fault Diagnosis of Ball Bearing Elements: A Generic Procedure based on Time-Frequency Analysis. *Meas. Sci. Rev.* **2019**, *19*, 185–194. [CrossRef]
22. Sun, Y.; Zhang, Y.G. New Developments in Fault Analysis Based on Dynamical Perspective. *IETE J. Res.* **2016**, *62*, 500–506. [CrossRef]
23. Krolczyk, G.M.; Krolczyk, J.B.; Legutko, S.; Hunjet, A. Effect of the disc processing technology on the vibration level of the chipper during operations. *Teh. Vjesn.* **2014**, *21*, 447–450.
24. Irfan, M.; Saad, N.; Ibrahim, R.; Asirvadam, V.S. Condition monitoring of induction motors via instantaneous power analysis. *J. Intell. Manuf.* **2017**, *28*, 1259–1267. [CrossRef]
25. Pandiyan, V.; Caesarendra, W.; Tjahjowidodo, T.; Tan, H.H. In-process tool condition monitoring in compliant abrasive belt grinding process using support vector machine and genetic algorithm. *J. Manuf. Process.* **2018**, *31*, 199–213. [CrossRef]
26. Zmarzly, D.; Boczar, T.; Fracz, P.; Borucki, S. High Voltage Power Transformer Diagnostics using Vibroacoustic Method. In Proceedings of the 2014 IEEE International Power Modulator and High Voltage Conference (IPMHVC), Santa Fe, NM, USA, 1–5 June 2014; pp. 561–564.
27. Kowalczyk, M.; Przewlocka, D.; Kryjak, T. Real-time implementation of contextual image processing operations for 4K video stream in Zynq UltraScale plus MPSoC. In Proceedings of the 2018 Conference on Design and Architectures for Signal and Image Processing (DASIP), 9–12 October 2018; pp. 37–42.
28. Kryjak, T.; Komorkiewicz, M.; Gorgon, M. Real-time Implementation of Foreground Object Detection From a Moving Camera Using the ViBE Algorithm. *Comput. Sci. Inf. Syst.* **2014**, *11*, 1617–1637. [CrossRef]
29. Kurtasz, P.; Boczar, T.; Witkowski, P.; Lorenc, M. The application of the multicomparative algorithm for classifying acoustic signals coming from partial discharges. *Prz. Elektrotech.* **2010**, *86*, 125–127.
30. Boczar, T.; Lorenc, M. The application of the descriptive statistics for recognizing electrical discharge forms registered by the acoustic emission method. *Prz. Elektrotech.* **2008**, *84*, 6–9.
31. Jablonski, M.; Tylek, P.; Walczyk, J.; Tadeusiewicz, R.; Pilat, A. Colour-Based Binary Discrimination of Scarified Quercus Robur Acorns under Varying Illumination. *Sensors* **2016**, *16*, 1319. [CrossRef]

32. Jaworek-Korjakowska, J.; Kleczek, P.; Tadeusiewicz, R. Detection and Classification of Pigment Network in Dermoscopic Color Images as One of the 7-Point Checklist Criteria. In *Recent Developments and Achievements in Biocybern. Biomed. Eng. 2018, Book Series: Advances in Intelligent Systems and Computing, Proceedings of the 20th Polish Conference on Biocybernetics and Biomedical Engineering, Kraków, Poland, 20–22 September 2017*; Springer: Berlin/Heidelberg, Germany, 2018; Volume 647, pp. 174–181. [CrossRef]
33. Glowacz, A.; Glowacz, Z. Recognition of images of finger skin with application of histogram, image filtration and K-NN classifier. *Biocybern. Biomed. Eng.* **2016**, *36*, 95–101. [CrossRef]
34. Kantoch, E. Recognition of Sedentary Behavior by Machine Learning Analysis of Wearable Sensors during Activities of Daily Living for Telemedical Assessment of Cardiovascular Risk. *Sensors* **2018**, *18*, 3219. [CrossRef]
35. Proniewska, K. Data mining with Random Forests as a methodology for biomedical signal classification. *Bio-Algorithms Med-Syst.* **2016**, *12*, 89–92. [CrossRef]
36. Proniewska, K.; Malinowski, K.; Pociask, E.; Proniewski, B. Classification of Sleep Disordered Breathing in the Evaluation of Acoustic Sound in Correlation with the ECG Signal. In Proceedings of the 2014 Computing in Cardiology Conference 2014 (CinC), Cambridge, MA, USA, 7–10 September 2014; Volume 41, pp. 153–156.
37. Jiang, Y.; Law, M.K.; Chen, Z.Y.; Mak, P.I.; Martins, R.P. Algebraic Series-Parallel-Based Switched-Capacitor DC-DC Boost Converter With Wide Input Voltage Range and Enhanced Power Density. *IEEE J. Solid-State Circuits* **2019**, *54*, 3118–3134. [CrossRef]
38. Mohey, A.M.; Ibrahim, S.A.; Hafez, I.M.; Kim, H. Design Optimization for Low-Power Reconfigurable Switched-Capacitor DC-DC Voltage Converter. *IEEE Trans. Circuits Syst. I-Regul. Pap.* **2019**, *66*, 4079–4092. [CrossRef]
39. Xie, F.Y.; Wu, B.C.; Liu, T.T. A Ripple Reduction Method for Switched-Capacitor DC-DC Voltage Converter Using Fully Digital Resistance Modulation. *IEEE Trans. Circuits Syst. I-Regul. Pap.* **2019**, *66*, 3631–3641. [CrossRef]
40. Zeng, T.; Wu, Z.; He, L.Z. Bridge modular switched-capacitor DC-DC converter with soft switching operation and multilevel voltage-gain range. *IEEJ Trans. Electr. Electron. Eng.* **2019**, *14*, 1399–1408. [CrossRef]
41. Kumar, M.; Ramesh, S. Design and Implementation of Three-Winding Coupled Inductor and Switched Capacitor-Based DC-DC Converter Fed PV-TDVR. *J. Circuits Syst. Comput.* **2019**, *28*. [CrossRef]

electronics

MDPI

Article

Numerical Laplace Inversion Method for Through-Silicon Via (TSV) Noise Coupling in 3D-IC Design

Khaoula Ait Belaid [1,*], Hassan Belahrach [1,2] and Hassan Ayad [1]

1 Faculty of Science and Technology, Cadi Ayyad University, Marrakesh 40000, Morocco
2 Electrical Engineering Department, Royal School of Aeronautics, Marrakesh 40000, Morocco
* Correspondence: aitbelaid.khaoula@gmail.com

Received: 20 August 2019; Accepted: 29 August 2019; Published: 10 September 2019

Abstract: Typical 3D integrated circuit structures based on through-silicon vias (TSVs) are complicated to study and analyze. Therefore, it seems important to find some methods to investigate them. In this paper, a method is proposed to model and compute the time-domain coupling noise in 3D Integrated Circuit (3D-IC) based on TSVs. It is based on the numerical inversion Laplace transform (NILT) method and the chain matrices. The method is validated using some experimental results and the Pspice and Matlab tools. The results confirm the effectiveness of the proposed technique and the noise is analyzed in several cases. It is found that TSV noise coupling is affected by different factors such as source characteristics, horizontal interconnections, and the type of Inputs and Outputs (I/O) drivers.

Keywords: 3D-IC design; NILT; TSV noise coupling; RDL; chain matrix; interconnect line

---

1. Introduction

Over the last four decades, silicon semiconductor technology has advanced at exponential rates in terms of performance and productivity [1,2]. Analysis of the fundamentals, materials, devices, circuits, and system limits discloses that silicon technology still has colossal potential for achieving terascale integration (TSI) of a significant number of transistors per chip. Such large-scale integration is feasible by assuming the development and bulk economic production of metal-oxide-semiconductor double-gate field-effect transistors. The development of interconnect lines for these transistors is a major challenge for the realization of nanoelectronics for TSI. Employing systems with high performance requires using two approaches. The first consists of reducing the size of the transistors, to enhance IC reduction technologies, and assembling ICs on the same chip (SoC) [3]. The second consists of developing high-performance technologies for interconnections between chips (SiP). For proper functioning, the area occupied by interconnections, which sometimes exceeds that occupied by the main functional blocks or chips, as well as their lengths must be reduced. However, since the interconnections are required in electronic systems, the number of interconnections cannot be decreased adversely to the area which can be reduced using 3D technology based on vertical interconnections.

Three-dimensional technology is acknowledged as an effective solution to overcome the challenges of miniaturization and distribution density. It combines More Moore and More than Moore, which offers many benefits. Some advantages of this technology are power efficiency, performance enhancement, cost reduction, and modular design [4–6]. Three-dimensional technology allows vertical stacking of chips through vertical interconnections like Through-Silicon-Via. Three-dimensional architectures contain different elements, such as through-silicon vias (TSVs), the substrate, redistribution layers (RDLs), and active circuits, which makes them difficult to model and study. To model these structures, each element is modeled using lumped circuits, and the entire model is then constructed by combining these element models in an appropriate manner.

Several papers have discussed the issue of modeling TSVs. In [7,8], the authors proposed a methodology based on Radio Frequency (RF) characterizations and simulations, leading to a frequency-dependent analytical model including the metal-oxide-semiconductor (MOS) effect of high ratio TSVs. The authors of [9] gave an accurate electrical model of TSVs considering metal-oxide-semiconductor (MOS) capacitance effects. The MOS capacitance accurately solved Poisson's equation in cylindrical coordinates. Another compact wideband equivalent circuit model for electrical modeling of TSVs has been presented in [10]. In another previous work [11], the Resistance, Inductance and Capacitance (RLC) parameters of TSVs were modeled as a function of physical parameters and material characteristics. The RLC model is applied to predict the resistance, inductance, and capacitance of small-geometry TSV architectures. TSV impedance can also be extracted using a fully analytical and physical model in addition to Green's function in high frequency [12]. All these previous works have given models of one TSV without considering general multi-TSV architectures. Thus, in [3,13,14] a TSV noise coupling model and TSV-to-active circuit have been proposed based on a three-dimensional transmission line matrix method (3D-TLM). Using this method, the noise transfer functions in the frequency domain from TSV-to-TSV and TSV-to-active circuit can be estimated. Other analytical models, for vias and traces, have been proposed in [15]. Vias are modeled using an analytical formulation for the parallel-plate impedance and capacitive elements, whereas the trace-via transitions are described by modal decomposition. All these proposed models are validated against full-wave methods and measurements up to 40 GHz. An efficient method to model TSV interconnections is proposed in [16]. This technique is based on solving Maxwell's equation in integral form, the method uses a small number of global modal basis functions and can be much faster than discretization-based integral-equation methods. The models proposed in the literature differ; indeed, some models contain the depletion capacitance, TSV resistance, and TSV inductance, others neglect these elements, especially for frequencies below 20 GHs [3,13,14].

The TSV capacitance depends on both the oxide capacitance and the depletion capacitance [17]. As the TSV gate bias increases, the depletion region capacitance starts to increase, and it acts in series with oxide capacitance. Hence, a TSV capacitor, C_{TSV}, is modeled with a series connection of the oxide capacitors and a depletion region capacitor [18]. The width of the depletion region is calculated for every geometrical variation by means of the exact Poisson's equation for an average TSV voltage of 0.5 V, and modeled as an area where the substrate has no free charge carriers [19]. Consequently, an increasing average TSV voltage increases its isolation from the substrate [20]. Thus, a power V_{dd}-TSV generally draws less E-field lines than a ground GND-TSV. However, the influence of the depletion region can be neglected [19].

RDLs have an important role in TSV packaging applications, they are used to connect various elements in 3D-IC and to redistribute the signals between dies. Therefore, different works have proposed several models for these interconnections. In [3,21], the authors gave analytic RLGC equations for the equivalent circuit model of a single-ended signal RDL to estimate the electrical characteristics. For the substrate, which has a distribution nature, its model can be extracted from numerical techniques mentioned in [22,23]. By combining each partial model, the global model of 3D structures is obtained.

One of the 3D-architecture challenges is to avoid noise coupling, which is a significant problem and causes serious effects. This noise degrades system performance and makes it more sensitive. It can also be transmitted directly to an active circuit through the substrate; therefore, the signal and power are corrupted, the system reliability is reduced, and the bit error rate is increased [24,25].

The investigation of the noise coupling in 3D architecture based on TSVs is mainly done in the frequency domain. Yet, as far as we know, no technique has been proposed to compute these noises in the time domain. Hence, the objective of this paper is to propose a method to compute noise coupling in 3D-IC in the time domain. It is necessary to obtain the wave forms of these noises in the time domain in order to analyze them, since the transition effects can be better observed in the time domain. Time-domain noise coupling was obtained by the NILT method and chain matrices. First, the method was applied to three different structures. Then, the TSV coupling noise was analyzed, for each structure, to deduce how the coupling between the horizontal interconnections affects it. Simulations in Pspice were done to validate the method.

The rest of the paper is organized as follows. The NILT method in addition to a chain matrix of many studied circuits are explained in Section 2. The results and simulations are analyzed in Section 3. The conclusions are drawn in the last section.

2. Calculation of Time-Domain TSV Noise Coupling in 3D-IC Design with NILT

The use of the Laplace transform method has simplified the solution of transients on transmission lines (TL), of transients of dynamic systems, and other problems in electrical engineering. However, some difficulties appear when transforming solutions to the time domain. This makes researchers concerned to find accurate and precise numerical methods. One of these numerical methods is the numerical inverse Laplace transform (NILT) method, which can be used in cases when, for instance, the transform is a transcendental, irrational or some other complex function; then finding the solution in its analytical form is difficult and sometimes impossible [26,27].

The NILT method has been used in several works. In [28], NILT methods were selected to evaluate their performance for dealing with solution transportation in the subsurface under uniform or radial flow conditions. The authors of [29] evaluate and compare some numerical algorithms of the NILT method for the inversion accuracy of some fractional order differential equation solutions. In [30–35] the multidimensional NILT method has been explained in detail for electrical circuits.

In this paper, we were interested in 1D-NILT. Thus, a one-dimensional Laplace transform of a function $f(t)$, with; $t \geq 0$, is defined as:

$$F(s) = \int_{0}^{\infty} f(t)e^{-st}dt \tag{1}$$

Under the assumption $|f(t)| \leq Me^{\alpha t}$, M is real positive, α is a minimal abscissa of convergence, and $F(s)$ is defined on a region $\{s \in C : \text{Re}[s] > \alpha\}$, with $s = c + j\Omega$, c is defined as an abscissa of convergence, $\Omega = \frac{2\pi}{\tau}$ as the generalized frequency step, and τ forms a region of the solutions $t \in [0\ \tau]$.

The original function can be given using the Bromwich integral [36]:

$$f(t) = \frac{1}{2\pi j} \int_{c-j\infty}^{c+j\infty} F(s).e^{st}ds \tag{2}$$

By using a rectangular rule of integration as mentioned in [30], Equation (3) is found.

$$\tilde{f}(t) = \frac{\exp(ct)}{\tau} \sum_{n=0}^{\infty} F(s)\exp(jn\Omega t) \tag{3}$$

As explained in [30], by substituting $s = c + jn\Omega$ into Equation (1), if the obtained function has integration ranges split into infinite numbers of steps of the length τ, $F(s)$ could be written as:

$$F_n = F(c + jn\Omega) = \sum_{l=0}^{\infty} \int_{l\tau}^{(l+1)\tau} g(t)\ \exp(-jn\Omega t)\ dt \tag{4}$$

$g(t)$ is an exponentially damped object function. Then for $t \in [l\tau, \tau(l+1)]$, the functions $g_l(t)$ and $F(s)$ are given by:

$$g_l(t) = f(t)\exp(-ct) \tag{5}$$

$$F(c + jn\Omega) = \tau \sum_{l=0}^{\infty} C_{l,n} \tag{6}$$

where:

$$C_{l,n} = \frac{1}{\tau} \int_{l\tau}^{(l+1)\tau} g_l(t)\ \exp(-jn\Omega t)dt \tag{7}$$

Applying complex Fourier series to Equation (5), $g_l(t)$ could be found as:

$$g_l(t) = \sum_{n=-\infty}^{+\infty} C_{l,n} \exp(jn\Omega t) \tag{8}$$

Moreover, by substituting Equation (6) into Equation (3) and considering Equation (8), it is found that the approximate original function exponentially damped could be expressed as the infinite sum of the newly defined periodical function, Equation (5).

By exploiting all the previous equations, $\tilde{f}(t)$ is obtained and the absolute error $\varepsilon(t) = \tilde{f}(t) - f(t)$ can be computed.

$$\tilde{f}(t) = f(t) + \sum_{l=1}^{\infty} f(l\tau + t).\exp(-cl\tau) \tag{9}$$

A limiting absolute error is determined as $\varepsilon_M(t) \geq \varepsilon(t)$, then $|f(t)| \leq Me^{\alpha t}$, so a limiting relative error δ_M could also be controlled, and a path of integration from a required limit relative error could be chosen using Equation (10).

$$c = \alpha - \frac{1}{\tau}\ln\left(1 - \frac{1}{1+\delta_M}\right) \approx \alpha - \frac{1}{\tau}\ln(\delta_M) \tag{10}$$

This formula is valid, with a relative error achieved by the NILT $\tilde{f}(t)$, if infinite numbers of terms are used in series, and is a suitable technique for accelerating a convergence and for achieving the convergence of infinite series in a suitable way. Equation (3) can be rewritten using FFT and IFFT algorithms for an effective computation. Based on the experience of the authors of [31], the quotient-difference (q-d) algorithm of Rutishanser seems to give errors rather close to δ_M predicted by Equation (10), while considering a relatively small number of additional terms.

While considering a discrete variable in the original domain, $t_k = kT$, where T is a sampling period, $\tilde{f}(t)$ could be expressed as:

$$\tilde{f}_k = \frac{\exp(ckT)}{\tau} \sum_{n=-\infty}^{\infty} \tilde{F}\left(c + jn\frac{2\pi}{\tau}\right) \exp\left(j2\pi\frac{nkT}{\tau}\right) \tag{11}$$

The above stated formula could be decomposed as:

$$\tilde{f}_k = C_k\left[\sum_{n=0}^{N-1} \tilde{F}^{(-n)} z_{-k}^n + \sum_{n=0}^{\infty} \tilde{G}^{(-n)} z_{-k}^n + \sum_{n=0}^{N-1} \tilde{F}^{(n)} z_k^n + \sum_{n=0}^{\infty} \tilde{G}^{(n)} z_k^n - \tilde{F}^{(0)}\right] \tag{12}$$

where $N = 2^k$, k integer, $\tilde{F}^{(\pm n)} = \tilde{F}(c - jn\Omega)$, $\tilde{G}^{(\pm n)} = \tilde{F}^{(\pm N \pm n)}$, $z_{\pm k} = \exp\left(\pm j\frac{2\pi kT}{\tau}\right)$, and $C_k = \frac{\exp(ckT)}{\tau}$, while $\tau = NT$, $\forall k$, and $z_{\pm k}^N = \exp(\pm j2\pi k) = 1$.

In Equation (12), the first and the third sum are evaluated using the FFT and IFFT algorithms, respectively, while other parts, which present the infinite sum, are used as the input data in the q-d algorithm that uses a very small number of necessary additional terms, as explained in [24]. The computing region should be chosen as: $O_{cal} = (0, t_{cal})$, where $t_{cal} = \left(\frac{N}{2} - 1\right).T$.

Time-domain noise coupling could be easily obtained by the explained method in 3D technology based on TSVs.

In order to compute the noise coupling, different circuits were treated. The first structure is illustrated in Figure 1. This figure represents a basic structure of the TSV–TSV noise coupling [3]. It is composed of two signal TSVs, two ground TSVs, and is terminated by I/O drivers. The simplified lumped circuit model of this structure is given in Figure 2, where $C_{TSV\text{-}equiv}$ is the total equivalent TSV capacitance, $R_{sub\text{-}equiv}$ is the substrate resistance, and $C_{sub\text{-}equiv}$ is the substrate capacitance. In this simplified model, proposed in [3], the TSV resistance (R_{TSV}), the TSV inductance (L_{TSV}), and the depletion region are neglected, but in our work R_{TSV} and L_{TSV} are kept. In the study just mentioned, the authors assume that their effects appear in frequencies above 12 GHz. To consider the effect of the depletion region, which is modeled by a capacitance, it is enough to add its value to the TSV capacitance. The I/O drivers can be modeled as a resistor for the output driver and as a capacitor for the input driver that represents the MOS gate capacitance. The I/O drivers are presented by the impedances Z_1, Z_2, Z_3, and Z_4. To apply the NILT method, the conceptual structure can be modeled with a T-matrix, as illustrated in the figure. The entire matrix of the circuit is the product of T_1, T_2, and T_3, as defined below.

$$\begin{pmatrix} V_1 \\ I_1 \end{pmatrix} = [T] \begin{pmatrix} V_2 \\ -I_2 \end{pmatrix} \tag{13}$$

where:

$$[T] = [T_4]\,[T_1]\,[T_2]\,[T_3]\,[T_4] \tag{14}$$

$$[T_1] = \begin{bmatrix} 1 & 0 \\ 1/\left(Z_2 + \frac{R_{tsv}}{2} + s\frac{L_{tsv}}{2}\right) & 1 \end{bmatrix} \tag{15}$$

$$[T_2] = \begin{bmatrix} 1 & Z_{eq} \\ 0 & 1 \end{bmatrix} \tag{16}$$

$$[T_3] = \begin{bmatrix} 1 & 0 \\ 1/\left(Z_3 + \frac{R_{tsv}}{2}\right) + s\frac{L_{tsv}}{2} & 1 \end{bmatrix} \tag{17}$$

$$[T_4] = \begin{bmatrix} 1 & R_{tsv} + sL_{tsv} \\ 0 & 1 \end{bmatrix} \tag{18}$$

$$Z_{eq} = \frac{2}{2C_{TSV-equiv}s} + \frac{R_{sub-equiv}}{1 + R_{sub-equiv}C_{sub-equiv}s} \tag{19}$$

Observing the circuit, Equations (14) and (15) are found:

$$V_{in}(s) = Z_1(s)I_1(s) + V_1(s) \tag{20}$$

$$V_2(s) = -Z_4(s)I_2(s) \tag{21}$$

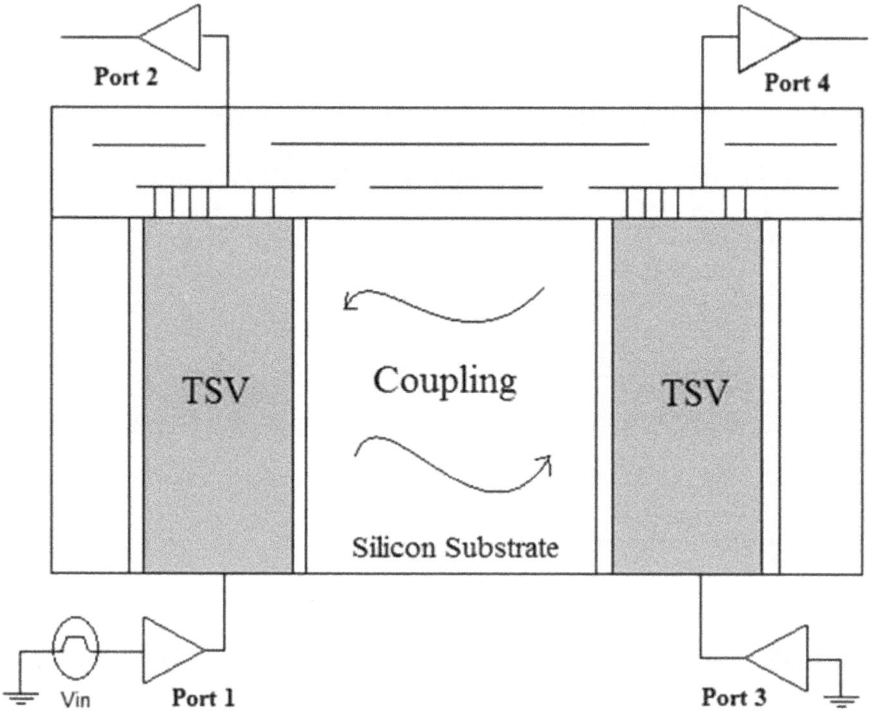

Figure 1. The through-silicon via (TSV)–TSV noise coupling structure with I/O termination.

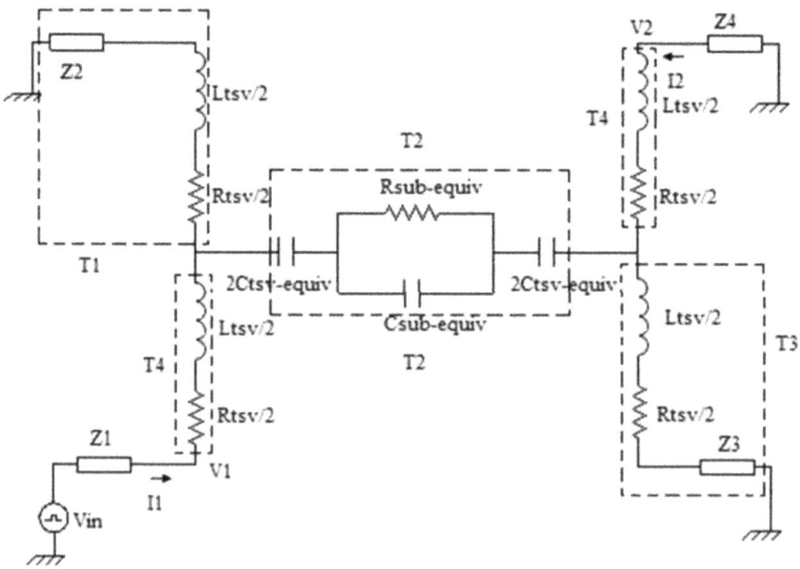

Figure 2. Lumped circuit model of TSV–TSV noise coupling.

By exploiting Equations (13)–(15), the noise V_2 could be expressed in the frequency domain according to V_{in}, then the NILT method can be applied, by replacing F(s) by V_2(s) in previous equations, to find the noise in the time domain. The voltage source V_{in} is a periodic trapezoidal signal switching expressed by Equation (16).

$$V_{in}(s) = \sum_{n=0}^{\infty} \exp(-snT).E(s) \tag{22}$$

where T is the period and E(s) represents the trapeze shape.

Then, while $\frac{1}{1-x} = \sum_{n=0}^{\infty} x^n$, Equation (16) could be written as:

$$V_{in}(s) = \frac{1}{1-\exp(-Ts)} E(s) \tag{23}$$

The second analyzed structure is given in Figure 3. It represents the conceptual view of TSV–active circuit noise coupling. The equivalent circuit model of this structure is similar to that in Figure 2, except that the capacity on the right is eliminated [3]. Consequently, the calculation was also done in the same way.

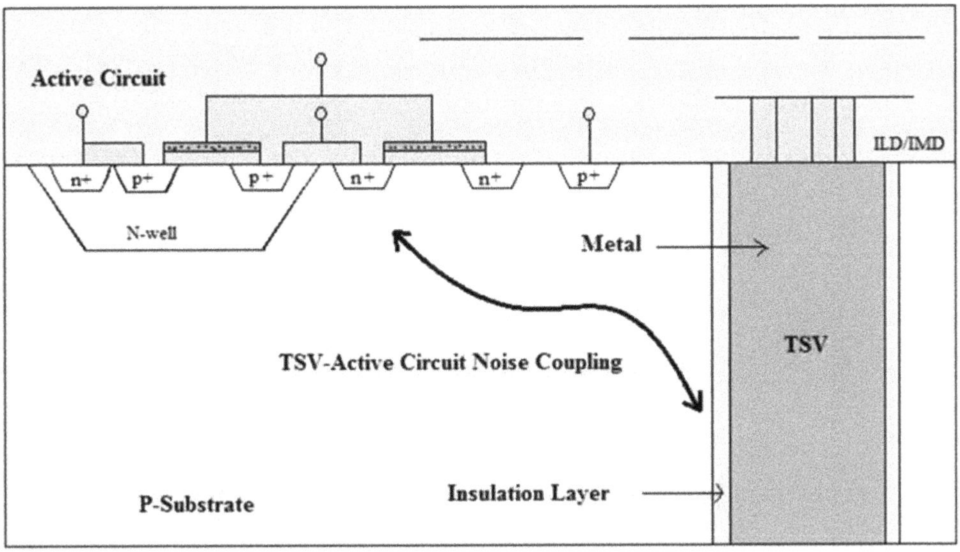

Figure 3. The conceptual view of TSV–active circuit noise coupling.

Because of the diversity of electronic devices, and the presence of many stacked dies in 3D technology, the second studied circuit contains two stacked dies with two interconnect lines. The concerned structure is presented in Figure 4. First, the noise coupling was calculated without taking into consideration the coupling between the two interconnect lines, only the coupling between the TSVs in each level was considered. This conceptual structure is modeled by a lumped circuit, as given in Figure 5.

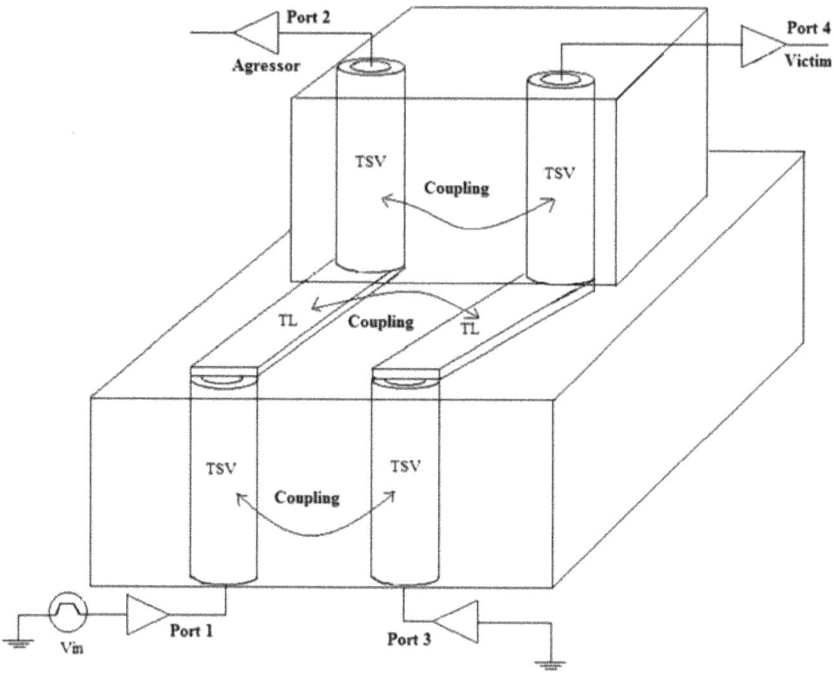

Figure 4. The conceptual view of a TSV noise coupling structure with interconnect lines and I/O drivers.

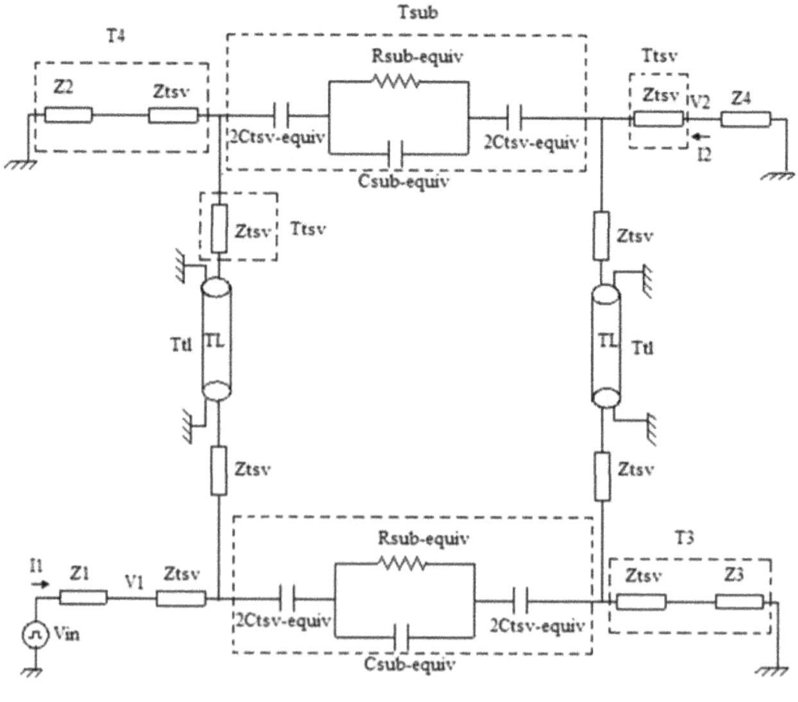

Figure 5. The equivalent circuit model of TSV noise coupling with interconnect line.

The electrical schema presented in Figure 5 is composed of a lumped circuit model of TSV–TSV noise coupling in each die, two interconnect lines to distribute signals between dies, and I/O drivers modeled by Z_1, Z_2, Z_3, and Z_4.

As explained above, before applying the NILT method, the global T-matrix of the circuit must be found. The matrices T_{sub}, T_{tsv}, T_{tl}, T_3, and T_4 were used. First, T_1 and T_2 were calculated using Equations (18) and (19), respectively, then a transformation to Y_1 and Y_2 of T_1 and T_2, respectively, was made. This transformation was performed to find the global Y_g of the circuit without Z_1, Z_4, and Z_{tsv} near Z_1 and Z_4. Then another transformation from Y_g to T_g was performed. When finding T_g, it is multiplied by T_{tsv} on the left and right sides, and by using Equations (13), (14), and (21) V_2 is found according to V_{in}.

$$[T_1] = [T_{sub}].[T_3].[T_{tsv}].[T_{tl}].[T_{tsv}] \tag{24}$$

$$[T_2] = [T_{tsv}].[T_{tl}].[T_{tsv}].[T_4].[T_{sub}] \tag{25}$$

$$[Y_g] = [Y_1] + [Y_2] \tag{26}$$

$$V_2 = -Z_4.I_2 \tag{27}$$

where:

$$[T_{tl}] = \begin{bmatrix} \cos(\beta l) & jZ_0 \sin(\beta l) \\ j\sin(\beta l)/Z_0 & \cos(\beta l) \end{bmatrix} \tag{28}$$

where β is the propagation constant, l and Z_0 are the length and the characteristic impedance, respectively, of the interconnect line, and:

$$[T_{sub}] = \begin{bmatrix} 1 & Zeq \\ 0 & 1 \end{bmatrix} \tag{29}$$

$$[T_{tsv}] = \begin{bmatrix} 1 & Z_{tsv} \\ 0 & 1 \end{bmatrix} \tag{30}$$

$$[T_4] = \begin{bmatrix} 1 & 0 \\ \frac{1}{Z_4+Z_{tsv}} & 1 \end{bmatrix} \tag{31}$$

$$[T_3] = \begin{bmatrix} 1 & 0 \\ \frac{1}{Z_{tsv}+Z_3} & 1 \end{bmatrix} \tag{32}$$

To consider the coupling between the interconnect lines, the conceptual structure presented in Figure 4 is modeled by the lumped circuit model shown in Figure 6. In the schema, the interconnect lines are presented by the equivalent circuit model of RDL [21]. As already explained above, to apply the NILT method, the total T-matrix of the circuit was calculated and then the noise V_n according to V_{in} was found.

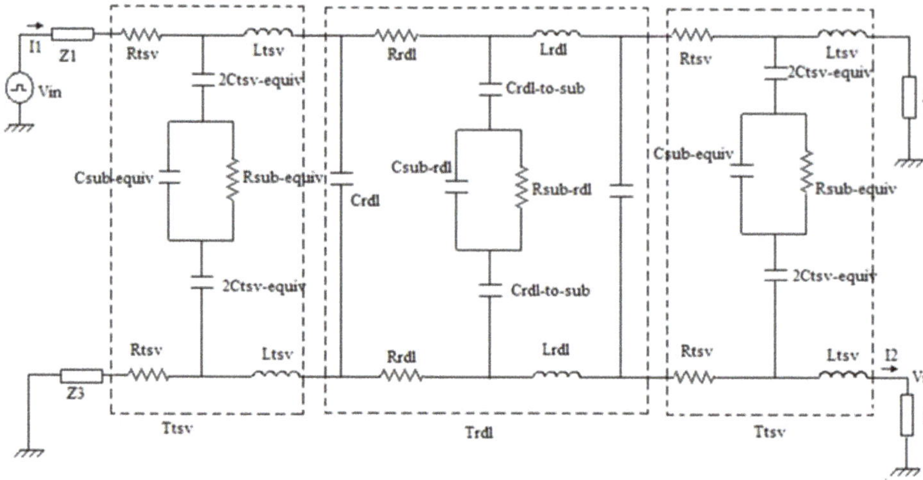

Figure 6. The equivalent circuit model of TSV noise coupling with redistribution layers (RDLs).

First, the total T-matrix, T_g, was computed as in Equation (23), then a transformation to Y_g was done to find the equivalent circuit of Figure 7. Hence, exploiting this figure and Equations (24)–(26), the noise V_n was calculated according to V_{in}.

$$\left[T_g\right] = [T_{tsv}].[T_{rdl}].[T_{tsv}] \tag{33}$$

$$\begin{pmatrix} I_1 \\ I_2 \end{pmatrix} = \begin{bmatrix} Y_{11} & Y_{12} \\ Y_{21} & Y_{22} \end{bmatrix} \begin{pmatrix} V_1 \\ V_2 \end{pmatrix} \tag{34}$$

$$V_{in} = (Z_1 + Z_3)\, I_1 + V_1 \tag{35}$$

$$V_2 + \left(\frac{Z_4 + Z_2}{Z_4}\right) V_n = 0 \tag{36}$$

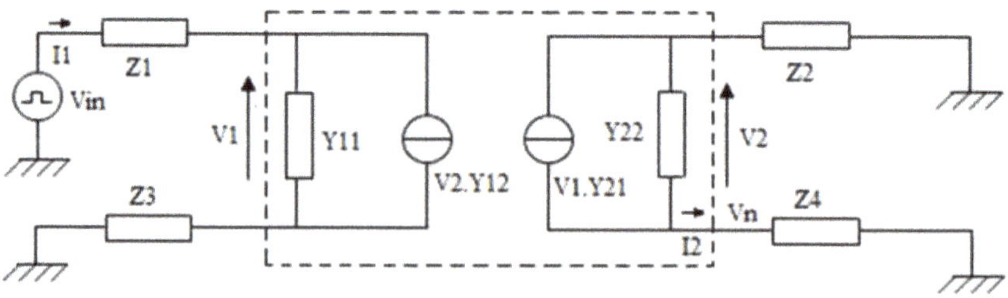

Figure 7. The admittance equivalent circuit of TSV noise coupling with RDLs.

The total admittance of all previous circuits could also be calculated, as mentioned in [37], before applying the NILT method.

The proposed method can be summarized in the diagram of Figure 8.

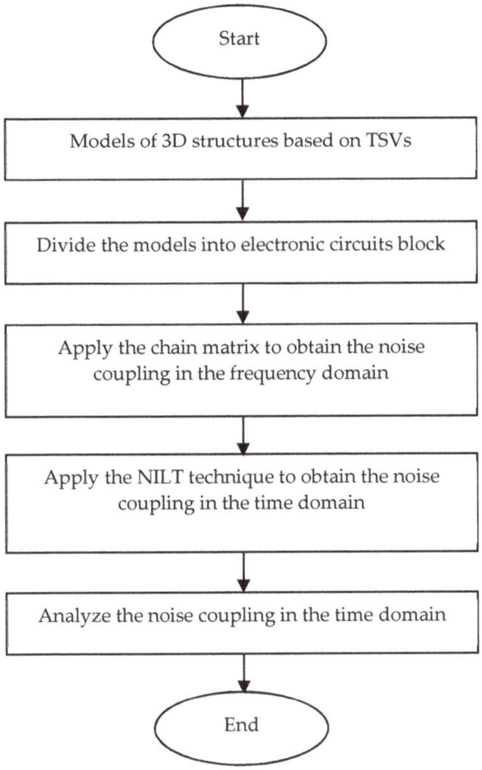

Figure 8. Block diagram of the proposed method.

3. Results and Discussions

In order to evaluate the effectiveness of the proposed method, simulation tests of the previous circuits were carried out. Simulations were performed with the Matlab and Pspice tools for all schemes, while the experimental tests of circuits 1 and 2 were taken from [13]. To take the measurements, the test vehicle in Figure 1 was fabricated using the Hynix via-last TSV process. The TSV circuit elements were calculated using the TLM-3D method; when the TSV diameter is 33 μm, the TSV pitch is 250 μm, the TSV dioxide thickness is 0.52 μm, and the TSV height is 105.2 μm. The RDL parameters were calculated using the method cited in [21]. Lumped circuit element values are listed in Tables 1–3. The accuracy and efficiency of the computing method were validated by simulations in Pspice and the measurements of [13].

Table 1. Lumped circuit elements of TSV–TSV noise coupling.

Component	Value
$C_{tsv\text{-}equi}$	201.3 fF
R_{tsv}	0.001 Ω
L_{tsv}	20.7 pH
$R_{sub\text{-}equi}$	928.5 Ω
$C_{sub\text{-}equiv}$	11.2 fF

Table 2. Lumped circuit elements of TSV–active circuit noise coupling.

Component	Value
$C_{tsv\text{-}equiv}$	817.5 fF
R_{tsv}	0.001 Ω
Ltsv	20.7 pH
$R_{sub\text{-}equiv}$	879.5 Ω
$C_{sub\text{-}equiv}$	12 fF

Table 3. Lumped circuit elements of the RDL.

Length of the Line	Component	Value
l_{RDL} = 200 μm	R_{rdl}	0.00672 Ω
	L_{rdl}	0.1664 nH
	C_{rdl}	7.66 fF
	$C_{rdl\text{-}to\text{-}sub}$	364.65 fF
	$C_{sub\text{-}rdl}$	0.13 fF
	$R_{sub\text{-}rdl}$	836.12 fF
l_{RDL} = 500 μm	R_{rdl}	0.0168 Ω
	L_{rdl}	0.42 nH
	C_{rdl}	19.15 fF
	$C_{rdl\text{-}to\text{-}sub}$	911.64 fF
	$C_{sub\text{-}rdl}$	0.33 fF
	$R_{sub\text{-}rdl}$	334.44 Ω

3.1. Validation of the Proposed Method

In order to verify the validity of the proposed method, it was applied first to the TSV–TSV and TSV–active circuit noise coupling circuits. The simulated waveforms of the electrical models of Figures 2 and 3 are shown in Figures 9–11. A trapezoidal signal switching from 0 to 1.8 V with a rising/falling time of 40 ps and a source resistance of 50 Ω at frequencies 100 MHz and 1 GHz is used. For a first test, Z_1, Z_2, Z_3, and Z_4 were replaced by resistances of 50 Ω.

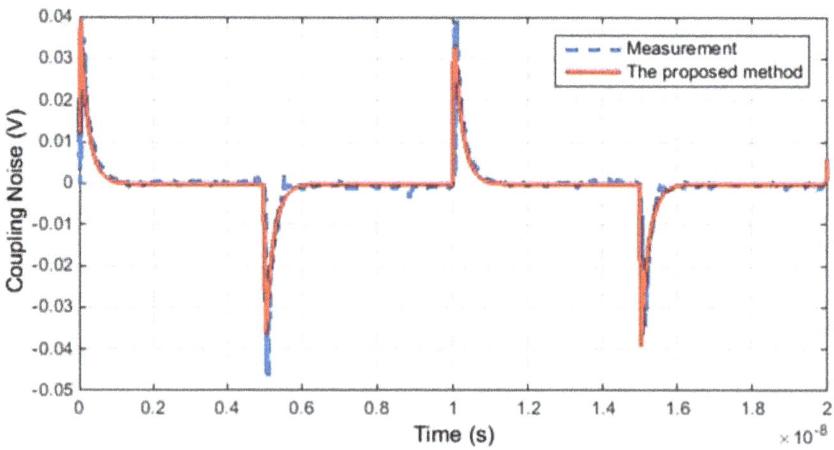

Figure 9. The proposed method and measured coupling of the TSV–TSV test vehicle (the input clock frequency is 100 MHz).

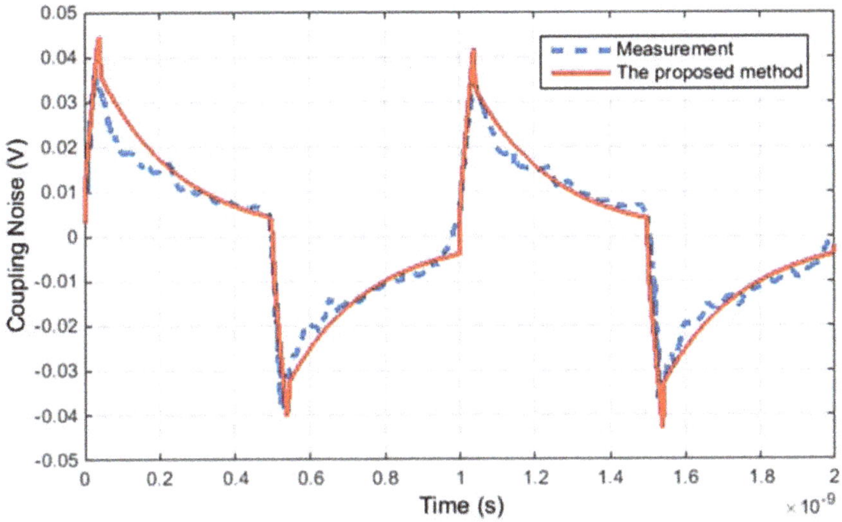

Figure 10. The proposed method and measured coupling of the TSV–TSV test vehicle (the input clock frequency at port 1 is 1 GHz).

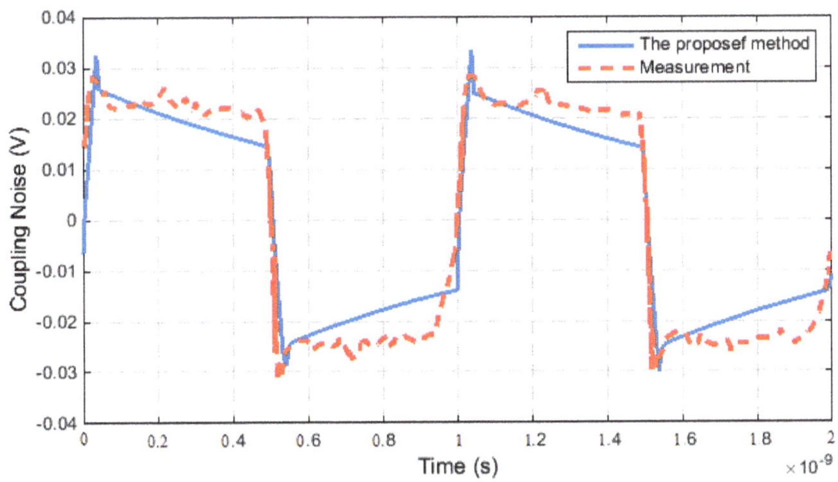

Figure 11. The proposed method and measured coupling noise of the TSV–active circuit (the input clock frequency at port 1 is 1 GHz).

Based on the results reported in the figures, it can be seen that the proposed method is in good agreement with the experiments. By analyzing these results, one can see that the proposed method is valid.

3.2. Time-Domain Analysis of the Coupling Noise with I/O Drivers Load

In Figures 9–11, the TSV coupling noise was computed based on the assumption that all TSVs are terminated with 50 Ω. However, TSVs are usually terminated with I/O drivers; therefore, the TSV I/O terminations must be considered as mentioned before. For the analysis, Z_2 and Z_4 were replaced by a capacitance of 10 fF. Figure 11 depicts the TSV–TSV noise coupling for a trapezoidal signal switching from 0 to 1 V and from 0 to 1.8 V. The results show that the coupling noise increases when Z_2 and Z_4 are replaced by the capacitances. The peak-to-peak coupling noise increases from 80 mV (Figure 10) to 170 mV (Figure 12). The peak-to-peak coupling noise increases from 170 mV to 310 mV when the source changes from 1 V to 1.8 V. These results imply that the type of termination and the source significantly affects the coupling noise. The TSV I/O buffer size also influences TSV noise coupling and must be considered.

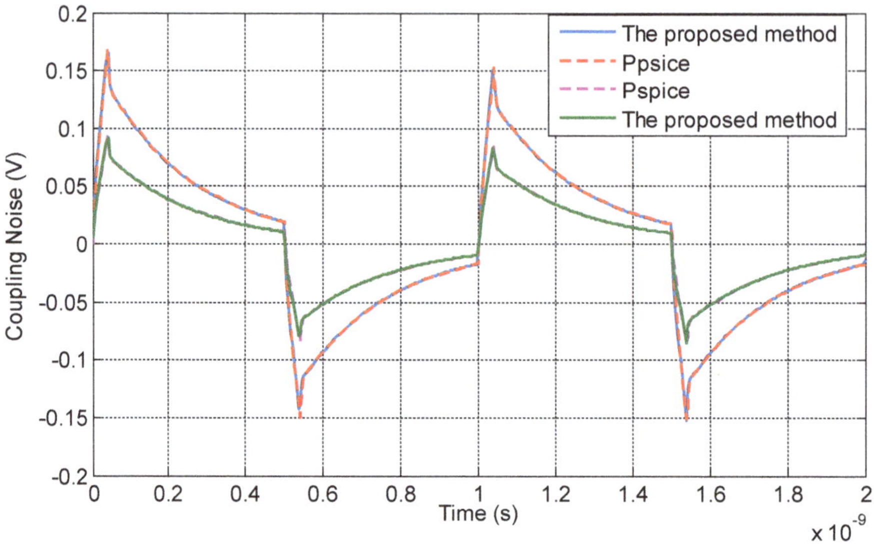

Figure 12. The proposed method and Pspice simulation of the coupling noise of TSV–TSV (V_{in} = 1 V and 1.8 V).

The RDL redistributes the signals to connect I/Os or power/ground when two different dies with via-last processed TSVs are integrated vertically. Therefore, for advanced 3D-IC design, analyzing TSV noise coupling with RDLs is very important.

The results found for the circuit presented in Figure 5 are illustrated in Figures 13–16 separately for l_{RDL} = 200 μm and l_{RDL} = 500 μm. These results present the TSV noise coupling without the coupling among the RDLs. A trapezoidal signal switching from 0 to 1.8 V with a rising/falling time of 10 ps and a source resistance of 50 Ω at frequency 1 GHz was used, Z_1 and Z_3 were replaced by resistances of 50 Ω, and Z_2 and Z_4 were replaced by capacitances of 10 fF.

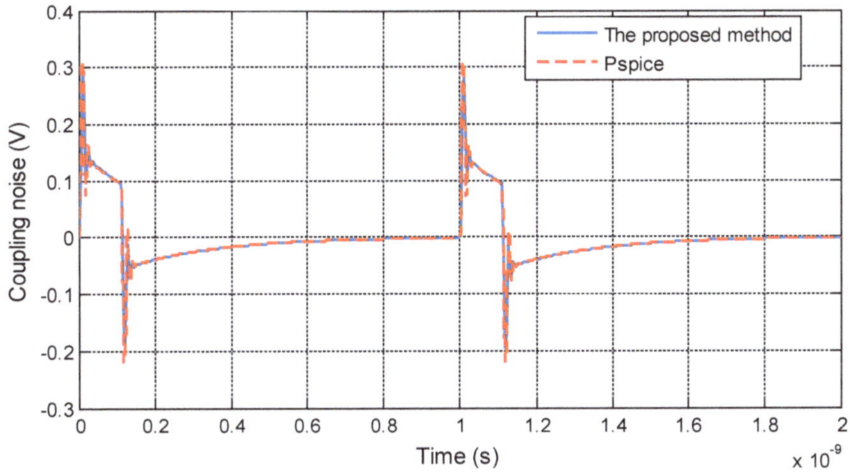

Figure 13. The proposed method and Pspice simulation of the TSV–TSV coupling noise with uncoupled RDLs (l_{RDL} = 200 μm) at port 4.

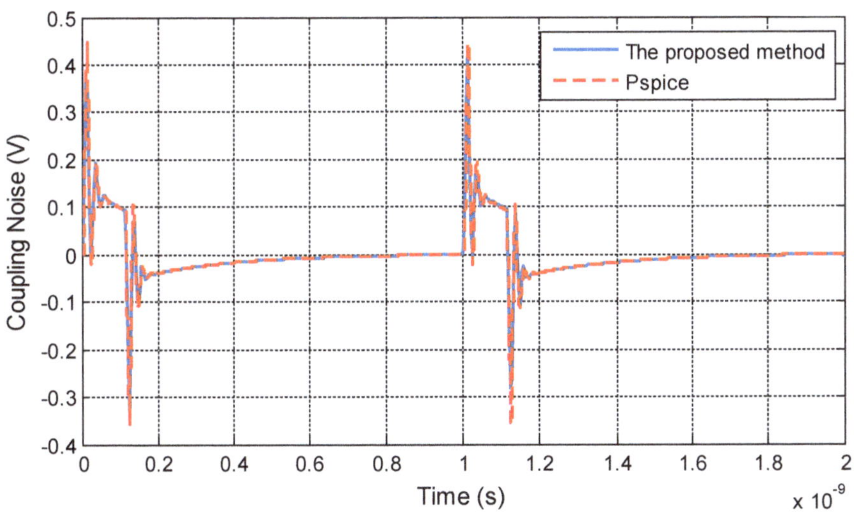

Figure 14. The proposed method and Pspice simulation of the TSV–TSV coupling noise with uncoupled RDLs (l_{RDL} = 500 μm) at port 4.

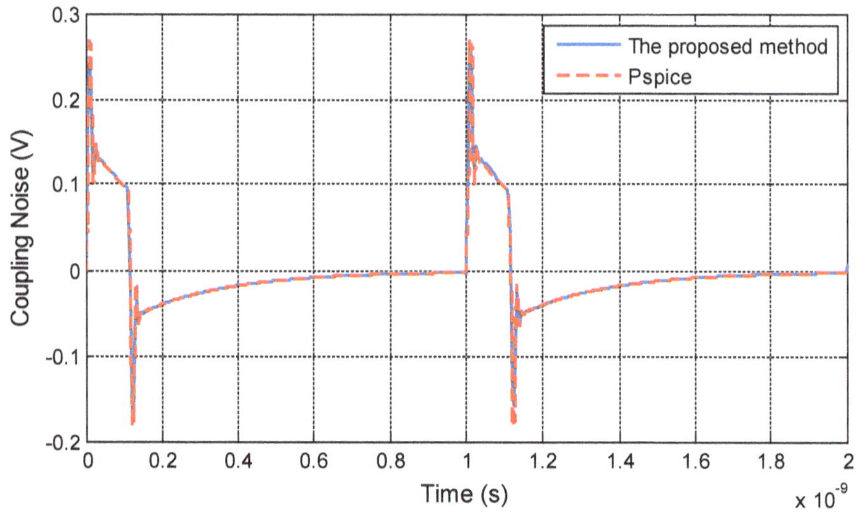

Figure 15. The proposed method and Pspice simulation of the TSV–TSV coupling noise with uncoupled RDLs (l_{RDL} = 200 μm) at port 3.

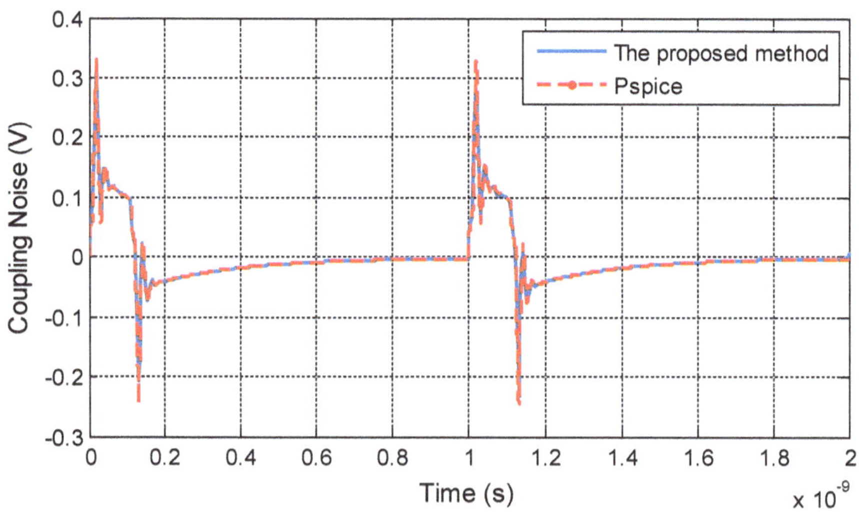

Figure 16. The proposed method and Pspice simulation of the TSV–TSV coupling noise with uncoupled RDLs (l_{RDL} = 500 μm) at port 3.

It is observed that the coupling noise spreads on the stacked dies through used interconnections. The peak-to-peak coupling noise increases from 50 mV to 80 mV when the length of the interconnect line (RDL) changes. It is also observed that both ports 3 and 4, which represent, respectively, the input and the output drivers, are affected by the coupling noise. By analyzing the obtained results, the presence of horizontal interconnections can add the coupling noise.

In high frequencies, coupling among the horizontal interconnections cannot be neglected. Indeed, a study including the coupling between the RDLs was done. The obtained results based on Figure 6 are depicted in Figures 17–19.

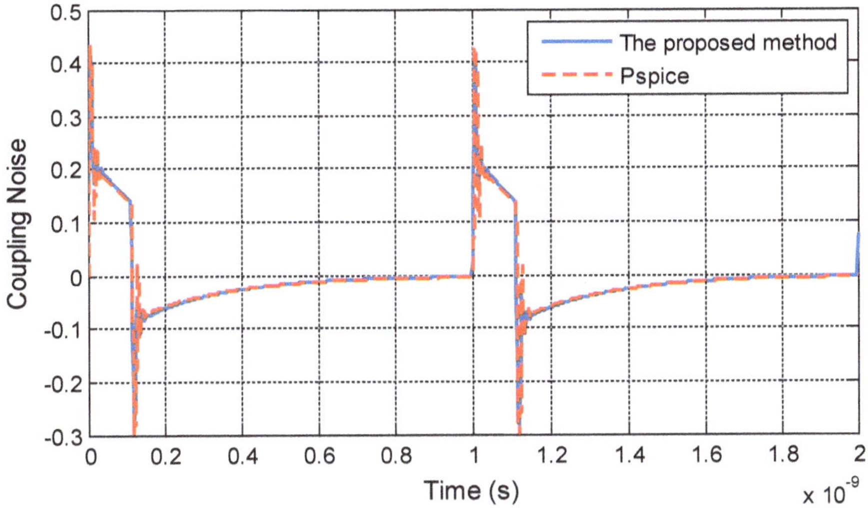

Figure 17. The proposed method and Pspice simulation of the TSV–TSV coupling noise with coupled RDLs (l_{RDL} = 200 μm and t_r= 10 ps) at port 4.

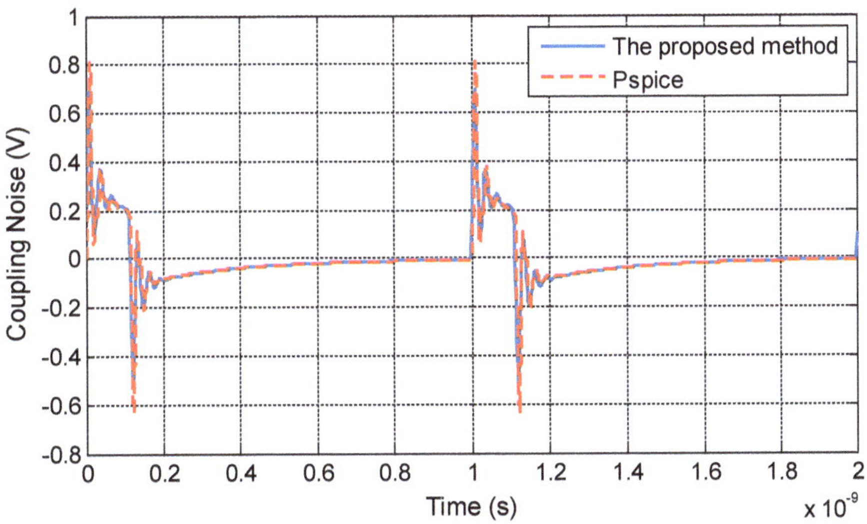

Figure 18. The proposed method and Pspice simulation of the TSV–TSV coupling noise with coupled RDLs (l_{RDL} = 500 μm and t_r = 10 ps) at port 4.

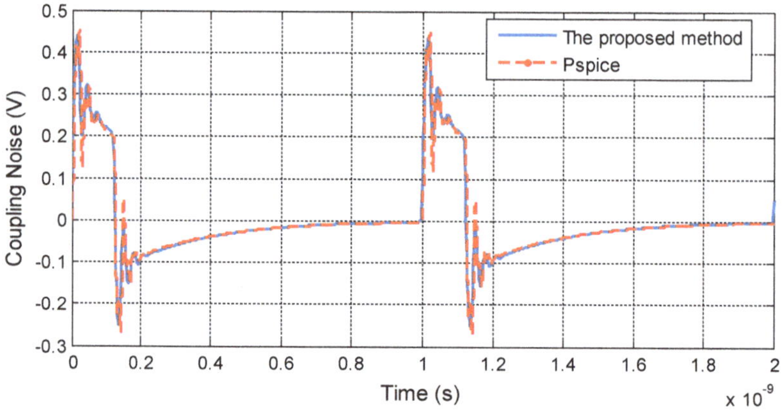

Figure 19. The proposed method and Pspice simulation of the TSV–TSV coupling noise with RDL (l_{RDL} = 500 μm and t_r = 20 ps) at port 4.

The simulations were done for different RDL lengths and several rise/fall time values. The noise was studied only at port 4.

Observing Figures 13 and 17, the peak-to-peak coupling noise increases when the coupling between RDLs is added. In addition, comparing the results of Figures 17 and 18, the peak-to-peak coupling noise increases when the RDL length increases. Simulation results of these case studies imply that, when the RDL length increases, the effect of the substrate elements among RDLs increases, and R_{RDL} and L_{RDL} change. Thus, the losses from the RDL are significant.

In a similar manner to the previous analysis, the effect of the rise/fall time variation is depicted in Figures 18–20. The results show that, as t_r increases from 10 ps to 20 ps and from 20 ps to 50 ps, pick-to-pick coupling noise decreases, respectively, from 1400 mV to 700 mV and from 700 mV to 550 mV. As a result, the rise/fall time is one of the most important factors that affect the TSV–TSV noise coupling in 3D-IC design.

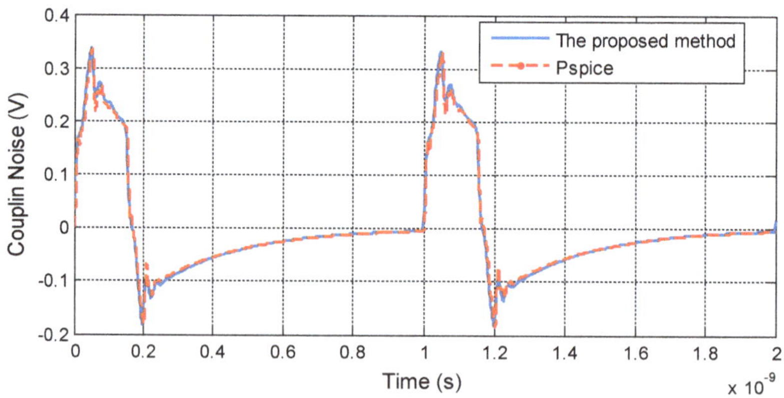

Figure 20. The proposed method and Pspice simulation of the TSV–TSV coupling noise with RDL (l_{RDL} = 500 μm and t_r = 50 ps) at port 4.

In summary, the method proposed to compute the coupling noise was validated using measurements and the Pspice and Matlab tools. Then, the time-domain analysis for several factors that must be considered was done.

4. Conclusions

In this paper, a method to compute the time-domain coupling noise in 3D-IC design has been proposed and explained in detail. The proposed method is based on 1D-NILT and chain matrices. It is effective and simple to apply. The used technique was validated using measurements of [13] and the Pspice tool.

The advantage of the proposed method is to compute the coupling noises of 3D structures based on TSVs, since transition phenomena are better observed in the time domain and not in the frequency domain.

A time domain analysis was done using several factors, such as different types of I/O drivers, the coupling between the horizontal interconnections, and the rise/fall time of the source. It was found that the type and the size of the TSV I/O buffer significantly influence the coupling noise. In addition, the presence of coupling between horizontal interconnections increases the noise at components of the 3D structures. These noises must be taken into consideration and must be minimized.

Author Contributions: Data curation, K.A.B.; Investigation, K.A.B. and H.B.; Methodology, K.A.B. and H.B.; Resources, K.A.B.; Supervision, H.B. and H.A.; Validation, H.B.; Writing—original draft, K.A.B.; Writing—review and editing, K.A.B. and H.B.

Funding: This research received no external funding.

Conflicts of Interest: The authors declare no conflict of interest.

References

1. Meindl, J.D.; Chen, Q.; Davis, J.A. Limits on Silicon Nanoelectronics for Terascale Integration. *Comput. Sci.* **2001**, *293*, 2044–2049. [CrossRef] [PubMed]
2. Borkar, S. Design Challenges of Technology Scaling. *IEEE Micro.* **1991**, *19*, 23–29. [CrossRef]
3. Lee, M.; Pak, J.S.; Kim, J. *Electrical Design of through Silicon Via*; Springer: Dordrecht, The Netherlands; Heidelberg, Germany; New York, NY, USA; London, UK, 2014.
4. Koester, S.J.; Young, A.M.; Yu, R.R.; Purushothaman, S.; Chen, K.-N.; La Tulipe, D.C.; Rana, N.; Shi, L.; Wordeman, M.R.; Sprogis, E.J. Wafer-level 3D integration technology. *IBM J. Res. Dev.* **2008**, *52*, 583–597. [CrossRef]
5. Knicherbacker, J.U.; Andry, P.S.; Dang, B.; Horton, R.R.; Interrante, M.J.; Patel, C.S.; Polastre, R.J.; Sakuma, K.; Sirdeshmukh, R.; Sprogis, E.J.; et al. Three-dimensional silicon integration. *IBM J. Res. Dev.* **2008**, *52*, 553–569. [CrossRef]
6. Garrou, P.E. Wafer-level 3D integration moving forward. *Semicond. Int.* **2006**, *29*, 12–17.
7. Cadix, L.; Fuchs, C.; Rousseau, M.; LeDuc, P.; Chaabouni, H.; Thuaire, A.; Brocard, M.; Valentian, A.; Farcy, A.; Bermond, C.; et al. Integration and frequency dependent parametric modeling of Through Silicon Via involved in high density 3D chip stacking. *ECS Trans.* **2010**, *33*, 1–21.
8. Ryu, C.; Lee, J.; Lee, H.; Lee, K.; Oh, T.; Kim, J. High Frequency Electrical Model of through Wafer Via for 3D Stacked Chip Packaging. In Proceedings of the ESTC, Dresden, Germany, 5–7 September 2006; pp. 215–220.
9. Bandyopadhyay, T.; Han, K.J.; Chung, D.; Chatterjee, R.; Swaminathan, M.; Tummala, R. Rigorous Electrical Modeling of TSVs with MOS Capacitance Effects. *IEEE Trans. Compon. Packag. Manuf. Technol.* **2011**, *1*, 893–903. [CrossRef]
10. Liu, E.X.; Li, E.P.; Ewe, W.B.; Lee, H.M.; Lim, T.G.; Gao, S. Compact Wideband Equivalent Circuit Model for Electrical Modeling of TSV. *IEEE Trans. Microw. Theory Tech.* **2011**, *59*, 1454–1460.
11. Katti, G.; Stucchi, M.; Meyer, K.D.; Dehaene, W. Electrical Modeling and Characterization of TSV for 3D-ICs. *IEEE Trans. Electron Devices* **2010**, *57*, 256–262. [CrossRef]
12. Xu, C.; Kourkoulos, V.; Suaya, R.; Banerjee, K. A Fully Analytical Model for the Series Impedance of TSV with Consideration of Substrate Effects and Coupling with Horizontal Interconnects. *IEEE Trans. Electron Devices* **2011**, *58*, 3529–3540. [CrossRef]
13. Cho, J.; Song, E.; Yoon, K.; Pak, J.S.; Kim, J.; Lee, W.; Song, T.; Kim, K.; Lee, J.; Lee, H.; et al. Modeling and Analysis of TSV Noise Coupling and Suppressing Using a Guard Ring. *IEEE Trans. Compon. Packag. Manuf. Technol.* **2011**, *1*, 220–233. [CrossRef]

14. Lim, J.; Cho, J.; Jung, D.H.; Kim, J.J.; Choi, S.; Kim, D.H.; Lee, M.; Kim, J. Modeling and analysis of TSV noise coupling effects on RFLC-VCO and shielding structures in 3D-IC. *IEEE Trans. Electromagn. Compat.* **2018**, *60*, 1939–1947. [CrossRef]
15. Rimolo-Donadio, R.; Gu, X.; Kwark, Y.; Ritter, M.; Archambeault, B.; De Paulis, F.; Zhang, Y.; Fan, J.; Bruns, H.-D.; Schuster, C. Physics-based via and trace models for efficient link simulation on multilayer structures up to 40 GHz. *IEEE Trans. Microw. Theory Tech.* **2009**, *57*, 2072–2083. [CrossRef]
16. Han, K.J.; Swaminathan, M.; Bandyopadhyay, T. Electromagnetic modeling of through-silicon via (TSV) interconnections using cylindrical modal basis functions. *IEEE Trans. Adv. Packag.* **2010**, *33*, 804–817. [CrossRef]
17. Beanato, G.; Gharibdoust, K.; Cevrero, A.; De Micheli, G.; Leblebici, Y. Design and analysis of jitter-aware low-power and high-speed TSV link for 3D-ICs. *Microelectron. J.* **2016**, *48*, 50–59. [CrossRef]
18. Attarzadeh, H.; Lim, S.K.; Ytterdal, T. Design and Analysis of a Stochastic Flash Analog-to-Digital Converter in 3D-IC technology for integration with ultrasound transducer array. *Microelectron. J.* **2016**, *48*, 39–49. [CrossRef]
19. Bamberg, L.; Najafi, A.; García-Ortiz, A. Edge effects on the TSV array capacitances and their performance influence. *Integration* **2018**, *61*, 1–10. [CrossRef]
20. Xu, C.; Li, H.; Suaya, R.; Banerjee, K. Compact AC modeling and performance analysis of through-silicon vias in 3D-ICs. *IEEE Trans. Electron. Devices* **2010**, *57*, 3405–3417. [CrossRef]
21. Kim, J.; Pak, J.S.; Cho, J.; Song, E.; Cho, J.; Kim, H.; Song, T.; Lee, J.; Lee, H.; Park, K.; et al. High-Frequency scalable electrical model and analysis of Through Silicon Via (TSV). *IEEE Trans. Compon. Packag. Manuf. Technol.* **2011**, *1*, 181–195.
22. Kerns, K.J.; Wemple, I.L.; Yang, A.T. Stable and efficient reduction of substrate model networks using congruence transforms. In Proceedings of the IEEE/ACM International Conference Computer-Aided Design, San Jose, CA, USA, 5–9 November 1995; pp. 207–214.
23. Verghese, N.K.; Allstot, D.J.; Masui, S. Rapid simulation of substrate coupling effects in mixed-mode ICs. In Proceedings of the IEEE Custom Integrated Circuits Conference, San Diego, CA, USA, 9–12 May 1993; pp. 18.3.1–18.3.4.
24. Shim, Y.; Park, J.; Kim, J.; Song, E.; Yoo, J.; Pak, J.; Kim, J. Modeling and analysis of simultaneous switching noise coupling for a CMOS negative-feedback operational amplifier in system-in-package. *IEEE Trans. Electromagn. Compat.* **2009**, *51*, 763–773. [CrossRef]
25. Koo, K.; Shim, Y.; Yoon, C.; Kim, J.; Yoo, J.; Pak, J.S.; Kim, J. Modeling and analysis of power supply noise imbalance on ultra-high frequency differential low noise amplifiers in a system-in-package. *IEEE Trans. Adv. Packag.* **2010**, *33*, 602–616. [CrossRef]
26. Li, J.; Farquharson, C.G.; Hu, X. Three effective inverse Laplace transform algorithms for computing time-domain electromagnetic responses. *Geophysics* **2016**, *81*, E113–E128. [CrossRef]
27. Liu, L.; Xue, D.; Zhang, S. Closed-loop time response analysis of irrational fractional-order systems with numerical Laplace transform technique. *Appl. Math. Comput.* **2019**, *350*, 133–152. [CrossRef]
28. Wan, Q.; Zhan, H. On different numerical inverse Laplace methods for solute transport. *Adv. Water Resour.* **2015**, *75*, 80–92.
29. Dingfelder, B.; Weideman, J.A.C. An improved Talbot method for numerical Laplace transform inversion. *Numer. Algorithms* **2015**, *68*, 167–183. [CrossRef]
30. Brancik, L. Numerical Inversion of Two-Dimensional Laplace Transforms Based on Partial Inversions. In Proceedings of the 17th international Conference Radioelektronika, Brno, Czech Republic, 24–25 April 2007.
31. Brancik, L. Technique of 3D NILT based on complex Fourier Series and Quotient-Difference Algorithms. In Proceedings of the 17th IEEE International Conference on Electronics Circuits and Systems, Athens, Greece, 12–15 December 2010.
32. Brancik, L. Matlab Simulation of Nonlinear Electrical Networks via Volterra Series Expansion and Multidimensional NILT. In Proceedings of the 2017 Progress in Electromagnetic Research Symposium-Fall, Singapore, 19–22 November 2017.
33. Chakarothai, J. Novel FDTD Scheme for Analysis of Frequency-Dependent Medium Using Fast Inverse Laplace Transform and Prony's Method. *IEEE Trans. Antennas Propag.* **2018**. [CrossRef]
34. Sheng, H.; Li, Y.; Chen, Y. Application of numerical inverse Laplace transform algorithms in fractional calculus. *J. Frankl. Inst.* **2011**, *348*, 315–330. [CrossRef]

35. Brancik, L. Error Analysis at Numerical Inversion of Multidimensional Laplace Transforms Based on Complex Fourier Series Approximation. *IEICE Trans. Fundam. Electron. Commun. Comput. Sci.* **2011**, *E94-A*, 999–1001. [CrossRef]
36. Al-Zubaidi, N.; Brančík, L. Convergence Acceleration Techniques for Proposed Numerical Inverse Laplace Transform Method. In Proceedings of the 24th Telecommunication form TELFOR, Belgrade, Serbia, 22–23 November 2016.
37. Ait-Belaid, K.; Belahrach, H.; Ayad, H. Investigation and Analysis of the Simultaneous Switching Noise in Power Distribution Network with Multi-Power Supplies of High-Speed CMOS Circuits. *Act. Passiv. Electron. Compon.* **2017**. [CrossRef]

electronics

MDPI

Article

Investigation of Induced Charge Mechanism on a Rod Electrode

Jiming Li [1], Jingyu Li [1], Xuezhen Cheng [1,*] and Guojin Feng [2,*]

1 College of Electrical Engineering and Automation, Shandong University of Science and Technology, Qingdao 266590, China

2 Department of Mechanical, Aerospace and Civil Engineering, Brunel University London, Middlesex UB8 3PH, UK

* Correspondence: zhenxc6411@163.com (X.C.); Guojin.Feng@brunel.ac.uk (G.F.); Tel.: +86-135-0532-4619 (X.C.); +44(0)1223-940276 (G.F.)

Received: 23 May 2019; Accepted: 30 August 2019; Published: 1 September 2019

Abstract: Rod electrodes based on an electrostatic induction mechanism are widely used in various industrial applications, but the analytic solution of an induced charge mechanism on a metal rod electrode has not yet been systematically established. In this paper, the theoretical model of the induced charge on a rod electrode is obtained through the method of images. Then, the properties of the rod electrode under the action of the point charge are studied, including the induced charge density distribution on the rod electrode, the amount of the induced charge with different diameters and lengths of the electrode, and the effective space region induced by the electrode. On this basis, a theoretical model of the induced current on a rod electrode is established, which is used to study the induced current properties by a moving point charge. It is found that both the magnitude and bandwidth of the induced current increase with the increased point charge velocity. Finally, three experimental studies are conducted, and the experimental results show good consistency with the analysis of the theoretical model, verifying the correctness, and accuracy of the model. In addition, the induced charge mechanism studied in this paper can act as an effective basis for the rod electrode sensor design in terms of the optimal radius and length.

Keywords: rod electrode; electrostatic induction; method of images; induced charge; induced current

1. Introduction

Dust pollution is a common issue in industrial and mining enterprises, such as coal mining, iron mining, etc. [1,2]. Dust moves together with the ventilation air and settles on walls and equipment. Dust in the ventilation air can have negative effects on the working conditions, posing a risk to the health of workers. Many measures have been taken to improve the air quality [3,4]. Real-time and accurate detection of the dust concentration is a basic guarantee to ensure an effective dust removal [5,6].

Based on the electrostatic induction phenomenon, electrodes are widely used to measure the electric charges carried on solid particles, mass flow rate, concentration, volume loading, mean flow velocity, and other electrical and mechanical parameters in two-phase and multiphase gas–solid flows such as those in pneumatic conveyances, in the air, etc. [7]. When charged particles move close to or away from a metal electrode, a charge with the opposite polarity to the dust particles is induced on the surface of the electrode. Hence, the physical parameters of the charged particles can be acquired by measuring the characteristics of the induced charge on the electrode. In practical applications, metal electrodes are fabricated in different shapes to acquire different measurement parameters in various environments. Typical electrodes mainly include the ring, curved, square, and rod electrodes and arrays composed of electrodes of different shapes.

The ring electrode has received wide attention by researchers because of its noninvasive characteristics. It is an electrode form with relatively mature theory and applications. Weinheimer [8] derived a charge numerical solution on the surface of ring electrodes induced by the point charge, which was applied to the measurement of meteorological precipitation charge. Yan, Gajewski, and Woodhead used a correlation method to measure velocity after studying the sensing mechanism, spatial sensitivity, and spatial filtering effect of noninvasive ring induction electrodes [9–11]. In recent years, ring charge-sensing sensors have been widely utilized in the measurement of dilute phase/dense phase of gas–solid two-phase flow parameters [12–14]. At the same time, some modern signal processing algorithms have further improved the measurement accuracy [15–19]. For example, Wang et al. [20] improved the measurement accuracy by applying the wavelet transform to the multiphase flow parameters. Considering that the signal measured by the ring electrode is an average feature over the entire cross-section, Zhang, Yang, and Dong's research found that the arc-shaped sensing electrode had an advantage in particle velocity and concentration distribution measurement in the monitoring of particle motion in gas–solid fluidized beds [21–23]. Qian also combined arc-shaped electrodes with digital images for the measurement of biomass–coal particles in fuel-injection pipelines [24]. Liu and Yao obtained the characteristics of square electrodes through theoretical and simulation studies, and utilized them on a square pneumatic conveying pipeline [25,26]. Compared with the above two models, Zhang used the method of images to obtain a simple analytical solution of the square electrode [27]. With the increasing complexity of the measurement environment, combinations of multistatic sensors and even electrostatic sensor arrays are applied to acquire different parameters of various pipelines [28,29].

Rod sensors based on the electrostatic induction mechanism have been widely used in industrial fields [7,30], mainly due to their simpler installation compared to other types of sensors, its working principle, and its applications in the real world are shown as Figure 1. Since they can accurately reflect pollutant emissions such as the steel mill flue gas, they have been widely applied for the detection of particulate matter concentration and dust in dust collector bags. Although the noninvasiveness is a great advantage of the ring electrodes, it usually takes the form of a spool piece installed in line with the pipe, which leads to an expensive and challenging installation. Moreover, signals collected by the ring electrodes are an overall result of the induced signals, making it difficult to detect local flow regimens. In comparison to the ring electrodes, the installation of rod electrodes is easier, since they only need a suitable drilling hole at any position of the pipe, making it possible to detect local flow regimes [30]. Therefore, rod electrodes are widely used in industrial areas and some research work has been performed. Shao compared the advantages and disadvantages of the ring electrode and rod electrode electrostatic sensors in measuring the pulverized coal speed [31,32], and proved that both types of electrodes achieved the same measurement accuracy in the coal dust measurement.

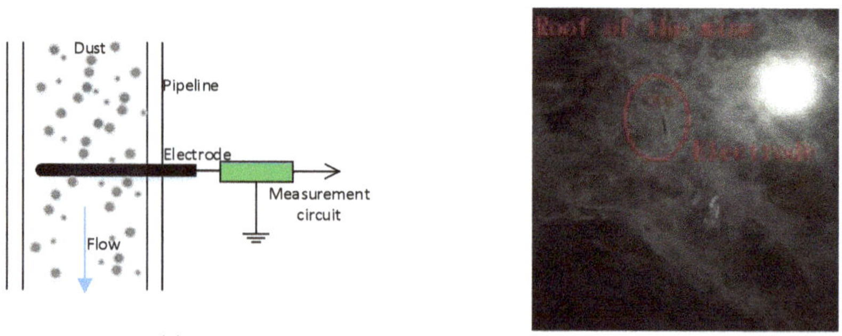

(a) **(b)**

Figure 1. *Cont.*

(c) (d)

Figure 1. Working principle figure of the electrode and its application: (**a**) Schematic figure; (**b**) in the iron mining company; (**c**) in the laboratory; and (**d**) in the steel mill company figure.

Electrostatic induction is a basic physical phenomenon in which the opposite induced charge is induced on the conductor if the point charge is close to the conductor. It is a qualitative conclusion. However, the analytic solution of the amount of induced charge when the point charge is close to the conductor has always been a difficult problem. It is a physical boundary value problem, which is difficult to represent with an analytical solution. The common method is to solve the Poisson equation through numerical calculations to get the amount of induced charge on the electrode. Krabicka studied the characteristics of electrostatic charges on rod electrodes by the finite element analysis (FEA) method and obtained an approximate solution [30]. However, this method requires remodeling of rod sensing electrodes of different lengths and diameters, which increases the modeling time. The simulation of large sensing electrodes takes too much time and the simulation accuracy is heavily dependent on the simulation software such as COMSOL. Therefore, this method is limited in practical application. Chen [33] established a mathematical model of rod electrodes by a theoretical derivation, but the fact that the induction conductor is a metal conductor was neglected, and the metal conductor was modeled as an insulator.

Analytic solutions of rod electrodes can be used to optimize the sensor design or interpret how particles at various locations influence the signal; for example, the bandwidth and amplitude. In order to establish a simple and easy-to-use mathematical model of rod electrodes, the method of images and the symmetry of rod electrodes are employed in this paper. The mathematical formula of the amount of charge induced by the point charge on the sensing rod electrode is obtained, and the equation is then used to study the physical properties of the sensing electrode under the action of the point charge. Based on this model, the distribution characteristics of the induced charge on the surface of the electrode and the influence of the electrode length on the induced charge density are studied, the induced charge of the induced electrode is simulated when the point charge moves in different directions, the amount of the induced current on the sensing electrode is studied, the spectral characteristics and the influence of the general measurement model of the induced current are analyzed, the experimental model is established, and the validity and accuracy of the model are verified by experiments.

2. Induced Charge Model on Rod Electrode by Point Charge

The method of images is an indirect method to solve the electrostatic field problem by applying the uniqueness theorem. The electrostatic method can be used to treat the actual partitioned uniform medium as uniform, and replace the actual complex charge distribution on the boundary with a simple charge distribution on the virtual closed setting boundary of the research field for calculation. According to the uniqueness theorem, this result is correct as long as the electric field generated by the imaginary charge together with the actual charge within the boundary satisfies a given boundary condition [34].

In this paper, the relationship between the rod electrode and point charge q is established by the method of images, which mainly models the relationship of the induced charge by the point charge q

with two basic parameters of the rod electrode, length, and radius. The steps of the proposed method to model the induced charge on a rod electrode is shown in Figure 2. The following subsections explains the steps in detail step by step.

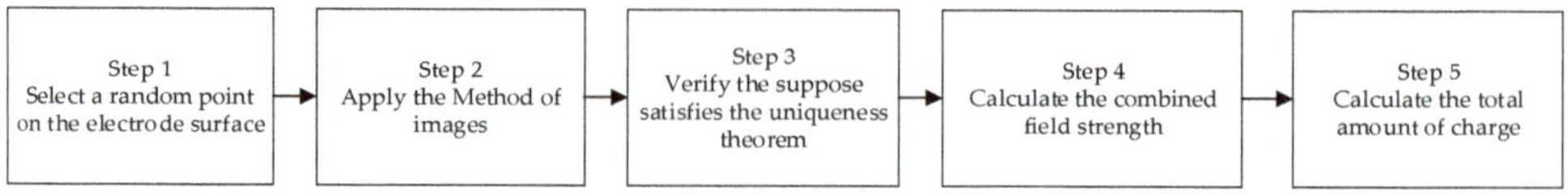

Figure 2. Steps of the proposed method to model the induced charge on a rode electrode.

2.1. *Step 1 Select A Random Point On the Electrode Surface*

To model the induced current on a rod electrode by a point charge, a cylindrical coordinate system is established, as shown Figure 3a, with the origin point O being the center of the electrode. Point C is the intersection of Oq and the electrode's surface, point A is one point on the surface of the electrode, and plane OAC is perpendicular to line L. Suppose θ is the angle between line OA and line OC, the coordinate of point A can be written as $(r,\ \theta,\ 0)$. Point charge q is in plane OAC and is located on the extension line of OC, and the distance to C is h and the distance to point A is h_1. Point B is a random point on the electrode surface and x is the vertical distance from B to plane OAC, hence the coordinates of point B are $(r,\ \theta, x)$.

2.2. *Step 2 Apply the Method of Images*

In addition to the electric field caused by point charge q in the electrolyte, the effect of the induced charge on the rod electrode should also be considered. However, the electric field distribution on the sensing electrode is unknown, which is the problem that this paper needs to solve. As shown in Figure 3b, it is assumed that an infinitely large plane S is tangent to the metal rod electrode and the intersection between the rod electrode and plane S forms a straight line L. The electric field intensity E and the electric field line distribution generated on infinite plane S can be obtained by the method of images, as shown in Figure 3c. Note that these electric field lines are perpendicular to the surface of the rod electrode. It is assumed that electric field E on line L of infinite plane S formed by point charge q is electric field E on the line of the rod electrode. Then, the image method is assumed to have a point charge at q′ with an equivalent charge (−q) to point charge q.

2.3. *Step 3 Verify the Suppose Satisfies the Uniqueness Theorem*

After removing the electrode and conductive plate S, and the following conditions are still satisfied:

(1) In addition to the position where the point charge is located, $\nabla^2\varphi = 0$ is satisfied everywhere, and φ is the potential.
(2) Taking infinity as the reference point, the potential at the interface between the medium and conductor L is zero.
(3) The direction of the electric field line is unchanged, which is the direction of the vertical electrode surface pointing to the axis.

These conditions satisfy the uniqueness theorem.

2.4. *Step 4 Calculate the Combined Field Strength*

By applying the method of images, we can obtain the electric field intensity E of point B. A cross-section view of plane OAC is illustrated in Figure 3d; the projection of point B is point A and the projection of plane S is line S'. The connecting line between point charge q and image charge q' intersects with S' at point D, the distance between point A and point D is l, and the distance between point q and point D is h'; d is the distance between point charge q and B.

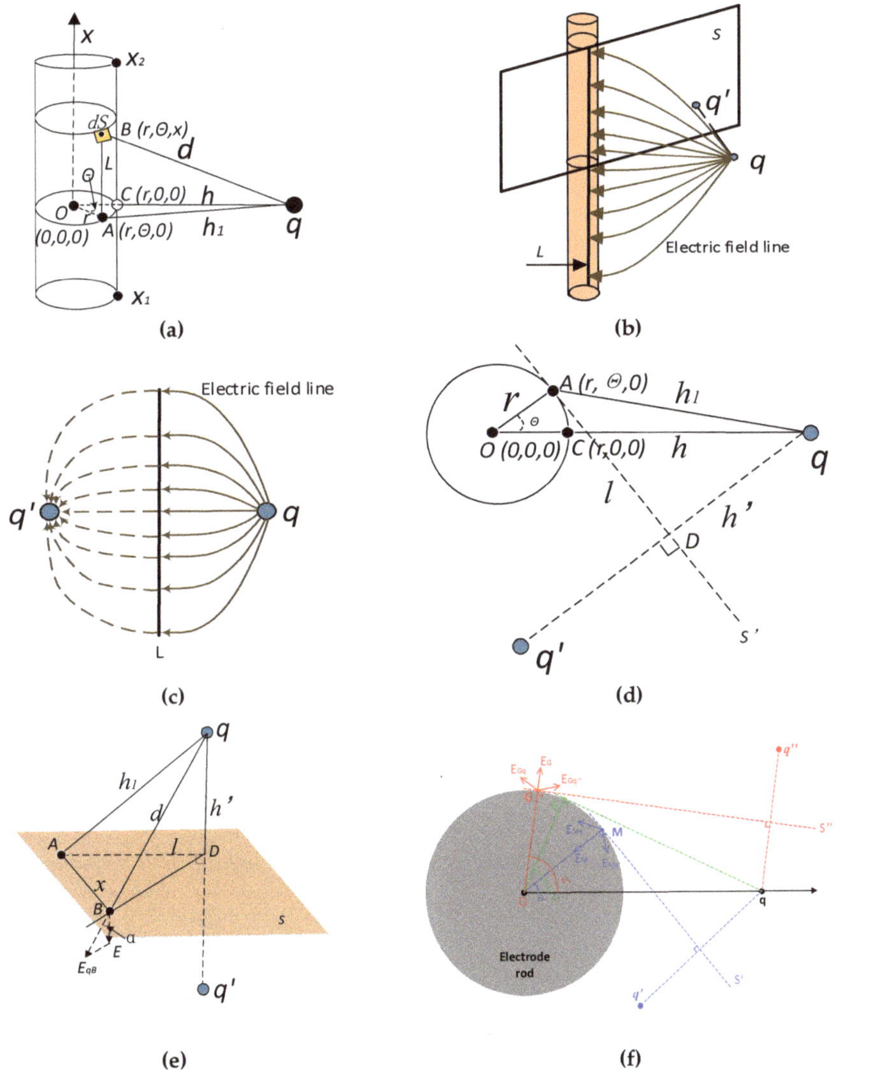

Figure 3. Rod electrode induced charge model: (**a**) Point charge q and its image; (**b**) electric field line distribution; (**c**) coordinate system; (**d**) projection of point *B* on plane *OAC*; (**e**) calculation of the field strength of *B* using the image method; and (**f**) determining the range of θ.

The mathematical expressions of h', l, and h_1 and the range of values of θ can be obtained, as follows:

$$h' = \left(h - \left(\frac{r}{cos\theta} - r\right)\right)cos\theta \tag{1}$$

$$l = rtan\theta + h'tan\theta \tag{2}$$

$$h_1 = \left(h'^2 + l^2\right)^{1/2} \tag{3}$$

$$d = \left(h_1{}^2 + x^2\right)^{1/2}. \tag{4}$$

To determine the range of θ, suppose point charge q carries a positive charge and is located at a position with minimum distance h to the surface of the electrode, as shown in Figure 3f. F is at the intersection of the tangential line through point charge q, and the angle θ_2 is equal to $acos(r/(r+h))$.

For any point M at a position that satisfies the condition $\theta_1 < acos(r/(r+h))$, its electric field strength can be analyzed as follows. According to the method of images, the imaginary charge is q', carrying a negative charge. E_{Mq} is the electric field strength generated by point charge q, and $E_{Mq'}$ is the electric field strength generated by point charge q'. Their synthetic electric field of strength is E_M, which points to the center point O. The orientation of E_M conforms with the electric field on the metal being perpendicular to the metal surface. As is known, if the charge of q is positive, the induced charge on the metal surface should be negative. This further confirms the correctness of the orientation of E_M.

For any point G at a position that satisfies the condition $\theta_3 > acos(r/(r+h))$, its electric field strength can be analyzed similarly. According to the method of images, the imaginary charge is q'', also carrying a negative charge. E_{Gq} is the electric field strength generated by point charge q, and $E_{Gq''}$ is the electric field strength generated by point charge q''. The synthetic field of strength is E_G, which points to a direction off the center O and is perpendicular to the metal surface. This indicates that the charge at point G is positive, which does not comply with the induced charge being negative. Hence, such a point should not have an induced charge generated from point charge q.

According to the above analysis, θ should not be larger than $acos(r/(r+h))$. Using the same method, we can get that the minimum value of θ is not smaller than $-acos(r/(r+h))$.

In summary, the range of θ can be determined as the following when the method of images is used:

$$\theta = \left[-acos\left(\frac{r}{r+h}\right), acos\left(\frac{r}{r+h}\right)\right] \tag{5}$$

It can be seen from Figure 3e that field strength E_{qB} is the summation result of point charge q and its image q'. E_{qB} can be expressed as

$$E_{qB} = 2\frac{q}{4\pi\varepsilon_0 d^2} \tag{6}$$

and E can be expressed as

$$E = E_{qB}cos\alpha. \tag{7}$$

Electric field strength E of point B based on the method of images can be obtained as

$$E = \frac{qh'}{2\pi\varepsilon_0 d^{3/2}} \tag{8}$$

where ε_0 is the vacuum permittivity.

2.5. Step 5 Calculate the Total Amount of Charge

The charge density ρ of point B is obtained as

$$\rho = -\varepsilon_0 E = \frac{-qh'}{2\pi d^{3/2}} \tag{9}$$

The dS marked in orange on the surface of the electrode contains point B, as in Figure 3c; the area of dS is expressed as

$$dS = rd\theta dx \tag{10}$$

When $d\theta$ and dx are infinite to zero, the charge density of dS will be the density of point B. The total amount of the induced charge on the metal electrode is obtained:

$$Q = \oint \rho dS = \int_{x_1}^{x_2} \int_{-acos(\frac{r}{r+h})}^{acos(\frac{r}{r+h})} \frac{-qh'r}{2\pi d^{3/2}} d\theta dx \tag{11}$$

where x_1, x_2 are the x-coordinates of the ends of the metal electrode.

By changing the order of the integration, we get

$$Q = \int_{-acos(\frac{r}{r+h})}^{acos(\frac{r}{r+h})} \int_{x_1}^{x_2} \frac{-qh'r}{2\pi d^{3/2}} dx d\theta. \tag{12}$$

The result is obtained as

$$Q = \int_{-acos(\frac{r}{r+h})}^{acos(\frac{r}{r+h})} \left(\frac{-qrh'x_2}{2\pi h_1{}^2 (h_1{}^2 + x_2{}^2)^{1/2}} - \frac{-qrh'x_1}{2\pi h_1{}^2 (h_1{}^2 + x_1{}^2)^{1/2}} \right) d\theta. \tag{13}$$

Let the charge distribution function along with x_1, x_2 and θ be $F(x_1, x_2, \theta)$:

$$F(x_1, x_2, \theta) = \frac{-qrh'x_2}{2\pi h_1{}^2 (h_1{}^2 + x_2{}^2)^{1/2}} - \frac{-qrh'x_1}{2\pi h_1{}^2 (h_1{}^2 + x_1{}^2)^{1/2}}. \tag{14}$$

When point charge q moves infinitely away from the electrode, the amount of induced charge is $\lim_{h \to \infty} Q = 0$. This means the amount of the induced charge generated by the point charge at infinity on the rod-shaped metal electrode is zero. From the analysis of Section 3.2.1, it can been seen that when point charge q is infinitely close to the electrode, the amount of the induced charge is $\lim_{h \to 0} Q = -q$; that is, the amount of the induced charge generated on the rod-shaped metal electrode is $-q$ when the point charge is infinitely close to the metal electrode.

3. Characteristics of Induced Charge on a Rod Electrode

In this section, the basic characteristics of the induced charge on a rod electrode are analyzed by utilizing the model obtained in Section 2.

3.1. Induced Charge Distribution under the Effect of Point Charge

The induced charge distribution with θ is studied. Four cases of h are studied, as shown in Figure 4a, where h is selected as 0.05 m, 0.10 m, 0.15 m, and 0.20 m. The diameter of the electrode is $r = 0.005$ m, the length of the rod electrode is 0.50 m, and the two ends of the electrode are $x_1 = -0.25$ m and $x_2 = 0.25$ m. Charge q carries a charge quantity of −1 C. Figure 4b shows the charge distribution with θ. It can be observed that the charge decreases as distance h increases and the charge reaches a maximum value when $\theta = 0$.

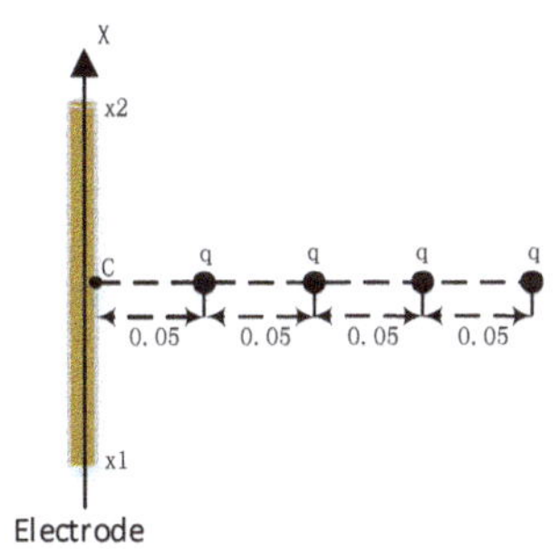

(a)

Figure 4. *Cont.*

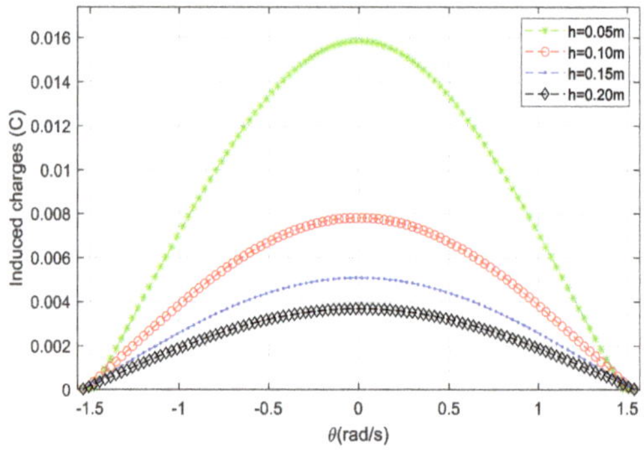

(b)

Figure 4. Induced charge density along θ: (**a**) Illustration of the point charge position, and (**b**) induced charge density.

The variation of the induced charge along (θ, x) on the electrode is shown in Figure 5. The induced charge reaches its maximum value at the point $(r, 0,\ 0)$. If the point charge is closer to the electrode, the charge distribution is more concentrated. When the sensing range x is smaller, the induced charge is greater. This indicates that the closer the point charge is to the rod electrode, the more charge is induced in the small sensing region, and the charge distribution tends to gradually decrease when the distance from the point $(r, 0,\ 0)$ increases.

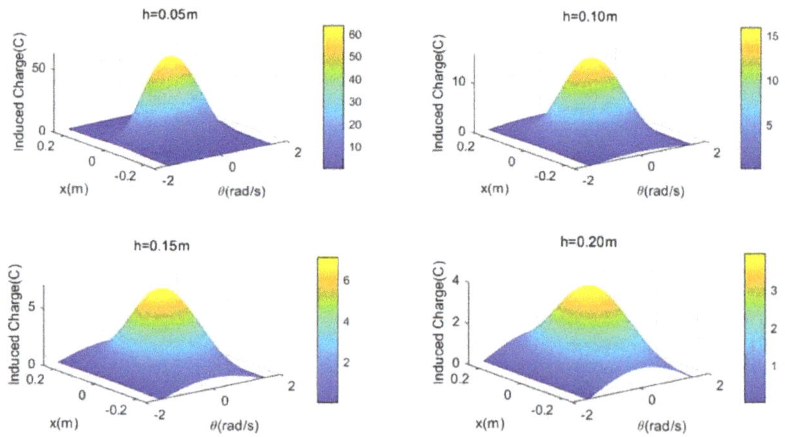

Figure 5. Distribution of the induced charge along (θ, x).

3.2. *Quantity of Charge Induced by Moving Point Charge*

3.2.1. Effect of Distance between Electrode and Point Charge

As shown in Figure 6a, the point charge q carries -1 C charge and moves at a constant velocity of 5 m/s along the direction parallel to the pipeline and perpendicular to the electrode's axis. The length of the electrode is 0.1 m and the radius of electrode r is 0.005 m.

Suppose the distance between the electrode and point charge is h. We performed simulations on five distances, as presented in Figure 6b, showing the cross-section of the pipeline that contains the electrode. The distances between the surface of the electrode and the five positions of A, B, C, D, E are 0.1 m, 0.2 m, 0.3 m, 0.4 m, and 0.5 m, respectively. The variations of the amount of charge induced on the electrode by these five simulations are shown in Figure 6c. Here, we define the distance from the point charge before it passes through points A, B, C, D, E as negative and the distance after as positive. It can be observed that the amount of charge decreases when the distance between the electrode and point charge increases. This applies when the distance between the electrode and point charge is within the range of −0.5 m to 0.5 m, and the amount of the induced charge can be neglected when it falls outside of this range Therefore, if there are charged particles uniformly distributed around the electrode, it can be considered that the induced charge on the electrode is generated by charges within a cylindrical section with a radius of 0.5 m.

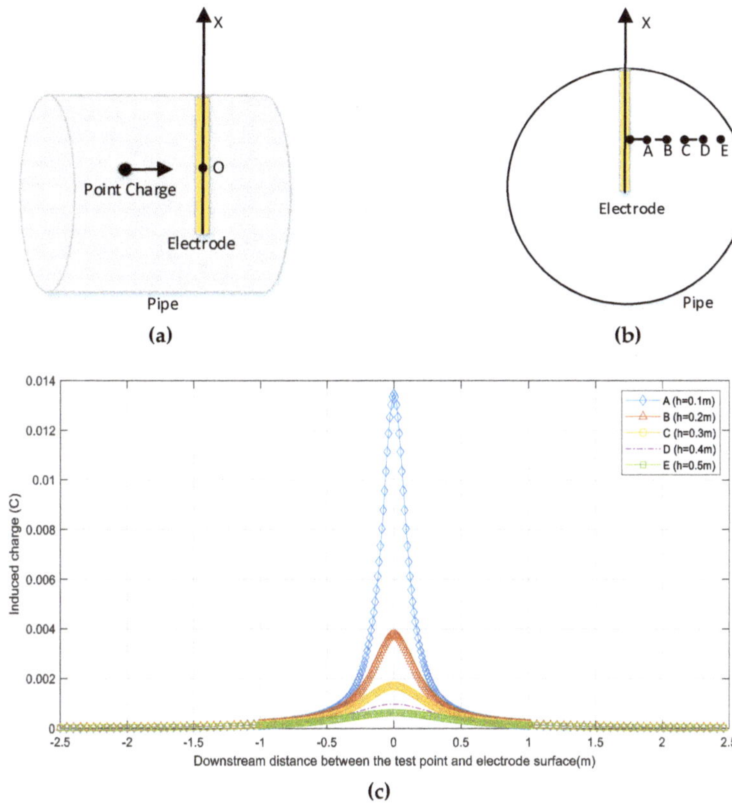

Figure 6. Locations of point charges A, B, C, D, E in the pipeline: (**a**) Point charge moving in the pipeline; (**b**) point charge distance from the electrode; and (**c**) induced charge of the point charge moving perpendicular to the electrode's axis.

As shown in Figure 7a, the point charge q moves from the surface of the electrode to point A and then to point B. The distance between point A and the surface of the electrode is 0.5 m and the distance between point B and the electrode is 0.5 m, the amount of charge Q is −1 C, the radius of rod electrode r is 0.005 m, and the length of the electrode is 0.5 m. As shown in Figure 7b, the closer the point charge is to the electrode, the larger the induced charge. With increased distance from the point charge, the induced charge decreases rapidly. When the distance between the point charge and the electrode reaches a certain value, the change of the induced charge is very small. The trend in Figure 7b can also be observed in Figure 6c. It is also consistent with the conclusion that when the charged particles are evenly distributed around the electrode, the charge on the electrode is induced by a cylindrical region around the electrode with a radius of 0.5 m.

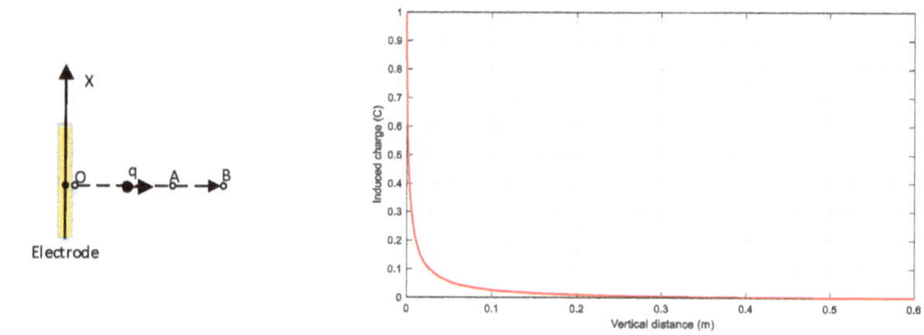

(a) **(b)**

Figure 7. Point charge away from the electrode: (**a**) Charge moving perpendicular to the electrode; and (**b**) induced charge change.

As shown in Figure 8a, the point charge passes through the points F, G, H, I with a constant velocity v = 5 m/s in a direction perpendicular to the electrode's axis. The distribution of F, G, H, I parallel to the length of the electrode is fixed, the amount of the point charge is $Q = -1$ C, the radius of the rod electrode is $r = 0.005$ m, and the vertical distance between F, G, H, I, and the surface of the electrode is $h = 0.05$ m. The distribution of the induced charge on the electrode is shown in Figure 8b. The amount of the induced charge increases gradually as the point charge gradually approaches the electrode. With the charge point away from the electrode, the amount of the induced charge gradually decreases. The induced charge at point I is very small, thus it can be concluded that the induced charge by the point charge being farther than I can be ignored when the charged particles are evenly distributed around the electrode.

The case where the point charge moves in a direction parallel to the axis of the electrode, as shown in Figure 9a, is considered, and the point charge moves from point A to point B and then to point C. The distance from point A to point C is 1.2 m. The charge of the point charge is $Q = -1$ C, the radius of the rod electrode is $r = 0.005$ m, and the vertical distance of points A, B, and C from the surface of the electrode is $h = 0.05$ m. The amount of the induced charge is shown in Figure 9b. The amount of the induced charge increases as the length of the electrode increases; the amount of the induced charge on the electrode in the range near the intermediate position of the electrode is larger, and the amount on the electrode becomes smaller. If the point charge is outside the range of the electrode length +0.15 m, the contribution to the amount of the induced charge will be negligible.

From the above analysis, it is concluded that the charged particles capable of inducing a charge on the rod electrode are mostly distributed in the shadow range, as shown in Figure 10. Note that a circular conveying pipe is considered in Figure 10 and the charged particles are uniformly distributed.

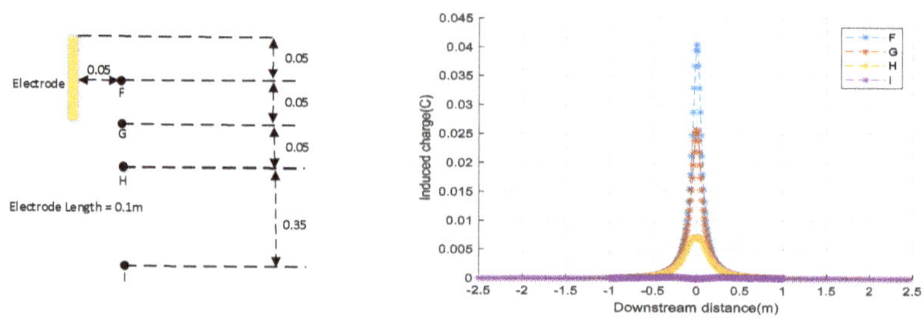

(a) **(b)**

Figure 8. Locations of point charges F, G, H, I in the pipeline: (**a**) Position of the point charge and electrode; and (**b**) amount of the induced charge on the electrode as distance changes.

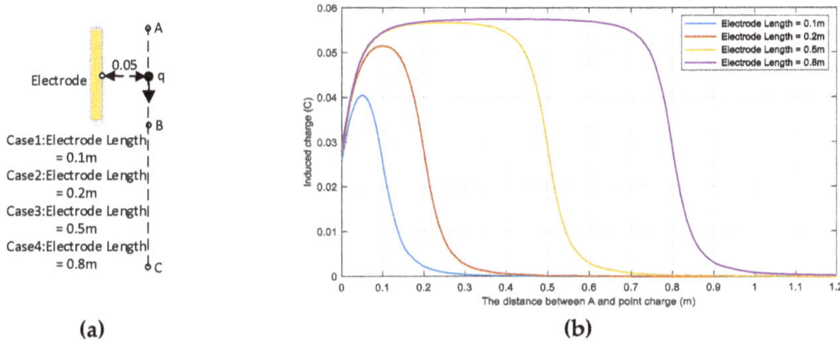

Figure 9. The point charge moves in a direction parallel to the axis of the electrode: (**a**) State of moving the point charge parallel to the electrode; and (**b**) amount of charge induced by moving the point charge parallel to the electrode.

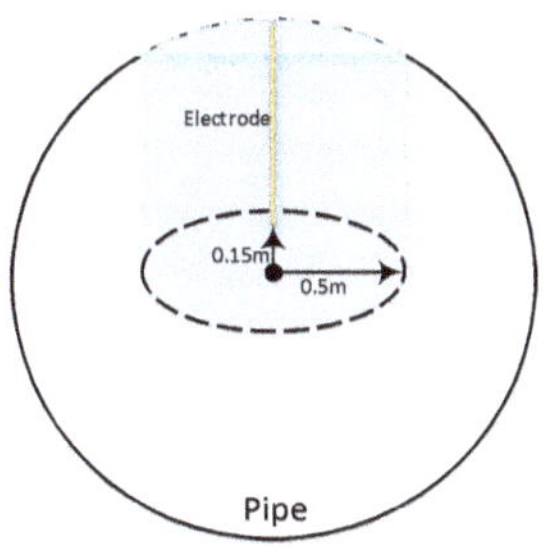

Figure 10. Active induced range of the rod electrode.

3.2.2. Effect of Electrode Length

The length of the rod electrode is an important parameter that needs to be determined in the sensor design stage. As shown in Figure 11a, four cases of the electrode length are chosen for the simulation study: 0.8 m, 0.4 m, 0.2 m, and 0.1 m. The point charge carries charge $Q = -1$ C and travels with constant velocity $v = 5$ m/s passing through point A in the direction perpendicular to the electrode. The radius of the rod electrode is $r = 0.005$ m, and the distance from point A to the surface of the electrode is $h = 0.05$ m. The change of the induced charge for the four cases is presented in Figure 11b. It can be observed that as the length of the electrode increases, the amount of the induced charge increases, but the increase rate decreases with the increase of the electrode. It can be concluded that the longer the electrode, the more charge can be induced. However, when the electrode length increases to a certain extent, its growth does not cause a significant change in the amount of the induced charge.

As shown in Figure 12a, the length of the electrode varies from 0 m to 4 m and the distance from the surface of the electrode is $h = 0.1$ m. The changed amount of the induced charge on the electrode is shown in Figure 12b. The charge of the point charge is $Q = -1$ C, and the radius of the rod electrode is $r = 0.005$ m. As shown in Figure 13b, once the length of the electrode reaches 1 m or more, the increase rate of the induced change becomes very small. In other words, increasing the electrode length contributes very little to the amount of the induced charge when the length reaches 1 m.

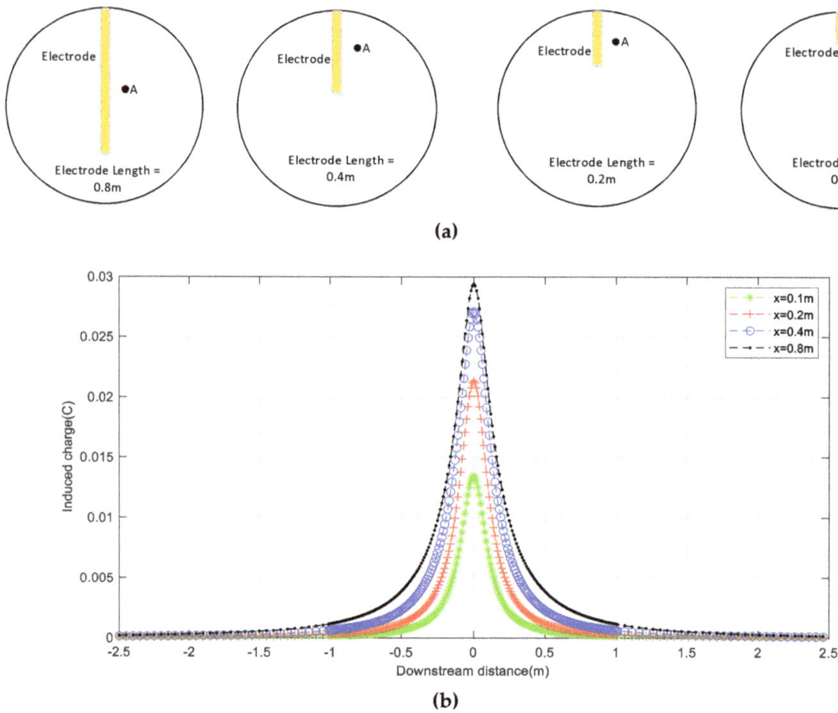

Figure 11. (**a**) Electrodes of different lengths and positions of the point charge; (**b**) induced charge distribution on electrodes of different lengths.

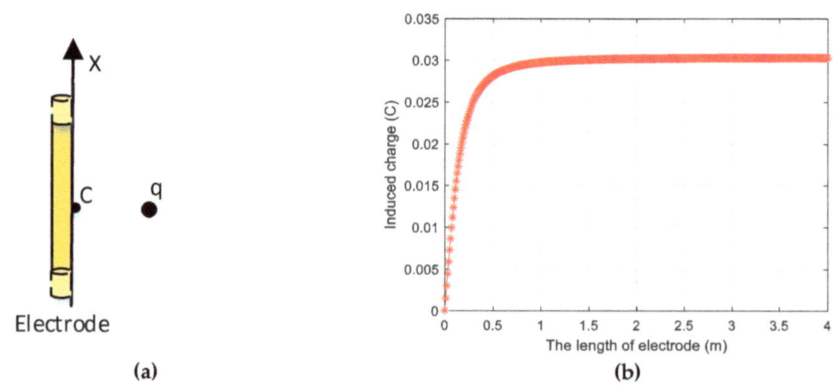

Figure 12. Effect of the electrode length change on the induced charge: (**a**) Increasing the length of the sensing electrode; and (**b**) variation of the amount of the induced charge with the electrode length.

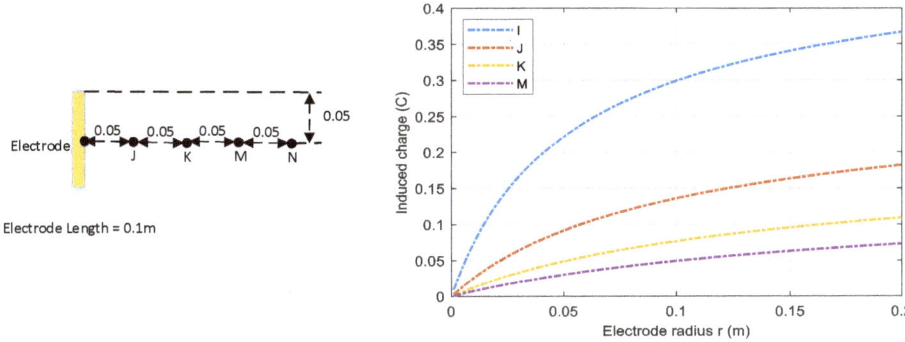

(a) (b)

Figure 13. Effect of the electrode radius: (**a**) Positions of different point charges and (**b**) variation of the induced charge with the electrode radius under the influence of the point charge at different positions.

3.2.3. Effect of Electrode Radius

As shown in Figure 13a, the distance of four points J, K, M, and N from the surface of the electrode is 0.05 m, 0.10 m, 0.15 m, and 0.20 m, respectively. The charge amount of the point charge is $Q = -1$ C, and the length of the rod electrode is 0.1 m. The radius varies from 0.001 m to 0.2 m. The amount of the induced charge on the electrode is shown in Figure 13b. As the radius of the electrode increases, the amount of the induced charge increases. This means that electrodes with a larger radius can induce more charge, which is beneficial to the design of the detection circuit in later stages. However, the radius of the electrode generally cannot be made too large to adapt to the actual situation in practice.

4. Induced Current on a Rod Electrode by Moving Point Charge

As shown in Figure 14, point charge q moves with velocity $v(t)$, assuming that its position at time *zero* is (x_0, y_0, z_0), then its position at time t is (x_t, y_t, z_t).

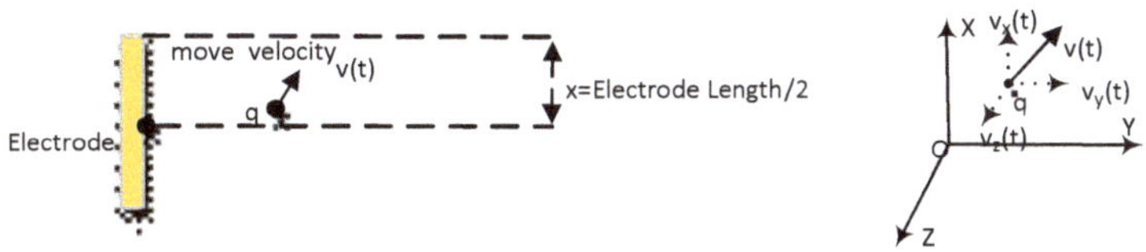

Figure 14. Moving point charge model.

x_t, y_t, and z_t can be expressed as

$$x_t = x_0 + \int_0^t v_x(t)dt \tag{15}$$

$$y_t = y_0 + \int_0^t v_y(t)dt \tag{16}$$

$$z_t = z_0 + \int_0^t v_z(t)dt. \tag{17}$$

h in Equation (1) can be expressed as

$$h = \left((y_t + r)^2 + z_t^2\right)^{1/2} - r. \tag{18}$$

Equation (13) can be expressed as

$$Q(t) = \int_{-acos(\frac{r}{r+h})}^{acos(\frac{r}{r+h})} \left(\frac{-qrh'(x_2 - x_t)}{2\pi h_1^2 \left(h_1^2 + (x_2 - x_t)^2\right)^{\frac{1}{2}}} - \frac{-qrh'(x_1 - x_t)}{2\pi h_1^2 \left(h_1^2 + (x_1 - x_t)^2\right)^{\frac{1}{2}}} \right) d\theta. \tag{19}$$

The induced current can be calculated by

$$I(t) = \frac{dQ(t)}{dt}. \tag{20}$$

4.1. Simulation for Induced Current

As shown in Figure 14, suppose that point charge q moves from $(0,\ y_0,\ z_0)$ to $(0,\ y_0,\ -z_0)$ with a different constant velocity v of 2 m/s, 5 m/s, 10 m/s, and 20 m/s. The charge of the point charge is $Q = -1$ C, $y_0 = 0.1$ m, the length of the electrode is 0.1 m, and the radius of the electrode is $r = 0.005$ m. As shown in Figure 15a, the induced current increases with the increased speed, and the range of the induced current is from −0.5 m to 0.5 m, which is consistent with the previous analysis in Section 3.2.1.

Figure 15b shows the variation of the maximum induced current with the velocity. Note that the maximum induced current is proportional to the velocity v_z of the point charge.

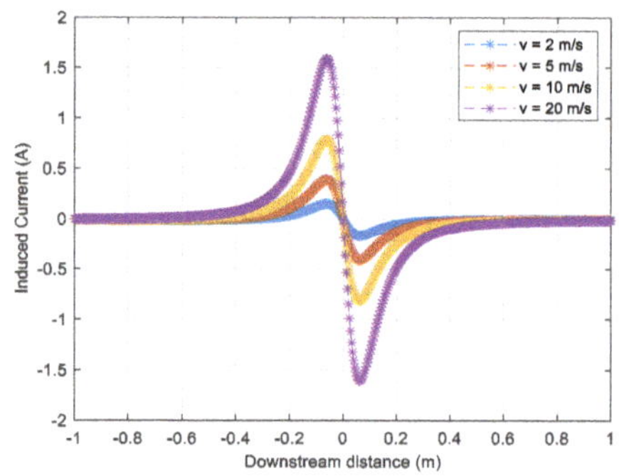

(a)

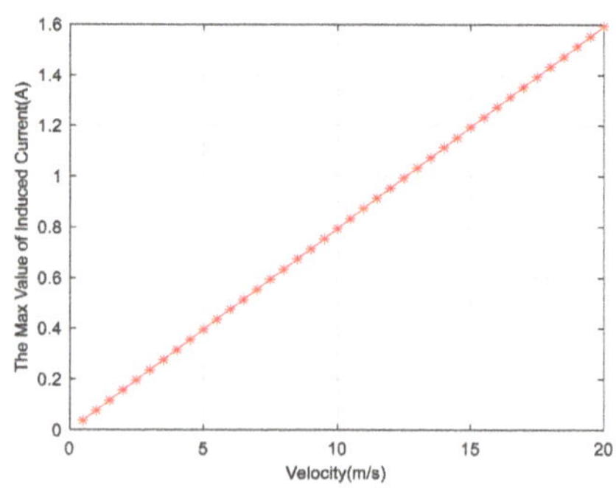

(b)

Figure 15. (**a**) Induced current under the effect of the point charge moving at different velocities, and (**b**) variation of the maximum induced current with the velocity change.

4.2. Spectrum of Induced Current

Four velocity cases (2 m/s, 5 m/s, 14.3 m/s, and 20 m/s) are selected to study the spectrum of the induced current. As shown in Figure 14, point charge q carries a charge of −1 C and moves from position $(0,\ y_0, z_0)$ to position $(0,\ y_0, -z_0)$ at velocity v, where $z_0 = 1$ m and $y_0 = 0.0015$ m. The length of the electrode is 0.5 m and the radius is $r = 0.005$ m. The results of the spectrum analysis are shown in Figure 16. It can be seen that the spectrum of the induced current spreads wider as the speed increases. In other words, as the charge moves faster, more abundant frequency components are introduced. This places a requirement on the design of the induced current measurement circuit. In order to capture the signals induced by faster-moving particles, it is necessary to design an acquisition circuit with a sufficiently wide frequency response, which will be discussed in Section 4.4.

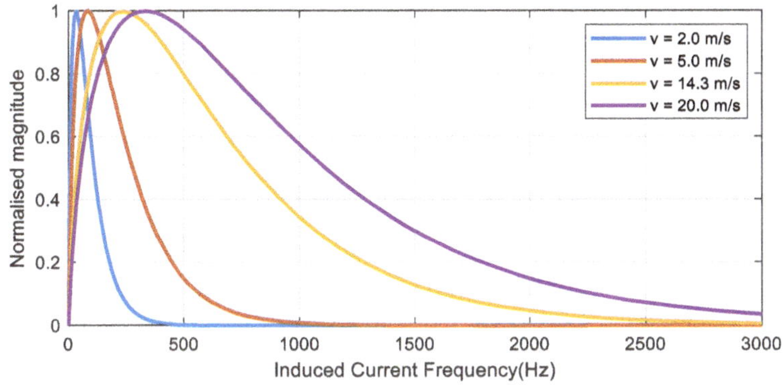

Figure 16. Spectrum analysis of the induced current under different moving speeds.

4.3. Analysis of the Variation of the Induced Current Spectrum over the Effective Range

To study the variation of the induced current spectrum in the range shown in Figure 10, we take velocity $v = 20$ m/s as an example. As shown in Figure 14, point charge q moves from $(0,\ y_0, z_0)$ to $(0,\ y_0, -z_0)$ at constant velocity v parallel to the z-axis direction, where $z_0 = 1$ m, the charge of the point charge is $Q = -1$ C, y_0 changes from 0.001 m to 0.5 m, the length of the electrode is 0.5 m, and the radius is $r = 0.005$ m. The vertical axis of Figure 17 represents the maximum frequency component. The result indicates that the higher the frequency of the induced current induced by particles closer to the electrode, the lower the frequency of the current induced by particles far from the electrode. In order to acquire high-frequency signals around the electrodes, the detection circuit requires a wide band-pass.

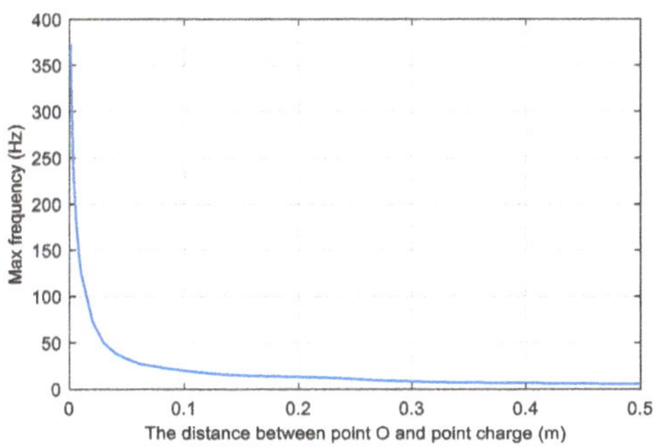

Figure 17. Variation of the main frequency and point charge position of the induced current spectrum under the action of the point charge moving at 20 m/s.

4.4. Measurement Circuit Analysis

An electrode typically has a high impedance output, hence a charge amplifier is typically added to match the impedance with the input of the data acquisition circuit. A typical charge amplifier is presented in Figure 18a. After passing through the circuit, the induced current signal is converted to a voltage signal. According to the virtual short and virtual break principle of the operational amplifier, it can be obtained that

$$I(t) = I_1(t) + I_2(t) \tag{21}$$

$$I_1(t) = -C\frac{dU(t)}{dt} \tag{22}$$

$$U(t) = -I_2(t)R. \tag{23}$$

By performing the Fourier transform on Equations (21)–(23), the response of the circuit can be derived as:

$$\frac{U(\omega)}{I(\omega)} = -\frac{R}{1 + j\omega CR} \tag{24}$$

The two most important features of the measurement circuit are its amplification and frequency characteristics. The amplification function is used to amplify the weak induced current signal to the extent that it can be acquired by an analogue to a digital converter. To study the spectral influence of the measurement circuit over the sensing circuit, its frequency characteristics are examined. From Equation (24), it can be concluded that the amplification factor of the output signal is determined by the resistor R and the cutoff frequency of the output signal $U(t)$ is determined by the capacitor C. The frequency responses by different R and C values are calculated and shown in Figure 18b. It can be observed that the cutoff frequency decreases with the increased C.

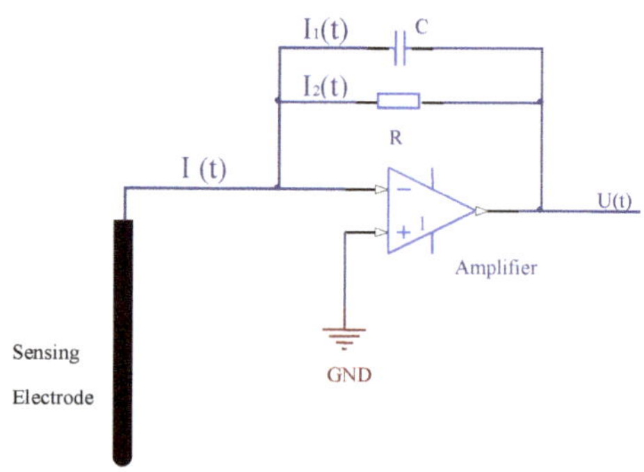

(a)

Figure 18. *Cont.*

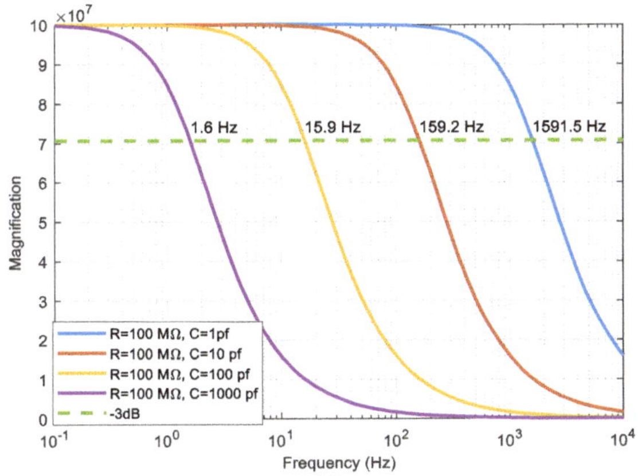

(b)

Figure 18. (**a**) Schematic of the measurement circuit, and (**b**) amplitude response of the circuit under different parameter settings.

5. Experimental Study and Discussion

From the simulation study in Section 4, we can understand the signals of the induced current by a moving point charge. In this section, an experimental system is set up and three experiments are performed to validate the established model.

5.1. Experimental Setup

The schematic of the experimental system is shown in Figure 19a. A rod electrode is placed on a supporting holder parallel to the ground, and a funnel is placed on top of the electrode. During the experiment, a charged ball is released from the funnel toward the ground. By adjusting the distance *Height1* between supporting rods 1 and 2, the moving speed of the charged ball can be adjusted. The distance *Height2* between supporting rod 2 and the ground can decide the final speed of the charging ball. Note that supporting rod 2 is made of insulating material to hold the rod electrode. In addition, supporting rods 1 and 2 are adjustable to a full 360°, hence the horizontal distance between the small ball and the rod electrode can be adjusted by changing the angle between supporting rods 1 and 2; that is, parameter h in Equation (19).

A picture of the rod electrode employed in this paper is shown in Figure 19b. The rod electrode is made of a 316L stainless steel, with a radius of 0.005 m and a length of 0.5 m. Note that the surface is coated with a Teflon insulation 0.3 mm thick to ensure that the current on the electrode is entirely induced by the electrostatic induction.

The output of the rod electrode was connected to a charge amplifier for amplification, as shown in Figure 18a, with $R = 100\ \mathrm{M\Omega}$ and $C = 10$ pf. According to the analysis in Section 4.4, the cutoff frequency was set at 159.2 Hz with such a setting. In the experiment, the maximum moving speed of the ball was less than 6 m/s, so the signal of interest could be appropriately magnified based on the analysis in Figure 16. Then, the signal was acquired by a data acquisition board and sent to the computer for further analysis. From the analysis in Section 4, we understand that the highest frequency of the signal is below 3500 Hz, so the sampling rate of the data acquisition board was set at 10 kHz to satisfy the Nyquist sampling theorem, i.e., the sampling frequency should be more than twice the maximum frequency.

The experiments are conducted in an environment with room temperature (about 20–25 °C) and normal pressure (about 101 kPa).

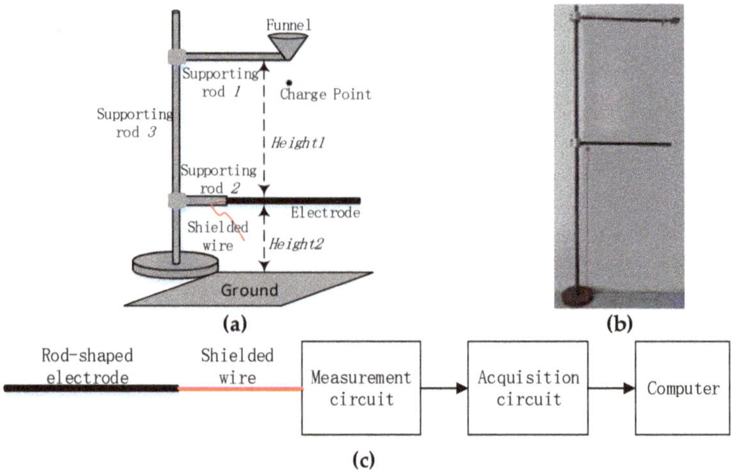

Figure 19. Experimental setup: (**a**) Schematic of the experimental system; (**b**) picture of the rod electrode employed; and (**c**) schematic of the signal connection.

5.2. Results and Discussion

Three situations were analyzed in this study, and in all these experiments, the gravity acceleration g was assumed to be 9.8 $\mathrm{m/s^2}$ and the ball fell from the funnel at an initial velocity of 0 m/s, hence the velocity of the ball at time t can be calculated by $v(t) = gt$.

Experiment 1: $Height1 = 0.57$ m, $Height2 = 0.90$ m, $h = 0.33$ m, ball carries a charge of $q = 486$ nC $v(t) = gt$. Hence, the velocity of the ball reached 3.34 m/s when passing by the rod electrode and reached 5.37 m/s when it landed on the ground. The measured signal by the rod electrode and its spectrum are shown in Figure 20a,b, respectively. As can be observed, there exists a DC component in the measured signal in Figure 20a. This DC component is from the data acquisition board, not from the true signal, as the data acquisition board has an added offset of 2.5 V to the input signal, converting the input voltage of −2.5 V to 2.5 V to 0 to 5 V. Hence, this DC component should be removed before further processing. It can be observed that the signal interfered with a severe 50 Hz power line signal and the signal of interest mainly falls within 20 Hz. With a proper low-pass digital filter, the power line frequency can easily be removed. The filtered signal and its spectrum are presented in Figure 20c,d, respectively. In Figure 20c, the voltage increases at the beginning when the ball moves nearer to the rod electrode, indicating that the rate of increase of the induced charge is increasing. Then, the voltage decreases to zero, indicating that the rate of increase of the induced charge is decreasing, and the induced charge reaches the maximum value when the ball is nearest to the electrode, at which point the voltage output is equal to zero.

As the ball moves away from the electrode, the voltage becomes negative, indicating that the induced charge on the electrode becomes smaller. The smaller the voltage, the larger the loss rate. When the loss rate becomes the smallest, then it becomes larger, indicating that the loss rate of the induced charge becomes smaller. At the point where the voltage reaches zero again, the induced charge returns to zero. This is in agreement with the fact that when a charged ball moves to an electrode, the electrode will induce charges, and the induced charges will be reduced when the charged ball moves away from the electrode.

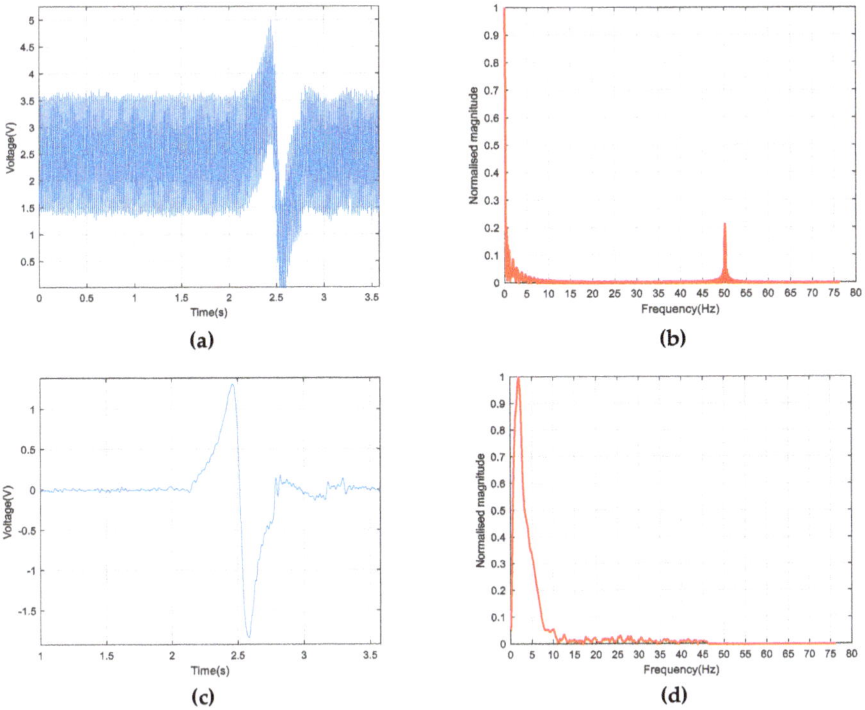

Figure 20. Results of experiment 1: (**a**) Measurement signal waveform; (**b**) spectrum of the measurement signal waveform; (**c**) filtered signal; and (**d**) spectrum of the filtered signal.

By using Equation (19), the induced current change with the distance between the ball and the funnel is calculated and presented in Figure 21a. The simulated signal output after the charge amplifier is shown in Figure 21b. It can be observed that this signal shows good similarity with the experimental output signal (Figure 20c) in both the shape and amplitude. The induced current increases to a maximum value with the ball moving close to the electrode, and then decreases to zero when the ball is closest to the electrode, corresponding with Figure 15a. The validity of the model is verified.

By resetting the time when the ball leaves the funnel at zero and converting the time to distance in Figure 20c, the theoretical output and measured signal are plotted together in Figure 22a and their spectra are compared in Figure 22b. It shows that these two signals have a high similarity in trend and their values are very close to each other, verifying the correctness of the model we established. Specifically, when the charged ball moved with a low velocity (distance *Height 1* is less than 0.8 m or velocity is less than 4 m/s), the theoretical and measured waveforms agreed very well. When the charged ball moved with a high velocity, i.e., velocity greater than 4 m/s, the measured signal was smaller than the theoretical one. A possible reason for this phenomenon could be that the charged ball experiences more resistance from the air with the increased velocity, which reduces the velocity of the ball to a level lower than theoretical calculations and hence less current is induced on the rod electrode.

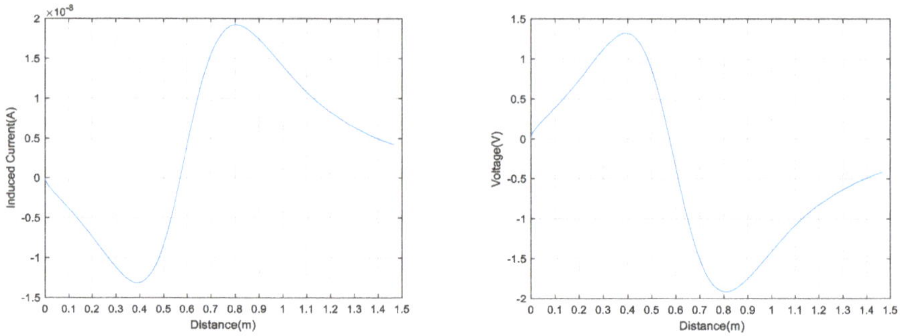

(a) **(b)**

Figure 21. (**a**) Theoretical induced current, and (**b**) theoretical output after the charge amplifier.

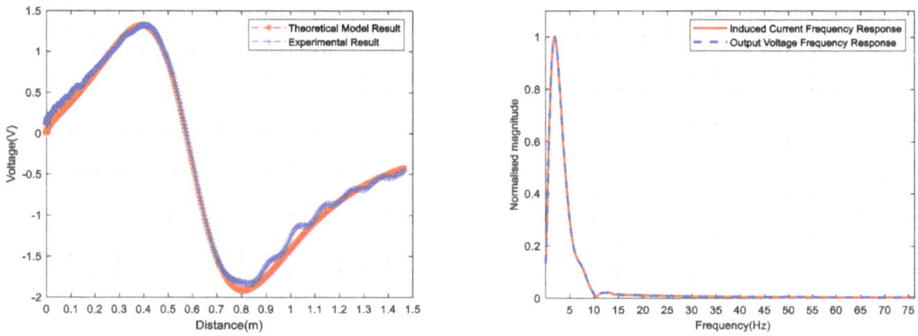

(a) **(b)**

Figure 22. Comparison of the theoretical and measured signal comparison: (**a**) Time domain waveform, and (**b**) spectrum.

Experiment 2: $Height1 = 0.80$ m, $Height2 = 0.90$ m, $h = 0.17$ m, *ball 1* carried a charge of 37.13 nC and *ball 2* carried a charge of 57.22 nC. Hence, the velocity of the ball reached 3.96 m/s when it passed by the rod electrode and reached 5.72 m/s when it landed on the ground.

The theoretical and measured output voltages are shown in Figure 23a. As can be observed, the waveforms show similar trends as those in Experiment 1. The induced current increased accordingly with the increased velocity of the ball, which is reflected by the maximum and minimum amplitude of the induced current. By comparing Figure 23a,c, it can be seen that the induced voltage increased significantly, mainly caused by an increased charge carried by the ball. This can be explained by Equation (19), which shows that the induced current is proportional to the amount of charge if other conditions remain unchanged. Taking the maximum value, minimum value, and measured value at four similar distances in Figure 23a,c, the results are summarized in Table 1. It can be observed that the induced current is approximately proportional to the amount of charge.

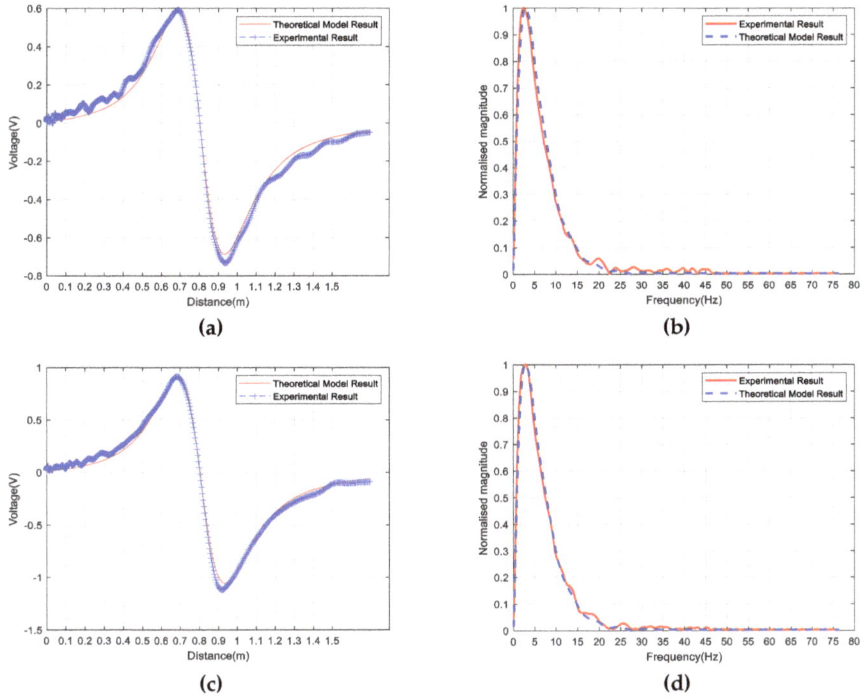

Figure 23. Results of experiment 2: (**a**) Comparison of the theoretical value with the measured voltage of charged *ball 1*; (**b**) comparison of the theoretical spectrum with the measured voltage spectrum of charged *ball 1*; (**c**) comparison of the theoretical value with the measured voltage of charged *ball 2*; and (**d**) comparison of the theoretical spectrum with the measured voltage spectrum of charged *ball 2*.

Table 1. Comparison of the experimental results between charged *ball 1* and charged *ball 2*.

	Charged Ball 1	Charged Ball 2	Ratio
Amount of charge (nC)	37.13	57.22	0.6490
Maximum measured value (V)	0.5909	0.9107	0.6488
Minimum measured value (V)	−0.7333	−1.1190	0.6553
Measured value at 0.5 m (V)	0.2788	0.4299	0.6485
Measured value at 0.6 m (V)	0.4794	0.6864	0.6984
Measured value at 1.1 m (V)	−0.4138	−0.6335	0.6531
Measured value at 1.2 m (V)	−0.2881	−0.4112	0.7006

Experiment 3: $Height1 = 0.45$ m, $Height2 = 0.90$ m, $h = 0.18$ m, ball carried a charge of 134 nC. Hence, the velocity of the ball reached 2.97 m/s when it passed by the electrode and reached 5.14 m/s when it landed on the ground.

The comparison of the output voltage between the theoretical and experimental waveforms is shown in Figure 24a and the spectrum comparison is shown in Figure 24b. By comparing Figures 22a and 24b, it can be seen that the high-frequency component decreases when h increases, which is consistent with the analysis in Section 4.3.

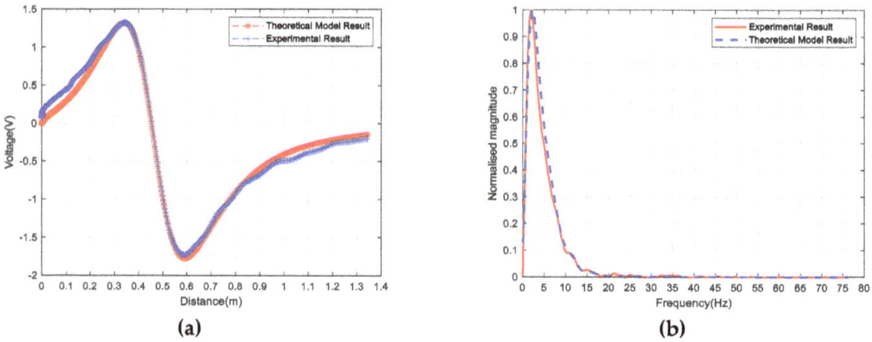

Figure 24. Results of experiment 3: (**a**) Comparison of the theoretical value with the measured voltage; and (**b**) comparison of theoretical spectrum with the measured spectrum of voltage.

Through three experiments with different scenario settings, it can be observed that the established model matches well with the experimental measurements, verifying the correctness and accuracy of the model. In addition, the following observations are also verified:

(1) The induced current decreased significantly with the increase of h, which can be observed in Figures 22b and 24a, and this is in line with the result in Figure 7b.
(2) The high-frequency components of the induced current are reduced with the increased h, which can be seen by comparing Figures 22a and 24b, and this is consistent with Figure 17.
(3) The high-frequency components increase with the increased velocity, which can be observed by comparing Figures 23b and 24b.
(4) When the velocity and the h value are fixed, the change in the amount of the charge on the ball does not influence the spectral characteristics of the induced current signal, which can be observed in Figure 23b,d.

6. Conclusions

In this paper, a theoretical model of the induced charge on a rod electrode by a point charge is established by the method of images and the accuracy of the model is verified by three experimental studies. The following conclusions are drawn based on the study.

1. The amount of the induced charge on the rod electrode are mainly determined by the following factors: The distance between the point charge and the electrode, the radius and the length of the electrode.
2. The general model of the relationship between the induced current and velocity is established and the spectrum of the induced current is studied. The induced current increases with the increase of the point charge's velocity and the maximum value of the induced current is linearly proportional to the point charge's velocity. The faster the velocity of the point charge, the wider the spectrum of the induced current. The further the point charge from the rod electrode, the narrower the spectrum of the induced current on the rod electrode.
3. For the measurement circuit, its amplification ratio and pass-band width are determined by the feedback resistance and feedback capacitance, respectively. With the increase of the feedback resistance, the amplification factor of the circuit increases. With the increase of the feedback capacitance, the pass-band width of the measurement circuit becomes narrower.

The phenomena observed in the experiments, which are well explained by the established model, would be not so easy to explain with just the qualitative analysis. This model provides a basis for more complex studies in the future; for example, the sum result of all the particles around the electrode can been studied with this model. In addition, the influence of the length and radius of the rod electrode on the induced charge are also studied, providing theoretical support for the rod electrode sensor development in the future.

Author Contributions: Conceptualization, J.L. (Jiming Li); Data curation, J.L. (Jiming Li) and J.L. (Jingyu Li); Formal analysis, J.L. (Jiming Li); Funding acquisition, X.C.; Investigation, J.L. (Jiming Li); Methodology, J.L. (Jiming Li); Project administration, X.C.; Resources, X.C.; Software, J.L. (Jiming Li) and G.F.; Supervision, X.C. and G.F.; Validation, J.L. (Jiming Li), J.L. (Jingyu Li), and G.F.; Writing—original draft, J.L. (Jiming Li); Writing—review and editing, J.L. (Jiming Li), X.C., and G.F.

Funding: This research received no external funding.

Acknowledgments: The work was supported by the National Natural Science Foundation of China (No. 61503224), the Shandong Natural Science Foundation of China (No. ZR2017MF048), the Graduate Education Quality Improvement Plan Construction of Shandong Province of China (No. 2016050), the Qingdao Minsheng Science and Technology Plan Project (No. 17-3-3-88-nsh), and the Shandong University of Science and Technology Postgraduate Innovation Program (SDKDYC190249).

Conflicts of Interest: All authors declare no conflict of interest.

References

1. Lebecki, K.; Małachowski, M.; Sołtysiak, T. Continuous dust monitoring in headings in underground coal mines. *J. Sustain. Min.* **2016**, *15*, 125–132. [CrossRef]
2. Brodny, J.; Tutak, M. Exposure to Harmful Dusts on Fully Powered Longwall Coal Mines in Poland. *Int. J. Environ. Res. Public Health* **2018**, *15*, 1846. [CrossRef] [PubMed]
3. Jeong, W.; Jeon, S.; Jeong, D. Advanced Backstepping Trajectory Control for Skid-Steered Duct-Cleaning Mobile Platforms. *Electronics* **2019**, *8*, 401. [CrossRef]
4. Mataloto, B.; Ferreira, J.C.; Cruz, N. LoBEMS—IoT for Building and Energy Management Systems. *Electronics* **2019**, *8*, 763. [CrossRef]
5. Marques, G.; Pitarma, R. A Cost-Effective Air Quality Supervision Solution for Enhanced Living Environments through the Internet of Things. *Electronics* **2019**, *8*, 170. [CrossRef]
6. Arroyo, P.; Lozano, J.; Suárez, J.I. Evolution of Wireless Sensor Network for Air Quality Measurements. *Electronics* **2018**, *7*, 342. [CrossRef]
7. Gajewski, J. Electrostatic Nonintrusive Method for Measuring the Electric Charge, Mass Flow Rate, and Velocity of Particulates in the Two-Phase Gas–Solid Pipe Flows—Its Only or as Many as 50 Years of Historical Evolution. *IEEE Trans. Ind. Appl.* **2008**, *44*, 1418–1430. [CrossRef]
8. Weinheimer, A.J. The charge induced on a conducting cylinder by a point charge and its application to the measurement of charge on precipitation. *J. Atmos. Ocean. Technol.* **1988**, *5*, 298–304. [CrossRef]
9. Woodhead, S.R.; Amadi-Echendu, J.E. Solid phase velocity measurement utilizing electrostatic sensors and cross correlation signal processing. In Proceedings of the 1995 IEEE Instrumentation and Measurement Technology Conference, Waltham, MA, USA, 23–26 April 1995; pp. 774–777.
10. Gajewski, J.B. Non-contact electrostatic flow probes for measuring the flow rate and charge in the two-phase gas–solids flows. *Chem. Eng. Sci.* **2006**, *61*, 2262–2270. [CrossRef]
11. Yan, Y.; Byrne, B.; Woodhead, S.; Coulthard, J. Velocity measurement of pneumatically conveyed solids using electrodynamic sensors. *Meas. Sci. Technol.* **1995**, *6*, 515–537. [CrossRef]
12. Qian, X.; Yan, Y.; Huang, X.; Hu, Y. Measurement of the Mass Flow and Velocity Distributions of Pulverized Fuel in Primary Air Pipes Using Electrostatic Sensing Techniques. *IEEE Trans. Instrum. Meas.* **2017**, *66*, 944–952. [CrossRef]
13. Li, L.; Hu, H.; Qin, Y.; Tang, K. Digital Approach to Rotational Speed Measurement Using an Electrostatic Sensor. *Sensors* **2019**, *19*, 2540. [CrossRef] [PubMed]
14. Xu, C.; Wang, S.; Tang, G.; Yang, D.; Zhou, B. Sensing characteristics of electrostatic inductive sensor for flow parameters measurement of pneumatically conveyed particles. *J. Electrost.* **2007**, *65*, 582–592. [CrossRef]
15. Li, J.; Ma, X.; Zhao, M.; Cheng, X. A Novel MFDFA Algorithm and Its Application to Analysis of Harmonic Multifractal Features. *Electronics* **2019**, *8*, 209. [CrossRef]
16. S´wirad, S.; Wydrzynski, D.; Nieslony, P.; Krolczyk, G.M. Influence of hydrostatic burnishing strategy on the surface topography of martensitic steel. *Measurement* **2019**, *138*, 590–601. [CrossRef]

17. Osornio-Rios, R.A.; Antonino-Daviu, J.A.; Romero-Troncoso, R.d.J. Recent Industrial Applications of Infrared Thermography: A Review. *IEEE Trans. Ind. Inf.* **2019**, *15*, 615–625. [CrossRef]
18. Mia, M.; Królczyk, G.; Maruda, R.; Wojciechowski, S. Intelligent Optimization of Hard-Turning Parameters Using Evolutionary Algorithms for Smart Manufacturing. *Materials* **2019**, *12*, 879. [CrossRef] [PubMed]
19. Glowacz, A. Fault Detection of Electric Impact Drills and Coffee Grinders Using Acoustic Signals. *Sensors* **2019**, *19*, 269. [CrossRef]
20. Wang, C.; Zhan, N.; Jia, L.; Zhang, J.; Li, Y. DWT-based adaptive decomposition method of electrostatic signal for dilute phase gas-solid two-phase flow measuring. *Powder Technol.* **2018**, *329*, 199–206. [CrossRef]
21. Zhang, W.; Cheng, X.; Hu, Y.; Yan, Y. Measurement of moisture content in a fluidized bed dryer using an electrostatic sensor array. *Powder Technol.* **2018**, *325*, 49–57. [CrossRef]
22. Yang, Y.; Zhang, Q.; Zi, C.; Huang, Z.; Zhang, W.; Liao, Z.; Wang, J.; Yang, Y.; Yan, Y.; Han, G. Monitoring of particle motions in gas-solid fluidized beds by electrostatic sensors. *Powder Technol.* **2017**, *308*, 461–471. [CrossRef]
23. Dong, K.; Zhang, Q.; Huang, Z.; Liao, Z.; Wang, J.; Yang, Y. Experimental investigation of electrostatic effect on bubble behaviors in gas-solid fluidized bed. *AIChE J.* **2015**, *61*, 1160–1171. [CrossRef]
24. Qian, X.; Yan, Y.; Wang, L.; Shao, J. An integrated multi-channel electrostatic sensing and digital imaging system for the on-line measurement of biomass–coal particles in fuel injection pipelines. *Fuel* **2015**, *151*, 2–10. [CrossRef]
25. Liu, S.; Chen, Q.; Wang, H.G.; Jiang, F.; Ismail, I.; Yang, W.Q. Electrical capacitance tomography for gas–solids flow measurement for circulating fluidized beds. *Flow Meas. Instrum.* **2005**, *16*, 135–144.
26. Yao, J.; Zhao, Y.L.; Fairweather, M. Numerical simulation of turbulent flow through a straight square duct. *Appl. Therm. Eng.* **2015**, *91*, 800–811. [CrossRef]
27. Zhang, S.; Yan, Y.; Qian, X.; Hu, Y. Mathematical Modeling and Experimental Evaluation of Electrostatic Sensor Arrays for the Flow Measurement of Fine Particles in a Square-Shaped Pipe. *IEEE Sens. J.* **2016**, *16*, 8531–8541.
28. Wang, L.; Yan, Y.; Hu, Y.; Qian, X. Rotational Speed Measurement Using Single and Dual Electrostatic Sensors. *IEEE Sens. J.* **2015**, *15*, 1784–1793.
29. Coombes, J.R.; Yan, Y. Experimental investigations into the flow characteristics of pneumatically conveyed biomass particles using an electrostatic sensor array. *Fuel* **2015**, *151*, 11–20. [CrossRef]
30. Krabicka, J.; Yan, Y. Finite-Element Modeling of Electrostatic Sensors for the Flow Measurement of Particles in Pneumatic Pipelines. *IEEE Trans. Instrum. Meas.* **2009**, *58*, 2730–2736. [CrossRef]
31. Shao, J.; Krabicka, J.; Yan, Y. Velocity Measurement of Pneumatically Conveyed Particles Using Intrusive Electrostatic Sensors. *IEEE Trans. Instrum. Meas.* **2010**, *59*, 1477–1484. [CrossRef]
32. Shao, J.; Krabicka, J.; Yan, Y. Comparative study of electrostatic sensors with circular and probe electrodes for velocity measurement of pulverized coal. *IEEE Trans. Instrum. Meas.* **2010**, *59*, 1477–1484.
33. Chen, J.-G.; Wu, F.-X.; Wang, J. Dust concentration detection technology of charge induction method. *J. Chin. Coal Soc.* **2015**, *40*, 713–718.
34. Griffiths, D.J. *Introduction to Electrodynamics*, 4th ed.; Addison-Wesley: Reading, MA, USA, 2012; pp. 124–126.

electronics

MDPI

Article

A Novel Image-Restoration Method Based on High-Order Total Variation Regularization Term

Jianhong Xiang, Pengfei Ye, Linyu Wang * and Mingqi He

College of Information and Communication Engineering, Harbin Engineering University, Harbin 150001, China
* Correspondence: wangly6853@163.com; Tel.: +86-1380-457-1419

Received: 25 June 2019; Accepted: 1 August 2019; Published: 5 August 2019

Abstract: This paper presents two new models for solving image the deblurring problem in the presence of impulse noise. One involves a high-order total variation (TV) regularizer term in the corrected total variation L1 (CTVL1) model and is named high-order corrected TVL1 (HOCTVL1). This new model can not only suppress the defects of the staircase effect, but also improve the quality of image restoration. In most cases, the regularization parameter in the model is a fixed value, which may influence processing results. Aiming at this problem, the spatially adapted regularization parameter selection scheme is involved in HOCTVL1 model, and spatially adapted HOCTVL1 (SAHOCTVL1) model is proposed. When dealing with corrupted images, the regularization parameter in SAHOCTVL1 model can be updated automatically. Many numerical experiments are conducted in this paper and the results show that the two models can significantly improve the effects both in visual quality and signal-to-noise ratio (SNR) at the expense of a small increase in computational time. Compared to HOCTVL1 model, SAHOCTVL1 model can restore more texture details, though it may take more time.

Keywords: image restoration; impulse noise; ADMM; HOCTVL1; spatially adapted regularization parameter

1. Introduction

In the field of electronics and information, signal processing is a hot research topic, and as a special signal, the study of image has attracted the attention of scholars all over the world [1–3]. In image processing, image restoration is one of the most important issues and this issue has received extensive attention in the past few decades [4–11]. Image restoration is a technology that uses degraded images and some prior information to restore and reconstruct clear images, to improve image quality. At present, this technology has been widely used in many fields, such as medical imaging [12,13], astronomical imaging [14,15], remote sensing image [16,17], and so on. In this paper, the problem of image deblurring under impulse noise is considered. Normally, camera shake, and relative motion between the target and the imaging device may cause image blurring; while in digital storage and image transmission, impulse noise may be generated.

In the process of image deblurring under impulse noise, the main task is to find the unknown true image $\boldsymbol{x} \in R^{n^2}$ from the observed image $\boldsymbol{f} \in R^{n^2}$ defined by

$$f = N_{imp}(\boldsymbol{K}\boldsymbol{x}) \tag{1}$$

where N_{imp} denote the process of image degradation by impulse noise, and $\boldsymbol{K} \in R^{n^2 \times n^2}$ is a blurring operator. It is known that when $\boldsymbol{K}$ is unknown, the model deals with blind restoration, and when $\boldsymbol{K}$ is known, it deals with image denoising.

There are two main types of impulse noise: salt-and-pepper noise and random-valued noise. Suppose the dynamic range of x to be $[d_{\min}, d_{\max}]$, for all $(i,j) \in \Omega = \{1,2,\ldots,n\} \times \{1,2,\ldots,n\}$, the $x_{i,j}$ is the gray value of an image x at location (i,j), and $d_{\min} \leq f_{i,j} \leq d_{\max}$. For 8-bit images, $d_{\min} = 0$ and $d_{\max} = 255$. Then for salt-and-pepper noise, the noisy version f at pixel location (i,j) is defined as

$$f_{i,j} = \begin{cases} d_{\min}, & \text{with probability } \frac{s}{2}, \\ d_{\max}, & \text{with probability } \frac{s}{2}, \\ x_{i,j}, & \text{with probability } 1-s, \end{cases} \tag{2}$$

where s is the noise level of the salt-and-pepper noise.

For random-valued noise, the noisy version f at pixel location (i,j) is defined as

$$f_{i,j} = \begin{cases} d_{i,j}, & \text{with probability } r, \\ x_{i,j}, & \text{with probability } 1-r, \end{cases} \tag{3}$$

where $d_{i,j}$ is uniformly distributed in $d_{\min} \leq f_{i,j} \leq d_{\max}$ and r is the noise level of random-valued noise. It is clear that compared with salt-and-pepper noise, the random-valued noise is more difficult to remove since it can be arbitrary number in $d_{\min} \leq f_{i,j} \leq d_{\max}$.

For image-restoration problem contaminated by impulse noise, the widely used model is composed of data fidelity term measured by ℓ_1 norm and the TV regularization term, which is called TVL1 model [18–20]. TVL1 model can effectively preserve image boundary information and eliminate the influence of outliers, so it is especially effective to deal with non-Gaussian additive noise such as impulse noise. Now, it has been widely and successfully applied in medical image and computer vision.

However, TVL1 model has its own shortcomings, which makes it ineffective in dealing with high-level noise, such as 90% salt-and-pepper noise and 70% random-valued noise [21]. In recent years, a large number of scholars have devoted themselves to this research, and a lot of algorithms have been proposed [22–27]. In 2009, Cai et al. [22] proposed a two-phase method, and in the first phase, damaged pixels of the contaminated image were explored, then in the second phase, undamaged pixels were used to restore images. Numerical experiments show that the two-phase method is superior to TVL1 model, it can handle as high as 90% salt-and-pepper noise, and as high as 55% random-valued noise, while it cannot perform effectively when the level of random-valued noise is higher than 55%. Similarly, considering the problem that TVL1 model may deviate from the data-acquisition model and the prior model, especially for high levels of noise, Bai et al. [23] introduced an adaptive correction procedure in TVL1 model and proposed a new model called the corrected TVL1 (CTVL1) model. The main idea is to improve the sparsity of the TVL1 model by introducing an adaptive correction procedure. The CTVL1 method also uses two steps to restore the corrupted image, the first step generates an initial estimator by solving the TVL1 model, and the second step generates a higher accuracy recovery from the initial estimator which is generated by TVL1 model. Meanwhile, for higher salt-and-pepper noise and higher random-valued noise, by repeating the correction step for several times, high levels of noise can be removed very well. Numerical experiments show that the CTVL1 model can remove salt-and-pepper noise as high as 90%, and remove random-valued noise as high as 70%, which is superior to the two-phase method.

Similar to CTVL1 method, Gu et al. [24] combined TV regularization with smoothly clipped absolute deviation (SCAD) penalty for data fitting, and proposed TVSCAD model; Gong et al. [25] used minimax concave penalty (MCP) in combination with TV regularization for data fitting, and he proposed TV-MCP model. Numerical experiments show that both TVSCAD model and TV-MCP model can achieve better effects than two-phase method. But compared with CTVL1 method, their contributions mainly focus on the convergence rate, and they did not improve much in terms of impulse noise removal.

However, as is described in [28], TV norm may transform the smooth area to piecewise constants, the so-called staircase effect. To overcome this deficiency, the efficient way is to replace the TV norm by a high-order TV norm [29]. In particular, second-order TV regularization schemes are widely studied for overcoming the staircase effects while preserving the edges well in the restored image. In [30], Si Wang proposed a combined total variation and high-order total variation model to restore blurred images corrupted by impulse noise or mixed Gaussian plus impulse noise. In [27], based on Chen and Cheng's an effective TV-based Poissonian image deblurring model, Jun Liu introduced an extra high-order total variation (HTV) regularization term to this model, which can effectively remove Poisson noise, and its effect is better than Chen and Cheng's model. In [31], Gang Liu combined the TV regularizer and the high-order TV regularizer term, and proposed HTVL1 model, which can better remove the impulse noise contrast to TVL1 model. However, since TVL1 model has its own defects, the restoration of HTVL1 model is limited. Besides, the author did not consider the removal effect of random-valued noise.

In this paper, we continue to study the problem of image deblurring under impulse noise. The main contributions of this paper include: (1) Combining high-order TV regularizer term with CTVL1 model, a new model named high-order corrected TVL1(HOCTVL1) model is proposed and the alternating direction method of multipliers (ADMM) is used to solve this new model. Compared with existing models, our model can get higher signal-to-noise (SNR) in dealing with image deblurring under impulse noise. (2) The spatially adapted regularization parameter is introduced into the HOCTVL1 model and SAHOCTVL1 model is proposed. Compared to HOCTVL1 model, SAHOCTVL1 model can further improve the effects of image restoration in some degree.

The rest of this paper is organized as follows. In Section 2, a brief review of related work is made. In Section 3, the presentation of HOCTVL1 model is discussed, and the HOCTVL1 algorithm is concluded. Section 4 introduces the spatially adapted regularization parameter selection scheme, and SAHOCTVL1 model is proposed. Numerical experiments are carried out in Section 5 and finally, the conclusion is presented in Section 6.

2. Brief Review of Related Work

For recovering the image corrupted by blur and impulse noise, the classic method is TVL1 model. Since a lot of literature [19,20,31–33] demonstrates that using L1-fidelity term for image restoration under impulse noise can achieve good effects, the TVL1 model is expressed as

$$\min_{x} TV(\boldsymbol{x}) + \lambda \|\boldsymbol{Kx} - \boldsymbol{f}\|_1 \tag{4}$$

where $\boldsymbol{f}$ is the observed image, $\boldsymbol{x}$ denotes the restoration image, $\boldsymbol{K}$ is a blur matrix, λ is a regularization parameter which is greater than zero, $TV(\boldsymbol{x})$ represents the discrete TV norm and is defined as $TV(\boldsymbol{x}) = \sum_{1\leq i,j\leq n} \left\|(\boldsymbol{Dx})_{i,j}\right\|$. Here, $\boldsymbol{D}$ denotes the discrete gradient operator (under periodic boundary conditions). The norm in $\left\|(\boldsymbol{Dx})_{i,j}\right\|$ can be taken as ℓ_1 norm or ℓ_2 norm. When the ℓ_2 norm is used, the resulting TV term is isotropic and when the ℓ_1 norm is used, the result is anisotropic. For more details about the TV norm, readers can refer to [18].

Since one of the unique characteristics of impulse noise is that an image corrupted with impulse noise still has intact pixels, the impulse noise can be modeled as sparse components, whereas the underlying intact image retains the original image characteristics [34]. Therefore, TVL1 model can efficiently remove abnormal value noise signals, and some points of the solution of the TVL1 model are close to the points of the original image. However, Nikolova [21] pointed out from the viewpoint of MAP that the solutions of the TVL1 model substantially deviate from both the data-acquisition model and the prior model, and Minru Bai [23] further pointed out that the TVL1 model does not perform well at the sparsity of $\boldsymbol{Kx} - \boldsymbol{f}$ and there are many biased estimates produced by the TVL1 model. To overcome this shortcoming, Bai et al. took a correction step to generate an estimator to obtain a better recovery performance.

Given a reasonable initial estimator $\widetilde{x} \in R^{n^2}$ generated by TVL1 model, let $\widetilde{z} = K\widetilde{x} - f$, then she established a model called CTVL1 model, which is defined as

$$\min_{x,z} \sum_{1 \le i,j \le n} \left\| (Dx)_{i,j} \right\| + \lambda(\|z\|_1 - \langle F(\widetilde{z}), z \rangle) \quad s.t. \quad z = Kx - f \tag{5}$$

Compared with TVL1 model, CTVL1 model is added a correction term $-\langle F(\widetilde{z}), z \rangle$, and $F : R^{n^2} \rightarrow R^{n^2}$ is an operator defined as

$$F(z) = \begin{cases} \phi(\frac{z}{\|z\|_\infty}), z \in R^{n^2} \setminus \{0\}, \\ 0, \qquad z = 0, \end{cases} \tag{6}$$

and the scalar function $\phi : R \rightarrow R$ takes the form

$$\phi(t) = \mathrm{sgn}(t)(1 + \varepsilon^\tau)\frac{|t|^\tau}{|t|^\tau + \varepsilon^\tau}, \; \forall t \in \mathrm{R}, \tag{7}$$

where $\tau > 0$ and $\varepsilon > 0$. Numerical results show that the CTVL1 model improves the sparsity of the data fidelity term $Kx - f$ greatly for the images deblurring under impulse noise, to achieve a good denoising effect.

Since TV regularization term may come into staircase effects, in the past few years, a lot of researchers have devoted to solving this problem, and they concluded that replacing the TV norm by a high-order TV norm can get a better effect. The majority of the high-order norms involve second-order differential operators because piecewise-vanishing second-order derivatives lead to piecewise-linear solutions that better fit smooth intensity changes [35]. The second-order TV norm is defined as

$$(D^2x)_{i,j} = ((Dx)_{i,j}^{x,x}, (Dx)_{i,j}^{x,y}, (Dx)_{i,j}^{y,x}, (Dx)_{i,j}^{y,y}) \tag{8}$$

where $(Dx)_{i,j}^{x,x}, (Dx)_{i,j}^{x,y}, (Dx)_{i,j}^{y,x}, (Dx)_{i,j}^{y,y}$ denote the second-order difference of the $((j-1)\,n + i)$th entry of the vector x. Here we just briefly mention the concept of second-order TV norm, for more details, readers can refer to [36].

Figure 1 shows the diagram of image restoration. In this paper, two models named HOCTVL1 model and SAHOCTVL1 model are proposed and are used to recover the corrupted images.

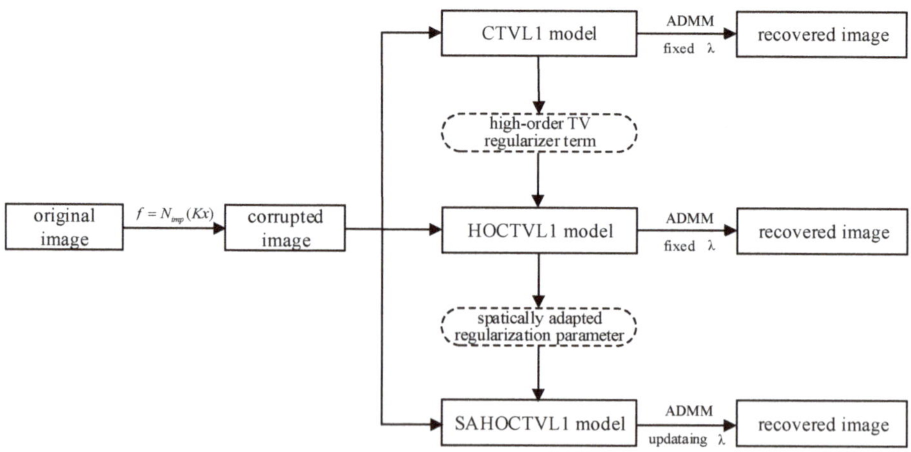

Figure 1. The block diagram of solving image deblurring problem under impulse noise.

3. New Method: The HOCTVL1 Algorithm

In this section, the HOCTVL1 model is proposed and the selection of $F(z)$ is talked about among CTVL1, SCAD and MCP models. Then ADMM is used to solve the proposed model and the HOCTVL1 algorithm is concluded.

3.1. Proposed New Model

Since the TV regularization norm $\left\|(Dx)_{i,j}\right\|$ can be taken as ℓ_1 norm or ℓ_2 norm, which is isotropic or anisotropic respectively. In [23–25,31], the authors all only consider the isotropic case, so in this paper, we will also only treat the isotropic case in detail, and the anisotropic case is similar to deal with. Based on this premise, the proposed high-order corrected TVL1 (HOCTVL1) model can be expressed as

$$\min_{x,z} \sum_{1\le i,j\le n} \alpha_{i,j}\left\|(Dx)_{i,j}\right\|_2 + (1-\alpha_{i,j})\left\|(D^2x)_{i,j}\right\|_2 + \lambda(\|z\|_1 - \langle F(\tilde{z}), z\rangle) \quad s.t. \quad z = Kx - f \tag{9}$$

where f is the corrupted image, x denotes the restoration image, K is a blur matrix, λ is a regularization parameter, $\left\|(Dx)_{i,j}\right\|_2$ denotes the first-order TV norm and $\left\|(D^2x)_{i,j}\right\|_2$ denotes the second-order TV norm, $-\langle F(\tilde{z}), z\rangle$ is a correction term, and $F(z)$ is an operator composed of a cluster of scalar functions.

$\alpha_{i,j}$ is a weighting parameter that discriminates the TV and second-order TV penalty, and there are several selection methods for the weighting parameter α. Here, the α in [31] is chosen since it can achieve better effects in experiments compared with the α in [37]. The α is expressed as

$$\alpha(i,j) = \begin{cases} 1, & \text{if } \left\|Dx_{i,j}^{k+1}\right\|_2 \ge c \\ \frac{1}{2}\cos\left(\frac{2\pi\left\|Dx_{i,j}^{k+1}\right\|_2}{c}\right) + \frac{1}{2}, & \text{else} \end{cases} \tag{10}$$

where c is a constant, and $0 \ll c < 1$.

About the selection of $F(z)$, in [24], Gu et al. used the SCAD penalty function for data fitting, and in [25], Gong et al. used the MCP penalty function for data fitting, the SCAD function $\xi(t)$ and MCP function $\varsigma(t)$ are described as

$$\xi(t) = \begin{cases} |t|, & \text{if } |t| \le \gamma_1 \\ \frac{-t^2+2\gamma_2|t|-\gamma_1^2}{2(\gamma_2-\gamma_1)}, & \text{if} \gamma_1 < |t| < \gamma_2 \\ \frac{\gamma_1+\gamma_2}{2}, & \text{if } |t| \ge \gamma_2 \end{cases} \tag{11}$$

$$\varsigma(t) = \begin{cases} \theta_1|t| - \frac{t^2}{2\theta_2}, & \text{if } |t| \le \theta_1\theta_2 \\ \frac{\theta_1^2\theta_2}{2}, & \text{if } |t| > \theta_1\theta_2 \end{cases} \tag{12}$$

where $\gamma_1, \gamma_2, \theta_1, \theta_2$ are all numbers greater than 0, and $0 \le t \le 1$.

It is easy to find that $\xi(t)$ and $\varsigma(t)$ are nonconvex and are difficult to solve. To solve this problem, Gu et al. adopted a difference of convex functions (DCA) algorithm to solve the nonconvex TVSCAD model. Similar to TVSCAD, Gong also adopted the DCA algorithm to solve the nonconvex TV-MCP model. The final processed functions $\varphi(t)$ in [24] and $\psi(t)$ [25] are respectively defined as

$$\varphi(t) = \begin{cases} 0, & \text{if } |t| \le \gamma_1 \\ \frac{t-\gamma_1 sgn(t)}{\gamma_2-\gamma_1}, & \text{if } \gamma_1 < |t| \le \gamma_2 \\ sgn(t), & \text{if } |t| > \gamma_2 \end{cases} \tag{13}$$

and

$$\psi(t) = \begin{cases} \frac{t}{\theta_2}, & \text{if } |t| \le \theta_1\theta_2 \\ \theta_1, & \text{if } |t| > \theta_1\theta_2 \end{cases} \tag{14}$$

Figure 2 shows the characteristics of $\phi(t)$, $\varphi(t)$ and $\psi(t)$. In TVL1 model, the regularization term is to enforce certain regularity conditions or prior constraints on the image, and the data fitting term penalizes the deviation of the observed data from the physical model. According to the analysis in Section 2 of [24], these three functions can all enforce less or even null data fitting and more regularization whenever $\boldsymbol{Kx}$ deviates significantly from $\boldsymbol{f}$. However, in the experimental simulation, it is found that the experimental results are almost the same no matter which function is selected. Therefore, in this paper, the scalar function still chooses $\phi(t)$.

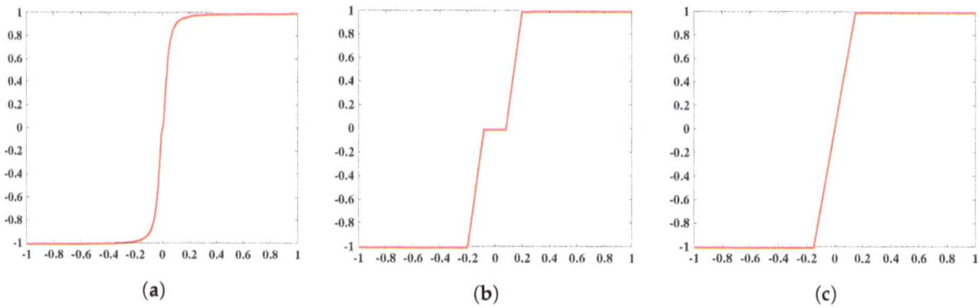

Figure 2. Plots of $\phi(t)$ in CTVL1, $\varphi(t)$ in TVSCAD and $\psi(t)$ in TV-MCP. (**a**) $\phi(t)$:$(\varepsilon^2, \tau) = (0.001, 2)$; (**b**) $\varphi(t)$:$(\gamma_1, \gamma_2) = (0.08, 0.2)$; (**c**) $\psi(t)$:$(\theta_1, \theta_2) = (1, 0.15)$.

3.2. The HOCTVL1 Algorithm

In this subsection, the solving procedure of the HOCTVL1 model by ADMM will be shown, and the HOCTVL1 algorithm will be concluded. About the details of ADMM, readers can refer to [38]. Firstly, let $y_{i,j} = (\boldsymbol{Dx})_{i,j}$, $w_{i,j} = (\boldsymbol{D}^2\boldsymbol{x})_{i,j}$, and $(i,j = 1,2,\ldots,n)$. Then Equation (9) can be rewritten as

$$\begin{aligned} &\min_{x,z} \sum_{1\le i,j\le n} \alpha_{i,j}\left\|(\boldsymbol{Dx})_{i,j}\right\|_2 + (1-\alpha_{i,j})\left\|(\boldsymbol{D}^2\boldsymbol{x})_{i,j}\right\|_2 \; + \lambda(\|\boldsymbol{z}\|_1 - \langle \boldsymbol{F}(\tilde{\boldsymbol{z}}), \boldsymbol{z}\rangle) \\ &s.t. \;\; y_{i,j} = \left\|(\boldsymbol{Dx})_{i,j}\right\|_2, \; w_{i,j} = \left\|(\boldsymbol{D}^2\boldsymbol{x})_{i,j}\right\|_2, \boldsymbol{z} = \boldsymbol{Kx} - \boldsymbol{f}, (i,j = 1,2,\ldots,n) \end{aligned} \tag{15}$$

where for each i, j, $y_{i,j} = ((y_1)_{i,j}, (y_2)_{i,j}) \in R^2$, $w_{i,j} = ((w_{11})_{i,j}, (w_{12})_{i,j}, (w_{21})_{i,j}, (w_{22})_{i,j}) \in R^4$, $\left\|y_{i,j}\right\|_2 = \sqrt{((y_1)_{i,j})^2 + ((y_2)_{i,j})^2}$, $\|w_{i,j}\|_2 = \sqrt{((w_{11})_{i,j})^2 + ((w_{12})_{i,j})^2 + ((w_{21})_{i,j})^2 + ((w_{22})_{i,j})^2}$.

Thus, the augmented Lagrangian function of Equation (15) is

$$\begin{aligned} L(\boldsymbol{x},\boldsymbol{y},\boldsymbol{z},\boldsymbol{w},\boldsymbol{\mu}_1,\boldsymbol{\mu}_2,\boldsymbol{\mu}_3) = &\sum_{i,j}\alpha_{i,j}\|y_{i,j}\|_2 - \boldsymbol{\mu}_1^T(\boldsymbol{y} - \boldsymbol{Dx}) + \frac{\beta_1}{2}\sum_{i,j}\left\|y_{i,j} - (\boldsymbol{Dx})_{i,j}\right\|_2^2 + \sum_{i,j}(1-\alpha_{i,j})\|w_{i,j}\|_2 \\ &- \boldsymbol{\mu}_2^T(\boldsymbol{w} - \boldsymbol{D}^2\boldsymbol{x}) + \frac{\beta_2}{2}\sum_{i,j}\left\|w_{i,j} - (\boldsymbol{D}^2\boldsymbol{x})_{i,j}\right\|_2^2 + \lambda(\|\boldsymbol{z}\|_1 - \langle \boldsymbol{F}(\tilde{\boldsymbol{z}}), \boldsymbol{z}\rangle) \\ &- \boldsymbol{\mu}_3^T(\boldsymbol{z} - (\boldsymbol{Kx} - \boldsymbol{f})) + \frac{\beta_3}{2}\|\boldsymbol{z} - (\boldsymbol{Kx} - \boldsymbol{f})\|_2^2 \end{aligned} \tag{16}$$

where $\mu_1 \in R^{2n^2}, \mu_3 \in R^{n^2}, \mu_2 \in R^{4n^2}$ are the Lagrangian multipliers, and $\beta_1, \beta_2, \beta_3 > 0$ are the penalty parameters. Then the ADMM for solving the model Equation (9) by updating x, y, w, z and λ as follows:

$$\begin{cases} y^{k+1} = \arg\min_y L(x^k, y, z^k, w^k, \mu_1^k, \mu_2^k, \mu_3^k) \\ w^{k+1} = \arg\min_w L(x^k, y^{k+1}, z^k, w, \mu_1^k, \mu_2^k, \mu_3^k) \\ z^{k+1} = \arg\min_z L(x^k, y^{k+1}, z, w^{k+1}, \mu_1^k, \mu_2^k, \mu_3^k) \\ x^{k+1} = \arg\min_x L(x, y^{k+1}, z^{k+1}, w^{k+1}, \mu_1^k, \mu_2^k, \mu_3^k) \\ \mu_1^{k+1} = \mu_1^k - \zeta\beta_1(y^{k+1} - Dx^{k+1}) \\ \mu_2^{k+1} = \mu_2^k - \zeta\beta_2(w^{k+1} - D^2x^{k+1}) \\ \mu_3^{k+1} = \mu_3^k - \zeta\beta_3(z^{k+1} - (Kx^{k+1} - f)) \end{cases} \tag{17}$$

where $\zeta > 0$ is the step length, and it can vary $(0, (\sqrt{5}+1)/2)$ [23]. For the y, w, z sub-problems, it is easy to get the scalar minimizer by using the soft thresholding, and for x sub-problem, it can be solved by fast Fourier transform (FFT) under periodic boundary conditions. Therefore, for y sub-problem, it can be obtained that

$$y_{i,j}^{k+1} = \max\left\{\left\|(Dx^k)_{i,j} + \frac{(\mu_1^k)_{i,j}}{\beta_1}\right\| - \frac{\alpha_{i,j}}{\beta_1}, 0\right\} \cdot \frac{(Dx^k)_{i,j} + (\mu_1^k)_{i,j}/\beta_1}{\left\|(Dx^k)_{i,j} + (\mu_1^k)_{i,j}/\beta_1\right\|_2} \tag{18}$$

here we assume the convention $0 \cdot (0/0) = 0, i, j = 1, 2, \ldots, n$. For the w sub-problem, there is

$$w_{i,j}^{k+1} = \max\left\{\left\|(D^2x^k)_{i,j} + \frac{(\mu_2^k)_{i,j}}{\beta_2}\right\| - \frac{1-\alpha_{i,j}}{\beta_2}, 0\right\} \cdot \frac{(D^2x^k)_{i,j} + (\mu_2^k)_{i,j}/\beta_2}{\left\|(D^2x^k)_{i,j} + (\mu_2^k)_{i,j}/\beta_2\right\|_2} \tag{19}$$

Taking account of z, there is

$$z^{k+1} = \max\left\{\left|Kx^k - f + \frac{\mu_3^k + \lambda F(\tilde{z})}{\beta_3}\right| - \frac{\lambda}{\beta_3}, 0\right\} \circ \operatorname{sgn}(Kx^k - f + \frac{\mu_3^k + \lambda F(\tilde{z})}{\beta_3}) \tag{20}$$

where $\circ$ denotes pointwise product. For x sub-problem, it can be solved by FFT and the result is shown as

$$x^{k+1} = \frac{D^T(\beta_1 y^{k+1} - \mu_1^k) + (D^2)^T(\beta_2 w^{k+1} - \mu_2^k) + K^T(\beta_3 z^{k+1} - \mu_3^k) + \beta_3 K^T f}{\beta_1 D^T D + \beta_2 (D^2)^T D^2 + \beta_3 K^T K} \tag{21}$$

Now, the HOCTVL1 Algorithm 1 can be concluded and is described as follows.

Algorithm 1: The HOCTVL1 algorithm

Input: $f, K, \lambda, \beta_1, \beta_2, \beta_3, \zeta, c, \delta_{tol}$, Maxiter

Initialization: $x^0 = f$, $\mu_1^0 = 0$, $\mu_2^0 = 0$, $\mu_3^0 = 0$, $k = 0$, $\alpha = 1$.

Step 1. Compute $y_{i,j}^{k+1}, w_{i,j}^{k+1}, z^{k+1}$ via Equations (18)–(20) respectively,

Step 2. Compute x^{k+1} by solving Equation (21),

Step 3. Update $\mu_1^{k+1}, \mu_2^{k+1}, \mu_3^{k+1}$, via Equation (17),

Step 4. Update $\alpha_{i,j}$ via Equation (10),

Step 5. If $k <$ Maxiter or $\left\|x^{k+1} - x^k\right\|_2 / \left\|x^k\right\|_2 > \delta_{tol}$, go to Step 1

Output: x^k.

4. SAHOCTVL1 Model

It is known that the regularization parameter λ controls the trade-off between the fidelity and the smoothness of the solution. Usually in most models, λ is a fixed value. In [39], Dong et al. developed a new automated spatially adapted regularization parameter selection method, and had a good effect on the Gaussian noise removal. In [27,40], the authors proposed a spatially adapted regularization parameter selection scheme for Poissonian image deblurring. In [31], the authors used the spatially adapted regularization parameter selection scheme for the impulse noise removal, while this method did not always show good results. In this section, the spatially adapted regularization parameter selection scheme described in [39] is adopted into the HOCTVL1 model, and SAHOCTVL1 algorithm is concluded.

Firstly, the SAHOCTVL1 model is defined as

$$\min_{x,z} \sum_{1\le i,j\le n} \left\|\alpha_{i,j}(\boldsymbol{D}\boldsymbol{x})_{i,j}\right\|_2 + (1-\alpha_{i,j})\left\|(\boldsymbol{D}^2\boldsymbol{x})_{i,j}\right\|_2 + (\|\boldsymbol{\lambda}\circ(\boldsymbol{z}-\boldsymbol{F}(\widetilde{\boldsymbol{z}})\circ\boldsymbol{z})\|_1 \;\; s.t. \;\; \boldsymbol{z}=\boldsymbol{K}\boldsymbol{x}-\boldsymbol{f} \tag{22}$$

where $\circ$ represents the pointwise product. Here, $\boldsymbol{\lambda}$ is a matrix as the same size of $\boldsymbol{f}$, and its all elements equal to one constant when we set its initial value.

As described in [39], the local window filter is defined as

$$\omega(a,b) = \begin{cases} \frac{1}{\omega^2}, & \text{if} \|b-a\|_\infty \le \frac{\omega}{2}, \\ 0, & \text{else}, \end{cases} \tag{23}$$

with $a \in \Omega$ fixed, and $\int_\Omega \omega(a,b)\mathrm{d}a\mathrm{d}b = 1$.

Let r represents the noise level, and ν represents the control constant for controlling the fidelity term. For the salt-and-pepper noise removal, we set

$$\nu = r/2 \tag{24}$$

and the $\boldsymbol{\lambda}$ updating rule is expressed as

$$(\widetilde{\lambda}^{p+1})_{i,j} = \eta\min((\widetilde{\lambda}^{p})_{i,j} + \tau\max(LEAVE_{i,j}-\nu,0),L) \tag{25}$$

$$(\widetilde{\lambda}^{p+1})_{i,j} = \frac{1}{\omega^2}\sum_{(s,t)\in\Omega^\omega_{i,j}} (\widetilde{\lambda}^{p+1})_{s,t} \tag{26}$$

where L is a large constant to ensure λ is finite, $1<\eta<2$, $\Omega^\omega_{i,j}$ is a local window with the center on (i,j), and $\tau = 2\|\lambda(:)\|_\infty/r$, $LEAVE_{i,j} = \frac{1}{\omega^2}\sum_{(s,t)\in\Omega^\omega_{i,j}} |\boldsymbol{K}\boldsymbol{x}-\boldsymbol{f}|_{s,t}$.

For the random-valued noise removal, the ν is defined as

$$\nu^\omega_{i,j} = \frac{1}{\omega^2}\sum_{(s,t)\in\Omega^\omega_{i,j}} r\cdot((\boldsymbol{K}\boldsymbol{x})^2_{s,t} - (\boldsymbol{K}\boldsymbol{x})_{s,t} + \frac{1}{2}) \tag{27}$$

Then the $\boldsymbol{\lambda}$ updating rule is expressed as

$$(\widetilde{\lambda}^{p+1})_{i,j} = \eta\min((\widetilde{\lambda}^{p})_{i,j} + \tau\cdot(LEAVE^\omega_{i,j}-\nu^\omega_{i,j})^+,L) \tag{28}$$

$$(\widetilde{\lambda}^{p+1})_{i,j} = \frac{1}{\omega^2}\sum_{(s,t)\in\Omega^\omega_{i,j}} (\widetilde{\lambda}^{p+1})_{s,t} \tag{29}$$

where the parameters η, L, τ, and $LEAVE_{i,j}$ are the same as before.

Now, the spatially adapted HOCTVL1 Algorithm 2 can be concluded.

Algorithm 2: Spatially adapted algorithm for solving the SAHOCTVL1 model

Input: $\boldsymbol{f}, \boldsymbol{K}, \boldsymbol{\lambda}, \beta_1, \beta_2, \beta_3, \zeta, c, \delta_{tol}$, Maxiter, r, ω, L.

Initialization: $x^0 = \boldsymbol{f}$, $\mu_1^0 = 0$, $\mu_2^0 = 0$, $\mu_3^0 = 0$, $k = 0$, $\alpha = 1$, $p = 0$.

Step 1. Solve the model Equation (22) by Algorithm 1, and get x^p,

Step 2. Update $\widetilde{\boldsymbol{\lambda}}^p$ via Equations (24)–(26) for salt-and-pepper noise,

Update $\widetilde{\boldsymbol{\lambda}}^p$ via Equations (27)–(29) for random-valued noise,

Step 3. stop or set $p = p + 1$ and return to Step 1.

Output: x^p.

5. Numerical Results

In this section, numerical results will be presented to illustrate the efficiency of the proposed models. Firstly the HOCTVL1 model is compared with TVL1 [20], HTVL1 [31], CTVL1 [23], then four state-of-the-art methods are selected for comparisons and the methods include LpTV-ADMM [26], the Adaptive Outlier Pursuit (AOP) method [41], the Penalty Decomposition Algorithm (PDA) [42], L0TV-PADMM [43]. It should be noted that we all only use HOCTVL1 model in these tests for comparison. In the last subsection, the efficiency of SAHOCTVL1 model will be compared with HOCTVL1 separately. In this section, the convergence of HOCTVL1 model is analyzed too. The test images are mainly: Lena, camera, pepper, boat, which are shown in Figure 3. In the experiments, for ease of comparison, we only consider "Gaussian" blurring kernel, since the model is also suitable for other blurring kernels. Besides, signal-to-noise ratio (SNR) is used to evaluate the quality of restoration, which is defined as

$$SNR(\boldsymbol{x}) = 10\log_{10}\frac{\|\widehat{\boldsymbol{x}} - E(\widehat{\boldsymbol{x}})\|_2^2}{\|\widehat{\boldsymbol{x}} - \boldsymbol{x}\|_2^2} \tag{30}$$

where $\widehat{x}$ and x denote the original and restored image respectively, and $E(\widehat{x})$ represents the mean of $\widehat{x}$. To evaluate the convergence rate, the running time of every algorithm is considered. For fairness, the stop criterion is the same among the algorithms mentioned in the experiments, which is expressed as

$$\frac{\left\|\boldsymbol{x}^{k+1} - \boldsymbol{x}^k\right\|_2^2}{\|\boldsymbol{x}^k\|_2^2} < \delta_{tol} \tag{31}$$

All experiments are operated under the Windows 10 and MATLAB R2018a with the platform Lenovo of Intel (R) Core (TM) i5-4200M CPU@2.50GHz 2.50 GHz made in Beijing, China.

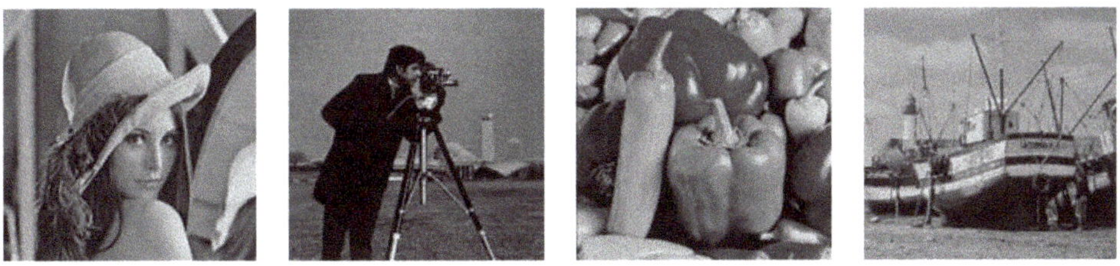

(a) (b) (c) (d)

Figure 3. Original images. First column: image name. Second column: image size. (**a**) Lena: 512 × 512; (**b**) camera: 256 × 256; (**c**) pepper: 512 × 512; (**d**) boat: 512 × 512.

5.1. Parameter Setting

In this subsection, we mainly define the values of some parameters in the experiments. In [23], the author set $(\beta_1, \beta_2) = (5, 350)$ for TVL1 model, and $(\beta_1, \beta_2) = (5, 20{,}000)$, $\zeta = 1.618$, $\tau = 2$ for CTVL1 model. In [31], the author set $(\beta_1, \beta_2, \beta_3) = (5, 10, 1000)$, $\eta = 1.1$, $c = 0.1$, the local window size $\omega = 21$ for HTVL1 model. In this paper, firstly we decide the value of β_3. Choose camera as test image and add salt-and-pepper noise with noise level 30%, the blurring kernel is Gaussian (hsize = 7, standard deviation = 5). When setting λ= 500 (not the most appropriate λ) and varying β_3 from 500 to 30,000, the trend of SNR of the restored image is shown in Figure 4. From Figure 4, it can be seen that with the increasing of β_3, SNR is also increasing, and when $\beta_3 > 20,000$, the trend keeps stable. In order to obtain good numerical results, in this paper, we set $\beta_3 = 25,000$.

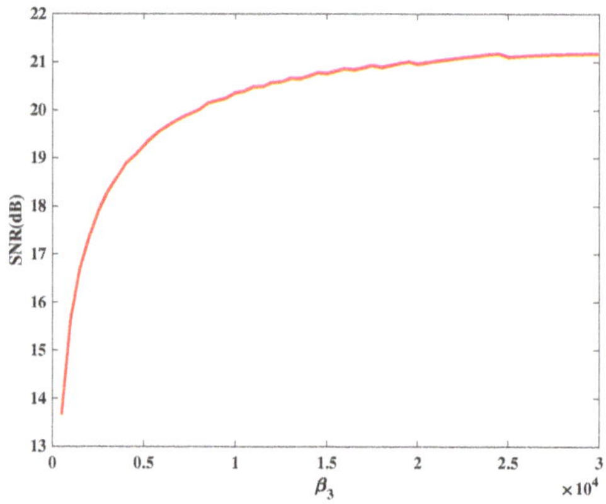

Figure 4. The SNR for results with different β_3.

Then considering the selection of λ, it is often a troublesome thing. In most cases, scholars obtain the appropriate λ through experience or a lot of attempts. In [24], Gong defined a selection scheme of λ based on numerical experiments, which is expressed as follows.

$$\lambda = \frac{c\lambda^*}{1-r} \tag{32}$$

where λ^* denotes the "best" λ found for TVL1 model, c is a constant and r represents the noise level. It means that we still need to struggle for the λ of TVL1 model. Through a large number of simulation experiments, we find that the difficulty in selecting λ mainly lies in the initial value of λ when the noise level is 10%, and as the noise increases, the value of λ decreases. Figure 5 shows the results of HOCTVL1 model and CTVL1 model corrupted by 10% salt-and-pepper noise with different λ and it can be seen that the appropriate λ for HOCTVL1 is almost the same with the λ for CTVL1. Therefore, similarly, we can adopt the λ for CTVL1 model in our HOCTVL1 model, and we set $\lambda = 800$ for impulse noise with noise level 10%.

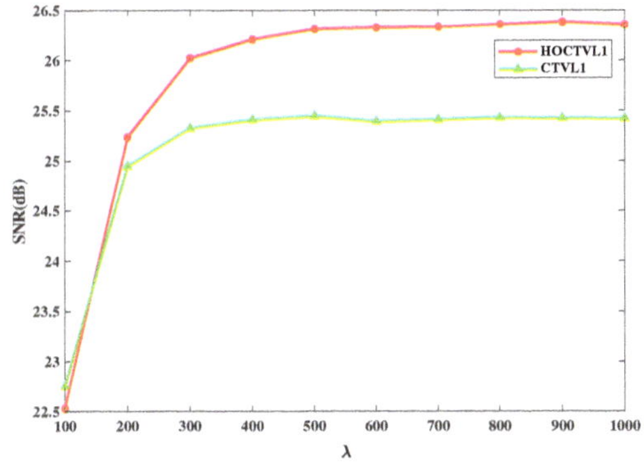

Figure 5. The SNR of HOCTVL1 and CTVL1 with different λ.

The size of local window ω is a factor that may influence the noise removal effect of SAHOCTVL1 model. In [33], the author illustrated through experiments that it can reduce more noise and recover more details when $\omega \geq 11$. Here, we also make an experiment. We choose camera as test image, and add 30% salt-and-pepper noise. When varying ω from 3 to 31, the SNR of the restored image is shown in Figure 6. From Figure 6, generally speaking, SNR does not change much, and when $\omega \geq 13$, SNR tends to be stable. Therefore, in this paper, we still set $\omega = 21$ and the other parameters are the same as mentioned before.

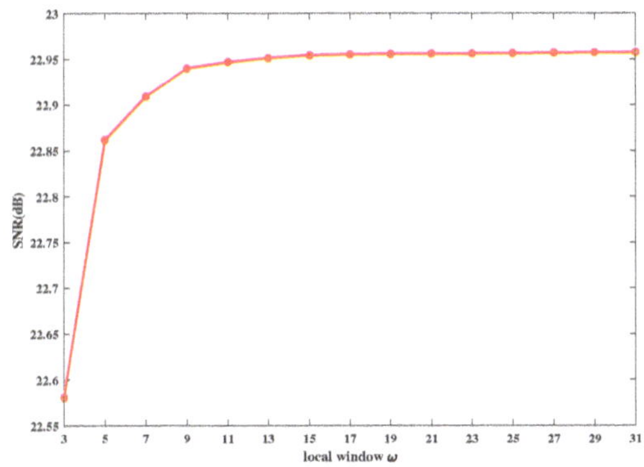

Figure 6. The SNR for results with different ω.

5.2. Convergence Analysis of HOCTVL1 Model

In this subsection, the convergence of HOCTVL1 algorithm will be analyzed. In [23], the author has proved that the CTVL1 model can converge to an optimal solution and its dual. Let $\left\{y^k, w^k, z^k, x^k, \mu_1^k, \mu_2^k, \mu_3^k\right\}$ be the iterative sequence generated by the ADMM approach, and set $Q_1(y) = \sum_{1\leq i,j\leq n} \|y_{i,j}\|_2$, $Q_2(w) = \sum_{1\leq i,j\leq n} \|w_{i,j}\|_2$, $Q_3(z) = \lambda(\|z\|_1 - \langle F(\tilde{z}), z\rangle)$. It is obvious that $Q_1 : R^{2n^2} \rightarrow R$, $Q_2 : R^{4n^2} \rightarrow R$, and $Q_3 : R^{n^2} \rightarrow R$ are closed proper convex functions. Then according to the subsection 4.3 of [23], it is easy for us to obtain the convergence result of HOCTVL1 model. Here, we verify the convergence property of HOCTVL1 model from another point of view. We observe the changes of SNR and $F(z)$ with the iterations by considering the camera image of size of 256×256 corrupted by Gaussian blur (hsize = 15, standard deviation = 5) and 50% salt-and-pepper noise, $\lambda = 500$, as are shown in Figure 7.

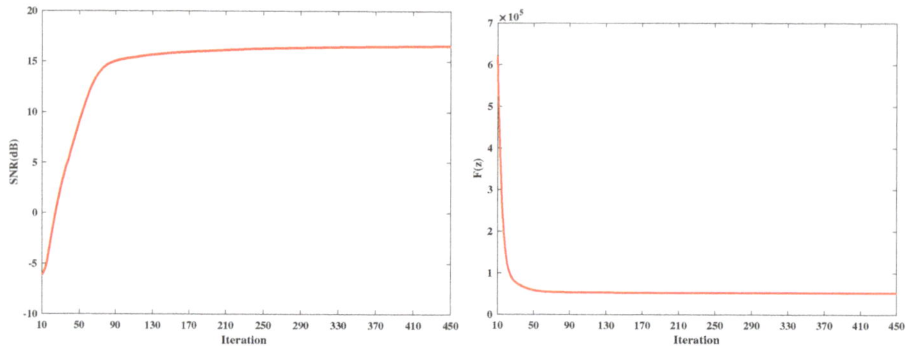

(a) (b)

Figure 7. Changes of SNR and $F(z)$ with the iterations for Camera image corrupted by Gaussian blur and 50% salt-and-pepper noise. (**a**) SNR; (**b**) $F(z)$.

It can be seen that the SNR value (or the function $\boldsymbol{F}(\boldsymbol{z})$) increases (or decreases) monotonically, which can demonstrate the convexity of the model. Besides, it can be seen that after 130th iteration, the SNR values keep stable, and the function remains unchanged after 70th iteration, which means that the model has converged to an optimal solution.

5.3. Comparisons of TVL1, HTVL1 and CTVL1 Models

In this subsection, some experiments are made to illustrate the superiority of HOCTVL1 model in removing the impulse noise and overcoming the staircase effects. By comparing the restoration effect of TVL1, HTVL1 and CTVL1 models, the superiority of our model is further illustrated. For CTVL1 model and HOCTVL1 model, the results of TVL1 model are used as the initial value, and the initial value is same. The blurring kernel is Gaussian (hsize = 15, standard deviation = 5). Next, the experiments will be carried out from three aspects: (1) Deblurring image under salt-and-pepper noise. (2) Deblurring image under random-valued noise. (3) Analysis of convergence rate.

5.3.1. For Salt-and-Pepper Noise

Firstly, the visual comparisons of Lena image corrupted by Gaussian blur and salt-and-pepper noise with noise levels 30%, 50%, 70% are carried out, and the results are shown in Figures 8–10 respectively. The unit of SNR value is dB.

(**a**) Corruption: 30% (**b**) SNR: 15.42 (**c**) SNR: 16.21 (**d**) SNR: 18.53 (**e**) SNR: 19.37

Figure 8. Comparisons of TVL1, HTVL1, CTVL1 and HOCTVL1 model on the Lena image corrupted by Gaussian blur and salt-and-pepper noise with noise level 30%. (**a**) Corrupted image. (**b**) TVL1 model. (**c**) HTVL1 model. (**d**) CTVL1 model. (**e**) HOCTVL1 model.

(**a**) Corruption: 50% (**b**) SNR: 14.82 (**c**) SNR: 15.14 (**d**) SNR: 18.01 (**e**) SNR: 18.61

Figure 9. Comparisons of TVL1, HTVL1, CTVL1 and HOCTVL1 model on the Lena image corrupted by Gaussian blur and salt-and-pepper noise with noise level 50%. (**a**) Corrupted image. (**b**) TVL1 model. (**c**) HTVL1 model. (**d**) CTVL1 model. (**e**) HOCTVL1 model.

(**a**) Corruption: 70% (**b**) SNR: 12.39 (**c**) SNR: 13.59 (**d**) SNR: 17.43 (**e**) SNR: 17.71

Figure 10. Comparisons of TVL1, HTVL1, CTVL1 and HOCTVL1 model on the Lena image corrupted by Gaussian blur and salt-and-pepper noise with noise level 70%. (**a**) Corrupted image. (**b**) TVL1 model. (**c**) HTVL1 model. (**d**) CTVL1 model. (**e**) HOCTVL1 model.

From Figures 8–10, it can be found that for noise levels 30% and 50%, four models can all remove the salt-and-pepper noise effectively, but the quality of the restored images is different. The restored image by HOCTVL1 model is closer to the original image and its SNR is the highest. For noise level 70%, the restored image by TVL1 model is not clear. The restored image by HTVL1 model is clearer than the restored image by TVL1 model, but there are some noise points in the image that have not been removed, while both CTVL1 model and HOCTVL1 model can get a good restored result. By comparing the results of TVL1 model and HTVL1 model, the image quality can be indeed improved by introducing the high-order TV regularizer term. However, since the poor performance of TVL1 model, the results of HTVL1 is not very good. Since an adaptive correction procedure is introduced in CTVL1 model, which can greatly enhance the effect of image deblurring. While we combine the CTVL1 model with second-order TV regularizer term, which can further improve the effect of image deblurring. Both for removing the salt-and-pepper noise with noise level from 30% to 70%, HOCTVL1 model can have a great performance. It cannot only provide a very good visual effect, but also achieve a higher SNR. Especially for noise level 30%, the SNR value of the restored image by HOCTVL1 model is more than 1 dB higher than that by CTVL1 model and for noise level 50%, the SNR value obtained by HOCTVL1 model is also about 0.6 dB higher than CTVL1 model.

For noise level 90%, HOCTVL1 model can also use the step correction to improve the removal effect, as shown in Figure 11. Figure 11 shows the results of CTVL1 model and HOCTVL1 model during five correction steps. It can be seen that after several correction steps, two models both can improve the effect of image deblurring though the effect of TVL1 model is worse. However, from first correction step to fifth correction step, the SNR of our HOCTVL1 model is always higher than CTVL1 model. After first correction step, the SNR of recovered image by HOCTVL1 model is about 0.6 dB higher than CTVL1 model. Meanwhile, after three correction steps, the SNR of the restored image keeps stable, and the noise is eliminated, which shows that the correction efficiency of HOCTVL1 model is very high.

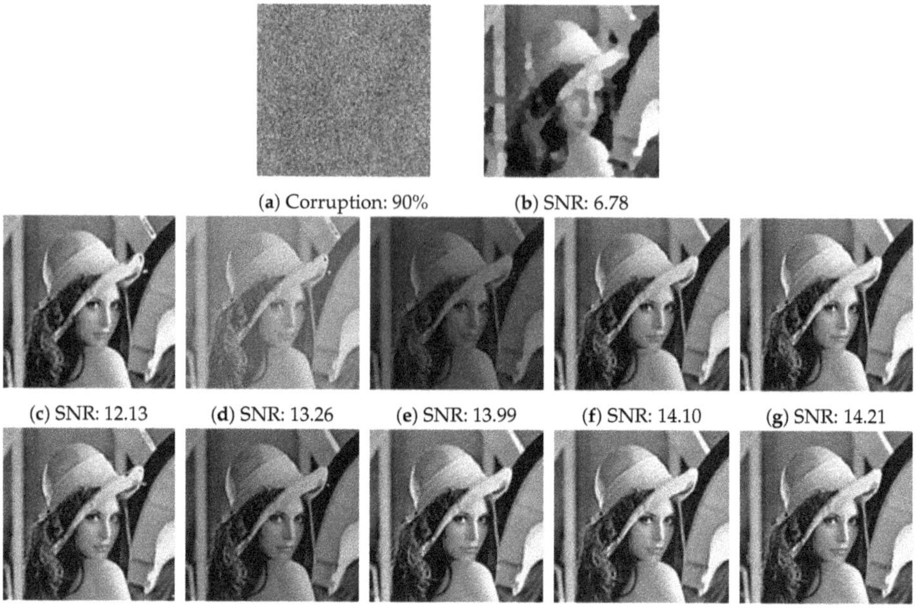

(a) Corruption: 90% (b) SNR: 6.78

(c) SNR: 12.13 (d) SNR: 13.26 (e) SNR: 13.99 (f) SNR: 14.10 (g) SNR: 14.21

(h) SNR: 12.77 (i) SNR: 14.19 (j) SNR: 14.50 (k) SNR: 14.51 (l) SNR: 14.53

Figure 11. Restored images of TVL1, CTVL1 and HOCTVL1 models on the Lena image corrupted by Gaussian blur and salt-and-pepper noise with noise level 90%. (**a**) Corrupted image. (**b**) TVL1 model. (**c–g**) first correction step to fifth correction step of CTVL1 model. (**h–l**) first correction step to fifth correction step of HOCTVL1 model.

Table 1 shows the results of the four models for restoring the corrupted images with noise levels 10%, 30%, 50%, 70% and 90%. The test images are what are shown in Figure 3. It should be noted that the values of CTVL1 and HOCTVL1 model for noise level 90% are the SNR of the restored images after first correction step. From Table 1, it can be seen that compared with other three models, HOCTVL1 model can achieve a higher SNR value. It can also be seen that there is a great improvement in HOCTVL1 model compared with TVL1 and HTVL1 model no matter what the noise level is. Compared with CTVL1 model, there is about 1 dB higher in restoring the Lena, pepper and boat images when noise levels are 10% and 30%. Even for recovering camera image, the SNR of the restored image by HOCTVL1 model is still at least 0.5 dB higher than CTVL1 model when noise levels are 10% and 30%. For noise level 90%, the SNR of our HOCTVL1 model is the highest among the four model, though there is only a slight improvement in the camera image.

Table 1. SNR of four different models for test images corrupted by Gaussian blur and salt-and-pepper noise.

Image	Noise Level	SNR(dB)			
		TVL1	HTVL1	CTVL1	HOCTVL1
Lena	10%	15.68	16.83	18.93	19.79
	50%	14.82	15.14	18.01	18.60
	70%	12.39	13.59	17.43	17.71
	90%	7.32	6.77	12.13	12.77
Camera	10%	12.95	13.64	16.27	16.92
	30%	12.30	12.99	15.73	16.21
	50%	11.48	11.79	14.79	15.28
	70%	10.26	10.27	12.44	12.54
	90%	5.73	5.56	8.08	8.20
Pepper	10%	19.40	21.65	25.08	26.00
	30%	18.96	21.01	24.24	25.43
	50%	17.98	19.31	23.65	24.63
	70%	16.14	16.48	22.77	23.25
	90%	7.90	7.12	13.16	14.75
Boat	10%	15.26	16.54	19.03	20.02
	30%	14.78	16.01	18.58	19.52
	50%	13.63	14.58	18.08	18.69
	70%	12.30	12.51	17.26	17.69
	90%	6.79	6.13	10.99	11.36

5.3.2. For Random-Valued Noise

For the sake of testing the performance of HOCTVL1 model in removing random-valued noise, we also carry out a series of experiments. Figures 12 and 13 show the visual comparisons of Lena image corrupted by Gaussian blur and random-valued with noise levels 30% and 50%. It can be seen that for noise level 30%, both four models can effectively restore the corrupted image, but the image restored by TVL1 is still somewhat blurred, the images recovered by other three models are clearer and the recovered image by HOCTVL1 model is closer to the original image and its SNR is the highest. For noise level 50%, there is a little noise in Figure 13b,c, which shows that TVL1 and HTVL1 models cannot completely remove the 50% random-valued noise though HTVL1 model has a better performance. While CTVL1 model and HOCTVL1 model can remove noise very well, and the recovered image by HOCTVL1 model is clearer than that by CTVL1 model, which illustrates that high-order regularizer term can effectively restrain the staircase effects. Meanwhile, the value of SNR of HOCTVL1 model also illustrates this point.

(**a**) Corruption: 30% (**b**) SNR: 15.41 (**c**) SNR: 16.25 (**d**) SNR: 18.48 (**e**) SNR: 19.32

Figure 12. Comparisons of TVL1, HTVL1, CTVL1 and HOCTVL1 model on the Lena image corrupted by Gaussian blur and random-valued noise with noise level 30%. (**a**) Corrupted image. (**b**) TVL1 model. (**c**) HTVL1 model. (**d**) CTVL1 model. (**e**) HOCTVL1 model.

(a) Corruption: 50% (b) SNR: 14.24 (c) SNR: 14.95 (d) SNR: 17.94 (e) SNR: 18.62

Figure 13. Comparisons of TVL1, HTVL1, CTVL1 and HOCTVL1 model on the Lena image corrupted by Gaussian blur and random-valued noise with noise level 50%. (**a**) Corrupted image. (**b**) TVL1 model. (**c**) HTVL1 model. (**d**) CTVL1 model. (**e**) HOCTVL1 model.

Figure 14 shows the results of TVL1, five correction steps of CTVL1 model and eighth correction steps of HOCTVL1 model for removing the random-valued noise with noise level 70%. Though the performance of TVL1 model is not good, after several correction steps, both CTVL1 model and HOCTVL1 model can improve the restoration effect. Meanwhile, it can be found that the SNR of the new model is always higher than CTVL1 model after the same correction step. Besides, after the eighth correction step, there are only a few noise points in the image and the restored image is very clear, which shows the superiority of HOCTVL1 model.

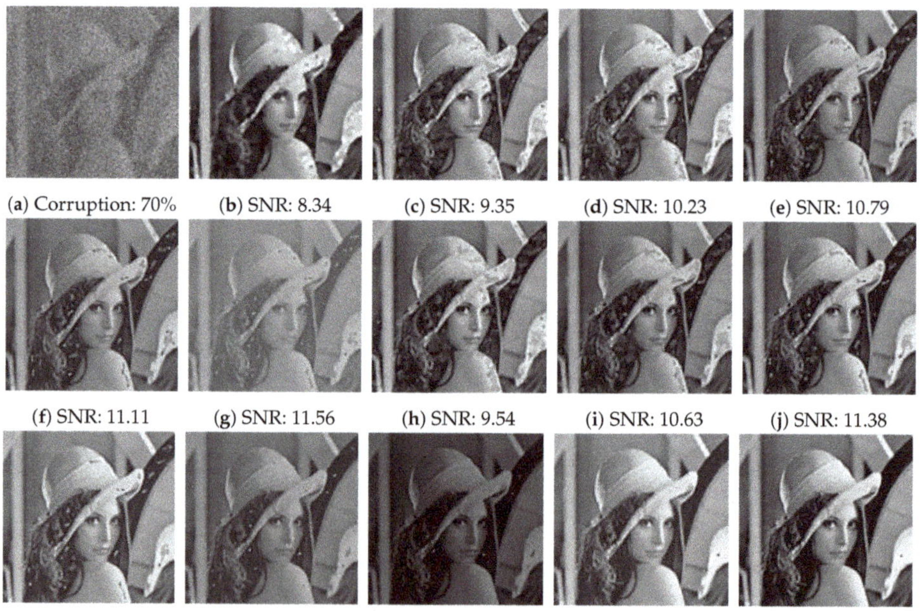

(a) Corruption: 70% (b) SNR: 8.34 (c) SNR: 9.35 (d) SNR: 10.23 (e) SNR: 10.79

(f) SNR: 11.11 (g) SNR: 11.56 (h) SNR: 9.54 (i) SNR: 10.63 (j) SNR: 11.38

(k) SNR: 11.95 (l) SNR: 13.08 (m) SNR: 13.89 (n) SNR: 14.55 (o) SNR: 14.97

Figure 14. Restored images of TVL1, CTVL1 and HOCTVL1 model on the Lena image corrupted by Gaussian blur and random-valued noise with noise level 70%. (**a**) Corrupted image. (**b**) TVL1 model. (**c–g**) first correction step to fifth correction step of CTVL1 model. (**h–o**) first correction step to eighth correction step of HOCTVL1 model.

Table 2 shows the results of the four models for restoring the corrupted images with noise levels 10%, 30%, 50%, 70%. The values of CTVL1 and HOCTVL1 models for noise level 70% in Table 2 are also the SNR of the restored images after first correction step. The values show the superiority of our model for removing the random-valued noise. Different to removing salt-and-pepper noise, there is only a little improvement compared to CTVL1 model for removing noise as high as 70%. But when the noise level is lower than 70%, the improvement is remarkable.

Table 2. SNR of four different models for test images corrupted by Gaussian blur and random-valued noise.

Image	Noise Level	SNR(dB)			
		TVL1	HTVL1	CTVL1	HOCTVL1
Lena	10%	15.70	16.82	18.87	19.68
	30%	15.41	16.25	18.48	19.32
	50%	14.24	14.95	17.94	18.62
	70%	8.34	8.28	9.35	9.54
Camera	10%	12.97	13.64	16.22	16.76
	30%	12.18	12.88	15.80	16.20
	50%	9.35	9.37	11.20	11.46
	70%	3.30	3.60	3.63	3.79
Pepper	10%	19.41	21.57	25.00	25.95
	30%	18.94	20.88	24.24	25.35
	50%	16.23	16.88	20.88	21.40
	70%	7.58	7.12	8.63	8.80
Boat	10%	15.26	16.48	19.01	19.90
	30%	14.76	15.94	18.55	19.47
	50%	12.90	13.62	17.06	18.31
	70%	6.57	7.06	7.88	7.98

5.3.3. Analysis of Convergence Rate

Now, we analyze the convergence rate of four models. We choose Lena as the test image, and use the running time to evaluate the convergence rate. Figure 15 shows the time that four models spend restoring the corrupted Lena image under impulse noise with different level. It can be seen that when dealing with the same noise, TVL1 and CTVL1 model cost relatively less time. Because of the combination of high-order TV regularizer term, which increases the computational complexity of the algorithm, HTVL1 and HOCTVL1 models consume more time compared to TVL1 and CTVL1 model. While HOCTVL1 model can effectively reduce the staircase effect and restore more details, it is worthwhile taking more time. Figure 16 shows the change of $F(z)$ with the iteration number when dealing with the salt-and-pepper noise with noise levels 30% and 50%. It can be seen that the convergence rate of HOCTVL1 model is slower than CTVL1 model, and the iteration number is about twice as much as that of CTVL1 model.

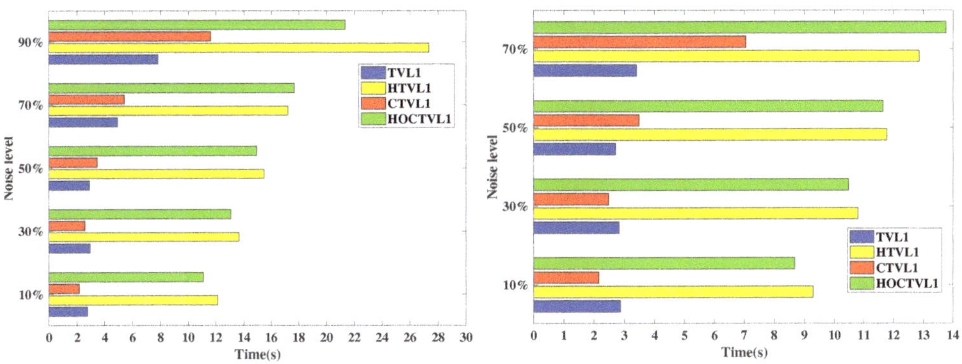

(a) (b)

Figure 15. Running time of four different models for Lena corrupted by Gaussian blur and impulse noise. (**a**) Salt-and-pepper noise; (**b**) Random-valued noise.

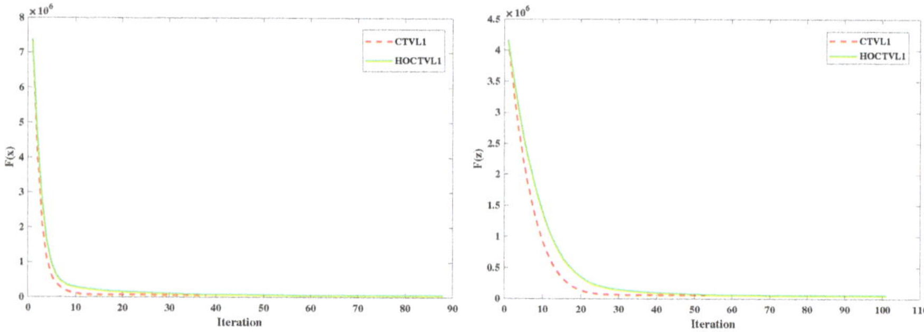

(**a**) (**b**)

Figure 16. Change of $F(z)$ value with Iteration number. (**a**) Corruption: 30%; (**b**) Corruption: 50%.

5.4. Comparisons of Some Other Methods

In this subsection, we compare the effect of the HOCTVL1 model with some other methods for image deblurring under impulse noise, mainly include: LpTV-ADMM [26], AOP [41], PDA [42] and L0TV-PADMM [43]. Since in [23], the author has shown the superiority of CTVL1 model by numerical experiments compared with two-phase method, in this subsection, we do not consider two-phase method. In this experiment, for ease of comparison, we choose "Gaussian" blurring kernel with hsize =9 and standard deviation =7, which is same with [43] and the parameter settings of these methods also obey to the related papers and readers can refer to them for details.

Firstly, we show the visual results of the pepper image corrupted by salt-and-pepper noise and random-valued noise with noise level 50% respectively, as are shown in Figures 17 and 18.

(**a**) Corruption: 50% (**b**) SNR: 14.33 (**c**) SNR: 16.97

(**d**) SNR: 15.95 (**e**) SNR: 26.62 (**f**) SNR: 30.64

Figure 17. Recovered images on Pepper image corrupted by salt-and-pepper noise with noise level 50%. (**a**) Corrupted image. (**b**) Lp-ADMM. (**c**) AOP. (**d**) PDA. (**e**) L0TV-PADMM. (**f**) HOCTVL1.

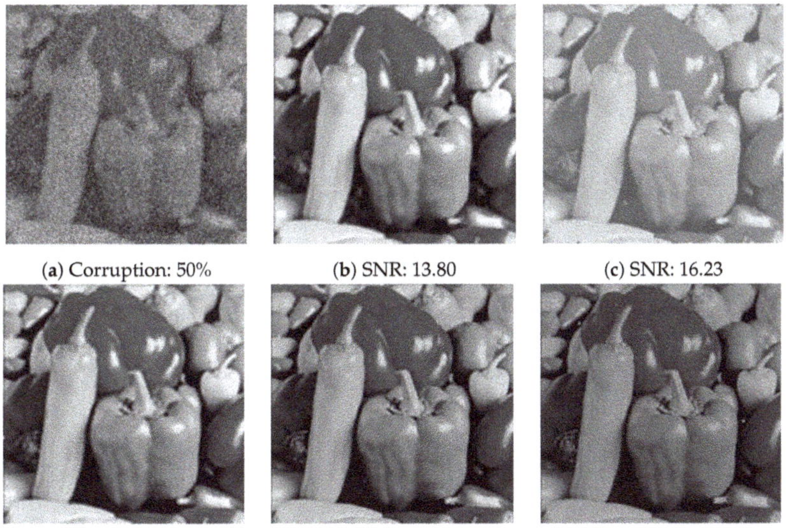

(**a**) Corruption: 50% (**b**) SNR: 13.80 (**c**) SNR: 16.23

(**d**) SNR: 15.45 (**e**) SNR: 20.82 (**f**) SNR: 22.94

Figure 18. Recovered images on Pepper image corrupted by random-valued noise with noise level 50%. (**a**) Corrupted image. (**b**) Lp-ADMM. (**c**) AOP. (**d**) PDA. (**e**) L0TV-PADMM. (**f**) HOCTVL1.

From Figures 17 and 18, it can be seen that the restoration effect of HOCTVL1 model is very remarkable. It is obvious that HOCTVL1 model has the highest SNR, followed by PDA and AOP methods, and the Lp-ADMM method has the lowest SNR. When removing 50% salt-and-pepper noise, as shown in Figure 17, compared with Lp-ADMM method, the SNR of (f) is more than twice as (b). Compared with L0-PADMM method, the SNR of our model is also 4 dB higher. When removing 50% random-valued noise, compared with L0-PADMM method, our model has only about 2 dB improvement in SNR, which is less than that in removing salt-and-pepper noise. But similarly, compared with Lp-ADMM method, our model has more than 100% improvement.

Table 3 shows the results of the five methods for restoring the corrupted images by impulse noise with different noise level, respectively. The value on the left of "/" represents the result after the first correction step and the value on the right of "/" represents the result after multi-correction steps. For removing salt-and-pepper noise, it is obvious that Lp-ADMM and PDA methods perform poorly, and when noise level is 90%, Lp-ADMM method has the worst effect. When the noise level varies from 30% to 70%, HOCTVL1 model has the highest SNR in most cases, except the SNR of L0TV-PADMM when dealing with the camera image with noise level 70%. It can also be seen that L0TV-PADMM method has the best restoration effect and its SNR value is higher than our model when the noise level is 90%. For dealing with random-valued noise, it can be seen that our model has the highest SNR when noise level varies from 30% to 50%. Similarly, for noise level 70%, L0TV-PADMM method has certain advantages; however, it can be seen from Lena and boat images that our model can achieve a higher SNR than L0TV-PADMM method after multi-correction steps.

Table 3. SNR of five methods for test images corrupted by Gaussian blur and impulse noise.

Image	Algorithm	Salt-and-Pepper Noise				Random-Valued Noise		
		30%	50%	70%	90%	30%	50%	70%
Lena	Lp-ADMM	14.18	12.72	8.25	2.21	14.24	12.79	7.21
	AOP	14.75	14.34	13.79	13.18	14.49	14.03	5.56
	PDA	14.22	13.75	12.96	9.12	14.07	13.41	11.77
	L0TV-PADMM	19.41	18.62	17.46	15.02	18.14	16.92	15.31
	HOCTVL1	23.66	22.19	18.52	11.05/13.82	23.49	19.44	9.27/17.24
Camera	Lp-ADMM	9.46	8.37	3.92	−0.43	9.43	7.54	2.68
	AOP	11.32	10.63	9.52	8.37	11.08	8.65	2.57
	PDA	9.91	9.57	8.98	5.86	9.64	8.92	6.54
	L0TV-PADMM	15.57	14.32	12.63	9.74	13.16	12.35	10.53
	HOCTVL1	22.27	17.89	12.55	6.22/6.46	21.67	14.79	3.65/9.51
Pepper	Lp-ADMM	16.63	14.33	7.80	1.92	16.75	13.80	6.52
	AOP	17.33	16.97	16.31	15.32	17.24	16.23	5.74
	PDA	16.53	15.95	14.91	9.54	16.30	15.45	13.10
	L0TV-PADMM	26.18	24.62	21.55	17.53	22.66	20.82	18.00
	HOCTVL1	32.07	30.64	22.58	9.80/11.25	30.34	22.94	8.06/16.09
Boat	Lp-ADMM	12.28	11.13	7.16	1.99	12.22	10.95	6.18
	AOP	13.70	12.80	12.24	11.43	13.11	12.51	5.43
	PDA	12.38	11.91	11.18	8.29	12.32	11.76	10.43
	L0TV-PADMM	19.45	18.12	16.58	13.26	17.38	15.76	14.04
	HOCTVL1	24.21	22.55	18.15	9.25/10.85	23.84	18.46	7.91/15.45

Figure 19 shows the running time of the five methods for restoring the corrupted pepper image. When dealing with 90% salt-and-pepper noise and 70% random-valued noise, HOCTVL1 model needs multi-correction steps, which costs a lot time, here we only show the time of removing salt-and-pepper noise as high as 70% and random-valued noise as high as 50%. It can be seen that compared to other four methods, our HOCTVL1 model spends the least time whatever the noise level is. It can also be found that the other four methods take several times as much time as our model, which illustrates the advantages of our model.

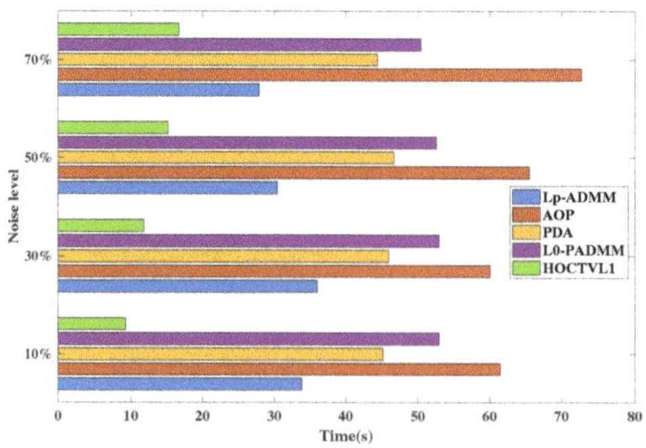

(a)

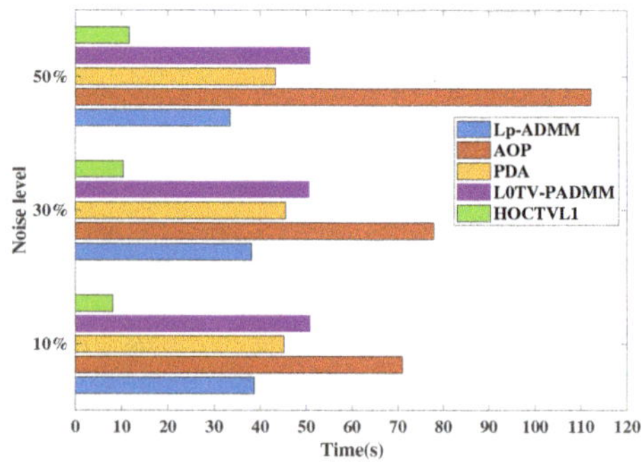

(b)

Figure 19. Running time of five methods for pepper corrupted by Gaussian blur and impulse noise. (**a**) Salt-and-pepper noise; (**b**) Random-valued noise.

5.5. Comparisons between SAHOCTVL1 Model and HOCTVL1 Model

In this subsection, the restoration effect of SAHOCTVL1 model will be analyzed. We choose camera and Lena as test images, we use "Gaussian" blurring with hsize = 15 and standard deviation = 5, and we set the spatially adapted iteration number $p \leq 3$. Since we have shown the superiority of HOCTVL1 model by many simulation experiments, here we only compare the effect of SAHOCTVL1 model and HOCTVL1 model on image restoration. Meanwhile, since we have shown the huge advantage of HOCTVL1 model compared to HTVL1 model, while the effect of SAHTVL1 model is similar to HTVL1 model, therefore we do not consider SAHTVL1 model in [31] either. In this subsection we do not consider the 90% salt-and-pepper noise and 70% random-valued noise since the multi-correction steps take much time. We evaluate the results by SNR and running time, and the restoration results are shown in Table 4.

Table 4. Comparisons of SAHOCTVL1 and HOCTVL1 model for Cameraman and Lena corrupted by Gaussian blur and impulse noise.

Image	Noise Type	Noise Level	SNR(dB)		Time(s)	
			HOCTVL1	SAHOCTVL1	HOCTVL1	SAHOCTVL1
Camera	SP	10%	16.85	16.98	2.32	6.76
		30%	16.18	16.51	2.67	8.20
		50%	15.33	15.72	3.02	9.32
		70%	12.50	12.57	5.15	15.47
	RV	10%	16.73	16.78	2.06	6.07
		30%	16.22	16.34	2.30	7.08
		50%	11.39	11.99	3.02	8.74
Lena	SP	10%	19.78	19.82	11.19	31.45
		30%	19.35	19.44	13.97	39.27
		50%	18.63	18.78	16.13	47.12
		70%	17.72	17.85	18.82	55.57
	RV	10%	19.68	19.72	9.18	30.60
		30%	19.29	19.32	11.39	36.77
		50%	18.56	18.68	12.07	41.35

SP denotes salt-and-pepper noise and RV denotes random-valued noise.

From Table 4, as is shown, generally speaking, SAHOCTVL1 model can achieve at least the same effect as HOCTVL1 model. For camera image, when noise is 30% and 50% salt-and-pepper noise, the SNR of SAHOCTVL1 model is about 0.4 dB higher than HOCTVL1 model and when noise is 50% random-valued noise, there is 0.6 dB higher than HOCTVL1 model. For Lena image, the advantage of SAHOCTVL1 model is little, and the SNR improves by about 0.1 dB when noise levels are 50% and 70%. Meanwhile, because we set $p \leq 3$, which makes the algorithm run 3 times, making the time SAHOCTVL1 model takes be about 3 times as much as that HOCTVL1 model takes. But we think it is worthwhile to obtain high SNR at the expense of running time.

6. Conclusions

This paper gives a contribution to solving the problem of image deblurring under impulse noise, and two models named HOCTVL1 model and SAHOCTVL1 model are proposed. Benefitting from the merits of high-order TV regularizer term and spatially adapted regularization parameter selection scheme, both models perform well in recovering the corrupted images. A great quantity of experiments is carried out to show the superiority of the two models. Compared to CTVL1 model, HOCTVL1 model can achieve better visual effects and higher SNR values. When dealing with salt-and-pepper noise with noise level less than 90% and random-valued noise with noise level less than 70%, there is about 0.5$\sim$1 dB improvement. When dealing with 90% salt-and-pepper noise and 70% random-valued noise, multi-correction steps are used to improve the restoration quality. HOCTVL1 model outperforms CTVL1 model both in the SNR value of each correction step and the number of correction steps they need. Compared to four other state-of-the-art methods, HOCTVL1 model always outperforms Lp-ADMM, AOP, and PDA methods. Compared to L0TV-PADMM method, HOCTVL1 model performs well in most cases and it can achieve about 1∼4 dB improvement. When dealing with 90% salt-and-pepper noise and 70% random-valued noise, L0TV-PADMM method performs well, while after several correction steps, HOCTVL1 model can obtain higher SNR value in some cases and HOCTVL1 model takes less time. In the last experiment, the comparisons of HOCTVL1 model and SAHOCTVL1 model are conducted and the results show that SAHOCTVL1 can achieve about 0.1∼0.6 dB improvement compared to HOCTVL1 model. However, it takes about three times as long as HOCTVL1 model, which is a problem that needs to be optimized in a future study.

Author Contributions: All authors have made great contributions to the work. J.X., P.Y., L.W. and M.H. conceived and designed the experiments; P.Y. performed the experiments and analyzed the data; J.X. and P.Y. gave insightful suggestions for the work; P.Y. wrote the paper.

Funding: This paper is supported by National Key Laboratory of Communication Anti-jamming191 Technology (614210202030217), Heilongjiang Education and Teaching Reform Project Funding in 2017 (SJGY20170515 and SJGY20170514), project support for Free Exploration of basic Scientific Research Business expenses in Central Colleges and Universities.

Conflicts of Interest: The authors declare no conflict of interest.

References

1. Tiwari, K.A.; Raisutis, R.; Tumsys, O. Defect Estimation in Non-Destructive Testing of Composites by Ultrasonic Guided Waves and Image Processing. *Electronics* **2019**, *8*, 315. [CrossRef]
2. Turajlic, E. Adaptive Block-Based Approach to Image Noise Level Estimation in the SVD Domain. *Electronics* **2018**, *7*, 397. [CrossRef]
3. Chervyakov, N.; Lyakhov, P.; Kaplun, D. Analysis of the Quantization Noise in Discrete Wavelet Transform Filters for Image Processing. *Electronics* **2018**, *7*, 135. [CrossRef]
4. Xu, J.; Tai, X.; Wang, L. A two-level domain decomposition method for image restoration. *Inverse Probl.* **2017**, *4*, 523–545. [CrossRef]
5. Guo, Z.; Sun, Y.; Jian, M.; Zhang, X. Deep Residual Network with Sparse Feedback for Image Restoration. *Appl. Sci.* **2018**, *8*, 2417. [CrossRef]
6. Orgiela, L.; Tadeusiewicz, R.; Ogiela, M.R. Cognitive analysis in diagnostic DSS-type IT systems. In Proceedings of the Artificial Intelligence and Soft Computing, Zakopane, Poland, 12–16 June 2006; pp. 962–971.
7. Simons, T.; Lee, D.J. Jet Features: Hardware-Friendly, Learned Convolutional Kernels for High-Speed Image Classification. *Electronics* **2019**, *8*, 588. [CrossRef]
8. Sun, G.; Leng, J.; Huang, T.J.I.A. An Efficient Sparse Optimization Algorithm for Weighted ℓ_0 Shearlet-Based Method for Image Deblurring. *IEEE Access* **2017**, *5*, 3085–3094. [CrossRef]
9. Xiang, J.H.; Yue, H.H.; Yin, X.J. A Reweighted Symmetric Smoothed Function Approximating L0 Norm Regularized Sparse Reconstruction Method. *Symmetry* **2018**, *10*, 583. [CrossRef]
10. Wang, L.Y.; Yin, X.J.; Yue, H.H. A Regularized Weighted Smoothed L0 Norm Minimization Method for Underdetermined Blind Source Separation. *Sensors* **2018**, *18*, 4260. [CrossRef]
11. Ma, X.; Hu, S.; Liu, S. Remote Sensing Image Fusion Based on Sparse Representation and Guided Filtering. *Electronics* **2019**, *8*, 303. [CrossRef]
12. Kittisuwan, P. Speckle Noise Reduction of Medical Imaging via Logistic Density in Redundant Wavelet Domain. *Int. J. Artif. Intell. Tools* **2018**, *27*, 1850006. [CrossRef]
13. Zhang, Y.D.; Zhang, Y.; Dong, Z.C.; Yuan, T.F.; Han, L.X.; Yang, M.; Carlo, C.; Lu, H.M. Advanced Signal Processing Methods In Medical Imaging. *IEEE Access* **2018**, *6*, 61812–61818. [CrossRef]
14. Vorontsov, S.; Jefferies, S. A new approach to blind deconvolution of astronomical images. *Inverse Probl.* **2017**, *33*, 055004. [CrossRef]
15. Shi, X.; Rui, G.; Yi, Z. Astronomical image restoration using variational Bayesian blind deconvolution. *J. Syst. Eng. Electron.* **2017**, *28*, 1236–1247.

16. Chen, S.; Sun, T.; Yang, F. An improved optimum-path forest clustering algorithm for remote sensing image segmentation. *Comput. Geosci.* **2018**, *112*, 38–46. [CrossRef]
17. Yong, Y.; Wan, W.; Huang, S.; Yuan, F.; Yang, S.; Yue, Q.J.I.A. Remote Sensing Image Fusion Based on Adaptive IHS and Multiscale Guided Filter. *IEEE Access* **2017**, *4*, 4573–4582. [CrossRef]
18. Guo, X.; Fang, L.; Ng, M.K. A Fast $\ell 1$-TV Algorithm for Image Restoration. *SIAM J. Sci. Comput.* **2009**, *31*, 2322–2341. [CrossRef]
19. Goldstein, T.; Osher, S. The Split Bregman method for L1 regularized problems. *SIAM J. Sci. Comput.* **2009**, *2*, 323–343. [CrossRef]
20. Yang, J.F.; Zhang, Y.; Yin, W.T. An Efficient Tvl1 Algorithm For Deblurring Multichannel Images Corrupted By Impulsive Noise. *SIAM J. Sci. Comput.* **2009**, *31*, 2842–2865. [CrossRef]
21. Nikolova, M. Model distortions in Bayesian map reconstruction. *Inverse Probl. Imaging* **2007**, *1*, 399–422. [CrossRef]
22. Cai, J.F.; Chan, R.H.; Nikolova, M. Fast Two-Phase Image Deblurring Under Impulse Noise. *J. Math. Imaging Vis.* **2008**, *2*, 187–204. [CrossRef]
23. Bai, M.; Zhang, X.; Shao, Q. Adaptive correction procedure for TVL1 image deblurring under impulse noise. *Inverse Probl.* **2016**, *32*, 085004. [CrossRef]
24. Yang, J.; Gu, G.; Jiang, S. A TVSCAD approach for image deblurring with impulsive noise. *Inverse Probl.* **2017**, *33*, 125008.
25. Minru, B.; Shihuan, G. TV-MCP: A New Method for Image Restoration in the Presence of Impulse Noise. *J. Hunan Nat. Sci.* **2018**, *45*, 126–130.
26. Xu, Z.; Chang, X.; Xu, F.; Zhang, H. L1/2 regularization: A thresholding representation theory and a fast solver. *IEEE Trans. Neural Netw. Learn. Syst.* **2012**, *23*, 1013–1027.
27. Liu, J.; Huang, T.Z.; Lv, X.G.; Si, W. High-order total variation-based Poissonian image deconvolution with spatially adapted regularization parameter. *Appl. Math. Model.* **2017**, *45*, 516–529. [CrossRef]
28. Chambolle, A.; Lions, P.L. Image recovery via total variation minimization and related problems. *Numer. Math.* **1997**, *76*, 167–188. [CrossRef]
29. Chan, T.; Marquina, A.; Mulet, P. High-Order Total Variation-Based Image Restoration. *SIAM J. Sci. Comput.* **2000**, *22*, 503–516. [CrossRef]
30. Wang, S.; Huang, T.Z.; Zhao, X.L.; Liu, J. An Alternating Direction Method for Mixed Gaussian Plus Impulse Noise Removal. *Abstract Appl. Anal.* **2013**, *2*, 233–255. [CrossRef]
31. Liu, G.; Huang, T.Z.; Liu, J. High-order TVL1-based images restoration and spatially adapted regularization parameter selection. *Comput. Math. Appl.* **2014**, *67*, 2015–2026. [CrossRef]
32. Clason, C.; Jin, B.; Kunisch, K. A Duality-Based Splitting Method for ℓ^1-TV Image Restoration with Automatic Regularization Parameter Choice. *SIAM J. Sci. Comput.* **2010**, *32*, 1484–1505. [CrossRef]
33. Hintermüller, M.; Rinconcamacho, M.M. Expected absolute value estimators for a spatially adapted regularization parameter choice rule in L1-TV-based image restoration. *Inverse Probl.* **2010**, *26*, 085005. [CrossRef]
34. Jin, K.; Ye, J. Sparse and Low-Rank Decomposition of a Hankel Structured Matrix for Impulse Noise Removal. *IEEE Trans. Image Process.* **2018**, 27, 1448–1461. [CrossRef]
35. Stamatios, L.; Aurélien, B.; Michael, U. Hessian-based norm regularization for image restoration with biomedical applications. *IEEE Trans. Image Process.* **2012**, *21*, 983–995.
36. Wu, C.; Tai, X.C. Augmented Lagrangian Method, Dual Methods, and Split Bregman Iteration for ROF, Vectorial TV, and High Order Models. *SIAM J. Imaging Sci.* **2012**, *3*, 300–339. [CrossRef]
37. Lysaker, M.; Tai, X.C. Iterative Image Restoration Combining Total Variation Minimization and a Second-Order Functional. *Int. J. Comput.* **2006**, *66*, 5–18. [CrossRef]
38. Ghadimi, E.; Teixeira, A.; Shames, I.; Johansson, M. Optimal Parameter Selection for the Alternating Direction Method of Multipliers (ADMM): Quadratic Problems. *IEEE Trans. Autom. Control.* **2015**, *60*, 644–658. [CrossRef]
39. Dong, Y.; Rincon-Camacho, M.M. Automated Regularization Parameter Selection in Multi-Scale Total Variation Models for Image Restoration. *J. Math. Imaging Vis.* **2011**, *40*, 82–104. [CrossRef]
40. Chen, D.Q.; Cheng, L.Z. Spatially adapted regularization parameter selection based on the local discrepancy function for Poissonian image deblurring. *Inverse Probl.* **2012**, *28*, 015004. [CrossRef]

41. Ming, Y. Restoration of Images Corrupted by Impulse Noise and Mixed Gaussian Impulse Noise using Blind Inpainting. *SIAM J. Imaging Sci.* **2013**, *6*, 1227–1245.
42. Lu, Z.; Yong, Z. Sparse Approximation via Penalty Decomposition Methods. *SIAM J. Optim.* **2012**, *23*, 2448–2478.
43. Yuan, G.; Ghanem, B. ℓ_0TV: A Sparse Optimization Method for Impulse Noise Image Restoration. *IEEE Trans. Pattern Anal. Mach. Intell.* **2019**, *41*, 352–364. [CrossRef] [PubMed]

electronics

Article

A n-out-of-n Sharing Digital Image Scheme by Using Color Palette

Ching-Nung Yang [1], Qin-Dong Sun [2], Yan-Xiao Liu [2,*] and Ci-Ming Wu [1]

[1] Department of CSIE, National Dong Hwa University, Hualien 97401, Taiwan
[2] Department of Computer Science and Engineering, Xi'an University of Technology, Xi'an 710021, China
* Correspondence: liuyanxiao@xaut.edu.cn; Tel.: +86-029-8231-2231

Received: 15 April 2019; Accepted: 13 July 2019; Published: 17 July 2019

Abstract: A secret image sharing (SIS) scheme inserts a secret message into shadow images in a way that if shadow images are combined in a specific way, the secret image can be recovered. A 2-out-of-2 sharing digital image scheme (SDIS) adopts a color palette to share a digital color secret image into two shadow images, and the secret image can be recovered from two shadow images, while any one shadow image has no information about the secret image. This 2-out-of-2 SDIS may keep the shadow size small because by using a color palette, and thus has advantage of reducing storage. However, the previous works on SDIS are just 2-out-of-2 scheme and have limited functions. In this paper, we take the lead to study a general n-out-of-n SDIS which can be applied on more than two shadow. The proposed SDIS is implemented on the basis of 2-out-of-2 SDIS. Our main contribution has the higher contrast of binary meaningful shadow and the larger region in color shadows revealing cover image when compared with previous 2-out-of-2 SDISs. Meanwhile, our SDIS is resistant to colluder attack.

Keywords: secret image sharing; digital image; n-out-of-n scheme; color palette; colluder attack

1. Introduction

A secret image sharing (SIS) scheme inserts a secret message into shadow images in a way that if shadow images are combined in a specific way, the secret image can be recovered. A SIS scheme is usually referred to by a threshold (k, n) SIS, where $k \leq n$, and can insert a secret image into n shadow images (referred to as shadows). In a (k, n)-SIS, we may recover the secret image by using any k shadows, but cannot recover the secret image from $(k-1)$ or fewer shadows. There are various types of SIS. Here, we give a brief survey for three major types of SIS schemes: the visual cryptography scheme (VC), the polynomial-based SIS (PSIS), and the bit-wise Boolean-operation based SIS.

The so-called VC [1–6] has a novel stacking-to-see property such that the involved participants can easily stack shadows to visually decode the secret through the human eye. This property makes VC applicable in many scenarios. Although VC has the ease of decoding, it has poor visual quality of reconstructed image. Another SIS adopts $(k-1)$-degree polynomial like Shamir's secret sharing [7] to design (k, n)-PSIS [8–15]. There are two major differences between VC and PSIS: the quality of recovered image and the decoding method. Unlike VC provided with the poor visual quality, the recovered secret image of PSIS is distortion-less. However, the decoding of VC only needs stacking operation but PSIS uses the computation of Lagrange interpolation to recover secret image. Some SIS schemes are based on Boolean operations [16–20]. Note: the stacking operation of VC, strictly speaking, is also a Boolean OR operation. However, this OR operation of VC is pixel-wise operation, which applied on black-and-white dots. However, Boolean operation in [16–20] is bit-wise operations, and can obtain a high-quality secret image (a distortion-less image like PSIS scheme). Besides, using -wise Boolean has much lower complexity when compared with Lagrange interpolation.

Recently, Wei et al. use the bit-wise XOR operation to design a $(2,2)$ sharing digital image scheme (SDIS) [17] to share a 256-color (or true color) digital image. Wei et al.'s $(2,2)$-SDIS is also a type of (k,n)-SIS where $k=n=2$. Wei et al.'s $(2,2)$-SDIS is the first SIS scheme using a 256-color palette. This color palette has 256 colors, where each color is composed of red (R), green (G), and blue (B) color planes. Each color and is chosen from a palette of 16,777,216(=2^{24}) colors (24 bits: each color plane has 8 bits). In VGA cards, 256 on-screen colors are chosen from a color palette, and these colors are most visible to the human eye and meanwhile conserve a bandwidth. When using a color palette, each pixel is represented by a color index in a 256-color color palette. Consider an example, a 256×256-pixel image. The file size is $256 \times 256 \times 1$ bytes (color indices) $+256 \times 3$ bytes (color palette) = 66,304 bytes, but is $256 \times 256 \times 3$ = 196,608 bytes for using 24-bit true color format. Thus, the file size of a color image can be kept small when represented by a color palette. Because Wei et al.'s $(2,2)$-SDIS is based on color palette, and thus it has the advantage of reducing storage.

However, there are three weaknesses in Wei et al.'s SDIS: the incorrect assignment of color palette data for the color index 255, the erroneous recovery in secret image, and the partial region in shadow revealing the cover image. In [19], Yang et al. address these weaknesses and propose a new $(2,2)$-SDIS. Both Wei et al.'s $(2,2)$-SDIS and Yang et al.'s $(2,2)$-SDIS are simple 2-out-of-2 scheme and have limited applications. In this paper, we take the lead to study a general (n,n)-SDIS, which can be applied on any $n \geq 3$. The main weakness of Wei et al.'s $(2,2)$-SDIS is the incorrect assignment of color palette data for some color indices, and this is tackled by using a complicated approach, partitioned sets, in Yang et al.'s $(2,2)$-SDIS. In the proposed (n,n)-SDIS, because of the number of shadows more than two, i.e., $n \geq 3$, a simple approach reducing Hamming weigh of a temporary block is adopted to easily solve this weakness. In addition, performance of our (n,n)-SDIS are enhanced when compared with the previous $(2,2)$-SDIS. The rest of this paper is organized as follows. Section 2 reviews Wei et al.'s $(2,2)$-SDIS and Yang et al.'s $(2,2)$-SDIS. The proposed (n,n)-SDIS is presented in Section 3. Also, an approach of enhancing visual quality of color meaningful shadow is introduced. A very extreme attack, the $(n-1)$-colluder attack, on the proposed (n,n)-SDIS is discussed in Section 4. The experiment, discussion and comparison are in Section 5. Finally, Section 6 concludes the paper.

2. Preliminaries

Notations in this paper and their descriptions are listed in Table 1. These notations are used throughout the whole paper to describe all the schemes, Wei et al.'s $(2,2)$-SDIS [17], Yang et al.'s $(2,2)$-SDIS [19], and the proposed (n,n)-SDIS.

In [17], Wei et al. first proposed a simple $(2,2)$-SDIS to insert a 256-color digital image SI into two binary noise-like shadows (NS_1 and NS_2). In Wei et al.'s $(2,2)$-SDIS, every 9-bit block B, i.e., $b_1 - b_9$, is obtained from the 256-color secret image SI and the color palette CP. Afterwards, the block B is subdivided into two blocks $B^{(1)}$ and $B^{(2)}$ on shadow 1 NS_1 and shadow 2 NS_2, respectively, by using XOR operation. As shown in Figure 1, $B = B^{(1)} \oplus B^{(2)}$, where each bit $b_i = b_i^{(1)} \oplus b_i^{(2)}$, $1 \leq i \leq 9$. Both shadow blocks of $B^{(1)}$ and $B^{(2)}$ are $\boxed{Y}$ blocks. Accomplish all blocks until all pixels in SI and the data in CP are processed. Because every pixel in SI is represented as a block, shadow sizes are nine times expanded. The first 8 bits $b_1 - b_8$ in B represents a color index, and the ninth bit b_9 in every block of NS_1 (i.e., the bit $b_9^{(1)}$) is collected to covey the CP information. Therefore, from the XOR-ed results $NS_1 \oplus NS_2$ we may obtain color indices and the CP to recover SI. There are other two types of shadows for Wei et al.'s $(2,2)$-SDIS. Noise-like shadows (NS_1, NS_2) can be extended to two binary meaningful shadows (BS_1, BS_2) and two color meaningful shadows (CS_1, CS_2), on which binary cover image BCI and color cover image CCI can be, respectively, visually viewed. In addition, Wei et al.'s $(2,2)$-SDIS can also be extended to directly insert a true color SI without using CP.

Table 1. Notations and Descriptions.

Notation	Description
CP	a 256-color color palette
SI	a secret image with the size with the size $(M \times N)$ pixels
CCI, BCI	binary (black-and-white) over image and color cover image with the size $(M \times N)$ pixels
NS_i	n noise-like shadows with the size $(3M \times 3N)$ (respectively, $(5M \times 5N)$) subpixels for 256-color (respectively, true color) secret image, where $i = 1, 2, ..., n$
BS_i	binary meaningful shadows with the size $(3M \times 3N)$ (respectively, $(5M \times 5N)$) subpixels for 256-color (respectively, true color) secret image
CS_i	color meaningful shadows with the size $(3M \times 3N)$ (respectively, $(5M \times 5N)$) subpixels for 256-color (respectively, true color) secret image
B	a 3×3-subpixel block B including 8-bit color index $b_1 - b_8$ and one bit b_9 (Note: the bit b_9 in B is collected to covey the CP information for the proposed (n, n)-SDIS)
B_r	a 3×3-subpixel block B_r including the first three 8-tuples, $(r_1 - r_8)$, $(g_1 - g_8)$, and $(bl_1 - bl_8)$, are used to represent R, G and B color planes, and the other one bit in B_r is p_9.
$B^{(i)}$	a 3×3-pixel block on shadow i, where $i = 1, 2, ..., n$, including 8-bit $b_1^i - b_8^i$ and one bit b_9^i. (Note: the ninth bit in every block $B^{(1)}$ (i.e., $b_9^{(1)}$) of NS_1 is collected to covey the CP information for Wei et al.'s $(2, 2)$-SDIS and Yang et al.'s $(2, 2)$-SDIS)
$xByW$	x black subpixels and y white subpixels in a block
$\boxed{X}, \boxed{Y}$	$\boxed{X}$ and $\boxed{Y}$ blocks have $6B3W$ and $5B4W$ subpixels, respectively
$H(\bullet)$	Hamming weight function, the number of "1" in a binary vector
$W(\bullet)$	Operation of Wei et al.'s $(2, 2)$-SDIS, i.e., $W(B) = B^{(1)} \oplus B^{(2)}$ where both are $\boxed{Y}$ blocks
$Y(\bullet)$	Operation of Yang et al.'s $(2, 2)$-SDIS, i.e., $Y(B) = B^{(1)} \oplus B^{(2)}$ where one is $\boxed{X}$ block and the other is $\boxed{Y}$ block

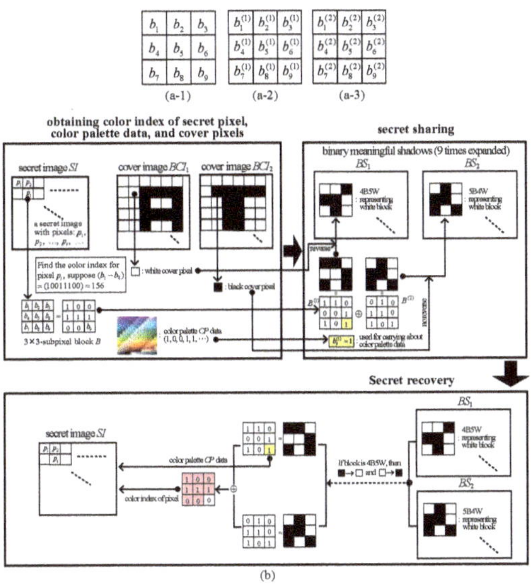

Figure 1. Blocks of $(2, 2)$-SDIS: (**a**) secret block B, shadow blocks $B^{(1)}$ and $B^{(2)}$ (**b**) diagrammatical representation of Wei et al.'s $(2, 2)$-SDIS with binary meaningful shadows.

For more clearly describing Wei et al.'s (2,2)-SDIS, Figure 1b illustrates diagrammatical representation of Wei et al.'s (2,2)-SDIS with binary meaningful shadows, which includes three processes: (i) obtaining color indices of secret pixels, color palette data, and cover pixels, (ii) secret sharing, and (iii) secret recovery. Consider a secret pixel pi with a color index $(b_1, b_2, ..., b_8) = (10011100) = 156$, and we may have $(b_1^{(1)}, b_2^{(1)}, ..., b_8^{(1)}) = (110001100)$ with $b_9^{(1)} = 1$ for carrying about CP data (suppose we embed "1" for this time), and $(b_1^{(2)}, b_2^{(2)}, ..., b_8^{(2)}) = (010110101)$ with $b_9^{(2)} = 1$. Then, we have $(b_1^{(1)}, b_2^{(1)}, ..., b_8^{(1)}) \oplus (b_1^{(2)}, b_2^{(2)}, ..., b_8^{(2)}) = (b_1, b_2, ..., b_8)$. Meantime, both blocks $B^{(1)} = (b_1^{(1)}, b_2^{(1)}, ..., b_9^{(1)})$ and $B^{(2)} = (b_1^{(2)}, b_2^{(2)}, ..., b_9^{(2)})$ are $5B4W$ blocks. For the corresponding position of this secret pixel pi, the cover pixels of BCI_1 and BCI_2 are white and black, respectively. We reverse the shadow $B^{(1)} = (b_1^{(1)}, b_2^{(1)}, ..., b_9^{(1)}) = (110001101)$ block to (001110010) $(4W5B)$ to represent the white color pixel in BCI_1, and we do not change $B^{(2)} = (b_1^{(2)}, b_2^{(2)}, ..., b_9^{(2)}) = (010110101)$ $(5B4W)$ to represent the black color pixel in BCI_2. In secret recovery, the color index can be easily derived from the exclusive OR result from $(b_1^{(1)}, b_2^{(1)}, ..., b_8^{(1)}) \oplus (b_1^{(2)}, b_2^{(2)}, ..., b_8^{(2)})$. In addition, the CP data can be obtained from every $b_9^{(1)}$ in BS_1.

However, Wei et al's (2,2)-SDIS has some weaknesses. For the color index 255, it has a problem with embedding the data of color palette. In addition, Wei et al.'s (2,2)-SDIS with color meaningful shadows cannot correctly extract the block data for white cover pixels, and this will cause erroneous recovery in the secret image. Moreover, Wei et al.'s SDIS uses $\boxed{Y}$ blocks on both shadows. Five black dots in a block B may not sufficiently demonstrate the visual quality of meaningful shadows.

It is obvious that more black subpixels in every block may enhance the visual quality of meaningful shadows BS_1 and BS_2, and CS_1 and CS_2. Accordingly, in [19], Yang et al. adopted $\boxed{X}$ block and $\boxed{Y}$ block half and half on blocks $B^{(1)}$ and $B^{(2)}$, such that the average number of black subpixels in $B^{(1)}$ and $B^{(2)}$ is enhanced from 5 to 5.5. This enhancement improved the visual quality of meaningful shadows. Meanwhile, Yang et al.'s (2,2)-SDIS also solved the other two weaknesses of Wei et al.'s (2,2)-SDIS.

3. Motivation and Design Concept

As described in Section 2, there are three weaknesses in Wei et al.'s SDIS: (1) the incorrect assignment of the color palette data for the color index 255, (2) the partial regions in meaningful shadows showing the content of the cover image, and (3) the erroneous recovery in secret image if the cover pixel is white in color meaningful shadows. Yang et al.'s (2,2)-SDIS already tackled these weaknesses.

By delving into these three weaknesses, we can see that the third weakness is a minor weakness caused from an intrinsic nature of color. A trivial approach in [19], using a near white color pixel instead of white pixels in cover image, is very efficient in addressing this weakness. Therefore, the approach can be still adopted in the proposed (n,n)-SDIS for solving this minor weakness. Our contribution is not just the extension from 2-out-of-2 scheme to n-out-of-n scheme. The proposed (n,n)-SDIS, where $n \geq 3$, has better solutions for other two major weaknesses. Because the number of shadows is more than two, we can easily solve the first weaknesses (note: the detail will be described in Section 3). However, Yang et al.'s (2,2)-SDIS uses a very complicated approach by partitioned sets to solve this weakness. For the second weakness, our (n,n)-SDIS uses $\boxed{X}$ blocks in most shadows This approach has large average black subpixels in shadow blocks to enhance visual qualities of meaningful shadows. In addition, the proposed (n,n)-SDIS embeds the CP information in b_9 but both (2,2)-SDISs [17,19] use $b_9^{(1)}$ in shadow block $B^{(1)}$. The bit b_9 obtained from the XOR-ed result B is more securely protected than the bit $b_9^{(1)}$ in one shadow block $B^{(1)}$.

A secret block $B = (b_1 ... b_9)$ has 8 bits $(b_1 ... b_8)$ to represent a color index, and one bit b_9 for representing the data of color palette CP. Together with CP, this color index can represent a pixel in secret image SI. All 9-bit blocks are obtained from the secret image SI and the color palate CP. Suppose that T is a 9-bit temporary block. Equations (1) and (2) are main statements in this paper, on which we can design the proposed (n,n)-SDIS. As shown in Equation (1), we may randomly generate

$(n-2)$ $\boxed{X}$ blocks $B^{(i_j)}, 1 \le j \le n-2$, and then determine a temporary block T via these $(n-2)$ blocks and the block B (see upper equation in Equation (1)). The content of T is provisional. Afterwards, T is divided into two blocks $\{B^{(j_1)}, B^{(j_2)}\}$ where $\{j_1, j_2\} = \{1, 2, ..., n\} - \{i_1, ..., i_{n-2}\}$. Using lower equation in Equation (1), we may insert T into two blocks based on Wei et al.'s $(2,2)$-SDIS or Yang et al.'s $(2,2)$-SDIS, which is dependent on the Hamming weigh of block T. In next subsection, we prove that lower equation in Equation (1) can be successfully accomplished. Via Equation (1), we can derive $B = B^{(1)} \oplus B^{(2)} \oplus ... \oplus B^{(n)}$ in Equation (2).

$$\begin{cases} T = B \oplus \overbrace{B^{(i_1)} \oplus ... \oplus B^{(i_{n-2})}}^{(n-2)\text{ random } \boxed{X} \text{ blocks}} \\ T = \oplus \overbrace{B^{(j_1)} \oplus B^{(j_2)}}^{\text{other two blocks}} \end{cases} \tag{1}$$

$$\begin{cases} T = B \oplus B^{(i_1)} \oplus ... \oplus B^{(i_{n-2})} \\ \Rightarrow B = T \oplus B^{(i_1)} \oplus ... \oplus B^{(i_{n-2})} \\ \Rightarrow B = B^{(j_1)} \oplus B^{(j_2)} \oplus B^{(i_1)} \oplus ... \oplus B^{(i_{n-2})} \\ \Rightarrow B = B^{(1)} \oplus ... \oplus B^{(n)}, (\{j_1, j_2\} \bigcup \{i_1, ..., i_{n-2}\} = \{1, ..., n\}) \end{cases} \tag{2}$$

Equation (2) implies that the block B can be subdivide into n shadow blocks $B^{(1)}, B^{(2)}, ..., B^{(n)}$, and meanwhile can be recovered from $B = B^{(1)} \oplus ... \oplus B^{(n)}$. All the n shadows in the proposed (n,n)-SDIS are illustrated in Figure 2. The operation of lower equation in Equation (1) using Wei et al.' $(2,2)$-SDIS is shown in Figure 2a, and using Yang et al.'s $(2,2)$-SDIS is shown in Figure 2b.

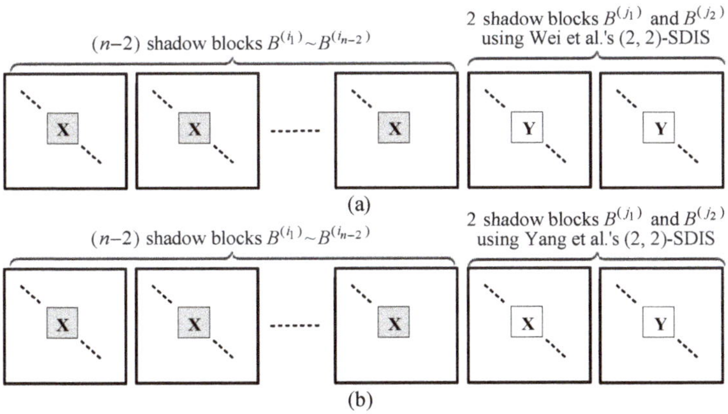

Figure 2. Shadows of the proposed (n,n)-SDIS: (**a**) using Wei et al.'s $(2,2)$-SDIS for $B^{(j_1)}$ and $B^{(j_2)}$ (**b**) using Yang et al.'s $(2,2)$-SDIS for $B^{(j_1)}$ and $B^{(j_2)}$.

Moreover, in [17], the authors claimed that the $(2,2)$-SDIS has a novel application to cover the transmission of confidential images. For example, as a supplementary aid to existing symmetric cryptography standards like DES which requires a pre-shared key, the $(2,2)$-SDIS remains a safe and less risky means for key distribution. Because the prosed scheme is extended from 2-out-of-2 to n-out-of-n, it implies that our (n,n)-SDIS can be applied on a group key distribution, which includes n members in this group. Besides the application in key distribution, the proposed scheme can be also applied to protection of secret image among multiple users. For instance, the colorful image of traffic or medical information are confidential, and our scheme provides a secure and high efficiency approach to safely keeping such image among n users, only all n users are able to recover the image with high quality.

Finally, in a shadow $NS_i, 1 \leq i \leq n$ there are $\boxed{X}$ blocks with percentage of $\frac{n-1.5}{n}(=\frac{1}{2}\times\frac{n-1}{n}+\frac{1}{2}\times\frac{n-2}{n})$, and $\boxed{Y}$ blocks with percentage of $\frac{1.5}{n}(=\frac{1}{2}\times\frac{1}{n}+\frac{1}{2}\times\frac{2}{n})$, respectively. The more $\boxed{X}$ blocks have the large number of black subpixels and may enhance visual qualities of meaningful shadows, and these percentages have more effective performance for large n.

4. The Proposed (n, n)-SDIS

4.1. Sharing and Recovering Algorithms

A block diagram of the proposed (n, n)-SDIS is illustrated in Figure 3. Shadows $NS_1 - NS_n$ are noise-like, which is the same as Boolean-operation based SIS [18]. For the proposed (n, n)-SDIS, we can complement the blocks for the corresponding white cover pixels to generate binary meaningful shadows $(BS_1 - BS_n)$ from noise-like shadows $(NS_1 - NS_n)$, i.e., $6B3W$ (or $5B4W$) for black color and $3B6W$ (or $4B5W$) for white color. However, the scheme in [18] does dot has this property. On the other hand, to implement color meaningful shadows (CS_1, CS_n), the 1s in blocks are replaced with the color of the corresponding cover pixel, and leave 0s blank. Therefore, we only describe how to generate noise-like shadows, and how to recover the secret image and color palette from n noise-like shadows.

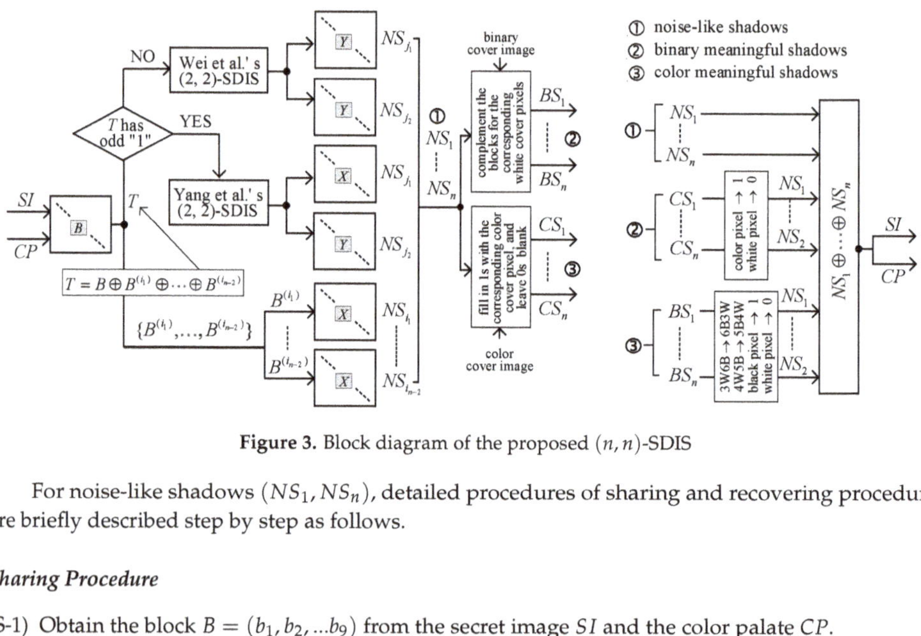

Figure 3. Block diagram of the proposed (n, n)-SDIS

For noise-like shadows (NS_1, NS_n), detailed procedures of sharing and recovering procedures are briefly described step by step as follows.

Sharing Procedure

(S-1) Obtain the block $B = (b_1, b_2, ...b_9)$ from the secret image SI and the color palate CP.

(S-2) Randomly generate $(n-2)$ $\boxed{X}$ blocks $B^{(i_1)}, B^{(i_2)}, ..., B^{(i_{n-2})}$.

(S-3) By $(n-2)$ random blocks and the block B, calculate the temporary block T via $T = B \oplus B^{(i_1)} \oplus ... \oplus B^{(i_{n-2})}$.

(S-4) If $H(T)$ is 9, we reduce its Hamming weight to $H(T) = 7$ via modifying any one shadow block of $\{B^{(i_1)}, ..., B^{(i_{n-2})}\}$.

/* (1) In Lemma 1, we prove that the reduction of Hamming weight can always be accomplished (2) After step (S-4), the Hamming weight distribution is $0 \leq H(T) \leq 8$ */.

(S-5) If $H(T)$ is odd ($H(T) = 1, 3, 5, 7$) then construct two other shadows $B^{(j_1)}, B^{(j_2)}$ by $Y(T) = \{B^{(j_1)}, B^{(j_2)}\}$; else by $W(T) = \{B^{(j_1)}, B^{(j_2)}\}$, where $\{j_1, j_2\} \bigcup \{i_1, ...i_{n-2}\} = \{1, 2, ..., n\}$.

/* In Lemma 2, we prove that $\{B^{(j_1)}, B^{(j_2)}\}$ can be obtained from $Y(T)$ for odd $H(T)$, and from $W(T)$ for even $H(T)$. */

(S-6) Process all the blocks, and output shadow blocks $B^{(1)}...B^{(n)}$ on n noise-like shadows $NS_1 - NS_n$, respectively.

Recovering procedure:

(S-1) Obtain B by XOR-ing $(B^{(1)} \oplus ... \oplus B^{(n)})$ via from n noise-like shadows $NS_1 - NS_n$.

/* Theorem 1, demonstrates that we can obtain the original block from $B = (B^{(1)} \oplus ... \oplus B^{(n)})$ */

(S-2) Recover the color index $(b_1 - b_8)$ and the data of color palette b_9, respectively, from B.

(S-3) Repeat the above until all blocks in $NS_1 \oplus ... \oplus NS_n$ are processed, and finally SI and CP can be recovered.

4.2. Extension of (n,n)-SDIS to Share True Color Secret Image

Same as $(2,2)$-SDIS and VC in [5], the proposed (n,n)-SDIS can be used to share a true color image. To share a true color secret image, we use a 25-subpixel block B_r, which the first three 8-tuples, $r_1, ..., r_8$, $g_1, ..., g_8$, and $bl_1, ..., bl_8$, are used to represent R, G and B color planes. The other one bit in B_r is p_9. This 25-subpixel block B_r is shown in Figure 4a. Because we share R, G and B colors directly, we do not need to use the bit p_9 to covey any information. Thus, this bit p_9 could be abandoned, or used as authentication bits to provide authentication capability like VC in [6] and PSIS in [10]. Collect $(x_1...x_8)$, where $x \in \{r, g, bl\}$, and append the bit p_9 to form red, green, and blue shadow blocks B_x where $x \in \{r, g, bl\}$ as shown in Figure 4b.

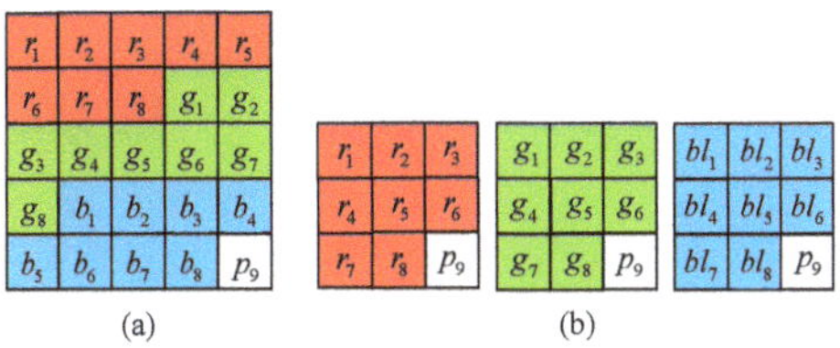

Figure 4. Blocks for sharing true clor image: (**a**) 25-bit B_T (**b**) 9-bit B_r, B_g, B_{bl}.

Detailed procedures of the proposed (n,n)-SDIS for sharing and recovering true color image are briefly described step by step as follows.

Sharing procedure:

(S′-1) Obtain 24-bit true color $r_1, ..., r_8$, $g_1, ..., g_8$, and $bl_1, ..., bl_8$ from the secret image SI, and random generate a bit p_9 to form a 25-bit block B_r, as shown in Figure 4a.

/* Parity bit p_9 is not used to covey any information, and thus it can be randomly generated */

(S′-2) Subdivide the true color block B_T to red, green, and blue shadow blocks B_r, B_g, B_{bl}.

(S′-3) Using B_r, B_g, B_{bl} as 9-bit block B in (S-1), respectively, to generate n shadow blocks $B_r^{(i)}, B_g^{(i)}, B_{bl}^{(i)}$, where $1 \le i \le n$, through (S-1) (S-6).

(S′-4) Collect every first 8 bits in $B_r^{(i)}, B_g^{(i)}, B_{bl}^{(i)}$, and append a black subpixel in the 25-th subpixel to generate a 25-bit shadow block $B^{(i)}$, where $1 \le i \le n$.

/* Because we do not use the 25-th bit p_9 in the XOR-ed result B_T to convey any information, we can use black subpixel in 25-th subpixel for all shadow blocks to enhance the number of black subpixels. */

(S′-5) Process all the blocks, and output blocks $B^{(1)} - B^{(n)}$ on n noise-like shadows $NS_1 - NS_n$, respectively.

Recovering procedure:

(R′-1) Obtain every 25-bit block B_T by XOR-ing $(B^{(1)} \oplus B^{(2)} \oplus ... \oplus B^{(n)})$ via XOR-ing n noise-like shadows $NS_1 - NS_n$.

(R$'$-2) Recover a true color from the first 24 bits in B_T, i.e., $r_1, ..., r_8$, $g_1, ..., g_8$, and $bl_1, ..., bl_8$.
(R$'$-3) Repeat the above until all blocks in $(NS_1 \oplus NS_2 ... \oplus NS_n)$ are processed, and finally a true color SI is obtained.

4.3. Enhancing Visual Quality of Color Meaningful Shadow

Consider sharing 256-color (respectively, true color) SI, noise-like shadows $NS_i, 1 \leq i \leq n$, are $3M \times 3N$ (respectively, $5M \times 5N$) times expanded. Based on noise-like shadow NS_i, we can fill in 1s in shadow blocks with the color of the corresponding cover pixel in CCI, and leave 0s blank to generate color meaningful shadow CS_i. Consider the case sharing 256-color SI. As shown in Figure 5a, there is a pixel with a blue color $\boxed{C}$ in CCI. Suppose that the block $B^{(i)}$ at corresponding position for this pixel in NS_i is $(b_1^{(i)} ... b_9^{(i)}) = (110101101)$ (see Figure 5b), and this block $B^{(i)}$ is a $\boxed{X}$ block with $6B3W$ sub-pixels (see see Figure 5c). By putting the blue cover pixel $\boxed{C}$ into all black sub-pixels in Figure 5c, we have color meaningful shadow $CS_{(i)}$ in Figure 5d. Noise-like shadow and color meaningful shadow have the same size $3M \times 3N$ subpixels and 9 times expanded when compared with CCI.

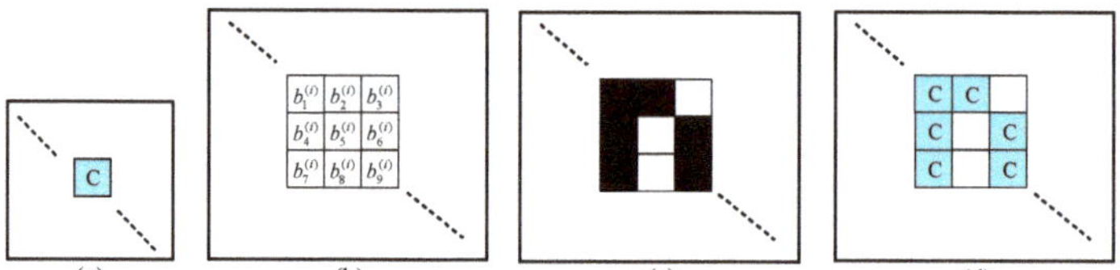

Figure 5. Block patterns: (**a**) a pixel with a color in CCI (**b**) the corresponding block $B^{(i)}$ in NS_i (**c**) the corresponding $6B3W$ block in NS_i (**d**) the corresponding block in CS_i

As shown in Figure 5d, the color at 1s in a block are the same. This is because SI and CCI have the same size with $M \times N$ pixels. To enhance visual quality of CS_i, we use a large color cover image CCI' with $3M \times 3N$ pixels (note: the original CCI has only $M \times N$ pixels). Obviously, this larger CCI' has the high resolution than CCI. As shown in Figure 6, our new approach uses a large CCI' (see Figure 6a). By putting the color pixels in to into all 1s of $B^{(1)}$ in Figure 6b, we have the CS_i' in Figure 6c. Because the color meaningful shadow CS_i' has more colors, and will have the high resolution. By the same argument, this approach can also be applied to sharing true color SI.

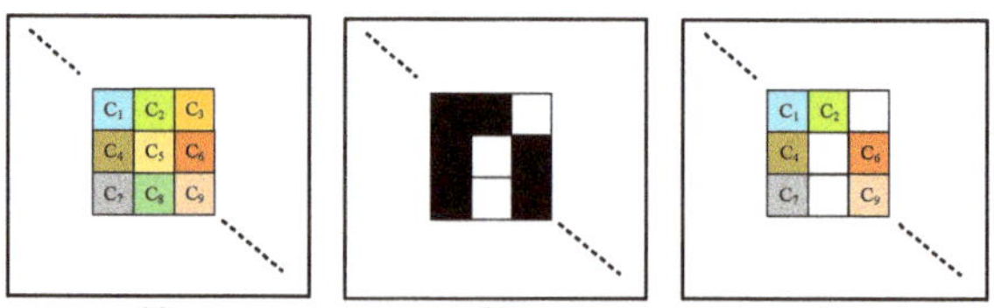

Figure 6. Block patterns: (**a**) 9 color pixels with color $C_1 - C_9$ in CCI' (**b**) the corresponding block $B^{(i)}$ in NS_i (**c**) the corresponding color block in CS_i'.

5. Theorem and Security Analysis

5.1. Main Theorems and Examples

Lemma 1. *Suppose that the block T in Equation (1) is all-1 block, i.e., $H(T) = 9$. We may change any two positions (one is $1 \to 0$ and the other is $0 \to 1$) in any one block $B^{(i_j)}, 1 \leq j \leq n-2$, such that the equation $B = T \oplus B^{(i_1)} \oplus ... \oplus B^{(i_{n-2})}$ holds, and $H(T)$ is reduced from 9 to 7. Meanwhile, all $(n-2)$ blocks $B^{(i_j)}, 1 \leq j \leq n-2$, are still $\boxed{X}$ blocks.*

Proof. As shown in Equation (1), all these $(n-2)$ blocks $B^{(i_1)} - B^{(i_{n-2})}$ are $\boxed{X}$ blocks. We choose one block $B^{(i_j)}$, and modify any two positions of $1 \to 0$ and $0 \to 1$. This modification will change the 1 in the block T to 0 at these two chosen modified positions. After that, $H(T)$ is reduced to $9-2=7$. Meanwhile, because we change two positions by $1 \to 0$ and $0 \to 1$, respectively, the Hamming weight $H(B^{(i_j)})$ is unchanged, and this shadow block $B^{(i_j)}$ is still a $\boxed{X}$ block. □

Lemma 2. *Blocks $B^{(j_1)}, B^{(j_2)}$ in step (S-5) can be obtained from $Y(T)$ for odd $H(T)$, and from $W(T)$ for even $H(T)$.*

Proof. Let X_1 be $\boxed{X}$ block, and both Y_1 and Y_2 be $\boxed{Y}$ blocks. We first prove that the possible Hamming weights of (Y_1, Y_2) are $0,2,4,6,8$, and the possible Hamming weights of (X_1, Y_2) are $1,3,5,7$. Because both blocks Y_1 and Y_2 have the same Hamming weight 5, the number of positions of $1 \to 0$ and $0 \to 1$ crossing from vectors Y_1 to Y_2 should be the same. Suppose that this number is y. Therefore, the (Y_1, Y_2) has the following form (see Equation (3)), where $0 \le y \le 4$. Obviously, the Hamming weight of $(Y_1 Y_2)$ in Equation (3) is $2y$, and thus $H(Y_1 Y_2)$ may be $0,2,4,6,8$. □

$$\begin{cases} Y_1 = \overbrace{1...1}^{y} \quad \overbrace{0...0}^{y} \quad \overbrace{1...1}^{5-y} \quad \overbrace{0...0}^{5-y} \\ \qquad\ \ \downarrow \qquad\ \ \uparrow \qquad\ \ \updownarrow \qquad\ \ \updownarrow \\ Y_2 = 0...0 \quad 1...1 \quad 1...1 \quad 0...0 \\ Y_1 \oplus Y_2 = 1...1 \quad 1...1 \quad 0...0 \quad 0...0 \end{cases} \tag{3}$$

Consider the XOR-ed block $(X_1 \oplus Y_2)$. Because blocks X_1 and Y_2 have Hamming weights 6 and 5, respectively. The number of positions of $1 \to 0$ and $0 \to 1$ crossing from vectors X_1 to Y_2 should differ with one. Suppose that the number crossing from vectors X_1 to Y_2 of $1 \to 0$ is $x+1$, and the number of $0 \to 1$ is x. Therefore, the $(X_1 \oplus Y_2)$ has the following form (see Equation (4)), where $0 \le x \le 3$. The Hamming weight of $(X_1 Y_2)$ in Equation (4) is $(2x+1)$, and thus $H(X_1 \oplus Y_2)$ may be $1,3,5,7$.

$$\begin{cases} X_1 = \overbrace{1...1}^{x+1} \quad \overbrace{0...0}^{x} \quad \overbrace{1...1}^{5-x} \quad \overbrace{0...0}^{3-x} \\ \qquad\ \ \downarrow \qquad\ \ \uparrow \qquad\ \ \updownarrow \qquad\ \ \updownarrow \\ Y_2 = 0...0 \quad 1...1 \quad 1...1 \quad 0...0 \\ X_1 \oplus Y_2 = 1...1 \quad 1...1 \quad 0...0 \quad 0...0 \end{cases} \tag{4}$$

Because Wei et al.'s $(2,2)$-SDIS uses two $\boxed{Y}$ blocks (say Y_1 and Y_2), therefore using Wei et al.'s $(2,2)$-SDIS has $H(Y_1 \oplus Y_2)$ with even values $0,2,4,6,8$. On the other hand, there are one $\boxed{X}$ block and one $\boxed{Y}$ block (say X_1 and Y_2) when using Yang et al.'s $(2,2)$-SDIS. Thus, using Yang et al.'s $(2,2)$-SDIS has $H(X_1 \oplus Y_2)$ with odd values $1,3,5,7$. Finally, the above implies that $\{B^{(j_1)}, B^{(j_2)}\}$ can be obtained from $Y(T)$ for odd $H(T) = 1,3,5,7$, and can be obtained from $W(T)$ for even $H(T) = 0,2,4,6,8$.

The following theorem shows that the proposed (n,n)-SDIS is a n-out-of-n sharing scheme that we can recover SI and CP from n noise-like shadows $(NS_1 - NS_n)$, and cannot obtain SI and CP from $(n-1)$ or fewer shadows.

Theorem 1. *The proposed (n,n)-SDIS is n-out-of-n sharing scheme that the XOR-ed result of n shadow blocks can represent* 0 255 *color indices and the data of color palette.*

Proof. We first prove that sharing procedure can successfully generate n shadow blocks $B^{(i)}, 1 \le i \le n$. Suppose that a block $B = (\overbrace{b_1...b_8}^{colorindex}, \overbrace{b_9}^{colorpalette})$ is composed of 8-bit color index (0 255) and 1-bit data of color palette, which are obtained from SI and CP. By Equation (1), we first randomly generate $(n-2)$ $\boxed{X}$ blocks $B^{(i_j)}, 1 \le j \le n-2$, and then calculate the temporary block T via $T = B \oplus B^{(i_1)} \oplus ... \oplus B^{(i_{n-2})}$. After step (S-4), the Hamming weight distribution of $H(T)$ is 0 8 (see Lemma 1). By Lemma 2, we can obtain $\{B^{j_1}, B^{(j_2)}\}$ from $Y(T)$ (respectively, $W(T)$) for odd $(1,3,5,7)$ (respectively, even $(0,2,4,6,8)$ $H(T)$. Finally, we have n shadow blocks $\{B^1, ..., B^n\}$. Process all the blocks, and we can generate n noise-like shadows.

Next, we consider the recovery. As shown in Equation (2), we can recover the original block $B = (b_1...b_9)$ from $B = B^{(1)} \oplus ... \oplus B^{(n)}$. Therefore, we can determine the color index $(b_1...b_8)$ and the data of color palette b_9. After obtaining all blocks, we can recover SI and CP. Because of $B = B^{(1)} \oplus ... \oplus B^{(n)}$, it is obvious that we cannot recover the original block B via $(n-1)$ or fewer shadow blocks. □

Let the ratio of average number of black subpixels in a block (i.e., the regions in shadows showing the content of cover image) for Wei et al.'s $(2,2)$-SDIS, Yang et al.'s $(2,2)$-SDIS, and the proposed (n,n)-SDIS be R_W, R_Y, R_P. In addition, let the contrasts of binary meaningful shadows for Wei et al.'s $(2,2)$-SDIS, Yang et al.'s $(2,2)$-SDIS, and the proposed (n,n)-SDIS be C_W, C_Y, C_P. The following theorem demonstrates $R_W \leq R_Y \leq R_P$ and $C_W \leq C_Y \leq C_P$.

Theorem 2. *The ratio of average numbers of black subpixels in a 9-bit block for Wei et al.'s $(2,2)$-SDIS, Yang et al.'s $(2,2)$-SDIS, and the proposed (n,n)-SDIS are $R_W = \frac{5}{9}, R_Y = \frac{5.5}{9}, R_P = \frac{6-1.5/n}{9}$ where $R_W < R_Y < R_P$. The contrasts of binary meaningful shadows for Wei et al.'s $(2,2)$-SDIS, Yang et al.'s $(2,2)$-SDIS, and the proposed (n,n)-SDIS are $C_W = \frac{1}{9}, C_Y = \frac{2}{9}, C_P = \frac{3-3/n}{9}$, where $C_W < C_Y < C_P$.*

Proof. Wei et al.'s $(2,2)$-SDIS has all $\boxed{Y}$ blocks on both shadows, and thus $R_W = \frac{5}{9}$. On the other hand, both shadows of Yang et al.'s $(2,2)$-SDIS are composed of $\boxed{X}$ and $\boxed{Y}$ blocks half and half. Therefore, we have $R_Y = \frac{(6+5)/2}{9} = \frac{5.5}{9}$. For the proposed (n,n)-SDIS, Step (S-5) implies that Yang et al.'s $(2,2)$-SDIS and Wei et al.'s $(2,2)$-SDIS are evenly used in the proposed (n,n)-SDIS. This is because the Hamming weights $H(T)$ are odd and even half and half. Therefore, the value of R_P is derived as follows.

$$\begin{cases} R_P = \frac{1}{2} \times \overbrace{\frac{((n-2)\times 6 + 2\times 5)/n}{9}}^{usingWeiet.al(2,2)-SDIS} + \\ \frac{1}{2} \times \overbrace{\frac{((n-1)\times 6 + 1\times 5)/n}{9}}^{usingWeiet.al(2,2)-SDIS} \\ = \frac{3-1/n}{9} + \frac{3-0.5/n}{9} = \frac{6-1.5/n}{9} \end{cases} \tag{5}$$

It is obvious that $R_P \geq \frac{5.5}{9}$ with equality for $n = 3$. From these values $R_W = \frac{5}{9}, R_Y = \frac{5.5}{9}, R_P = \frac{6-1.5/n}{9}$, we have $R_W < R_Y < R_P$

The contrast is the difference of blackness for black block and white block. In binary meaningful shadows $BS_1 - BS_n$, we complement the blocks for the corresponding white cover pixels to generate white shadow blocks. Thus, if the number of black subpixels in a black shadow block is n_B, then the he number black subpixels in a white shadow block is $9 - n_B$. Thus, we have $C_W = \frac{5-(9-5)}{9} = \frac{1}{9}, C_Y = \frac{5.5-(9-5.5)}{9} = \frac{2}{9}, C_P = \frac{6-1.5/n-(9-6+1.5/n)}{9} = \frac{3-3/n}{9}$. It is obvious that $C_P \geq \frac{2}{9}$ with equality for $n = 3$. From these values $C_W = \frac{1}{9}, C_Y = \frac{2}{9}, C_P = \frac{3-3/n}{9}$ we have $C_W < C_Y \leq C_P$. □

An illustrative example gives a quick understanding for the proposed (n,n)-SDIS.

Example 1. *Share and recover the following information $(c,d) = (176,0)$ and $(49,1)$, where c is the color index and d is the data of color palette, by the proposed $(4,4)$-SDIS.*

Given $(c,d) = (176,0)$, we have the block $B = (\overbrace{b_1...b_8}^{c}, \overbrace{b_9}^{d}) = (\overbrace{10110000}^{176}, \overbrace{0}^{0})_2$. By step (S-2), we randomly generate two $\boxed{X}$ blocks (say $B^{(1)}, B^{(2)}$). Suppose that these two random blocks are

$B^{(1)} = (101110110)$ and $B^{(2)} = (111101001)$ with $H(B^{(1)}) = H(B^{(2)}) = 6$, and then we obtain the temporary block T via the following equation.

$$\begin{cases} T = B \oplus B^{(1)} \oplus B^{(2)} \\ = (101100000) \oplus (101110110) \oplus (111101001) \\ = (111111111) \end{cases} \tag{6}$$

Because of $H(T) = 9$, we should modify any two positions (one is $1 \to 0$ and the other is $0 \to 1$) in one block (say $B^{(2)}$), to reduce $H(T)$ from 9 to 7. For example, we may modify $B^{(2)}$ as (111101100). Finally, we have $T = (111111010)$ with $H(T) = 7$, and meanwhile the new block $B^{(2)} = (111101100)$ is still a $\boxed{X}$ block. Since $H(T) = 7$ is odd, we apply Yang et al.'s (2,2)-SDIS to obtain $Y(111111010) = 7$, which can be determined from Equation (7). Finally, all four shadow blocks are $B^{(1)} = (101110110), B^{(2)} = (111101100), B^{(3)} = (110110101), B^{(4)} = (001001111)$, where $B^{(1)}, B^{(2)}, B^{(3)}$ are $\boxed{X}$ blocks, and $B^{(4)}$ is $\boxed{Y}$ block.

$$\begin{cases} B^{(3)} = (110110101) : \boxed{X} \\ \oplus \qquad \downarrow\downarrow\uparrow\downarrow\downarrow\uparrow\updownarrow\uparrow\updownarrow \\ B^{(4)} = (001001111) : \boxed{Y} \\ T = (111111010) \end{cases} \tag{7}$$

Consider another case $(c, d) = (49, 1)$. We have the block $B = (\overbrace{001100011}^{49}, \overbrace{1}^{1})_2$. From step (S-2), we randomly select two $\boxed{X}$ blocks (say $B^{(1)}, B^{(2)}$). Suppose that these two random blocks are $B^{(1)} = (011111001)$ and $B^{(2)} = (110011011)$, and then we obtain the temporary block T via Equation (8).

$$\begin{cases} T = B \oplus B^{(1)} \oplus B^{(2)} \\ = (001100011) \oplus (011111001) \oplus (110011011) \\ = (100000001) \end{cases} \tag{8}$$

Since $H(T) = 2$ is even, we apply Wei et al.'s (2,2)-SDIS to obtain $Y(100000001) = (B^{(3)}, B^{(4)})$, which can be determined from Equation (9). Finally, all four blocks are $B^{(1)} = (011111001), B^{(2)} = (110011011), B^{(3)} = (110101010), B^{(4)} = (010101011)$, where $B^{(1)}, B^{(2)}$ are $\boxed{X}$ blocks, and $B^{(3)}, B^{(4)}$ are $\boxed{Y}$ blocks.

$$\begin{cases} B^{(3)} = (110101010) : \boxed{Y} \\ \oplus \qquad \downarrow\updownarrow\updownarrow\updownarrow\updownarrow\updownarrow\updownarrow\updownarrow\uparrow \\ B^{(4)} = (010101011) : \boxed{Y} \\ T = (100000001) \end{cases} \tag{9}$$

For recovery, consider the case: $B^{(1)} = (101110110), B^{(2)} = (111101100), B^{(3)} = (110110101), B^{(4)} = (001001111)$. The XOR-ed result is $B = B^{(1)} \oplus B^{(2)} \oplus B^{(3)} \oplus B^{(4)} = (101100000)$, and thus $(c, d) = (176, 0)$. For the other case: $B^{(1)} = (011111001), B^{(2)} = (110011001), B^{(3)} = (110101010), B^{(4)} = (010101011)$, the XOR-ed result is $B = B^{(1)} \oplus B^{(2)} \oplus B^{(3)} \oplus B^{(4)} = (101100000)$. Therefore, we have $(c, d) = (49, 1)$.

Let R_P' be the ratio of average numbers of black subpixels in a 25-bit shadow block for the proposed (n, n)-SDIS sharing true color image. The following theorem demonstrates $R_P' > R_P$, i.e., the meaningful shadows of sharing true color secret image have the better visual quality than those of sharing 256-color secret image.

Theorem 3. *The ratio of average numbers of black subpixels in a* 25*-bit block for the proposed* (n, n)*-SDIS sharing true color image is* $R_P' = \frac{17}{25} - \frac{0.16}{n}$*, where* $R_P' > R_P$.

Proof. If the blocks $B_r^{(i)}, B_g^{(i)}, B_{bl}^{(i)}$ are $\boxed{X}$ (respectively, $\boxed{Y}$) blocks, then the first 8 bits in $B_r^{(i)}, B_g^{(i)}, B_{bl}^{(i)}$ has 6 black subpixels with $\frac{C_8^6}{C_9^6}$ percentage and 5 black subpixels with $\frac{C_8^5}{C_9^6}$ percentage (respectively, 5 black subpixels with $\frac{C_8^5}{C_9^5}$ percentage and 4 black subpixels with $\frac{C_8^4}{C_9^5}$ percentage). The average number of black pixels for the first 8 bits in $B_r^{(i)}, B_g^{(i)}, B_{bl}^{(i)}$ is $6 \times \frac{C_8^6}{C_9^6} + 5 \times \frac{C_8^5}{C_9^6} = \frac{16}{3}$ for $\boxed{X}$ blocks, and is $5 \times \frac{C_8^5}{C_9^5} + 4 \times \frac{C_8^4}{C_9^5} = \frac{40}{3}$ for $\boxed{X}$ blocks. Therefore, the average number of black subpixels in every 8 bits in the first 24 bits of $B^{(i)}$ is $\frac{16n-4}{3n}$, as derived below.

$$\begin{cases} \frac{1}{2} \times \overbrace{((n-2) \times 16/3 + 2 \times 40/9)/n}^{usingWeiet.al(2,2)-SDIS} + \\ \frac{1}{2} \times \overbrace{((n-1) \times 16/3 + 1 \times 40/9)/n}^{usingYanget.al(2,2)-SDIS} \\ = \frac{24n-8}{9n} + \frac{24n-4}{9n} = \frac{16n-4}{3n} \end{cases} \tag{10}$$

Because the 25-th bit in shadow block is always 1, and thus the value of R_P' is determined as $R_P' = \frac{3\times(16n-4)/3n+1}{25} = \frac{17}{25} - \frac{0.16}{n}$. The following equation implies $R_P' > R_P$.

$$\begin{cases} R_P' = \frac{17}{25} - \frac{0.16}{n} > \frac{6}{9} - \frac{0.16}{n} > \frac{6}{9} - \frac{1.5/9}{n} \\ \frac{6-1.5/n}{9} = R_P \end{cases} \tag{11}$$

□

5.2. Security Analysis: The $(n-1)$-Colluder Attack

Here, we consider an attack way that $(n-1)$ participants collude together and want to figure out SI and CP. The $(n-1)$-colluder attack is a very extreme attack for the proposed (n,n)-SDIS, because it needs $(n-1)$ participants jointly providing their shadows for guessing SI and CP. We first discuss the $(n-1)$-colluder attack on Wei et al.'s $(2,2)$-SDIS and Yang et al.'s $(2,2)$-SDIS. Suppose that Participant 1 wants to predict SI and CP from his own shadow NS_1. Because the color palette CP information is conveyed by the ninth bit $b_9^{(1)}$ of every block on NS_1. Therefore, the CP can be completely obtained from NS_1. Even though Participant 1 has the color palette CP, but he cannot obtain the information about color index. An attacker has $\frac{1}{256} \approx 0.004$ probability to figure out the correct color index $(b_1 ... b_8)$ of block B without any shadow. This value of $\frac{1}{256}$ is a brute-force probability, which tries all possible 256 colors in the color palette. However, for the $(n-1)$-colluder attack, Participant 1 has $B^{(1)}$. By cryptanalytic attacks relying on knowing one shadow (the first eight bit of $B^{(1)}$), Participant 1 may guess the color index. Let the successful probability to recover the block B for Wei et al.'s $(2,2)$-SDIS and Yang et al.'s $(2,2)$-SDIS be P_W and P_Y, respectively, when collecting one shadow. Because both shadow blocks of Wei et al.'s $(2,2)$-SDIS are all $\boxed{Y}$ blocks $(5B4W)$, obviously P_W is $\frac{1}{C_9^5} = \frac{1}{126} \approx 0.008$. On the other hand, shadow blocks of Yang et al.'s $(2,2)$-SDIS are evenly composed of $\boxed{X}$ blocks and $\boxed{Y}$ blocks. Thus, $P_Y = \frac{1/C_9^6 + 1/C_9^5}{2} = \frac{1/84 + 1/126}{2} \approx 0.01$. Both probabilities 0.08 and 0.01 are higher than the brute-force probability 0.004. However, these probabilities 0.08 and 0.01 are still practically secure for guessing 256 colors.

Let the successful probability to recover the block B for $(n-1)$-colluder attack, for the proposed (n,n)-SDIS, be P_P. In the following theorem, we theoretically prove $P_P = \frac{1}{C_9^6} - \frac{3}{2n} \times (\frac{1}{C_9^6} - \frac{1}{C_9^5})$.

Theorem 4. *The successful probability to recover the block B in the proposed (n,n)-SDIS for $(n-1)$-colluder attack is $P_P = \frac{1}{C_9^6} - \frac{3}{2n} \times (\frac{1}{C_9^6} - \frac{1}{C_9^5})$, where $P_W \le P_Y \le P_P$.*

Proof. Suppose that there are $(n-1)$ shadows (say $B^{(1)} - B^{(n-1)}$) for reconstruction, on which we may guess the type of shadow block in the corresponding position of $B^{(n)}$. The block $B^{(n)}$ has $\boxed{X}$ block and $\boxed{Y}$ block with $\frac{2n-3}{2n}$ probability and $\frac{3}{2n}$ probability, respectively, which are derived below.

$$\begin{cases} \frac{1}{2} \times \overbrace{\frac{C_2^2 \cdot C_{n-2}^1}{C_n^{n-1}}}^{Weietal's(2,2)SDIS} + \frac{1}{2} \times \overbrace{\frac{C_1^1 \cdot C_{n-1}^1}{C_n^{n-1}}}^{Yanget.al's(2,2)-SDIS} = \frac{2n-3}{2n} \\ (B^{(n)} : \boxed{X}) \\ \frac{1}{2} \times \overbrace{\frac{C_2^1 \cdot C_{n-2}^{n-2}}{C_n^{n-1}}}^{Weietal's(2,2)SDIS} + \frac{1}{2} \times \overbrace{\frac{C_1^1 \cdot C_{n-1}^{n-1}}{C_n^{n-1}}}^{Yanget.al's(2,2)-SDIS} = \frac{3}{2n} \\ (B^{(n)} : \boxed{Y}) \end{cases} \tag{12}$$

If $B^{(n)}$ is $\boxed{X}$ block (respectively, $\boxed{Y}$ block), there is $\frac{1}{C_9^6}$ (respectively, $\frac{1}{C_9^5}$) probability to guess the correct color index $(b_1...b_8)$, which is better than brute-force probability $\frac{1}{256}$. Thus, P_P is calculated as follows.

$$P_P = \overbrace{\frac{2n-3}{2n} \times \frac{1}{C_9^6}}^{\boxed{X}block} + \overbrace{\frac{3}{2n} \times \frac{1}{C_9^5}}^{\boxed{Y}block} = \frac{1}{C_9^6} - \frac{3}{2n} \times \left(\frac{1}{C_9^6} - \frac{1}{C_9^5}\right) \tag{13}$$

□

Since $P_W = \frac{1}{C_9^5}$ and $P_Y = \frac{1/C_9^5 + 1/C_9^6}{2}$, we have $P_W < P_Y$. About P_Y and P_P, the relation is derived as follows.

$$\begin{cases} P_P = \frac{1}{C_9^6} - \frac{3}{2n} \times \left(\frac{1}{C_9^6} - \frac{5}{9}\right) = \frac{2n-3}{2n} \times \frac{1}{C_9^6} + \frac{3}{2n} \times \frac{1}{C_9^5} \\ = \frac{n}{2n} \times \frac{1}{C_9^6} + \left(\frac{n-3}{2n} \times \frac{1}{C_9^6} + \frac{3}{2n} \times \frac{1}{C_9^5}\right) \\ \geq \frac{n}{2n} \times \frac{1}{C_9^6} + \left(\frac{n-3}{2n} \times \frac{1}{C_9^5} + \frac{3}{2n} \times \frac{1}{C_9^5}\right) \\ = \frac{1/C_9^5 + 1/C_9^6}{2} = P_Y \end{cases} \tag{14}$$

For $n = 3$, the value of P_P is $P_P = \frac{1/C_9^5 + 1/C_9^6}{2} = P_Y$, and P_P approaches to $\frac{1}{C_9^6}$ for large n. In fact, the value of $\frac{1}{C_9^6} = \frac{1}{84} \approx 0.012$ is almost the same as $P_y \approx 0.01$. For this extreme case, the $(n-1)$-colluder attack, the security of the proposed (n,n)-SDIS is close to that of Yang et al.'s $(2,2)$-SDIS. By the same argument, for other cases collecting $(n-2)$ or shadows, the possible combination of collected shadows is more difficult to analyze compared with collecting $(n-1)$ shadows, and even less than the brute-force probability.

In the proposed (n,n)-SDIS, the color palette information is conveyed by b_9 (the ninth bit in B), but not the ninth bit $b_9^{(1)}$ of the block $B^{(1)}$ in NS_1. Therefore, the color palette CP may be obtained from only one shadow for Wei et al.'s $(2,2)$-SDIS and Yang et al.'s $(2,2)$-SDIS. Even though an attacker has the CP information, he still cannot obtain the secret image SI. For the proposed (n,n)-SDIS, the color palette information in B is securely protected and only can be determined from XOR-ing n blocks $B^{(1)} \oplus ... \oplus B^{(n)}$. This makes the cryptanalysis is more difficult for the proposed (n,n)-SDIS. The following theorem demonstrates the successful probability P_C to recover a correct color in CP for the proposed (n,n)-SDIS when collecting $(n-1)$ shadows.

Theorem 5. *The successful probability to recover a correct color in CP for the proposed (n,n)-SDIS when collecting $(n-1)$ shadows is $P_C = (\frac{2}{3} - \frac{1/6}{n})^{24}$.*

Proof. Each color information in CP is encapsulated in 24 blocks, which every block should be derived from $B = B^{(1)} \oplus ... \oplus B^{(n)}$. If colluders have $(n-1)$ shadows (say $NS_1 - NS_{n-1}$), for a block B, they have the XOR-ed result $B' = B^{(1)} \oplus ... \oplus B^{(n-1)}$, and can guess that the shadow block $B^{(n)}$ is $\boxed{X}$ block and $\boxed{Y}$ block with $\frac{2n-3}{2n}$ probability and $\frac{3}{2n}$ probability, respectively. For $\boxed{X}$ block, it implies that we have $\frac{6}{9}$ probability that the bit b_9 is the complementary bit b_9' in B'. On the other hand, we have $\frac{5}{9}$ probability that the bit b_9 is the complementary bit b_9' in B' for $\boxed{Y}$ block. Therefore, the average probability of guessing b_9 is derived as $\overbrace{\frac{2n-3}{2n} \times \frac{6}{9}}^{\boxed{X}\,block} + \overbrace{\frac{3}{2n} \times \frac{5}{9}}^{\boxed{Y}\,block} = \frac{2}{3} - \frac{1/6}{n}$ Note: every block has one-bit color palette information, and a true color is represented by 24-bit $R, G,$ and B color planes. Because colluders can guess the bit b_9 with $\frac{2}{3} - \frac{1/6}{n}$ probability, P_C is $(\frac{2}{3} - \frac{1/6}{n})^{24}$. □

Therefore, the value $P_C = (\frac{2}{3} - \frac{1/6}{n})^{24}$ is less than $(\frac{2}{3})^{24} \approx 5.94 \times 10^{-5}$, and this implies that the color palette cannot be recovered under $(n-1)$-colluder attack.

6. Evaluation and Comparisons

6.1. Experimental Results

Seven experiments (Experiments $A - H$) are conducted to evaluate the proposed (n, n)-SDIS from various aspects: (A) noise-like shadows NS_1, NS_2, NS_3 sharing 256-color image for $(3, 3)$-SDIS (B) binary meaningful shadows BS_1, BS_2, BS_3 sharing 256-color image for $(3,3)$-SDIS (C) color meaningful shadows CS_1, CS_2, CS_3 sharing 256-color image for $(3,3)$-SDIS (D) color meaningful shadows CS_1, CS_2, CS_3 sharing true color image for $(3,3)$-SDIS (E) binary meaningful shadows $(NS_1 - NS_4)$ and color meaningful shadows $(CS_1 - CS_4)$ for $(4,4)$-SDIS (F) binary meaningful shadows $(NS_1 - NS_5)$ and color meaningful shadows $(CS_1 - CS_5)$ for $(5,5)$-SDIS (G) color meaningful shadows CS_1', CS_2', CS_3' sharing 256-color image for $(3,3)$-SDIS by the approach of enhancing visual quality.

Experiments $A - D$ are the $(3,3)$-SDIS. Experiment A has noise-like shadows, and other four experiments are meaningful shadows. Experiments D demonstrates sharing true color secret image. Experiments E and F demonstrate binary and color meaningful shadows for $(4,4)$-SDIS and $(5,5)$-SDIS, respectively. In Experiment G, we redo Experiment C to enhance the visual quality of color meaningful shadows by using the approach in Figure 6.

In all experiments, five binary cover images $BCI_1 - BCI_5$ with black-and-white printed texts $\boxed{A}, \boxed{B}, \boxed{C}, \boxed{D}, \boxed{E}$, and five color cover images $CCI_1 - CCI_5$ with photos of birds are used. In addition, two secret images SI_1 (Lena: 256-color image), SI_2 (Kaleidoscope: true color image) are used. All these images $BCI_1 - BCI_5$ (see Figure 7), $CCI_1 - CCI_5$ (see Figure 8), and SI_1, SI_2 (see Figure 9) are 256×256 pixels.

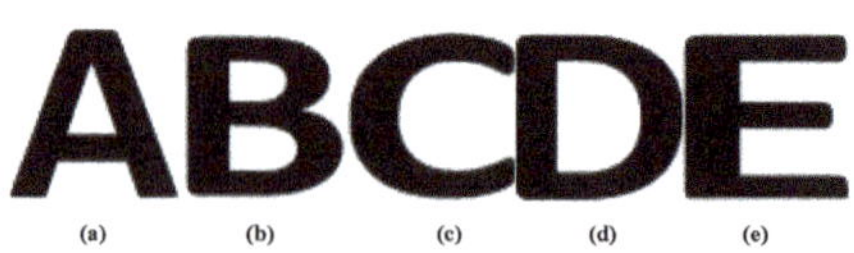

Figure 7. Five color cover images with photos of birds: (**a**) BCI_1 (**b**) BCI_2 (**c**) BCI_3 (**d**) BCI_4 (**e**) BCI_5.

Figure 8. Five color cover images with photos of birds: (**a**) CCI_1 (**b**) CCI_2 (**c**) CCI_3 (**d**) CCI_4 (**e**) CCI_5.

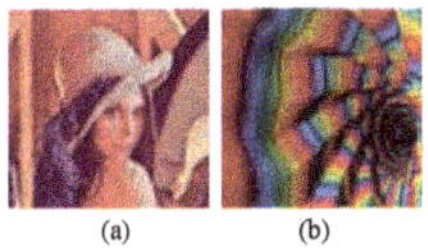

Figure 9. Two secret images: (a) SI_1: 256-color Lena (b) SI_2: true color Kaleidoscope.

Because shadows may be 9 or 25 times expanded in experiments, for demonstrating all the images in a single page, the shadow images in experiments are not correctly proportional.

Experiment A. Three noise-like shadows $NS_1 - NS_3$ of the proposed $(3,3)$-SDIS sharing a 256-color secret image are demonstrated.

The secret image SI_1: 256-color Lena in Figure 9a is used to test the proposed $(3,3)$-SDIS. Each noise-like shadow has $\frac{2n-3}{2n} = \frac{6-3}{6} = 50\%$ $\boxed{X}$ blocks and $\frac{3}{2n} = \frac{3}{6} = 50\%$ $\boxed{Y}$ blocks, which are the same as Yang et al.'s $(2,2)$-SDIS. As shown in Figure 10, three noise-like shadows are expanded to 768×768 pixels. Via recovering procedure, we can recover the original 256-color secret image Lena.

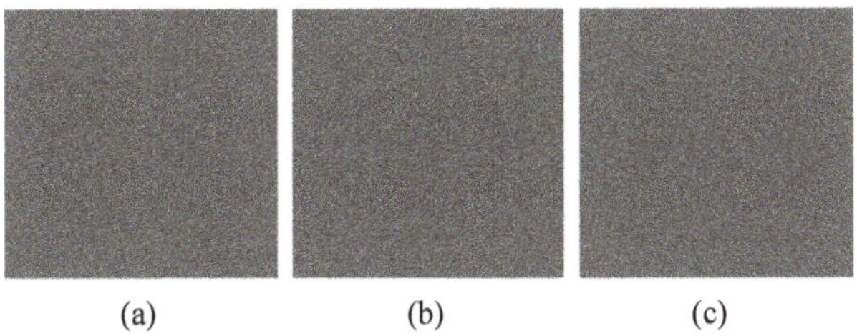

Figure 10. Noise-like shadows of the proposed $(3,3)$-SDIS: (a) NS_1 (b) NS_2 (c) NS_3.

Experiment B. Three binary meaningful shadows $BS_1 - BS_3$ of the proposed $(3,3)$-SDIS sharing a 256-color secret image are demonstrated.

By revering (respectively, unchanging) the color of subpixels in a block of $B^{(1)}$, $B^{(2)}$, and $B^{(3)}$ on NS_1, NS_2 and NS_3 in **Experiment A** to represent the white (respectively, black) color in $BCI_1 - BCI_3$ (A, B, and C in Figure 7a–c). The proposed $(3,3)$-SDIS has the contrast $C_P = \frac{3-(3/n)}{9} = \frac{3-(3/3)}{9} = \frac{2}{9}$ (see Theorem 2). It is observed that the printed-texts A, B, and C are revealed indeed on $BCI_1 - BCI_3$, with the size of 768×768 pixels (see Figure 11a–c). Consider recovery. We first transfer the $3B6W$ block and $4B5W$ block to $6B3W$ block and $5B4W$ block, respectively. Afterwards, via the recovering procedure, we may recover the 256-color secret image Lena.

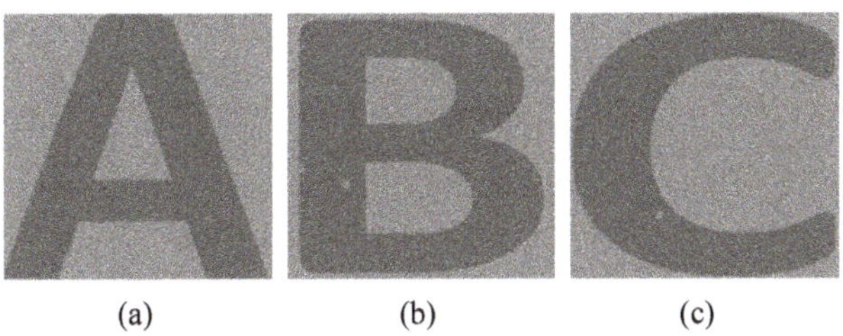

Figure 11. Binary meaningful shadows of the proposed $(3,3)$-SDIS: (a) BS_1 (b) BS_2 (c) BS_3.

Experiment C. Three color meaningful shadows $CS_1 - CS_3$ of the proposed $(3,3)$-SDIS sharing a 256-color secret image are demonstrated.

By adopting the color pixels in $CCI_1 - CCI_3$ into black subpixels in blocks $B^{(1)}, B^{(2)}$, and $B^{(3)}$ on NS_1, NS_2 and NS_3, respectively, we generate three color meaningful shadows $CS_1 - CS_3$ with the size of 768×768 pixels. Each color meaningful shadow has $R_P = \frac{6-(1.5/n)}{9} = \frac{6-(1.5/3)}{9} = \frac{5.5}{9}$. As shown in Figure 12a–c, it is observed that the images of three photos of birds in Figure 8a–c are revealed on $CS_1 - CS_3$. Consider recovery. We first transfer the color subpixel in every block to "1"s and white subpixel to "0". Afterwards, via the recovering procedure, we may recover the 256-color secret image Lena.

Figure 12. Color meaningful shadows of the proposed (3,3)-SDIS: (**a**) CS_1 (**b**) CS_2 (**c**) CS_3.

Experiment D. Three color meaningful shadows $CS_1 - CS_3$ of the proposed (3,3)-SDIS sharing a true color secret image are demonstrated.

The secret image SI_2: true color Kaleidoscope is used to test the proposed (3,3)-SDIS sharing a true color secret image. For a secret pixel, we use the information of R, G, and B color planes to form a 25-bit block. By adopting the color pixels in $CCI_1 - CCI_3$ into three 25-subixle shadow blocks, we can generate three color meaningful shadows $CS_1 - CS_3$ with the size of 1280×1280 pixels (25 times expanded). Each color meaningful shadow has $R'_P = \frac{17}{25} - \frac{0.16}{n} = 0.627$ (see Theorem 3) larger than $R_P = \frac{5.5}{9} = 0.611$ in Experiment C, to show the content of cover image. As shown in Figure 13a–c, it is observed that the images $CCI_1 - CCI_3$ are revealed on $CS_1 - CS_3$. Via the recovering procedure, we may recover the true color secret image Kaleidoscope.

Experiment E. Four binary meaningful shadows $BS_1 - BS_4$ and four color meaningful shadows $CS_1 - CS_4$ of the proposed (4,4)-SDIS sharing a 256-color secret image are demonstrated.

Four binary cover images printed-texts in Figure 7a–d, and four color cover images $CCI_1 - CCI_4$ in Figure 8a–d are used. Finally, four binary meaningful shadows $BS_1 - BS_4$, and four color meaningful shadows $CS_1 - CS_4$ are illustrated in Figure 14a,b, respectively. All these shadows have the sizes of 768×768 pixels. Binary meaningful shadows of (4,4)-SDIS have $C_P = \frac{3-(3/n)}{9} = \frac{3-(3/4)}{9} = \frac{2.25}{9}$, and color meaningful shadows of (4,4)-SDIS have $R_P = \frac{6-(1.5/n)}{9} = \frac{6-(1.5/4)}{9} = \frac{5.625}{9}$. Both values are greater than $\frac{2}{9}$ (Experiment B) and $\frac{5.5}{9}$ (Experiment C), respectively.

Experiment F. Five binary meaningful shadows $BS_1 - BS_5$ and five color meaningful shadows $CS_1 - CS_5$ of the proposed (5,5)-SDIS sharing a 256-color secret image are demonstrated.

Five color cover images printed-texts in Figure 7a–e, and five color cover images $CCI_1 - CCI_5$ in Figure 8a–e are used. Finally, fiver binary meaningful shadows $BS_1 - BS_5$, and five color meaningful shadows $CS_1 - CS_5$ are illustrated in Figure 15a,b, respectively. All these shadows have the sizes of 768×768 pixels. Binary meaningful shadows of (5,5)-SDIS have $C_P = \frac{3-(3/n)}{9} = \frac{3-3/5}{9} = \frac{2.4}{9}$, and color meaningful shadows of (5,5)-SDIS have $R_P = \frac{6-(1.5/n)}{9} = \frac{6-1.5/5}{9} = \frac{5.7}{9}$. Both values are better than those of (3,3)-SDIS.

Experiment G. Redo Experiment C, but use the approach of enhancing visual quality of color meaningful shadows. Three $CS'_1 - CS'_3$ are demonstrated.

In Experiment C, three 256×256-pixel color cover images $CCI_1 - CCI_3$ in Figure 8a–c are used. To enhance the visual quality of $CS_1 - CS_3$, we use another three 768×768-pixel $CCI'_1 - CCI'_3$, which has high resolution. These three images $CCI_1' - CCI_3'$ are omitted here for brevity. By using the approach in Figure 6, we use color pixels in $CCI_1' - CCI_3'$ into black subpixels in blocks $B^{(1)}$, $B^{(2)}$, and $B^{(3)}$ on NS_1, NS_2 and NS_3, respectively to generate three color meaningful shadows $CS_1' - CS_3'$ with the size of 768×768 pixels. As shown in Figure 16a–c, it is observed that Figure 16 has better visual quality than Figure 12. However, the photos $CCI_1 - CCI_3$ used in this experiment may not clearly demonstrate the enhancement. Here, we use a cover image, a colorful centered fractal, for testing. Figure 17(a-1,b-1) shows two color meaningful shadows using the original one and new enhancement, respectively. For clear observation, cropped image areas of Figure 17(a-1,b-1) are shown in Figure 17(a-2,b-2). Visual inspection of cropped image areas in Figure 17(a-2,b-2) reveals that the original method spoils some edges and fine details in shadow images. Our enhancement has clear color sharpness, especially the clearness of edges.

For fairer comparison, we adopt visual quality assessment, the structural similarity (SSIM) index, and the feature similarity (FSIM) index to compare Figure 17(a-1) and Figure 17 (b-1). Let the original image be a colorful centered fractal with the size $768 * 768$ pixels. According to the image quality assessment from Laboratory for Computational Vision in New York University (please refer to https://www.cns.nyu.edu/~lcv/ssim/#usage), to calculate SSIM and FSIM for color images, it would be better to convert the color image to gray image with the formula $0.2989R + 0.5870G + 0.1140B$, and then calculate its SSIM and FSIM. Finally, SSIM and FSIM of Figure 17(a-1) are 0.2532 and 0.8400, and SSIM and FSIM of Figure 17(b-1) are 0.3300 and 0.8498, respectively. These values of SSIM and FSIM demonstrate a consistency with the performance in Figure 17(a-2,b-2).

(a) (b) (c)

Figure 13. Color meaningful shadows of the proposed (3,3)-SDIS sharing a true color secret image: (**a**) CS_1 (**b**) CS_2 (**c**) CS_3.

(a-1) (a-2) (a-3) (a-4)

(b-1) (b-2) (b-3) (b-4)

Figure 14. Binary snd color meaningful shadows of the proposed: (**a**) $BS_1 - BS_4$ (**b**) $CS_1 - CS_4$.

Figure 15. Binary snd color meaningful shadows of the proposed: (**a**) $BS_1 - BS_5$ (**b**) $CS_1 - CS_5$.

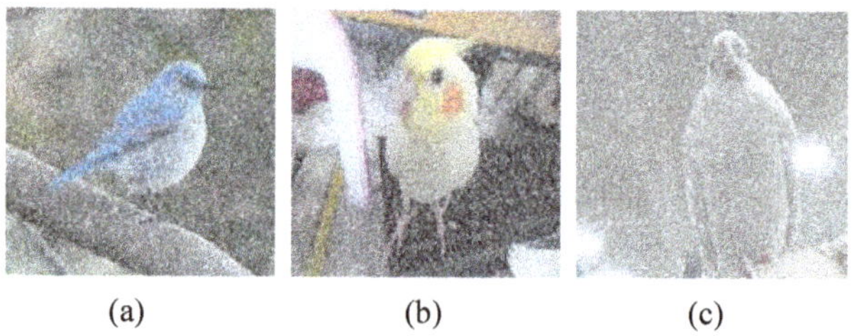

Figure 16. Color meaningful shadows of (3,3)-SDIS by the approach of enhancing visual quality: (**a**) CS_1' (**b**) CS_2' (**c**) CS_3'.

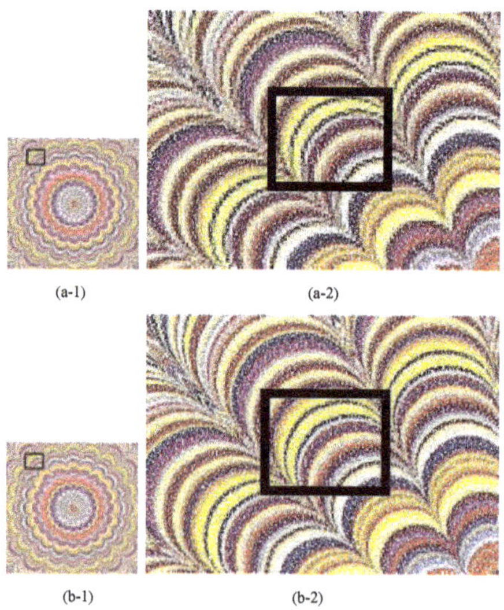

Figure 17. Color meaningful shadows and enlarged parts of cropped image area for (3,3)-SDIS: (**a**) using the original method (**b**) using the approach of enhancing visual quality.

6.2. Discussion and Comparison

6.2.1. Enhancing R_P

In step (S-2), we first randomly generate $(n-2)\boxed{X}$ blocks $B^{(i_1)}, B^{(i_2)}, ..., B^{(i_{n-2})}$. Afterwards, in step (S-5), we evenly use Wei et al.'s (2,2)-SDIS and Yang et al.'s (2,2)-SDIS to generate two other shadows $B^{(j_1)}, B^{(j_2)}$, where $\{j_1, j_2\} = \{1...n\} - \{i_1...i_{n-2}\}$. Finally, R_P is $\frac{6-(1.5/n)}{9}$ (see Equation (5)). In fact, we may further enhance R_P by using $\boxed{W}$ block instead of $\boxed{X}$ block to generate $(n-2)$

$B^{(i_1)}, B^{(i_2)}, ..., B^{(i_{n-2})}$, where $\boxed{W}$ block may be $7B2W$ or $8B1W$. When using $\boxed{W} = 6B3W$, the approach changes back to the original (n, n)-SDIS. By this approach, the R_P is enhanced to $\frac{7-3.5/n}{9}$ and $\frac{8-5.5/n}{9}$ for $\boxed{W} = 7B2W$ and $\boxed{W} = 8B1W$, as derived in Equations (15) and (16), respectively.

$$\begin{cases} R_P = \frac{1}{2} \times \overbrace{\frac{((n-2)\times 7 + 2\times 5)/n}{9}}^{Weietal's(2,2)SDIS} + \\ \frac{1}{2} \times \overbrace{\frac{((n-2)\times 7 + 1\times 5 + 1\times 6)/n}{9}}^{Yanget.al's(2,2)-SDIS} \\ = \frac{3.5-(2/n)}{9} + \frac{3.5-(1.5/n)}{9} = \frac{7-(3.5/n)}{9} \end{cases} \quad (15)$$

$$\begin{cases} R_P = \frac{1}{2} \times \overbrace{\frac{((n-2)\times 8 + 2\times 5)/n}{9}}^{Weietal's(2,2)SDIS} \\ +\frac{1}{2} \times \overbrace{\frac{((n-2)\times 8 + 1\times 5 + 1\times 6)/n}{9}}^{Yanget.al's(2,2)-SDIS} \\ = \frac{4-(3/n)}{9} + \frac{4-(2.5/n)}{9} = \frac{8-(5.5/n)}{9} \end{cases} \quad (16)$$

Consider $(n-1)$-colluder attack for the case using $\boxed{W}$ block with Hamming weight w. The following theorem demonstrates the successful probability to recover the block B under $(n-1)$-colluder attack.

Theorem 6. *When using $\boxed{W}$ block in the proposed (n, n)-SDIS, the successful probability to recover the block B for $(n-1)$-colluder attack is $R_P = \frac{2n-4}{2n} \times \frac{1}{C_9^w} + \frac{1}{2n} \times \frac{1}{C_9^6} + \frac{3}{2n} \times \frac{1}{C_9^5}$.*

Proof. Suppose that using $\boxed{W}$ block with Hamming weight w in step (S-2). Consider the case that colluders already have $(n-1)$ shadows (say $B^{(1)} - B^{(n-1)}$) for reconstruction. Based on these $(n-1)$ shadows, colluders may guess the type of shadow block $B^{(n)}$ in the other shadow, The block $B^{(n)}$ has $\boxed{W}$ block, $\boxed{X}$ block and $\boxed{Y}$ block with $\frac{2n-4}{2n}$ probability, $\frac{1}{2n}$ probability and $\frac{3}{2n}$ probability, respectively, which are derived below. Note: if $\boxed{W}$ is $6B3W$ Equation (17) is reduced to Equation (12).

$$\begin{cases} \frac{1}{2} \times \overbrace{\frac{C_2^2 \times C_{n-1}^1}{C_n^{n-1}}}^{Weietal's(2,2)SDIS} + \frac{1}{2} \times \overbrace{\frac{C_1^1 \times C_1^1 \times C_{n-2}^1}{C_n^{n-1}}}^{Yanget.al's(2,2)-SDIS} \\ = \frac{2n-4}{2n} (B^{(n)} is \boxed{W} block) \\ \frac{1}{2} \times \overset{0}{+} \frac{1}{2} \times \overbrace{\frac{C_1^1 \times C_1^1 \times C_{n-2}^{n-2}}{C_n^{n-1}}}^{Yanget.al's(2,2)-SDIS} = \frac{1}{2n} (B^{(n)} is \boxed{X} block) \\ \frac{1}{2} \times \overbrace{\frac{C_2^1 \times C_{n-2}^{n-2}}{C_n^{n-1}}}^{Weietal's(2,2)SDIS} + \frac{1}{2} \times \overbrace{\frac{C_1^1 \times C_1^1 \times C_{n-2}^{n-2}}{C_n^{n-1}}}^{Yanget.al's(2,2)-SDIS} = \\ \frac{3}{2n} (B^{(n)} is \boxed{Y} block) \end{cases} \quad (17)$$

There is probability $\frac{1}{C_9^w}, \frac{1}{C_9^5}, \frac{1}{C_9^6}$ to guess the correct block B when $B^{(n)}$ is $\boxed{W}$ block, $\boxed{X}$ block, and $\boxed{Y}$ block, respectively. Therefore, the P_P is calculated as follows.

$$P_P = \overbrace{\frac{2n-4}{2n} \times \frac{1}{C_9^w}}^{\boxed{W}} + \overbrace{\frac{1}{2n} \times \frac{1}{C_9^5}}^{\boxed{X}} + \overbrace{\frac{3}{2n} \times \frac{1}{C_9^5}}^{\boxed{Y}} \quad (18)$$

□

The value of P_P is $\frac{1}{C_9^7} - \frac{0.038}{n}$ and $\frac{1}{C_9^8} - \frac{0.204}{n}$ for $w = 7$ and 8. The values are about $\frac{1}{C_9^7} = \frac{1}{36}$ and $\frac{1}{C_9^8} = \frac{1}{9}$, respectively, for large n. Even though these values are larger than $P_P = \frac{1}{C_9^6} - \frac{3}{2n}(\frac{1}{C_9^6} - \frac{1}{C_9^5})$ for using $\boxed{W}$ block in step (S-2), it is still practically secure for applications. This is because our CP information is protected in the XOR-ed result, but not conveyed on $b_9^{(1)}$ in $B^{(1)}$ like (22)-SDIS [17,19]. For example, when using $8B1W$ as $\boxed{W}$ block. If colluders have $(n-1)$ shadows (say $NS_1 - NS_{n-1}$), for a block B, they have the XOR-ed result $B' = B^{(1)} \oplus ... \oplus B^{(n-1)}$, and can guess the shadow block $B^{(n)}$ is $\boxed{W}$ block with a very high probability for large n (note: $\frac{2n-4}{2n} \rightarrow 1$ for large n). It implies that there is about $\frac{8}{9}$ probability that the bit b_9 in B is the complementary bit b_9' of B'. By using the same argument in proof of Theorem 5, for this case, the successful probability to recover a correct color in CP is $P_C = (\frac{8}{9})^{24} \simeq 0.059$. Therefore, we cannot get the correct CP back. Although colluders may recover the first 8 bits $(b_1 - b_8)$ in B, i.e., a color index by complementing the first 8 bits $(b_1' - b_8')$ in B' with $\frac{1}{9}$ probability. This probability of guessing a color index is larger than the brute-force probability $\frac{1}{256}$. However, colluders do not have the correct CP, and thus they cannot recover the original SI. Obviously, it is more difficult to apply $(n-1)$-colluder attack on using $7B2W$ as $\boxed{W}$ block, because P_C is $(\frac{7}{9})^{24} \simeq 0.0024$. This is why we claim that using $\boxed{W}$ block is still practically secure for applications.

To demonstrate the above phenomenon, we use $8B1W$ as $\boxed{W}$ block in the proposed (5,5)-SDIS. Five color meaning shadows using color cover images $CCI_1 - CCI_5$ in Figure 8a–e are illustrated in Figure 18a, where the approach of enhancing visual quality in Section 4.3 is also adopted. Figure 18b are the 256-color SI (Lena), and its corresponding CP. The recovered 256-color secret image SI' and the color palette CP' are shown in Figure 18c. It is observed that these five color meaning shadows in Figure 18a have high resolutions with $R_P = \frac{8-5.5/n}{9} = 0.767$ for $n = 5$, which have better visual qualities than those in Figure 15b. From, Figure 18c, there is not any secret information of CP and SI leaked for $(n-1)$-colluder attack.

Figure 18. The proposed (5,5)-SDIS using $8B1W$ block (**a**) five color meaningful shadows (**b**) 256-color SI and its corresponding CP (**c**) the recovered 256-color SI^1 and color palette CP' under $(n-1)$-colluder attack.

6.2.2. Comparison

We extend (2,2)-SDIS to the proposed (n,n)-SDIS. Because the percentage of $\boxed{X}$ block is greater than 50%, the resolution of binary and color meaningful shadows are enhanced. Note: Yang et al.'s (2,2)-SDIS uses $\boxed{X}$ block and $\boxed{Y}$ block half and half, while Wei et al.'s (2,2)-SDIS only uses $\boxed{Y}$ blocks. On the other hands, Wei et al.'s (2,2)-SDIS has the incorrect assignment of color palette data for the color index 255. This problem comes from from all-1 9-bit vector. In [19], Yang et al. adopted a complicated approach using partitioned sets to address this problem. In the proposed (n,n)-SDIS, the number of shadows of (n,n)-SDIS is more than two, i.e., $n \geq 3$. Thus, we can easily adopt a simple approach by reducing $H(T)$ to $H(T) = 7$ in step (S-4) via modifying any one shadow block to solve this problem. Meantime, as described in Section 5.1, we may enhance R_P and simultaneously retain the practical security by using $\boxed{W}$ block.

As shown in Table 2, a complete comparison is given among Wei et al.'s (2,2)-SDIS, Yang et al.'s (2,2)-SDIS, and the proposed (n,n)-SDIS. The comparison includes the structure of block, percentages of blocks, the region in color meaningful shadows revealing cover image, the contrast of binary meaningful shadows, enhancing R_P, the embedding of color palette data, where to embed color palette data, enhancing visual quality of color meaningful shadows, encoding/decoding complexity, and the security. About the security, although the successful probability to recover B under $(n-1)$-colluder attack $P_P = \frac{1}{C_9^6} - \frac{3}{2n}(\frac{1}{C_9^6} - \frac{1}{C_9^5}) \simeq \frac{1}{C_9^6} = 0.012$ for large n is larger than $P_W = \frac{1}{C_9^5} = 0.008$ and $P_Y = \frac{1/C_9^6 + 1/C_9^5}{2} = 0.01$. This value is still practical secure for practical application. Besides, the CP of the proposed (n,n)-SDIS cannot be obtained under $(n-1)$-colluder attack, but the CP of $(2,2)$-SDIS can be obtained from only one shadow. Based on this observation, the proposed (n,n)-SDIS is much securer than $(2,2)$-SDIS.

Table 2. Comparison of Three SDIS Schemes.

		Wei et al.'s $(2,2)$-SDIS	Yang et al.'s $(2,2)$-SDIS	The Proposed (n,n)-SDIS
number of shadows		2	2	$n \geq 3$
structure of block		$\boxed{Y}$ block	$\boxed{X}$ and $\boxed{Y}$ blocks	$\boxed{X}$ and $\boxed{Y}$ blocks
percentage of block		$\boxed{Y}$:100%	$\boxed{X}$:50%, $\boxed{Y}$:50%	$\boxed{X}:\frac{2n-3}{2n}$, $\boxed{Y}:\frac{3}{2n}$
region in color shadows revealing cover image		$R_W = \frac{5}{9}$	$R_Y = \frac{5.5}{9}$	$R_P = \frac{6-1.5/n}{9}$
			$R_W < R_Y \leq R_P$	
contrast of binary meaningful shadows		$C_W = \frac{1}{9}$	$C_Y = \frac{2}{9}$	$C_P = \frac{3-3/n}{9}$
			$C_W < C_Y \leq C_P$	
enhancement of R_P		No	No	Yes
embedding the data of color palette data		having a problem for the color index 255	using partitioned sets for some color indices	using a simple approach by reducing Hamming weight
where to embed color palette data		the bit $b_9^{(1)}$ in $B^{(1)}$	the bit $b_9^{(1)}$ in $B^{(1)}$	the bit b_9 in the XOR-ed B
enhancing visual quality of color meaningful shadows		No	No	Yes
encoding/decoding complexity		XOR operation	XOR operation; lookup table	XOR operation
security	probability to recover B under $(n-1)$-colluder	$P_W = \frac{1}{C_9^5}$	$P_Y = \frac{1/C_9^5 + 1/C_9^6}{2}$	$P_P = \frac{1}{C_9^6} - \frac{3}{2n}(\frac{1}{C_9^6} - \frac{1}{C_9^5})$
			$P_W < P_Y \leq P_P$	
	probability to obtain CP under $(n-1)$-colluder	CP can be obtained from only one shadow	CP can be obtained from only one shadow	$P_C = (\frac{2}{3} - \frac{1/6}{n})^{24}$

7. Conclusions

In this paper, we discussed the general (n,n)-SDIS, which can be applied to any $n \geq 3$. The proposed (n,n)-SDIS is skilfully implemented on basis of previous $(2,2)$-SDIS. Our main contribution is theoretically to prove the proposed (n,n)-SDIS being able to resist $(n-1)$ colluder attack. Meanwhile, the contrast of binary meaningful shadow and the region in color shadows revealing cover image are both enhanced. The main weakness of Wei et al.'s $(2,2)$-SDIS is the incorrect assignment of color palette data for some color indices, and this is tackled by using partitioned sets in Yang et al.'s $(2,2)$-SDIS. In the proposed (n,n)-SDIS, because of the number of shadows more than two, i.e., $n \geq 3$, a simple approach reducing Hamming weigh of a temporary block can be adopted to easily solve this weakness. Since the proposed (n,n)-SDIS is based on color palette and resistant to $(n-1)$-colluder attack, and also enhances the visual quality of meaningful shadows, it is suitable for modern visual communication applications where features such as secure transmission, storage sensitive, and high-quality image reconstruction are required.

Author Contributions: Designing scheme, writing—original draft preparation, C.-N.Y.; security analysis, Q.-D.S.; editing and responding to reviewer, Y.-X.L.; experiment, C.-M.W.

Funding: Ministry of Science and Technology, under Grant MOST 107-2221-E-259-007, 108-2221-E-259-009-MY2; Natural Science Foundation of China under Grant No. 61502384,61571360,61872289.

Conflicts of Interest: The authors declare there is no conflicts of interest regarding the publication of this paper

References

1. Naor, M.; Shamir, A. Visual cryptography. In *Advances in Cryptology-EUROCRYPT'94;* LNCS 950; Springer: Berlin/Heidelberg, Germany, 1995; pp. 1–12.
2. Shyu, S.J.; Jiang, H.W. General constructions for threshold multiple-secret visual cryptography Schemes. *IEEE Trans. Inf. Forensics Secur.* **2013**, *8*, 733–743. [CrossRef]
3. Yang, C.N.; Wu, C.C.; Lin, Y.C. *k* out of *n* region-based progressive visual cryptography. *IEEE Trans. Circuits Syst. Video Technol.* **2017**. [CrossRef]
4. Karolin, M.; Meyyappan, T.; Thamarai, S.M. Encryption and decryption of color images using visual cryptography. *Int. J. Pure Appl. Math.* **2018**, *118*, 277–281.
5. Kansal, I.; Kasana, S.S. Sharing two true colour images using $(3, 3)$-extended visual cryptography technique. *J. Mod. Opt.* **2018**, *65*, 1949–1959.
6. Yang, C.N.; Wu, F.H.; Peng, S.L. Enhancing multi-factor cheating prevention in visual cryptography based minimum (k, n)-connected graph. *J. Vis. Commun. Image Represent.* **2018**, *55*, 660–676. [CrossRef]
7. Shamir, A. How to share a secret. *Commun. Assoc. Comput. Mach.* **1979**, *22*, 612–613. [CrossRef]
8. Thien, C.C.; Lin, J.C. Secret image sharing. *Comput. Graph.* **2002**, *26*, 765–770. [CrossRef]
9. Liu, Y.X.; Yang, C.N.; Wu, C.M.; Sun, Q.D.; Bi, W. Threshold changeable secret image sharing scheme based on interpolation polynomial. *Multimed. Tools Appl.* **2019**, *78*, 18653–18667. [CrossRef]
10. Yang, C.N.; Chen, T.S.; Yu, K.H.; Wang, C.C. Improvements of image sharing with steganography and authentication. *J. Syst. Softw.* **2007**, *80*, 1070–1076. [CrossRef]
11. Pakniat, N.; Noroozi, M.; Eslami, Z. Secret image sharing scheme with hierarchical threshold access structure. *J. Vis. Commun. Image Represent.* **2014**, *25*, 1093–1101. [CrossRef]
12. Liu, Y.X.; Zhang, Y.Z.; Yang, C.N. Reducing file size and time complexity in secret sharing based document protection. *Math. Biosci. Eng.* **2019**, *16*, 4802–4817. [CrossRef]
13. Kanso, A.; Ghebleh, M. An efficient lossless secret sharing scheme for medical images. *J. Vis. Commun. Image Represent.* **2018**, *56*, 245–255. [CrossRef]
14. Li, P.; Liu, Z.; Yang, C.N. A construction method of (t, k, n)-essential secret image sharing scheme. *Signal Process. Image Commun.* **2018**, *65*, 210–220. [CrossRef]
15. Wu, X.; Yang, C.N.; Zhuang, Y.T.; Hsu, S.C. Improving recovered image quality in secret Image sharing by simple modular arithmetic. *Signal Process. Image Commun.* **2018**, *66*, 42–49. [CrossRef]
16. Lukac, R.; Plataniotis, K.N. Bit-level based secret sharing for image encryption. *Pattern Recognit.* **2005**, *38*, 767–772. [CrossRef]
17. Wei, S.C.; Hou, Y.C.; Lu, Y.C. A technique for sharing a digital image. *Comput. Stand. Interfaces* **2015**, *40*, 53–61. [CrossRef]
18. Yang, C.N.; Chen, C.H.; Cai, S.R. Enhanced Boolean-based multi secret image sharing scheme. *J. Syst. Softw.* **2016**, *116*, 22–34. [CrossRef]
19. Yang, C.N.; Wu, C.H.; Yeh, Z.X.; Wang, D.; Kim, C. A new sharing digital image scheme with clearer shadow images. *Comput. Stand. Interfaces* **2017**, *51*, 118–131. [CrossRef]
20. Liu, Y.X.; Yang, C.N.; Wu, S.Y.; Chou, Y.S. Progressive (k, n) secret image sharing schemes based on Boolean operations and covering codes. *Signal Process. Image Commun.* **2018**, *66*, 77–86. [CrossRef]

electronics

MDPI

Article

A Low-Cost, High-Precision Method for Ripple Voltage Measurement Using a DAC and Comparators

Jincheng Liu [1], Jiguang Yue [1], Li Wang [1,*], Chenhao Wu [1] and Feng Lyu [2]

[1] College of Electronic and Information Engineering, Tongji University, No. 4800, Cao'an Highway, Shanghai 201804, China; 1730755@tongji.edu.cn (J.L.); yuejiguang@tongji.edu.cn (J.Y.); 1152448@tongji.edu.cn (C.W.)

[2] School of Ocean and Earth Science, Tongji University, No. 1239, Siping Road, Shanghai 200092, China; lf@tongji.edu.cn

* Correspondence: 2015wangli@tongji.edu.cn; Tel.: +86-021-6598-9241

Received: 1 May 2019; Accepted: 23 May 2019; Published: 27 May 2019

Abstract: As the core of electronic system, the switched-mode power supply (SMPS) will lead to serious accidents and catastrophes if it suddenly fails. According to the related research, the monitoring of ripple can acquire the health degree of SMPS indirectly. To realize low-cost, high-precision, and automatic ripple measurement, this paper proposes a new ripple voltage (peak-to-peak value) measuring scheme, utilizing a DAC and two high-speed comparators. Within this scheme, the DC component of SMPS output is blocked by a high-pass filter (HPF). Then, the filtered signal and the reference voltage from a DAC together compose the input of a high-speed comparator. Finally, output pulses of the comparator are captured by a microcontroller unit (MCU), which readjusts the output of the DAC by calculation, and this process is repeated until the DAC output is exactly equal to the peak (or valley) value of ripple. Moreover, in order to accelerate the measurement process, a peak estimation method is specially designed to calculate the output ripple peak (or valley) value of buck topology through merely two measurements. Then the binary search method is utilized to obtain a more exact value on the basis of estimative results. Additionally, an analysis of the measurement error of this ripple measurement system is executed, which shows that the theoretical error is less than 0.5% where the ripple value is larger than 500 mV. Furthermore, appropriate components are selected, and a prototype is manufactured to verify the validity of the proposed theory.

Keywords: ripple voltage measurement; DAC; comparator; peak-ripple estimation; binary search; low-cost

1. Introduction

Switched-mode power supply (SMPS) is widely applied due to the advantages of low power consumption and high efficiency [1]. However, SMPS may lead to serious accidents and catastrophes if it suddenly fails. Among the components used in SMPS, aluminum electrolytic capacitors (AECs) have the shortest life and are the most common source of failure [2–6]. During the period of usage, the performance of AECs will degrade continuously until the whole power system fails. A decreased capacitance value and increased equivalent series resistance (ESR) is the major feature of the AECs' degradation [7,8]. Therefore, the health degree of the SMPS can be obtained by the real-time monitoring of AEC parameters (capacitance or ESR), which also contributes to preventive maintenance and the indication of future failure occurrences.

According to differences in data acquisition, the current research achievement of using ripple to evaluate the health state of AECs can be summarized as the indirect calculation of ripple through simulation software [9–11], the offline computing from sampled data [12], AEC parameter calculation based on the mathematical model of a particular topology and some specific measurement value (voltage or current) [13–19], and direct measurement of ripple through high-cost oscilloscopes, data acquisition cards, or high-speed analog-to-digital converters (ADCs) [20–23]. Nevertheless, the developed methods are rarely adopted in practical industry applications due to the increased cost, complexity, and other relevant issues [8,24]. In practice, the method using specific topology mathematical models to indirectly obtain the ripple (or ESR) presents low generality and a measurement precision that highly depends on the accuracy of components. The use of a high-speed ADC or oscilloscope is restricted in the laboratory and hard to popularize because of the high cost of data acquisition.

Compared to other health-monitoring methods, calculating the ESR with ripple values has an intuitive feature and does not require complicated calculations. Figure 1 shows the principle of ripple generation in a common buck topology. That is, the high-frequency pulse-width modulation (PWM) signal causes voltage fluctuations of the same frequency, which cannot be completely filtered by the subsequent filter circuit, thus generating high-frequency ripple at the output. Due to the high frequency of the PWM, the impedance of the output AEC is almost equal to the impedance of its ESR at this frequency. Therefore, when the load is constant, the relationship between the ripple value and the ESR is approximately proportional, which indicates that the remaining life of the AECs can be easily obtained from ripple. However, in order to accurately measure high-frequency ripple, high-speed ADCs with sampling rates far exceeding the switching frequency must be used, which makes the relevant research not able to be applied to industrial applications such as built-in test (BIT) due to the high cost of data acquisition. Thus, it is of significance to focus on the technology of common low-cost, high-precision measurement of ripple wave.

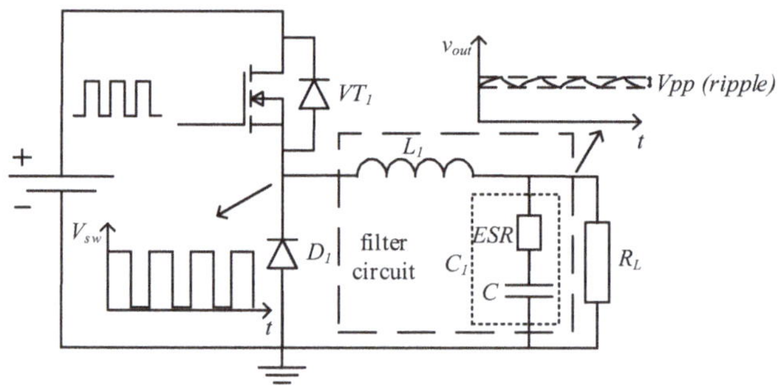

Figure 1. The occurrence of ripple in switched-mode power supply (SMPS) (taking buck topology as an example).

When built-in test (BIT) is required for health monitoring based on ripple measurement, researchers always turn to other low-cost instruments. Therefore, several special ripple measurement schemes have been designed. Some studies focus on root mean square (RMS) measurements, one of which is to measure ripple by AC millivoltmeter [25]. Le provides another scheme based on RMS-to-DC chips [26]. Since the ripple voltage is not a standardized sinusoidal signal, the deteriorate degree of its spike could not be purely obtained by RMS measurement, thus the schemes above have not been widely applied.

Since ripple voltage could be measured by obtaining both peak and valley values, there exist various approaches to acquire the peak value. The well-known basic peak value detector is a solution convenient to implement; however, it is limited by bandwidth and is vulnerable to transient changes in tested SMPS. Some related studies have made improvements regarding. Jerry proposed an analogy peak measurement circuit, but it could not obtain the accurate value of ripple and only determined whether there was ripple beyond a fixed reference value [27]. Zhou offered an implementation of ripple testing by specific peak detecting chips, but its low bandwidth increases measurement error [28].

Smith proposed an improved peak measurement method by adding a high-speed comparator with open-drain output. It has the property of ameliorative accuracy at high frequency without diode, but it fails to acquire the valley voltage [29]. Ren proposed a method to measure the peak-to-peak value of a periodic signal [30]. In this scheme, the input periodic signal and a reference signal from a digital-to-analog converter (DAC) constitute both inputs of the comparator, whose output is captured by an MCU used for adjusting the DAC value by a binary search method. After multiple operating cycles, the DAC output could gradually approximate to the peak or valley values of the input signal. But it is only designed for power frequency AC measurement; in this case, the convergence is too slow. The authors did not conduct either simulation or practical experiments.

Under constant load, the SMPS output ripple can be regarded as a periodic signal. This paper proposes some improvements and realizes low-cost peak-to-peak ripple measurement. The contributions of this study include: the design of the ripple measurement system, the analysis of the measurement error, and the proposal of an optimization algorithm with a higher convergence speed according to the shape characteristics of the ripple signal. The experimental results indicate that the measurement error of this designed system is less than 0.5% with 20 ms consumption where the ripple value is larger than 500 mV, which completely satisfies the requirements of engineering applications and scientific research.

2. Ripple Measurement Scheme and Error Analysis

The ripple value of SMPS could be acquired by an expensive high-speed ADC. However, the high price makes it difficult to apply in BIT. This study provides a new scheme to cut down the expense by replacing the ADC with two high-speed comparators and adding a DAC to generate the reference voltage (one DAC is capable of measuring the peak and valley of a ripple at different moments and one high speed comparator is also enough with utilizing an analog switch). This method ensures accuracy with cost reduction. The designed ripple measurement system consists of four modules: high-pass filter (HPF) circuit, peak measurement, valley measurement, and MCU feedbacks. The block diagram of ripple measurement and the inner signals of peak measurement are shown in Figure 2. And the waveforms of key nodes are shown in Figure 3.

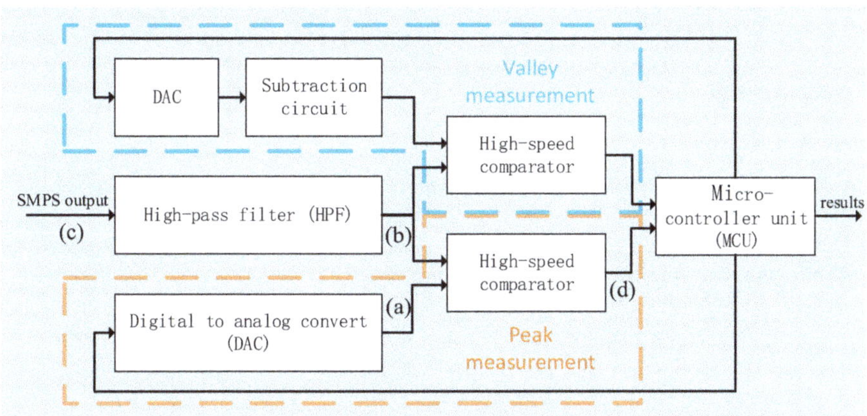

Figure 2. Block diagram of peak measurement and some inner signals.

Figure 3c shows that the input signal originating from the output of the SMPS contains a high DC bias and an AC ripple. In the first step, an HPF is used to filter out the DC component. Ideally, the high-frequency ripple signal is hardly affected, while the DC voltage is completely eliminated. The filtered ripple voltage is shown in Figure 3b.

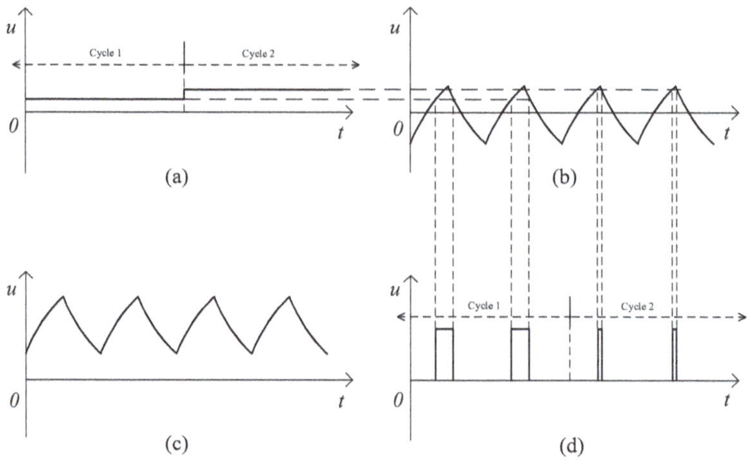

Figure 3. Waveform of key nodes in Figure 2. (**a**) The DAC output, with constant voltage in each feedback cycle. (**b**) The HPF output, only maintaining an AC ripple. (**c**) The SMPS output, containing a high DC bias and an AC ripple. (**d**) The output of comparator, providing a periodic pulse in most cases.

The workflow of the measurement system is somewhat similar to the ADC with a successive approximation register (SAR). Each measurement requires several feedback cycles for convergence. During a certain feedback cycle, as illustrated in Figure 3a, the DAC output voltage is constant and combines with the ripple signal to compose the input of the high-speed comparator. When the DAC output is lower than the peak, the high-speed comparator provides a periodic pulse, as shown in Figure 3d, otherwise it remains low (this situation is not shown in Figure 3). Once any periodic pulse is captured by the MCU, it increases the DAC output to approximate the peak. Otherwise, it drops the DAC value. The MCU repeats the above procedure until it cannot capture the pulse exactly, when the output of the DAC is highly equal to the peak voltage of the ripple.

As shown in the block diagram, the measurement system mainly includes an HPF, high-speed comparator, and DAC (with its reference source). The circuit design and the error analysis of each part are introduced below.

2.1. High-Pass Filter (HPF)

The HPF is used to block the DC component of SMPS. However, before the DC-blocking capacitor C_1 has been charged, the high DC voltage may damage the comparator and MCU. Therefore, the protection diode D_1 should be added. In order to provide a discharge path for C_1 after the input switches off, the resistor R_1 should be placed in the circuit. The modified HPF diagram is shown in Figure 4a.

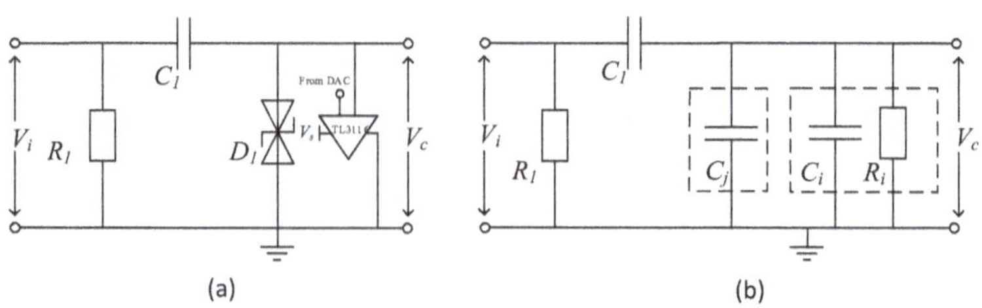

Figure 4. HPF diagram, (**a**) the actual components, (**b**) the parasitic parameters of components.

By setting the component values properly, the input DC element can be completely filtered out, while the AC element can be fully reserved. However, as Figure 4 demonstrated, the parasitic parameters such as the input impedance of the subsequent high-speed comparator and the parasitic capacitance of the protection diode would introduce a small attenuation to the input signal. Since the impedance of R_i is much larger than that of C_j and C_i, the HPF transfer function would be approximately expressed as follow

$$v_c = \frac{\frac{1}{jw(C_i+C_j)}}{\frac{1}{jwC_1} + \frac{1}{jw(C_i+C_j)}} \cdot v_i, \tag{1}$$

$$v_c = \frac{C_1}{C_1 + C_i + C_j} \cdot v_i, \tag{2}$$

$$v_c = (1 + \delta_{hpf}) \cdot v_i, \tag{3}$$

where

$$\delta_{hpf} = -\frac{C_i + C_j}{C_1 + C_i + C_j}. \tag{4}$$

The high-pass filter δ_{hpf} is independent of frequency, hence the peak value of the input voltage v_{ip} and the peak value of the filtered signal have similar expressions compared to Equation (3)

$$v_{cp} = (1 + \delta_{hpf}) \cdot v_{ip}. \tag{5}$$

Since the exact values of C_j and C_i are difficult to obtain, the system error of the HPF circuit is hard to correct and is added in δ_{hpf}, which get its maximal value when C_1 turns to minimal and C_i and C_j to maximal.

2.2. High-Speed Comparator Circuit

The measurement error caused by the comparator mainly includes two aspects: one part is affected by the frequency of the input signal and the bandwith of the comparator, and the other is derived from the manufacturing characteristics of different input transistors of comparators,which may not be exactly matched. By choosing the high-speed comparator with a cut-off frequency much higher than the ripples', the error caused by the ripple frequency can be ignored. Then the latter component, which is represented by the input offset voltage v_{offset}, can be considered as the dominated error.

The two input signals of the comparator come from the DAC and HPF, respectively. Each filtered signal peak v_{cp} corresponds to a threshold point. If the DAC output crosses the point, the comparator output flips.

$$v_{th} = v_{cp} + v_{offset} \tag{6}$$

$$v_{th} = (1 + \delta_{cmp}) \cdot v_{cp}, \tag{7}$$

where

$$\delta_{cmp} = \frac{v_{offset}}{v_c}. \tag{8}$$

2.3. DAC Circuit

The digital signal from the MCU is converted into analog quantity by a DAC, which requires a input reference voltage (v_{ref}) to operate properly. Given a digital signal M, the ideal output voltage of a DAC is

$$v_{dac\_ideal} = \frac{M}{2^N} \cdot v_{ref}. \tag{9}$$

Since M is a discrete value, v_{dac_ideal} may not be totally equal to v_{th}. Given the assumption that

$$v_{th} = \frac{m}{2^N} \cdot v_{ref}, \tag{10}$$

then M is the integral part of m, and hence it generates quantization error δ_q.

$$M = [m] = m \cdot (1 + \delta_q), \tag{11}$$

where

$$|\delta_q| < \frac{1}{m}. \tag{12}$$

The conversion accuracy of the DAC circuit is affected by a series of factors. For example, the input reference voltage (v_{ref}) is generated by a reference source with a subtle error. The DAC characteristics such as integral nonlinearity and DAC differential nonlinearity also reduce the accuracy. Generally, the DAC error is 2 or 3 times its least significant bit (LSB). Let N stand for the number of DAC bits and K stand for the LSBs of the DAC error. The DAC output v_{dac} would be derived as follows:

$$v_{dac} = \frac{M + K}{2^N} \cdot (v_{ref} + \Delta v_{ref}), \tag{13}$$

$$v_{dac} = (1 + \delta_{dac})(1 + \delta_{ref}) \cdot \frac{M}{2^N} \cdot v_{ref}, \tag{14}$$

where

$$\delta_{dac} = \frac{K}{M}, \tag{15}$$

$$\delta_{ref} = \frac{\Delta v_{ref}}{v_{ref}}. \tag{16}$$

2.4. Total Error

When the digital signal M meet the threshold point, the measure ripple v_m could be calculated by:

$$v_m = \frac{M}{2^N} \cdot v_{ref} \tag{17}$$

On the basis of Equations (4), (8), (12), (15), and (16),

$$v_m = (1 + \delta_{hpf})(1 + \delta_{cmp})(1 + \delta_q)(1 + \delta_{dac})(1 + \delta_{ref}) \cdot v_{ip}. \tag{18}$$

Since $\delta_{hpf}, \delta_{cmp}, \delta_q, \delta_{dac}$, and δ_{ref} are much smaller than 1, the total error could be approximately expressed as

$$v_m = (1 + \delta)v_{ip}, \tag{19}$$

where

$$\delta \approx \delta_q + \delta_{dac} + \delta_{ref} + \delta_{cmp} + \delta_{hpf}. \tag{20}$$

In the valley measurement, in order to generate a negative reference voltage, the DAC output is processed by a subtractor that contains an operational amplifier (OPA). In this study, the error of a high-precision OPA with an offset voltage is much smaller than the error of the selected high-speed comparator, hence the subtractor error could be negligible.

Referring to Equations (8) and (12), the total measurement error is related to the ripple amplitude. The larger the ripple value is, the smaller the measurement error. In the case that the ripple is too small, a special amplifying circuit is required, which is not mentioned in this paper.

Consider the traditional measurement scheme, i.e., using a high-speed ADC to directly sample the HPF output signal. Similar to the derivation of Equation (20), the measurement error of a high-speed ADC scheme can be described as follows:

$$\delta_{adc} = \delta_{hpf} + \delta_q + \delta_{adc} + \delta_{ref}, \tag{21}$$

where δ_{hpf}, δ_q, δ_{adc}, and δ_{ref} have similar expressions to Equations (4), (12), (15), and (16), respectively. But high-speed ADC is generally with 8 or 10 bits, which is less than the slow-speed DAC, leading to a greater quantization error δ_q and ADC internal error δ_{adc}. The exact value of the errors will be further discussed in the section on experiment verification.

3. Ripple Waveform Analysis

The measurement error of this design is determined by the components, and the measurement speed can be improved by using some prior information. In some applications, such as predicting the remaining life of SMPS by ripple value, the information about the topology of SMPS and the ripple waveform characteristics are already known, which can contribute to the feedback cycle reduction. This paper takes the typical buck converter topology as an example. Its output ripple can be approximated as a triangular wave in continuous current mode (CCM). Then the output signal of an HPF can be described as follows:

$$v_c = \begin{cases} v_{min} + k_1 \cdot t & 0 \le t < DT \\ v_{max} + k_2(t - DT) & DT \le t < T, \end{cases} \tag{22}$$

where v_{min} and v_{max} are the minimum and maximum value of v_c, respectively, T is the switching period of the SMPS to be measured, D is the duty ratio, and k_1 and k_2 are the rising and falling slopes of the approximated triangular wave, respectively. Since v_c has only AC components, the triangular waves are symmetrical, and the ripple signal is periodic, we can obtain following expressions:

$$v_{min} = -v_{max}, \tag{23}$$

$$v_c(0) = v_c(T) = v_{min}, \tag{24}$$

$$v_c(DT) = v_{max}. \tag{25}$$

Referring to Equations (22)–(25), we can obtain

$$\frac{k_1}{k_2} = \frac{D-1}{D}. \tag{26}$$

Moreover, the mathematical expression of the comparator output is as follows:

$$y_c = \begin{cases} 1 & v^+ > v^- \\ 0 & v^- > v^+ , \end{cases} \tag{27}$$

where v^+ and v^- are the comparator's noninverting and inverting inputs, respectively. The comparator can produce a square wave under the condition that Equation (28) is satisfied. Otherwise, it either remains high or remains low.

$$v_{min} \le v_{dac} \le v_{max}. \tag{28}$$

The square wave in a single cycle is described as follows:

$$y_c = \begin{cases} 0 & 0 < t < \frac{v_{dac} - v_{min}}{k_1} \\ 1 & \frac{v_{dac} - v_{min}}{k_1} < t < \frac{v_{dac} - v_{max}}{k_2} + DT \\ 0 & \frac{v_{dac} - v_{max}}{k_2} + DT < t < T. \end{cases} \tag{29}$$

In combination with Equations (22)–(29), the high duration t_h of the output signal is derived as follows:

$$t_h = \frac{(k_1 - k_2)(v_{dac} - v_{max})}{k_1 k_2}. \tag{30}$$

If the DAC's two output values v_{dac1} and v_{dac2} both satisfy Equation (28) and their corresponding high-level durations are t_{h1} and t_{h2}, they correspond to the relationship

$$\frac{t_{k1}}{t_{k2}} = \frac{v_{dac1} - v_{max}}{v_{dac2} - v_{max}}, \tag{31}$$

$$v_{max} = \frac{v_{dac} t_{h2} - v_{dac2} t_{h1}}{t_{h2} - t_{h1}}. \tag{32}$$

Equation (32) means only two cycles are required for the peak measurement of buck topology, but it is not recommended that it be directly applied because it enlarges the measurement error. The practical measurement algorithm will be discussed in the next section. The principle of valley measurement is similar to peak measurement, and the details will not be described here.

4. Algorithm Design

In order to converge the DAC output to the ripple peak rapidly and accurately, the MCU should capture the output pulses of the high-speed comparator and adjust the DAC output according to a certain method. For example, binary search is an available algorithm that updates one DAC bit at each iteration to narrow the search region. Hence, the iterative cycles are equal to the DAC bits, that is, the higher DAC accuracy it achieves, the greater number of iterations it costs. The required time for an iteration contains three parts: pulse capturing, instructions transmission, and setting time of DAC output. The first is the dominant cost, which is always several dozen times higher than the ripple period for reducing the random error. Therefore, a faster measurement can be realized by reducing the cycle number or cutting the waiting time in each iteration. According to Equation (32), only two measurements are necessary to calculate the peak voltage. However, there is an estimated error since the ripple is not completely equivalent to the triangular wave and the pulse width measurement also introduces quantization errors. Therefore, a new algorithm based on the triangular wave approximation and the binary search is proposed here. The algorithm not only improves the convergence speed but also guarantees measurement accuracy. Firstly, the algorithm estimates the approximate peak value by two measurements. Secondly, the algorithm determines the upper and lower bounds of the peak in the vicinity of the estimated value. Finally, the binary search is used to ensure the converge of DAC output to the peak of the ripple. The flowchart is shown in Figure 5, and the details of three steps are introduced below and shown in Figure 6 .

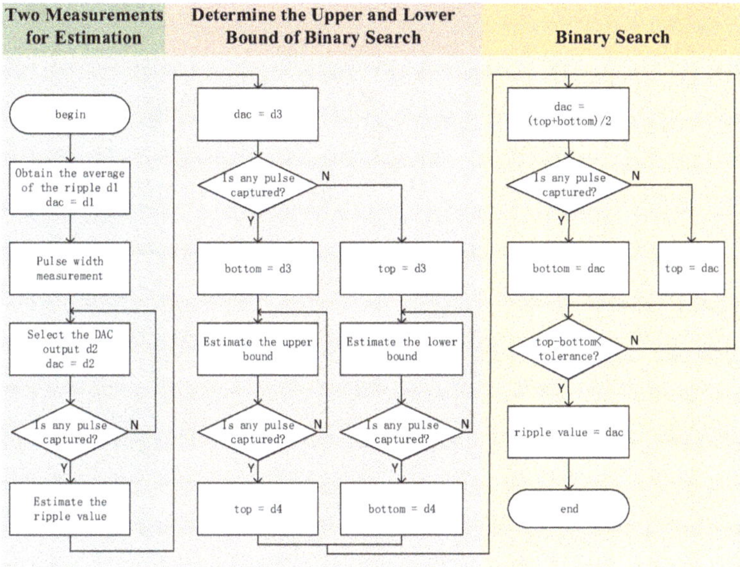

Figure 5. Flowchart of the proposed algorithm.

4.1. Step 1. Estimate the Peak Value by Two Measurements

According to Equation (32), two sets of measurement data that satisfy Equation (28) are required for peak estimation. The first set can be accessed by taking DAC output as the average of ripples (zero, in most cases). In order to obtain the second data set, an appropriate DAC output should be selected. If it is too small, the estimated error will be quite large; when it is too large, Equation (28) is not satisfied. Therefore, we take 1/8 of the scale range of DAC output in the initial attempt. If it is still larger than the peak value, then we continue to take another 1/8 of the last output until the pulse signal is captured. When the ripple is small, the convergence speed of the algorithm is three times of the traditional binary search. Then the MCU calculates the ripple peak after obtaining two sets of measurement data.

4.2. Step 2. Determine the Upper and Lower Bounds of Binary Search from the Estimated Value

The DAC output is set to the estimated value of step 1. If the estimated value is lower than the ripple peak, the MCU can receive pulses from the comparator briefly. At this time, the above estimation value can be set as the lower bound of the binary search. Otherwise, the upper bound is acquired. When the lower bound has already been determined, an empirical constant (for example, 50 mV) is added to the lower bound as the upper bound. If the value is still smaller than the peak, the MCU then sets the DAC output as the new lower bound and iterates the process until the MCU fails to capture the comparator output pulse. In general, the upper bound could be determined without repeated iterations. The upper bound is determined in the calculation process, which is similar to the case shown above in Figure 5.

4.3. Step 3. Determine the Peak Value based on Binary Search

When the lower and upper bounds are all determined, the binary search method can be applied to obtain the DAC output. The DAC output is firstly set to the average value of the two bounds. During each measurement, if the MCU receives feedback pulses, which means the current DAC output is smaller than the peak value of the ripple, the lower bound should be updated to the output of the DAC. Otherwise, the DAC output is too large and the upper bound should be updated. The above process is repeated until the distance of both bounds is smaller than the tolerance, then we eventually obtain the ripple voltage.

The differences in peak measurement between binary search and the proposed method are shown in Figure 6. Generally, Step 1, Step 2, and Step 3 take 2 or 3 cycles, 2 or 3 cycles, and 4 or 5 cycles, respectively. It is obvious that the required time will be significantly reduced. As previously mentioned, the waiting time in each cycle can be several dozen times of the period of a ripple to ensure a high measurement accuracy. Since the first several cycles only aim for a narrow search area rather than high accuracy, less time (one-quarter of an accuracy cycle) can be allocated in these cycles. As Figure 6 shows, when the shape of a waveform is used in ripple estimation, the measuring time is about three quarters of the binary search method. On this basis, it can achieve a shorter measurement time by cutting down the waiting time in the first several cycles because the results of the first several cycles are only used for peak estimation with lower accuracy requirement. The process of valley measurement is similar to peak measurement; therefore, the details are not described here.

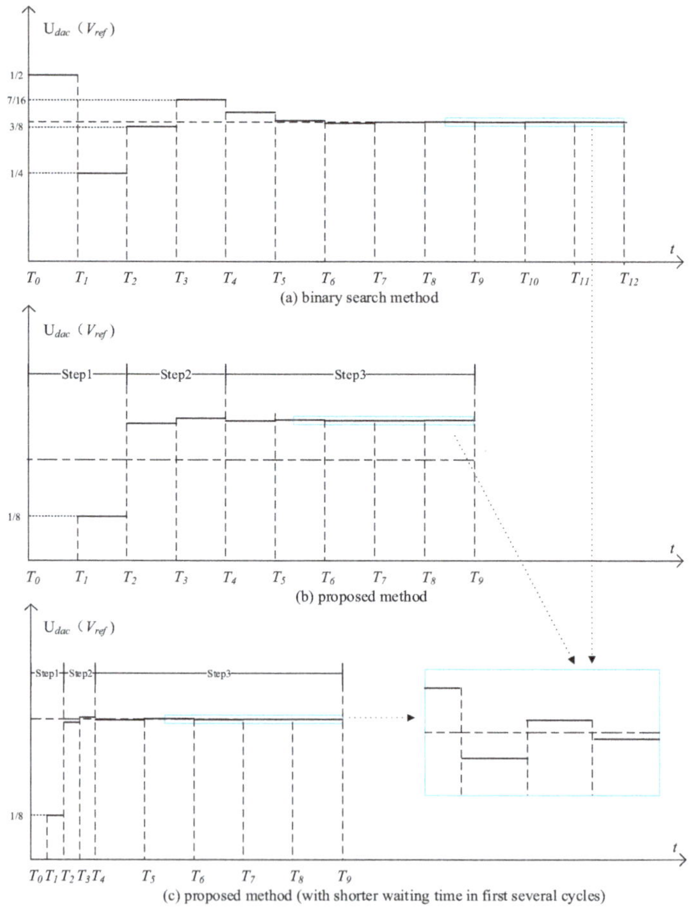

Figure 6. The differences between binary search and the proposed method. (**a**) Binary search method, the requiring measurement cycles are equal to the DAC bits. (**b**) Proposed method, reducing the measurement cycles by peak estimation (step 1) and bound determination (step 2). (**c**) Further reduction of measurement time by cutting the waiting time in first several cycles.

5. Experimental Verification

In order to prove the characteristics of the proposed scheme, a print circuit board (PCB) was designed for verification. The prototype is demonstrated in Figure 7 and the major components are described in Table 1. The theoretical error calculated by Equation (20) is shown in Table 2 and is compared with the error of traditional ADC measurement calculated by Equation (21) (assuming that the internal error of the 8-bit, 10-bit, and 12-bit ADC is 1, 2 and 3, respectively). The results are shown in Figure 8.

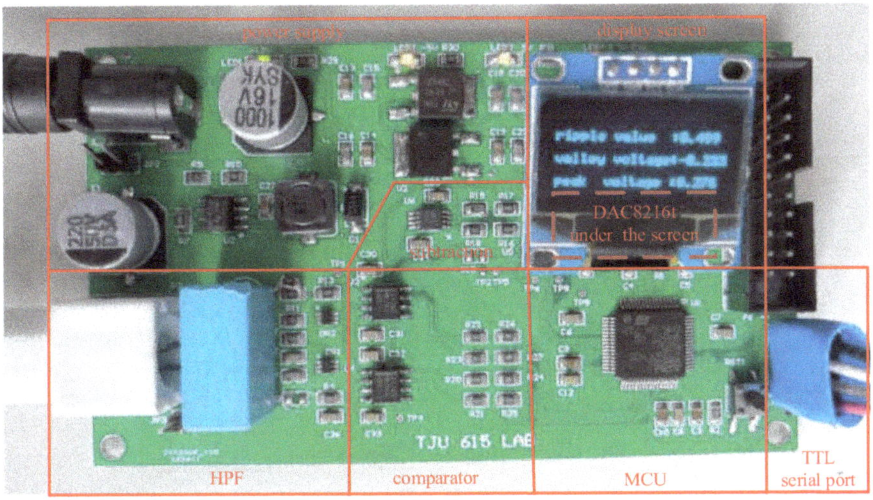

Figure 7. The prototype.

Table 1. The major components of the prototype.

Component	Part Name	Major Parameter
MCU	STM32F405RGT6	Frequency: 168 MHz
DAC	DAC8162t	14 bits, 2 channels
Reference source	REF2125	Accuracy: 0.05%
Operational amplifier (OPA)	OPA209	Input offset voltage: 150 μV
Resistor	SMD resistors	Accuracy: 0.05%
Protection diode	PESD3V3L5UV	Diode capacitance: 22 pF
Filter capacitor	Film capacitors	Value: 47 nF ± 20%
High-speed comparator	TL3116	Bandwidth: 100 MHz. Input offset voltage: 0.5 mV

Table 2. The theoretical measurement errors in different inputs.

Ripple Value (mV)	Percentage Error (%)	Absolute Error (mV)
20	4.19	0.84
50	1.78	0.89
100	0.97	0.97
200	0.57	1.14
500	0.33	1.65
1000	0.25	2.49
2000	0.21	4.17
4000	0.19	7.54

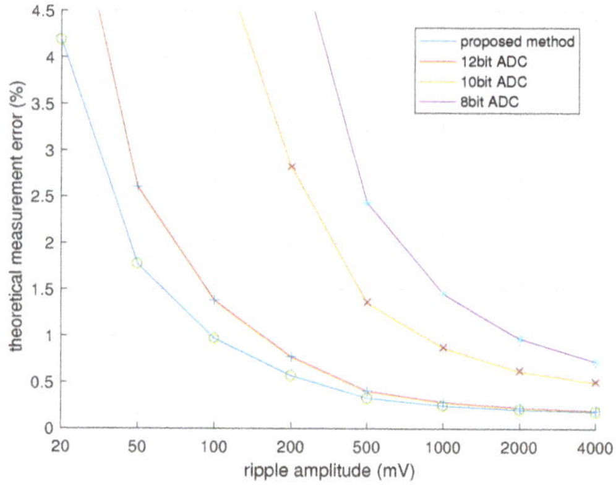

Figure 8. The theoretical measurement errors of different methods.

From Figure 8, we see that the error of the prototype based on proposed method is even smaller than that of the 12-bit high-speed ADC (for the measurement of small signals, an amplification circuit is necessary to lower quantization error and improve measurement precision, which is not considered in ADC error calculation). In fact, when the ripple amplitude is above 100 mV, the theoretical error of the prototype is under 1%, which has an advantage over the majority of oscilloscopes (vertical accuracy between 3–5%). Due to the lack of high-accuracy ripple measurement equipment, it is difficult to calibrate the actual measurement error of the prototype. Here, we use another alternative, i.e., utilizing a WAVESURFER10 oscilloscope with 1% DC vertical error to carry out the comparison experiment.

The experimental environment is shown in Figure 9. The DG1022U signal generator is used to generate triangular waves of different amplitude, frequency, and duty ratio. The peak-to-peak value is measured simultaneously by the WAVESERFER 10 oscilloscope and the prototype. The measurement results are displayed on the screen and sent to the computer for data storage. The differences in measurement results between the WAVESURFER 10 oscilloscope and the designed prototype are shown in Figures 10–12. Since the theoretical error of the selected oscilloscope is larger than the designed prototype, the vertical axis is the "absolute difference" instead of the "measurement error".

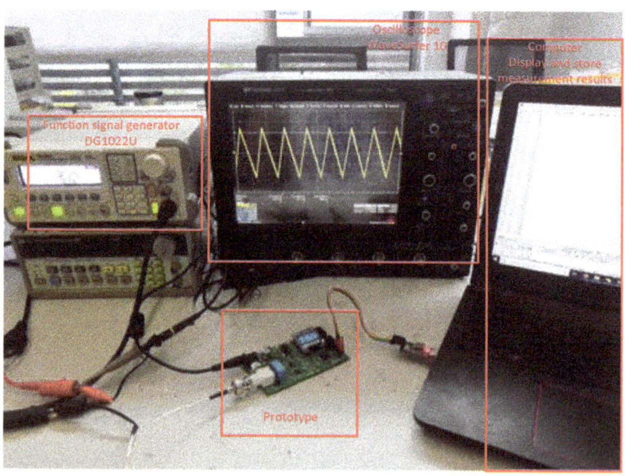

Figure 9. The experimental environment.

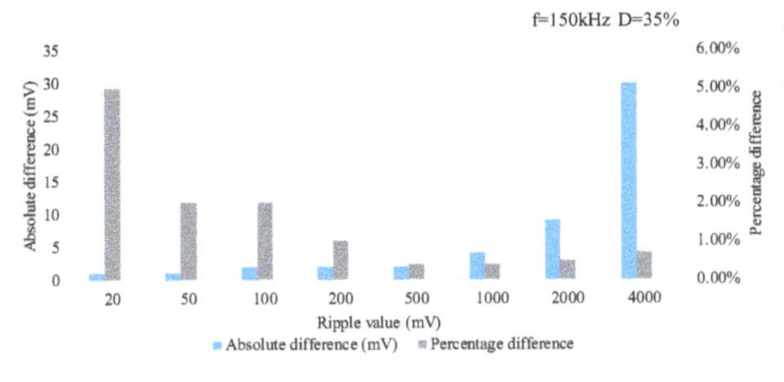

Figure 10. The measurement differences for different amplitudes.

Figure 10 demonstrates that the absolute differences in measurement between the two methods are 1 mV (5%) and 1 mV (2%), respectively, for the 20 mV and 50 mV ripples, which is in accordance with the theoretical calculation shown in Table 2 (with consideration of the oscilloscope error and resolution, for example, the minimum resolution of oscilloscope is 1 mV). The absolute difference in measurement results remains 2 mV (2%) in response to the input of a 100 mV ripple, which possibly is the maximum error from the test equipment and oscilloscope simultaneously, along with opposite error symbol. Of course, it is also possible that the actual measurement error at this point is greater than the theoretical value. When the ripple is over 100 mV, the result of the two instruments is vicinal (the measurement difference is below 1% of the ripple).

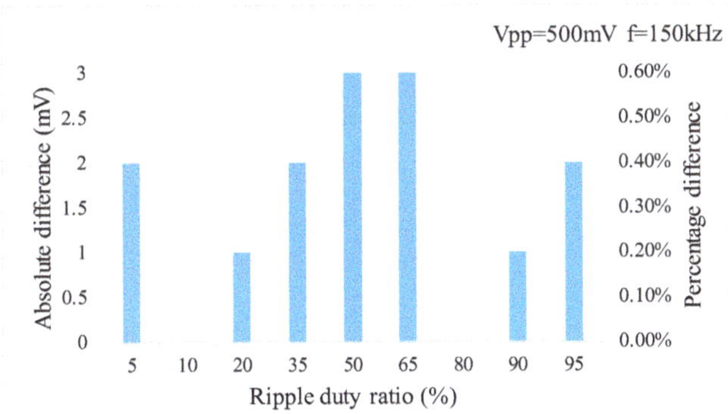

Figure 11. The measurement differences for different duty ratios.

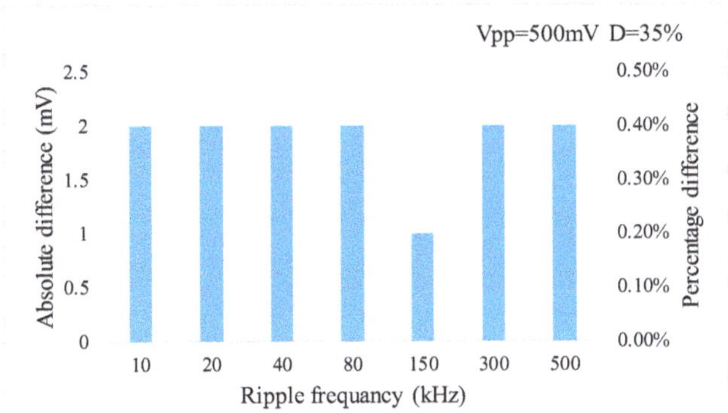

Figure 12. The measurement differences for different frequencies.

Figure 11 shows a slight difference between absolute errors in measurement results for different duty ratios (less than the resolution ratio of the oscilloscope in this range, 3 mV). Figure 12 illustrates the capability of the proposed measurement system for ripples up to 500 kHz.

The above experiments support the following conclusions:

(1) When the ripple amplitude is under 100 mV, the measurement error of the prototype is consistent with the theoretical value and smaller than the measurement error of the general oscilloscope (5%).
(2) When the ripple amplitude is larger than 100 mV, the measurement error of the prototype is also not smaller than the professional oscilloscope (1%).
(3) The measurement precision of the proposed system is hardly affected by the variation of both frequency and duty ratio, which indicates that the designed system can be utilized in extensive field applications .

Since the period of the ripple is usually less than 0.1 ms, the measurement period was set to 3 ms in the experiment. Experimental results show that if a triangular wave is used as the input signal, the ordinary binary research method takes 40 ms, the improved method takes 28 ms, and if the waiting time of the first few cycles is reduced, it only takes 20 ms. In addition, the above methods have the same measurement error. The experimental results are consistent with the theoretical analysis (Figure 6), and the results indicate that the measurement speed can be doubled using the improved method, which is useful in some data-driven applications.

6. Conclusions

Ripple acts as one of the crucial parameters of SMPS, reflecting the operating health status. This manuscript presents a new ripple measuring scheme that utilizes inexpensive a low-speed DAC and high-speed comparators, instead of costly high-speed ADCs, and is also characterized by low cost, high precision, portability, and automation. The operating details and error sources are described, and a new advanced strategy for ripple measurement cycle reduction is proposed. Both the theory and experiment show the designed measurement can be utilized in extensive field applications, including the measurement of ripple under different amplitudes, frequencies, and duty ratios.

Due to the limited cost, the health management of SMPS is mainly realized by monitoring the voltage and current of the system. In this paper, the ripple value can be obtained at a low cost, so as to promote the implementation of relevant studies on the health management of SMPS using ripple. This is of great significance for detecting potential problems in power system operation and preventing sudden failure accidents. Moreover, the high precision and automation of the proposed method enables it to be widely applied for the quality testing of SMPS and other engineering fields.

7. Patents

Patents for the research results of this paper have been applied for in China, the patent number is 201810330485.9 and is currently in the publicity period.

Author Contributions: Conceptualization, J.L. and J.Y.; Data curation, J.L.; Funding acquisition, F.L.; Resources, L.W.; Software, J.L.; Writing—original draft, C.W.; Writing—review & editing, C.W.

Funding: This research was funded by the Shanghai Science and Technology Innovation Action Plan under Grant No. 16DZ1205000.

Acknowledgments: The authors would like to acknowledge the support from the School of Ocean and Earth Science in Tongji University under the project "Research on the power system of large-scale scientific cabled seafloor observatory networks". The authors also wish to express their sincere thanks to the editors and anonymous reviewers for their valuable opinions and contributions.

Conflicts of Interest: The authors declare no conflict of interest. The funders had no role in the design of the study; in the collection, analyses, or interpretation of data; in the writing of the manuscript, or in the decision to publish the results.

References

1. Farjah, E.; Givi, H.; Ghanbari, T. Application of an efficient Rogowski coil sensor for switch fault diagnosis and capacitor ESR monitoring in nonisolated single-switch DC–DC converters. *IEEE Trans. Power Electron.* **2016**, *32*, 1442–1456. [CrossRef]
2. Lahyani, A.; Venet, P.; Grellet, G. Failure prediction of electrolytic capacitors during operation of a switchmode power supply. *IEEE Trans. Power Electron.* **1998**, *13* 1199–1207. [CrossRef]
3. Nakao, H.; Yonezawa, Y.; Sugawara, T. Online evaluation method of electrolytic capacitor degradation for digitally controlled SMPS failure prediction. *IEEE Trans. Power Electron.* **2018**, *33* 2552–2558. [CrossRef]
4. Bhambra, J.K.; Perinpanayagam, S.; Taurand, C.; Peyrat, S. Health monitoring of POL converter using digital PWM controller. In Proceedings of the 8th IEEE Symposium on Diagnostics for Electrical Machines, Power Electronics & Drives, Bologna, Italy, 5–8 September 2011.
5. Hofmeister, J.P.; Wagoner, R.S.; Taurand, C.; Goodman, D.L. Prognostic health management (PHM) of electrical systems using condition-based data for anomaly and prognostic reasoning. *Chem. Eng. Trans.* **2013**. Available online: https://www.ridgetopgroup.com/wp-content/uploads/2015/07/236_Paper_final_Milan.pdf (accessed on 30 April 2019)
6. Lifeng, W.; Yinyu, D.; Shihong, Z. Effect of Electrolytic Capacitors on the Life of SMPS. *J. Converg. Inf. Technol.* **2011**, *6*, 491–499.
7. Gao, J.; Huang, D.; Lu, J. Online Output Capacitor Monitor for Buck DC-DC Converter. In Proceedings of the 2018 Prognostics and System Health Management Conference (PHM-Chongqing), Chongqing, China, 26–28 October 2018.
8. Gao, J.; Huang, D.; Lu, J. Online Health Monitoring of the Electrolytic Capacitor in DC-DC Converters. In Proceedings of the 2017 International Conference on Computer Systems, Electronics and Control (ICCSEC), Dalian, China, 25–27 December 2017.
9. Wang, G.; Guan, Y.; Zhang, J.; Wu, L.; Zheng, X.; Pan, W. ESR estimation method for DC-DC converters based on improved EMD algorithm. In Proceedings of the IEEE 2012 Prognostics and System Health Management Conference (PHM-2012 Beijing), Beijing, China, 23–25 May 2012.
10. Leite, V.; Teixeira, H.; Cardoso, A.J.; Araujo, R. A simple ESR identification methodology for electrolytic capacitors condition monitoring. In Proceedings of the 20th International Congress and Exhibition on Condition Monitoring and Diagnostic Engineering Management, Faro, Portugal, 10–13 June 2007.
11. Anusree, R.; Sreelekshmi, R.S.; Nair, M.G. Study & Simulation For Determining the Age of Electrolytic Capacitor Using ESR. In Proceedings of the 2018 IEEE Distributed Computing, VLSI, Electrical Circuits and Robotics (DISCOVER), Mangalore (Mangaluru), India, 13–14 August 2018.
12. Imam, A.M.; Habetler, T.G.; Harley, R.G. Condition monitoring of electrolytic capacitor in power electronic circuits using adaptive filter modeling. In Proceedings of the 2005 IEEE 36th Power Electronics Specialists Conference, Recife, Brazil, 16 June 2005.
13. Li, H.; Yao, K.; Zhou, X.; Yang, F.; Zhang, J. A novel online ESR and C identification method for output capacitor of flyback converter. In Proceedings of the 2016 IEEE Energy Conversion Congress and Exposition (ECCE), Milwaukee, WI, USA, 18–22 September 2016.
14. Kai, Y.; Wenbin, H.; Weijie, T.; Jianguo, L.; Jingcheng, C. A novel online ESR and C identification method for output capacitor of buck converter. In Proceedings of the 2014 IEEE Energy Conversion Congress and Exposition (ECCE), Pittsburgh, PA, USA, 14–18 September 2014.
15. Liu, L.; Guan, Y.; Wu, M.; Wu, L. Failure prediction of electrolytic capacitors in switching-mode power converters. In Proceedings of the IEEE 2012 Prognostics and System Health Management Conference (PHM-2012 Beijing), Beijing, China, 23–25 May 2012.
16. Khandebharad, A.R.; Dhumale, R.B.; Lokhande, S.S. Electrolytic capacitor online failure detection and life prediction methodology. *Int. J. Res. Eng. Technol.* **2015**, *4*, 636–641.
17. Ahmad, M.W.; Agarwal, N.; Anand, S. Online monitoring technique for aluminum electrolytic capacitor in solar PV-based DC system. *IEEE Trans. Ind. Electron.* **2016**, *63*, 7059–7066. [CrossRef]
18. Ahmad, M.W.; Agarwal, N.; Kumar, P.N.; Anand, S. Low-frequency impedance monitoring and corresponding failure criteria for aluminum electrolytic capacitors. *IEEE Trans. Ind. Electron.* **2017**, *64*, 5657–5666. [CrossRef]

19. Hannonen, J.; Honkanen, J.; Ström, J.P.; Kärkkäinen, T.; Räisänen, S.; Silventoinen, P. Capacitor aging detection in a DC–DC converter output stage. *IEEE Trans. Ind. Appl.* **2016**, *52*, 3224–3233. [CrossRef]
20. Buiatti, G.M.; Martin-Ramos, J.A.; Garcia, C.H.R.; Amaral, A.M.; Cardoso, A.J.M. An online and noninvasive technique for the condition monitoring of capacitors in boost converters. *IEEE Trans. Instrum. Meas.* **2010**, *59*, 2134–2143. [CrossRef]
21. Amaral, A.M.R.; Cardoso, A.M. A simple offline technique for evaluating the condition of aluminum–electrolytic–capacitors. *IEEE Trans. Ind. Electron.* **2009**, *56*, 3230–3237. [CrossRef]
22. Laadjal, K.; Sahraoui, M.; Cardoso, A.J.M.; Amaral, A.M.R. Online Estimation of Aluminum Electrolytic-Capacitor Parameters Using a Modified Prony's Method. *IEEE Trans. Ind. Appl.* **2018**, *54*, 4764–4774. [CrossRef]
23. Ren, L.; Gong, C.; Zhao, Y. An Online ESR Estimation Method for Output Capacitor of Boost Converter. *IEEE Trans. Power Electron.* **2019**. [CrossRef]
24. Soliman, H.; Wang, H.; Blaabjerg, F. A review of the condition monitoring of capacitors in power electronic converters. *IEEE Trans. Ind. Appl.* **2016**, *52*, 4976–4989. [CrossRef]
25. Xu, B.; Huang, Y. Thinking of the Problem of Using AC Voltmeter to Measure Ripple Voltage of Stable Voltage Power. *Sci. Technol. Inf.* **2008**, *36*, 393.
26. Le, J.; Yao, E.; Zhang, M. Design of measurement circuit on true RMS for DC power ripple based on AD637. *Res. Des. Power Technol.* **2014**, *38*, 1926–1929.
27. Jerry, O. Measuring Power Supply Ripple Voltage. *Electron. Test* **1981**. Available online: https://www.engineeringvillage.com/search/doc/abstract.url?&pageType=quickSearch&usageZone=resultslist&usageOrigin=searchresults&searchtype=Quick&SEARCHID=f9181e92963e4fb8b9714bfb04db25bc&DOCINDEX=1&ignore_docid=inspec_base801791136&database=8388611&format=quickSearchAbstractFormat&tagscope=&displayPagination=yes (accessed on 30 April 2019).
28. Zhou, J.; Wang, K. Design and implementation of DC voltage and ripple tester based on STC89C52. *Mod. Electron. Tech.* **2012**, *35*, 138–140.
29. Anthony, H.S.; Scitech, B.; Wang, K. Inexpensive peak detector requires few component. *EDN* **2005**, *50*, 88–92.
30. Ren, L.; Tong, Z.; Na, X. A method on the measurement of a repeated signal's peak value. *Electr. Meas. Instrum.* **2001**, *38*, 24–26.

electronics

MDPI

Article

Noise Reduction for High-Accuracy Automatic Calibration of Resolver Signals via DWT-SVD Based Filter

Meishan Guo and Zhong Wu *

School of Instrumentation and Optoelectronic Engineering, Beihang University, Beijing 100191, China; msguo@buaa.edu.cn

* Correspondence: wuzhong@buaa.edu.cn; Tel.: +86-10-8233-9703

Received: 16 April 2019; Accepted: 6 May 2019; Published: 9 May 2019

Abstract: High-accuracy calibration of resolver signals is the key to improve its angular measurement accuracy. However, inductive harmonics, residual excitation components, and random noise in signals dramatically restrict the further improvement of calibration accuracy. Aiming to reduce these unexpected noises, a filter based on discrete wavelet transform (DWT) and singular value decomposition (SVD) is designed in this paper. Firstly, the signal was decomposed into a time-frequency domain by DWT and several groups of coefficients were obtained. Next, the SVD operation of a Hankel matrix created from the coefficients was made. Afterwards, the noises were attenuated by reconstructing the signal with a few selected singular values. Compared with a conventional low-pass filter, this method can almost only preserve the fundamental and DC components of the signal because of the multi-resolution characteristic of DWT and the good correspondence between the singular value and frequency. Therefore, the calibration accuracy of the imperfect characteristics could be improved effectively. Simulation and experimental results demonstrated the effectiveness of the proposed method.

Keywords: resolver; discrete wavelet transform; singular value decomposition; automatic calibration; noise reduction

---

1. Introduction

The accurate information of the motor angular position is desired in high-performance servo control systems. Due to the simple structure, strong robustness, and adaptability to various harsh environments [1], resolvers have attracted great attention as shaft angle sensors in servo control applications such as antennas, radars, steering engines, and industrial robots.

Generally, a complete angular measurement system consists of a resolver and a Resolver-Digital Converter (RDC). In the software-based RDCs, the output signals of the resolver are transformed into sinusoidal and cosinusoidal envelopes with respect to the shaft angle after detection. Next, the angular position and velocity are obtained from the demodulation of envelopes [2]. However, there are usually some mechanical and electrical errors in a resolver. The former are caused by the manufacturing tolerance, assembled mismatch, and deformation. The latter result from winding nonlinearity, circuit asymmetry, and excitation signal distortion. Because of these errors, the envelopes contain five nonideal characteristics, such as amplitude imbalances, DC offsets, and imperfect quadrature [3], all of which seriously affect the accuracy of demodulation. Therefore, it is necessary to calibrate and correct the imperfect parameters in the resolver envelope signals.

As the calibration of the resolver signals is equivalent to the parameter estimation of non-orthogonal sinusoidal pair signals, approaches have been widely reported in recent years including a look-up table, optimization, observer, neural network, etc. An offline look-up table was constructed in Reference [4] to compensate the imperfectness in encoder signals. However, a trade-off has to be made between a larger table and increased sensitivity to noise. Heydemann [5] firstly proposed optimization approach by establishing a quadratic equation of five unknown parameters and obtained the optimal numerical solution by employing the least square method. Based on this, many literatures have presented improved methods [6,7]. However, the nonlinearity equation has multiple roots and lacks the ability to escape from local optimization if the initial iteration values are selected as unreasonable. To solve this issue, an adaptive estimator was given in Reference [8] that tracks the imperfect parameters of a characteristic ellipse formed by resolver signals. An automatic calibration algorithm based on state observer was introduced in Reference [9]. However, the strong coupling between parameters and the angular velocity in the mathematical model was undesired because the improvement of the calibration accuracy depended on the angular frequency. Therefore, an improved algorithm based on two-step gradient estimators was presented to decouple them [10]. Owing to the more accurate information of angular velocity, the calibration accuracy was further improved. Besides, signal flow network and deep learning algorithm in Reference [11] were introduced to ensure the independence of the variables.

However, the above methods are based on simplified models. The direct influence of inductive harmonics, residual excitation components, and random noise on the calibration accuracy was ignored. Since resolver windings are unevenly distributed and not exactly sinusoidal or cosinusoidal functions with respect to angular position, the output signals always contain harmonics [3]. Moreover, residual excitation components and random noise appear because of the excitation signal distortion and electrical errors from the conditioning circuit. These noises seriously limit the further improvement of the calibration accuracy no matter which method above is used.

Several studies on noise reduction have concerned themselves with improving the calibration accuracy. Common methods include mathematical modeling, filters, and phase-locked loop. Lara et al. [12] utilized a higher order approximation to describe harmonics but had a slight convergence deviate. The smaller the deviation was, the more complex model needed to be established. Shang et al. [13] analyzed the harmonics by Fourier transform and weakened the 3rd harmonic through adding a corresponding harmonic in the shape function of the rotor structure. Obviously, it required a special rotor structure. Similarly, the error profile curve with respect to the angle was described by Fourier series [14]. However, it was not an automatic calibration. Finite Impulse Response filter was applied in a self-tuning circuit [15], which reduced noise but had an inherent time delay and phase distortion. An adaptive phase-locked loop proposed in Reference [16] was able to filter noise online to a certain extent. However, the continuous calibration increased the unnecessary delay with the errors supposed constant in a short time. Another novel RDC algorithm performed in a frequency domain was studied in Reference [17]. Since the detection was unrequired and only the carrier frequency component was utilized to estimate parameters, it was preferable to suppress the disturbances outside of the carrier frequency. However, the amplitude imbalances were out of consideration.

In order to achieve high-accuracy calibration of the imperfect parameters, it is important to reduce the three types of noises. Some image noising methods are worth learning and using for reference. The discrete wavelet transform (DWT) has been widely used to signal or image denoising. Because of the characteristic of multi-resolution, DWT can distinguish noise and useful information to different frequency bands [18–20]. But the conventional wavelet threshold denosing method [21] is difficult to flexibly select a reasonable threshold and has little effectiveness in noise reduction near the fundamental wave. Moreover, nonlocal self-similarity prior learning [22], convolutional neural network [23], and singular value decomposition (SVD) [24] are also used in image denoising. Guo et al. [24] used a few large singular values and corresponding singular vectors to estimate the image and reduce noise. Recently, because of the multi-resolution characteristic of DWT and the good correspondence between the singular value and frequency, the cooperation between DWT and SVD [25,26] in the time-frequency domain has attracted the attention of researchers. At present, several different combinations have been adopted in image watermarking [27], image contrast enhancement [28], image compression and denoising [29], and the feature extraction of signals [30].

Aiming to reduce the noises and obtain the high calibration accuracy of resolver signals, a DWT-SVD based filter in time-frequency domain is designed in this paper. Since this method is able to reduce inductive harmonics, residual excitation components, and random noise in resolver signals with only the fundamental and DC components being retained, the calibration accuracy can be improved effectively. Simulation and experimental results verify the effectiveness of the proposed method.

This paper is organized as follows: The calibration principle of resolver signals is introduced and the problem of noises is formulated in Section 2. Section 3 presents the designed DWT-SVD based filter and describes the filtering processing in detail. To verify the effectiveness of the method, simulation and experimental results are analyzed in Section 4. Finally, the concluding remarks are given in Section 5.

2. Calibration Principle and Problem Formulation of Resolver

As shown in Figure 1, in a software-based RDC, when the rotor winding of resolver is excited with a high frequency voltage, the two spatially orthogonal windings on the stator will produce amplitude modulation signals which have sinusoidal and cosinusoidal envelopes with respect to shaft angle. Then the envelopes are obtained from detection. Finally, owing to the mathematical properties of trigonometric function, the angular position θ and velocity ω are calculated from envelopes by phase-locked loop, arctangent or other demodulation algorithms.

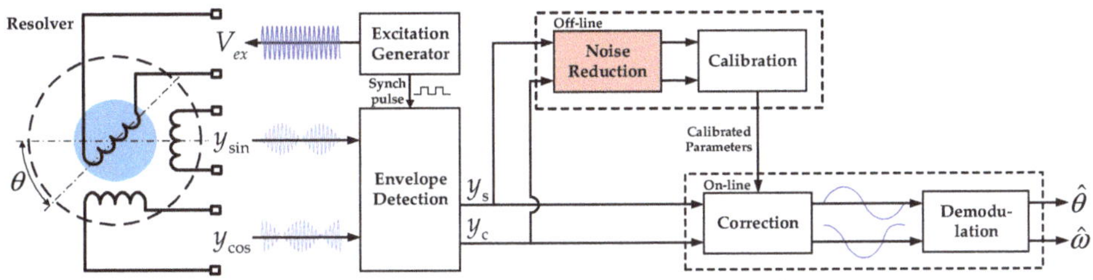

Figure 1. Schematic block diagrams of a resolver and RDC.

In practice, the resolver signals after detection are always disturbed by imperfect characteristics. The amplitude imbalances and DC offsets result from the eccentric rotor, unequal winding, and asymmetric circuit. The imperfect quadrature arises when the space angle of two coils on stator are not exactly equal to $\pi/2$. Therefore, the envelopes should be described as

$$\begin{cases} y_s = a_{s1}\sin\theta + a_{s0} \\ y_c = a_{c1}\cos(\theta+\beta) + a_{c0} \end{cases} \tag{1}$$

where a_{s1} and a_{c1} are the amplitudes, a_{s0} and a_{c0} are the offsets, β represents the imperfect quadrature. Obviously, it is necessary to calibrate the envelopes and correct (1) to the standard form of sine and cosine functions before demodulation.

The calibration of resolver signals is a process of estimating the five imperfect parameters of non-orthogonal sinusoidal pair signals. These estimation methods have been widely reported in recent years. By using a look-up table, optimization, observer, neural network or other estimation algorithm, the imperfect parameters can be estimated to correct and reduce demodulation error. Thereafter, the signals can be calibrated by substituting the estimated value into the following equation:

$$\begin{cases} \hat{y}_s = (y_s - \hat{a}_{s0})/\hat{a}_s = \sin\theta \\ \hat{y}_c = (y_c - \hat{a}_{c0})/\hat{a}_c\cos\beta + (y_s - \hat{a}_{s0})tan\beta/\hat{a}_s = \cos\theta \end{cases} . \tag{2}$$

Unfortunately, most calibration algorithms are based on simplified models and ignore the noises like harmonics, residual excitation components, and random noise in envelopes, all of which seriously affect the calibration of the resolver. The harmonic distortion arises when the unevenly distributed windings are not exactly sinusoidal or cosinusoidal shaped with respect to the angular position. The residual excitation components and random noise exist due to the electrical errors from conditioning circuit. Hence, the Equation (1) can be rewritten in the following manner:

$$\begin{cases} y_s = a_{s0} + a_{s1}\sin\theta + \sum\limits_{n=2}^{\infty} a_{sn}\sin n\theta + d_s \\ y_c = a_{c0} + a_{c1}\cos(\theta+\beta) + \sum\limits_{n=2}^{\infty} a_{cn}\cos n\theta + d_c \end{cases} \tag{3}$$

where n is the harmonic order, a_{sn} and a_{cn} represent the amplitudes of the nth harmonic, d_s and d_c are random noise.

As shown in Figure 1, aiming at suppressing noises and improving calibration accuracy, several methods including mathematical modeling and low-pass filter have been used recently. However, the mathematical modeling method makes an inevitable deviation and is pretty complex. The low-pass filter has an inherent phase distortion and cannot attenuate the noises in the passband. Therefore, it is still a serious problem to filter the noises without phase distortion and preserve the fundamental and DC component only.

3. Design of DWT-SVD Based Filter

In order to reduce the three types of noises in resolver signals without phase distortion and preserve the fundamental and DC component only, a DWT-SVD based filter is designed in this paper. As shown in Figure 2, this method is divided into 3 steps: (1) Decompose the resolver envelopes into several groups of coefficients corresponding to different frequency bands through DWT; (2) Process the coefficients by SVD to filter noise; (3) The filtered envelopes are reconstructed with the processed coefficients. Since the procedure of the sinusoidal pair signals are identical, the following only considers the sinusoidal envelope y_s in Equation (3).

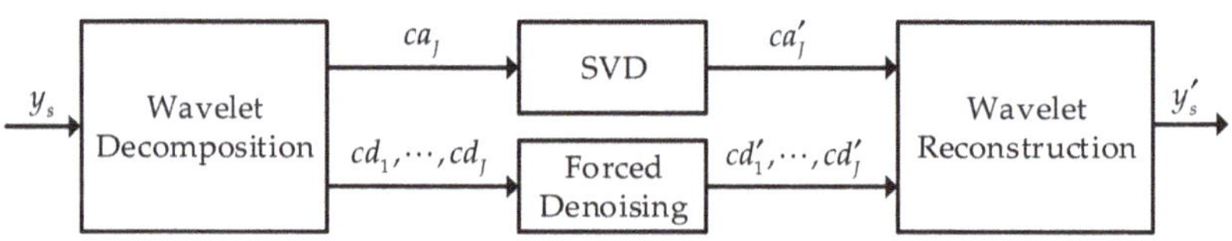

Figure 2. Block diagram of the proposed filter.

3.1. Signal Decomposition

The first step is to decompose the signal into approximation coefficients and detail coefficients through J-layer DWT. Actually, the essence of DWT can be regarded as a process of utilizing a set of high-pass and low-pass filters on the signal. Furthermore, the high-pass and low-pass filters depend on the selected wavelet base function. Thus, the approximation coefficients ca which represented low frequency information and detail coefficients cd which represented high frequency information are obtained. In this method, the Mallat algorithm is employed to achieve J-layer DWT. The coefficients ca and cd of each layer are calculated as follows:

$$\begin{cases} ca_j(k) = \sum\limits_{n} h(n-2k)ca_{j-1}(n) \\ cd_j(k) = \sum\limits_{n} g(n-2k)ca_{j-1}(n) \end{cases}, j = 1,2,3,\cdots,J \tag{4}$$

where h and g represent the impulse responses of low-pass filter and high-pass filter, respectively, when $j=1$, ca_{j-1} represents the envelope signal of resolver.

The procedure of multi-layer decomposition is shown in Figure 3. Assuming the sampling frequency f_s satisfies the Nyquist Sampling Theorem and the total layer is J, the spectrum of the signal is limited in $(0 \sim f_s/2)$ according to the normalized frequency band. Due to the multi-resolution characteristic of DWT, the frequency band of cd_1, cd_2 and cd_3, respectively, are $(f_s/4 \sim f_s/2)$, $(f_s/8 \sim f_s/4)$, $(f_s/16 \sim f_s/8)$. And, more remarkably, ca_3 is in the low frequency band $(0 \sim f_s/16)$ which contains the fundamental and DC components of resolver envelope. If the layer J is too small, the data length of ca_J would be overmuch and then increase the computational complexity of SVD. Otherwise, the useful information would leak into the detail coefficients. Therefore, selecting the layer reasonably would directly determine whether the ca_J includes a fundamental wave. Moreover, it is important to make the detail coefficients possess harmonics and residual excitation components as far as possible.

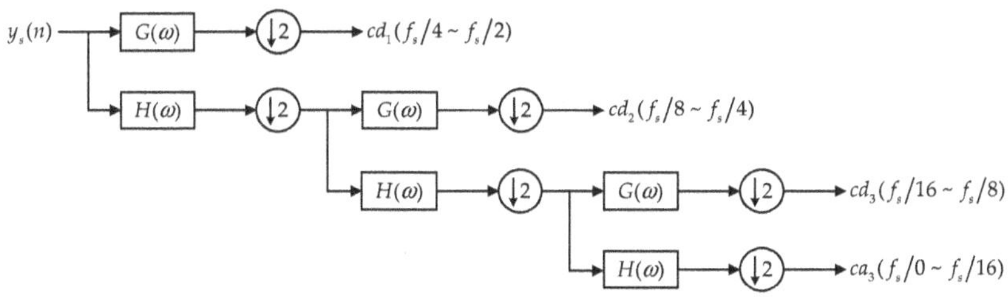

Figure 3. Scheme of wavelet decomposition.

3.2. Coefficient Processing

The second step is to analyze the approximation coefficient ca_J and detail coefficients from cd_1 to cd_J. Since the detail coefficients contain residual excitation components and some harmonics with so little useful information, they can be addressed by forced noise reduction. The coefficient ca_J, which involves the fundamental wave, is still affected by noises, such as random noise and harmonics. Therefore, SVD is employed to reduce these noises.

The SVD of a matrix $\boldsymbol{H} \in \mathbf{R}^{m \times n}$ is defined as the following equation:

$$\boldsymbol{H} = \boldsymbol{USV}^T \tag{5}$$

where $\boldsymbol{U} \in \mathbf{R}^{m \times m}$ and $\boldsymbol{V} \in \mathbf{R}^{n \times n}$ are orthogonal matrices. The diagonal matrix $\boldsymbol{S} \in \mathbf{R}^{m \times n}$ can be given by

$$\boldsymbol{S} = (diag(\sigma_1, \sigma_2, \cdots, \sigma_p), \boldsymbol{O}) \tag{6}$$

where $p = \min(m,\ n)$ is the number of singular values, and $\sigma_i (i = 1, 2, \cdots p)$ represent the singular values of matrix $\boldsymbol{H}$ which satisfy $\sigma_1 \geq \sigma_2 \geq \cdots \geq \sigma_p > 0$.

As $ca_J = (x_1, x_2, \cdots, x_N)$ is a one-dimension data, a Hankel matrix $\boldsymbol{H}$ needs to be construct when processing ca_J by SVD. The matrix can be expressed as

$$\boldsymbol{H} = \begin{bmatrix} x_1 & x_2 & \cdots & x_{N-n+2} \\ x_2 & x_3 & \cdots & x_{N-n+2} \\ \vdots & \vdots & \ddots & \vdots \\ x_{N-n+1} & x_{N-n+2} & \cdots & x_N \end{bmatrix}_{m \times n} \tag{7}$$

where n satisfies $1 < n < N$ and $m = N - n + 1$. From Equation (7) each row vector in the Hankel matrix lags only one data behind the previous row vector, which means the adjacent row vectors are highly correlated with useful information and independent of noises. Therefore, the fundamental and DC components of the signal which contain the main energy will be concentrated in a few large singular values. Due to the good correspondence between the singular value and frequency, the first two maximum values represent the fundamental wave, and the value which represents the DC component can be selected from test. According to the principle, the modified ca'_J can be calculated from Equation (5) by using only three singular values.

3.3. Signal Reconstruction

The last step is reconstruction. The procedure of multi-layer wavelet reconstruction is shown in Figure 4. The formula of reconstruction is given by

$$ca'_{j-1}(n) = \sum_k h^*(n-2k)ca'_j(k) + \sum_k g^*(n-2k)cd'_j(k). \tag{8}$$

Since the detail coefficients are forced to be zero, the envelope signal of resolver is reconstructed with the modified approximation coefficient ca_J. Finally, the signals of resolver after noise reduction are obtained.

From the above description, it can be seen that the filter can reduce the harmonics, residual excitation components, and random noise and extract the fundamental and DC components of resolver envelopes without phase distortion.

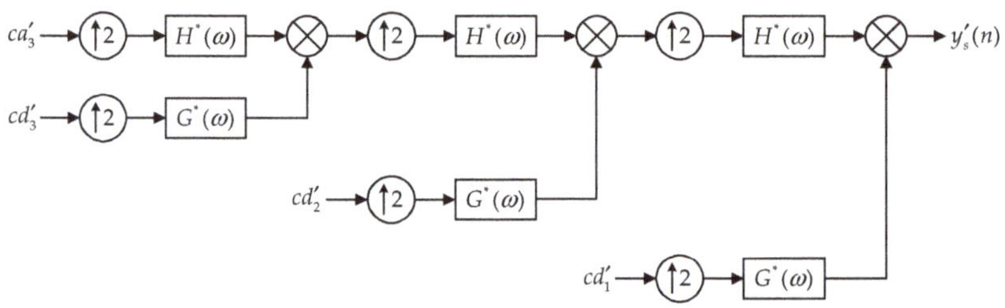

Figure 4. Scheme of wavelet reconstruction.

4. Simulation and Experimental Results

Aiming to evaluate the performance of the proposed method, the spectrums of signals are compared among the following four groups both in simulation and experiment.

Group 1: The original signals;

Group 2: The signals denoised by the low-pass Butterworth filter;

Group 3: The signals denoised by the DWT based filter;

Group 4: The signals denoised by the DWT-SVD based filter.

Next, in order to verify the influence of the filter on the calibration accuracy, the imperfect parameters of the above signals are estimated by an automatic calibration algorithm based on two-step gradient estimators in Reference [10]. The simulation and experimental results are analyzed as follows.

4.1. Simulation Results

In the simulation, sinusoidal pair signals are generated to simulate the envelopes of resolver. The angular frequency ω is 2π rad/s. The imperfect parameters are set as $a_{s1} = 1.8370$ V, $a_{s0} = 0.1365$ V, $a_{c1} = 1.9520$ V, $a_{c0} = 0.1452$ V and $\beta = 1.2°$. The harmonics are shown in Table 1. In addition, the residual excitation components are 0.0010 V and 0.0011 V, respectively, with the frequency being 10 kHz. The SNR of signals is 35 dB by adding Gaussian white noise. The simulation is proceeded by using MATLAB.

Table 1. Harmonics in signals.

Order	2	3	4	5
y_s (V)	0.0255	0.0130	0.0078	0.0032
y_c (V)	0.0243	0.0128	0.0082	0.0025

In the DWT-SVD based filter (Group 4), a biorthogonal wavelet basis function "bior 5.5" is chosen. Since the biorthogonal wavelet has a linear phase, the signals can be completely reconstructed without phase distortion. Whereby, the layer of wavelet decomposition is 4. As comparisons, the low-pass Butterworth filter in Group 2 is designed with no more than 0.1 dB of ripple in a passband from 0 to 3 Hz, and at least 30 dB of attenuation in the stopband. The DWT based filter in Group 3 is designed by using 6-layer wavelet decomposition and reconstruction to reduce the high-frequency noise.

The calibration method in Reference [10] is constructed as

$$\begin{cases} \dot{x} = \xi + y_s \\ \dot{\xi} = -\lambda_0 x - \lambda_1(\xi + y_s) \\ \dot{\eta} = k[x^2(\lambda_0 - \hat{\omega}^2) + (\lambda_1 x + y_s)(\xi + y_s)] \\ \hat{\omega}^2 = \eta - kxy_s \\ \dot{\hat{a}}_0 = \gamma(y_s - \hat{y}_s) \\ \dot{\hat{a}}_1 = \gamma \sin \hat{\omega}t(y_s - \hat{y}_s) \\ \dot{\hat{a}}_2 = \gamma \cos \hat{\omega}t(y_s - \hat{y}_s) \\ \hat{y}_s = \hat{a}_0 + \hat{a}_1 \sin \hat{\omega}t + \hat{a}_2 \cos \hat{\omega}t \end{cases} \tag{9}$$

where the estimator gains are chosen as $k = 100, \lambda_0 = 15, \lambda_1 = 15, \gamma = 0.8$. The angular velocity $\hat{\omega}$ is estimated by the first four equations. Then, the amplitude $\hat{a}_{s1} = \sqrt{\hat{a}_1^2 + \hat{a}_2^2}$, DC offset $\hat{a}_{s0} = \hat{a}_0$ and phase $\hat{\phi}_s = \tan^{-1}(\hat{a}_2/\hat{a}_1)$ of y_s can be estimated by the rest of equations. Since the procedure of y_c is same as y_s, phase shift is calculated by $\hat{\beta} = \hat{\phi}_s - \hat{\phi}_c$.

The results are analyzed as follows:

(1) As shown in Figure 5, the detail coefficients $cd_1 \sim cd_4$ of y_s reflect noises with no useful information. In contrast, the approximation coefficient ca_4 contains the information of fundamental and DC components with a few harmonics and noises. Thus the decomposition can be understood as a pre-filter. Then SVD operation of a Hankel matrix created from ca_4 is made. The singular values are given in Table 2. It is obvious that the 1st and 2nd singular values represent the fundamental wave and the 3rd reflects the DC components. Therefore, y_s can be finally reconstructed from the new ca_4' which is calculated by the three singular values.

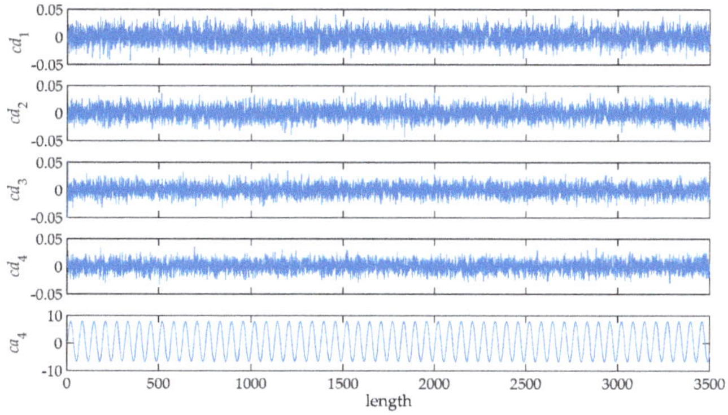

Figure 5. The first 3500 data of approximation coefficient and detail coefficients of y_s in simulation.

Table 2. Partial singular values of approximation coefficient ca_4.

Number	1	2	3	4	5
Value	6896.7	6893.6	1026.7	95.2	95.1

(2) The performance of the filter can be verified from spectral analysis. As shown in Figure 6, the spectrum of the original signal includes harmonics and noises. However, the low-pass filter is unable to reduce noises in the passband and results in a slight amplitude attenuation of fundamental wave. The DWT based filter has no effect on fundamental wave but is unable to suppress low-order harmonics. Unlike these filters, it is showed obviously in Figure 6d that the DWT-SVD filter retains almost only the fundamental and DC components.

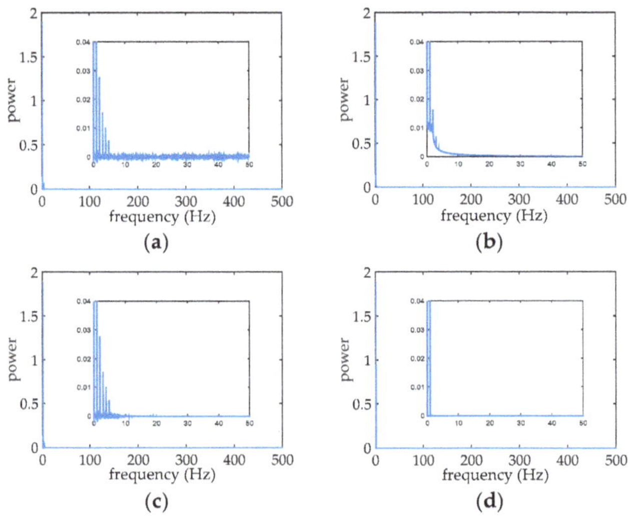

Figure 6. Spectrums of y_s in simulation (**a**) before the filter; (**b**) after the Butterworth filter; (**c**) after the DWT based filter and (**d**) after the designed filter.

(3) By the calibration algorithm in Reference [10], the estimations of the angular frequency and five imperfect parameters in Groups 1–4 are given in Figures 7–10, respectively. And Table 3 shows the estimated results calculated by means of the data and the standard deviations (STD) in the range of 40–50 s. From Figures 7–10, the steady-state error of Group 4 is smaller than that of the other groups. Compared with the preset values in Table 3, the accuracy of ω after the designed filter reaches 10^{-5} rad/s, while the accuracies of the other groups are 10^{-3} rad/s, 10^{-4} rad/s and 10^{-4} rad/s, respectively. The accuracy of amplitudes after the designed filter reaches to 10^{-4} rad/s, while the others are 10^{-3} rad/s and Group 2 has a slight attenuation. Moreover, the STD is reduced at least two orders of magnitude more than the other groups. It is worth noting that the designed filter leads to a high-accuracy phase due to the phase undistorted characteristic, while the low-pass filter causes a phase shift. Therefore, the DWT-SVD filter apparently improves the calibration accuracy and is more stable than other ways.

Table 3. Results of calibration in simulation.

Parameters		ω(rad/s)	a_s(V)	a_c(V)	a_{s0}(V)	a_{c0}(V)	β(°)
Preset values		6.283185	1.83700	1.95200	0.13650	0.14520	1.2000
Calibrated directly	Estimates	6.285769	1.83893	1.95372	0.13643	0.14632	1.2019
	STD	1.20×10^{-2}	1.25×10^{-3}	1.26×10^{-3}	1.63×10^{-3}	1.63×10^{-3}	3.02×10^{-2}
After the Butterworth filter	Estimates	6.283505	1.83577	1.95035	0.13644	0.14631	1.2017
	STD	5.20×10^{-3}	7.32×10^{-4}	7.00×10^{-4}	7.54×10^{-4}	6.95×10^{-4}	1.48×10^{-2}
After the DWT	Estimates	6.283640	1.83741	1.95210	0.13644	0.14632	1.2019
	STD	1.01×10^{-2}	1.22×10^{-3}	1.23×10^{-3}	1.25×10^{-3}	1.18×10^{-3}	2.55×10^{-2}
After the designed filter	Estimates	6.283179	1.83691	1.95187	0.13648	0.14623	1.2002
	STD	2.49×10^{-6}	1.11×10^{-5}	2.28×10^{-5}	4.59×10^{-6}	3.44×10^{-6}	4.23×10^{-4}

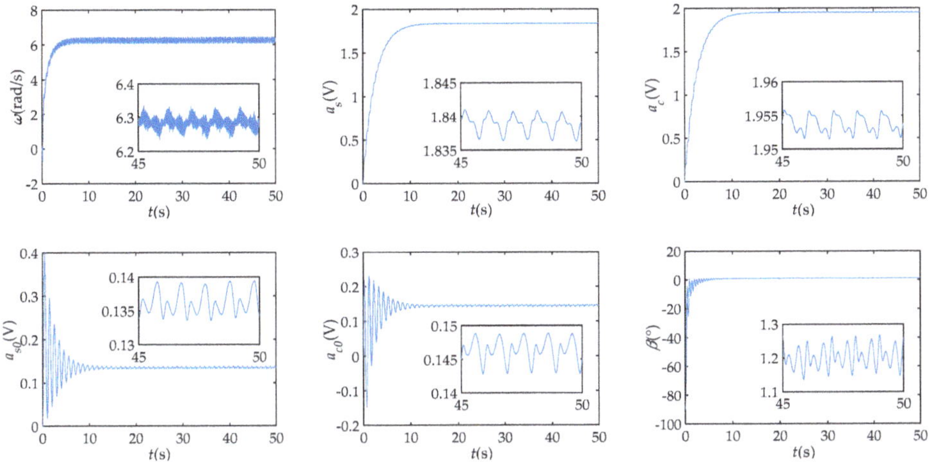

Figure 7. Estimations of angular velocity and imperfect parameters in simulation before filter (Group 1).

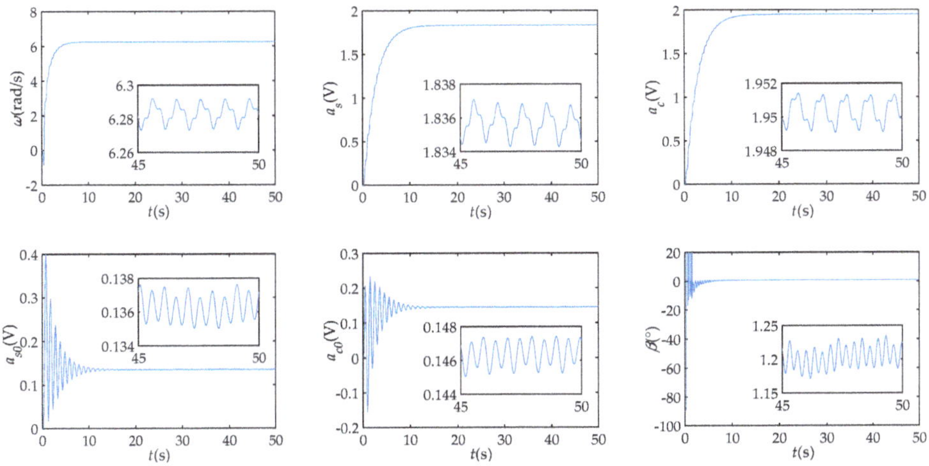

Figure 8. Estimations of angular velocity and imperfect parameters in simulation after the Butterworth filter (Group 2).

4.2. Experimental Results

The experimental platform is shown in Figure 11. A control board drives a permanent magnet synchronous motor (PMSM) and a resolver (Infranor, Zurich, Switzerland). The parameters of PMSM and resolver are given in Table 4. In this experiment, PMSM is driven to rotate at $\omega = 2\pi$ rad/s and the resolver measures its angular position. After envelope detection circuits, the envelops of resolver output signals are uploaded to the upper computer through USB. Then the envelops are denoised and calibrated in the upper computer.

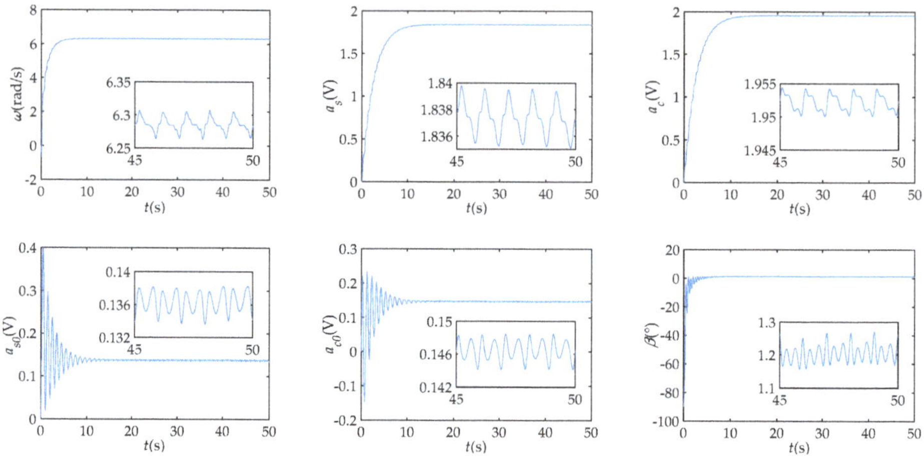

Figure 9. Estimations of angular velocity and imperfect parameters in simulation after the DWT (Group 3).

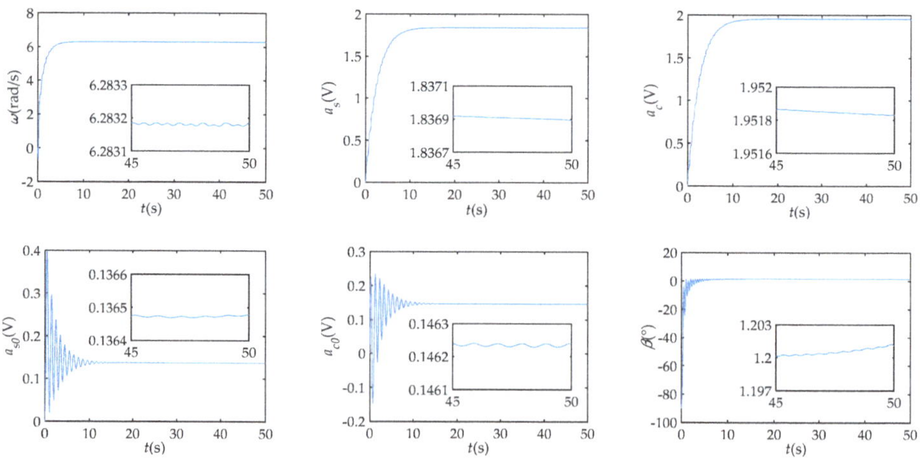

Figure 10. Estimations of angular velocity and imperfect parameters in simulation after the designed filter (Group 4).

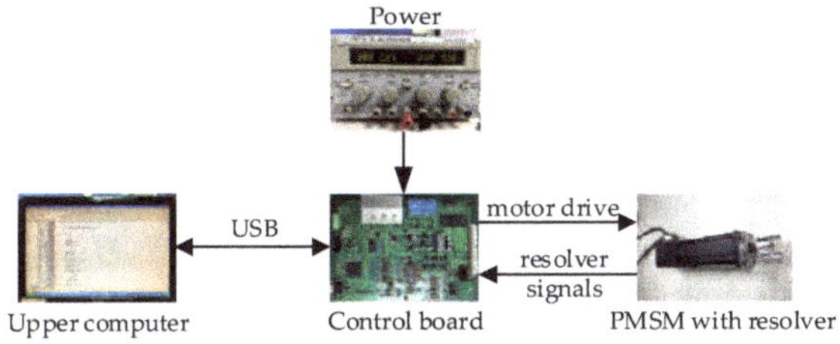

Figure 11. Experimental platform.

Table 4. PMSM and resolver parameters.

PMSM		**Resolver**	
Pole pairs	2	Pole pairs	1
Rated voltage	110 V(AC)	Input voltage	5 V ± 0.2 V (AC)
Rated speed	3000 r/min	Input frequency	10 kHz
Torque constant	0.15 Nm/A	Output voltage	>2 V
Phase resistance	8 Ω	Transformer ratio	0.5 ± 5%
Phase inductance	10 mH	Electrical error	≤ 10′

In this experiment, the parameters of four groups are set the same as in the simulation. The results are analyzed as follows:

(1) The coefficients and singular values of y_s calculated from the DWT-SVD based filter are given in Figure 12 and Table 5. From Figure 12, the approximation coefficient ca_4 has already pre-filtered the residual excitation components and most of the random noise. Next, according to a rigorous test, the 1st and 2nd singular values in Table 5 reflect the fundamental wave and the 5th value reflects the DC components. Finally, the signal can be reconstructed by the three singular values and corresponding singular vectors.

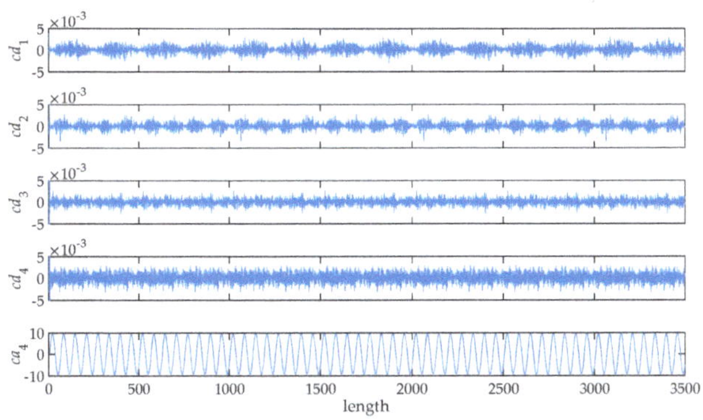

Figure 12. The first 3500 data of approximation coefficient and detail coefficients of y_s in experiment.

Table 5. Partial singular values of approximation coefficient ca_4 in experiment.

Number	**1**	**2**	**3**	**4**	**5**	**6**	**7**
value	9418.9	9410.4	19.8	19.7	11.7	9.3	9.2

(2) The spectrums in Figure 13 also verify the performance of the designed filter. As shown in Figure 13a, the spectrum of the original signal contains harmonics and random noise. However, the spectrum in Figure 13b shows that the low-pass filter attenuates the DC component seriously

and cannot reduce noise in the passband. The spectrum in Figure 13c shows that the DWT-based filter is unable to suppress low-order harmonics although it can reduce the high-frequency noise. Compared with Groups 1–3, the DWT-SVD filter in Group 4 preserves almost only the fundamental and DC components.

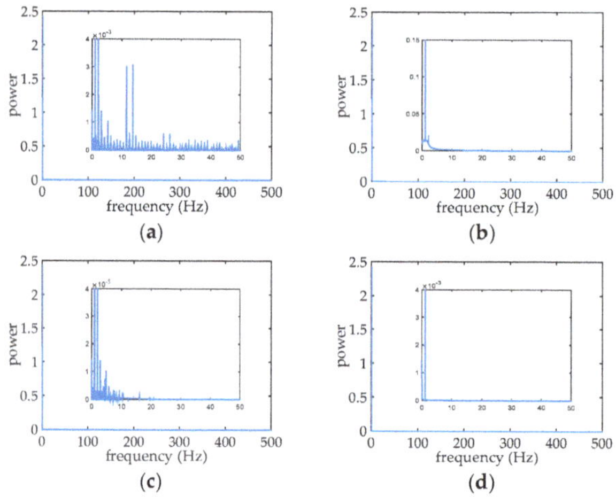

Figure 13. Spectrums of y_s in experiment (**a**) before filter, (**b**) after the Butterworth filter; (**c**) after the DWT based filter and (**d**) after the designed filter.

(3) As show in Figures 14–17 and Table 6, the estimations of the angular frequency ω and five imperfect parameters a_{s1}, a_{c1}, a_{s0}, a_{c0} and β in Groups 1–4 are carried out by the calibration algorithm in [10], respectively. From Figures 14–17, the steady-state errors in Groups 1 and 3 are in the same order of magnitude while in Group 2 is smaller, since the harmonics in Group 2 is weaker than Groups 1 and 3. Compared with them, Group 4 has the smallest steady-state error among the four groups because the proposed method can suppress harmonics effectively. In order to further verify the effectiveness of the proposed method, Table 6 gives the STDs of estimated parameters, which are calculated from the data in the range of 40–50 s. The STD is an important index to compare the four groups while the true values are unknown. From Table 6, it is obvious that Group 4 has the smallest STDs which are reduced at least two orders of magnitude than others.

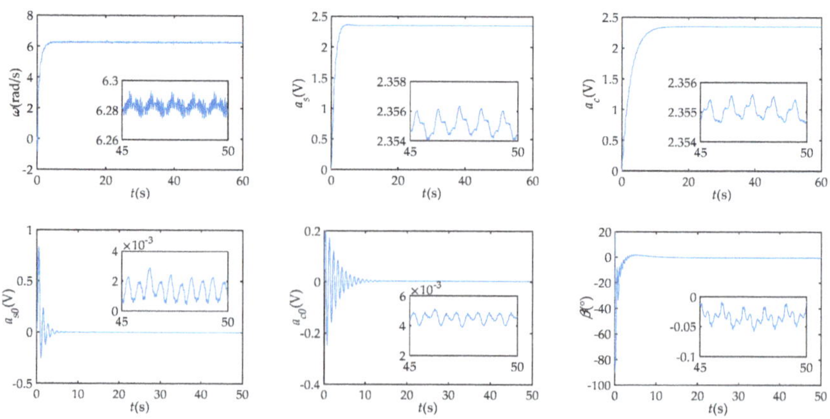

Figure 14. Estimations of angular velocity and imperfect parameters before filter (Group 1).

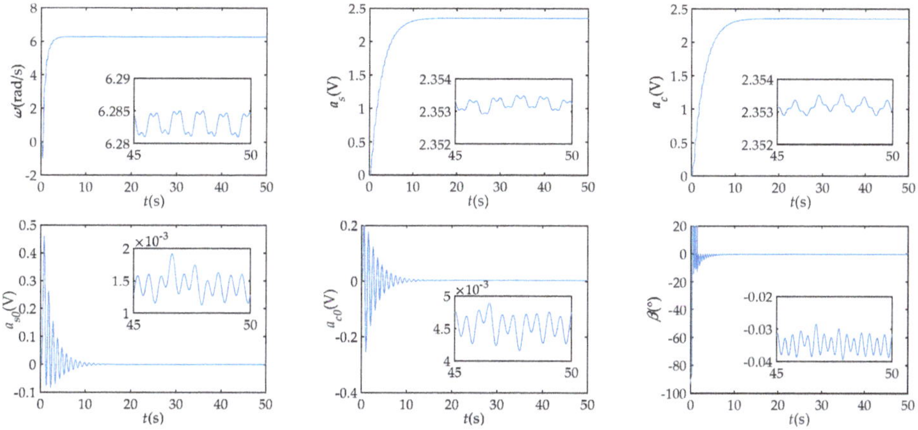

Figure 15. Estimations of angular velocity and imperfect parameters after the Butterworth filter (Group 2).

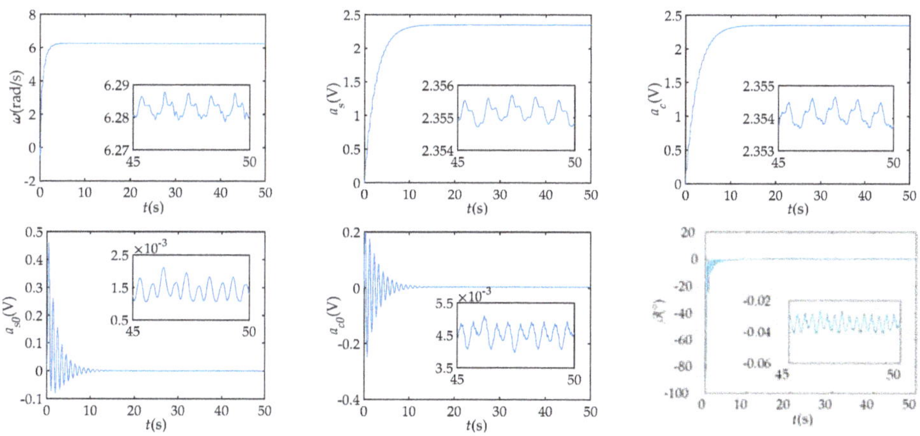

Figure 16. Estimations of angular velocity and imperfect parameters after the DWT (Group 3).

Table 6. Results of calibration in experiment.

Parameters		ω(rad/s)	a_s(V)	a_c(V)	a_{s0}(V)	a_{c0}(V)	β(°)
Calibrated directly	Estimates	6.28288	2.3551	2.3550	1.446×10^{-3}	4.548×10^{-3}	−0.03450
	STD	2.88×10^{-3}	2.54×10^{-4}	2.56×10^{-4}	2.85×10^{-4}	2.86×10^{-4}	3.99×10^{-3}
After the Butterworth filter	Estimates	6.28304	2.3532	2.3532	1.445×10^{-3}	4.545×10^{-3}	−0.03462
	STD	1.80×10^{-3}	1.54×10^{-4}	1.57×10^{-4}	1.70×10^{-4}	1.66×10^{-4}	2.44×10^{-3}
After the DWT	Estimates	6.28299	2.3552	2.3541	1.447×10^{-3}	4.547×10^{-3}	−0.03448
	STD	2.19×10^{-3}	2.48×10^{-4}	2.53×10^{-4}	2.56×10^{-4}	2.57×10^{-4}	3.83×10^{-3}
After the designed filter	Estimates	6.28318	2.3553	2.3532	1.416×10^{-3}	4.544×10^{-3}	−0.03436
	STD	3.30×10^{-6}	7.99×10^{-6}	1.07×10^{-7}	2.48×10^{-6}	2.54×10^{-6}	4.16×10^{-5}

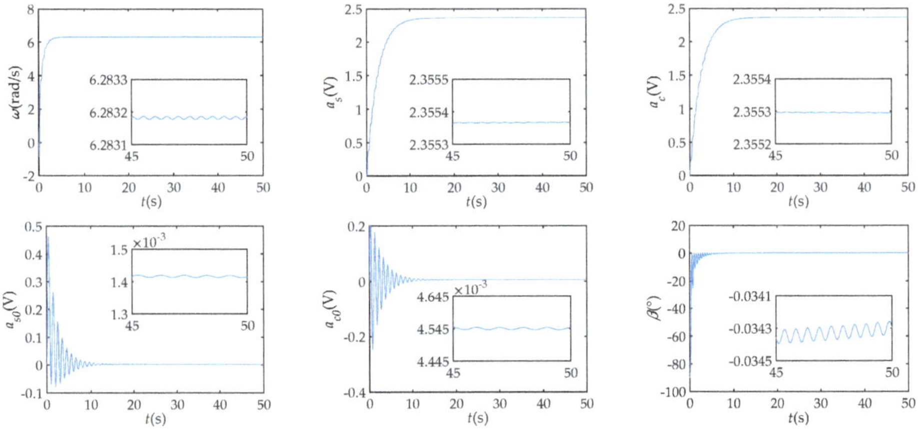

Figure 17. Estimations of angular velocity and imperfect parameters after the DWT-SVD based filter (Group 4).

5. Conclusions

In order to improve the calibration accuracy of the resolver signals, a DWT-SVD based filter was designed in this paper to reduce the noises. Most of the noises in the resolver, such as the inductive harmonics, residual excitation components, and random noise were taken into account. Firstly, the resolver signals were decomposed to the approximation coefficient and detail coefficients by DWT. The decomposition pre-filtered the residual excitation components and part of the noises. Next, the singular values of approximation coefficient were calculated. Finally, the signals were reconstructed by a few selected singular values to suppress harmonics and preserve almost only the fundamental and DC components. Because of the multi-resolution characteristic of DWT and the good correspondence between the singular value and frequency, this method is favorable to dramatically reduce the noises. Therefore, the proposed filter improved the calibration accuracy of the nonideal parameters, such as amplitude deviations, DC offsets, and imperfect quadrature in resolvers. The effectiveness of the designed filter was verified by simulation and experimental results.

Author Contributions: Conceptualization, Z.W.; methodology, M.G.; validation, M.G.; writing—Original draft preparation, M.G. and Z.W.

Funding: This work was partially supported by The National Natural Science Foundation of China (10772011).

Conflicts of Interest: The authors declare no conflict of interest.

References

1. Alipour-Sarabi, R.; Nasiri-Gheidari, Z.; Tootoonchian, F.; Oraee, H. Performance analysis of concentrated wound-rotor resolver for its applications in high pole number permanent magnet motors. *IEEE Trans. Ind. Electron.* **2017**, *17*, 7877–7885. [CrossRef]
2. Liu, H.; Wu, Z. Demodulation of angular position and velocity from resolver signals via Chebyshev filter-based type III phase locked loop. *Electronics* **2018**, *7*, 354. [CrossRef]
3. Hanselman, D.C. Resolver signal requirements for high accuracy resolver-to-digital conversion. *IEEE Trans. Ind. Electron.* **1991**, *37*, 556–561. [CrossRef]
4. Tan, K.K.; Zhou, H.X.; Lee, T.H. New interpolation method for quadrature encoder signals. *IEEE Trans. Instrum. Meas.* **2002**, *51*, 1073–1079. [CrossRef]
5. Heydemann, P.L.M. Determination and correction of quadrature fringe measurement errors in interferometers. *Appl. Opt.* **1981**, *20*, 3382–3384. [CrossRef] [PubMed]

6. Balemi, S. Automatic calibration of sinusoidal encoder signals. In Proceedings of the 16th Triennial World Congress, Prague, Czech Republic, 3–8 July 2005; pp. 1189–1195.
7. Hoang, H.V.; Jeon, W.J. Signal compensation and extraction of high resolution position for sinusoidal magnetic encoders. In Proceedings of the International Conference on Control, Automation and Systems, Seoul, Korea, 17–20 October 2007; pp. 1368–1373.
8. Hoseinnezhad, R.; Bab-Hadiashar, A.; Harding, P. Calibration of resolver sensors in electromechanical braking systems: A modified recursive weighted least-squares approach. *IEEE Trans. Ind. Electron.* **2007**, *54*, 1052–1060. [CrossRef]
9. Zhang, J.; Wu, Z. Automatic calibration of resolver signals via state observers. *Meas. Sci. Technol.* **2014**, *25*, 095008. [CrossRef]
10. Wu, Z.; Li, Y. High-accuracy automatic of resolver signals via two-step gradient estimators. *IEEE Sens. J.* **2018**, *18*, 2883–2891. [CrossRef]
11. Gao, Z.; Zhou, B.; Hou, B.; Li, C.; Wei, Q.; Zhang, R. Self-calibration of angular position sensors by signal flow networks. *Sensors* **2018**, *18*, 2513. [CrossRef]
12. Lara, J.; Chandra, A. Position error compensation in quadrature analog magnetic encoders through an iterative optimization algorithm. In Proceedings of the Industrial Electronics Society IECON 2014—40th Annual Conference of the IEEE, Dallas, TX, USA, 29 October–1 November 2014; pp. 3043–3048.
13. Wang, H.; Shang, J.; Li, Y.; Xu, Y. The finite element analysis and parameter optimization of the axial flux variable-reluctance resolver with short pitch distributed winding. *Int. J. Appl. Electromagn. Mech.* **2014**, *45*, 441–447. [CrossRef]
14. Kaul, S.K.; Tickoo, A.K.; Koul, R.; Kumar, N. Improving the accuracy of low-cost resolver-based encoders using harmonic analysis. *Nucl. Instrum. Methods Phys. Res.* **2007**, *586*, 345–355. [CrossRef]
15. Faber, J. Self-calibration and noise reduction of resolver sensor in servo drive application. In Proceedings of the 2012 Elektro of the IEEE, Rajeck Teplice, Slovakia, 21–22 May 2012; pp. 174–178.
16. Sarma, S.; Venkateswaralu, A. Systematic error cancellations and fault detection of resolver angular sensors using a DSP based system. *Mechatronics* **2009**, *19*, 1303–1312. [CrossRef]
17. Zhu, M.; Wang, J.; Ding, L.; Zhu, Y. A Software based robust resolver-to-digital conversion method in designed in frequency domain. In Proceedings of the 2011 International Symposium on Computer Science and Society, Kota Kinabalu, Malaysia, 16–17 July 2011; pp. 244–247.
18. Minaee, S.; Abdolrashidi, A.A. Highly accurate palmprint recognition using statistical and wavelet features. In Proceedings of the 2015 IEEE Signal Processing and Signal Processing Education Workshop, Salt Lake City, UT, USA, 9–12 August 2015.
19. Huang, Z.; Li, W.; Wang, J.; Zhang, T. Face recognition based on pixel-level and feature-level fusion of the top-level's wavelet sub-bands. *Inf. Fusion* **2015**, *22*, 95–104. [CrossRef]
20. Minaee, S.; Abdolrashidi, A. On The power of joint wavelet-DCT features for multispectral palmprint recognition. In Proceedings of the 2015 49th Asilomar Conference on Signals, Systems and Computer, Pacific Grove, CA, USA, 8–11 November 2015; pp. 1593–1597.
21. Xu, X.; Luo, M.; Tan, Z.; Pei, R. Echo signal extraction method of laser radar based on improved singular value decomposition and wavelet threshold denoising. *Infrared Phys. Technol.* **2018**, *92*, 327–335. [CrossRef]
22. Xu, J.; Zhang, L.; Zuo, W.; Zhang, D.; Feng, X. Patch group based nonlocal self-similarity prior learning for image denoising. In Proceedings of the 2015 IEEE International Conference on Computer Vision, Santiago, Chile, 7–13 December 2015; pp. 244–252.
23. Zhang, K.; Zuo, W.; Zhang, L. FFDNet: Toward a fast and flexible solution for CNN based image denoising. *IEEE Trans. Image Process.* **2018**, *27*, 4608–4622. [CrossRef]
24. Guo, Q.; Zhang, C.; Zhang, Y.; Liu, H. An efficient SVD-based method for image denoising. *IEEE Trans. Circuits Syst. Video Technol.* **2016**, *26*, 868–880. [CrossRef]
25. Zhao, X.; Ye, B.; Chen, T. The relationship between non-zero singular values and frequencies and its application to signal decomposition. *Acta Electron. Sin.* **2017**, *45*, 2008–2018.
26. Paul, J.S.; Reddy, M.R.; Kumar, V.J. A transform domain SVD filter for suppression of muscle noise artefacts in exercise ECG's. *IEEE Trans. Biomed. Eng.* **2000**, *47*, 654–663. [CrossRef] [PubMed]
27. Bhatnagar, G.; Raman, B. A new robust reference watermarking scheme based on DWT-SVD. *Comput. Stand. Interfaces* **2009**, *31*, 1002–1013. [CrossRef]

28. Kallel, F.; Hamida, A.B. A new adaptive gamma correction based algorithm using DWT-SVD for non-contrast CT image enhancement. *IEEE Trans. Nanobiosci.* **2017**, *16*, 666–675. [CrossRef]
29. Kumar, M.; Vaish, A. An efficient encryption-then-compression technique for encrypted images using SVD. *Digit. Signal Prog.* **2017**, *60*, 81–89. [CrossRef]
30. Jiang, Y.; Tang, B.; Qin, Y.; Liu, W. Feature extraction method of wind turbine based on adaptive Morlet wavelet and SVD. *Renew. Energy* **2011**, *36*, 2146–2153. [CrossRef]

electronics

MDPI

Article

HPEFT for Hierarchical Heterogeneous Multi-DAG in a Multigroup Scan UPA System

Yuzhong Li [1,2], **Wenming Tang** [1] **and Guixiong Liu** [1,*]

[1] School of Mechanical & Automotive Engineering, South China University of Technology, Guangzhou 510641, China; melyz@mail.scut.edu.cn (Y.L.); tang.wm@mail.scut.edu.cn (W.T.)

[2] School of Information Engineering, Huizhou Economic and Polytechnic College, Huizhou 516057, China

[*] Correspondence: megxliu@scut.edu.cn; Tel.: +86-020-8711-0568

Received: 2 April 2019; Accepted: 29 April 2019; Published: 5 May 2019

Abstract: Multidirected acyclic graph (DAG) workflow scheduling is a key problem in the heterogeneous distributed environment in the distributed computing field. A hierarchical heterogeneous multi-DAG workflow problem (HHMDP) was proposed based on the different signal processing workflows produced by different grouping and scanning modes and their hierarchical processing in specific functional signal processing modules in a multigroup scan ultrasonic phased array (UPA) system. A heterogeneous predecessor earliest finish time (HPEFT) algorithm with predecessor pointer adjustment was proposed based on the improved heterogeneous earliest finish time (HEFT) algorithm. The experimental results denote that HPEFT reduces the makespan, ratio of the idle time slot (RITS), and missed deadline rate (MDR) by 3.87–57.68%, 0–6.53%, and 13–58%, respectively, and increases relative relaxation with respect to the deadline (RLD) by 2.27–8.58%, improving the frame rate and resource utilization and reducing the probability of exceeding the real-time period. The multigroup UPA instrument architecture in multi-DAG signal processing flow was also provided. By simulating and verifying the scheduling algorithm, the architecture and the HPEFT algorithm is proved to coordinate the order of each group of signal processing tasks for improving the instrument performance.

Keywords: hierarchical heterogeneous multi-DAG workflow; multigroup scan; ultrasonic phased array; heterogeneous earliest finish time

1. Introduction

Ultrasonic phased array (UPA) systems with a large number of elements can achieve multigroup scanning, increase the scanning flexibility, and enhance the resolution and contrast of the resulting images. Hierarchical signal processing flow that accounts for the constraints of a directed acyclic graph (DAG) can be realized by adding a bus and an arbitrator to the hardware architecture. In a distributed software environment, multigroup UPA scans can use different scanning modes in different groupings; thus, several different signal processing methods can be implemented. These processes must also be hierarchically processed using the heterogenous shared resources according to the priority constraints. The priority constraints between the tasks in each group are represented using the DAG diagrams, and each shared resource can only perform the specified signal processing tasks because of functional constraints. Further, the priority constraints related to the multigroup tasks combine with the functional constraints on shared resources to form a hierarchical heterogeneous multi-DAG workflow problem (HHMDP). To address this problem, a scheduling algorithm is required to coordinate task processing with the heterogeneous shared resources so that the various signal processing steps involved in the distributed UPA instruments can be executed in an orderly manner.

With the rapid development of information technology and the increasing complexity of the associated problems, distributed computing resources with high performance are required to complete the computing tasks subject to the deadline constraints. Further, distributed resource scheduling for DAG tasks has been the subject of several previous researches and has developed into a mature method. More than 100 scheduling algorithms have been developed based on the homogeneous or heterogeneous distributed environments, the structural differences of the scheduling tasks, and the different optimization objectives [1]. Among these algorithms, the DAG scheduling model and the heterogeneous earliest finish time (HEFT) algorithm proposed by Topcuoglu [2] in 2002 have been extensively adopted. These are the models and methods that have been recently employed in distributed systems for performing tasks such as grid and cloud computing.

An existing study has denoted that the DAG scheduling problem is a nondeterministic polynomial complete problem [3]. Further, popular task scheduling algorithms used to obtain the makespan can be classified into the following two categories: static scheduling and dynamic scheduling algorithms [4]. Among the static scheduling algorithms, the list scheduling algorithms that employ heuristic techniques have been proved to produce the most efficient scheduling, and their complexities, associated with the number of involved tasks involved, are generally quadratic [5]. In addition, the list scheduling algorithm is fast, easy to implement, and has wider applicability than that of other scheduling algorithms. Some of the most famous list scheduling algorithms are HEFT [2], predicted earliest finish time (PEFT) [3], heterogeneous critical parent tree (HCPT) [6], high-performance task scheduling (HPS) [7], and performance effective task scheduling (PETS) [8]. HEFT is known to produce a scheduling length that can be compared with that produced by other scheduling algorithms exhibiting a lower time complexity. Further, the HEFT algorithm can be enhanced by either introducing a new mechanism for calculating the task priority or adding a new mechanism (such as prior attributes) to improve the processor selection [9].

The idle time slot of a single DAG task is large, and its resource utilization is low owing to the data transmission delay between the tasks and the imbalanced DAG structure. To effectively improve these shortcomings, Honig [10] and H. Zhao [11] initially proposed a scheduling algorithm in 2006 for multi-DAGs sharing a set of distributed resources. A DAG task can use idle slots generated by other DAG tasks scheduled on the same distributed resource group. Further, strategies to minimize makespan and scheduling fairness were also proposed. Although Yu et al. [12] made some improvements in this regard, the multi-DAG scheduling algorithm proposed in their study did not consider the deadline constraints. By considering the earliest deadline first (EDF) [13] algorithm—an application of the sequential scheduling algorithm based on the deadline—Stavrinides [14] proposed the usage of the deadline of each DAG for determining the priority of the multi-DAG task scheduling and the usage of the time slots for accurate calculation of the DAG tasks. This approach is suitable for DAGs with similar DAG structures and sizes. However, the deadline is insufficient to respond to the emergency of DAG because the number of DAG tasks is different. Furthermore, the relative strictness of Multi-DAGs with deadlines (MDRS) [15] scheduling strategy improves the fairness of scheduling according to the relative strictness of each DAG. However, before the scheduling tasks are selected, the HEFT algorithm is used to preschedule the remaining tasks for each DAG after which the relative duration of each DAG is calculated, and the most urgent task in the DAG is scheduled for execution. Therefore, each DAG can decentralize hybrid scheduling before the deadline, thereby improving resource utilization. However, some DAGs would be discarded when resource shortage is observed. Tian et al. [16] improved the existing fair scheduling algorithm by solving the fairness problem that is encountered while scheduling multiple DAG workflows that have the same priority but are submitted at different times. In addition, Xu et al. [17] proposed a cooperative scheduling algorithm to further improve the utilization of the computing resources for the workflow in a distributed heterogeneous environment exhibiting better performance in terms of throughput, time slot wastage, fairness, and time complexity when compared with those exhibited by MDRS, EDF, and fairness algorithms.

For ultrasonic phased array and DAG workflow scheduling, Tang et al. [17] studied ultrasonic bus transmission scheduling using the MFBSS algorithm to schedule between FIFOs, so that the utilization rate of transmission channels was not less than 92%. Li et al. [18], based on time division multiplexing, proposed an IBF algorithm for focus and delay module scheduling, which increased the maximum completion time by 8.76 to 21.48%, reduced resource consumption by 30 to 40%. Li et al. [19] also proposed SSPA algorithm for heterogeneous signal processing. Compared with the FCFS algorithm and SPT algorithm, the SSPA algorithm improves bandwidth utilization by 9.72% and reduces maximum completion time by 11%. Anwar and Deng [20] proposed a novel Hybrid Bio-inspired Metaheuristic for Multiobjective Optimization (HBMMO) algorithm based on a nondominant sorting strategy for the workflow scheduling problem, which decrease makespan, execution cost, and inefficient utilization of the virtual machines (VMs). Miao et al. [21] investigates H_∞ consensus problem of heterogeneous multiagent systems under Markov switching topologies; consensus algorithms with communication time delay via output were also proposed. By applying stochastic stability analysis, model transformation techniques and graph theory, sufficient conditions of mean square consensus and H_∞ consensus are obtained, respectively. Drozdov [22] address image processing workflow scheduling problems on a multicore digital signal processor cluster. They proposed Pessimistic Heterogeneous Earliest Finish Time scheduling algorithm for Ligo and Montage applications and presented its better performance than others. Feng et al. [23] studied gene function prediction, which includes the hierarchical multilabel classification (HMC) task, and proposed an algorithm for solving this problem based on the Gene Ontology (GO), the hierarchy of which is a directed acyclic graph (DAG). The algorithm has better performance compared with true path rule and CLUS-HMC algorithm on eight yeast data sets annotated by the GO.

However, these scheduling algorithms do not consider the challenge in performing heterogeneous resource processing tasks at different layers using different resources, i.e., some processors can only handle specific tasks, and these tasks or transactions can be sequentially executed if a single DAG can be approximated as a distributed permutation flow shop scheduling problem in a distributed environment and if the processor selection phase can be considered to be hierarchical. Thus, in general, there exists little research and few achievements related to the multi-DAG hierarchical scheduling problem and its application in the signal processing scheduling of ultrasonic systems even though many related problems require urgent attention. This study uses the critical path method by considering the hierarchical and specified function constraints of the shared resources to propose an improved method for HEFT that incorporates predecessor pointer adjustment (PPA), which can be referred to as the heterogeneous predecessor earliest finish time (HPEFT). This method can coordinate the different signal processing steps to satisfy the priority constraints represented by the multi-DAG model under the multigroup scanning architecture of a UPA system with multiple layers and shared resources for performing the specified functions. This coordination reduces the maximum completion time (makespan) and improves the utilization rate of the shared resources, improving the real-time frame rate and reducing the energy consumption.

The remainder of the manuscript is organized as follows. Section 2 discusses the hierarchical heterogeneous multi-DAG workflow scheduling problem and presents both the previous and proposed algorithms along with some basic examples; Section 3 discusses the experimental settings, problem generators, parameters, the experiments performed herein, and the results; Section 4 presents the signal processing scheduling optimization strategies for a multigroup scan UPA system; and Section 5 summarizes this study and discusses future work.

2. A Hierarchical Heterogeneous Multi-DAG Workflow Problem (HHMDP) and HPEFT Algorithm

2.1. Problem Description

Suppose there are multiple workflows that can be modeled as DAG $\{G_k, k \in 1, 2 \ldots K\}$. Each DAG node, ki, represents a task, T_{ki}, and each edge represents the sequential relation between two different nodes. Thus, it can be said that ki exhibits a hierarchical structure. The nodes belonging to the first layer have no predecessor nodes, and they have start time ST_{ki}. The tasks in the first layer must wait until the specified ST_{ki} before they can be executed by the processor. Further, the tasks in the final layer have no successor nodes. All the nodes must be dispatched to a collection of heterogeneous shared resources M = $\{\{M_j{}^l, j = 1, 2 \ldots Q_l\}, l = 1, 2, 3 \ldots$ L}; the set of shared resources for each layer of $M_j{}^l$ contains several shared resources that can execute the corresponding layer nodes. The shared resources in a layer can only perform tasks corresponding to the nodes in the same layer, and the nodes in the same layer to be executed by the shared resource must belong to the same layer. Further, the nodes in the same DAG have a sequential relation. With the exception of the first and the final layer nodes, the remaining nodes in a DAG must have predecessor and successor nodes and cannot be isolated from other nodes. A node can have multiple predecessor and successor nodes, and its predecessor nodes can only be in the upper layer, whereas its successor nodes can only be in the subsequent layer. However, a node cannot be both the predecessor and successor node of another node simultaneously. All the tasks must be sequentially processed according to the DAG graph, and the computing tasks of the upper node must be processed before the current node can be processed. For any given node, the processing time P_{ki} for the computing tasks is determined by the data length D_{ki} in the node, as follows

$$P_{ki} = D_{ki}/SD_j + C_j. \tag{1}$$

The difference in D_{ki} and the shared resource speed, SD_j, can change the execution time of a given node $P_{ki,j}$. C_j is the delay that is required for the current processor to run. The delay is considered to be generally small. The system used in the present research was interconnected by buses with the following conditions. The tasks of the same node cannot be executed twice on the same shared resource, and the predecessor or successor nodes of any given node cannot be executed on the same shared resource because the hierarchical structure is in different sets of shared resources. Therefore, in the problem that is considered herein, the delay is not equivalent to the communication time, as observed in the case of a classical HEFT algorithm. Hence, we propose a strategy, which ensures that each node produces C_j in any processor, with the communication time for shared resource switching observed to be zero.

Each shared resource in the system can be simultaneously executed and communicated. The scheduling problem is to minimize the makespan. Furthermore, the DAG actual finish time (DAFT) after the scheduling of the algorithm represents the makespan after all the nodes in a DAG are executed.

$$\text{makespan} = \max\{\text{DAFT}(ki), \text{ where } ki \text{ is the latest execution node in DAG}\} \tag{2}$$

The maximum number of layers of tasks (node) from the top to the bottom of the DAG and the number of layers of shared resources is equal to L. Further, the longest path from the top to the bottom denotes the critical path of the DAG. *ki* at the earliest start time (EST) of the shared resource M_j is given as follows

$$\text{EST}(T_{ki}, M_j) = \max\left\{\text{Tavailable}\left(M_j\right), \max_{T_{ks} \in \text{Pred}(T_{ki})}\{\text{AFT}(T_{ks})\}, ST_{ki}\right\} \tag{3}$$

Tavailable (M_j) is the time when the shared resource M_j is ready for performing new tasks. For the top (first) level nodes, all the processors have not yet performed node tasks, and there are no previous nodes; however, the processors have the EST, ST_{ki}, for node tasks. In such a situation, EST becomes equal to ST_{ki}. The earliest finish time (EFT) is the earliest time when task *ki* can be processed using an assigned shared resource.

$$\text{EFT}(T_{ki}, M_j) = \text{EST}(T_{ki}, M_j) + P_{ki,j} \tag{4}$$

The actual start time (AST) refers to the real time when a given node task begins executing after a DAG task is scheduled. In this algorithm, AST is considered to be equal to EST. The actual finish time (AFT) refers to the actual completion time of a node after task scheduling. Table 1 presents the symbols used in this study.

Table 1. Table of symbols.

Symbol	Description
PR (*ku*,*kv*)	Precedent relation between node *ku* and *kv*.
Pred (*ki*)	Direct predecessor of node *ki*.
Sucd (*ki*)	Direct successor of node *ki*.
Tavailable (M_j)	Time required to issue a new task in the shared resource M_j.
Shed (M_j)	Node set for scheduling the shared resource M_j
Layer (*ki*)	Node *ki* layer
Layer (M_j)	Shared resource M_j layer
NumMac (*ki*)	The number of shared resources on the same layer as node *ki*
PAFT (*ki*)	The maximum actual completion time of the previous nodes (predecessor tasks)

2.2. HPEFT Algorithm

A previous study [2] demonstrated that HEFT can be used to obtain the critical path of scheduling and generate upward rank (Rank_u) with respect to the critical path. However, the problems presented in this study are different from the ones available in the literature on HEFT using PEFT, given as follows [4]; (1) multi-DAG scheduling; (2) each DAG has a different start time ST_{ki}; (3) the communication consumption between the shared resources M_j is 0; however, considering the different latency of each processor, the shared resources can be classified as P_{ki}; and (4) the shared resources and DAG are hierarchical.

When compared with the HEFT algorithm, the proposed algorithm can satisfy the requirements of hierarchical scheduling. In addition, hierarchical scheduling, where any two layers are connected by edges, contains a prioritized set of nodes. The current node is scheduled to execute after the execution of all the predecessor nodes, thereby making the scheduling compact. As depicted in Figure 1, the tasks in any two DAGs are assumed to be tasks A1 and A2; here, they are the tasks of DAG A, and the tasks of B1 and B2 are the tasks of DAG B. If A1 and B1 belong to the same layer and are scheduled on the same processor, B1 and B2 are the tasks of DAG B. A2 and B2 belong to the same layer and are scheduled on the same processor. A1 is scheduled on the shared resource M1 before B1. In the shared resource M2, the task A2 is scheduled first and is followed by B2. The completion time is observed to be short. The higher the ratio of processors to tasks in a given layer, the shorter will be the maximum completion time in the layer. The highest execution efficiency can be achieved when the number of processors and tasks in a given layer is the same.

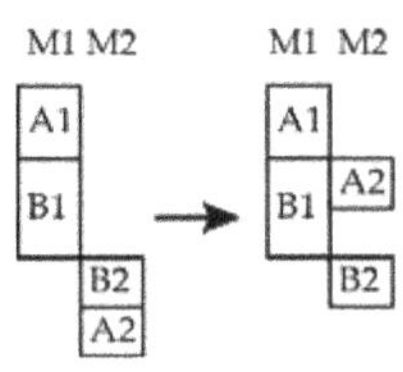

Figure 1. Demonstration of compact scheduling.

The proposed algorithm can be given as follows:

In stage 1, the upward weights are calculated as

$$\begin{aligned} \text{Rank}_\text{p}(ki) &= \overline{P_{ki}} + \min_{ks\in \text{sucd}(ki)} \left[\text{Rank}_\text{p}(ks)\right] \\ \overline{P_{ki}} &= \sum_{Layer(ki)=j} P_{ki,j}/\text{NumMac}(ki) \end{aligned} \tag{5}$$

Communication delay, i.e., the average of the execution time of $\overline{P_{ki}}$ in all the machines at the same layer of *ki*, has been incorporated into $P_{ki,j}$. *ks* denotes the direct successor of *ki*. $P_{ki,j}$ denotes the processing time of *ki* in the shared resource M_j; Rank_p denotes the upward rank. If the node is at the final level, $\text{Rank}_\text{p}(ki) = \overline{P_j}$, NumMac (*ki*) denotes the shared resource in the same layer as node *ki*. The average execution time of all the previous tasks is related to the shared resource NumMac (*ki*) at that layer.

Stage 2 involves the selection of tasks with the highest Rank_p in the list. According to the maximum completion time of all the scheduled predecessor tasks, the available slots are searched, and the location of the shared resource M_j with the earliest time slots is allocated to the available (M_j) task nodes.

In stage 3, after completing all the scheduled tasks in stage 1 and 2, the time slots among the scheduled resources are researched according to each shared resource, as denoted in Equation (6).

$$M_{j\_\text{PPA}} \in \{M_j, \text{AFT}(ks) - \text{AST}(kf) > \text{sum}(P_{ki,\,j})\}, ks, kf, ki \in \text{shed}(M_j) \quad (6)$$

If a shared resource M_{j_PPA} has a time slot in the scheduled task, we can find the maximum actual completion time of their predecessor tasks (PAFT) for tasks arranged by the previous DAG, as denoted in Equation (7); next, we calculate the PPA table and reschedule M_{j_PPA} accordingly.

$$\text{PAFT}(ki) = \max[\text{AFT}(ks)], ks \in \text{Pred}(ki) \quad (7)$$

2.3. Example Demonstration

The three DAGs and their node constraints are depicted in Figure 2; the dashed arrow denotes the start time of the tasks. A1–A6 belong to G1 (DAG A), B1–B6 belong to G2 (DAG B), C1–C6 belong to G3 (DAG C), and M1–M5 denote a set of shared resources. The hierarchy of the shared resources and task nodes can also be observed. The priority constraints between the tasks in each DAG are represented by the solid arrowhead lines. For example, the successor tasks of A1 are A4, and the successor tasks of C3 are C4. The specific functional constraints (layering) and the task layering of all the shared resources are also depicted in Figure 2, i.e., A1, A2, B1, B2, and C1. C1 can be executed by the shared resources M1, M2, and M3, which belong to Layer 1, or by other layers.

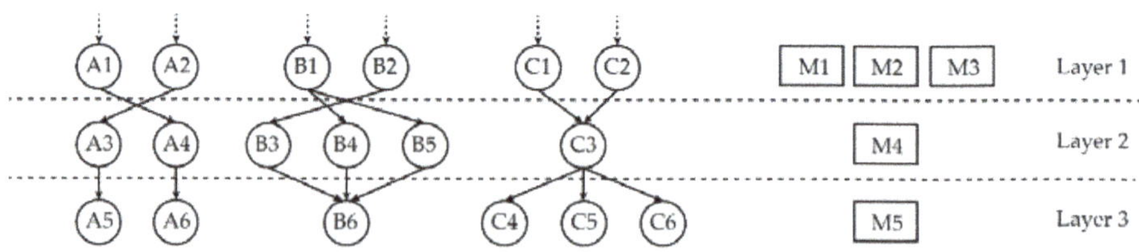

Figure 2. Hierarchical diagram of the multi-DAG task nodes and shared resources.

Table 2 is a hierarchical table of shared resources presenting the serial number of shared resources and their corresponding layers. Table 3 presents the data length of each task node, which can be calculated using Equation (1) to obtain the execution time $P_{ki,j}$. Table 4 presents SD_j and C_j of the shared resources. Table 5 denotes the ST_{ki} values. For a start node $ki_\text{start} = \{ki, \text{Pred}(ki) = \emptyset\}$ without predecessor tasks, there exists a corresponding start time. Tables 6–8 denote that after calculating the processing time ($P_{ki,j}$), all the corresponding "-" lines indicate that node *ki* cannot be executed on the shared resource M_j. Therefore, the specific functional constraints (hierarchical relations) can also be observed from these tables.

Table 2. Shared resources of each layer.

Layer	Shared Resource		
1	M1	M2	M3
2	M4	-	-
3	M5	-	-

Table 3. Data length (D_{ki}) of DAGs.

DAG	1	2	3	4	5	6
A	2	1	2	2	1	1
B	1	3	1	4	3	1
C	2	3	2	4	1	2

Table 4. Speed (SD_j) and delay (C_j) of the shared resources.

M_j	1	2	3	4	5
Speed	1	1	1	2	1
Delay	1	1	2	1	1

Table 5. Start time (ST_{ki}) for each DAG.

DAG	1	2
A	3	2
B	2	4
C	3	4

Table 6. Processing time with priority relation (PR) in DAG A.

	A1	A2	A3	A4	A5	A6
M1	3	2	-	-	-	-
M2	4	2	-	-	-	-
M3	3	3	-	-	-	-
M4	-	-	2	2	-	-
M5	-	-	-	-	2	2

Table 7. Processing time with priority relation (PR) in DAG B.

	B1	B2	B3	B4	B5	B6
M1	2	4	-	-	-	-
M2	2	4	-	-	-	-
M3	3	5	-	-	-	-
M4	-	-	2	3	3	-
M5	-	-	-	-	-	2

Table 8. Processing time with priority relation (PR) in DAG C.

	C1	C2	C3	C4	C5	C6
M1	3	4	-	-	-	-
M2	3	4	-	-	-	-
M3	4	5	-	-	-	-
M4	-	-	2	-	-	-
M5	-	-	-	5	2	3

Figure 3a–c denotes the schematics of three scheduling algorithms: shortest processing time (SPT), round-robin (R-R), and HEFT. No constraint connection is depicted on the graph for clarity. The constraint relation between the tasks is shown on the edges of Figure 2.

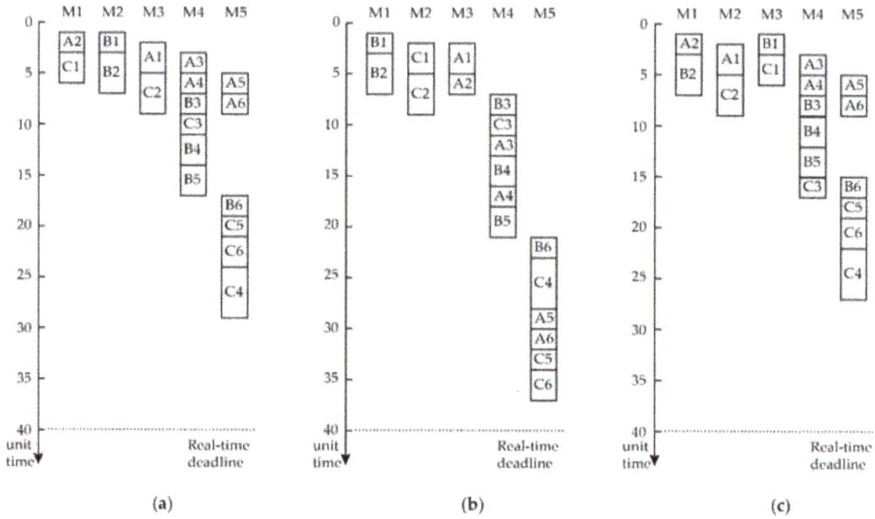

Figure 3. Scheduling examples of three different scheduling algorithms: (**a**) SPT; (**b**) round-robin; and (**c**) heterogeneous earliest finish time (HEFT).

The SPT algorithm sorts all the DAG tasks on the layer from small to large according to the processing time; further, the algorithm schedules the tasks according to the earliest time slot of the shared resources that they possess until all the task nodes in all the layers complete task scheduling.

The R-R algorithm can be used to schedule the tasks in the DAG according to the serial number of the DAG and to allocate tasks to the earliest time slot of shared resources in the corresponding layer.

The associated communication time $C_{ks,ki} = 0$. In addition, for the end node, Rank_p can be calculated using Equation (8), whereas, for other nodes, Rank_p can be calculated using Equation (5).

$$\text{Rank}_\text{p}(ki)|_{\text{Sucd}(ki)=\varnothing} = \sum P_{ki,j}/NumMac(M_j) \quad \text{s.t. Layer}(j) = \text{Layer}(ki) \tag{8}$$

Here, NumMac (M_j) denotes the number of matching shared resources, i.e., Layer (j) = Layer (ki).

Table 9 presents the PPA table, and Figure 4 depicts the adjustment method using a PPA table diagram. The graph shown herein reveals that the HPEFT algorithm is arranged according to Rank_p (ki); B5 is scheduled next to B6, subsequently followed by C5, C6, and C4. However, this approach is not optimal. According to the algorithm proposed in the previous section, C5, C6, and C4 are not directly connected with C3 scheduling. Hence, there is a time slot between A6 and B6. Therefore, according to Equation (6), all node tasks in the shared resource M5 can evaluate their PPA table according to the tasks that were scheduled in M5 in the previous stage with respect to the PPA table presented in Table 9.

Table 9. The predecessor pointer adjustment (PPA) table.

Node(*ki*)	**A5**	**A6**	**C4**	**C5**	**C6**	**B6**
PAFT (*ki*)	5	7	11	11	11	17

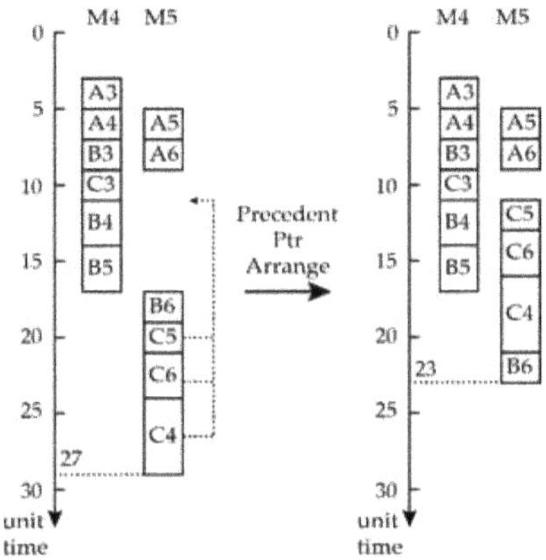

Figure 4. Predecessor pointer adjustment (PPA).

As depicted in Figure 4, the first step is to arrange A5 and A6 at time 5 and 7; schedule C5, C6, and C4 at time 11; and finally schedule B6 to C4, which can reduce the makespan from 27 to 23.

2.4. Description of the HPEFT Algorithm

Algorithm 1 HPEFT

Input: DAG group $G = \cup_{k=1}^{K}\{G_k\}$, resource group $M = \cup_{j=1}^{Q}\{M_j\}$, processing time matrix of tasks in a shared resource set $\boldsymbol{P} = \cup_{k=1,i=1,j=1}^{k=L,i=N_l,j=Q}\{P_{ki,j}\}$, DAG deadline, constraint matrix $\boldsymbol{E} = \cup_{i=1,s=1,k=1}^{i=N_k,s=N_k,k=K}\{e_{i,j,k}\}$, and start time matrix $\boldsymbol{ST} = \cup_{k=1}^{k=K}\{ST_{ki}\}$

Output: Scheduling List

HPEFT ()

1. Unscheduled DAG list *unscheTasks*←T_{ki} (k = 1,2 … K, i = 1, 2 … N_k)
2. Calculate Rank_p of each T_{ki} in *unscheTasks*, and arrange them in the ascending order
3. Using Equation (1), the data length D_i, shared resource speed SD_j, and delay C_j are substituted, and the processing time matrix $\boldsymbol{P}$ is obtained.
4. For l = 1, 2 … L
5. Find whether all the tasks performed at this *unscheLayerTasks*←{T_{ki}, Layer(ki) ∈ l} in unschTasks.
6. WHILE (*unscheLayerTasks* ≠ ∅)
7. Sort all the tasks T_{ki} in *unscheLayerTasks* according to Rank_p (T_{ki}), and find the current minimum T_{ki} as T_{ki_urgent}.
8. Using Equation (3), find an EST (T_{ki},M_j) suitable for scheduling.
9. Assign T_{ki_urgent} to the Scheduling List HPEFT ().
10. Delete the T_{ki_urgent} Task from *unscheLayerTasks*.
11. END WHILE
12. END FOR
13. Find the idle time slot in the M_j Gantt chart using Equation (6).
14. If M_j has an idle time slot, all the tasks scheduled to M_j should be returned to rearrange(j)←{T_{ki}, T_{ki} ∈ Shed (M_j)}; then, clear the M_j scheduling table, i.e., the scheduling list.
15. According to their predecessor AFT, calculate the corresponding PAFT(ki) in the ascending order for the PPA table.
16. WHILE (rearrange(j) ≠ ∅)
17. Schedule the minimum PAFT(ki) of task, $T_{ki\_rearrange\_urgent}$, in the PPA table to M_j ∈ Tavailable(M_j) in the scheduling list; if overlay exists, postpone the other tasks for $T_{ki\_rearrange\_urgent}$.
18. Delete $T_{ki\_rearrange\_urgent}$.
19. END WHILE
20. Makespan is calculated using Equation (2), RITS is calculated using Equation (9), and RLBD is calculated using Equation (10).
21. Return the *scheduling list*

End Procedure

The HPEFT flowchart is depicted in Figure 5.

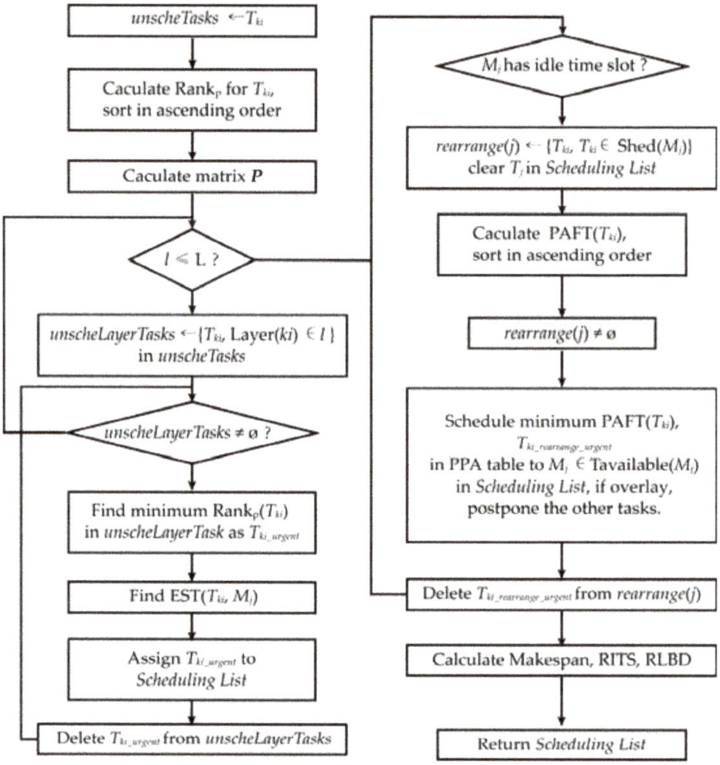

Figure 5. Flowchart of HPEFT.

2.5. Time Complexity

Assuming that each DAG has N nodes and that the number of computing resources is Q, the worst-case time complexity of the HEFT algorithm is computed to be $O(N^2Q)$ [2]. If there are K such DAGs that need to be simultaneously scheduled, the time complexity of HPEFT becomes $O(K^2N^2Q)$. HPEFT needs to adjust the order of scheduling in shared resource with time slots and obey the compact scheduling rule because it can find all the time slots and sort PPA table in stage 3. The order of time complexity increases the algorithm's complexity; therefore, the complexity of the algorithm is $O(K^2N^2Q^2)$.

3. Experimental Result and Analysis

3.1. Parameter Setting and Test Data Generation

The main parameters of the test sample data are the total number of tasks N_k, the number of layers L, task range TR, the total number of shared resources Q, number of DAGs k, uniform deadline time and start time ST_{ki}, shared resource speed SD_j, and delay C_j.

The tasks and their hierarchical generation are as follows: select k-th DAG tasks and randomly distribute them in each layer. At least one task node is required in each layer, and the task number is distributed from the top to the bottom in an orderly manner. Further, the next DAG can be selected until all the DAGs have completed the tasks and the allocation of layers. The shared resource hierarchy is generated as follows; select all the shared resources Q; then, randomly allocate these shared resources to L layers, with each layer having at least one shared resource from top to bottom according to the serial number from small to large. Further, the start time ST_{ki} and execution time $P_{ki,j}$ are generated as follows; first, according to the task serial number, K × N_k, TR is randomly generated from [1, TR]; furthermore, the nodes without previous tasks also generate ST_{ki} equally and randomly using the same task number, with a length of [1, 1.2 × TR]; finally, SD_j is randomly and evenly generated according to the length in the range [1, 0.2 × TR]. C_j of the shared resources is generated according to the scope [1,2]. $P_{ki,j}$ of each task is calculated using Equation (1). The priority relation matrix PR (*ku*, *kv*) can be generated as follows; find all the tasks in layer L and layer L + 1. The number of tasks in these layers is recorded as $N_{k,\,l}$ and $N_{k,\,l\,+\,1}$. According to $N_{k,\,l}$ and $N_{k,\,l\,+\,1}$, construct a diagonal unit matrix with $N_{k,\,l}$ as the number of rows and $N_{k\,+\,l,\,1}$ as the number of columns. If the columns are not sufficient, duplicate until the elements of the matrix are filled. If $N_{k,\,l} \geq N_{k\,+\,l,\,1}$, randomly scramble the rows; if $N_{k,\,l} < N_{k\,+\,l,\,1}$, randomly scramble the columns, and increase the row (column) element 1 with a probability of 0.1 to the rows; assign the matrix to the corresponding position of the priority relation constraint matrix PR as depicted in Figure 6. The diagonal unit matrix can be used to ensure that the upper and lower tasks have nodes connected to their proper edges and to prevent the generation of the isolated nodes. The row (column) elements are increased with a probability of 0.1 to ensure full coverage of the test set as far as possible. The test case data can be obtained after generating the test case.

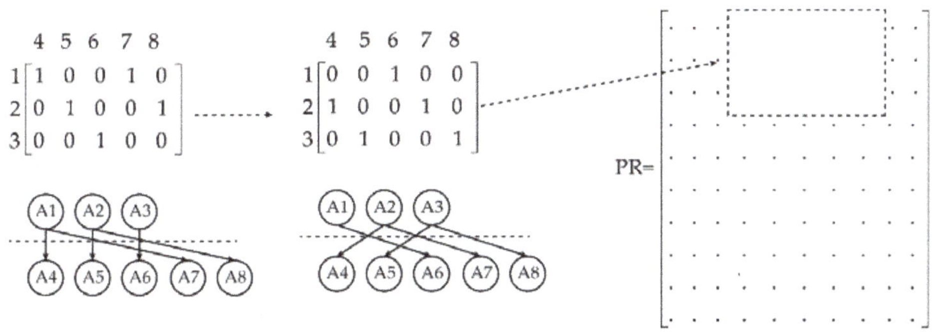

Figure 6. Diagram of the test case generation process.

3.2. Definition of the Performance Evaluation Indices

3.2.1. Ratio of the Idle Time Slot (RITS)

RITS is obtained by dividing the total length of the scheduling node task in each shared resource according to the difference between the actual EST and the actual latest end time for all the tasks in the shared resource. After adding these times, the percentages of the ratios of the idle time slots, as shown in Equation (9), are subtracted from 1. Equation (9) denotes the mathematical definition of RITS.

$$\mathrm{RITS} = 1 - \sum_{j} \frac{\sum\limits_{ki \in Shed(M_j)} P_{ki,j}}{\{\max[AFT(ki)] - \min[AST(ki)]\}} \% \tag{9}$$

with the presence of hierarchical limited resources, the rate of idle slots generated by the multi-DAG scheduling algorithm determines the percentage of time wasted by all the shared resources after applying scheduling. The larger this value, the more will be the wasted time because of the hierarchical arrangement of the shared resources.

3.2.2. Relative Laxity with Respect to the Deadline (RLD)

RLD denotes the sum of the differences between the maximum completion time and the deadlines for each shared resource, representing the overall scheduling performance while using all the shared resources. Equation (10) defines RLD, which indicates the number of time slot intervals between the maximum completion time and the specified deadline.

$$\text{RLD} = \text{sum}\{\text{Deadline} - \text{makespan}\ (M_j)\},\ \text{makespan}(M_j) = \max(\text{AFT}(ki)) \\ \text{subject to } ki \text{ scheduling in } M_j \tag{10}$$

3.3. *Experimental Analysis*

This section presents the performance of the algorithm using four experiments. The experimental settings are as follows. All the tasks ranged from 1 to 10 time units, with the deadline of time units being represented by (Number of tasks) × (Number of layers) × (Range of tasks). Other experimental parameters were set as presented in Table 10.

Table 10. Test parameter setting.

Test No.	Variable	Tasks (per DAG)	Share Rsources	DAGs	Layers
Test 1	Number of tasks	10–80 step 10	5	6	3
Test 2	Number of shared resources	30	5–12 step 1	6	3
Test 3	Number of DAGs	30	20	2–9 step 1	3
Test 4	Number of layers	30	5	3	2–9 step 1

Test 1 verifies the effect of the number of tasks on the algorithm. When the number of DAGs and the sharing of resources and layers are similar, heavier and increased number of tasks will result in better scheduling performance. As can be observed from Figure 7a, the makespan of each algorithm increases as the number of tasks increases. Among the algorithms considered herein, HPEFT exhibited the minimum makespan with increasing number of tasks; its advantage of the maximum completion time is considerably pronounced. Figure 7b depicts the RITS. HPEFT exhibited more compact scheduling compared with the other algorithms. SPT, HEFT, and HPEFT exhibited more idle slots with respect to the deadline. As can be observed from Figure 7c, the RLD of HPEFT increases as the number of tasks increased when compared with other algorithms, implying that the more the number of tasks, the better will be the scheduling performance for the same deadline. For the elapsed time of algorithm in test 1, although the HPEFT time increased when compared with that of other algorithms, the average increase in time is 1.9, 2.3, 1.1, and 1.4 times that of SPT, R-R, HEFT, and PEFT, respectively. For the maximum task condition in test 1, the number of tasks is 90 × 3 = 270, and the increase in time is 65 ms, as presented in Table 11.

Table 11. Elapsed time of test 1.

Algorithm	SPT	R-R	HEFT	PEFT	HPEFT
Elapsed Time (ms) [1]	31.1	25.8	56.7	42.7	65.0

[1] Task number was 90 per DAG and 270 in total; other parameter settings are the same as test 1.

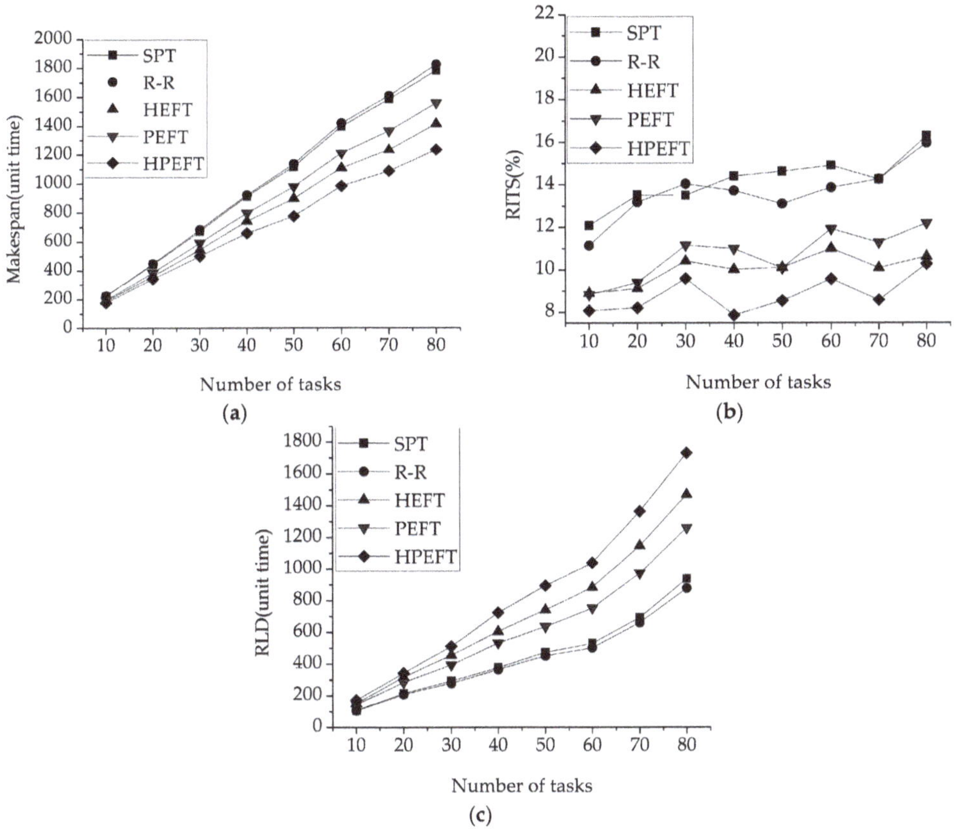

Figure 7. Variable number of tasks: (**a**) makespan, (**b**) RITS, and (**c**) RLD.

Test 2 presents the effect of the number of shared resources on the algorithm. In the case of the same number of tasks and the same number of layers, the lower the number of shared resources, and the heavier the scheduling task will be. Figure 8a denotes the relation between the number of shared resources and makespan. It can be observed from the figure that the HPEFT algorithm exhibited smaller makespan than that exhibited by the other algorithms, and the lower the number of shared resources, the greater the advantage of HPEFT will be. Figure 8b shows that RLD increases with an increase in the number of shared resources, and HPEFT has a slight advantage over other algorithms. The greater the number of shared resources, the earlier the completion of each scheduled task; accordingly, more idle time slots with deadlines are observed. However, for RITS, the effect of HPEFT in test 2 is 0.14–1.22% more than that of HFET, and there is no obvious advantage.

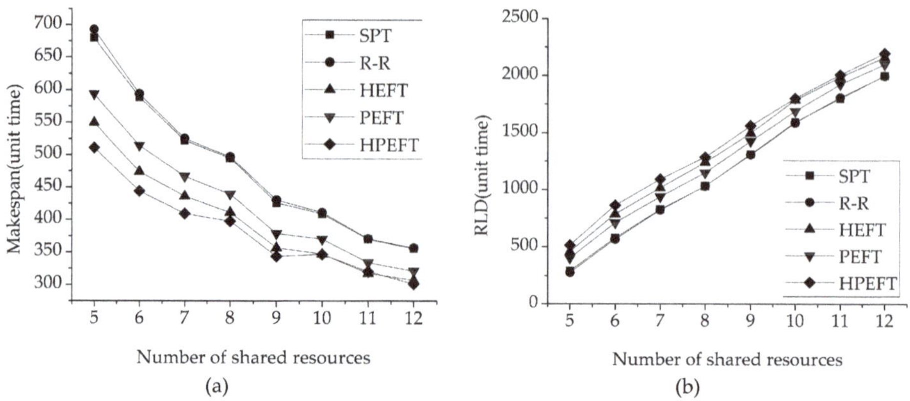

Figure 8. Variable number of shared resources: (**a**) makespan and (**b**) RLD.

The number of DAGs is an important factor affecting the multi-DAG workflow scheduling. Figure 9a shows that HPEFT finishes execution in lesser time than that required by the other algorithms as the number of DAGs increases. As the number of tasks increases, this advantage will become obvious. Figure 9b shows that as the number of DAGs increases, the RITS of HPEFT has the minimum value, whereas the time slot utilization improves. Figure 9c shows that HPEFT has advantages with respect to RLD, and the greater the number of DAGs, the greater the advantage.

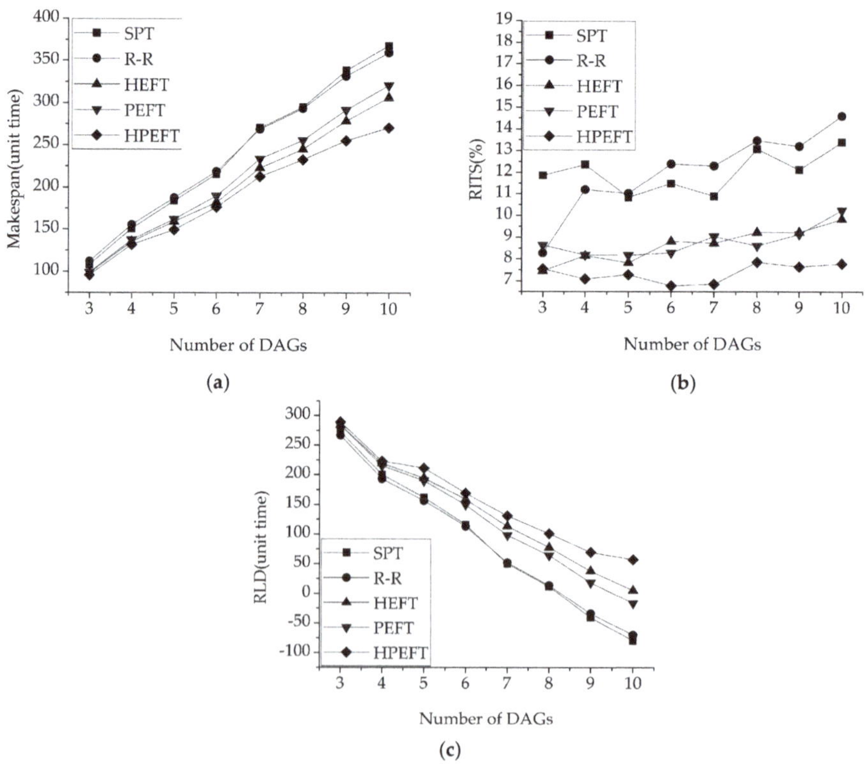

Figure 9. Variable number of DAGs: (**a**) makespan, (**b**) RITS, and (**c**) RLD.

Test 3 also compared the algorithms in terms of the missed deadline rate (MDR). MDR is defined as the number of times that a deadline was missed when 100 randomly generated scheduling problems were solved. Table 12 presents the MDRs for the three most serious cases in test 3. The MDR was decreased by 13 to 30% using the HPEFT method, according to the data presented in the sixth column.

Table 12. MDRs of test 3.

Number of DAGs [1]	SPT	R-R	HEFT	PEFT	HPEFT
8	38%	35%	14%	19%	13%
9	63%	60%	21%	32%	19%
10	88%	78%	43%	54%	30%

[1] The three most serious conditions in test 3.

In test 4, the number of layers refers to the number of layers of the DAG and tasks. The higher is the number of layers, the longer the precedence relation. The interaction between multi-DAG and the layers increases the complexity of the problem and tests the scheduling performance of the algorithm.

Figure 10a denotes that HPEFT always has the smallest makespan. The higher the number of layers, the greater the advantages of HPEFT with respect to the makespan. Figure 10b shows that RLD has an advantage in HPEFT. The RITS of HPEFT is 0.327–1.722% larger than that of HEFT, and there is no obvious difference.

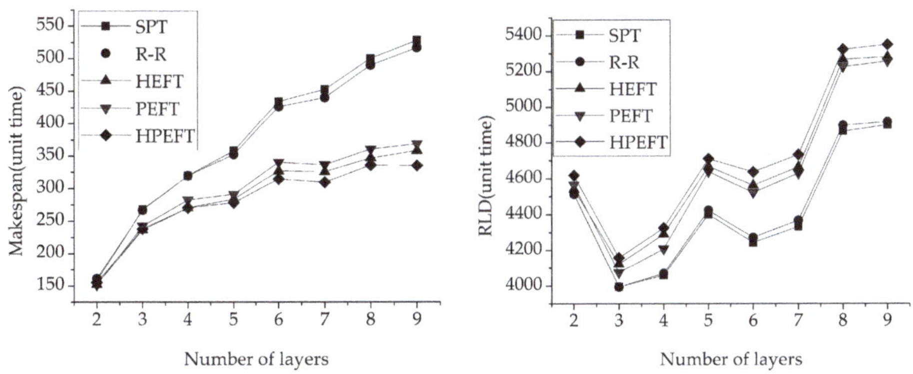

(a) (b)

Figure 10. Variable number of layers: (**a**) makespan and (**b**) RLD.

Test 5 was run to study the statistical characteristics of the HPEFT algorithm in terms of the makespan. Eighty tasks were scheduled in the test; the other parameters are the same as those in test 1, and 1000 calculations were run. Figure 11a is a boxplot of the makespan results. As can be observed from the figure, when compared with other algorithms, the upper and lower quartile of the results were lower than those of the other algorithms, and the interquartile range (IQR) was not considerably different. Figure 11b gives a 95% confidence interval (CI) plot. As depicted in the figure, the average result from the HPEFT is smaller than that obtained from other algorithms; however, the difference in the CI is also small.

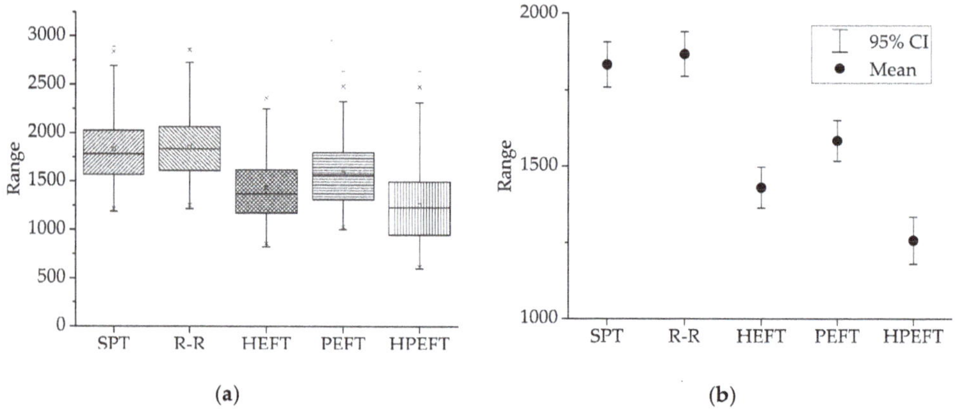

Figure 11. Statistical plots: (**a**) box plot and (**b**) 95% CI plot.

Table 13 summarizes the indicators recorded from tests 1 to 4. The makespan decreases by 3.87 to 57.68%; RITS decreases by 0 to 6.53%; RLD increases by at least 2.27 to 8.58% because of the different deadlines; and the elapsed time increases by 42.14 to 63.62%. Although the elapsed time increases, we can observe from Table 11 that the difference in computing time is acceptable.

Table 13. Percentage differences between various indicators and their worst values.

	Makespan Difference	RITS Difference	RLD Difference [1]	Elapsed Time Difference
Test 1	27.60–47.61%	0–6.53%	39.22–51.62%	42.14–63.62%
Test 2	18.18–35.64%	0–5.53%	9.05–46.46%	46.71–47.60%
Test 3	17.02–35.88%	4.32–6.82%	7.81–239.71%	33.86–49.48%
Test 4	3.87–57.68%	0–2.77%	2.27–8.58%	45.74–48.91%

[1] Due to different deadlines, consider the smallest value.

4. Optimization of the Signal Processing Scheduling Process for a Multigroup Scan UPA System Based on HPEFT

Figure 12 depicts the architecture of a multigroup scan UPA system using the TFM method [24]. After the acquisition, multigroup scan ultrasound signals are sent to the on-chip memories (OCMs) of the FPGA chip. The shared signal processing modules, such as Hilbert transform and FIR noise reduction, are connected to the Avalon-MM bus of the system. The scheduling control module reads the signals by writing the OCMs in the control and status register and interrupt request (IRQ) control chips, and the signals are sent to the corresponding signal processing module with respect to the DAG tasks. After completing DAG processing, the signal is sent to the DDR3 buffer controlled by the DDR controller from which signals are sent directly to a PCI-E bus controller using the Scatter–Gather DMA through the Avalon-ST bus. The PCI-E controller receives the signal of the Scatter–Gather DMA. After all DAG tasks are processed, all signals are sent to the PC through the PCI-E PHY physical terminal.

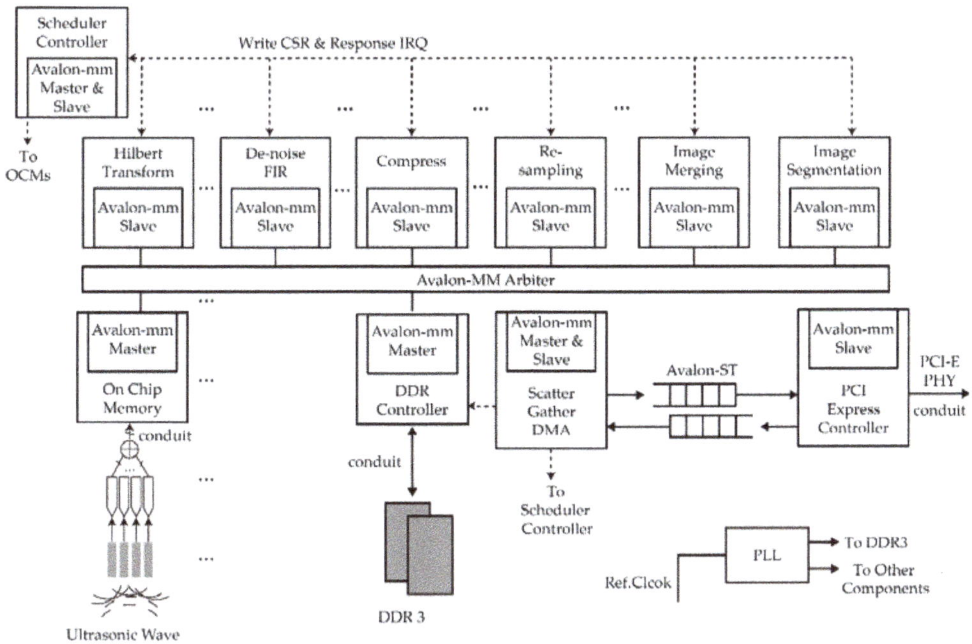

Figure 12. Architecture of the multigroup-scan ultrasound TFM system.

A virtual example of multiple DAG scheduling for a multigroup-scan UPA system is depicted in Figure 13.

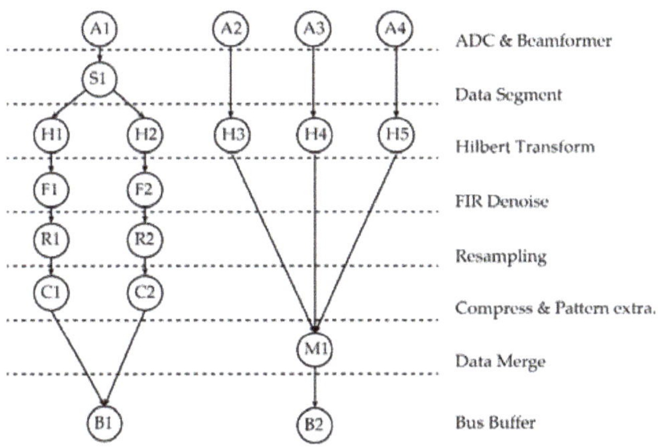

Figure 13. An example of two DAG scans.

The first DAG is the graph formed after a set of piezoelectric chips acquires the segment data and then applies the focusing delays, the Hilbert transform, FIR denoising, resampling, compression & pattern extraction and data merge. Further, the data is sent to the bus buffer. The second DAG expresses the functions obtained when the signals collected by the three sets of piezoelectric wafers are subjected to focusing delays; only then is the Hilbert transform performed along with data merging, and the data are finally sent to the bus buffer.

To simplify the experiment, the process modules in a single layer (performing the same special function) are considered to be homogeneous, and the ADC and beam-forming steps are considered to be the start time of DAG. Table 14 denotes the processing time of the signal processing modules, Table 15 presents the number of signal processing modules used to facilitate the calculation in the FPGA by considering k = 1024. The time unit is a single clock cycle in the FPGA and is 10 ns (100 MHz) in these experiments.

Table 14. Processing time required for performing tasks.

Task symbol	**A1**	**A2**	**A3**	**A4**	**S1**	**H1**	**H2**	**H3**	**H4**	**H5**
Proc. time [1]	1040k	1030k	1032k	1036k	6k	6k	6k	6k	6k	6k
Task symbol	**F1**	**F2**	**R1**	**R2**	**M1**	**C1**	**C2**	**B1**	**B2**	-
Proc. time [1]	12k	8k	6k	6k	6k	10k	8k	16k	16k	-

[1] All time unit is clock cycle, 1k = 1024.

Table 15. Number of signal processing modules.

SPM [1]	AD [1]	DS [1]	HT [1]	FD [1]	RS [1]	CP [1]	DM [1]	BB [1]
Number	4	1	2	2	2	2	1	2

[1] SPM, AD, DS, HT, FD, RS, CP, DM, BB refer to signal processing module, ADC & beamformer, data segment, Hilbert transform, FIR de-noise, resampling, compression & pattern extract, data merge and bus buffer, respectively.

Figure 14 provides a Gantt chart of the whole system as scheduled by HPEFT. In this figure, we can observe the effect of hierarchical scheduling to address the functional constraints.

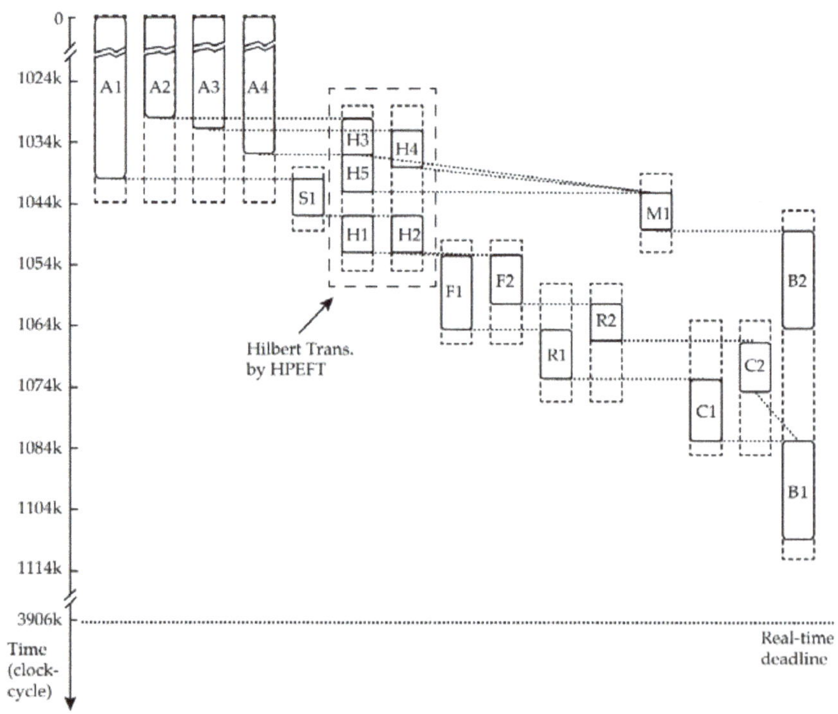

Figure 14. Gantt chart after scheduling multiple groups of scanning tasks by HPEFT.

To clearly denote the effect of scheduling, we selected the Hilbert transform as part of the overall system scheduling to verify the results of the algorithm after the simulation of the FPGA scheduler. We used Hilbert transform tasks H1–H5 to illustrate the scheduling situation and generate the simulation results from Modelsim 10.2 SE (Mentor Co., Ltd., Wilsonville, OR, USA). In this case, two DAGs arrive at the Hilbert transform tasks H1–H5 after handling tasks A1, A2, A3, A4, and S1. Tasks H1–H5 are ranked as in Table 16 by the HEFT algorithm; therefore, the scheduling order of the HEFT algorithm is H1, H2, H3, H4, H5, and the simulation results are presented in Figure 15a. The PPA table obtained by the HPEFT algorithm is depicted in Table 17. The scheduling orders are H3, H5, H1 in Hilbert Transform module 1 and H4, H2 in Hilbert Transform module 2. The simulation results are denoted in Figure 15b.

Table 16. Task Rank_p of tasks in Hilbert transforms.

Task [1]	**H1**	**H2**	**H3**	**H4**	**H5**
Rank_p	48	48	16	16	16

[1] H1 and H2 have the same $P_{ki,j}$ and C_j, and H3, H4, and H5 have the same $P_{ki,j}$ and C_j.

Table 17. PPA table for tasks in Hilbert transforms.

Task [1]	**H3**	**H4**	**H5**	**H1**	**H2**
PAFT(Hx)	6	8	12	18	18

[1] H1 and H2 have the same $P_{ki,j}$ and C_j, and H3, H4, and H5 have the same $P_{ki,j}$ and C_j.

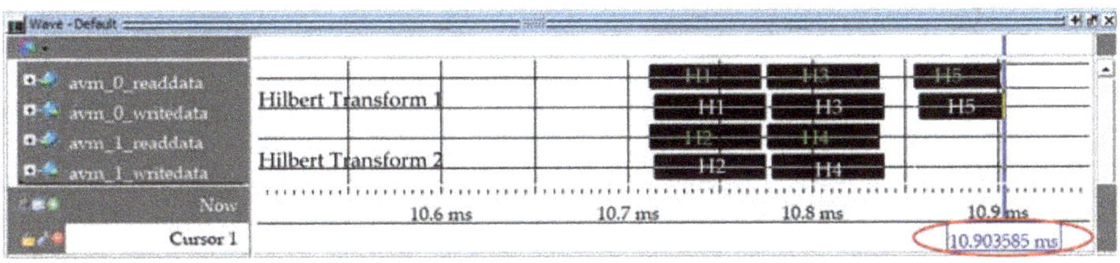

(a)

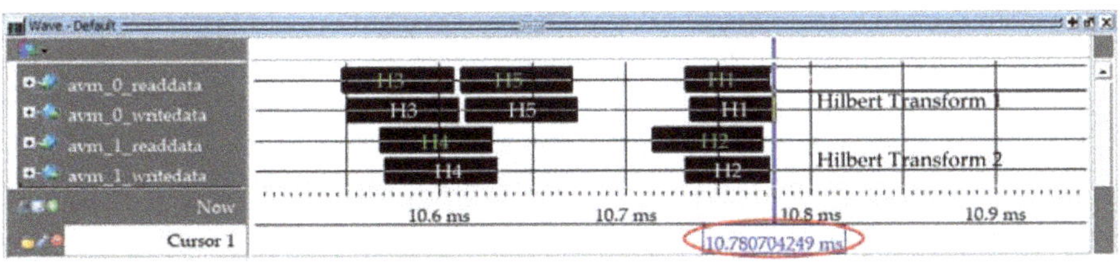

(b)

Figure 15. Hilbert transform scheduling simulation in ModelSim: (**a**) HEFT and (**b**) HPEFT.

The experimental parameters are set as follows: the processing times of all the tasks (H1-H5) $P_{ki,j}$ are 6k clock cycles, and the Hilbert transform shared resources number is two. There are four scanning groups; each group has 32 elements with 16k sample depth; therefore, all the ADC & beam-forming times are 1024k clock cycles [25]. As depicted in Figure 15a,b, the completion time of all the schedules using HPEFT is approximately 10.78 ms, whereas that of the schedules employing HEFT is 10.90 ms. If 1 ms is given to the remaining signal processing modules, the frame periods, as shown in Figure 15a,b, will be 11.90 ms and 11.78 ms, respectively. Therefore, using the HPEFT algorithm, the frame period was increased by 1% in this experiment. If ADC and beam-forming require less time, the increase in frame period will be made obvious by scheduling.

In our experiment verification environment, Figure 16 shows the experiment circuit board and Signaltap II (Intel Corporation, Santa Clara, CA., USA) diagram with the small-scale local experiment.

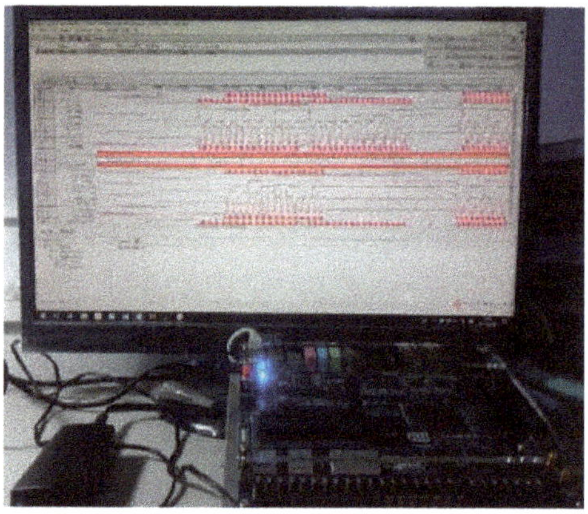

Figure 16. Experiment circuit board and Signaltap II diagram.

5. Conclusions

Based on the existing multi-DAG resource scheduling algorithms, this study proposes a deadline, constraint, multi-DAG, sharing-limited HHDMP scheduling problem and proposes an HPEFT algorithm for solving it. This algorithm inherits the advantages of both the HEFT algorithms for calculating the upward rank for critical paths, and it is improved for performing hierarchical tasks and for obtaining shared resources. Based on the characteristics of the hierarchical resources, wherein the DAG predecessors and successors must be compact, a stage 3 PPA algorithm was proposed. After stage 1 and 2 scheduling, PPA can find a large time slot to make the same DAG task of the same shared resource schedule compact, shortening the time of the multi-layer resource scheduling problem. This study also adopted two indicators with respect to the hierarchical scheduling problem: RITS and RLD. When compared with several classical algorithms, such as SPT, R-R, HEFT, and PEFT, the experimental results denote that the makespan of the proposed algorithm was reduced by 5 to 16%, RITS was reduced by 0 to 6.53%, RLD was increased by 2.27 to 8.58%, and MDR was decreased by 13 to 58%. Even so, the algorithm still exhibits some limitations. First, when the number of shared resources and layers increases, the RITS index of the HPEFT algorithm shows no clear advantage over that of HEFT. Second, the time complexity is increased, and the computing time increases by approximately 50%. Third, in the experiments that were not presented above, the PPA method can significantly increase the scheduling imbalance between DAGs. An example of a multigroup scanning UPA system based on the Altera Qsys architecture was also presented, and the HPEFT algorithm scheduling was verified in this architecture by scheduling the Hilbert transform tasks in two DAGs.

In future works, we intend to focus on selecting the initial resources for the algorithm with respect to different types of ultrasound scanning; the relation between the shared resources in each layer and the successors, predecessors, and number of tasks in the layer, and studying the layer delay of shared resources. More complex and numerous signal processing modules based on FPGA will have to be tested in the future to verify the effectiveness of the scheduling algorithms discussed in this study.

Author Contributions: Y.L., W.T., and G.L. conceived the concept of the study; Y.L. performed the experiments; Y.L. and W.T. designed the system model; and Y.L. wrote the study.

Funding: This work was financially supported by the National Key Foundation for Exploring Scientific Instrument (2013YQ230575) and Guangzhou Science and Technology Plan Project (201509010008).

Acknowledgments: We thank Guangzhou Doppler Electronic Technology Co., Ltd. (www.cndoppler.cn) for technical support.

Conflicts of Interest: The authors declare no conflicts of interest.

References

1. Byun, E.J.; Choi, S.J.; Baik, M.S.; Gil, J.; Park, C.; Hwang, C. MJSA: Markov job scheduler based on availability in desktop grid computing environment. *Future Gener. Comput. Syst.* **2007**, *23*, 616–622. [CrossRef]
2. Topcuoglu, H.; Hariri, S.; Wu, M.Y. Performance-effective and low-complexity task scheduling for heterogeneous computing. *IEEE Trans. Parallel Distrib. Syst.* **2002**, *13*, 260–274. [CrossRef]
3. Ullman, J.D. NP-complete scheduling problems. *J. Comput. Syst. Sci.* **1975**, *10*, 384–393. [CrossRef]
4. Arabnejad, H. List Based Task Scheduling Algorithms on Heterogeneous Systems—An Overview. Available online: https://paginas.fe.up.pt/~{}prodei/dsie12/papers/paper_30.pdf (accessed on 5 June 2013).
5. Sakellariou, R.; Zhao, H. A hybrid heuristic for DAG scheduling on heterogeneous systems. In Proceedings of the 18th International Parallel and Distributed Processing Symposium, Santa Fe, NM, USA, 26–30 April 2004; p. 111.
6. Hagras, T.; Janecek, J. A simple scheduling heuristic for heterogeneous computing environments. In Proceedings of the Second International Conference on Parallel and Distributed Computing, Ljubljana, Slovenia, 13–14 October 2003; IEEE Computer Society: Washington, DC, USA, 2003; pp. 104–110.
7. Ilavarasan, E.; Thambidurai, P.; Mahilmannan, R. High performance task scheduling algorithm for heterogeneous computing system. In *International Conference on Algorithms and Architectures for Parallel Processing*; Springer: Berlin/Heidelberg, Germany, 2005; pp. 193–203.
8. Ilavarasan, E.; Thambidurai, P.; Mahilmannan, R. Performance effective task scheduling algorithm for heterogeneous computing system. *J. Comput. Sci.* **2007**, *3*, 28–38.
9. Bittencourt, L.F.; Sakellariou, R.; Madeira, E.R.M. DAG scheduling using a lookahead variant of the heterogeneous earliest finish time algorithm. In Proceedings of the 18th Euromicro International Conference on Parallel, Distributed and Network-Based Processing, Pisa, Italy, 17–19 February 2010; IEEE: Piscataway, NJ, USA, 2010; pp. 27–34.
10. Honig, U.; Schiffmann, W. A meta-algorithm for scheduling multiple DAGs in homogeneous system environments. In Proceedings of the eighteenth IASTED International Conference on Parallel and Distributed Computing and Systems, Dallas, TX, USA, 16–18 November 2006.
11. Zhao, H.; Sakellariou, R. Scheduling multiple DAGs onto heterogeneous systems. In Proceedings of the 20th IEEE International Parallel & Distributed Processing Symposium, Rhodes Island, Greece, 25–29 April 2006; p. 14.
12. Yu, Z.; Shi, W. A planner-guided scheduling strategy for multiple workflow applications. In Proceedings of the 2008 International Conference on Parallel Processing, Portland, OR, USA, 8–12 September 2018; pp. 1–8.
13. Baker, T.P. An analysis of EDF schedulability on a multiprocessor. *IEEE Trans. Parallel Distrib. Syst.* **2005**, *16*, 760–768. [CrossRef]
14. Stavrinides, G.L.; Karatza, H.D. Scheduling multiple task graphs with end-to-end deadlines in distributed real-time systems utilizing imprecise computations. *J. Syst. Softw.* **2010**, *83*, 1004–1014. [CrossRef]
15. Tian, G.; Xiao, C.; Xie, J. Scheduling and fair cost-optimizing methods for concurrent multiple DAGs with deadline sharing resources. *Chin. J. Comput.* **2014**, *37*, 1607–1619.
16. Tian, G. Research several problems of scheduling multiple DAGs sharing resources. Ph.D. Thesis, Beijing University of Technology, Beijing, China, 2014.
17. Xu, X.; Xiao, C.; Tian, G.; Sun, T. Expansion slot backfill scheduling for concurrent workflows with deadline on heterogeneous resources. *Clust. Comput.* **2017**, *20*, 471–483. [CrossRef]
18. Tang, W.; Liu, G.; Li, Y.; Tan, D. An Improved Scheduling Algorithm for Data Transmission in Ultrasonic Phased Arrays with Multigroup Ultrasonic Sensors. *Sensors* **2017**, *17*, 2355. [CrossRef] [PubMed]
19. Li, Y.; Tang, W.; Liu, G. Improved scheduling algorithmfor signal processing in asynchronous distributed ultrasonic total-focusing method system. *PLoS ONE* **2019**, *14*, 906.
20. Anwar, N.; Deng, H. A Hybrid Metaheuristic for Multi-Objective Scientific Workflow Scheduling in a Cloud Environment. *Appl. Sci.* **2018**, *8*, 538. [CrossRef]

21. Miao, G.; Li, G.; Li, T.; Liu, Y. H_∞ Consensus Control for Heterogeneous Multi-Agent via Output under Markov Switching Topologies. *Electronics* **2018**, *7*, 453. [CrossRef]
22. Drozdov, A.Y.; Tchernykh, A.; Novikov, S.V.; Vladislavlev, V.E.; Rivera-Rodriguez, R. PHEFT: Pessimistic Image Processing Workflow Scheduling for DSP Clusters. *Algorithms* **2018**, *11*, 76. [CrossRef]
23. Feng, S.; Fu, P.; Zheng, W. A Hierarchical Multi-Label Classification Algorithm for Gene Function Prediction. *Algorithms* **2017**, *10*, 138. [CrossRef]
24. Holmes, C.; Drinkwater, B.W.; Wilcox, P.D. Post-processing of the full matrix of ultrasonic transmit-receive array data for non-destructive evaluation. *NDT E Int.* **2005**, *38*, 701–711. [CrossRef]
25. Li, Y.; Tang, W.; Liu, G. Improved bound fit algorithm for fine delay scheduling in a multigroup scan of ultrasonic phased arrays. *Sensors* **2019**, *19*, 906. [CrossRef] [PubMed]

electronics

MDPI

Article

Rolling 3D Laplacian Pyramid Video Fusion

Rade Pavlović [1] and Vladimir Petrović [2,*]

[1] Ministry of Defence, Military Technical Institute, Ratka Resanovica 1, 11000 Belgrade, Serbia; rade_pav@yahoo.com
[2] Faculty of Technical Sciences, University of Novi Sad, Trg Dositeja Obradovica 6, 21000 Novi Sad, Serbia
* Correspondence: vladimir.petrovic@uns.ac.rs; Tel.: +381-65-2017677

Received: 25 January 2019; Accepted: 1 April 2019; Published: 19 April 2019

Abstract: In this paper, we present a novel algorithm for video fusion of multi-sensor sequences applicable to real-time night vision systems. We employ the Laplacian pyramid fusion of a block of successive frames to add temporal robustness to the fused result. For the fusion rule, we first group high and low frequency levels of the decomposed frames in the block from both input sensor sequences. Then, we define local space-time energy measure to guide the selection based fusion process in a manner that achieves spatio-temporal stability. We demonstrate our approach on several well-known multi-sensor video fusion examples with varying contents and target appearance and show its advantage over conventional video fusion approaches. Computational complexity of the proposed methods is kept low by the use of simple linear filtering that can be easily parallelised for implementation on general-purpose graphics processing units (GPUs).

Keywords: image fusion; multi-sensor fusion; night vision

1. Introduction

Multi-sensor night-vision systems use multiple sensors based on different physical phenomena to monitor the same scene. This eliminates reliability deficiencies of individual sensors, and leads to a reliable scene representation in all conditions. For example, combinations of thermal infrared (IR) sensors and visible range cameras can operate in both day and nighttime.

Additional sensors however, mean more data to process as well as display to human observers who cannot effectively monitor multiple video streams simultaneously [1]. Some form of coordination of all data sources is necessary. These problems can be solved by using multi-sensor data fusion methods [1–57], which combine multiple image or video signals into a single, fused output signal. These algorithms significantly reduce the amount of raw data with ideally, minimal loss of information, which is a reliable path to follow when dealing with information fusion from several sensors.

Video signal processing used in many fields of vision and algorithms for video fusion that combine two or more video streams into a single fused stream are developing rapidly. The main goal is a better computational efficiency with equivalent or even improved fusion performance. The use of real-time image or video fusion is important in military, civil aviation and medical applications. The requirements for video, also known as dynamic fusion are broadly similar to those of static image fusion. Given that fusion is a significant data reduction process, it is necessary to preserve as much useful information as possible from the input videos while avoiding distortions in the fused signal. An additional requirement, specific to video fusion is the temporal stability of the fused result, which means a temporally consistent fused output despite the dynamically changing scene content. Finally, video fusion algorithms are generally supposed to work in real-time, which means a fusion rate of at least 25 frames per second, or indeed up to 60 for real-time head-up-display applications [6,24].

There are many methods to achieve for image and video fusion, but the field is dominated by multi-resolution and multi-scale methods [1–11]. The multi-resolution analysis decomposes image signals, or frames in case of video, into pyramid representations containing sub-band signals of decreasing resolution, where each sub-band is a part of the original spectrum. Larger structures in the scene are represented in lower frequency sub-bands, while finer details are in high frequency sub-bands. Fusing multi-resolution pyramids rather than complete image signals, provides greater flexibility when choosing relevant information for fused image, allowing the selection of spatially overlapping features from different inputs, if they occupy different scale ranges. The most common multi-resolution techniques are the Laplacian pyramid (LAP) [25,27], ROLP or Contrast pyramid [26,45], Discrete wavelet transform (DWT) [46–48], Shift invariant discrete wavelet (SIDWT) [21], bilateral filter [11], guided filter [12,13], Shearlet Transform [3], Nonsubsampled contourlet transform [14] etc.

2. Video Fusion

Video fusion algorithms can be classified into three basic categories [15]. First, are static image fusion algorithms, developed over the last 30 years, where fusion is performed frame by frame to form the fused video sequence. The most popular and widely used algorithms are the Laplacian pyramid fusion [25,27] and Wavelet transform [46,47]. Further to these classic algorithms, new multi-scale techniques have more recently been proposed based on the static fusion using Curvelets [50], Ridgelets [51], Contourlets [14], Shearlet [3] as well as the Dual tree complex wavelet transform (DTCWT) [48]. The static fusion methods for video fusion are generally less computationally demanding, but since they ignore the temporally varying component of the available scene information, they can result in temporally unstable fused sequences exhibiting blinking effect distortions that affect the perceived fused video quality [15,24].

In the second category are fusion algorithms that take the temporal, as well as spatial component of the data into account. Most common techniques use some of the static image methods or modified static image fusion method with additional calculation of temporal factors such as optical flow [22], motion detection or motion compensation [15]. These algorithms compare pixel or pixel block change through frames, forming the selection decisions for fused pixels in sequence. These "real" video fusion methods achieve better results than static fusion applied dynamically, but these methods, depending on the used technique and its complexity, can generally jeopardize real-time operation. The most popular algorithms in this category are Optical flow [22], and Discrete wavelet transform with motion compensation [15]. The algorithm in [53] periodically calculates the background over a specific period T (T = 4 s) by taking the most repetitive pixel value. The background is refreshed every T/4. That way the background image fusion is also executed every T/4, while the moving object fusion is calculated for each frame using the Laplace pyramid fusion [27].

Finally, the third category is made up of so-called 3D algorithms [54–59]. These algorithms represent an extension of the conventional static image fusion algorithms into 3D space. The most important aspect of these algorithms is that they cannot be used in real-time applications, even though they provide better results than the algorithms described above. It should also be taken into consideration that video signals are not a simple 3D extension of 2D static images; and motion information needs to be considered very carefully. Computational demands, as well as memory consumption are, in this case, way above the requirements of algorithms from the first two groups. In the 3D Laplace pyramid fusion [54], the Gaussian pyramid decomposition is performed in three dimensions using identical 1×5 1D Gaussian filter response (with values: [1 4 6 4 1]/16). The condition for this type of pyramid decomposition is that the length of the sequence is greater than 2^{N+1}, where N is the number of pyramid levels. Similar to the 2D filtering situation, where each next level is obtained by decimation with factor 2, in the 3D case the number of frames is also decreased with factor 2 (Figure 1). The equivalent 3D Laplacian pyramid of a sequence is obtained in the same way as in the 2D case, using the Gaussian pyramid expansion and subtraction. The 3D pyramid fusion can then be performed using the same conventional methods of pyramid fusion used in image fusion. The final fused sequence is formed by reconstructing the 3D Laplace pyramid (Figure 1). Other methods of the static image fusion extended to the 3D fusion in this manner are 3D DWT [54], 3D DT CWT [55,56] and 3D Curvelets [16,17]. A related, advanced 3D fusion approach used to additionally achieve noise reduction is polyfusion [59], which performs the Laplace pyramid fusion of different 2D sections of the 3D pyramid (e.g., spatial only sections or spatio-dynamic sections involving lateral pyramid side (Figure 2). The final fused sequence is obtained by fusing these two fusion results, while taking care of the dynamic value range.

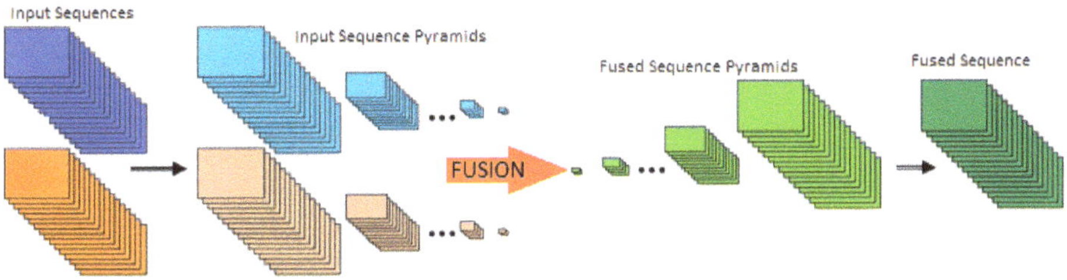

Figure 1. 3D Gaussian pyramid decomposition.

Figure 2. Polyview representation of mean opinion score (MSO1) sequence.

Figure 3 shows a multi-sensor view, in this case IR and TV images, of the same scene. The IR image clearly shows a human figure but not the general structure of the scene [57,58], while it is not immediately detectable in the TV image. Figure 4 shows a fused image using the Laplacian pyramid fusion [27]. Laplacian fusion robustly transfers important objects from the IR image and preserves structures from the TV image.

Figure 3. TV and IR sequences of the same scene.

Figure 4. Fused image using the Laplacian pyramid.

3. Dynamic Laplacian Rolling-Pyramid Fusion

Video fusion methods mentioned above take into account the temporal data component and give better results than standard frames by frame methods, but they are time-consuming and for higher video resolutions cannot be used in real-time. These methods require the fusion of already existing multi-resolution methods, decomposing more than one frame for calculating the fusion current-frame coefficient and additional temporal parameters (motion detection, temporal filters), which significantly increases their computational complexity.

Therefore, a new approach for video sequence fusion is required that would not only alleviate identified shortcomings of current methods but also introduce spatio-temporal stability into the fusion process. Furthermore, it must be computationally efficient to allow real-time fusion of two multi-sensor streams with a maximum latency of no more than a single frame period. Both subjective tests and objective measures comparisons of still image fusion methods have shown that the Laplacian pyramid fusion provides optimal or near optimal fusion results in terms of both of the subjective impression of the fused results and objective fusion performance as measured with a range of objective fusion metrics. Furthermore, this is achieved with a lower complexity in comparison to algorithms that give similar results [6]. In [6] 18 different fusion methods [10,27–43] are compared using nine objective fusion performance metrics and computational complexity evaluations. The analysis concluded that out of the real-time capable fusion algorithms, the Laplacian fusion performs best for the majority of metrics.

For these reasons, the Laplacian pyramid approach could solve existing problems in video fusion while being suitable for real-time operation. In order to reduce processing time and process the temporal information properly, it is necessary to reduce the number of frames to be processed. The approach however must facilitate robust selection input structures from input pyramids, which critically affects the fused result.

The proposed algorithm broadly follows the conventional strategy of decomposing the input streams into pyramid representations, which are then fused using a spatio-temporal pyramid fusion approach and finally reconstructed into the fused sequence. The adopted approach uses a modified version of the multi-dimensional Laplacian pyramid to decompose the video sequence. Specifically, it maintains a rolling buffer version of the 3D pyramid constructed from the 2D Laplacian pyramids of three successive frames only, current and two previous frames, to fuse each frame. The advantage in complexity of this algorithm in comparison with existing fusion methods is the fact that for the fusion of one frame only one frame needs to be decomposed into its pyramid, while the two other frames used in the 3D pyramid are taken from memory (previous frame pyramids). Furthermore, the pyramid fusion is performed on one 2D frame pyramid only and only one fused frame needs to be reconstructed from a 2D representation. All this results in a significantly faster operation. Additionally, there is no need for further processing such as motion detection or background subtraction.

The dynamic pyramid fusion, as mentioned above, is applied to the whole rolling 3D pyramid but only to fuse the central frame. Specifically, only the central frame of the fused pyramid is constructed from equivalent frames in the rolling input pyramids. For this purpose, only values from these input frames are used to construct the fused value at each location, while previous and next frames serve to determine their respective importance and combination factors (Figure 5).

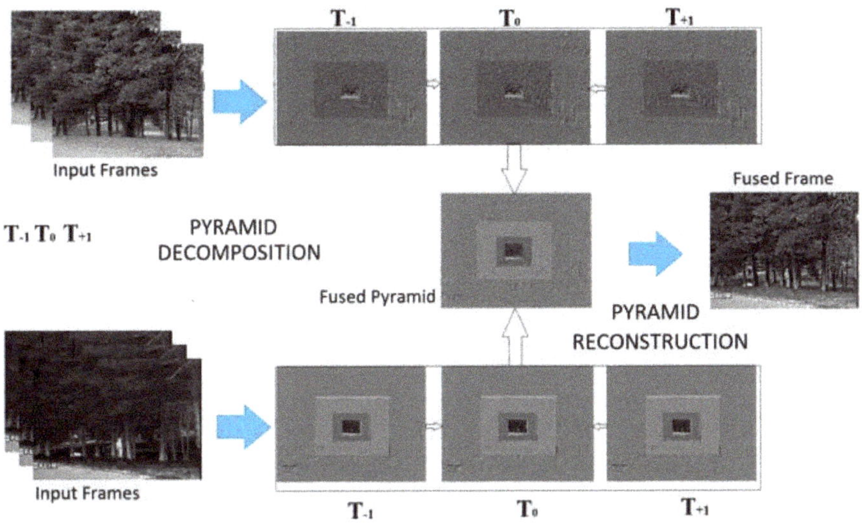

Figure 5. Block diagram of the video fusion method based on spatio-temporal Laplacian pyramid.

The first step is to group high and low frequency levels of the pyramid of all three frames from both input sensor sequences. The fusion rule for low frequency details is a spatio-temporal selection rule based on central pixel neighbourhood energy. The neighbourhood evaluation space is thus $M \times N \times T$, where M, N are window dimensions, and T is the number of frames in our rolling pyramid (in our case we use simply $M = N = T = 3$). Even though this neighbourhood seems small both spatially and temporally, it is in fact enough as will be shown to achieve temporal stability.

Low-frequency coefficients of the fused Laplacian pyramid are obtained by:

$$F_{L0}^{k}(m,n) = \omega_{Va}^{k}(m,n)L_{0}^{Va,k}(m,n) + \omega_{Vb}^{k}(m,n)L_{0}^{Vb,k}(m,n) \tag{1}$$

where $L_{0}^{Va,k}(m,n)$ and $L_{0}^{Vb,k}(m,n)$ are low-frequency Laplacian pyramid coefficients of the current frame k in the input video sequences V_a and V_b at position $(m,\ n)$ and $\omega_{Va}^{k}(m,n)$ and $\omega_{Vb}^{k}(m,n)$ are the local weight coefficients that represent the energy of the pixel environment in a spatial-temporal domain. Low-frequency coefficients represent the lowest level of the pyramid in which the main energy and larger structures of the frame are contained. It means that the weight coefficients for fusing the low-frequency coefficients of the Laplacian pyramid are determined from:

$$\omega_{Va}^{k}(m,n) = \frac{E_{Va}^{k}(m,n) + \varepsilon}{E_{Va}^{k}(m,n) + E_{Vb}^{k}(m,n) + \varepsilon} \tag{2}$$

$$\omega_{Vb}^{k}(m,n) = 1 - \omega_{Va}^{k}(m,n) \tag{3}$$

where ε is a small positive constant, to prevent division with 0, set throughout to 10^{-6}. The local spatio-temporal energy E of a central pixel at m, n in frame t and is determined as the total amount of high-frequncy activity, measured through square of local pyramid coeffcient magnitude, in its immediate, $3 \times 3 \times 3$ spatio-temporal neighbourhood according to:

$$E^k_{\{Va,Vb\}}(m,n,t) = \sum_{m=-M/2}^{M/2} \sum_{n=-N/2}^{N/2} \sum_{\eta=-T/2}^{T/2} \left| L_0^{\{Va,Vb\}}(x+m, y+n, t+\eta) \right|^2 \quad (4)$$

where $\{V_a, V_b\}$ signifies the spatial energy computed for video V_a and V_b in turn, for the sake of brevity. Interesting locations around the salient static and moving structures, that we want to preserve in the fused sequence, will have significant pyramid coefficients $L_i^V(m,n,t)$ leading to high local energy estimates. The next step is to fuse the coefficients of the Laplacian Pyramid $L_i^{Va}(m,n,t)$ and $L_i^{Vb}(m,n,t)$ which represent higher frequencies and, therefore finer details in the incoming multi-sensory sequences. Similar to the fusion of large-scale structures, the spatio-temporal energy approach based on a local neighborhood of $M \times N \times T$ is also used here. The window size has been kept the same at 3.

It is an established practice in the fusion field that for fusing information of higher frequencies derived from multi-resolution decompositions, the choice of the maximum absolute pixel value from either of the inputs is a reliable method of maximizing contrast and preserving the most important input information. However, in our case, we have information from three successive frames, and using the local energy approach a local $3 \times 3 \times 3$ of pyramid pixels will be influenced by each coefficient eliminating the effects of noise and temporal flicker due to shift variance effects of the pyramid decomposition.

Comparing this approach to the simple select-max applied to central frame only, using the objective *DQ* video fusion performance measure [24] on a representative sequence illustrated in Figure 3, Figure 6 below, we see that the proposed approach improves fusion performance. However, although the increase in *DQ* is significant, there are still large oscillations through the frames. Figure 7 below shows successive frames obtained by the proposed dynamic fusion where flicker through sequences still causes temporal instability. This is also evident in the difference image obtained between these two frames, in the form of "halo" effects around the person and pixels that have a higher value, although there are no significant changes in the scene background. We appreciate that it is difficult to convey this type of dynamic effect on a still image and include this fused sequence in the Supplementary Material.

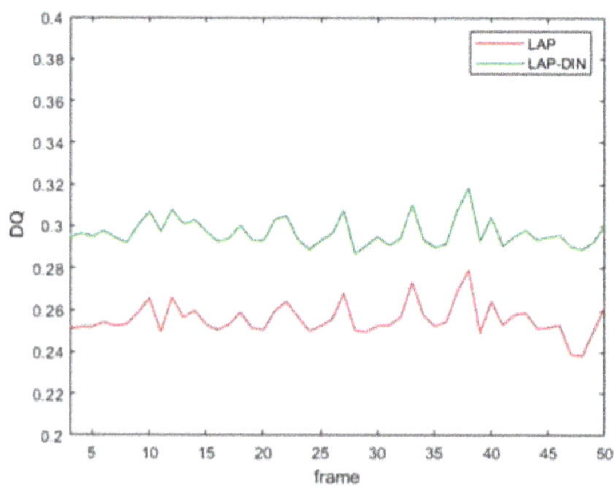

Figure 6. Video fusion performance of proposed local energy HF detail fusion (green) compared to conventional frame-by-frame select-max fusion (red) measured using objective fusion performance metric *DQ*.

Figure 7. Fused two successive frames (top images) and difference image obtained between these two frames (bottom image).

Temporally Stable Fusion

Temporal instability is often caused in areas where local pyramid energies of the input images are similar which in turn causes frequent changes of coefficient selection decisions between the inputs across space and time, causing source flicker. This behavior can be remedied through a more advanced fusion approach applied to higher frequency details. Specifically, we can use the spatio-temporal similarity index $S^k_{Vab}(m,n)$ to compare the input pyramid structures before deciding on the optimal fusion approach [15]. Similarity between inputs at each location is evaluated according to:

$$S^k_{Vab}(m,n) = \frac{2\sum_{m=-M/2}^{M/2}\sum_{n=-N/2}^{N/2}\sum_{\eta=-T/2}^{T/2}\left|L_i^{Va}(x+m,y+n,t+\eta)L_i^{Vb}(x+m,y+n,t+\eta)\right|+\varepsilon}{E^k_{Va}(m,n)+E^k_{Vb}(m,n)+\varepsilon} \quad (5)$$

S^k_{Vab} ranges between 0 and 1, where 1 signifies identical signals and values around 0 indicate very low input similarity. If S is small, below a threshold ξ, one of the inputs is usually dominant and the coefficient from the pyramid with higher local energy is taken for the fused pyramid. If similarity is high, we preserve both inputs in a weighted summation with weight coefficients based on their relative local energies.

$$F^k_{Li}(m,n) = \begin{cases} L_j^{Va,k}(m,n), \ S^k_{Vab}(m,n) < \xi \ and \ E_j^{Va}(m,n) \geq E_j^{Vb}(m,n) \\ L_j^{Vb,k}(m,n), \ S^k_{Vab}(m,n) < \xi \ and \ E_j^{Va}(m,n) < E_j^{Vb}(m,n) \\ \omega_j^{Va}(m,n)L_j^{Va,k}(m,n) + \omega_j^{Vb}(m,n)L_j^{Vb,k}(m,n) \end{cases} \quad (6)$$

To determine the optimal value of the similarity threshold ξ, we applied the proposed method on a set of six different multi-sensor sequences, varying ξ from 0 to 1 with a step of 0.05. When $\xi = 0$ resolves to a selection of coefficients with maximum local energy and 1 implies fusion using exclusively linear weighted combination of inputs. We measured the average fusion performance for each tested value of ξ using the dynamic fusion performance measure DQ [24]. The result of this analysis for a relevant subset of threshold values is shown in Figure 8 below, identifying that $\xi = 0.7$ gives optimal fusion performance.

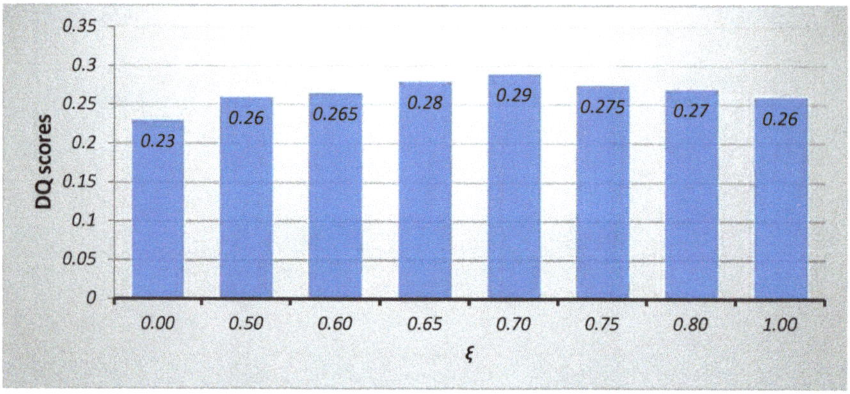

Figure 8. Results of objective measure *DQ* on proposed video fusion algorithm changing value of similarity threshold ξ from 0 to 1.

Figure 9 illustrates the effects of the proposed pyramid fusion approach compared to the static fusion. Pyramid fusion selection maps, static, left, and proposed right, for the frames shown in Figure 3 above (bright pixels are sourced from the visible range and dark ones from the thermal sequence with gray values showing split sourcing in the dynamic fusion case) show a significantly greater consistency in the proposed dynamic method. This directly affects spatio-temporal stability.

Figure 9. Pyramid fusion selection maps of the static Laplacian fusion (**left**) and proposed fusion method (**right**).

4. Results

Performance of the multi-sensor fusion is traditionally measured using subjective and objective measures. Subjective measures derived from collections of subjective scores provided by human observers on representative datasets, are generally considered to be the most reliable measures, since humans are the intended end users of fused video imagery in fields such as surveillance and night vision. Outputs of such subjective evaluation trials are human observer quality measures represented through mean opinion scores – MOS. MOS is a widely used method of subjective quality scores generalization, defined as a simple arithmetic mean of observers' score for a fused signal *i*:

$$MOS_i = \frac{1}{N_s} \sum_{1}^{N_s} SQ(n,i) \tag{7}$$

where $SQ(n,i)$ – subjective quality estimate of fused sequence *i* by the observer *n* while N_s is the total number of observers that took part in the trial.

Objective fusion metrics are algorithmic metrics providing a significantly more efficient fusion evaluation compared to subjective trials [60,61]. Even though an extensive field of still fusion objective metrics exists, these methods do not consider temporal data vital for video fusion. Video fusion metrics need to consider temporal stability implying that temporal changes in the fused signal can only be a result of changes in an input signal (any input) and not the result of a fusion algorithm. Furthermore, temporal consistency requires that changes in input sequences have to be represented in fused sequence without delay or contrast change. A direct video fusion metric *I* was proposed on these principles in [21] based on the calculation of common information in inter-frame-differences (IFDs), of the inputs and fused sequence.

DQ metric based on measuring preservation of spatial and temporal input information in the fused sequence was proposed to explicitly measure video fusion performance [24]. *DQ* measures the similarity of spatial and temporal gradient information between the inputs and the fused sequences (Figure 10). The evaluation is based on three consecutive frames of all three sequences with spatial information extracted from the current and temporal information from the other two, previous and following, frames using a robust temporal gradient approach. A perceptual gradient preservation model is then applied to evaluate information preservation at each location and time in the sequence. Spatial and temporal preservation estimates are then integrated into a single spatio-temporal information preservation estimate for each location and frame. These localized estimates are then pooled using local perceptual importance estimates into frame scores and then averaged into a single, complete sequence fusion performance score.

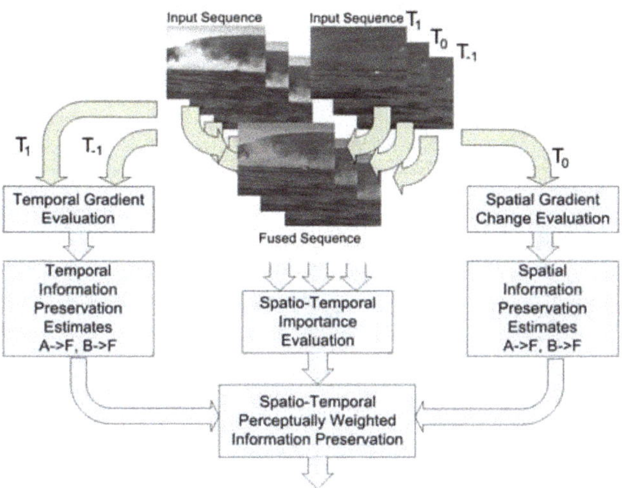

Figure 10. Dynamic fusion evaluation metric DQ.

We also used the objective video fusion quality metric Q_{ST} with the structural similarity (SSIM) index and the perception characteristics of human visual system (HVS) [62]. First, for each frame, two sub-indices, i.e., the spatial fusion quality index and the temporal fusion quality index, are defined by the weighted local SSIM indices. Second, for the current frame, an individual-frame fusion quality measure is obtained by integrating the above two sub-indices. Last, the global video fusion metric is constructed as the weighted average of all the individual-frame fusion quality measures. In addition, according to the perception characteristics of HVS, some local and global spatial–temporal information, such as local variance, pixel movement, global contrast, background motion and so on, is employed to define the weights in the metric Q_{ST}.

Finally, we also evaluate our fusion results with a non-reference objective image fusion metric FMI based on mutual information which calculates the amount of information conducted from the source images to the fused image [63]. The considered information is represented by image features like gradients or edges, which are often in the form of two-dimensional signals.

The performance of the proposed LAP-DIN method was evaluated on a database of dynamic multi-sensor imagery from six different scenarios, Figure 11. The compromises local sharpness for the sake of temporal stability and fewer spatial artifacts, which can be seen in the sharpest SIDWT method.

Figure 11. Database set for testing different fusion methods.

Figure 12 illustrates its performance alongside the Laplacian pyramid [27] and SIDWT fusion [21], image fusion methods with shift-invariance well suited to dynamic fusion, applied frame by frame. The proposed method is generally no less sharp than the other two methods, see left column, but in some examples the dynamic selection.

The left column shows the static Laplacian pyramid fusion [27], the middle–static SIDWT fusion [21], while the right proposed LAP-DIN fusion, all applied with the same decomposition depth of four. The proposed fusion provides clearer, higher contrast images than the other two methods. Further, a noise mitigation effect is also visible in the second row where the thermal image noise, is transferred into the fused signal by the two static methods, but not the LAP-DIN approach.

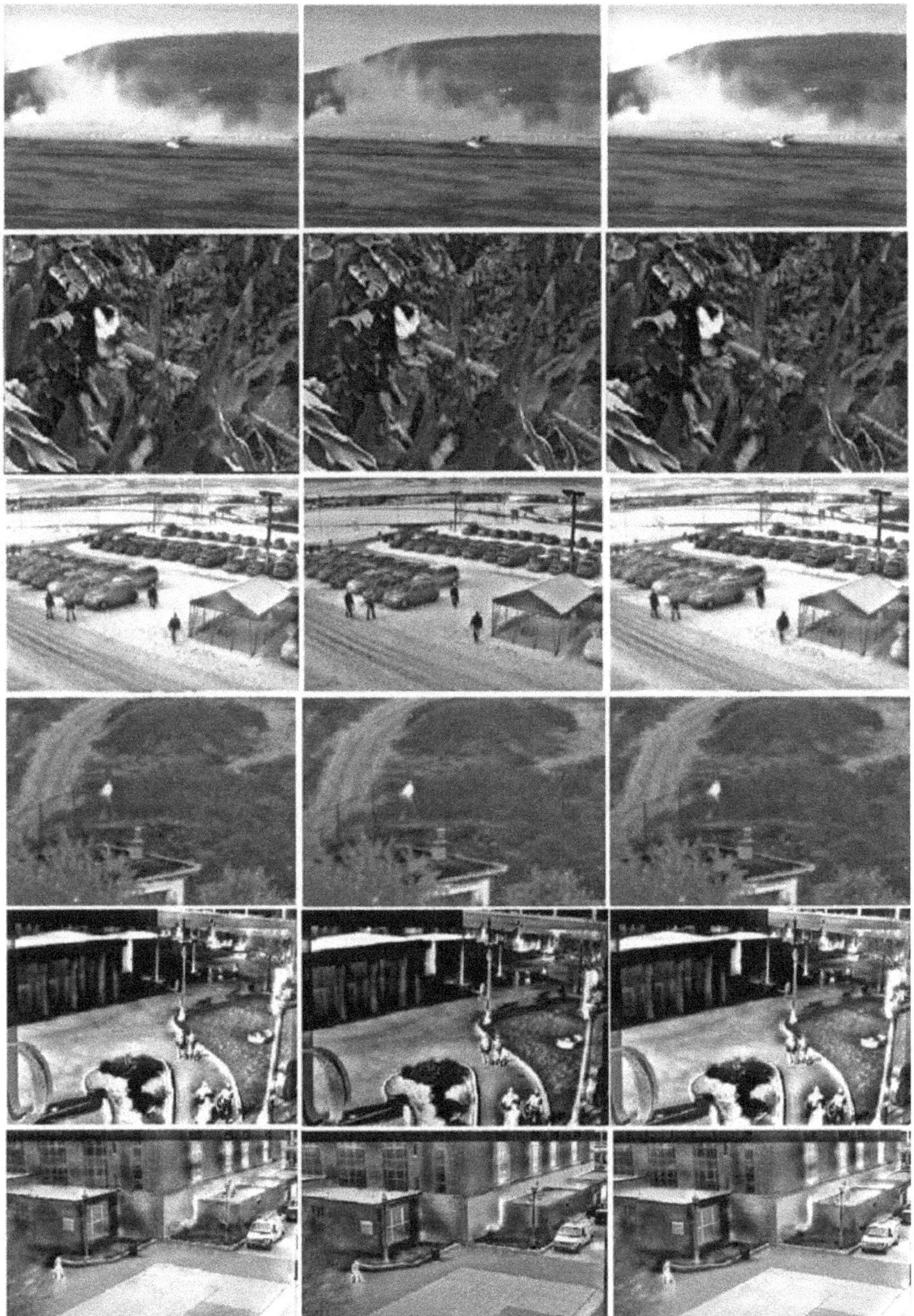

Figure 12. Fused images with Laplacian pyramid (**left column**), the middle Shift invariant discrete wavelet (SIDWT) (**middle column**) and proposed LAP-DIN fusion (**right column**).

4.1. Objective Evaluation

Objective performance evaluation was performed by the *DQ* and *I* metrics on the fused video obtained from our test database. *DQ* scores for the three methods considered first, shown in Figure 13 below and given for all sequences individually in Table 1, indicate that the LAP-DIN method clearly preserves spatial and temporal input information better overall and for all scenarios individually.

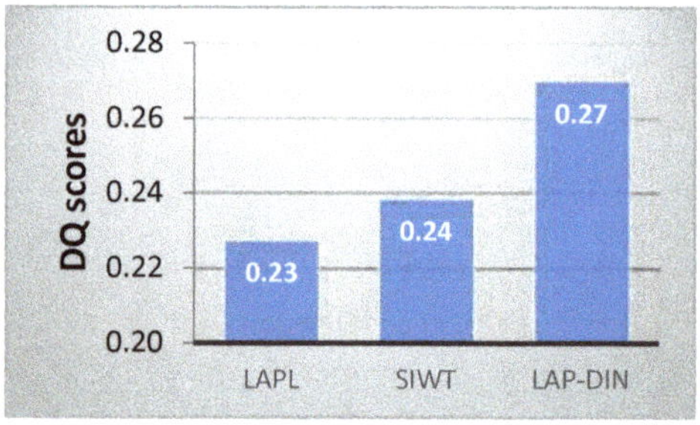

Figure 13. *DQ* fusion performance scores.

Table 1. Fusion performance scores for individual sequences.

	LAP	SIWT	LAP-DIN
Seq 1	0.23	0.23	0.26
Seq 2	0.26	0.25	0.30
Seq 3	0.20	0.22	0.23
Seq 4	0.26	0.26	0.29
Seq 5	0.23	0.26	0.28
Seq 6	0.19	0.21	0.23
Mean	0.23	0.24	0.27

As an indication of temporal stability of fusion scores, *DQ* values for the first 50 frames of sequence 1 are shown in Figure 14 below. LAP-DIN scores exhibits considerably less temporal variation 0.049 compared to 0.079 and 0.0076 for the LAP and SIDWT static algorithms respectively, on the same fused video section. The remaining score changes are the result of a significant scene movement.

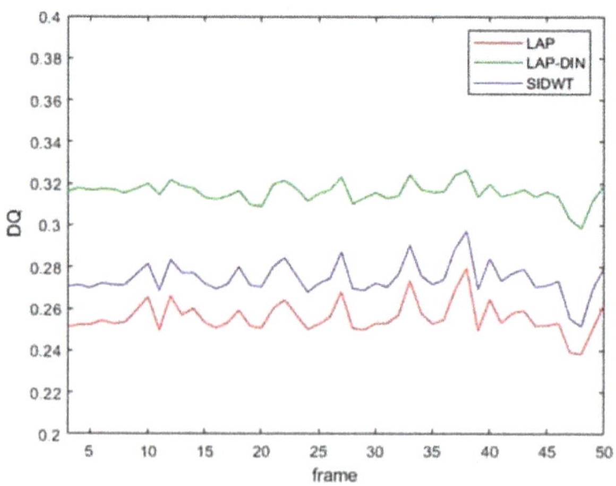

Figure 14. Temporal fusion performance stability.

In Figure 15 and Table 2, we compare *DQ* scores of the proposed method directly with those of the video fusion methods that explicitly deal with temporal information: MCDWT based on motion detection estimation and the discrete wavelet transformation [15] and the non-causal Laplacian 3D pyramid fusion method [54] not suitable for real-time operation. It indicates that the true 3D pyramid is the most successful video fusion technique, followed by the LAP-DIN method and the MCDWT, which is better than static methods.

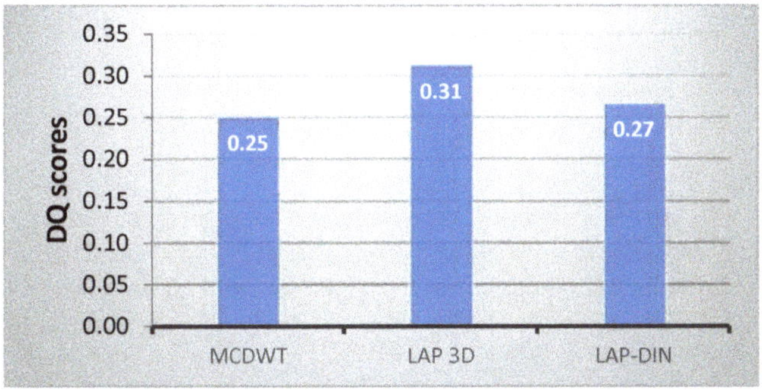

Figure 15. Results of objective measure *DQ* on three video fusion methods on database set.

Table 2. Results of objective measure *DQ* on dataset sequences separately.

	MCDWT	LAP 3D	LAP-DIN
Seq 1	0.24	0.32	0.26
Seq 2	0.26	0.37	0.30
Seq 3	0.23	0.30	0.23
Seq 4	0.22	0.30	0.23
Seq 5	0.26	0.30	0.28
Seq 6	0.30	0.28	0.29
Mean	0.25	0.31	0.27

These findings were confirmed by the *I* metric [21] as shown in Figure 16 below.

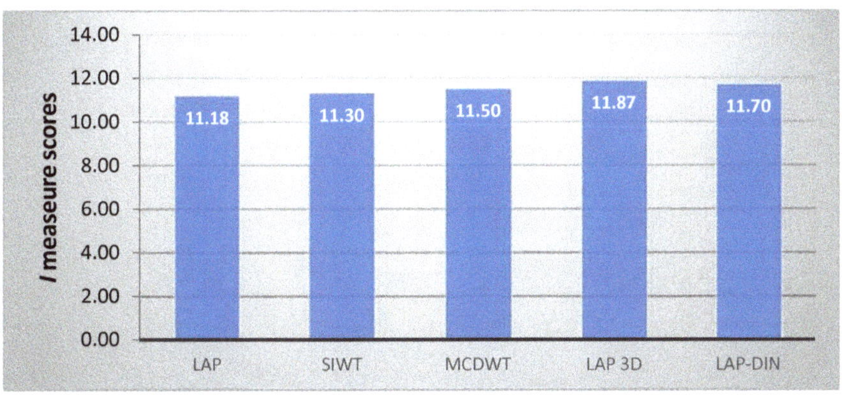

Figure 16. Comparing results of objective measure *I* on six fusion methods (static and dynamic) on database set.

Finally, Table 3 provides the results of the evaluation by four different objective video fusion performance metrics. All the metrics confirm the non-causal 3D Laplacian pyramid fusion as the most successful method, with the proposed method next best, with the exception of the FMI metric, which ranks the conventional Laplacian fusion second. FMI is a static image fusion metric and does not take into account dynamic effects in fused sequences.

Table 3. Results of four objective measures on dataset sequences.

	LAP	SIDWT	MCDWT	LAP 3D	LAP-DIN
DQ	0.23	0.24	0.25	0.31	0.27
I	11.18	11.30	11.50	11.87	11.70
Q_{ST}	0.87009	0.870064	0.871903	0.878615	0.876745
FMI	0.673145	0.641313	0.661495	0.692034	0.676197

4.2. Subjective Evaluation

The proposed video fusion method was also evaluated through formal subjective trials. Observers with general image and video processing research experience but no specific multi-sensor fusion experience were recruited to perform the test in a daylight office environment, until the subjective ratings converged. In all 10 observers completed the trial on six different fusion scenarios displayed in a sequence on a 27″ monitor using 1920 × 1080 (full HD) resolution. Participants freely adjusted their position relative to the display and had no time limit. They rated each fused sequence on a scale of 0 to 5, and were free to award equivalent grades (no forced choice).

Each observer was separately induced into the trial by performing an evaluation of two trial video sets which were not included in the analysis. They were explained the aim of the evaluation and various effects of video fusion. Each observer then evaluated the same number, six fused video sets. During the evaluation stage, the upper portion of the display showed the two input video streams and lower portion of the display showed three fused alternatives produced using different fusion algorithms. The order of the fusion methods altered randomly between video sets and observers to avoid positional bias. The sequence duration varied between six and 12 s. Each observer could replay the sequences, which replayed simultaneously, an unlimited number of times until they were satisfied with their assessment and moved onto the next video set. Trial time was not limited.

The first test compared the static Laplacian and SIDWT fusion methods applied frame by frame with the proposed LAP-DIN method. Subjective MOS scores for each method, shown in Figure 17 match the results of objective evaluation. The proposed dynamic method outperforms static ones which perform similarly.

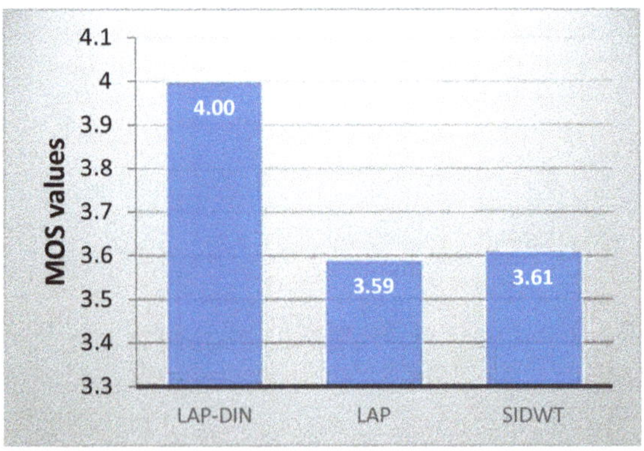

Figure 17. Subjective MOS scores of different fusion methods.

The second subjective trial, run in identical conditions on an identical dataset directly compared three true video/3D fusion methods: MCDWT [15], full 3D pyramid fusion [54] and proposed LAP-DIN method. The results, shown in Figure 18, again support objective metric findings and identify full 3D Laplace pyramid fusion, MOS = 4.1, as the best of the three, followed by proposed LAP-DIN and MCDWT.

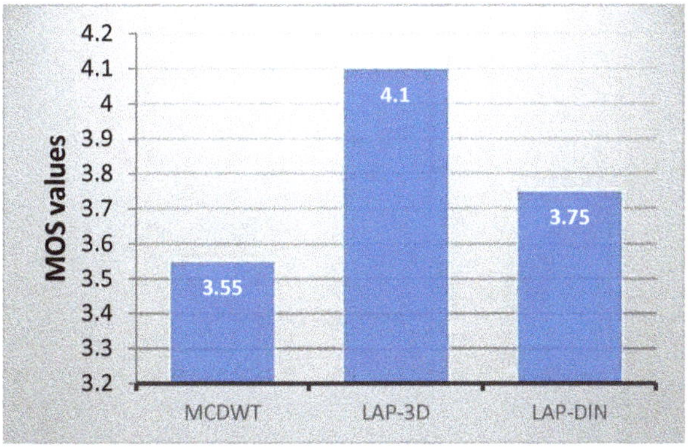

Figure 18. Subjective MOS scores of different video fusion methods.

This result underlines the well-known fact of the power of hindsight: Full 3D pyramid fusion requires knowledge of the entire signal well into the future and being in possession of all the facts we can more easily arrive at the optimal result. The proposed LAP-DIN fusion trades a single frame latency for a considerable improvement in performance on the fully causal frame-by-frame approach.

An interesting observation is the relative difference of the LAP-DIN MOS between the two trials run in identical conditions on identical data. It reflects the influence of other methods in the trial which generally performed better than those in the first trial, and undermines the value of absolute quality scores but also underlines the value of relative, or ranking scores produced by subjective trials.

4.3. Computational Complexity

Computational complexity, of vital importance in real-time operation, was evaluated for each method on video fusion at resolution of 640 × 480 pixels using the same i7 processor with 8GB of RAM. Results comparing their per-frame cost relative to the static Laplacian fusion are shown in Table 4. MCDWT is the most demanding due to motion estimation while LAP-DIN is the most efficient among dynamic methods and can be implemented to operate in real-time with 25 frames per second.

Table 4. Relative computational complexity of different video fusion methods.

	LAP	SIDWT	MCDWT	LAP-3D	LAP-DIN
Multiple of LAP	1	1.6	1.8	1.75	1.3

5. Conclusions

A new dynamic video fusion method is proposed based on the construction of a fused rolling-multiscale-Laplacian pyramid from equivalent input stream pyramids. The method uses a sophisticated local energy pyramid fusion rule that successfully transfers important structure information from the input video sequences into the fused, achieving considerable temporal stability and consistency. Furthermore, this is achieved with a significantly lower computational complexity compared to other dynamic fusion methods. Comprehensive assessment of the proposed method using subjective and objective evaluation on a number of well-known multi-sensor videos from multiple surveillance scenarios showed that the proposed method performs better than comparable causal video fusion methods. The results also indicate that extending the latency of the fusion process further could add further robustness to the fusion process and we intend to explore this performance-latency boundary in our further work.

Further work on the video fusion will include exploration of different methods to obtain a more compact description of spatio-temporal information. Also, we are planning to make a new database of multi-sensor sequences in different conditions and test the algorithm with subjective and objective tests.

Supplementary Materials: The following are available online at http://www.mdpi.com/2079-9292/8/4/447/s1, Video S1: Proposed LAP-DIN fusion video.

Author Contributions: Conceptualization, R.P. and V.P.; Methodology, R.P. and V.P.; Software, R.P.; Validation, R.P. and V.P.; Formal Analysis, R.P. and V.P.; Investigation, R.P. and V.P. Resources, R.P. and V.P.; Data Curation, R.P. and V.P.; Writing-Original Draft Preparation, R.P.; Writing-Review & Editing, V.P.; Visualization, R.P. and V.P.; Supervision, R.P. and V.P.

Funding: The APC was funded by Ministry of Defence, Military Technical Institute, Belgrade, Republic of Serbia.

Acknowledgments: The authors would like to thank the Ministry of Defence, Military Technical Institute, Belgrade, Republic of Serbia for material support used for experiments.

Conflicts of Interest: The authors declare no conflict of interest.

References

1. Pavlović, R.; Petrović, V. Objective evaluation and suppressing effects of noise in dynamic image fusion. *Sci. Tech. Rev.* **2014**, *64*, 21–29.
2. Du, Q.; Xu, H.; Ma, Y.; Huang, J.; Fan, F. Fusing infrared and visible images of different resolutions via total variation model. *Sensors* **2018**, *18*, 3827. [CrossRef] [PubMed]
3. Huang, Y.; Bi, D.; Wu, D. Infrared and visible image fusion based on different constraints in the non-subsampled shearlet transform domain. *Sensors* **2018**, *18*, 1169. [CrossRef]
4. Li, S.; Kang, X.; Fang, L.; Hu, J.; Yin, H. Pixel-level image fusion: A survey of the state of the art. *Inf. Fusion* **2017**, *33*, 100–112. [CrossRef]
5. Jin, X.; Jiang, Q.; Yao, S.; Zhou, D.; Nie, R.; Hai, J.; He, K. A survey of infrared and visual image fusion methods. *Infrared Phys. Technol.* **2017**, *85*, 478–501. [CrossRef]
6. Jiayi, M.A.; Yong, M.A.; Chang, L. Infrared and visible image fusion methods and applications: A survey. *Inf. Fusion* **2019**, *45*, 153–178.
7. Li, H.; Liu, L.; Huang, W.; Yue, C. An improved fusion algorithm for infrared and visible images based on multi-scale transform. *Infrared Phys. Technol.* **2016**, *74*, 28–37. [CrossRef]
8. Dogra, A.; Goyal, B.; Agrawal, S. From multi-scale decomposition to non-multi-scale decomposition methods: A comprehensive survey of image fusion techniques and its applications. *IEEE Access* **2017**, *5*, 16040–16067. [CrossRef]
9. Chang, L.; Feng, X.; Zhu, X.; Zhang, R.; He, R.; Xu, C. CT and MRI image fusion based on multiscale decomposition method and hybrid approach. *IET Image Process.* **2018**, *13*, 83–88. [CrossRef]
10. Bavirisetti, D.P.; Dhuli, R. Two-scale image fusion of visible and infrared images using saliency detection. *Infrared Phys. Technol.* **2016**, *76*, 52–64. [CrossRef]
11. Xing, C.; Wang, Z.; Meng, F.; Dong, C. Fusion of infrared and visible images with Gaussian smoothness and joint bilateral filtering iteration decomposition. *IET Comput. Vis.* **2018**, *13*, 44–52. [CrossRef]
12. Gan, W.; Wu, X.; Wu, W.; Yang, X.; Ren, C.; He, X.; Liu, K. Infrared and visible image fusion with the use of multi-scale edge-preserving decomposition and guided image filter. *Infrared Phys. Technol.* **2015**, *72*, 37–51. [CrossRef]
13. Toet, A.; Hogervorst, M.A. Multiscale image fusion through guided filtering. *Proc. SPIE* **2016**, *9997*, 99970J.
14. Cai, J.; Cheng, Q.; Peng, M.; Song, Y. Fusion of infrared and visible images based on nonsubsampled contourlet transform and sparse K-SVD dictionary learning. *Infrared Physics & Technology* **2017**, *82*, 85–95.
15. Liang, X.; Junping, D.; Zhenhong, Z. Infrared-visible video fusion based on motion-compensated wavelet transforms. *IET Image Process.* **2015**, *9*, 318–328.

16. Zhang, Q.; Yueling, C.; Long, W. Multisensor video fusion based on spatial–temporal salience detection. *Signal Process.* **2013**, *93*, 2485–2499. [CrossRef]
17. Zhang, Q.; Wang, Y.; Levine, M.D.; Yuan, X.; Wang, L. Multisensor video fusion based on higher order singular value decomposition. *Inf. Fusion* **2015**, *24*, 54–71. [CrossRef]
18. Gangapure, V.N.; Nanda, S.; Chowdhury, A.S. Superpixel-based causal multisensor video fusion. *IEEE Trans. Circuits Syst. Video Technol.* **2018**, *28*, 1263–1272. [CrossRef]
19. Hu, H.M.; Wu, J.; Li, B.; Guo, Q.; Zheng, J. An adaptive fusion algorithm for visible and infrared videos based on entropy and the cumulative distribution of gray levels. *IEEE Trans. Multimed.* **2017**, *19*, 2706–2719. [CrossRef]
20. Jiawei, W.; Hai-Miao, H.; Yuanyuan, G. A realtime fusion algorithm of visible and infrared videos based on spectrum characteristics. In Proceedings of the 2016 IEEE International Conference on Image Processing (ICIP); IEEE: Piscataway, NJ, USA, 2016; pp. 3369–3373.
21. Rockinger, O.; Fechner, T. Pixel-level image fusion: The case of image sequences. *Proc. SPIE* **1998**, *3374*, 378–388.
22. Li, J.; Nikolov, S.; Benton, C.; Scott-Samuel, N. Motion-based video fusion using optical flow information. In Proceedings of the 9th International Conference on Information Fusion, Florence, Italy, 10–13 July 2006; pp. 1–8.
23. Blum Rick, S.; Zheng, L. *Multi-Sensor Image Fusion and Its Applications*; CRC Press: Boca Raton, FL, USA, 2005.
24. Petrovic, V.; Cootes, T.; Pavlovic, R. Dynamic image fusion performance evaluation. In Proceedings of the 9th International Conference on Information Fusion, Québec, Canada, 9–12 July 2007; pp. 1–7.
25. Vanmali, V.; Gadre, V.M. Visible and nir image fusion using weight-map-guided laplacian–gaussian pyramid for improving scene visibility. *Sadhana* **2017**, *42*, 1063–1082.
26. Xu, H.; Wang, Y.; Wu, Y.; Qian, Y. Infrared and multi-type images fusion algorithm based on contrast pyramid transform. *Infrared Phys. Technol.* **2016**, *78*, 133–146. [CrossRef]
27. Burt, P.; Adelson, E. The Laplacian pyramid as a compact image code. *IEEE Trans. Commun.* **1983**, *COM-31*, 532–540. [CrossRef]
28. Chipman, L.J.; Orr, T.M.; Graham, L.N. Wavelets and image fusion. In Proceedings of the International Conference on Image Processing, Washington, DC, USA, 23–26 October 1995; pp. 248–251.
29. Adu, J.; Gan, J.; Wang, Y.; Huang, J. Image fusion based on nonsubsampled contourlet transform for infrared and visible light image. *Infrared Phys. Technol.* **2013**, *61*, 94–100. [CrossRef]
30. Naidu, V. Novel image fusion techniques using dct. *Int. J. Comput. Sci. Bus. Inf.* **2013**, *5*, 1–18.
31. Kumar, B.S. Image fusion based on pixel significance using cross bilateral filter. *Signal Image Video Process.* **2015**, *9*, 1193–1204. [CrossRef]
32. Zhou, Z.; Wang, B.; Li, S.; Dong, M. Perceptual fusion of infrared and visible images through a hybrid multi-scale decomposition with gaussian and bilateral filters. *Inf. Fusion* **2016**, *30*, 15–26. [CrossRef]
33. Li, S.; Kang, X.; Hu, J. Image fusion with guided filtering. *IEEE Trans. Image Process.* **2013**, *22*, 2864–2875.
34. Bavirisetti, D.P.; Dhuli, R. Fusion of infrared and visible sensor images based on anisotropic diffusion and karhunen-loeve transform. *IEEE Sens. J.* **2016**, *16*, 203–209. [CrossRef]
35. Liu, Y.; Wang, Z. Simultaneous image fusion and denoising with adaptive sparse representation. *IET Image Process.* **2014**, *9*, 347–357. [CrossRef]
36. Liu, Y.; Liu, S.; Wang, Z. A general framework for image fusion based on multi-scale transform and sparse representation. *Inf. Fusion* **2015**, *24*, 147–164. [CrossRef]
37. Qu, X.; Hu, C.; Yan, J. Image fusion algorithm based on orientation information motivated pulse coupled neural networks. In Proceedings of the World Congress on Intelligent Control and Automation, Chongqing, China, 25–27 June 2008; pp. 2437–2441.
38. Qu, X.B.; Yan, J.W.; Xiao, H.Z.; Zhu, Z.Q. Image fusion algorithm based on spatial frequency motivated pulse coupled neural networks in nonsubsampled contourlet transform domain. *Acta Autom. Sin.* **2008**, *34*, 1508–1514. [CrossRef]
39. Naidu, V. Hybrid ddct-pca based multi sensor image fusion. *J. Opt.* **2014**, *43*, 48–61. [CrossRef]
40. Bavirisetti, D.P.; Xiao, G.; Liu, G. Multi-sensor image fusion based on fourth order partial differential equations. In Proceedings of the 20th International Conference on Information Fusion, Xi'an, China, 10–13 July 2017; pp. 1–9.

41. Zhang, Y.; Zhang, L.; Bai, X.; Zhang, L. Infrared and visual image fusion through infrared feature extraction and visual information preservation. *Infrared Phys. Technol.* **2017**, *83*, 227–237. [CrossRef]
42. Zhang, X.; Ma, Y.; Fan, F.; Zhang, Y.; Huang, J. Infrared and visible image fusion via saliency analysis and local edge-preserving multi-scale decomposition. *JOSA A* **2017**, *34*, 1400–1410. [CrossRef]
43. Ma, J.; Chen, C.; Li, C.; Huang, J. Infrared and visible image fusion via gradient transfer and total variation minimization. *Inf. Fusion* **2016**, *31*, 100–109. [CrossRef]
44. Luo, X.; Wang, S.; Yuan, D. Weber-aware weighted mutual information evaluation for infrared–visible image fusion. *J. Appl. Remote Sens.* **2016**, *10*, 045004. [CrossRef]
45. Toet, A. Image fusion by a ratio of low-pass pyramid. *Pattern Recognit. Lett.* **1989**, *9*, 245–253. [CrossRef]
46. Li, H.; Manjunath, B.S.; Mitra, S.K. Multisensor image fusion using the wavelet transform. *Gr. Models Image Process.* **1995**, *57*, 235–245. [CrossRef]
47. Zhan, L.; Zhuang, Y.; Huang, L. Infrared and visible images fusion method based on discrete wavelet transform. *J. Comput.* **2017**, *28*, 57–71. [CrossRef]
48. Madheswari, K.; Venkateswaran, N. Swarm intelligence based optimisation in thermal image fusion using dual tree discrete wavelet transform. *Quant. Infrared Thermogr. J.* **2017**, *14*, 24–43. [CrossRef]
49. Liu, S.; Piao, Y.; Tahir, M. Research on fusion technology based on low-light visible image and infrared image. *Opt. Eng.* **2016**, *55*, 123104. [CrossRef]
50. Candès, E.; Demanet, L.; Donoho, D.; Ying, L. Fast discrete curvelet transforms. *Multiscale Model. Simul.* **2006**, *5*, 861–899. [CrossRef]
51. Do, M.N.; Vetterli, M. The finite ridgelet transform for image representation. *IEEE Trans. Image Process.* **2003**, *12*, 16–28. [CrossRef] [PubMed]
52. Zuo, Y.; Liu, J.; Bai, G.; Wang, X.; Sun, M. Airborne infrared and visible image fusion combined with region segmentation. *Sensors* **2017**, *17*, 11–27. [CrossRef] [PubMed]
53. Masini, A.; Branchitta, F.; Diani, M.; Corsini, G. Sight enhancement through video fusion in a surveillance system. In Proceedings of the 14th International Conference on Image Analysis and Processing, Modena, Italy, 10–14 September 2007; pp. 554–559.
54. Hill, R.; Achim, A.; Bull, D. Scalable video fusion. In Proceedings of the 2013 IEEE International Conference on Image Processing; IEEE: Piscataway, NJ, USA, 2013; pp. 1277–1281.
55. Wang, Y.; Wang, I.; Selesnick, I.; Vetro, A. Video coding using 3D dual-tree wavelet transform. *J. Image Video Process.* **2007**, *1*, 1–15.
56. Hill, R.; Achim, A.; Bull, D. Scalable fusion using a 3D dual tree wavelet transform. In Proceedings of the Sensor Signal Processing for Defence (SSPD 2011), London, UK, 27–29 September 2011; p. 35.
57. Hogervorst, M.A.; Toet, A. Improved Color Mapping Methods for Multiband Nighttime Image Fusion. *J. Imaging* **2017**, *3*, 36. [CrossRef]
58. Vlahovic, N.; Graovac, S. Sensibility Analysis of the Object Tracking Algorithms in Thermal Image. *Scientific Technical Review* **2017**, *67*, 13–20. [CrossRef]
59. Kai, Z.; Zhou, W. Polyview fusion: A strategy to enhance video denoising algorithms. *IEEE Trans. Image Process.* **2012**, *21*, 2324–2328.
60. Zheng, Y.; Blasch, E.; Liu, Z. *Multispectral Image Fusion and Colorization*; SPIE Press: Bellingham, WA, USA, 2018.
61. Petrović, V. Subjective tests for image fusion evaluation and objective metric validation. *Inf. Fusion* **2007**, *8*, 208–216. [CrossRef]
62. Zhang, Q.; Wang, L.; Li, H.; Ma, Z. Video fusion performance evaluation based on structural similarity and human visual perception. *Signal Process.* **2012**, *92*, 912–925. [CrossRef]
63. Haghighat, M.B.A.; Aghagolzadeh, A.; Seyedarabi, H. A non-reference image fusion metric based on mutual information of image features. *Comput. Electr. Eng.* **2011**, *37*, 744–756. [CrossRef]

electronics

MDPI

Article

A 2.5 Gbps, 10-Lane, Low-Power, LVDS Transceiver in 28 nm CMOS Technology

Xu Bai [1,2,*], Jianzhong Zhao [1], Shi Zuo [1,2] and Yumei Zhou [1,2]

[1] Smart Sensing R&D Centre, Institute of Microelectronics of Chinses Academy of Science, Beijing 100029, China; zhaojianzhong@ime.ac.cn (J.Z.); zuoshi@ime.ac.cn (S.Z.); ymzhou@ime.ac.cn (Y.Z.)

[2] Institute of Microelectronics, University of Chinese Academy of Sciences, 19A Yuquan Rd., Shijingshan District, Beijing 100049, China

* Correspondence: baixu@ime.ac.cn; Tel.: +86-152-2230-6300

Received: 16 January 2019; Accepted: 16 March 2019; Published: 22 March 2019

Abstract: This paper presents a 2.5 Gbps 10-lane low-power low voltage differential signaling (LVDS) transceiver for a high-speed serial interface. In the transmitter, a complementary MOS H-bridge output driver with a common mode feedback (CMFB) circuit was used to achieve a stipulated common mode voltage over process, voltage and temperature (PVT) variations. The receiver was composed of a pre-stage common mode voltage shifter and a rail-to-rail comparator. The common mode voltage shifter with an error amplifier shifted the common mode voltage of the input signal to the required range, thereby the following rail-to-rail comparator obtained the maximum transconductance to recover the signal. The chip was fabricated using SMIC 28 nm CMOS technology, and had an area of 1.46 mm^2. The measured results showed that the output swing of the transmitter was around 350 mV, with a root-mean-square (RMS) jitter of 3.65 ps@2.5 Gbps, and the power consumption of each lane was 16.51 mW under a 1.8 V power supply.

Keywords: LVDS; high-speed serial interface; transmitter; receiver; low-power

1. Introduction

While scaled CMOS technology continues to enhance on-chip operating speeds, the power dissipation also increases at the same time. This means that reducing power consumption is critical for battery-powered systems to extend battery life. Low voltage differential signaling (LVDS), as one of the data transmission standards, is now pervasive in communication networks and is used extensively in applications such as laptop computers [1], office imaging [2,3], and medical [4] and automotive [5,6] applications. It features a low-voltage swing (250–400 mV) and achieves a high data rate (up to several gigahertz per single pair) with less power dissipation. A typical LVDS serial link [7,8] point-to-point communication is shown in Figure 1, and involves a single transmitter (TX) and receiver (RX) pair. A current source (Is) is derived from the TX, and the output amplitude is formed by the current source flowing through the terminated resistor (R_T) to establish voltage in the input of RX. By changing the current direction, the same amplitude with the opposite polarity is created to generate the logic of zeros and ones. The simple termination, low-power, and low-noise characteristics have gradually made LVDS the technology of choice for gigabit-per-second serial transmission. In addition, the wide common mode input of LVDS makes its devices easily interoperable with other differential signaling technologies [9–11].

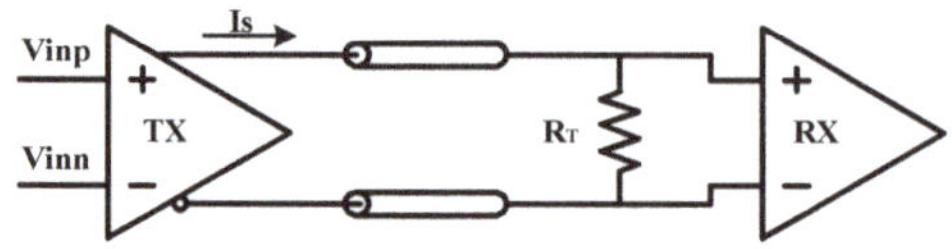

Figure 1. Low voltage differential signaling (LVDS) serial link communication block.

In general, the architecture of LVDS drivers is divided into fully-differential NMOS-only style [12], fully-differential PMOS-only style [13] and complementary MOS style [14–16]. As shown in Figure 2, all configurations consist of four MOS switches arranged in an H-bridge structure. The NMOS-only style LVDS driver, shown in Figure 2a, works well if the supply voltage (VDD) is 2.5 V or greater [17]. However, when the supply voltage is scaled down (1.8 V for 28 nm CMOS technology), it is not applicable, as there is not enough voltage headroom. According to the LVDS standard specifications [18], a 1.125–1.325 V common mode voltage range and 250–400 mV output swing of the output signals is required, which would cause the transistors (M1a and M2a) to cut off. To overcome the supply voltage headroom issues, PMOS-only (shown in Figure 2b) and complementary MOS (shown in Figure 2c) LVDS drivers need to be addressed. A benefit of PMOS-only style drivers is that they can work without the body effect. However, the inherent speed limitation in PMOS devices precludes their use in high speed data communication. To achieve the same speed as CMOS style drivers, the size of the transistors must be increased. Consequently, the area cost and power consumption will also increase. Comparing the above-mentioned LVDS drivers, the complementary MOS style driver is the optimum choice for LVDS transmission systems operating under low supply voltage, as it is not only compatible with the LVDS standard, but also faster than the other options.

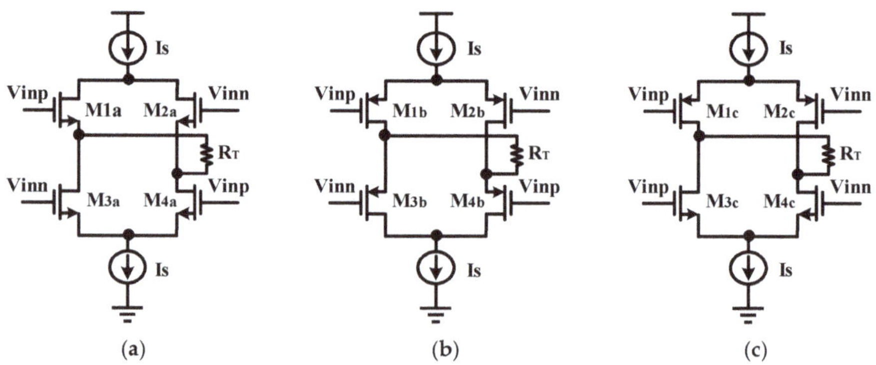

Figure 2. Simplistic circuit of LVDS output driver: (**a**) NMOS-only style; (**b**) PMOS-only style; (**c**) Complementary MOS style.

In this paper, a 2.5 Gbps 10-lane low-power LVDS transceiver is presented. The transceiver can operate at a data rate up to 2.5 Gbps, and is fully compatible with ANSI/TIA/EIA-644-A standards. The paper is organized as follows: Section 2 describes the architecture of the proposed LVDS transceiver, and presents some related simulation results. In Section 3, the measurement results are discussed. Finally, a summary and the conclusions are outlined in Section 4.

2. Architecture Design

The proposed 10-lane, low-power, LVDS transceiver is shown in Figure 3. Each lane is comprised of a receiver followed by a transmitter. It employs differential data transmission and the receiver is configured as a switched-polarity signal generator. The receiver is composed of a pre-stage common mode voltage (Vcm) shifter and a rail-to-rail comparator (COMP), while the transmitter includes a

CMOS H-bridge output driver with a common mode feedback (CMFB) circuit, a high-speed level shifter (LS) and pre-emphasis (PE) driver. In addition, two bandgap references (BGR) are embedded in the scheme to provide proper DC bias for receivers and transmitters, respectively. In the design, the differential data are firstly addressed by the receiver, then the transmitter deals with the data and sends them out in accordance with specified requirements. Therefore, only if both the receiver and transmitter are operated properly can the transmitted signals be output. The detailed implementation of the transceiver will be expatiated in the following sections.

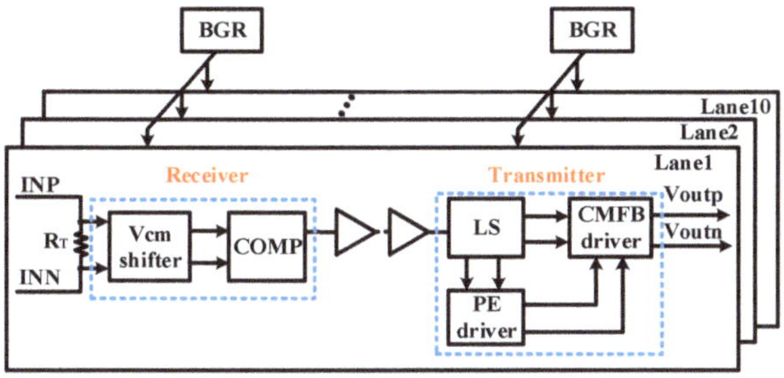

Figure 3. Simplistic circuit of 10-lane LVDS transceiver.

2.1. Receiver

According to LVDS specifications [18], a receiver is required to operate in a wide input common mode voltage range of 0.05–2.35 V. Therefore, with the 1.8 V supply voltage, the receiver firstly needs to achieve the common mode voltage conversion. Figure 4 shows the simplistic circuit of a pre-stage common mode voltage shifter, which includes a current regulator and an error amplifier. The error amplifier detects the common mode voltage difference between input data (INP and INN) and reference voltage (VREF) and amplifies the voltage difference to control the current regulator by injecting or extracting currents from resistors R1 and R2. As a result, voltage drops across R1 and R2 are generated, and the common mode voltage is shifted [19]. It is obvious that the shifted common mode voltage is affected by VREF. Thus, the value of VREF was set at 0.9 V for the following rail-to-rail comparator to obtain a higher gain.

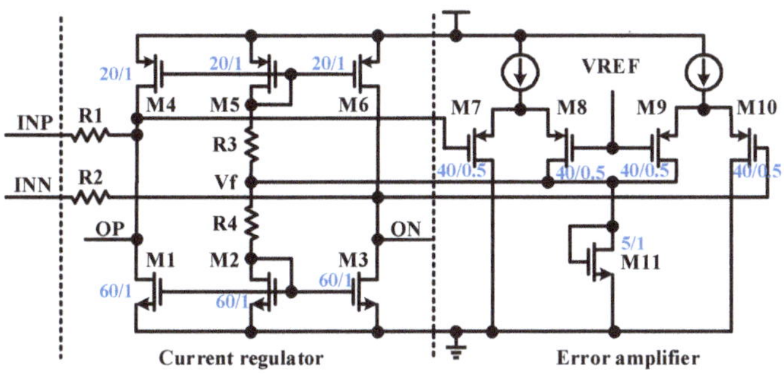

Figure 4. The input common mode voltage shifter.

A simple rail-to-rail comparator [20,21], as shown in Figure 5, was constructed as a composite of P and NMOS pairs. The amplifier with rail-to-rail input identifies the voltage difference from the input data (OP and ON) and converts them into currents through the input trans-conductor cell (M1–M4). After this, the currents are both mirrored and summed up at the node N1, before the data is reinstituted and reshaped by the last-stage shaping buffer.

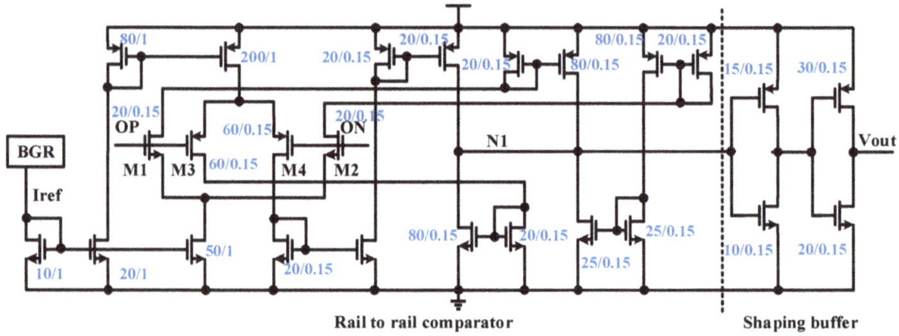

Figure 5. Schematic of the rail-to-rail comparator.

2.2. Transmitter

In this paper, the transmitter contained three parts: a high-speed level shifter, a pre-emphasis driver and an output driver. The high-speed level shifter [22,23] was introduced to achieve the different voltage domain conversion in the pre-stage of the transmitter, whose circuit is presented in Figure 6. A pair of NMOS devices (M3 and M4) receive the low-voltage input signals (Dp_L and Dn_L) and convert them into high-voltage signals through the positive feedback transistors (M1 and M2). Then, the buffer chain with several inverters reshapes the output signals under the high-voltage (VDDH) supply.

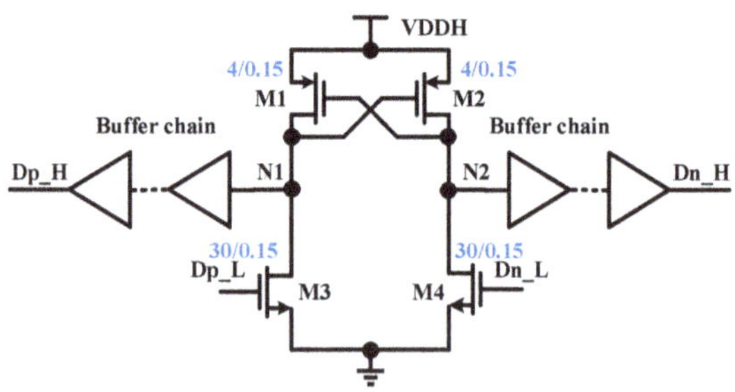

Figure 6. Simplified schematic level shifter.

Figure 7 shows the proposed transmitter output driver based on the CMOS H-bridge structure. As Figure 7 shows, the output stage of the driver uses the PMOS and NMOS configuration. A simple common mode feedback (CMFB) circuit [24,25], with transistors M5–M8, is used to stabilize the output common mode voltage (Vcm), and is less dependent on PVT. The two differential output voltages (Voutp and Voutn) are averaged to form a common mode voltage (Vcm) by two resistors (R1 and R2), which is compared with the designed reference common mode voltage (Vbg). The difference is then amplified and converted into the common mode current to adjust the common mode voltage (Vcm). In addition, an Rc and Cc pole-zero compensation network is exploited to obtain an adequate phase margin of CMFB under the conditions created by the PVT variations. Meanwhile, a cascade current mirror (M9–M12) is utilized to provide high precision current bias at a 1.8 V voltage supply.

In addition, a pre-emphasis driver with a simple pulse-width modulation (PWM) technique [26,27] is used in the transmitter to enhance signal integrity. A simplistic circuit of this pre-emphasis driver is presented in Figure 8. The pre-emphasis driver exploits the timing relationship between signals and delay signals to establish the signal-related pulse (UP and DN), which is only enabled at the rise and fall of the signal [28,29]. During the signal transition, the pre-emphasis driver adds a current to the output node, and also extracts the current from the output node by the UP and DN pulses, so that the rise and fall time is decreased. Figure 9 shows the eye diagram of the transmitter after the channel, which operates at 2.5 Gbps. Figure 9a presents the simulated results of the eye diagram without a pre-emphasis driver, while the simulated results of the eye diagram with a pre-emphasis driver are shown in Figure 9b. As shown, the pre-emphasis driver is not only able to shorten the rise time but also improves the amplitude of the output signal.

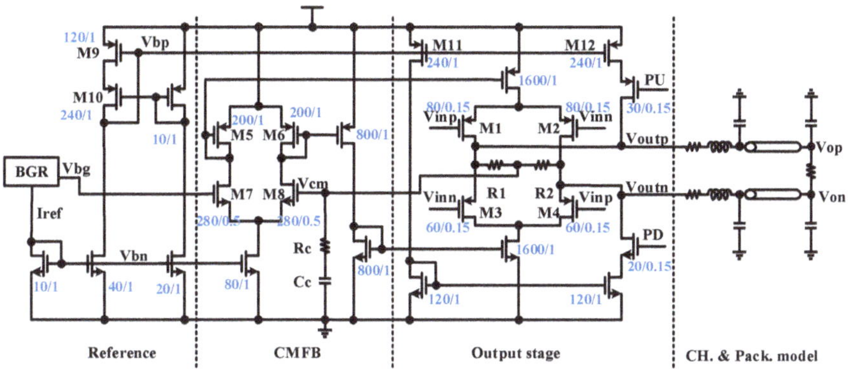

Figure 7. The architecture of the output driver.

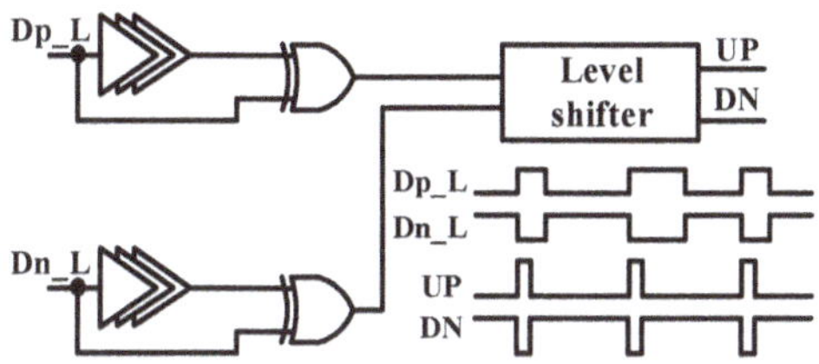

Figure 8. Simplified schematic of the pre-emphasis driver.

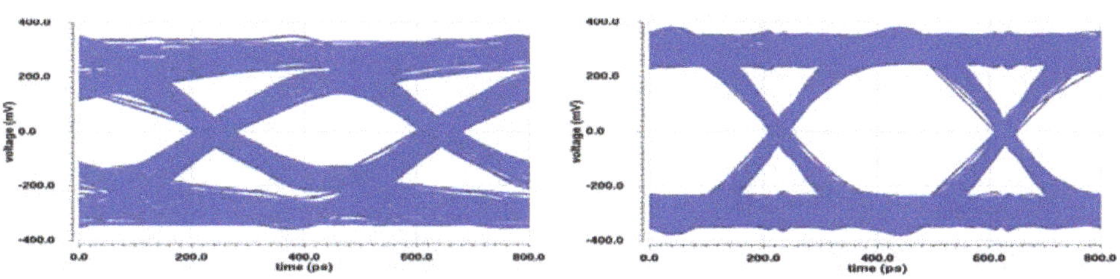

Figure 9. Simulated result of the eye diagram (**a**) without and (**b**) with the pre-emphasis driver.

3. Measured Result Analysis and Discussion

Figure 10 shows a chip microphotograph of the 10-lane LVDS transceiver. The entire chip was fabricated with SMIC 28 nm CMOS technology and the total area was 1.46 mm^2. The area of each TX/RX lane was 0.0333 mm^2, where TX and RX occupy 0.0306 mm^2 and 0.0027 mm^2, respectively. In multi-lane high-speed serial links, crosstalk and interference of lanes are important issues that deteriorate the performance of output signals. In this paper, two lanes of the transceiver shared supply voltage to improve the power integrity, and the BGR utilized a pair of individual supply voltages to provide the dependable DC bias for TX and RX, respectively. Plentiful on-chip decoupling capacitors were also inserted in the empty area to enhance signal integrity. These methods simply and effectively suppressed output jitter.

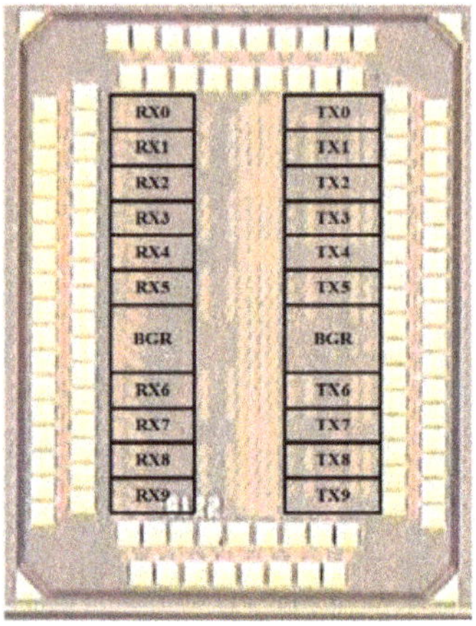

Figure 10. Microphotograph of the LVDS transceiver.

An Agilent pulse generator 81134A was used to produce 2^{31}-1 pseudorandom bit sequence (PRBS) data patterns to the receiver, while a Tektronix MSO71604C mixed signal oscilloscope was used to detect the differential output eye diagram of the transmitter. A 22-inch coupled micro-strip line on the testing PCB acted as the transmission channel, the channel loss of which is shown in Figure 11. The channel loss was 2.2 dB at 625 MHz, and 1.8 dB at 1.25 GHz.

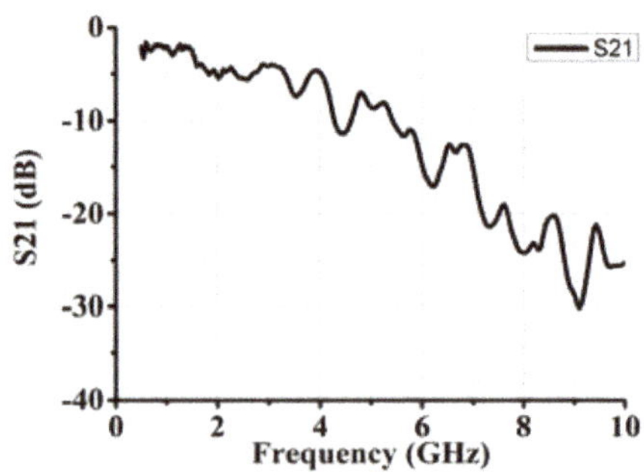

Figure 11. The frequency response of the 22-inch FR4 channel.

According to the measured results, the maximum data rate of the transceiver reached 2.5Gbps. Figure 12a,b shows the single lane of transmitter differential output eye diagrams with 2^{31}-1 PRBS patterns and data rates of 1.25 Gbps and 2.5 Gbps. Both output swings of the two operating data rates were around 350 mV, and the root-mean-square (RMS) jitters were 5.48 ps and 3.65 ps, respectively. Figure 12c,d show transmitter differential output eye diagrams of 1.25 Gbps and 2.5 Gbps for multi-lane transmission communication. Similarly, their output swings were around 350 mV, but their performance was degraded. This is due to the lane-to-lane interference of signals and power lines, which introduced higher deterministic jitter (DJ) that deteriorated the signal integrity of the output signals. The total power dissipation of the two operating data rates were 8.72 mW and 16.51 mW at a 1.8 V power supply for each lane.

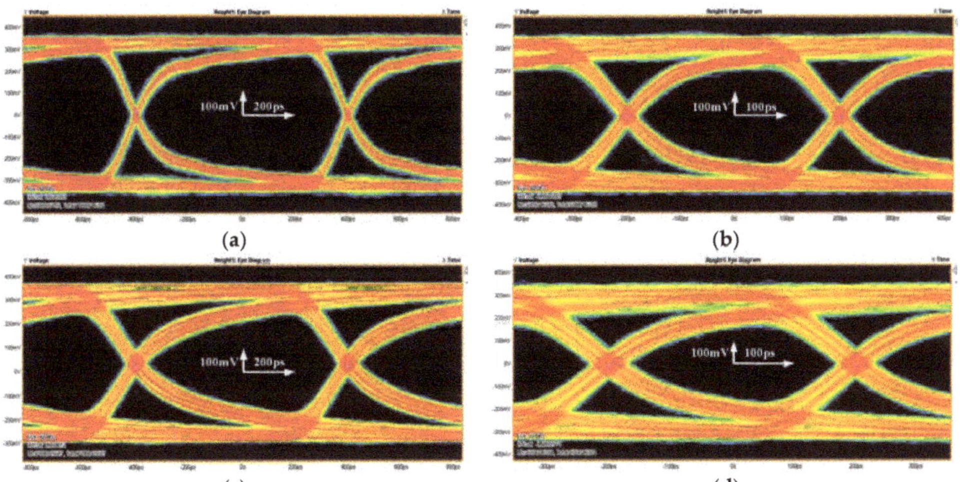

(a) (b) (c) (d)

Figure 12. Measured output eye diagrams for different data rates (**a**) 1.25 Gbps of single lane; (**b**) 2.5 Gbps of single lane (**c**) 1.25 Gbps of multi-lane (**d**) 2.5 Gbps of multi-lane.

Table 1 summarizes the comparison of the performance of the previously reported LVDS transmitters. This LVDS transmitter, based on a complementary MOS H-bridge, had excellent noise immunity performance, with an RMS jitter of 3.65 ps with a data rate up to 2.5 Gbps. The proposed LVDS transmitter also had superior power consumption performance of 16.51 mW at a data rate of 2.5 Gbps, with a figure of merit (FOM) of 6.6 mW/Gbps.

Table 1. Comparison with previous works.

Ref.	[9] *	[15] **	[30] *	[31] **	This Work **
Year	2016	2011	2014	2018	2019
Technology (nm)	28 CMOS	180 CMOS	40 CMOS	28 CMOS	28 CMOS
Supply voltage (V)	1.8/1	2.5	1.8/1	1.8/1	1.8/0.9
Output swing (mV)	350	313	320	348	350
Data rate (Gbps)	1	2	1	1	2.5
RMS jitter (ps)	2.2	7.65	4	9.8	3.65
Power(mW)	8.7	15.41	7	7.9	16.51
Area (mm^2)	0.009	0.061	0.0168	0.085	0.0306
FOM # (mW/Gbps)	8.7	7.705	7	7.9	6.60

*: Simulated result; **: Measured result; #: FOM = Power (mW)/Data rate (Gbps).

4. Conclusions

In this paper, a 2.5 Gbps, 10-lane, low-power, LVDS transceiver was presented. In the receiver, a pre-stage common mode voltage shifter was introduced to implement the common mode voltage conversion, and a rail-to-rail comparator embedded with a shaping buffer was utilized to recover the input signal. Compared with the characteristics of previous LVDS driver architectures, a complementary MOS LVDS driver using a CMFB circuit was exploited to provide the required output common mode voltage and differential output swing at 1.8 V supply voltage. In addition, a high-speed level shifter was designed for voltage domain conversion, and a pre-emphasis driver with PWM technique was employed to reduce the signal transition time. Further, the proposed LVDS transceiver was compatible with ANSI/TIA/EIA-644-A standards. The tranceiver is easy to interoperate with other differential signaling technologies, and can be embedded in other chips as an IP core, which makes it suitable for use in portable electronics. The whole circuit was fabricated with SMIC 28 nm CMOS technology, with a total chip area of 1.46 mm^2. The measured results show that the proposed low-power LVDS was able to be properly operated at 2.5 Gbps, with an RMS jitter of 3.65 ps and an FOM of 6.6 mW/Gbps.

Author Contributions: Conceptualization, X.B.; formal analysis, X.B. and J.Z.; data curation, X.B. and Y.Z.; writing—original draft preparation, X.B. and S.Z.; writing—review and editing, X.B. and S.Z.; visualization, X.B. and S.Z.; supervision, J.Z.; project administration, Y.Z.

Funding: This research was funded by National Science and Technology Major Project of China, grant number 2014ZX0302002.

Acknowledgments: The authors would like to thank the National Science and Technology Major Project of China for their support.

Conflicts of Interest: The authors declare no conflict of interest.

References

1. Park, J.; Chae, J.H.; Jeong, Y.U.; Lee, J.W.; Kim, S. A 2.1-Gb/s 12-channel transmitter with phase emphasis embedded serializer for 55-in UHD intra-panel interface. *IEEE J. Solid-State Circuits* **2018**, *53*, 2878–2888. [CrossRef]
2. Yousefzadeh, A.; Jabłoński, M.; Iakymchuk, T.; Linares-Barranco, A.; Rosado, A.; Plana, L.A.; Temple, S.; Serrano-Gotarredona, T.; Furber, S.B.; Linares-Barranco, B. On multiple AER handshaking channels over high-speed bit-serial bidirectional LVDS links with flow-control and clock-correction on commercial FPGAs for scalable neuromorphic systems. *IEEE Trans. Biomed. Circuits Syst.* **2017**, *11*, 1133–1147. [CrossRef] [PubMed]
3. Jiang, B.J.; Pan, Z.B.; Qiu, Y.H. Study on the key technologies of a high-speed CMOS camera. *Optik-Int. J. Light Electron Opt.* **2017**, *129*, 100–107. [CrossRef]
4. Shi, Z.; Tang, Z.A.; Feng, C.; Cai, H. Improvement to the signaling interface for CMOS pixel sensors. *Nuclear Instrum. Methods Phys. Res. Sect. A Accel. Spectrometers Detect. Assoc. Equip.* **2016**, *832*, 77–84. [CrossRef]
5. Jayshree; Verma, S.; Chatterjee, A. A methodology for designing LVDS interface system. In Proceedings of the IEEE Sixth International Symposium on Embedded Computing and System Design, Patna, India, 15–17 December 2016; pp. 284–288. [CrossRef]
6. Gilbert, A.; Mehmet, R.Y.; Jean-Michel, R. An integrated LVDS transmitter-receiver system with increased self-immunity to EMI in 0.18-μm CMOS. *IEEE Trans. Electromagn. Compat.* **2016**, *58*, 231–240. [CrossRef]
7. Gupta, H.S.; Hari, S.G.; Parmar, R.M.; Dave, R.K. High speed LVDS driver for serdes. In Proceedings of the IEEE International Conference on Emerging Trends in Electronic and Photonic Devices & Systems, Varanasi, India, 22–24 December 2009; pp. 92–95. [CrossRef]
8. Ning, H.W.; Zhen, G.Y.; Ren, Y.F. An optimal design of LVDS interface. In Proceedings of the IEEE International Conference on Computer Science and Network Technology, Harbin, China, 24–26 December 2011; pp. 2024–2026. [CrossRef]
9. Graceffe, G.A.; Gatti, U.; Calligaro, C. A 400 Mbps radiation hardened by design LVDS compliant driver and receiver. In Proceedings of the IEEE International Conference on Electronics, Circuits and Systems, Bordeaux, France, 11–14 December 2016; pp. 109–112. [CrossRef]
10. Sun, Z.Y.; Zhang, D.; Fang, W. A ASIC chip with pipeline ADCs for CCD sensor imaging system. *Sens. Actuators A Phys.* **2018**, *279*, 284–292. [CrossRef]
11. Xu, H.Y.; Wang, J.; Lai, J.M. Design of a power efficient self-adaptive LVDS driver. *IEICE Electron. Express* **2018**, *15*. [CrossRef]
12. Li, S.; Zhang, Q.; Zhao, X.; Liu, S.; Yuan, Z.; Zhang, X. Dynamic data transmission technology for expendable current profiler based on low-voltage differential signaling. *Geosci. Instrum. Methods Data Syst.* **2017**, *6*, 263–267. [CrossRef]
13. Marar, H.W.; Abugharbieh, K.; Al-Tamimi, A.K. A power efficient 3 Gbps 1.8 V PMOS-based LVDS output driver. In Proceedings of the 19th IEEE International Conference on Electronics, Circuits, and Systems, Seville, Spain, 9–12 December 2012; pp. 240–243. [CrossRef]

14. Traversi, G.; De Canio, F.; Liberali, V.; Stabile, A. Design of LVDS driver and receiver in 28 nm CMOS technology for associative memories. In Proceedings of the IEEE International Conference on Modern Circuits and Systems Technology, Thessaloniki, Greece, 4–6 May 2017; pp. 1–4. [CrossRef]
15. Lv, J.; Ju, H.; Yuan, L.; Zhao, J.; Zhang, F.; Wu, B.; Jiang, J.; Zhou, Y. A high speed low jitter LVDS output driver for serial links. *Analog Intergr. Circuits Signal Process.* **2011**, *68*, 387–395. [CrossRef]
16. Chen, M.; Silva-Martinez, J.; Nix, M.; Robinson, M.E. Low-voltage low-power LVDS drivers. *IEEE J. Solid-State Circuits* **2005**, *40*, 472–479. [CrossRef]
17. Lee, S.S.; Lee, L.; Kung, F.W.; Saad, A.; Tan, G.H. A fully integrated and high precision 350 mV amplitude regulated LVDS transmitter compensating PVT variations. *Microelectron. J.* **2018**, *81*, 192–199. [CrossRef]
18. Telecommunications Industry Association. *Electrical Characteristics of Low Voltage Differential Signaling (LVDS) Interface Circuits*; Standard ANSI/TIA/EIA-644-A (2001); Telecommunications Industry Association: Arlington, VA, USA, 1996.
19. Louis, L.; John, C.; Jeffrey, D. A continuous-time common-mode feedback circuit (CMFB) for high-impedance current-mode applications. *IEEE Trans. Circuit Syst. II Analog Digit. Signal Process.* **2000**, *47*, 363–369. [CrossRef]
20. Divide, M.; Gaetano, P.; Salvatore, P. A new compact low-power high-speed rail-to-rail Class-B buffer for LCD applications. *J. Disp. Technol.* **2010**, *6*, 184–190. [CrossRef]
21. Nagy, L.; Arbet, D.; Kovac, M.; Potocny, M.; Stopjakova, V. Design and performance analysis of ultra-low voltage rail-to-rail comparator in 130 nm CMOS technology. In Proceedings of the IEEE 21st International Symposium on Design and Diagnostics of Electronic Circuits and Systems, Budapest, Hungary, 25–27 April 2018; pp. 51–54. [CrossRef]
22. Priti, G.; Ujwala, G. Design of voltage level shifter for multi-supply voltage design. In Proceedings of the International Conference on Communication and Signal Processing, Melmaruvathur, India, 6–8 April 2016; pp. 853–857. [CrossRef]
23. Lanuzza, M.; Corsonello, P.; Perri, S. Low-Power Level Shifter for Multi-Supply Voltage Designs. *IEEE Trans. Circuits Syst. II Express Briefs* **2012**, *59*, 922–926. [CrossRef]
24. Ahmed, N.; Edgar, S.-S.; Jose, S.-M. A fully balanced pseudo-differential OTA with common mode feedback and inherent common mode detector. *IEEE J. Solid-State Circuits* **2003**, *38*, 663–668. [CrossRef]
25. Basu, J.; Mandal, P. Effect of switched-capacitor CMFB on the gain of fully differential OP-Amp for design of integrators. In Proceedings of the IEEE International Symposium on Circuits and System, Florence, Italy, 27–30 May 2018; pp. 1–5. [CrossRef]
26. Ševčík, B.; Brančík, L. Time-domain pre-emphasis technique based on pulse-width modulation scheme. In Proceedings of the IEEE International Conference on Telecommunications and Signal Processing, Budapest, Hungary, 18–20 August 2011; pp. 483–486. [CrossRef]
27. Gilbert, A.; Huang, H.Y. Equalization and pre-emphasis based LVDS transceiver. *Analog Intergr. Circ. Signal Process.* **2013**, *75*, 109–123. [CrossRef]
28. Jawed, S.A.; Asghar, A.; Khan, K.; Abbasi, S.; Naveed, M.; Siddiqi, Y.; Siddiqi, W. A configurable 2-Gbps LVDS transceiver in 150-nm CMOS with pre-emphasis, equalization, and slew rate control. *Int. J. Circuit Theory Appl.* **2017**, *45*, 1369–1381. [CrossRef]
29. Xu, Y.; Sun, T.Q.; Zhao, F.; Hu, C. A full-integrated LVDS transceiver in 0.5 μm CMOS technology. In Proceedings of the IEEE Conference on Industrial Electronics and Applications, Hangzhou, China, 9–11 June 2014; pp. 1672–1675. [CrossRef]
30. Ayyagari, R.; Gopal, K. Low power LVDS transmitter design and analysis. In Proceedings of the IEEE The Asia-Pacific Conference on Communication, Ishigaki, Japan, 1–3 October 2014; pp. 42–45. [CrossRef]
31. Traversi, G.; De Canio, F.; Liberali, V.; Stabile, A. Characterization of an LVDS link in 28 nm CMOS for multi-purpose pattern recognition. In Proceedings of the IEEE International Symposium on Circuits and Systems (ISCAS), Florence, Italy, 27–30 May 2018; pp. 43–47. [CrossRef]

electronics

MDPI

Article

A 2.6 GS/s 8-Bit Time-Interleaved SAR ADC in 55 nm CMOS Technology

Dong Wang [1,2], Xiaoge Zhu [1], Xuan Guo [1], Jian Luan [1], Lei Zhou [1], Danyu Wu [1], Huasen Liu [1,2], Jin Wu [1] and Xinyu Liu [1,*]

1 Institute of Microelectronics of the Chinese Academy of Sciences, Beijing 100029, China; wangdong@ime.ac.cn (Dong W.); zhuxiaoge1989@163.com (X.Z.); guoxuan@ime.ac.cn (X.G.); luanjian@ime.ac.cn (J.L.); zhoulei@ime.ac.cn (L.Z.); wudanyu@ime.ac.cn (Danyu W.); liuhuasen@ime.ac.cn (H.L.); wujin@ime.ac.cn (J.W.)

2 School of Microelectronics, University of Chinese Academy of Sciences, Beijing 100049, China

* Correspondence: xyliu@ime.ac.cn

Received: 9 February 2019; Accepted: 6 March 2019; Published: 8 March 2019

Abstract: This paper presents an eight-channel time-interleaved (TI) 2.6 GS/s 8-bit successive approximation register (SAR) analog-to-digital converter (ADC) prototype in a 55-nm complementary metal-oxide-semiconductor (CMOS) process. The channel-selection-embedded bootstrap switch is adopted to perform sampling times synchronization using the full-speed master clock to suppress the time skew between channels. Based on the segmented pre-quantization and bypass switching scheme, double alternate comparators clocked asynchronously with background offset calibration are utilized in sub-channel SAR ADC to achieve high speed and low power. Measurement results show that the signal-to-noise-and-distortion ratio (SNDR) of the ADC is above 38.2 dB up to 500 MHz input frequency and above 31.8 dB across the entire first Nyquist zone. The differential non-linearity (DNL) and integral non-linearity (INL) are +0.93/−0.85 LSB and +0.71/−0.91 LSB, respectively. The ADC consumes 60 mW from a 1.2 V supply, occupies an area of 400 μm × 550 μm, and exhibits a figure-of-merit (FoM) of 348 fJ/conversion-step.

Keywords: analog-to-digital converter; successive approximation register; direct sampling; time-interleaved; channel-selection-embedded bootstrap; segmented pre-quantization and bypass

1. Introduction

High-speed analog-to-digital converters (ADCs) with a moderate resolution of 6–8 bits while maintaining excellent power efficiency for longer battery life are highly demanded for applications such as 802.11 ad (WiGig) radio architectures and the next-generation mobile communication system (5G) [1]. Compared with flash and pipeline ADC, successive approximation register (SAR) ADC has superior energy efficiency and is more suitable for the aggressive downscaling of technology because of its primarily digital nature [2–4]. In order to overcome the speed limitation of a single ADC, time-interleaved (TI) architecture running multiple parallel ADCs is an attractive approach. In general, TI SAR ADC is the most feasible solution to realize both over GHz operation and medium resolution around 8-bit [5]. However, the inter-channel non-ideal factors like offset, gain mismatch, and time skew, will deteriorate the overall performance [6], and can be compensated for by either circuit improvement in the analog domain or special digital calibration. As digital calibration is often complex, it is preferred to minimize these mismatches using on-chip design optimization to relax calibration requirements, especially when the number of interleaved channels is not large.

This paper demonstrates a 2.6 GS/s 8-bit SAR ADC prototype with an eight-channel direct sampling TI architecture. In the sampling front-end, in order to suppress the time skew error among different channels, the channel-selection-embedded bootstrap switch is used as the sampling switch to ensure the uniformity of sampling times by the master clock. In the sub-channel SAR ADC design, segmented pre-quantization and a bypass switching scheme is employed to avoid unnecessary large capacitors switching, reducing power consumption and non-linearity. Double comparators clocked asynchronously in alternate mode are used to improve the conversion rate, with the background offset calibration function integrated on-chip. This ADC exhibits lower calibration complexity and achieves an acceptable efficiency in terms of area and power consumption.

2. Proposed TI SAR ADC Architecture

TI structures can generally be categorized as hierarchical sampling and direct sampling. In a hierarchical sampling structure, there are at least two sampling switches in series in each sub-channel. In contrast, a direct sampling architecture, wherein all parallel channels have individual sample/hold circuits, provides the shortest signal transmission path from the input to the sampling capacitors, and is very efficient for a small number of parallel channels (usually ≤8) [7].

In this design, a direct sampling architecture is adopted to implement the 2.6 GS/s 8-bit TI ADC prototype, which mainly consists of multi-phase clock generator (MPCG), sampling switch, eight-channel 325 MS/s 8-bit SAR ADC, and multiplexer (MUX), as depicted in Figure 1.

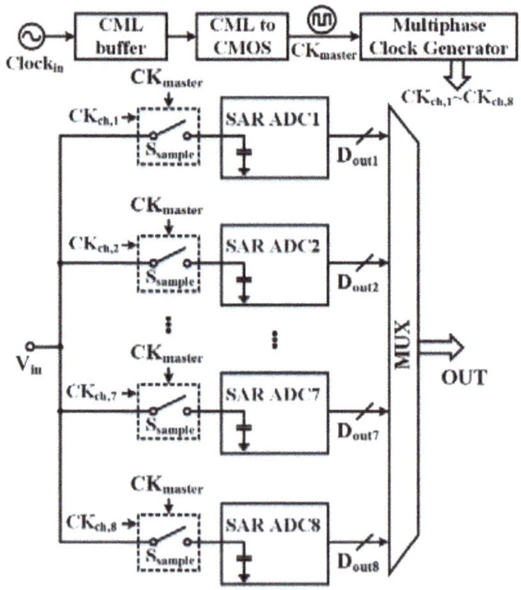

Figure 1. Eight-channel time-interleaved (TI) successive approximation register (SAR) analog-to-digital converter (ADC) architecture.

The current-mode logic (CML) sinusoidal input clock signal is buffered and then transformed to complementary metal-oxide-semiconductor (CMOS) level full-speed master clock CK_{master}. The multi-phase clock pulse signals $CK_{ch,1}$–$CK_{ch,8}$ are generated from CK_{master} using the cascaded D flip-flop (DFF) chain [8–10] and combinational logic circuits using synchronous frequency division, as shown in Figure 2a.

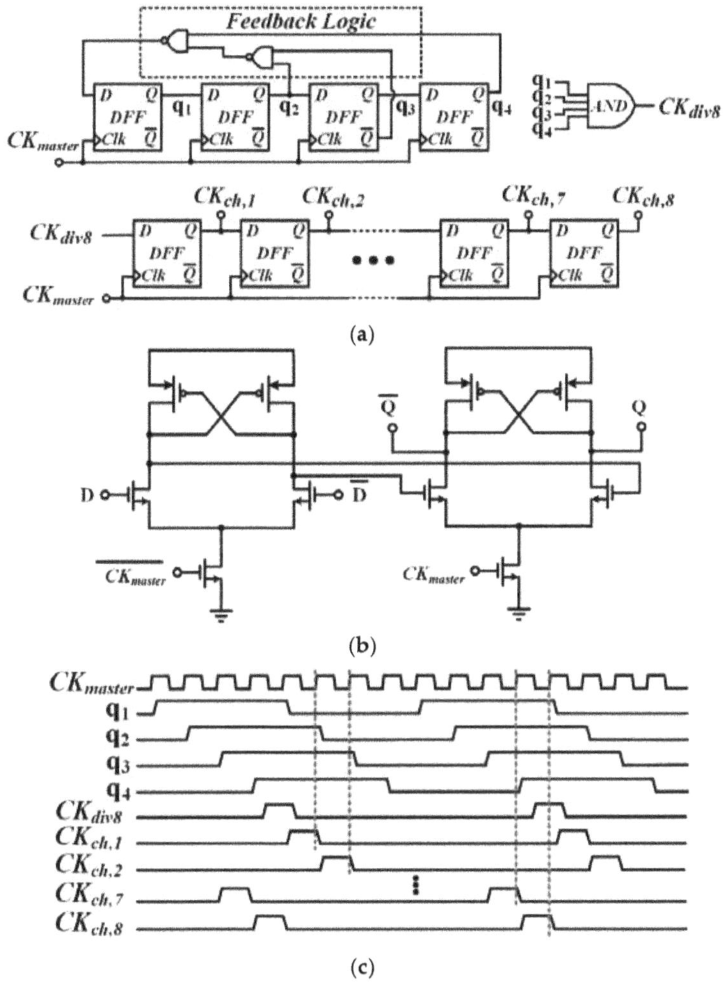

Figure 2. (**a**) Multi-phase clock generator (MPCG), (**b**) D flip-flop (DFF), and (**c**) timing diagram.

The DFF is constructed using two latches [11], as shown in Figure 2b. The internal signals q1, q2, q3 and q4 with a duty cycle of 50% are the divide-by-eight clocks of CK_{master} such that a clock pulse signal CK_{div8} with a duty cycle of 12.5% is obtained through AND logic operation. Since the initial state of the shift registers is uncertain, the feedback logic is added to activate self-starting such that the MPCG can automatically return from the non-ideal state to normal. The shift operation of CK_{div8} is executed to get $CK_{ch,1}$–$CK_{ch,8}$, which have definite phase sequence relationships and the delay between each other is one period of CK_{master}, as shown in Figure 2c. The non-overlapping sampling phases guarantee that only one sampling switch is turned on, thus reducing the channel load at the input. $CK_{ch,1}$–$CK_{ch,8}$ are not only used as the control signals for the sampling switches, but also initiate the conversion process of each channel.

As the input signal is sampled onto the capacitors array of respective sub-ADC sequentially for quantization, the sampling instants are controlled uniformly using CK_{master} to mitigate the time skew between channels. Finally, the multi-path digital outputs from the sub-ADCs array are aggregated into a one-way data stream using MUX.

3. Circuit Implementation Details

3.1. Channel-Selection-Embedded Bootstrap Switch

Time skew refers to the mismatch in the sampling instants among TI channels, which originated from the non-uniform sampling clock edges [12]. Some timing-skew, calibration-free techniques have been proposed to suppress the time skew by circuit design and layout [13–15]. We attempted to realize similar skew tolerance via circuit improvement in the analog domain for design simplicity rather than digital calibration with complex algorithms that will consume extra hardware and power. The channel-selection-embedded bootstrap switch [16] is utilized in the sampling front-end, so that the sampling instants of each channel are aligned to the master clock CK_{master}, while the corresponding TI clock signals $CK_{ch,i}$ (i = 1–8) are used to perform the channel selection, as described in Figure 3.

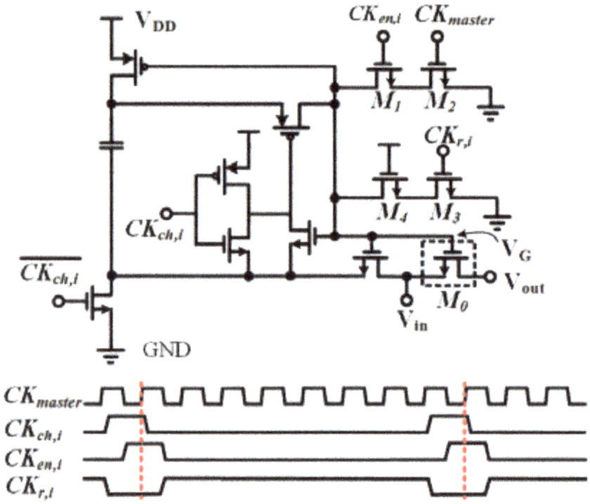

Figure 3. Channel-selection-embedded bootstrap switch.

When $CK_{ch,i}$ becomes high, the switching transistor M_0 is turned on, and the channel begins to track the input signal. $CK_{en,i}$ is a delayed version of $CK_{ch,i}$, used to prepare for sampling synchronization. When $CK_{en,i}$ is high, once the rising edge of CK_{master} comes, M_1 and M_2 are both turned on, and the gate voltage V_G of M_0 is released from V_{in} + V_{DD} to the ground level, thus the sampling instants are determined by the rising edge of CK_{master}. Then, the channel finishes the sampling process and enters the holding phase. As $CK_{en,i}$ goes low, the gate of M_0 is in floating state and vulnerable to the interference of other signals. M_3 from $CK_{r,i}$ provides a discharging path, and the V_G is fixed to the ground level, therefore avoiding the floating node at the holding phase. The inter-channel sampling synchronization that is determined uniformly by CK_{master} ensures the consistency of the sampling instants, and it is beneficial to suppress the time skew among channels.

3.2. Sub-ADC Design

3.2.1. Asynchronous Timing of Alternate Comparators

The sub-ADC is implemented with 325 MS/s 8-bit SAR ADC [17], including fundamental building blocks, such as capacitive digital-to-analog converter (CDAC), comparators and control logic, as illustrated in Figure 4a.

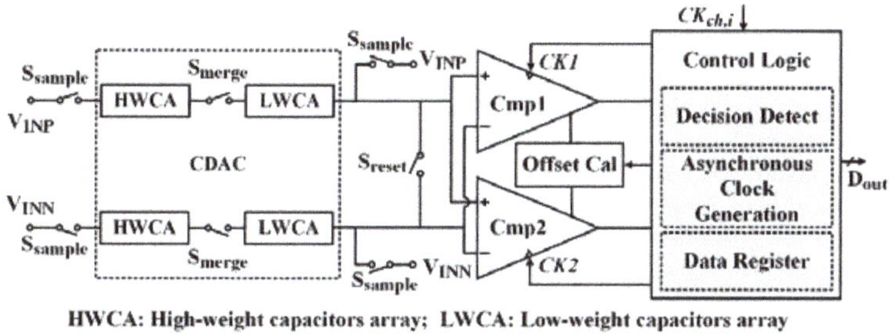

(a)

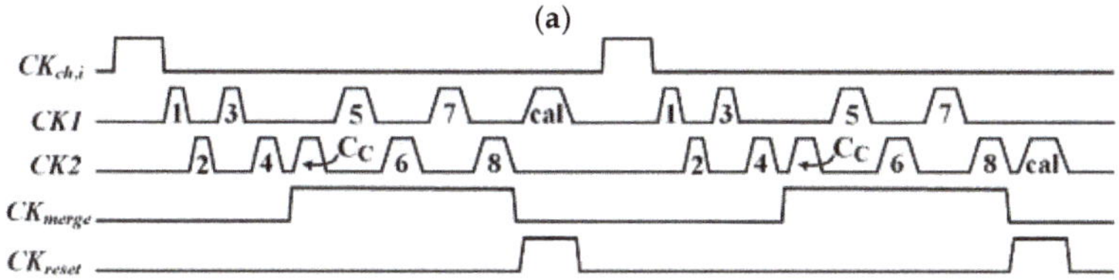

(b)

Figure 4. (**a**) SAR ADC overview, and (**b**) asynchronous timing diagram.

If only one comparator is used in the SAR ADC, the comparator needs to be fully reset to avoid the residual effect from the previous conversion process, so the overall conversion speed is slowed down. Double dynamic comparators with cross-coupled latches are used [18], and the preamplifier has been designed for moderate gain to reduce its offset, and more importantly, to limit the kick-back noise from the latch [19]. The two comparators are clocked asynchronously in alternate operation, as shown in Figure 4b. Under the control of *CK1* and *CK2*, when one comparator is in a comparison state, another comparator is reset, reducing the impact of reset time on the critical path, and thus improving the conversion speed [20]. Once the preceding comparison is over, a successful decision is detected as the trigger signal for the following comparison process. This asynchronous conversion process repeats like dominoes until all bits are resolved [21].

The pulse labeled "C_C" corresponds to the operation for the compensation capacitor in the CDAC. At the end of the conversion, the inputs of the comparators are shorted together using CK_{reset} for calibration purposes. With the pulse labeled "cal" the background offset calibration based on the charge pump principle in the analog domain for each comparator is carried out once every two cycles.

3.2.2. The Background Offset Calibration of Comparator

The diagram of comparator's background offset calibration is shown in Figure 5. An auxiliary differential pair is introduced to calibrate the offset voltage of the comparator. Only the generation circuit of V_{calp} is presented. Another calibration voltage V_{caln} can be generated in two ways, one is set to a constant voltage using a resistive voltage divider [20], and the other is similar to the generation of V_{calp}, except that the corresponding control voltage is different. In this design, the latter method is adopted. V_{calp} and V_{caln} change in opposite directions and jointly cancel the impact of the offset.

During the offset calibration phase of the comparator, the input signals V_{ip} and V_{in} of the main differential pair are shorted together. In the case of no offset, V_{ip} = V_{in}, then the output voltage of the preamplifier stage is equal, that is, V_A = V_B. Finally, the output signal of the comparator *OUTP* = *OUTN* = 0, and the corresponding complementary output signal $\overline{OUTP} = \overline{OUTN} = 1$. At this time, M5 and M8 are on, the upper capacitor C_P is charged to the power supply, and the lower C_P is discharged to the ground level. Both M6 and M7 are off, and the calibration voltage V_{calp} remains unchanged. There is no calibration effect yet.

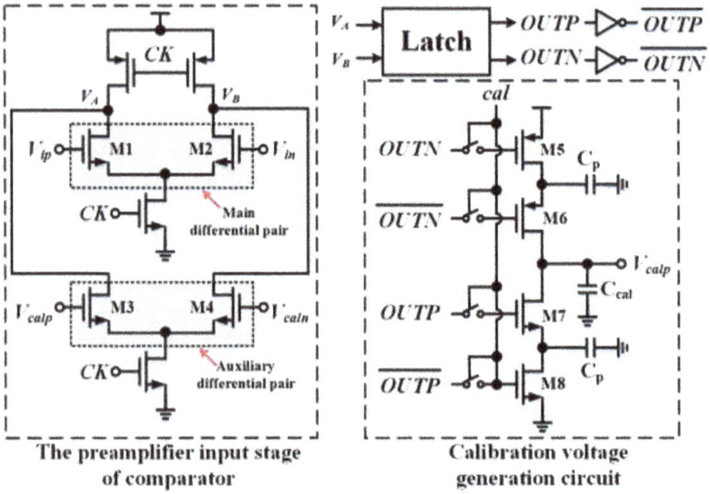

Figure 5. The diagram of comparator's background offset calibration.

If the offset exists, *OUTP* or *OUTN* will change. Usually, the output changes caused by offset can be equivalent to the input changes. Taking an offset output case as an example, if *OUTP* = 0 and *OUTN* = 1, this offset effect is the same as the situation when $V_{ip} < V_{in}$. Note that $V_A > V_B$ now. In the calibration voltage generation circuit, M6 is turned on, the voltage on C_p charges C_{cal}, and V_{calp} is pumped up. Then V_{calp} is fed back to the preamplifier input stage such that V_A decreases to approximate V_B. After a number of calibration cycles, once the decreasing V_A is equal to V_B, *OUTP* = *OUTN*, thereby realizing offset calibration. The calibration step size or accuracy is related to the capacitance values of C_p and C_{cal}. The parasitic capacitor can be used as C_p, which is usually small. The larger the calibration capacitor C_{cal}, the higher the calibration accuracy and the better the calibration effect. However, a large C_{cal} will affect the calibration settling time, and a trade-off between calibration accuracy and settling time is required to determine the value of C_{cal}.

3.2.3. Segmented Pre-Quantization and Bypass Switching Scheme

The control logic is used to generate internal asynchronous clocks, register the decision results of the comparators, and control the switching of the CDAC accordingly [20]. Several typical power-efficient switching sequences for CDAC, such as monotonic [22], splitting monotonic [23], and bypass switching techniques, have been proposed to improve the power efficiency. According to the reference [24], it was observed that the bypass method yields better results with less switching activity [25–27] because of the basic idea to skip the conversion steps for several significant bits when the signal is within a predefined window. Moreover, the skip operation reduces the error accumulation to improve the static performance. In this design, the CDAC is built with a segmented pre-quantization and bypass switching scheme [28], and the actual differential structure is displayed as single-ended for clarity in Figure 6.

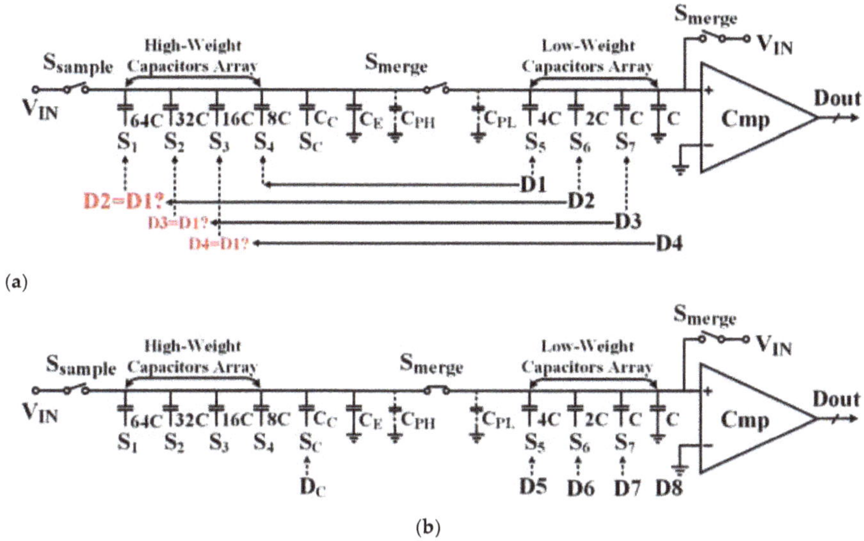

Figure 6. (**a**) Pre-quantization stage, and (**b**) global quantization stage in segmented pre-quantization and bypass switching scheme.

The capacitors array of CDAC is divided into two parts with high and low weight by the switch S_{merge}. To keep the comparator's input common-mode voltage constant, each capacitor is split into two identical small capacitors. The input signal is sampled onto the capacitors array via channel-selection-embedded bootstrap switches.

After sampling, all the switches are turned off, and only the low-weight capacitors array is connected to the comparator, equivalent to a 4-bit ADC. The comparator directly performs the first comparison without switching any capacitors to obtain the first digital code D1 [22], which is fed back to switch the minimum capacitor 8C in the high-weight capacitors array, providing an initial voltage. The subsequent digital codes D2–D4 are compared with D1, as Figure 6a shows. If one of the codes is the same as D1, that means the previous output of the CDAC is not enough, so the associated large capacitor is switched, contributing a corresponding weight output to the CDAC. In case the code is different from D1, it indicates that the last output of the CDAC is excessive, and the relevant large capacitor is bypassed, just maintaining the original state without switching. The monotonic procedure is either upward or downward to avoid unnecessary opposite direction switching of high-weight large capacitors, therefore reducing the power consumption and nonlinearity.

Once the high-weight capacitors are properly set, the switch S_{merge} is turned on, and the two arrays are merged. Meanwhile the low-weight capacitors array is reset to the initial condition. Then, the entire structure is changed back to 8-bit ADC, entering the residual quantization phase for the low 4-bit digital codes D5–D8, as shown in Figure 6b. In the whole conversion process, all quantization is done using the low-weight capacitors array, relaxing the settling constraints of the CDAC.

Ideally, the proportion of 4C in the low-weight capacitors array should be the same as that of 64C in the whole capacitors array, both being 1/2. However, the presence of parasitic capacitance changes the ratio. Since the total capacitance of high-weight capacitors array is much larger than that of low-weight capacitors array, even the same parasitic capacitance will occupy different proportions in different weight capacitor arrays, resulting in gain errors in CDAC output between high 4-bit coarse quantization and low 4-bit global quantization stages, so it is necessary to insert equilibrium capacitor (denoted as C_E) to balance the parasitic differences between the two capacitor arrays, as shown below:

$$\frac{C_L}{C_L + C_{PL}} = \frac{C_H}{C_H + C_{PH} + C_E} \tag{1}$$

C_H and C_L represent the total capacitance of high and low weight capacitors array, respectively; that is, C_H = 64C + 32C + 16C + 8C = 120C, C_L = 4C + 2C + C + C = 8C. C_{PH} and C_{PL} represent the parasitic capacitance respectively, which can be obtained from the layout parameters extraction.

In order to further solve the potential wrong conversion caused by inaccurate parameter extraction and manufacturing process variation, a compensation capacitor (denoted as C_C) with the weight of 4 is used to provide 1-bit redundancy (corresponding to digital code D_C), whose error correction range is up to 4/128 = 3.125%.

The gain error of TI SAR ADC mainly comes from the parasitic effect and capacitance mismatch of CDAC. When selecting a capacitor size, there are two main factors to consider: thermal noise (kT/C) and matching accuracy [9]. A compact and reasonable CDAC layout is deliberately designed by using full-custom metal-oxide-metal (MOM) capacitors [29] with the unit capacitance of 1.5 fF. Benefiting from 1-bit redundancy, intrinsic capacitor matching, and careful layout routing, the gain error can be minimized to a tolerant level [30].

4. Measured Results

The ADC prototype was manufactured in a 55-nm one-poly nine-metal (1P9M) CMOS process with a core area of 400 μm × 550 μm, and a large number of decoupling capacitors were filled inside the chip to keep the power supply voltage clean and stable. The die micrograph is shown in Figure 7. The static performance of differential non-linearity (DNL) and integral non-linearity (INL) is shown in Figure 8. The measured DNL and INL were +0.93/−0.85 LSB and +0.71/−0.91 LSB, respectively.

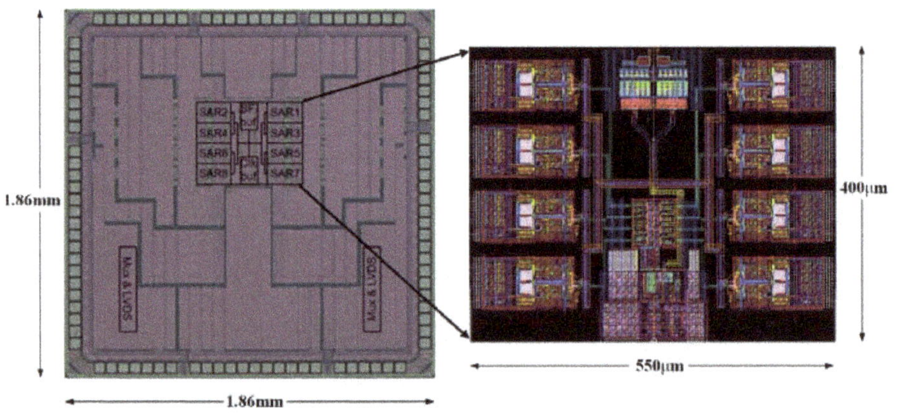

Figure 7. Die micrograph with layout view.

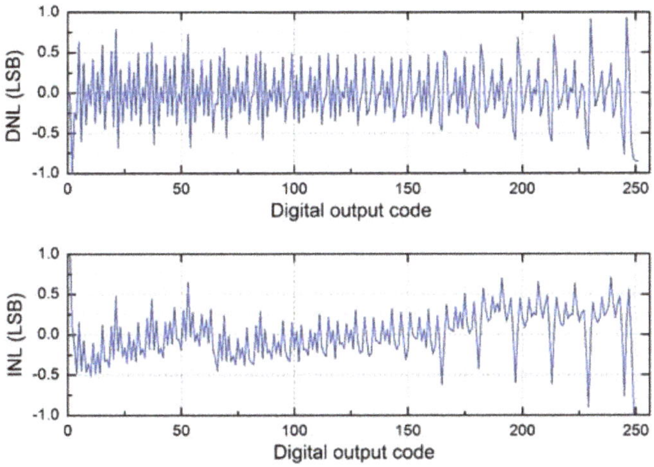

Figure 8. Measured differential non-linearity (DNL) and integral non-linearity (INL).

The output fast Fourier transform (FFT) spectrum is shown in Figure 9 at a 115 MHz input frequency and 2.6 GS/s, with an spurious-free dynamic range (SFDR) of 52.0 dB and signal-to-noise-and-distortion ratio (SNDR) of 41.52 dB. Figure 10 shows SNDR and SFDR versus input frequency at 2.6 GS/s. Within the input frequency range of 500 MHz, the SFDR was greater than 47.9 dB, the SNDR was greater than 38.2 dB, and the effective number of bits (ENOB) was greater than 6-bit. SFDR was above 40.3 dB and SNDR was above 31.8 dB in the first Nyquist zone. However, as the input frequency increased to the Nyquist frequency, the SNDR decreased by about 9 dB, which was much lower than the expected theoretical values. This result reveals that although the proposed method could suppress sample/hold circuit mismatch, the performance was not satisfactory in the high-frequency region due to other non-ideal factors, such as the master clock path mismatch, input signal path mismatch, and so on [5].

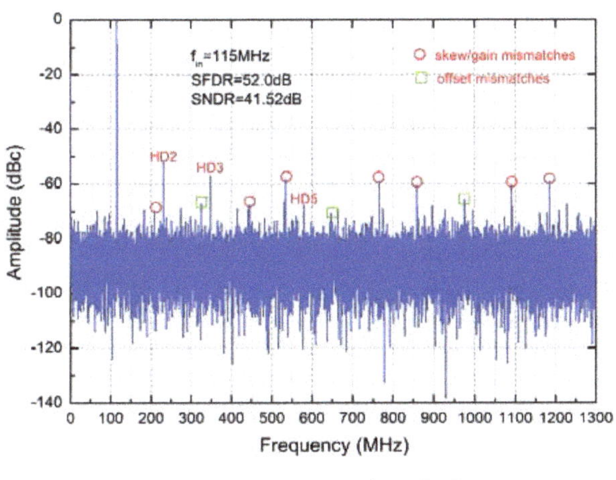

Figure 9. Output fast Fourier transform (FFT) spectrum.

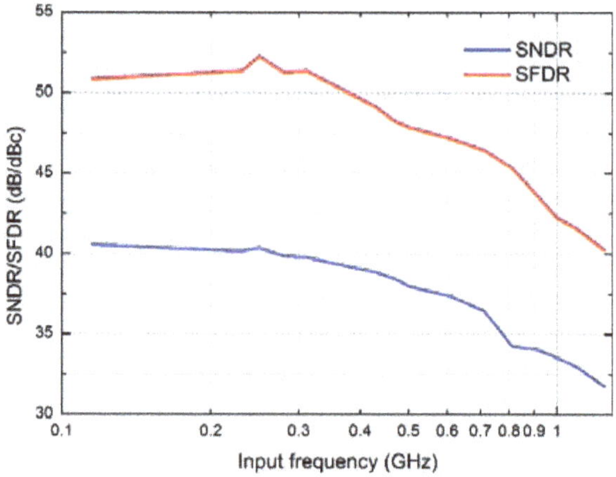

Figure 10. Signal-to-noise-and-distortion ratio (SNDR) and spurious-free dynamic range (SFDR) versus input frequency.

Based on the simulation results, the total power consumption of 60 mW at 1.2 V supply voltage was composed as follows: 12 mW for the clock generation module and 48 mW for the SAR ADCs array (that is, 6 mW/slice for every sub-ADC). The FoM calculated within the 500 MHz input frequency was 348 fJ/conversion-step. The performance summary is shown in Table 1.

Table 1. Performance summary.

Technology	55-nm 1P9M CMOS
Architecture	8-channel TI SAR
Sampling Rate	2.6 GS/s
Resolution	8-bit
Power	60 mW
Active Area	0.22 mm^2
DNL	+0.93/−0.85 LSB
INL	+0.71/−0.91 LSB
SFDR	≥50.94 dB (up to115 MHz) ≥47.9 dB (up to 500 MHz) ≥40.3 dB (up to Nyquist)
SNDR	≥40.54 dB (up to115 MHz) ≥38.2 dB (up to 500 MHz) ≥31.8 dB (up to Nyquist)
FoM [1]	348 fJ/conversion-step
Calibration Complexity	On-chip offset calibration only

[1] FoM = Power/(2^{ENOB} × Sampling frequency).

5. Conclusions

A 2.6 GS/s 8-bit SAR ADC prototype with eight-channel direct sampling TI architecture has been presented. The SNDR was above 38.2 dB up to 500 MHz input frequency and above 31.8 dB up to the Nyquist frequency. The DNL and INL were +0.93/−0.85 LSB and +0.71/−0.91 LSB, respectively. The ADC consumed 60 mW, occupied an area of 400 μm × 550 μm, and realized a FoM of 348 fJ/conversion-step. In general, this design is a beneficial attempt of time-skew, calibration-free technology, which achieves acceptable results in low and medium frequency, and provides a reference for related research and design. If the calibration for time skew is used for future work, better performance can be promised.

Author Contributions: D.W. (Dong Wang) designed the circuits, analyzed the measurement data, and wrote the manuscript. X.Z., J.L., X.G. and L.Z. assisted the circuits implementation and simulation. D.W. (Danyu Wu) and H.L. performed the chip test. J.W. contributed to the technical discussions and reviewed the manuscript. X.L. gave some valuable guidance and confirmed the final version of manuscript.

Funding: This research was supported by National Science and Technology Major Project of China, grant number 2016ZX03001002.

Conflicts of Interest: The authors declare no conflict of interest.

References

1. Lin, C.; Wei, Y.; Lee, T. A 10-bit 2.6-GS/s Time-Interleaved SAR ADC With a Digital-Mixing Timing-Skew Calibration Technique. *IEEE J. Solid-State Circuits* **2018**, *53*, 1508–1517. [CrossRef]
2. Harpe, P.J.A.; Zhou, C.; Bi, Y.; van der Meijis, N.P.; Wang, X.; Philips, K.; Dolmans, G.; de groot, H. A 26μW 8 bit 10 MS/s Asynchronous SAR ADC for Low Energy Radios. *IEEE J. Solid-State Circuits* **2011**, *46*, 1585–1595. [CrossRef]
3. Doris, K.; Janssen, E.; Nani, C.; Zanikopoulos, A.; van der Weide, G. A 480 mW 2.6 GS/s 10b Time-Interleaved ADC With 48.5 dB SNDR up to Nyquist in 65 nm CMOS. *IEEE J. Solid-State Circuits* **2011**, *46*, 2821–2833. [CrossRef]
4. Fredenburg, J.A.; Flynn, M.P. Statistical Analysis of ENOB and Yield in Binary Weighted ADCs and DACS With Random Element Mismatch. *IEEE Trans. Circuits Syst. Regul. Pap.* **2012**, *59*, 1396–1408. [CrossRef]
5. Miki, T.; Ozeki, T.; Naka, J. A 2-GS/s 8-bit Time-Interleaved SAR ADC for Millimeter-Wave Pulsed Radar Baseband SoC. *IEEE J. Solid-State Circuits* **2017**, *52*, 2712–2720. [CrossRef]
6. Xu, D.-G.; Pu, J.; Xu, S.-L.; Zhang, Z.-P.; Chen, K.-R.; Chen, Y.-Y.; Zhang, J.-A.; Wang, J.-A. A 10-bit 1.2 GS/s 45 mW time-interleaved SAR ADC with background calibration. *IEICE Electron. Express* **2018**, *15*, 1–12.
7. Kull, L.; Pliva, J.; Toifl, T.; Schmatz, M.; Francese, P.A.; Menolfi, C.; Brandli, M.; Kossel, M.; Morf, B.; Andersen, T.M.; et al. Implementation of Low-Power 6–8 b 30–90 GS/s Time-Interleaved ADCs With Optimized Input Bandwidth in 32 nm CMOS. *IEEE J. Solid-State Circuits* **2016**, *51*, 636–648. [CrossRef]
8. Stepanovic, D.; Nikolic, B. A 2.8 GS/s 44.6 mW Time-Interleaved ADC Achieving 50.9 dB SNDR and 3 dB Effective Resolution Bandwidth of 1.5 GHz in 65 nm CMOS. *IEEE J. Solid-State Circuits* **2013**, *48*, 971–982. [CrossRef]
9. Lee, S.; Chandrakasan, A.P.; Lee, H. A 1 GS/s 10b 18.9 mW Time-Interleaved SAR ADC With Background Timing Skew Calibration. *IEEE J. Solid-State Circuits* **2014**, *49*, 2846–2856. [CrossRef]
10. Lee, H.; Aurangozeb; Park, S.; Kim, J. A 6-bit 2.5-GS/s Time-Interleaved Analog-to-Digital Converter Using Resistor-Array Sharing Digital-to-Analog Converter. *IEEE Trans. Very Large Scale Integr. VLSI Syst.* **2015**, *23*, 2371–2383. [CrossRef]
11. Wei, H.; Zhang, P.; Sahoo, B.D.; Razavi, B. An 8 Bit 4 GS/s 120 mW CMOS ADC. *IEEE J. Solid-State Circuits* **2014**, *49*, 1751–1761. [CrossRef]
12. Zhu, Y.; Chan, C.-H.; Zheng, Z.-H.; Li, C.; Zhong, J.-Y.; Martins, R.P. A 0.19 mm2 10 b 2.3 GS/s 12-Way Time-Interleaved Pipelined-SAR ADC in 65-nm CMOS. *IEEE Trans. Circuits Syst. Regul. Pap.* **2018**, *65*, 3606–3616. [CrossRef]
13. Janssen, E.; Doris, K.; Zanikopoulos, A.; Murroni, A.; van der Weide, G.; Lin, Y.; Alvado, L.; Darthenay, F.; Fregeais, Y. An 11b 3.6 GS/s time-interleaved SAR ADC in 65nm CMOS. In Proceedings of the 2013 IEEE International Solid-State Circuits Conference Digest of Technical Papers, San Francisco, CA, USA, 17–21 February 2013; pp. 464–465.
14. Kundu, S.; Alpman, E.; Lu, H.-L.; Lakdawala, H.; Paramesh, J.; Jung, B.; Zur, S.; Gordon, E. A 1.2 V 2.64 GS/s 8 bit 39 mW Skew-Tolerant Time-interleaved SAR ADC in 40 nm Digital LP CMOS for 60 GHz WLAN. *IEEE Trans. Circuits Syst. Regul. Pap.* **2015**, *62*, 1929–1939. [CrossRef]
15. Lien, Y.-C. A 14.6 mW 12b 800 MS/s 4× time-interleaved pipelined SAR ADC achieving 60.8 dB SNDR with Nyquist input and sampling timing skew of 60fsrms without calibration. In Proceedings of the 2016 IEEE Symposium on VLSI Circuits (VLSI-Circuits), Honolulu, HI, USA, 15–17 June 2016; pp. 1–2.

16. Zhu, Y.; Chan, C.-H.; U, S.-P.; Martins, R.P. An 11b 900 MS/s time-interleaved sub-ranging pipelined-SAR ADC. In Proceedings of the ESSCIRC 2014—40th European Solid State Circuits Conference (ESSCIRC), Venice Lido, Italy, 22–26 September 2014. [CrossRef]
17. Zhu, X.; Zhou, L.; Wu, D.; Wu, J.; Liu, X. A 6 mW 325 MS/s 8 bit SAR ADC with background offset calibration. *IEICE Electron. Express* **2017**, *14*, 1–8. [CrossRef]
18. Kuo, B.-Y.; Chen, B.-W.; Tsai, C.-M. A 0.6 V, 1.3 GHz dynamic comparator with cross-coupled latches. In Proceedings of the VLSI Design, Automation and Test (VLSI-DAT), Hsinchu, Taiwan, 27–29 April 2015; pp. 1–4.
19. Fang, J.; Thirunakkarasu, S.; Yu, X.; Silva-Rivas, F.; Zhang, C.; Singor, F.; Abraham, J. A 5-GS/s 10-b 76-mW Time-Interleaved SAR ADC in 28 nm CMOS. *IEEE Trans. Circuits Syst. Regul. Pap.* **2017**, *64*, 1673–1683. [CrossRef]
20. Kull, L.; Toifl, T.; Schmatz, M.; Francese, P.A.; Menolfi, C.; Brandli, M.; Kossel, M.; Morf, T.; Andersen, T.M.; Leblebici, Y. A 3.1 mW 8b 1.2 GS/s Single-Channel Asynchronous SAR ADC With Alternate Comparators for Enhanced Speed in 32 nm Digital SOI CMOS. *IEEE J. Solid-State Circuits* **2013**, *48*, 3049–3058. [CrossRef]
21. Chen, S.M.; Brodersen, R.W. A 6-bit 600-MS/s 5.3-mW Asynchronous ADC in 0.13-μm CMOS. *IEEE J. Solid-State Circuits* **2006**, *41*, 2669–2680. [CrossRef]
22. Liu, C.-C.; Chang, S.-J.; Huang, G.-Y.; Lin, Y.Z. A 10-bit 50-MS/s SAR ADC With a Monotonic Capacitor Switching Procedure. *IEEE J. Solid-State Circuits* **2010**, *45*, 731–740. [CrossRef]
23. Liu, C.-C.; Chang, S.-J.; Huang, G.Y.; Lin, Y.-Z.; Huang, C.-M. A 1V 11fJ/conversion-step 10bit 10MS/s asynchronous SAR ADC in 0.18μm CMOS. In Proceedings of the 2010 Symposium on VLSI Circuits, Honolulu, HI, USA, 15–18 June 2010; pp. 241–242.
24. Santhanalakshmi, M.; Yasoda, K. Verilog-A implementation of energy-efficient SAR ADCs for biomedical application. In Proceedings of the 2015 19th International Symposium on VLSI Design and Test, Ahmedabad, India, 26–29 June 2015; pp. 1–6.
25. Huang, G.-Y.; Chang, S.-J.; Liu, C.-C.; Lin, Y.-Z. A 1-μW 10-bit 200-kS/s SAR ADC With a Bypass Window for Biomedical Applications. *IEEE J. Solid-State Circuits* **2012**, *47*, 2783–2795. [CrossRef]
26. Liu, Y.; Yuan, C.; Hung, Y.L.Y. A capacitor constructed bypass window switching scheme for energy-efficient SAR ADC. In Proceedings of the 2014 IEEE International Symposium on Circuits and Systems (ISCAS), Melbourne, VIC, Australia, 15–18 June 2014; pp. 1352–1355.
27. Wang, T.-Y.; Li, H.-Y.; Ma, Z.-Y.; Huang, Y.-J.; Peng, S.-Y. A Bypass-Switching SAR ADC With a Dynamic Proximity Comparator for Biomedical Applications. *IEEE J. Solid-State Circuits* **2018**, *53*, 1743–1754. [CrossRef]
28. Wang, X.; Zhou, X.; Li, Q. A High-Speed Energy-Efficient Segmented Prequantize and Bypass DAC for SAR ADCs. *IEEE Trans. Circuits Syst. Express Briefs* **2015**, *62*, 756–760. [CrossRef]
29. Abusleme, A.; Dragone, A.; Haller, G.; Murmann, B. Mismatch of lateral field metal-oxide-metal capacitors in 180 nm CMOS process. *Electron. Letters* **2012**, *48*, 286–287. [CrossRef]
30. Zhu, Y.; Chan, C.-H.; U, S.-P.; Martins, R.P. An 11b 450 MS/s Three-Way Time-Interleaved Subranging Pipelined-SAR ADC in 65 nm CMOS. *IEEE J. Solid-State Circuits* **2016**, *51*, 1223–1234. [CrossRef]

electronics

MDPI

Article

A Multispectral Backscattered Light Recorder of Insects' Wingbeats

Iraklis Rigakis [1], Ilyas Potamitis [2,*], Nicolaos-Alexandros Tatlas [1], Ioannis Livadaras [3] and Stavros Ntalampiras [4]

1. Department of Electrical and Electronics Engineering, University of West Attica, 12241 Athens, Greece; iraklis.rigakis@gmail.com (I.R.); ntatlas@uniwa.gr (N.-A.T.)
2. Technological Educational Institute of Crete, 71410 Heraklion, Greece
3. Institute of Molecular Biology and Biotechnology of the Foundation for Research and Technology Hellas (IMBB-FORTH), 71110 Heraklion, Greece; livadara@imbb.forth.gr
4. Department of Computer Science, University of Milan, 20133 Milan, Italy; stavros.ntalampiras@unimi.it

* Correspondence: potamitis@staff.teicrete.gr; Tel.: +30-28310-21900

Received: 10 January 2019; Accepted: 23 February 2019; Published: 2 March 2019

Abstract: Most reported optical recorders of the wingbeat of insects are based on the so-called extinction light, which is the variation of light in the receiver due to the cast shadow of the insect's wings and main body. In this type of recording devices, the emitter uses light and is placed opposite to the receiver, which is usually a single (or multiple) photodiode. In this work, we present a different kind of wingbeat sensor and its associated recorder that aims to extract a deeper representational signal of the wingbeat event and color characterization of the main body of the insect, namely: a) we record the backscattered light that is richer in harmonics than the extinction light, b) we use three different spectral bands, i.e., a multispectral approach that aims to grasp the melanization and microstructural and color features of the wing and body of the insects, and c) we average at the receiver's level the backscattered signal from many LEDs that illuminate the wingbeating insect from multiple orientations and thus offer a smoother and more complete signal than one based on a single snapshot. We present all the necessary details to reproduce the device and we analyze many insects of interest like the bee *Apis mellifera*, the wasp *Polistes gallicus*, and some insects whose wingbeating characteristics are pending in the current literature, like *Drosophila suzukii* and *Zaprionus*, another member of the drosophilidae family.

Keywords: Fresnel lens; wingbeat; insects; optoelectronics; bees; wasps; fruit flies; e-traps

---

1. Introduction

This work belongs to a broader context of applications that relate to automated insect surveillance of insects of economic and hygienic importance. In order to monitor the presence and density of insects, as well as design policies and apply measures, entomologists deploy a high number of traps that are currently checked manually [1]. Our goal is to automate the reporting procedure of sampled insect fauna without involving a human in the loop. To this end, we embed optoelectronic sensors in typical traps that, depending on the situation, count insects, discern sex and species of captured insects, report daily results, wirelessly, to remote servers, and update infestation maps, decision support systems and predictive analytics. To give some lucid examples of how things are already evolving in several application areas we are involved, we report the completed tasks as well as their technological readiness level (TRL) (EC-H2020 definition):

(a) In the context of the IoBee project (https://cordis.europa.eu/project/rcn/210011_en.html), e-gates applied in the entrance of beehives measure the bee traffic and discern the presence of a drone from a worker and inform for the case of an outgoing queen (TRL-9).

(b) In the context of the REMOSIS project (https://cordis.europa.eu/result/rcn/230808_en.html), optical counters of mosquitoes being sucked into commercial traps have been upgraded to sensors that discriminate sex and species (TRL-8).
(c) In the case of insect traps that are based on pheromones to attract targeted insects, such as the grain pitfall for stored-products insects, Picusan traps for the red palm weevil, Lindgren and funnel traps [2], an optical counter is incorporated, and the accuracy in insect counts relies on the effectiveness and specificity of the pheromone attractants (TRL-9).

Cases (a) and (c) are simple in the sense that they count insects that pass through specific constrictions and quantify their size based on the measured optical intensity variation between an emitter and a receiver. For these applications, we recommend a simple optical counter based on the extinction light. There are other cases, however, where there are no widely accepted pheromones that attract both sexes (entomologists are especially interested in female counts); therefore, a general food bait is used (i.e., in the case of some fruit flies or sent in the cases of mosquitoes). In such cases, we rely on the wingbeat of the incoming insect and the analysis of its frequency content to classify sex and/or species identity [3,4]. In Reference [5], we have demonstrated that a backscattered light signal originating from an insect is better that the extinction light provided by the same wingbeat event in terms of signal to noise ratio (SNR) and number of harmonics standing out of the noise floor. In this work, we elaborate on this finding and the new accomplishments are that we expand to a multispectral sensor configuration that integrates recordings from multiple orientations. The sensor aims to extract complementary information from the microstructural and melanization features of the wing and coloration of the main body of the species. The information contained in the samples of these recordings will provide complementary information and precise quantification of size on the difficult task of discerning morphologically similar insects whose wingbeat spectrum may overlap significantly in the frequency domain. Note that, although we present a stand-alone device as in References [6–8], it is designed in a way that is detachable from its base and its size is reconfigurable so that it can take different forms depending on the e-trap in which we are interested in embedding it.

There are currently two approaches based on optical technology (excluding camera-based vision): a) the e-traps [2–5], and b) the light detection and ranging (LIDAR) based approaches [9–12].

E-traps sample the insect fauna based on baits and usually aim at capturing target-specific insects. They try to locate the onset of an infestation, the correct timing to apply treatment, or to assess the after-treatment impact. LIDAR technology aims to characterize insects over larger distances and enable the mapping of densities and fluxes on very short time scales due to the large number of insect counts. E-traps can be made of low-cost elements such as LEDs, photodiodes, and acrylic lenses, and therefore, it is possible to deploy a large number of traps and still have a cost-effective monitoring plan. LIDAR technology is orders of magnitude more expensive and requires an external power supply. The techniques are suited for different kinds of studies.

The paper is organized as follows. First, the methodology to retrieve the optical signals and their frequency content is described in the "Materials and Methods" section. Then, we present experimental results based on recordings of different species of insects some of them never reported in the literature (i.e., *Zaprionus, Pollistes galicus*). Finally, we discuss, based on our experimental results, the possibilities of different spectral bands for gender and species identification.

2. Materials and Methods

It has been reported in LIDAR applications that near-infrared wavelengths (NIR), e.g., 808 nm, are affected by melanization, and that different spectral bands carry complementary information on the insect's main body and wings coloration [10–12]. We pursue this direction by developing a device (see Figure 1) that examines the possibility of extracting more information about the cast shadow based on backscattered light recordings of wingbeat events under different spectral bands. In this task, we use three different LEDs (one in the visible frequency range (450–700 nm), one at 810 nm, and one at 940 nm).

Figure 1. The multispectral sensor: (**left**) The disk in the middle is the Fresnel lens. The three LED's in each plate emit, in turn, visible light 450–700nm, and infrared at 810 nm and 940 nm. All eight LEDs of each spectral band are lit simultaneously for 20.8 microseconds (i.e., three circular arrays of eight LEDs each). All LEDs' supporting plates have a 68.5° orientation with respect to the Fresnel plane (**right**). The photodiode is placed at the focal point of the Fresnel lens.

Note that this configuration is naturally expandable to more wavelengths that are distinct, but in this work, we are constrained by the cost of the LEDs, their operational wavelength, and their capability to operate at high frequencies. We illuminate the wingbeating insect from different orientations and we average on a per wavelength basis to achieve a smoother signal. In brief, the main concept of the sensor is as follows: the central processing unit (CPU) turns on the three circular arrays of eight LED's successively, each having the same wavelength. The Fresnel lens focuses the backscattered light stemming from the wingbeating insect onto the photodiode. The photodiode directs its output to the demultiplexer that has three sample-and-hold circuits. The demultiplexer sends its output to a multichannel analog to digital converter (ADC) and the latter back to the CPU, and finally a wav-type recording to the secure digital (SD) card (see also Figure 2). In detail, the CPU (ST STM32L4R7 Microelectronics. 39, Chemin du Champ des Filles, Geneva, CH 1228, Switzerland) (Figure A1) produces the synchronization signals for all system units, receives the digital words from the analog to a multichannel analog to digital converter (ADC), and stores the signals to a three-channel 16 KHz 24-bit wav in the SD card of the recorder. We place the photodiode (TEMD5080X01, Vishay Intertechnology, Malvern, Pennsylvania, USA) at the focal point of the Fresnel lens (Fresnel Technologies Inc., 101 W. Morningside Drive Fort Worth, TX 76110, Part number: 3*). The three LED types are white: GW CS8PM1.PM and 810 nm, SFH4780S (both from Osram, Munich, Germany) and the 940 nm L1I0 (LUMILEDS, San Jose, CA 95131, USA) and emit for 20.8 µs. The ADC (Figure A2) is based on the AD7768-4 IC (Analog Devices, One Technology Way Norwood, MA, USA). The ADC receives the three analog outputs of the demultiplexer (Figure A3) and converts them to digital words. The output signal from the photodiode is amplified (Texas Instruments, Dallas, Texas 75266-0199, OPA380 transimpedance amplifier (Figure A4)) and then driven to the demultiplexer. It is worth mentioning that the feedback loop in Figure A4 ensures the possibility of operating the device in the presence of the sun and allows for considerable power saving in field operation. The demultiplexer (Texas Instruments analog switch TS12A44514 and OPAMP OPA4376 as a Sample & Hold amplifier) separates the photodiode's output to three different signals, one for each band, and drives it to the three-channel ADC. The LED drivers (Figure A5) produce consecutive pulses to three LED arrays (Figure A6) with constant current controlled by the CPU (see Figure 3 for the timing of operations). The CPU also controls the current level of each wavelength. It is based on Infineon (Am Campeon 1-15,

85579 Neubiberg, Germany) metal-oxide-semiconductor field-effect transistor IRF7341, on OPAMPs ADA4805 of Analog Devices (Norwood, MA 02062) and Texas Instruments TS12A44514 analog switches. The multichannel sigma-delta ADC converter AD7768-4 receives the three analog audio signals and sends the digital words to the CPU using the time division multiplexing (TDM) output.

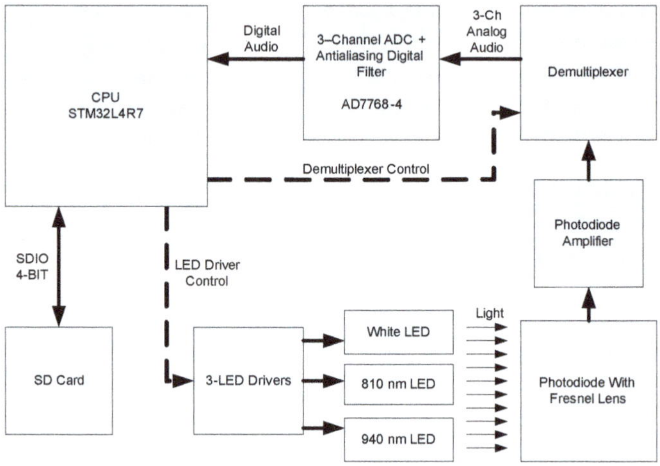

Figure 2. Block diagram of the multispectral device. The system is controlled from an STM32L4R7 ARM CPU of ST. The LED drivers produces the sequentially current pulses of each LED.

Powering of the LEDs is carried out through the TPS54302 IC of Texas Instruments (Figure A7 left) and the device state (trigger, SD card, power status) is indicated in the front LEDs of the device (Figure A7 right).

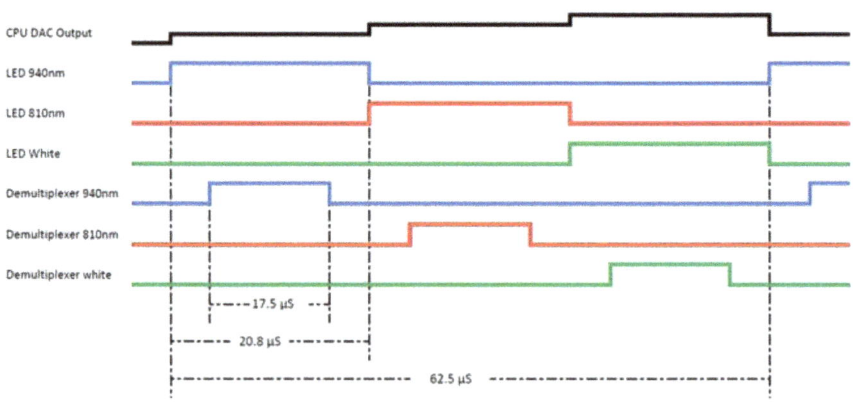

Figure 3. The CPU digital to analog converter (DAC) output defines the current level for each light pulse per spectral band. The signals LED 940 nm, LED 810 nm, and LED White initiate the output of the LED drivers for each spectral band. The signals: demultiplexer 940 nm, demultiplexer 810 nm, and demultiplexer-white initiate the corresponding sample and hold amplifier so that the photodiode is demultiplexed.

We show an internal picture of the completed device in Figure 4 and a cost analysis in Table A1 in Appendix A.

Figure 4. The multispectral recorder prototype. We circle and annotate the main components of the recorder.

Regarding the software, the embedded microprocessor runs a constantly looping program that processes data captured by the sensors. The board is programmed in C/C++. The digital audio output from the optoelectronic sensor is copied to six circular buffers. The first three buffers are used to monitor the backscattered signal's root-mean-square (RMS) using a window of 128 samples (16 ms in 8 kHz sampling rate). The other three circular buffers of 32 kwords (a word equals 32 bit in this processor) each store the recoding of each band. If any of these bands exceeds a common threshold, it triggers the recording process for all bands (see Figure 5). The recordings of the signal are coded in 24-bit resolution, at an 8 kHz sampling rate. The first 20% of the samples are drawn before and up to the triggering point and 80% after that point in order to ensure that the onset of a wingbeat event is not lost. Wingbeat events are short in time for fast flying insects and one cannot afford discarding any useful part of the signal such as the onset. The sampling frequency, window length, and triggering threshold are pre-stored in the SD-card of the system and the settings (i.e., sampling frequency, triggering level, and record length in samples) are read once from the SD card during powering-on. The software is written in C language using the IAR Embedded workbench. The programming of the flash memory was carried out using the ST-Link V2 programmer. The code initialization was done using the STM32CubeMX of ST. For programming the peripheral sub-components, such as the SD and ADC, we made use of the STM32 HAL drivers. The control signals and data transfers were done using the direct memory access (DMA) controller of STM32L4R7.

Regarding the insect specimens, we collected the insect species *Zaprionus* (Diptera: Drosophilidae), *Drosophila suzukii* (Diptera: Drosophilidae), and *Drosophila melanogaster* (Diptera: Drosophilidae) from the area Gouves, Chersonisoss Crete, March–April 2017. The insects have been transported to an entomological laboratory to breed and reproduce. Their diet contained sugar, yeast, agar, cornflower, and nipogen. Their breeding conditions had been kept constant at 25 °C, 60% relative humidity following a cycle of 14 h of light and 10 h of darkness. After breeding, the adult insects have been transferred to different cages 50 × 60 cm and the device has been inserted in turn into the cages.

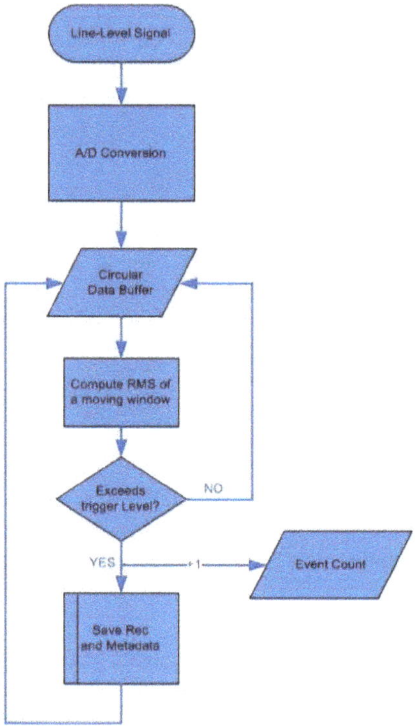

Figure 5. A diagram explaining the recording procedure. The signal is continuously stored in a cyclic buffer and the RMS value of a window of 128 samples is monitored. Once the RMS value exceeds a threshold, it triggers a recording session. A total of 20% of the samples are taken prior to the triggering point from the buffer and 80% after this event to ensure that the wingbeat onset is not missed. Metadata is composed of a timestamp, humidity, and temperature stamps that are passed to the filename.

3. Results

Figure 6 demonstrates the various ways we can use the device. One can tether an insect with an inverted tweezer inside the sensor's probe volume, or confine the insect in a Plexiglas box, or insert the sensor in an insectary cage that contains the insects of interest without the Plexiglas box.

In this work, we present multispectral recordings from various flying insects using the confinement into a transparent box. The reflected light relates to the refractive index of the wing membrane and the glittering of the insect [10–12]. Therefore, the intensity of light in the absence of an insect is theoretically zero and practically equal to the minimum light reflection stemming from the black termination plane. We performed many recording sessions for various insects and the main results of this experiment show that:

(a) Walking insects are efficiently detected due to the backscattered light from their body.
(b) We visually confirmed that all wingbeat events observed in the confinement box were registered, and the frequency content of the recording was clearly resolved.
(c) The wingbeat "signature" of insects in the spectral domain is consistent and repeatable with small interspecies variation.
(d) The signal to noise ratio is very high (30–35 dB) and the number of harmonics often exceeds 20 (see Figure 7a–d).
(e) One can see that the power spectral density (PSD) of the different spectral bands are not identical (see Figure 7a–d), and this is an encouraging observation as we aim at extracting complementary

information. We derive the PSD estimate of the discrete-time signals using Welch's averaged, modified periodogram method using a Hanning window of 512 samples, 50% overlap, and 512 samples Fast Fourier Transform (FFT) at 8 kHz sampling frequency. The spectral peaks, corresponding to the fundamental frequency and their harmonics reside on the same frequencies as expected since they relate to the same insect and wingbeat event. However, the details of the spectral signature, especially at high frequencies, are different for each spectral band.

(f) We can discern morphologically different insects, such as the bee, the wasp, and the fruit flies, as they have very different wingbeating frequencies (*Apis mellifera*: 190 Hz, *Polistes gallicus*: 124 Hz, *Zaprionus*: 220 Hz, *D. suzukii*: 250 Hz, *D. melanogaster*: 250 Hz) and distribution of power over harmonics. We took all measurements at the same temperature.

(g) The light intensity close to the DC frequencies can quantify the size of the insect. In terms of physical size and in descending order the insects are ranked as follows: bee–>wasp–>*Zaprionus*–>*D. suzukii*–>*D. melanogaster*. The PSD plots in Figure 7 follow the same ranking. Note that size classification is correct not only for the gross cases of a bee versus fruit flies, but also among fruit flies, paving the way for automatic discrimination of similar fruit flies.

(h) The spectral tilt in the PSD of large insects has a slope, whereas in small insects, it is more flat and we attribute this to their main body contribution.

(i) The current implementation is sensitive to the AC frequency of artificial light and further development is needed to make it noise immune.

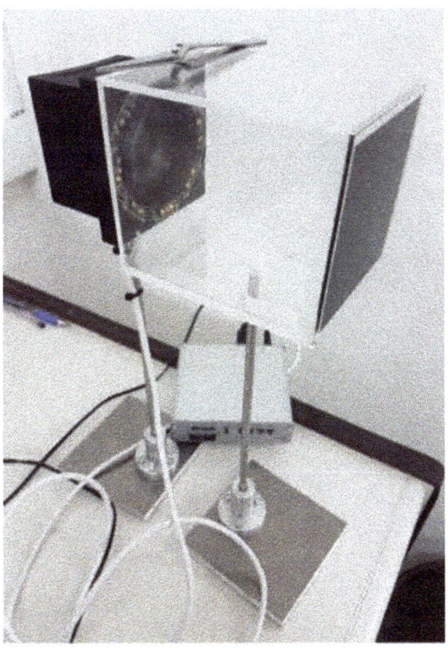

Figure 6. The three ways to use the suggested sensor and its associated recorder: by tethering an insect inside the probe volume of the sensor (e.g., by holding its legs with an inverted tweezer), by confining the insect in a transparent cage that is large enough to allow the insect to fly, or by inserting the sensor in an insectary cage containing a free-flying insect. Notice the black termination plane on the right of the Plexiglas (MAXiBLACK, Advanced Coating Products, Acktar Store LTD, Kiryat-Gat, Israel).

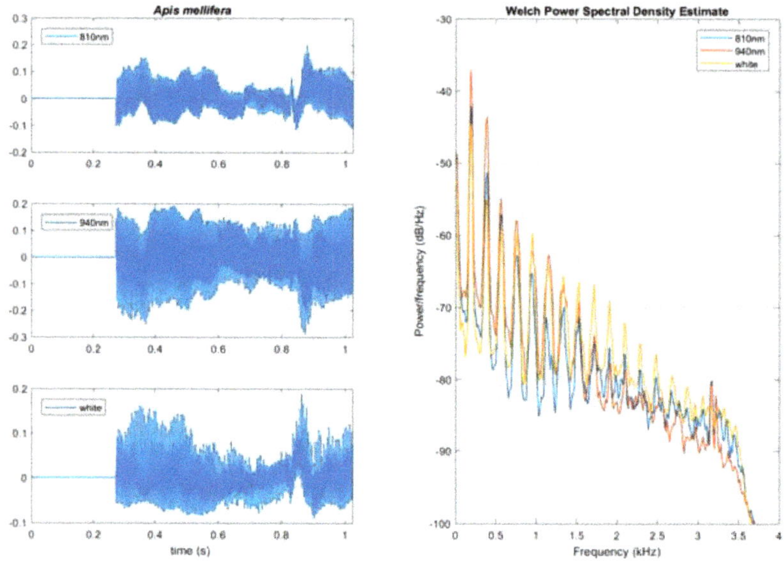

(**a**)

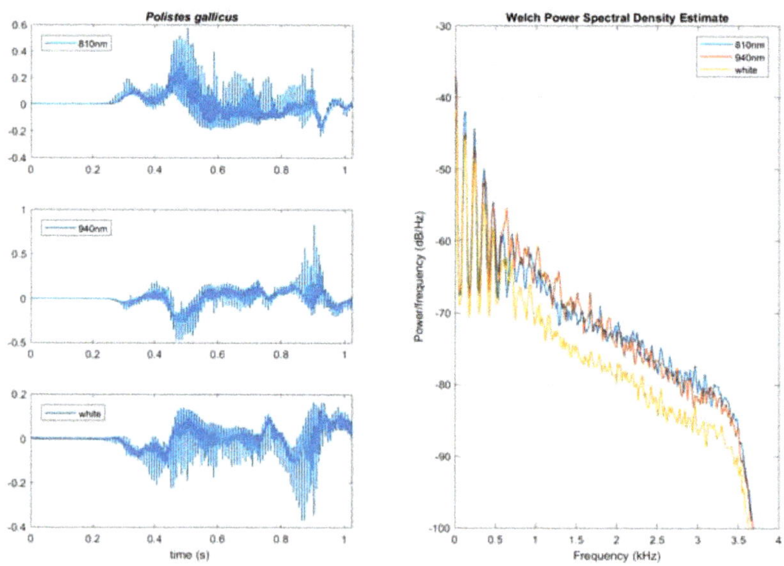

(**b**)

Figure 7. *Cont.*

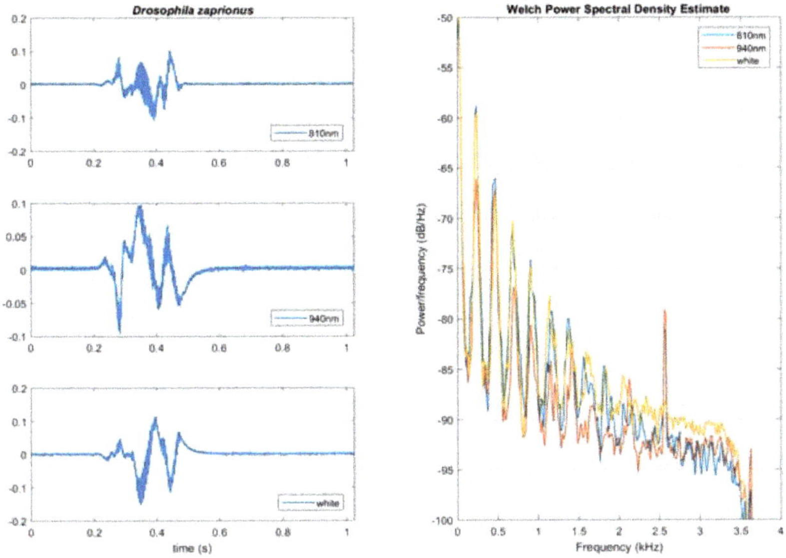

(**c**)

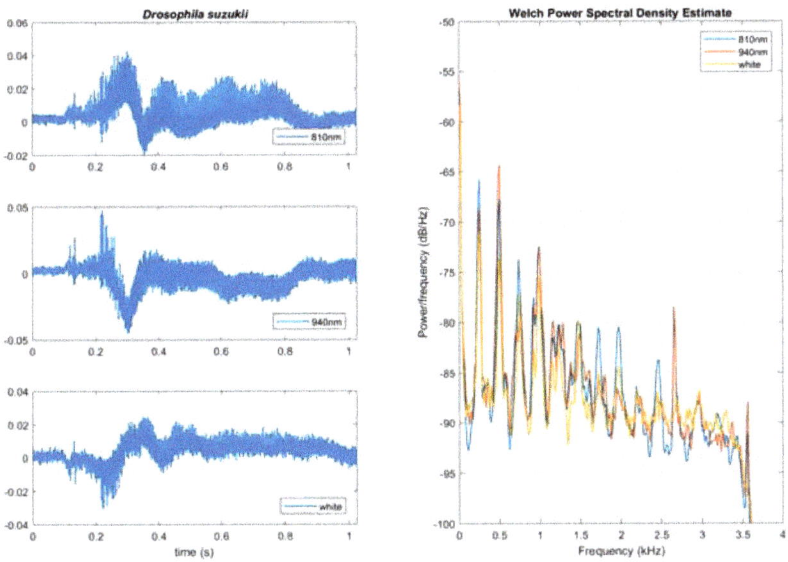

(**d**)

Figure 7. *Cont.*

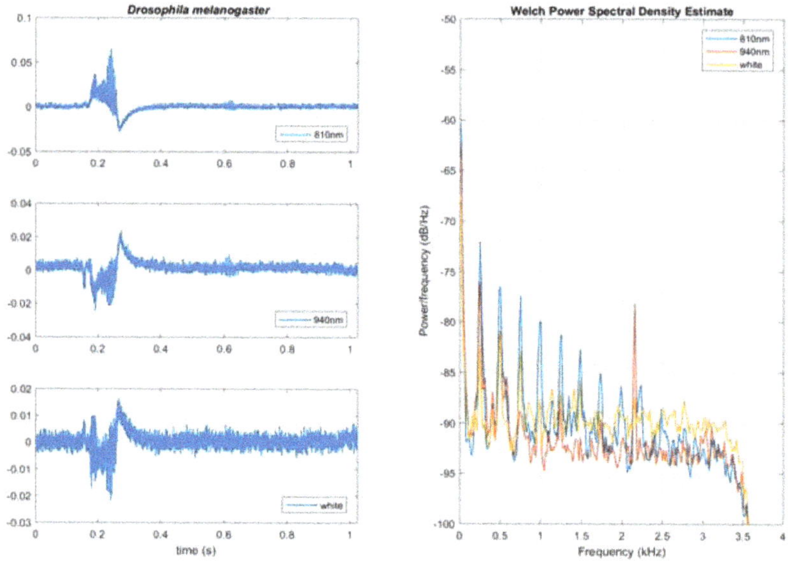

(e)

Figure 7. Optical wingbeat recording of backscattered light wings and main body using three different spectral bands (810 nm, 940 nm, white). Time domain signal at 8 kHz sampling rate for each band on the left column of each sub-figure and the Welch power spectral density on the right column. Recordings are treated as audio and the amplitude in y-axis is normalized between $[-1, 1]$. (**a**) The bee *Apis mellifera*, (**b**) the wasp *Polistes gallicus*, (**c**) the fruit fly *Zaprionus*, (**d**) the fruit fly *Drosophila suzukii*, and (**e**) the fruit fly *Drosophila melanogaster*.

4. Discussion

Undoubtedly, any sensor type related to the context of our application, such as microphones and vision cameras, have advantages and disadvantages [13,14]. Multispectral imaging has been suggested in a different context to our work in agriculture and entomology [15–17]. From our point of view, optical sensors are the suitable choice for use in electronic insect gates and automatized insect traps working in the field because they record intermittently, i.e., on per event basis, and only if their probe volume (that can be shaped with proper lenses) is interrupted in contrast to the continuous recording of microphones. Microphones receive continuous input from an uncontrolled and unknown number of audio sources in the field and are not generally suitable for field applications. The proposed multispectral sensors do not require the bandwidth of a vision camera and do not face the difficulty of a photograph of a pile of insects that are not easily discernable in detail. Fresnel lenses provide an affordable way to collimate light, and therefore, it is possible to effectively avoid interferences from the sun or diffuse light sources. Using the sensor presented in this work, walking insects (e.g., bees and wasps) were efficiently detected and their presence was registered in the power of low frequencies around the DC level. The power level of the received light was suitable to rank insects according to their size. The wingbeat event could be easily discerned from a walking event due to the harmonic structure of the power spectral density of the former and the flat spectrum of the latter. The wingbeat "signature" of all insects in the spectral domain was consistent and repeatable [18]. The signal to noise ratio of the backscattered light sensor was at 30–40 dB and often reached 20–30 harmonics. Multispectral signatures look richer than the ones provided by simple one-band sensors but their advantage on classification improvement needs to be clarified and quantified with large-scale experiments (see References [19–21] for related work).

Future work will focus on different wavelengths with an aim to discern mosquitoes that have had a blood meal, or are dyed with fluorescent dust or carry a marker gene in the context of the sterile insect technique. The work in Reference [22] demonstrated that is possible to discern with high accuracy whether a mosquito carries a virus load based on NIR spectroscopy. While our current application constraints do not allow us to reach this level of analysis, our ultimate goal is to finally embed this kind of sensor in commercial mosquito traps. With the advance of high rate RGB wavelength demultiplexers [23] and all-optical neural networks [24], we envision that the size of the sensors will become smaller and artificial intelligence tools will be embedded in smart traps that will provide a detailed analysis of incoming insects based on their back-scattered multi-spectral signature. A dispersed network of "e-flowers" like the one depicted in Figure 1, when deployed in the field, could unobtrusively sample the insect's fauna and report on insect densities that can be correlated to pollination studies (e.g., estimate insect counts and distribution of bees) or assess agricultural risks, e.g., due to aphids.

Author Contributions: Conceptualization, I.P.; methodology, I.P. and I.R.; software N.-A.T.; data curation, S.N.; writing—original draft preparation, I.P.; writing—review and editing, I.P. and N.-A.T.; visualization, S.N., I.L. bred the fruit flies recorded in this work. All authors have contributed have contributed equally to the work reported.

Funding: This research was funded by the European Commission H2020-EU.2. - PRIORITY 'Industrial leadership, H2020-EU.3. - PRIORITY' societal challenges, project IoBee: Beehive health IoT application to fight Honey Bee Colony Mortality, grant number 760342.

Acknowledgments: We thank Michael Weber [5] for fruitful discussions on the backscattered signal from insects.

Conflicts of Interest: The authors declare no conflict of interest. The funders had no role in the design of the study; in the collection, analyses, or interpretation of data; in the writing of the manuscript, or in the decision to publish the results.

Appendix A

Hereinafter, we present details schematics of the electronic device described in the text. Data supplemental to the main text is included in the directory of recordings.

Table A1. Cost break-down of the multispectral sensor and recorder (Euros).

Item		Manufacturer	Price
Photodiode 1X TEMD5080X01		Vishay Intertechnology, Malvern, USA	1.56E
LED GW CS8PM1.PM (8 pieces)		Osram, Munich, DE	10.48E
LED 810nm, SFH4780S (8 pieces)		Osram, Munich, DE	39.02E
LED 940nm, L1I0-094006000 (8 pieces)		LUMILEDS, San Jose, USA	23.68E
Fresnel lens, Part number: 3*		Fresnel Technologies Inc., USA	25E
CPU 1X STM32L4R7		Microelectronics. Geneva, Switzerland	12.67E
Four-channel ADC AD7768-4		Analog Devices	16.89E
ADC Drivers 3X THS4531		Texas Instruments	7.86E
Demultiplexer	1X OPA4376	Texas Instruments	3.27E
	1X TS12A44514	Texas Instruments	1.35E
LED Drivers	3X ADA4805	Analog Devices	11.52E
	1X TS12A44514	Texas Instruments	1.35E
	2X IRF7341	Infineon	3.12E
Power Supplies	3X LP2985-33	Texas Instruments	1.48E
	1X ADP7104	Analog Devices	3.58E
	1X TPS76901	Texas Instruments	0.74E
	1X TPS54302	Texas Instruments	1.65E
Passives	Resistors, Capacitors		10E

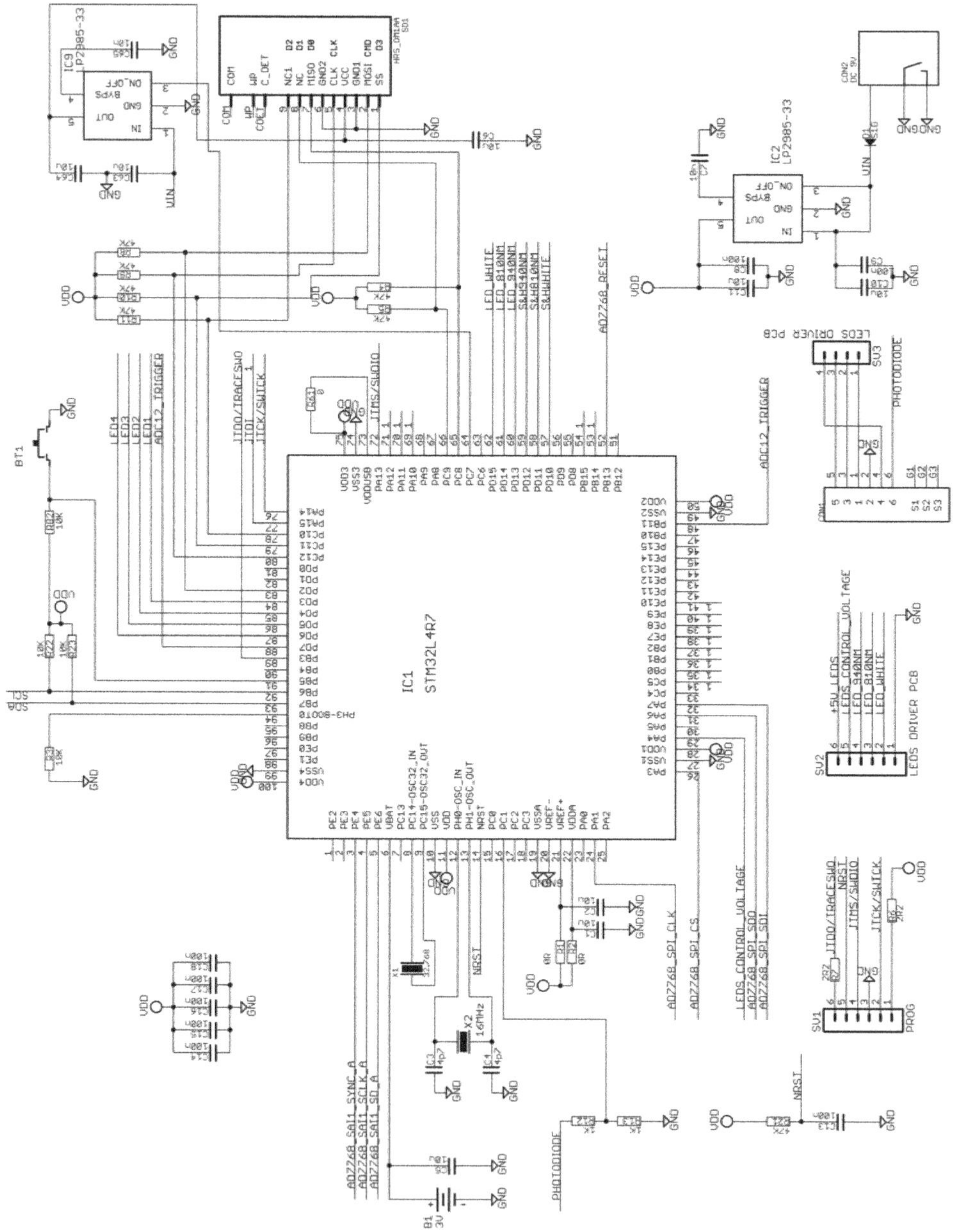

Figure A1. Microprocessor unit. The microprocessor STM32L4R7 operates at 3.3 V and is synchronized by the 16 MHz clock provided by the oscillator X2. It controls all functions of the recorder: storing at the SD card, controlling the ADC, and production of the synchronization signals. The internal real time clock is powered from battery B1 and is clocked by X1 (32.768 kHz crystal) that time-stamps detection events. The CPU also drives the LEDs and the user interface through the function button and the status LEDs, named LED1 to LED4. The connector SV1 is used for programming the CPU, and the SV2 and SV3 is used for the connection of LEDs driver PCB to main PCB. The connector CON1 is used for the connection of the sensor with the device.

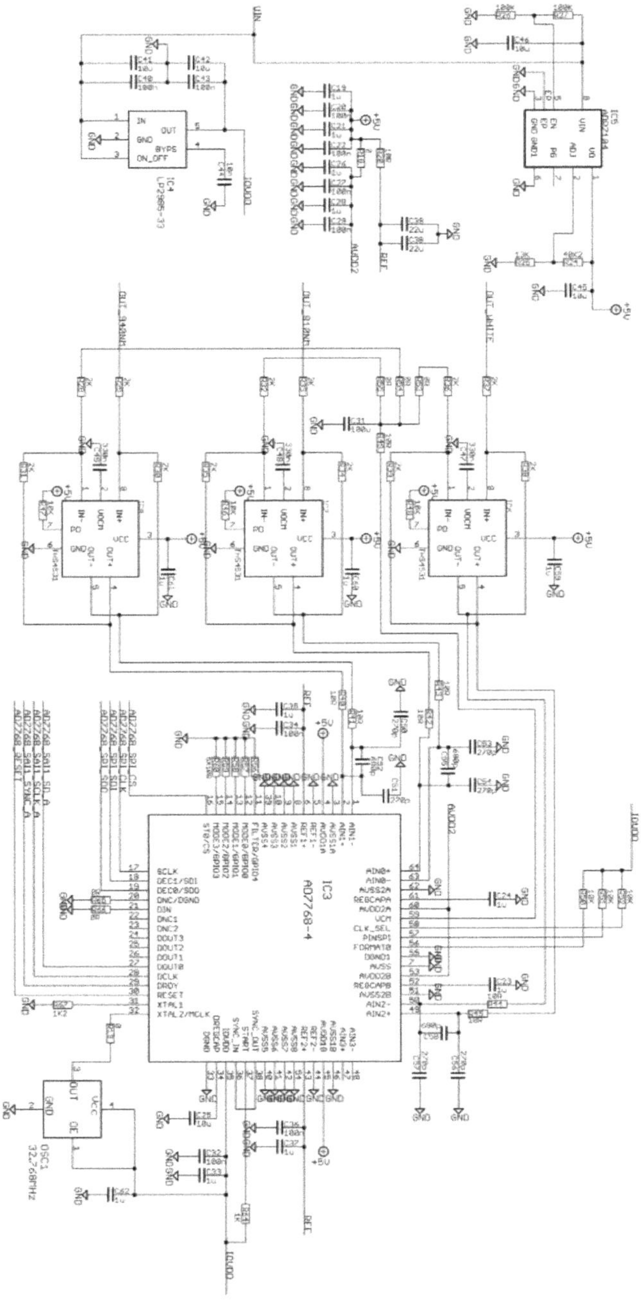

Figure A2. Three-channel ADC. The ADC is based on AD7768-4 (IC3) and converts the analogue audio signals to digital words. It is a four-channel 24-bit sigma delta ADC and we use the three of them. The three analog inputs of ADC are driven by three THS4531 differential amplifiers (ADC drivers). All functions of ADC (sampling rate, digital filtering, etc.) are controlled by the main CPU. The control interface is SPI and the digital audio data is transferred using the TDM protocol.

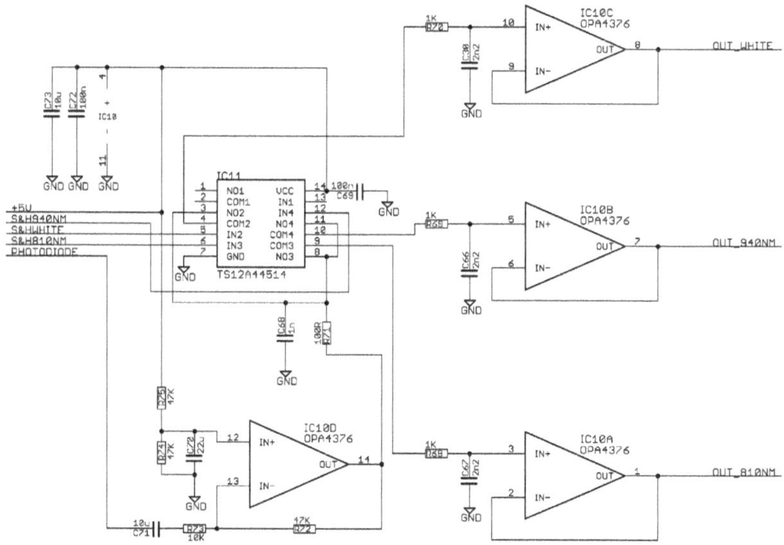

Figure A3. Demultiplexer. The demultiplexer receives the photodiode amplifier output and sends three dedicated analogue audio signals to the ADC.

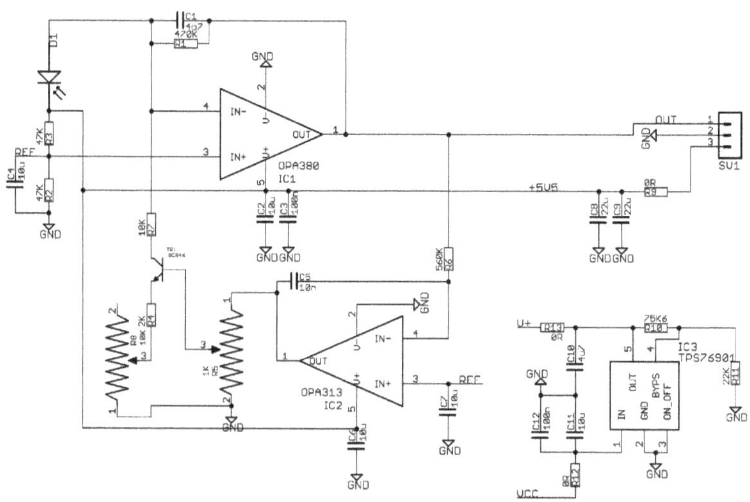

Figure A4. Photodiode receiver. It is based on the OPA380 transimpedance amplifier. The IC9 with the TR1 transistor functions as a feedback amplifier. Only the DC component of the IC8 output passes through IC9. The output of IC9 controls the conductivity of TR1, which in turn, subtracts the photodiode's DC current. Therefore, the input of IC8 is contains only the AC component of the current. This way, the sensor can function in the presence of sunlight. Without the feedback loop, it would not be possible to function properly due to the high amplification of the OPA380 (470K feedback resistor).

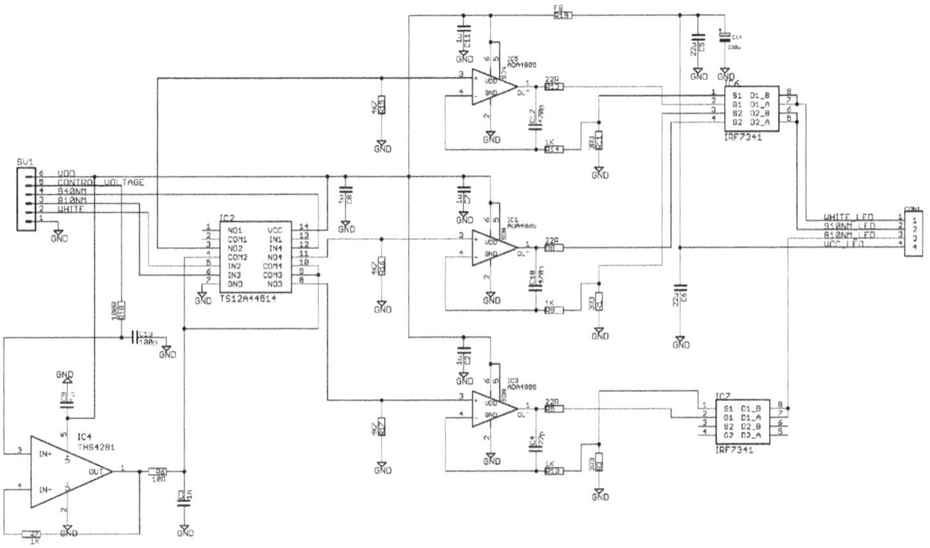

Figure A5. Three-channel LED driver. The led driver is based on IC1, IC3, IC5, IC6, and IC7. The voltage in the input of each ADA4805 controls the current of each LED array. This circuit is controlled by the CPU that is connected to SV1. The control voltage of the drivers is different for each spectral band to account for the different sensitivity of the photodiode responding to different spectral bands and is produced by the DAC of the main processor. Therefore, with the help of the analogue switch IC2 and the timing signals for the 940nm, 810nm, and white LEDs, the different voltages, in turn, give input to the ADA4805.

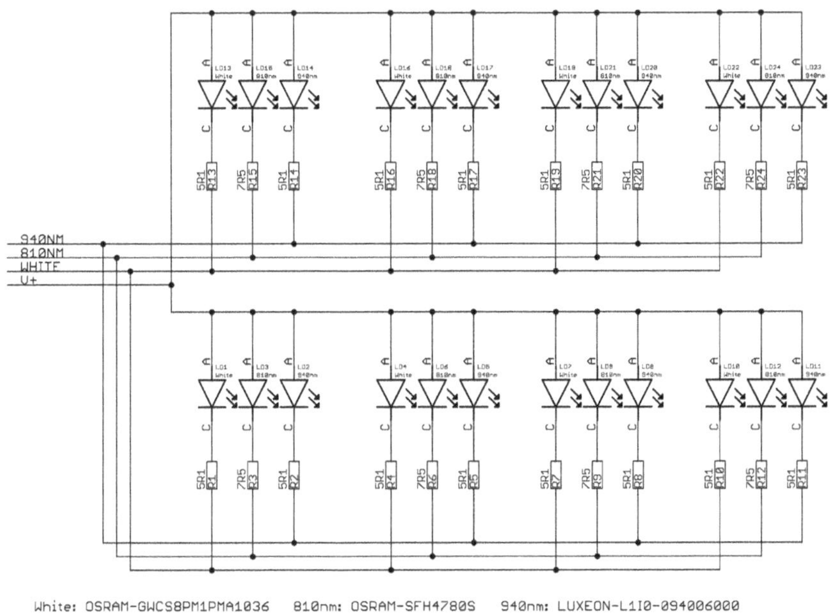

Figure A6. LEDs array. One LED per spectral band.

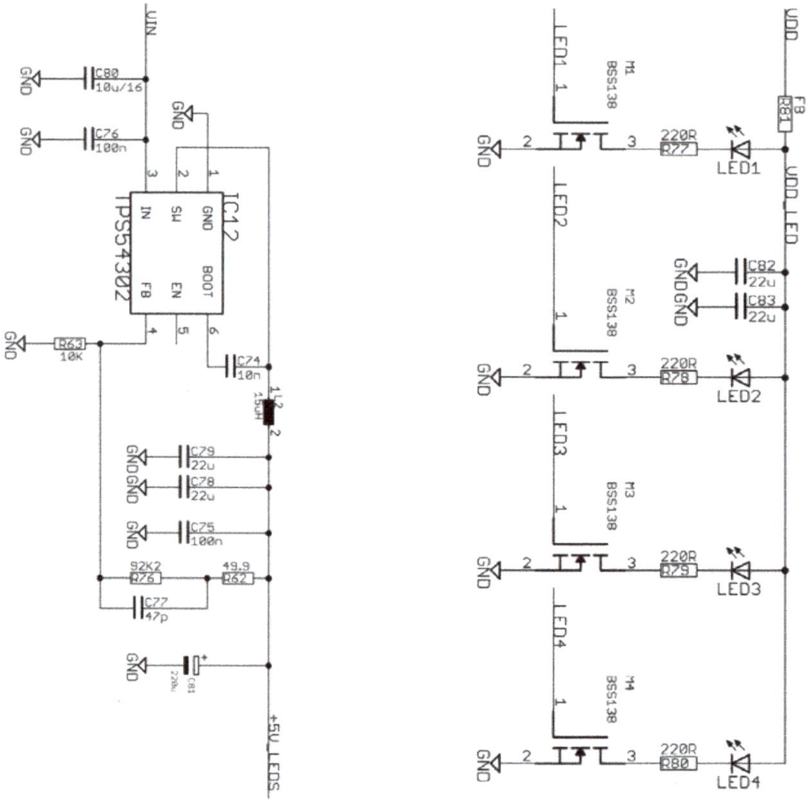

Figure A7. (**left**) Sensor LEDs' power supply. (**right**) Indicator LEDs. The LED arrays are powered with 5 V. The SMPS TPS54302 (IC12) produces this voltage. The indicator LEDs are controlled by the CPU through the MOSFETs M1, M2, M3, and M4.

References

1. Goldshtein, E.; Cohen, Y.; Hetzroni, A.; Gazit, Y.; Timar, D.; Rosenfeld, L.; Grinshpon, Y.; Hoffman, A.; Mizrach, A. Development of an automatic monitoring trap for the Mediterranean fruit fly (Ceratitis capitata) to optimize control applications frequency. *Comput. Electron. Agric.* **2017**, *139*, 115–125. [CrossRef]
2. Potamitis, I.; Eliopoulos, P.; Rigakis, I. Automated Remote Insect Surveillance at a Global Scale and the Internet of Things. *Robotics* **2017**, *6*, 19. [CrossRef]
3. Chen, Y.; Why, A.; Batista, G.; Mafra-Neto, A.; Keogh, E. Flying insect detection and classification with inexpensive sensors. *J. Vis. Exp.* **2014**, *92*, e52111. [CrossRef] [PubMed]
4. Mukundarajan, H.; Hol, F.J.H.; Castillo, E.A.; Newby, C.; Prakash, M. Using mobile phones as acoustic sensors for high-throughput mosquito surveillance. *Elife* **2017**, *6*, e27854. [CrossRef] [PubMed]
5. Potamitis, I.; Rigakis, I.; Vidakis, N.; Petousis, M.; Weber, M. Affordable Bimodal Optical Sensors to Spread the Use of Automated Insect Monitoring. *J. Sens.* **2018**, *2018*, 3949415. [CrossRef]
6. Potamitis, I.; Rigakis, I. Novel Noise-Robust Optoacoustic Sensors to Identify Insects Through Wingbeats. *IEEE Sens. J.* **2015**, *15*, 4621–4631. [CrossRef]
7. Raman, D.R.; Gerhardt, R.R.; Wilkerson, J.B. Detecting Insect Flight Sounds in the Field: Implications for Acoustical Counting of Mosquitoes. *Agric. Biosyst. Eng. Pub.* **2007**, *57*, 1481–1485.
8. Santos, D.A.A.; Teixeira, L.E.; Alberti, A.M.; Furtado, V.; Rodrigues, J.J.P.C. Sensitivity and Noise Evaluation of an Optoelectronic Sensor for Mosquitoes Monitoring. In Proceedings of the 2018 3rd International Conference on Smart and Sustainable Technologies (SpliTech), Split, Croatia, 26–29 June 2018; pp. 1–5.

9. Genoud, A.P.; Basistyy, R.; Williams, G.M.; Thomas, B.P. Optical remote sensing for monitoring flying mosquitoes, gender identification and discussion on species identification. *Appl. Phys. B* **2018**, *124*, 46. [CrossRef] [PubMed]
10. Gebru, A.; Jansson, S.; Ignell, R.; Kirkeby, C.; Prangsma, J.C.; Brydegaard, M. Multiband modulation spectroscopy for the determination of sex and species of mosquitoes in flight. *J. Biophotonics* **2018**, *11*, e201800014. [CrossRef] [PubMed]
11. Gebru, A.K.; Rohwer, E.G.; Neethling, P.; Brydegaard, M.S. Investigation of atmospheric insect wing-beat frequencies and iridescence features using a multispectral kHz remote detection system. *J. Appl. Remote Sens.* **2014**, *8*. [CrossRef]
12. Gebru, A.; Jansson, S.; Ignell, R.; Kirkeby, C.; Brydegaard, M. Multispectral polarimetric modulation spectroscopy for species and sex determination of malaria disease vectors. In Proceedings of the 2017 Conference on Lasers and Electro-Optics (CLEO), San Jose, CA, USA, 14–19 May 2017; pp. 1–2.
13. Chen, C.-P.; Chuang, C.-L.; Jiang, J.-A. Ecological Monitoring Using Wireless Sensor Networks—Overview, Challenges, and Opportunities. In *Smart Sensors, Measurement and Instrumentation; Book Section in Advancement in Sensing Technology*; Mukhopadhyay, S.C., Jayasundera, K.P., Eds.; Springer: Berlin/Heidelberg, Germany, 2013; pp. 1–21.
14. Prince, P.; Hill, A.; Piña Covarrubias, E.; Doncaster, P.; Snaddon, J.L.; Rogers, A. Deploying Acoustic Detection Algorithms on Low-Cost, Open-Source Acoustic Sensors for Environmental Monitoring. *Sensors* **2019**, *19*, 553. [CrossRef] [PubMed]
15. Fennell, J.; Veys, C.; Dingle, J.; Nwezeobi, J.; van Brunschot, S.; Colvin, J.; Grieve, B. A method for real-time classification of insect vectors of mosaic and brown streak disease in cassava plants for future implementation within a low-cost, handheld, in-field multispectral imaging sensor. *Plant Methods* **2018**, *14*, 82. [CrossRef] [PubMed]
16. Ma, T.; Kobori, H.; Katayama, N.; Tsuchikawa, S. Non-Destructive Inspection of Insects in Chocolate Using near Infrared Multispectral Imaging. *J. Near Infrared Spectrosc.* **2016**, *24*, 391–397. [CrossRef]
17. Vasas, V.; Hanley, D.; Kevan, P.G.; Chittka, L. Multispectral images of flowers reveal the adaptive significance of using long-wavelength-sensitive receptors for edge detection in bees. *J. Compar. Physiol.* **2017**, *203*, 301–311. [CrossRef] [PubMed]
18. Ellington, C. The novel aerodynamics of insect flight: Applications to micro-air vehicles. *J. Exp. Biol.* **1999**, *202*, 3439–3448. [PubMed]
19. Santos, D.A.; Rodrigues, J.J.; Furtado, V.; Saleem, K.; Korotaev, V. Automated electronic approaches for detecting disease vectors mosquitoes through the wing-beat frequency. *J. Clean. Prod.* **2009**, *217*, 767–775. [CrossRef]
20. Ouyang, T.H.; Yang, E.C.; Jiang, J.A.; Lin, T.T. Mosquito vector monitoring system based on optical wingbeat classification. *Comput. Electron. Agric.* **2015**, *118*, 47–55. [CrossRef]
21. Van Roy, J.; De Baerdemaeker, J.; Saeys, W.; De Ketelaere, B. Optical identification of bumblebee species: Effect of morphology on wingbeat frequency. *Comput. Electron. Agric.* **2014**, *109*, 94–100. [CrossRef]
22. Fernandes, J.N.; Dos Santos, L.M.; Chouin-Carneiro, T.; Pavan, M.G.; Garcia, G.A.; David, M.R.; Beier, J.C.; Dowell, F.E.; Maciel-de-Freitas, R.; Sikulu-Lord, M.T. Rapid, noninvasive detection of Zika virus in *Aedes aegypti* mosquitoes by near-infrared spectroscopy. *Sci. Adv.* **2018**, *4*, eaat0496. [CrossRef] [PubMed]
23. Dadabayev, R.; Shabairou, N.; Zalevsky, Z.; Malka, D. A visible light RGB wavelength demultiplexer based on silicon-nitride multicore PCF. *Opt. Laser Technol.* **2019**, *111*, 411–416. [CrossRef]
24. Shabairou, N.; Cohen, E.; Wagner, O.; Malka, D.; Zalevsky, Z. Color image identification and reconstruction using artificial neural networks on multimode fiber images: Towards an all-optical design. *Opt. Lett.* **2018**, *43*, 5603–5606. [CrossRef] [PubMed]

electronics

MDPI

Article

A Highly Robust Interface Circuit for Resistive Sensors

Emmanuel Gómez-Ramírez [1,*], L. A. Maeda-Nunez [1], Luis C. Álvarez-Simón [2] and F. G. Flores-García [3]

1. CONACyT—Tecnológico Nacional de México—Instituto Tecnológico de la Laguna, Torreón 27000, Mexico; luis.maeda@gmail.com
2. CONACyT—Universidad Autónoma del Estado de México, Ecatepec de Morelos 55020, Mexico; alvarez.simon.dr@gmail.com
3. Instituto Tecnológico de la Laguna, Torreón 27000, Coahuila, Mexico; francisco.floresgarcia@gmail.com

* Correspondence: egomez.ram@gmail.com or egomez@conacyt.gob.mx ; Tel.: +52-222-173-8794

Received: 17 January 2019; Accepted: 19 February 2019; Published: 28 February 2019

Abstract: The signal from a resistive sensor must be converted into a digital signal to be compatible with a computer through an interface circuit. Resistance-to-Period converter, used as interface, is preferred if the resistance variations are very large. This paper presents the structure of an interface circuit for resistive sensors that is highly robust to component and power supply variations. Robustness is achieved by using the ratiometric approach, thus complex circuits or highly accurate voltage references are not necessary. To validate the proposed approach, a prototype was implemented using discrete components. Measurements were carried out considering a variation of ±35% in the single supply voltage and a range from 1 kΩ to 1 MΩ.

Keywords: Resistance-to-Period converter; robust read-out circuits; ratiometric technique

1. Introduction

In recent years, the demand for sensors has increased in many areas, from medical and consumer electronics to automotive and industrial applications. In particular, resistive sensors are widely used in the industry for the measurement of displacement, strain, flow, force, pressure, temperature, light, weight, humidity, gas concentration and moisture, among others. Their resistance may vary from a few tens Ω such as thermistors, strain gauges, Resistance Temperature Detectors (RTD), piezo-resistive sensors, etc., to several MΩ such as various gas chemiresistive sensors, light dependent resistors (LDR), soil moisture, etc. To assess the resistive parameter, voltage may be applied across the sensor while the current is read or vice versa. Then, this voltage or current must be converted into a digital domain to be compatible with a computer, DAQ system, microcontroller or microprocessor using an interface.

According to [1], direct measurement of resistance changes can be done by two different ways: In the case of small resistance variations, circuits based on voltage dividers and Wheatstone bridges followed by precision differential or instrumentation amplifiers to reduce the offset voltage are used. This results in large and complex configurations and linearization techniques must be applied due to the intrinsic limitation in the dynamic range [2–4]. In contrast, quasi-digital converters, i.e., resistance-to-frequency [5,6], -period [7,8] or -duty-cycle [9] converters, are preferred if the resistance variations are very large. This converters not only provide a wider dynamic range but also simplify interfacing to digital systems, as no analog-to-digital converters (ADCs) are required [10,11]. In this way, the resistance is measured indirectly using a simple digital counter.

Most of the quasi-digital converters proposed in the literature focus on achieving a wide dynamic range (e.g., [12–16]), whereas other issues such as robustness and simplicity have not been fully addressed, as required for low-cost, low-power readout circuits for practical implementations.

Accuracy and robustness of most quasi-digital converters strongly depend on a bandgap voltage reference or on high performance building blocks, such as low-offset and high-speed comparators, resulting in higher complexity and/or power consumption [17,18]. In fact, the few papers on these converters found in the literature that consider robustness to voltage variations suffer from these disadvantages [19–22]. In this paper, the ratiometric concept is applied to achieve robustness to component and power supply variations without increasing complexity and without the use of bandgap circuits. The ratiometric approach is a common technique for measuring analog signals from sensors, to be conditioned in a digital signal to be compatible with a computer, DAQ system, microcontroller or microprocessor [23–26]. The advantage of the ratiometric approach is that it avoids the need for internal regulators, which reduces energy consumption and costs, but one disadvantage is that the dynamic range of the output depends on the supply voltage, and would be a major problem where the supply voltage decreases continuously, e.g., in autonomous portable equipment.

The aim of this study was to design a low-cost and versatile interface for resistive sensors, offering both wide operative range and robustness to component and power supply variations. The result is a simple, compact, low-cost, low-voltage and low-power solution.

The paper is organized as follows: after the introduction in Section 1, the proposed approach is described in Section 2 An implementation with discrete devices following the proposed approach is described in Section 3. Section 4 shows the measurement results of a prototype. Finally, the conclusion is provided in Section 5.

2. Circuit Description and Operation

The principle of the proposed resistive sensor interface is shown in Figure 1, which is implemented using a Resistance-to-Current (R-I) converter and a Current Controlled Oscillator (CCO). In the R-I converter, V_{bias} provides a constant voltage to biasing the sensor R_S and produces a current inversely proportional to the sensor resistance $I_{out}(R_S) = V_{bias}/R_S$. This current charges and discharges a capacitor C, generating a triangular signal $V_C(t)$ whose slope depends on the current. $V_C(t)$ is compared with a reference voltage V_{ref} to generate a square wave, $T_{out}(I_{out})$, whose period is proportional to the resistance variation.

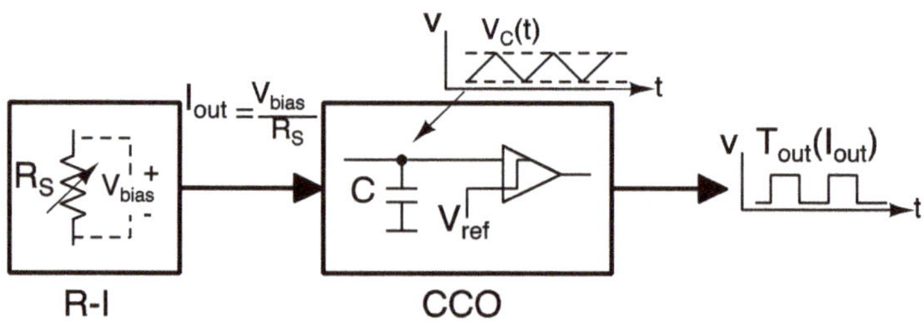

Figure 1. Block diagram of the proposed resistive sensor interface.

In Figure 2, an implementation of a Resistance-to-Current converter is shown. It is used to give an output current inversely proportional to the sensor resistance. The high open-loop gain of the amplifier will force the Gate of M1 to the required voltage such that V_{bias} appears across R_S. The current in R_S will be V_{bias}/R_S. This will flow only in the source of the transistor M1 and will be replied at its drain. Finally, it is mirrored by the current mirror with a gain of $1/\beta$. In this way, output current is equal to

$$I_{out} = \frac{I_{in}}{\beta} = \frac{V_{bias}}{R_S\beta}. \tag{1}$$

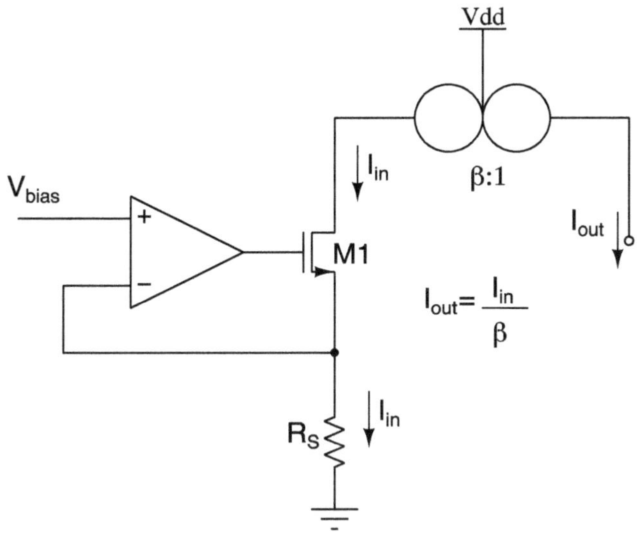

Figure 2. R-I converter.

The output current, I_{out}, will charge and discharge a capacitor C through a switch (SW1), which changes depending of the comparator output, as can be seen in Figure 3. The Voltage across the capacitor C, $V_C(t)$ is compared with a reference voltage V_H when it is charged and to another reference voltage V_L when it is discharged, being $V_H > V_L$. Each time $V_C(t)$ reaches V_H or V_L, the direction of the current through the capacitor is reversed. Therefore, the period of the output signal is given by

$$T_{out} = \frac{2C(V_H - V_L)}{I_{out}}, \tag{2}$$

where C is the capacitance from capacitor C.

Substituting Equation (1) into Equation (2),

$$T_{out} = \frac{2CR_S\beta(V_H - V_L)}{V_{bias}}. \tag{3}$$

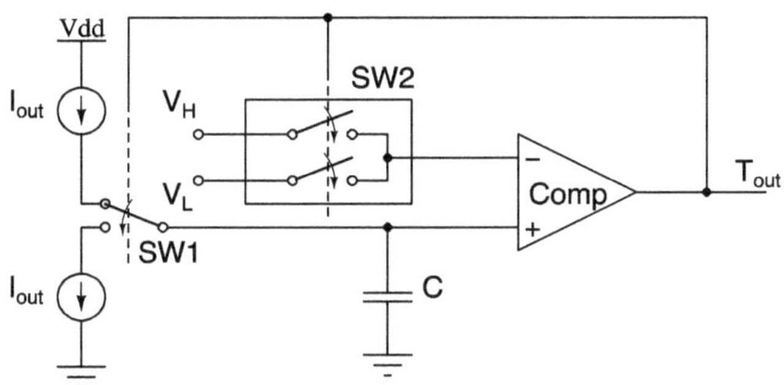

Figure 3. Implementation of the Current Controlled Oscillator.

To avoid the need for robust voltage reference, for V_{bias} implementation, the ratiometric approach is applied by using the same voltage reference for both the first block and the second block, i.e., $V_{bias} = (V_H - V_L)$. Thus, the period of the output signal is now given by

$$T_{out} = 2CR_S\beta. \tag{4}$$

In this way, the proposed architecture produces an output independent of any reference voltage, and no bandgap circuit is required. This reduces complexity, cost and power consumption of the system.

Minimal capacitance tolerance in commercial capacitors (ΔC) is 0.5%, so the maximum error is equal the same value. Additionally, error due to the delay in the comparator and switches is considered by adding a delay factor, t_d. To consider errors due to variations on the capacitor and delays, the period of the output signal of the proposed interface is given by:

$$T_{out} = 2(C \pm \Delta C)\beta R_S + t_d. \tag{5}$$

The switches that control the reference level (SW2) were implemented with a pair of complementary transistors. In this way, the change in the reference level is faster than the change in the integration sign, avoiding a deadlock situation, as shown in Figure 4. The change in the integration sign can be seen as a negative feedback mechanism, as the integrator output voltage changes backward. In turn, the change in the reference level is a positive feedback mechanism since it is the action that makes the whole system regenerative. In the ideal case, both actions occur at the same time, as depicted in Figure 4a; however, in real circuits, both feedbacks have a delay. When the output voltage of the integrator becomes higher than the high reference voltage, two actions are performed: switch the sign of the integrated constant and switch the reference voltage. If the reference voltage does not change faster than the sign of the integrated constant, the output voltage of the integrator will cross the high reference voltage again, leading to a deadlock state, as illustrated in Figure 4b. To avoid a deadlock situation, the positive feedback loop has to be stronger than the negative feedback mechanism to ensure that the transition to the next state takes place, as shown in Figure 4c [27].

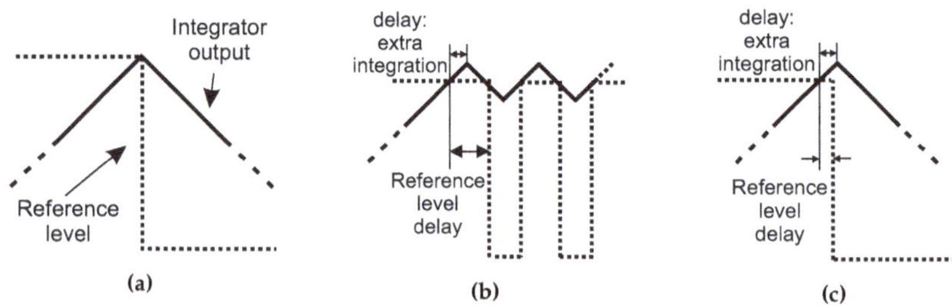

Figure 4. Transitions of both feedback loop. (**a**) Ideal case; (**b**) Deadlock situation; (**c**) No deadlock situation.

3. Proposed Circuit

This section presents an implementation designed with discrete devices to demonstrate the potential of the proposed architecture.

A variation of a linear voltage to current converter [28] shown in Figure 5 was used to implement the current mirror of Figure 2. Therefore, I_{out} is now

$$I_{out} = \frac{I_{in}R_1}{R_2}. \tag{6}$$

Thus,

$$\beta = \frac{R_2}{R_1}. \tag{7}$$

The maximum error in β is $\pm 1.4\%$ considering that commercial resistor have a minimum tolerance of 1%.

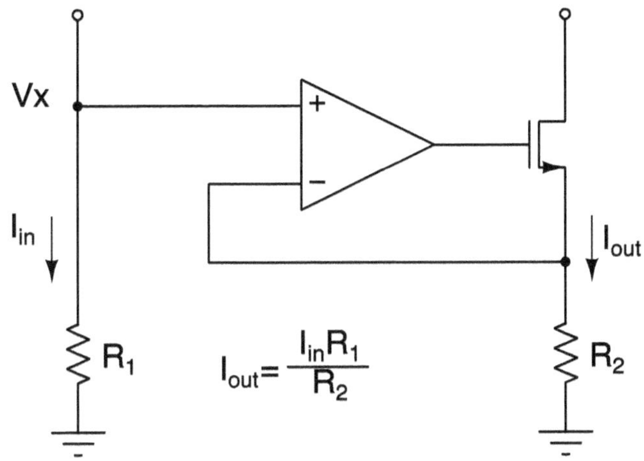

Figure 5. Current mirror implementation.

The complete resistive sensor interface is shown in Figure 6. The R-I converter consists of one resistor ladder (R_{ladder}), two amplifiers (A1 and A2) and two current mirrors (CM1 and CM2). The voltage across the resistive sensor R_S is set by amplifiers A1 and A2. The current $I(R_S)$ flowing through the sensor is driven to M1 and M2, and then current mirrors are used to improve accuracy of the copy. In this way, the current injected to or subtracted from the capacitor C remains almost independent of the variations in $V_C(t)$. The CCO consists of one comparator (comp), two switches (SW1 and SW2) and two NOT gates. The circuit that controls the current flow direction (SW1) was implemented using complementary transistors (M4 and M5). The switches that set the appropriate reference voltage level (V_H or V_L), were implemented with transmission gates to ensure that the change in the reference level is faster than the change in the integration sign.

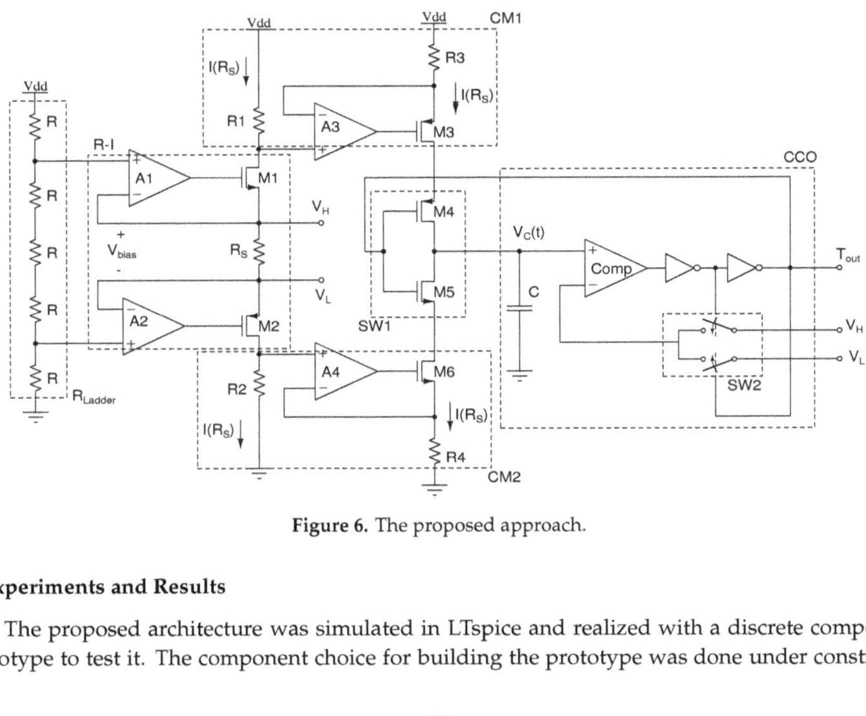

Figure 6. The proposed approach.

4. Experiments and Results

The proposed architecture was simulated in LTspice and realized with a discrete component prototype to test it. The component choice for building the prototype was done under constraints of compactness, single-supply voltage operation and low cost. The component selection includes quad-operational amplifiers (LM324), complementary enhancement mode field effect transistors (AO4614) and precision resistors of 1 kΩ. Both NOT gates were implemented using complementary transistors. Finally, a capacitor C of 100 nF (with 5% accuracy) was used. A picture of the realized printed circuit board (PCB) is shown in Figure 7. Optimizations could therefore be made to further improve the circuit compactness.

Figure 7. PCB from proposed circuit implementation.

For the experimental setup a DC Power Supply from BK PRECISION to power the circuit and a TDS1002 Mixed Signal Oscilloscope from Tektronix and a 5491A Digital Count-Multimeter from BK PRECISION to analyze the output signal were used. Experimental set-up is shown in Figure 8.

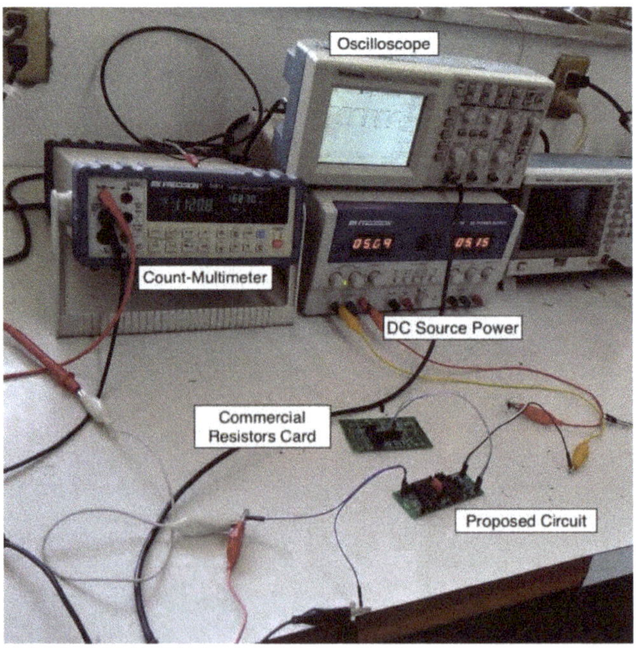

Figure 8. Experimental setup.

The converter was tested with sample resistors R_S (ranging from 1 kΩ to 100 MΩ with 1% accuracy) to emulate the sensor. Commercial resistors were used, whose values were previously determined with a 5491A Digital Multimeter from BK PRECISION. To measure output signal and to evaluate the system performance in terms of relative standard deviation and linearity, multiple-period averaging was applied. For each tested resistance value, 1000 consecutive measurements were taken.

A snapshot from oscilloscope with an input resistance R_S equal to 1 kΩ and its corresponding frequency ($1/T_{out}$) are shown in Figure 9. Both simulation and experimental data, as well as the linear Equation (5), using the weighted least square regression, are shown in Figure 10. The circuit was tested with Vdd equal to 5 V. A linear correlation coefficient of 0.9953 was obtained with respect to Equation (5). Vertical error bars show $\pm 2\sigma$.

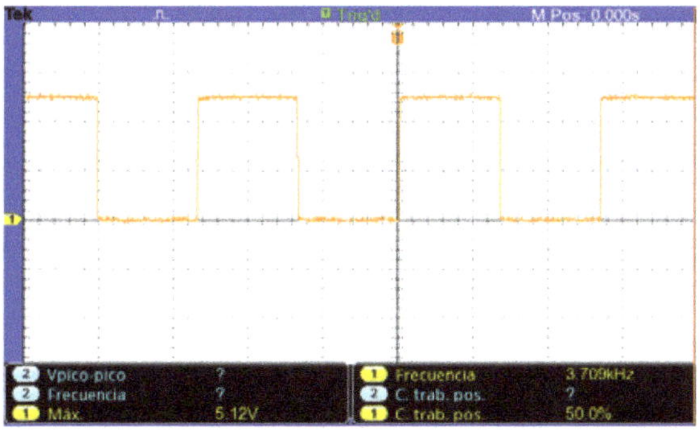

Figure 9. Oscilloscope snapshot for R_S equal to 1 kΩ.

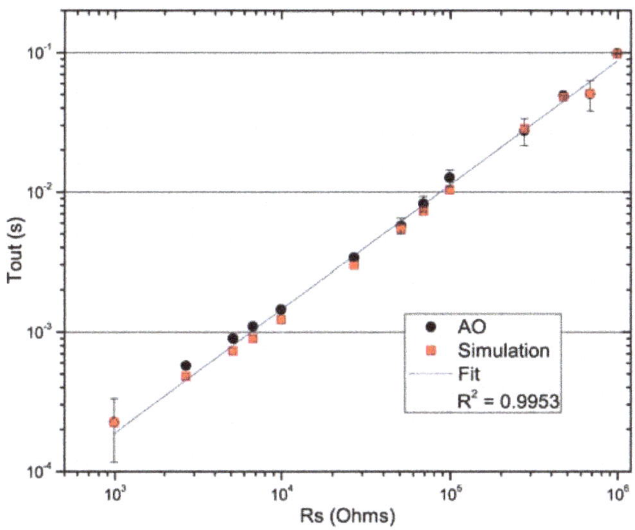

Figure 10. Period as function of the sample resistors. Simulation in solid red squares, experiment in solid black circles and linear Equation (5) in solid line.

F. Reverter et al. [29] presented an analysis of the effects of power supply interference on the output information in quasi-digital and modulating sensors, in which variations in power supply around 25, 50, and 100 mV and 0–5 times the central frequency were tested. In both cases, quasi-digital sensors have time-based outputs that are susceptible to power supply interferences. In this work, experimental measurements were carried out considering a variation of ±10% and of ±35% in the single supply voltage with a nominal value of 5 V, to demonstrate that dependence with the supply voltage (V_{DD}) was reduced when V_{bias} was matched with (V_H-V_L). In Figure 11, the experimental data from ±10% variation with a maximum error of ±2% is shown. The error was calculated using the nominal value. In the same way, Figure 12 shows a maximum standard deviation of ±10% with respect to the mean, for a sweep from 3.5 to 7.5 V in the Vdd (more than 50% of V_{dd} variation).

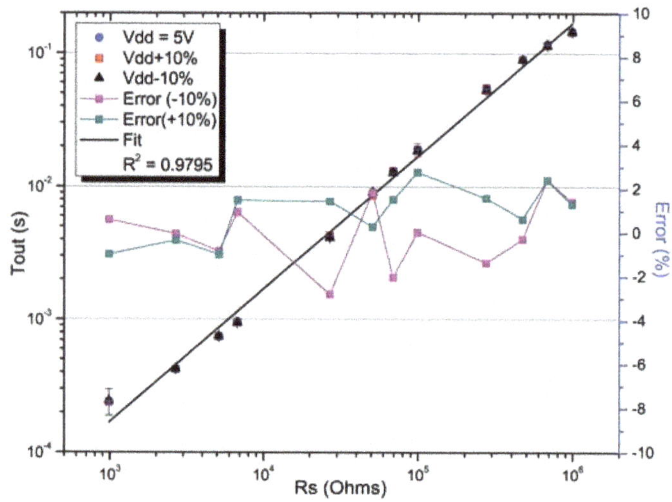

Figure 11. Period as function of the sample resistors. Experimental results with a power supply variation of ±10% from a nominal value of 5 V.

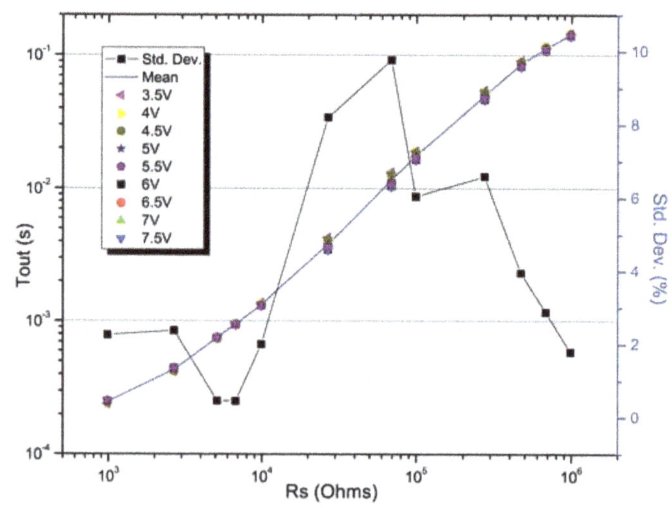

Figure 12. Period as function of the sample resistors. Experimental results with sweep from 3.5 to 7.5 V in the power supply.

To prove robustness to component variation, a second discrete component prototype was implemented. The complementary enhancement mode field effect transistors from the AO4614 family were replaced with the DMC4050 family and both NOT gates were implemented using a TTL7402. The operational amplifiers, resistors, and capacitor were kept with the same values. Figure 13 shows the experimental data from the comparison between the component replacement of the AO with the

DMC family. A correlation coefficient of 0.9956 was obtained. In the same way, another measurement was done where all operational amplifiers from the LM324 family were rotated, i.e., all the encapsulated opamps were changed by others of the same family. The rotation was done four times. Figure 14 shows four histograms and the mean value from the four measurements. Error from the histogram is less than ±1.5%. Thus, it was found that changes on the components did not represent a significant change in the response of the proposed circuit.

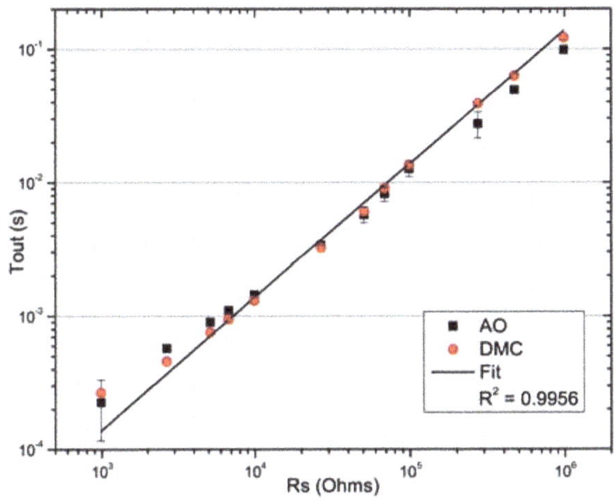

Figure 13. Experimental results from two discrete component prototypes: the first implemented with transistors from the AO4614 (AO, square) family and the second from the DMC4050 family (DMC, circle). Correlation coefficient equal to 0.9956.

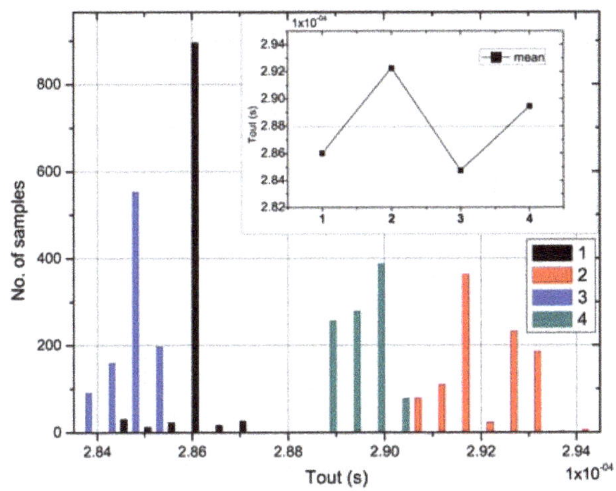

Figure 14. Histogram from the opamp variations with four different encapsulated. R_S = 2.7 KΩ.

Finally, Figure 15 shows the response of the proposed circuit for different values of capacitance C. It should be mentioned, that according to Equation (5), the smaller is the capacitance, the greater will be the error due to delay t_d.

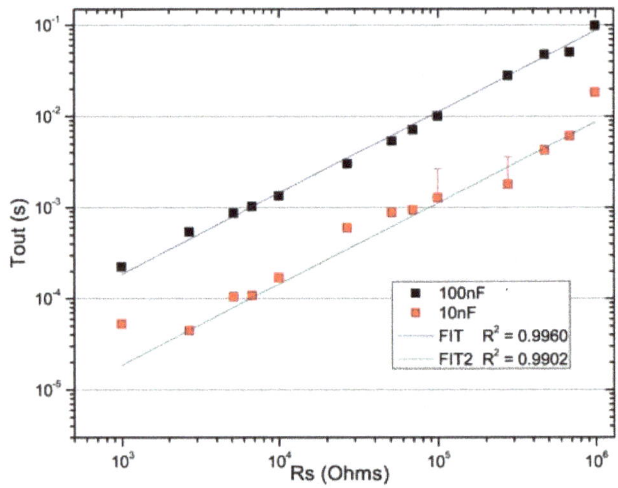

Figure 15. Measurement for different values of capacitance C.

5. Conclusions

A robust converter for obtaining a quasi-digital output directly from a resistive sensor is presented in this work. The proposed solution is based on the ratiometric approach, so the output becomes independent of any reference voltage. The square-wave output signal provides an easy way to process the sensor resistance allowing a simple data acquisition through low-cost digital processing systems.

Several works on the conditioning of resistive sensors have been reported in the literature, however, none has contemplated or shown the robustness to variations of the power supply and change of components according to the authors' knowledge. In this work, an architecture is shown that focuses on this robustness maintaining a wide dynamic range and precision.

Analysis of the proposed architecture for possible sources of errors indicates that most of the non-idealities introduce a gain error in β (in the current mirrors) and a delay t_d (by the comparator and switches), which can easily be canceled after calibration step.

Through several experimental measurements using two fabricated discrete components prototype, commercial Op-Amps and resistors emulating sensors, the results presented here establish the efficacy of the architecture proposed.

Based on the presented results, regarding its robustness and linearity with different variations, the proposed circuit is a suitable solution for portable sensor interfacing applications that have not been fully explored.

Author Contributions: Conceptualization, E.G.-R. and L.C.Á.-S.; formal analysis, E.G.-R., L.A.M.-N, and L.C.Á.-S.; methodology, E.G.-R., L.A.M.-N., L.C.Á.-S. and F.G.F.-G.; investigation, E.G.-R., L.A.M.-N. and L.C.Á.-S.; resources, E.G.-R., L.A.M.-N., L.C.Á.-S. and F.G.F.-G.; data curation, E.G.-R., L.A.M.-N. and L.C.Á.-S.; project administration, F.G.F.-G.; supervision, F.G.F.-G.; visualization, E.G.-R., L.A.M.-N., and L.C.Á.-S.; writing—original draft, E.G.-R., L.A.M.-N., and L.C.Á.-S.; and writing—review and editing, E.G.-R., L.A.M.-N., L.C.Á.-S. and F.G.F.-G.

Funding: This research was funded by Tecnológico Nacional de México.

Acknowledgments: The authors acknowledge CONACyT for the support received through the Cátedras CONACyT program.

Conflicts of Interest: The authors declare no conflict of interest.

Abbreviations

The following abbreviations are used in this manuscript:

LDR	Light Dependent Resistor
RTD	Resistance Temperature Detector
DAQ	Data Acquisition
R-I	Resistance-to-Current
CCO	Current Controlled Oscillator

References

1. Yurish, S.Y. Low-Power, Low-Voltage Resistance-to-Digital Converter for Sensing Applications. *Sens. Transducers* **2016**, *204*, 1–10.
2. Jain, V.; George, B. An efficient digitization scheme for resistive sensors interfaced through quarter bridge. In Proceedings of the 2017 Eleventh International Conference on Sensing Technology (ICST), Sydney, Australia, 4–6 December 2017; pp. 1–5.
3. Solar, H.; Beriain, A.; Jiménez-Irastorza, A.; Alvarado, U.; Berenguer, R.; Ortiz de Landaluce, M.; Cojocariu, M.; Martínez, C. A CMOS sensor signal conditioner for an automotive pressure sensor based on a piezo-resistive bridge transducer. In Proceedings of the Conference on Design of Circuits and Integrated Systems (DCIS), Granada, Spain, 23–25 November 2016; pp. 1–5.
4. Ramanathan Nagarajan, P.; George, B.; Kumar, V.J. An Improved Direct Digital Converter for Bridge-Connected Resistive Sensors. *IEEE Sens. J.* **2016**, *16*, 3679–3688. [CrossRef]
5. De Marcellis, A; Reig, C.; Cubells-Beltran, M. A Capacitance–to–Time Converter-Based Electronic Interface for Differential Capacitive Sensors. *Electronics* **2019**, *8*, 80. doi:10.3390/electronics8010080. [CrossRef]
6. Koay, K.C.; Chan, P.K. A 0.18-μm CMOS Voltage-to-Frequency Converter With Low Circuit Sensitivity. *IEEE Sens. J.* **2018**, *18*, 6245–6253 [CrossRef]
7. Hijazi, Z.; Grassi, M.; Caviglia, D.D.; Valle, M. Time-based calibration-less read-out circuit for interfacing wide range MOX gas sensors. *Integration* **2018**, *63*, 232–239. [CrossRef]
8. Sreenath, V.; Semeerali, K.; George, B. A Resistive Sensor Readout Circuit With Intrinsic Insensitivity to Circuit Parameters and Its Evaluation. *IEEE Trans. Instrum. Meas.* **2017**, *66*, 1719–1727. [CrossRef]
9. Lim, J.; Rezvanitabar, A.; Degertekin, F.L.; Ghovanloo, M. An Impulse Radio PWM-Based Wireless Data Acquisition Sensor Interface. *IEEE Sens. J.* **2019**, *19*, 603–614. [CrossRef]
10. Vooka, P.; George, B. Capacitance-to-digital converter for leaky capacitive sensors. *Electron. Lett.* **2016**, *52*, 456–458. [CrossRef]
11. Nowicki, M. A Modified Impedance-Frequency Converter for Inexpensive Inductive and Resistive Sensor Applications. *Sensors* **2019**, *19*, 121. doi:10.3390/s19010121. [CrossRef] [PubMed]
12. Hijazi, Z.; Grassi, M.; Caviglia, D.D.; Valle, M. 153dB Dynamic Range Calibration-Less Gas Sensor Interface Circuit with Quasi-Digital Output. In Proceedings of the 2017 New Generation of CAS (NGCAS), Genova, Genoa, 6–9 September 2017; pp. 109–112.
13. Ciciotti, F.; Buffa, C.; Gaggl, R.; Baschirotto, A. A programmable dynamic range and digital output rate oscillator-based readout interface for MEMS resistive and capacitive sensors. In Proceedings of the 2018 International Conference on IC Design & Technology (ICICDT), Otranto, Italy, 4–6 June 2018; pp. 41–44.
14. George, A.K.; Shim, W.; Je, M.; Lee, J. A 114-Af RMS- Resolution 46-Nf/10-MΩ-Range Digital-Intensive Reconfigurable RC-to-Digital Converter with Parasitic-Insensitive Femto-Farad Baseline Sensing. In Proceedings of the 2018 IEEE Symposium on VLSI Circuits, Honolulu, HI, USA, 18–22 June 2018; pp. 157–158.
15. Dai, S.; Perera, R.T.; Yang, Z.; Rosenstein, J.K. A 155-dB Dynamic Range Current Measurement Front End for Electrochemical Biosensing, *IEEE Trans. Biomed. Circ. Syst.* **2016**, *10*, 935–944. [CrossRef] [PubMed]
16. Chen, M.; Liu, Y.; Li, Z.; Xiao, J.; Chen, J. A High Dynamic Range CMOS Readout Chip for Electrochemical Sensors. *IEEE Sens. J.* **2016**, *16*, 3504–3513.

[CrossRef]

17. Ciciotti, F.; Baschirotto, A.; Buffa, C.; Gaggl, R. A MOX Gas Sensors Resistance-to-Digital CMOS Interface with 8-bits Resolution and 128dB Dynamic Range for Low-Power Consumer Applications. In Proceedings of the 2017 13th Conference on Ph.D. Research in Microelectronics and Electronics (PRIME), Giardini Naxos, Italy, 12–15 June 2017; pp. 21–24.
18. Ko, Y.; Kim, H.; Mun, Y.; Lee, B.; Kim, G.; Sul, W.; Lee, B.; Ko, H. 31.6 pJ/Conversion-step Energy-efficient 16-bit Successive Approximation Register Capacitance-to-digital Converter in a 0.18 µm CMOS Process. *Sens. Mater.* **2018**, *30*, 1765–1773. [CrossRef]
19. De Marcellis, A.; Depari, A.; Ferri, G.; Flammini, A.; Marioli, D.; Stornelli, V.; Taroni, A. A CMOS Integrable Oscillator-Based Front End for High-Dynamic-Range Resistive Sensors. *IEEE Trans. Instrum. Meas.* **2008**, *57*, 1596–1604. [CrossRef]
20. Gupta, R.; George, B. Resistance-to-digital converter designed for high power-line interference rejection capability. *IET Circ. Devices Syst.* **2017**, *11*, 446–451. [CrossRef]
21. Malcovati, P.; Grassi, M.; Baschirotto, A. Towards high-dynamic range CMOS integrated interface circuits for gas sensors. *Sens. Actuators B Chem.* **2013**, *179*, 301–312. [CrossRef]
22. Ferri, G.; Carlo, C.D.; Stornelli, V.; Marcellis, A.D.; Flammini, A.; Depari, A.; Jand, N. A single-chip integrated interfacing circuit for wide-range resistive gas sensor arrays. *Sens. Actuators B Chem.* **2009**, *143*, 218–225. [CrossRef]
23. Yu, Z.; Scherjon, C.; Mahsereci, Y.; Burghartz, J.N. A new CMOS stress sensor ratiometric readout for in-plane stress magnitude and angle detection. In Proceedings of the 2017 IEEE SENSORS, Glasgow, UK, 29 October–1 November 2017; pp. 1–3.
24. Ganesan, H.; George, B.; Aniruddhan, S.; Haneefa, S. A Dual Slope LVDT-to-Digital Converter. *IEEE Sens. J.* **2019**, *19*, 868–876. [CrossRef]
25. Amini, S.; Johns, D.A. A pseudo-differential charge balanced ratiometric readout system for capacitive inertial sensors. In Proceedings of the 2015 IEEE 58th International Midwest Symposium on Circuits and Systems (MWSCAS), Fort Collins, CO, USA, 2–5 August 2015; pp. 1–4.
26. Beriain, A.; Gutierrez, I.; Solar, H.; Berenguer, R. 0.5 V and 0.43 pJ/bit Capacitive Sensor Interface for Passive Wireless Sensor Systems. *Sensors* **2015**, *15*, 21554–21566. [CrossRef] [PubMed]
27. Westra, J.; Verhoeven, C.; Van Roermound, A. *Oscillators and Oscillator Systems: Classification, Analysis and Synthesis*; Springer: Berlin, Germany, 1999.
28. Wan, M.; Liao, W.; Dai, K.; Zou, X. A Nonlinearity-Compensated All-MOS Voltage-to-Current Converter. *IEEE Trans. Circ. Syst. II Express Briefs* **2016**, *63*, 156–160. [CrossRef]
29. Reverter, F.; Gasulla, M.; Pallás-Areny, R. Analysis of power supply interference effects on quasi-digital sensors. *Sens. Actuators A* **2005**, *119*, 187–195. [CrossRef]

electronics

MDPI

Article

A 13-bit 3-MS/s Asynchronous SAR ADC with a Passive Resistor Based Loop Delay Circuit

Hyungyu Ju and **Minjae Lee** *

The School of Electrical and Computer Science, Gwangju Institute of Science and Technology, Gwangju 61005, Korea; hgju@gist.ac.kr

* Correspondence: minjae@gist.ac.kr; Tel.: +82-062-715-2205

Received: 9 February 2019; Accepted: 26 February 2019; Published: 27 February 2019

Abstract: An asynchronous successive approximation register (SAR) ADC incorporates a passive resistor based delay cell to reduce power consumption and accommodate the SAR ADC with a reconfigurable sampling frequency or tapered bit period without repeated delay calibration. The ADC aims to have a sampling frequency of several MS/s. The proposed delay cell adopts resistance controlled delay architecture to generate a delay of nanoseconds with high linearity. The resistance controlled delay cell is based on a passive resistor instead of a MOS transistor using a triode region to avoid the nonlinear delay characteristic of active devices. From the analysis of the linearity of delay cell, the passive resistor based delay cell achieves a delay error of about 5 percent. The prototype ADC to validate the proposed passive resistor based delay cell is fabricated in 40 *nm* CMOS. The ADC occupies 0.054 mm^2 and achieves an SNDR of 57.4 dB under 67 μW power dissipation at a 1.1 V supply with a 3 MHz sampling frequency.

Keywords: asynchronous; delay cell; passive resistor; SAR ADC; loop delay circuit

1. Introduction

From low speed applications such as industrial monitoring, bio-medical and sensor node [1–3] to high speed applications such as high speed links and next generation communication systems [4,5], Successive Approximate Register (SAR) ADC is the most widely adopted ADC architecture owing to its low power operation from a simple operating principle. Moreover, an asynchronous architecture is also widely used to mitigate the requirements of the comparison time of the comparator in SAR ADC.

In the asynchronous SAR ADC, the time budget of the sampling clock is composed of an input sample time, comparator comparison time, capacitor DAC settling time, digital logic propagation delay, and the number of repetitions proportional to the resolution, as shown in Figure 1. While many studies have been conducted to reduce each timing budget to enhance the ADC performance, the loop delay that replaces the DAC settling time has been subjected to few methodological studies. Therefore, this paper focuses on digitally controlled delay generation methods with linear delay characteristics when the asynchronous SAR ADC has a sampling frequency of several MS/s. The linear delay characteristics can be applied to design the loop delay circuit of SAR ADC with reconfigurable sampling frequency or tapered bit periods [6,7]. For the SAR ADC of this paper, we designed an ADC with reconfigurable sampling frequency to obtain an adjustable frame rate in a touch screen panel (TSP) readout IC.

There are several ways to implement a digitally controlled variable delay cell. The popular method is a delay cell with shunt capacitors [8]. Assuming that the delay cell is a first-order RC circuit, this delay cell adjusts the capacitance depending on the digital input patterns. Similarly, a method adjusting the resistance in the first-order RC circuit was introduced in reference [9]. An MOS transistor array was used as a variable resistor. Another method used to generate a digitally controlled delay cell is a current starved architecture [10]. The delay cell adopts a tail current source, thus adjusting delay by digitally controlled current level. Using digital logic propagation delay is another way to implement a digitally controlled delay cell [11]. This method adjusts the delay by varying the number of logic gates.

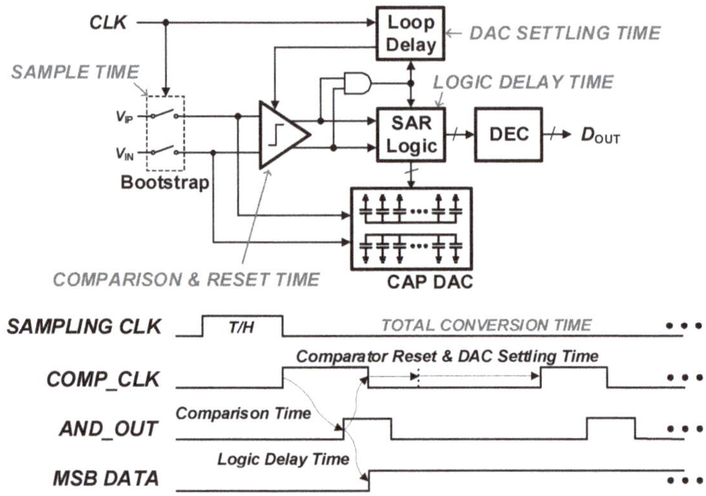

Figure 1. A structure and timing budget of the asynchronous SAR ADC.

When the delay cell is applied to the asynchronous SAR ADC, with sampling frequency of few MS/s and medium resolution, the delay cell should have a delay value of several tens of nanoseconds. Thus, the delay generation methods using the logic propagation delay and variable capacitance are inadequate due to power consumption in proportion to delay. In addition, the current staved structure requires an analog bias voltage, which leads to static current consumption by an additional bias generation circuit. In the case of using a MOS transistor as a variable resistor, the coding problem occurs due to the unpredictable circuit characteristic when the delay is generated by the equivalent resistance value.

Among the methods used to generate delay, the proposed delay cell adopts a resistive controlled delay cell to achieve high linearity with low power consumption. Thus, the proposed delay cell replaces the MOS transistor array to passive resistors to avoid the nonlinear delay characteristic and code dependent parasitic capacitance of the active device. The structure of the proposed delay cell is described in Section 2, along with the analysis of the linearity of the delay. The detailed circuit implementation of the proposed delay cell is represented in Section 3 with a delay calibration method. Finally, the simulation and measurement results are shown in Section 4, and we conclude the paper in Section 5.

2. Linearity of Delay Cell

Figure 2 shows a simple delay cell structure with a RC low-pass filter. When the input changes from high to low, the output voltage $V_{OUT}(t)$ can be expressed as a first order response given as

$$V_{out}(t) = V_{sup}\left(1 - e^{-t/\tau}\right) \tag{1}$$

where V_{sup} is supply voltage of delay cell. Then, we can solve the (1) for t, yielding

$$t = \tau \ln \frac{V_{sup}}{V_{sup} - V_{out}(t)} \tag{2}$$

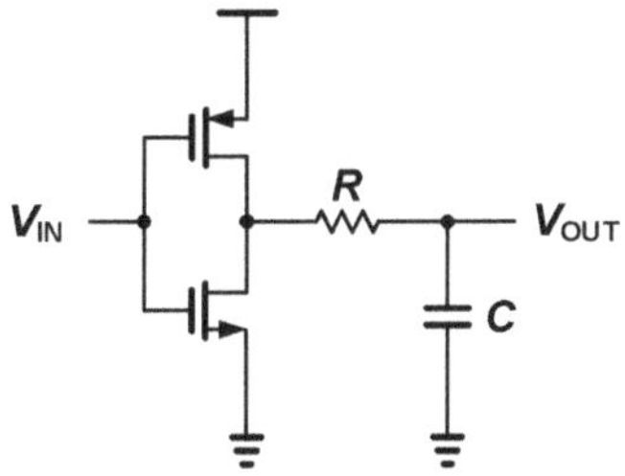

Figure 2. Simple delay cell using a passive resistor and capacitor.

Assuming that a logical signal is delivered to the next digital logic when the voltage settling is about 50 percent, the final value of $V_{out}(t)$ can be approximated to be 0.5 V_{sup}. Then, the logarithm term in (2) can be replaced by a constant coefficient α. Thus, the calculated delay time for the time constant can be expressed as

$$t = \alpha \cdot \tau = \alpha \cdot (R_{on.p} + R)C \tag{3}$$

where $R_{on.p}$ is the turn on resistance of the P-type MOS transistor. From (3), the resistance R, capacitance C and equivalent resistance of the MOS transistor $R_{on.p}$ are the factors that can adjust the delay time. However, adjusting C is difficult, because the power consumption of delay cell is proportional to C. In addition, using turn on resistance $R_{on.p}$ of the transistor is also not suitable due to the poor linearity of delay from the nonlinear characteristic of active devices. Therefore, adjusting R is the most reasonable approach in terms of power consumption and delay linearity. Then, we obtain a linear delay step according to the derivation of R which is given as

$$\Delta t = \alpha \cdot \Delta \tau = \alpha \cdot \Delta RC \tag{4}$$

It is important that the effect on $R_{on.p}$, which causes the nonlinearity, is eliminated in the resistive controlled delay cell.

The proposed digitally controlled delay cell is shown in Figure 3. The resistance is controlled by a switch with a small resistance value. C_p is the parasitic capacitance of the switch. As the switch is made of a pass transistor, the parasitic capacitance changes depending on whether the switches are on or off. $C_{p.tot}$ is sum of the parasitic capacitances for all pass transistors.

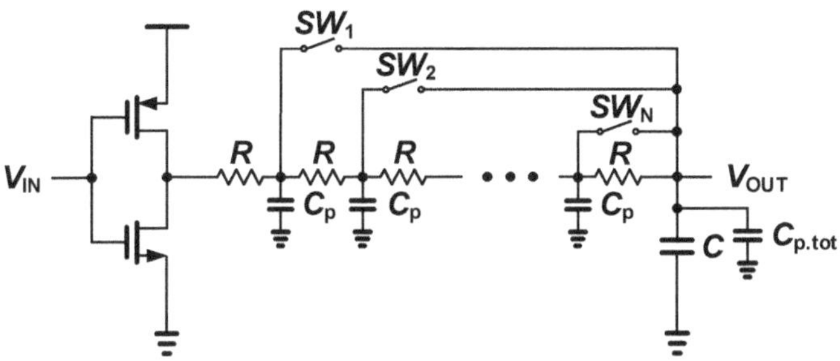

Figure 3. Proposed resistive controlled delay cell.

From (3) the delay time is a calculated by the product of time constant of the circuit and constant coefficient α. In other words, we can obtain the delay time by calculating the time constant of delay cell. To derive the time constant of entire RC network of proposed delay circuit, the Elmore delay model is employed.

Figure 4 represents the RC network model of proposed delay cell. $R_{on.ptr}$ is the turn on resistance of a switch made of pass transistor. $C_{p.on}$ and $C_{p.off}$ are parasitic capacitances of turn on and off switch, respectively. Among the switches, only one switch is turned on as the digital control input, so the internal RC network connected in parallel with the $R_{on.ptr}$ must be converted to a network, to which the Elmore delay model is applicable.

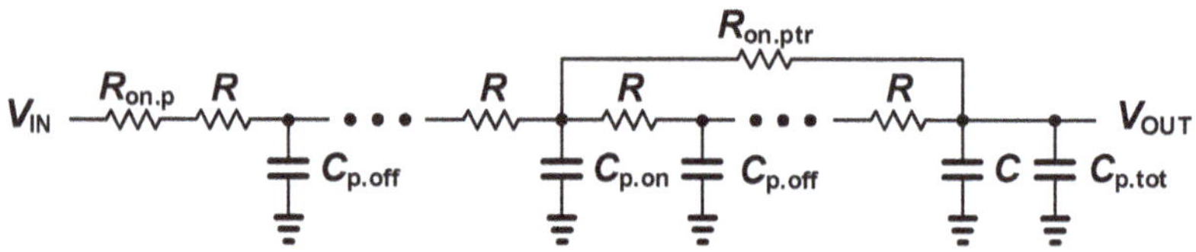

Figure 4. RC network model of proposed delay cell.

To simplify the internal RC network, Figure 5 describes a lumped PI-T transform method [12]. After calculating the time constant of PI and T model using Elmore delay model, two relations of the lumped PI-T transform are given as

$$R_P = 2R_T,\ C_T = 2C_P \tag{5}$$

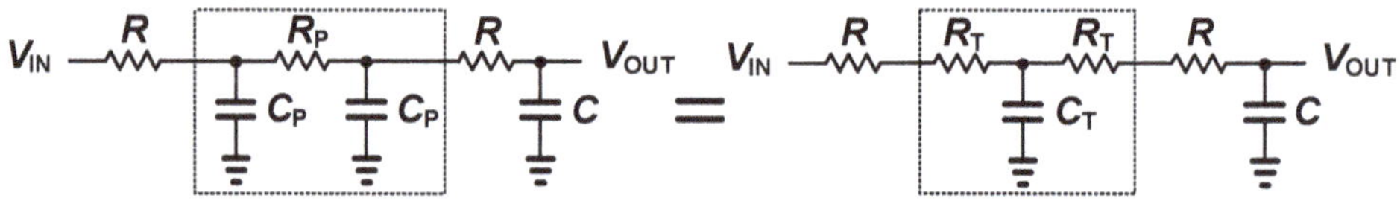

Figure 5. Lumped PI-T transform.

Using the lumped PI-T transform, the internal RC network can be simplified in the form of PI, and the parameters of the simplified PI model R_P and C_P are calculated as

$$R_P = kR,\ C_P = 0.5(k-1)C_{p.off} \tag{6}$$

where k is the number of R in the internal RC network.

Figure 6 shows the RC network model of proposed delay cell using (6). N represents the total number of switches. Assuming that $R_{on.ptr}$ is much smaller than R, the resistance of parallel resistors can be approximated as

$$(N-k+1)R \parallel R_{on.ptr} \approx R_{on.ptr} \tag{7}$$

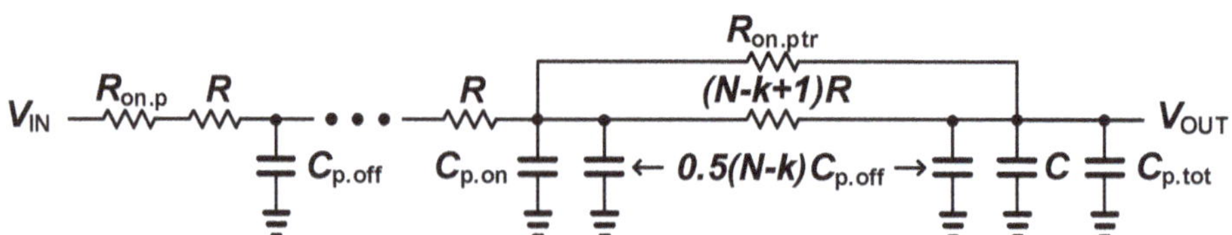

Figure 6. Equivalent RC network model of proposed delay cell.

Then, the normalized time constant equation of the proposed delay cell can be obtained as

$$\begin{aligned} \tau_k = g(k) + \left(R_{on.p} + kR\right)\left(C_{p.on} + 0.5(N-k)C_{p.off}\right) \\ +\left(R_{on.p} + kR + R_{on.ptr}\right)\left(C + C_{p.tot} + 0.5(N-k)C_{p.off}\right) \end{aligned} \tag{8}$$

where

$$g(k) = \begin{cases} 0 & , \quad k = 1 \\ \sum_{i=1}^{k-1} (R_{on.p} + iR)C_{p.off}, & k \geq 2 \end{cases}$$

Finally, the delay step can be obtained by the difference between the previous and current step of the normalized time constant from (4) and is derived as

$$\begin{aligned} \Delta t &= \alpha \cdot (\tau_k - \tau_{k-1}) \\ &= \alpha \cdot \left[R\left(C + C_{p.tot} + C_{p.on} + (N+1)C_{p.off} - 0.5kC_{p.off}\right) - 0.5R_{on.ptr}C_{p.off} \right] \end{aligned} \tag{9}$$

It is important to note that the nonlinearity of proposed delay cell is determined by the term multiplied by the value of k. If N is large enough, Equation (9) can be approximated as

$$\Delta t \cong \alpha \cdot \left[R\left(C + 2C_{p.tot} - 0.5kC_{p.off}\right) \right] \tag{10}$$

The term including $R_{on.ptr}$ is very small and hence negligible. If C is not in (10), then the delay linearity of the proposed structure has an error of approximately ±12.5 percent at maximum. In order to attain more linearity, the passive capacitor C is added at the cost of the increased power consumption.

3. Circuit Implementation

The implementation of loop delay circuit for asynchronous SAR ADC is presented in Figure 7. This circuit senses the comparison completion of the comparator and makes the comparator reset and operation clock repeatedly. The detailed circuit operation is as follows. When the START generated by inverting the sampling clock goes high, the loop is activated and the first comparison is made as the comparator clock COMP_CLK goes high. After the comparison of the comparator, the COMP_DONE indicating the end of the comparison becomes high and quickly resets COMP_CLK through the comparator reset path. As the comparator is reset, COMP_DONE is also reset. Then, the low state of COMP_DONE is propagated through DAC settling path, and COMP_CLK becomes high again after a propagation delay of delay cell. This SAR operation loop is repeated until the STOP goes high after all conversion cycles have ended.

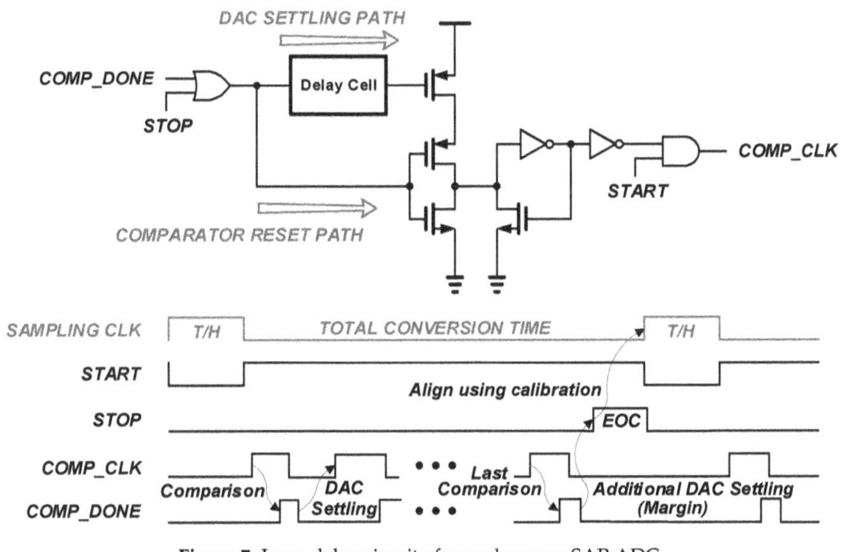

Figure 7. Loop delay circuit of asynchronous SAR ADC.

One feature of the loop delay circuit is that it is divided into the DAC settling path and the comparator reset path depending on the logical state of COMP_DONE. This two path operation has the advantage of using both fast and slow signal passing. However, the internal node of the delay cell can be in undesired states when the low state is applied to the delay cell input, because the input of the delay cell may change before the internal logical states by earlier high state input are entirely propagated. To prevent these uncertain states in the proposed delay cell, it includes reset switches as shown in Figure 8. The reset switches quickly reset the internal nodes of the delay cell when the comparator reset path is activated. The delay cell is controlled by 5-bit digital input, hence 32 resistors and switches consist of a proposed delay cell. R_{init}, representing the first passive resistance, is replaced by a lager resistance, instead of R, to increase the minimum delay. In addition, a Schmitt trigger is employed at the output stage to prevent any glitch caused by the supply fluctuations [13].

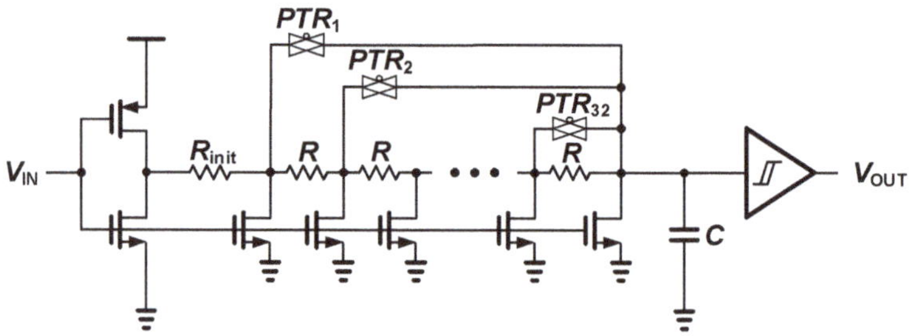

Figure 8. Proposed delay cell circuit including reset switch and Schmitt trigger.

The main drawback of using a passive resistor is that the resistance in the silicon process varies significantly with the process corner or temperature. Particularly, the delay variation is mostly dominated by process variation rather than temperature variation. To resolve this process variation, a foreground calibration is performed for the loop delay circuit. The delay is calibrated so that the STOP indicating the end of the conversion is aligned with the rising edge of next track and hold clock, as shown in Figure 7. Moreover, an additional DAC settling cycle is added to cover the temperature variation, hence securing a 1 cycle margin.

4. Simulation and Measurement Results

Figure 9a shows the calculated delay times from (8) with simulated delays from SPICE simulation according to the input digital codes. In the proposed delay cell, R_{init} and R are 15.1 $k\Omega$ and 1.9 $k\Omega$, respectively, at the nominal corner. $C_{p.off}$ and $C_{p.on}$ are 2.77 fF and 3.45 fF, respectively, and are extracted by the SPICE simulation. Because foreground calibration is performed to correct the delay mismatch from process variation, the delays in all corner conditions meet the 15 ns after calibration. The gray lines show the delays as temperature variations at a nominal corner. Compared to the corner variation, the delay variation from the temperature variation is very small, which is covered by the 1 cycle margin. The calculated delay values reflected the different α values of (3) as the temperature and corner. $R_{on.p}$ and $R_{on.ptr}$ also reflected the temperature and corner variation. Figure 9b shows the step delays from SPICE simulation of the proposed delay cell. The proposed delay cell achieves an error of about 5 percent, owing to the additional 155 fF of Metal-Oxide-Metal (MOM) capacitor C to improve linearity. Figure 10 shows simulation results of best and worst delay errors by component mismatch. From 200 samples of a Monte Carlo simulation, the best and worst delay error are 2.8 percent and 4.4 percent, respectively, with a mean of 3.6 percent and a standard derivation of 0.2 percent. Thus, the effect of mismatch does not significantly affect the delay linearity.

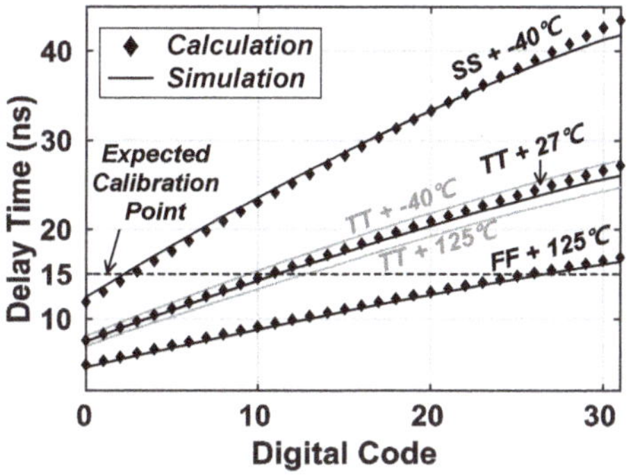

(a)

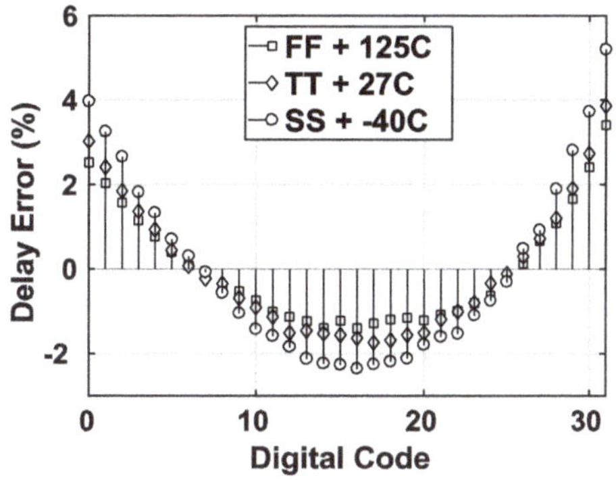

(b)

Figure 9. (**a**) Calculation and SPICE simulation results of delay versus the delay codes; (**b**) linearity of proposed delay cell from SPICE simulation.

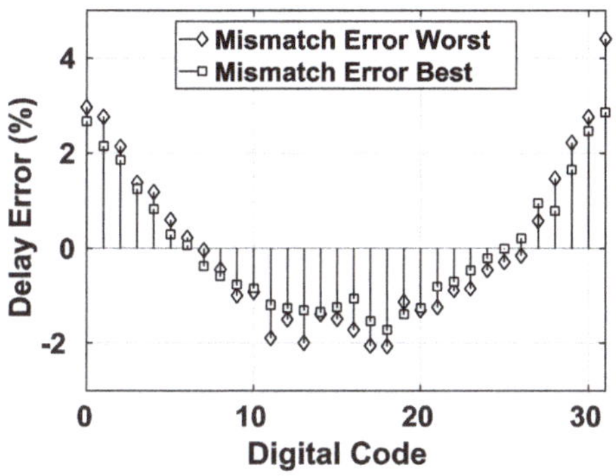

Figure 10. Worst and best linearity of proposed delay cell from Monte Carlo simulation.

Table 1 shows the comparison of delay cell structures with respect to current consumption and delay error. In the case of the current staved delay cell, we followed the design procedure of reference [10] with 30 fF of load capacitor. For fair comparison, the current consumption is measured with clock frequency of 20 MHz, and the each digital codes that adjusts the delay is set to 15ns. In terms of current consumption, the current starved delay cell has low current consumption. However, it has a 45 percent delay error, which is inadequate for tapered bit periods and reconfigurable sampling frequency. Contrarily, the shunt capacitor based delay cell has high linearity but consumes too much current. In case of the passive resistor based delay cell without C, the delay error is 8 percent with same current level of the current starved delay cell. To reduce delay error, the passive capacitor C is added in our application with additional 10 percent of current consumption. In the total power consumption of ADC, the passive resistor based delay cell consumes 28.4 μW.

Table 1. Comparison of delay cell structure.

Delay Generation Structure	Current (μA)	Delay Error (%)
Shunt Capacitor [8]	84	4
Current Starved [10]	13.2	45
Passive Resistor with C	14.6	5
Passive Resistor without C	13.2	8

Figure 11 shows the signal to noise and distortion ratio (SNDR) performance of prototype SAR ADC according to the delay codes. As the delay code increases, LSB conversions are not performed due to increased DAC settling delay at a given time, hence degrading SNDR performance. In order to use the proposed delay cell for the ADC with reconfigurable sampling frequency, two frequencies and their corresponding delay codes are required to obtain the relation equation. For example, measuring the sampling frequency of prototype SAR ADC at 2 MHz and 3 MHz from two delay calibrations, the corresponding delay codes from calibration are 0 and 11, respectively. Then, in the case of 2.5 MHz, the calculated delay code is 5.5 from the relation equation made with a two point calibration, and thus, delay code of 5 is applied, which is well matched to the measurement at a 2.5 MHz sampling frequency. The delay codes where the bits are skipped also show that the delay cell is sufficiently linear to be utilized for the ADC with a varied sampling frequency.

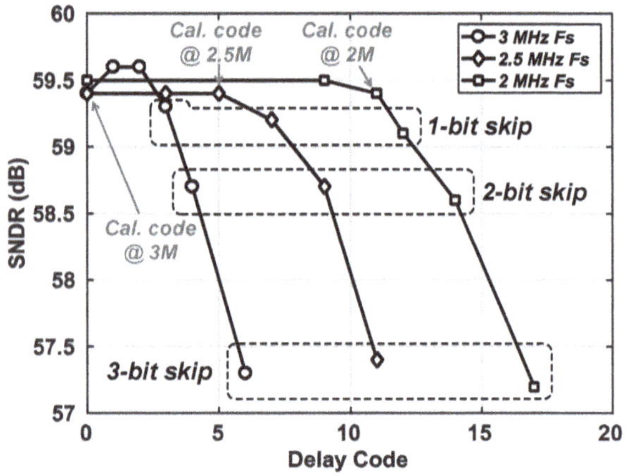

Figure 11. SNDR performance of SAR ADC as delay codes.

The prototype SAR ADC is fabricated in the 40 nm CMOS process. Figure 12 shows the die photograph of prototype SAR ADC, which occupies a core area measuring 670 um × 80 um. The delay cell occupies 35 um × 17 um, which is about 1 percent of total area. The prototype SAR ADC operates under 1.1 V supply voltage, consuming 67 μW at the 3 MHz sampling frequency. Figure 13 shows the differential nonlinearity (DNL) and integral nonlinearity (INL). The peak DNL are 2.9/-1 LSB, and the peak INL are 5.8/−10.5 LSB. Figure 14 shows an output spectrum from the prototype SAR ADC for near 100 kHz and 1.5 MHz. With a near 100 kHz input frequency, a measured spurious free dynamic range (SFDR) and SNDR are 68 dB and 59.4 dB, respectively. With a near 1.5 MHz input frequency, the prototype ADC achieves an SFDR of 65.2 dB and SNDR of 57.4 dB, yielding a FoM of 35.4 fJ/conversion-step. The overall performance of the prototype ADC is summarized in Table 2 and compared to references [14–17].

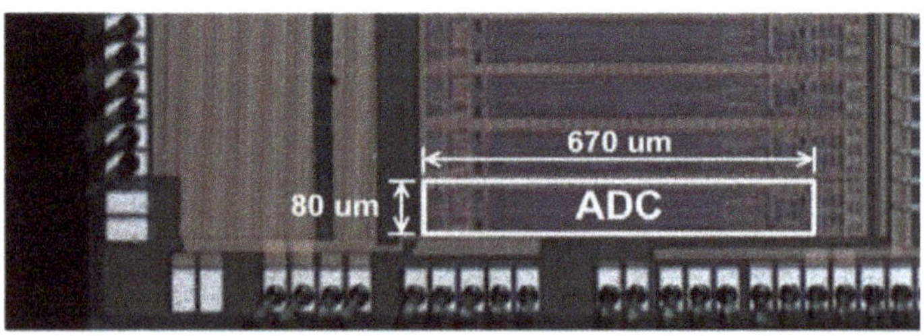

Figure 12. Die photograph.

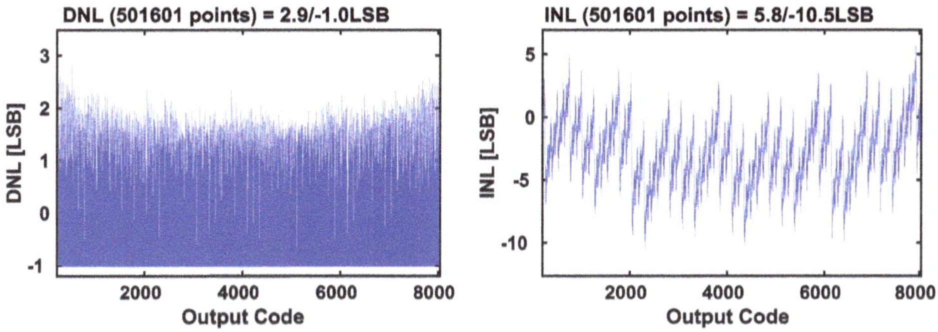

Figure 13. Static performance of prototype SAR ADC.

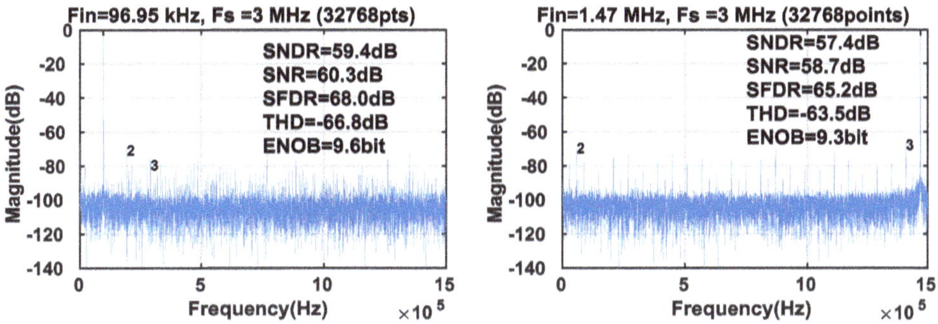

(a) (b)

Figure 14. Measured output spectrum of prototype SAR ADC with (**a**) a near 100 kHz sinusoidal input; (**b**) a near 1.5 MHz sinusoidal input.

Table 2. Performance summary and comparison.

Reference	[14]	[15]	[16]	[17]	This work
Technology (nm)	110	180	180	28	40
Resolution (bit)	10	10	12	12	13
Supply Voltage (V)	1.2	0.9	1.8	1	1.1
Sampling Rate (MS/s)	10	2	10	4	3
ENOB (bit)	8.6	9.07	10.82	10.1	9.3
FoM (fJ/conversion-step)	409	20.6	44.2	26	35.4
Power (μW)	1640	22.12	820	115	67
Area (mm^2)	0.25	0.21	0.359	0.016	0.054

5. Conclusions

In this paper, a 13 bit 3 MS/s asynchronous SAR ADC with a passive resistor based delay cell is presented. The proposed delay cell adopts passive resistors, which yields a delay error of less than 5 percent with reduced power consumption. The prototype SAR ADC achieves 57.4 dB of SNDR with 67 μW power dissipation, which converts to FoM of 35.4 fJ/conversion-step. This measurement shows that the scaling of delay codes can cope with the reconfigurable sampling frequency.

Author Contributions: Conceptualization, H.J.; Data curation, H.J.; Formal analysis, H.J.; Funding acquisition, M.L.; Investigation, H.J.; Methodology, H.J.; Project administration, M.L.; Supervision, M.L.; Visualization, H.J.; Writing—original draft, H.J.; Writing—review & editing, H.J.

Funding: This research was supported by the National Research Foundation of Korea Grant funded by the Korean Government (NRF-2016R1A2B4016544). The EDA tools were supported by IDEC, Korea.

Conflicts of Interest: The authors declare no conflict of interest.

References

1. Koppa, S.; Mohandesi, M.; John, E. An ultra-low power charge redistribution successive approximation register A/D converter for biomedical applications. *J. Low Power Electron.* **2016**, *12*, 385–393. [CrossRef] [PubMed]
2. Bai, W.; Zhu, Z. A 0.5-V 9.3-ENOB 68-nW 10-kS/s SAR ADC in 0.18-μm CMOS for biomedical applications. *Microelectron. J.* **2017**, *59*, 40–46. [CrossRef]
3. Fan, H.; Heidari, H.; Maloberti, F.; Li, D.; Hu, D.; Cen, Y. High resolution and linearity enhanced SAR ADC for wearable sensing systems. In Proceedings of the IEEE International Symposium on Circuits and Systems (ISCAS), Baltimore, MD, USA, 28–31 May 2017; pp. 1–4.
4. Frans, Y.; Shin, J.; Zhou, L.; Upadhyaya, P.; Im, J.; Kireev, V.; Borrelli, C. A 56-Gb/s PAM4 wireline transceiver using a 32-way time-interleaved SAR ADC in 16-nm FinFET. *IEEE J. Solid-State Circuits* **2017**, *52*, 1101–1110. [CrossRef]
5. Aryanfar, F.; Hossain, M. A quad channel 11-bit 1 GS/s 40 mW Collaborative ADC based enabling digital beamforming for 5G wireless. In Proceedings of the IEEE Radio Frequency Integrated Circuits Symposium (RFIC), Honolulu, HI, USA, 4–6 June 2017; pp. 120–123.
6. Shen, Y.; Zhu, Z.; Liu, S.; Yang, Y. A Reconfigurable 10-to-12-b 80-to-20-MS/s Bandwidth Scalable SAR ADC. *IEEE Trans. Circuits Syst. Regul. Pap.* **2018**, *65*, 51–60. [CrossRef]
7. Janke, D.; Monk, A.; Swindlehurst, E.; Layton, K.; Chiang, S.H.W. A 9-Bit 10-MHz 28-μW SAR ADC using Tapered Bit Periods and a Partially Interdigitated DAC. *IEEE Trans. Circuits Syst. Express Briefs* **2018**, *66*, 187–191. [CrossRef]
8. Ramazanoglu, S.; Batur, O.Z. Switched Capacitor Variable Delay Line. In Proceedings of the IEEE International Symposium on Circuits and Systems (ISCAS), Florence, Italy, 27–30 May 2018; pp. 1–5.
9. Saint-Laurent, M.; Swaminathan, M. A digitally adjustable resistor for path delay characterization in high-frequency microprocessors. In Proceedings of the IEEE Southwest Symposium on Mixed-Signal Design (SSMSD), Austin, TX, USA, 25–27 February 2001; pp. 61–64.
10. Maymandi-Nejad, M.; Sachdev, M. A digitally programmable delay element: Design and analysis. *IEEE Trans. Very Large Scale Integr. VLSI Syst.* **2003**, *11*, 871–878. [CrossRef]
11. Yang, R.J.; Liu, S.I. A 40–550 MHz harmonic-free all-digital delay-locked loop using a variable SAR algorithm. *IEEE J. Solid-State Circuits* **2007**, *42*, 361–373. [CrossRef]
12. Weste, N.H.; Harris, D. *CMOS VLSI Design: A Circuits and Systems Perspective*; Pearson Education: Chennai, India, 2015.
13. Melek, L.A.P.; da Silva, A.L.; Schneider, M.C.; Galup-Montoro, C. Analysis and design of the classical CMOS Schmitt trigger in subthreshold operation. *IEEE Trans. Circuits Syst. Regul. Pap.* **2017**, *64*, 869–878. [CrossRef]
14. Nam, S.P.; Kim, Y.M.; Hwang, D.H.; Kim, H.J.; An, T.J.; Park, J.S.; Lee, S.H. A 10b 1MS/s-to-10MS/s 0.11 um CMOS SAR ADC for analog TV applications. In Proceedings of the IEEE International SoC Design Conference (ISOCC), Jeju, Korea, 4–7 November 2012; pp. 124–127.
15. Zhu, Z.; Qiu, Z.; Liu, M.; Ding, R. A 6-to-10-Bit 0.5 V-to-0.9 V Reconfigurable 2 MS/s Power Scalable SAR ADC in 0.18 um CMOS. *IEEE Trans. Circuits Syst. Regul. Pap.* **2015**, *62*, 689–696. [CrossRef]
16. Liu, S.; Shen, Y.; Zhu, Z. A 12-bit 10 MS/s SAR ADC with high linearity and energy-efficient switching. *IEEE Trans. Circuits Syst. Regul. Pap.* **2016**, *63*, 1616–1627. [CrossRef]
17. Haenzsche, S.; Höppner, S.; Ellguth, G.; Schüffny, R. A 12 bit 4 MS/s SAR ADC with configurable redundancy in 28 nm CMOS technology. *IEEE Trans. Circuits Syst. Express Briefs* **2014**, *61*, 835–839. [CrossRef]

electronics

MDPI

Article

A Study of Movement Classification of the Lower Limb Based on up to 4-EMG Channels

Diana C. Toledo-Pérez [1,†], Miguel A. Martínez-Prado [2,†], Roberto A. Gómez-Loenzo [2,†], Wilfrido J. Paredes-García [2,†] and Juvenal Rodríguez-Reséndiz [2,*]

1 División de Investigación y Posgrado, Facultad de Informática, Universidad Autónoma de Querétaro (UAQ), Av. de las Ciencias S/N, Juriquilla, Querétaro C.P. 76230, Mexico; dtoledo16@alumnos.uaq.mx

2 División de Investigación y Posgrado, Facultad de Ingeniería, Universidad Autónoma de Querétaro (UAQ), Cerro de las Campanas, S/N, Col. Las Campanas, Querétaro C.P. 76010, Mexico; miguel.prado@uaq.mx (M.A.M.-P.); rob@uaq.mx (R.A.G.-L.); wparedes17@alumnos.uaq.mx (W.J.P.-G.)

* Correspondence: juvenal@uaq.edu.mx; Tel.: +52-442-192-1200

† These authors contributed equally to this work.

Received: 18 January 2019; Accepted: 22 February 2019; Published: 27 February 2019

Abstract: The number and position of sEMG electrodes have been studied extensively due to the need to improve the accuracy of the classification they carry out of the intention of movement. Nevertheless, increasing the number of channels used for this classification often increases their processing time as well. This research work contributes with a comparison of the classification accuracy based on the different number of sEMG signal channels (one to four) placed in the right lower limb of healthy subjects. The analysis is performed using Mean Absolute Values, Zero Crossings, Waveform Length, and Slope Sign Changes; these characteristics comprise the feature vector. The algorithm used for the classification is the Support Vector Machine after applying a Principal Component Analysis to the features. The results show that it is possible to reach more than 90% of classification accuracy by using 4 or 3 channels. Moreover, the difference obtained with 500 and 1000 samples, with 2, 3 and 4 channels, is not higher than 5%, which means that increasing the number of channels does not guarantee 100% precision in the classification.

Keywords: intention of movement classification; EMG-Signals; Support Vector Machines

1. Introduction

In recent decades, the use of signals obtained from the muscles has become popular due to its implementation in different applications such as health monitoring, assistive technology, and prosthetic control. This is due to the increase in technological advances in wearable electronics for the exploration of muscle signals.

When a muscle contraction or relaxation occurs, it generates an electrical potential that can be measured with an electromyographic sensor. There are two different approaches to place this kind of sensor—invasive and non-invasive methods. In the case of the former, the sensor is intramuscular; whereas in the latter, commonly called surface electromyography (sEMG), the sensor is placed on the skin surface; the former approach is the most common technique since it does not require surgical intervention.

To improve the classification accuracy, Oskoei, M. A. and Hu, H. [1] experimented with the quantity and type of characteristics used in the feature vector, while She et al. [2] varied the kernels utilized by the classifier.

Using another approach, Englehart K. and Hudgins B. [3] compared the effect in the accuracy due to the method used to obtain the features in frequency time, like Fast Fourier Transform, Wavelet Transform, and Wavelet Packet Transform. In general, the most common way used to improve this accuracy is to find some algorithm, variation or combination of these, to try to reach an accuracy of 100% [1,4–13].

Some researchers have also increased the number of channels used to classify; for example, Fukuda O. et al. [9] used six sensors, authors [14–17] used eight and Ceseracciu et al. [18] even used sixteen, but, none yields 100% accuracy in the classifications. In an effort to improve accuracy, some researchers have not only increased the number of channels but also the number of features employed; for example, Alizadeh et al. [19] increased both the number of features up to 28 and the EMG channels up to six.

The methods for signal analysis involve time-domain and frequency-domain features, time-frequency analysis methods, power spectrum density, and higher-order spectra [20]. For example, Pancholi S. and Joshi A. M. [21] combines two of them, time and frequency domain features, using nine features for the time domain and seven features for the frequency domain, that is 16 in total. In order to analyze the sEMG signals, this work only considers time-domain features, since they are easy to compute and do not require any transformation. Therefore, Mean Absolute Value (MAV), Zero Crossings (ZC), Waveform Length (WL) and Slope Sign Changes (SSC) are recommended characteristics to obtain a better classifier performance [2,3,8,13,16,22].

Increasing the number of channels for the classification introduces a dimensionality problem, which leads to lower classification performance [23]. Some tools can be used to analyze signals to improve the classification accuracy without increasing the number of them processed, e.g., the Empirical Mode Decomposition used only in a single-channel [24].

Aside from the techniques looking for ways to improve the accuracy in classification, other research works are focused on reducing the dimensionality problem such as Principal Component Analysis (PCA), Independent Component Analysis (ICA), Linear Discriminant Analysis (LDA), Canonical Correlation Analysis (CCA), among others.

Although the goal of PCA is usually to find out an optimal linear transformation which represents the original data and to reduce the dimensionality of the features vector [25–28], this research work only uses this method in order to achieve better accuracy, not to reduce the dimensionality of features.

Support Vector Machines (SVMs) are used for classification because they have a high potential for classifying signals in myoelectric control systems since they can recognize complex patterns [1].

However, in previous research, the difference in classification accuracy caused by increasing the number of channels or by varying the muscle from which the EMG signal is extracted has not been shown. This study offers the researchers the opportunity to decide whether the increase in resources used for processing is worthwhile or not.

In this article, sEMG signals were recorded on four opposite muscles on the lower limb and are used to compare the classification accuracy; there were four different stages with an increase in the number of signals in each stage that is, in a first step, only one signal was used, then two of them, then three and finally four signals in a final stage. The muscles selected to place the sensors on them were tibials anterioris (TA), gastrocnemius medials (GM), biceps femoris (BF) and vastus lateralis (VL), which presents a better movement signal [29].

This paper is organized as follows. Section 2 provides a brief background of the conventional techniques used for sEMG signals analysis and the most commonly used features, and describes the Support Vector Machine algorithm and PCA. Section 3 describes the experimental design and the analysis of sEMG signals. Comparison results from the SVM classifier varying the number of channels and their origin are presented in Section 4. Section 5 presents our concluding remarks.

2. Background

2.1. Analysis of sEMG Signals

Myoelectric control success depends highly on the classification accuracy. Classification methods and feature extraction are essential to attain high performance in the classification for pattern recognition [1].

Depending on the level of muscle contraction, sEMG signals vary in amplitude, variance, energy, and frequency. Given those measures, a variety of features is extracted from them for their analysis. As mentioned earlier, the most recommended in literature are MAV, ZC, SSC, and WL, and are described in the following paragraphs.

- MAV: It is the average of the N absolute values of the sEMG samples within a given time epoch, and is given by:

$$\text{MAV} = \frac{1}{N}\sum_{i=1}^{N}|x_i|. \tag{1}$$

- ZC: It is the number of times that the signal samples $\{x_i\}$ cross zero, whether it goes from a negative value to a positive one or the other way around, as in equation:

$$\text{ZC} = \sum_i f_{\text{ZC}}(x_i), \tag{2}$$

where

$$f_{\text{ZC}}(x_i) = \begin{cases} 1, & \text{if } x_i > 0 \quad \text{and} \quad x_{i+1} < 0 \\ & \text{or } x_i < 0 \quad \text{and} \quad x_{i+1} > 0, \\ 0, & \text{otherwise.} \end{cases} \tag{3}$$

- WL: It is the accumulated variation of a signal that can indicate the degree of signal oscillation and is given by equation:

$$\text{WL} = \sum_{i=1}^{N-1}|x_{i+1} - x_i|. \tag{4}$$

- SSC: It counts the number of times that the slope of the signal changes sign, which make necessary to evaluate where it is, where it was and where the signal goes. SSC is calculated with equation:

$$\text{SSC} = \sum_i f_{\text{SSC}}(x_i), \tag{5}$$

where

$$f_{\text{SSC}}(x_i) = \begin{cases} 1, & \text{if } x_i < x_{i+1} \quad \text{and} \quad x_i < x_{i-1} \\ & \text{or } x_i < x_{i+1} \quad \text{and} \quad x_i > x_{i-1}, \\ 0, & \text{otherwise.} \end{cases} \tag{6}$$

2.2. Principal Component Analysis

PCA is a statistical technique that performs a linear transformation from an original set of values into a smaller one of uncorrelated variables, which represents the most relevant information of the original set. Thus, the dimensionality of the original set is reduced or kept but never increased. The idea was conceived by K. Pearson [30] and later developed by Hotelling [31].

The PCA technique uses the covariance matrix from the original set (X) and the correlation between every one of these components, in such a way that a smaller Y output space is found, by representing the statistical information contained in X as it is described in equation:

$$Y = XC, \tag{7}$$

where C is the $m \times n$ matrix with the principal components selected, where $n < m$, which implies the dimensionality reduction from the original set. The procedure to determine C consists in constructing the covariance matrix, then compute the eigenvalues and eigenvectors to project the data matrix with these eigenvectors in decreasing magnitude order. Finally, it is only necessary to consider the desired information and to select the number of vectors that compose it.

2.3. Support Vector Machines

SVMs are commonly used as a classification algorithm for body movements, images, sounds, and other data. An SVM builds an optimum separation hyperplane in a feature space which is said to be of high dimension when the inputs are mapped using non-linear functions, to be able to distinguish between two or more object types. In 1995 this theory was introduced in [32].

In an SVM, the training algorithm is reformulated as a global and unique problem to solve using Quadratic Programming (QP) for input training data $(x_1, y_1), \ldots, (x_m, y_m) \in \mathbb{R}^N \times \{-1, +1\}$, where x_i corresponds to the input value and y_i the assigned value of the object type to which it belongs (also known as a class); if these data are not linearly separable, they are a mapped (non-linearly) by a kernel function $\varphi\colon \mathbb{R}^N \mapsto F$ into a characteristic space F. In this way, the obtained linear hyperplanes that separate the object types can be described as:

$$\omega \in \{x \mid \varphi(x) + b = 0\}, \qquad \omega \in \mathbb{R}^N, \quad b \in \mathbb{R} \tag{8}$$

Thus, by constructing an optimal hyperplane with the maximum value of the separation margin and a closed error ξ in the training of the algorithm, the QP problem is stated as:

$$\min_{w,b} \frac{1}{2}\|\omega\|^2 + C\sum_{i=1}^{m} \xi_i \tag{9}$$

The first term in a cost function that generates a maximum separation margin between classes, while the second one provides an upper bound for mistakes in the training data. Finally, the constant $C \in [0, \infty)$ creates a compensation between the number of poorly classified samples with a maximum margin.

Finally, the solution to the objective function proposed in Equation (9) can be obtained as mentioned in the previous paragraph using QP tools or methods such as those proposed by Pérez-Hernández [33].

3. Methods and Experimentation

3.1. Data Acquisition

The data in this research were acquired from eight healthy subjects, four females and four males. All subjects are aged between 23 and 32 years old, normally limbed and without muscle disorders.

The sensor system was placed on the skin over the muscles and comprised nine electrodes, eight of them, positioned in pairs, sensing the differential potential from muscles, and the last one used as a ground reference. Electrodes used were *Kendall Medi Trace 200* (Ag/AgCI circular bipolar electrodes, with 10 mm in diameter with an adhesive conducting gel). The sEMG signals were amplified almost 1000 times after passing through the INA114 amplifier. Then, the signals went through an analog 60 Hz notch filter to remove electric line interference, implemented with an operational amplifier. Later, an offset was applied to the signal to set a reference voltage of 1.67 V, because the ADC (Analog to Digital Converter) has a range from 0 to 3.3 V (Figure 1).

The signals were sampled with the aid of an STM32F103C8 microcontroller with a 12-bit ADC at a sampling frequency of 1000 Hz; each sample was packed as 2 bytes, which were sent to a PC, and then stored in an ASCII text format.

Each pair of sensors was placed according to the distances described in [29] with a 2.5 cm separation between them to obtain the best signal quality: For VL the best place is at 66% of the muscle length on the line from the anterior spina iliaca superior to the lateral side of the patella; for the TA is at 47.5% on the line between the tip of the fibula and the tip of the medial malleolus. The optimal electrode position in GM is at 38% of the muscle length from the medial side of the popliteal cavity to the medial side of the Achilles tendon insertion, starting from the Achilles tendon; and, for BF 50%, the position on the line between the ischial tuberosity and the lateral epicondyle of the tibia presents the best quality of the signal. For a thorough discussion of the relevant issues regarding electrode placement, refer to [34].

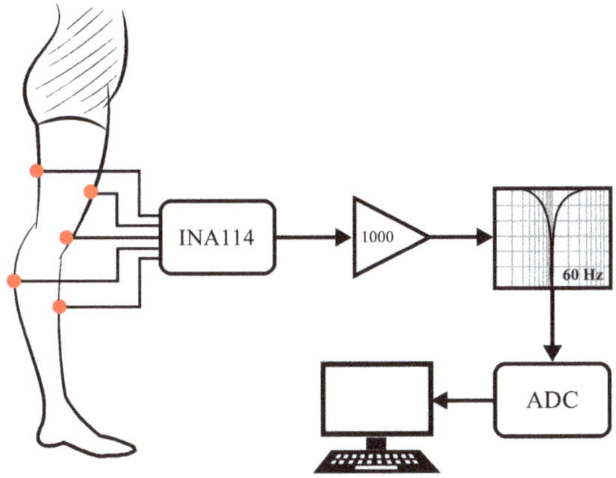

Figure 1. Basic experiment diagram.

3.2. Data Processing

For the data processing, MATLAB and the LIBSVM library version 3.2 were used in this work [35]. Although this library provides a module dedicated to applying a variety of kernels, such as linear, polynomial, RBF (Radial Basis Function), or sigmoid, none of them was required.

In software, two different digital filters were applied to remove undesirable noise from the collected sEMG signals. First a 60 Hz notch filter and then an elliptical bandpass filter between 10 and 500 Hz. The functions used were *filter*, *ellipord*, and *ellip*.

For the training process, first, the feature vector was built with MAV, ZC, WL, and SSC for windows of different sizes and for each of the channels individually; this was to make a comparison of the results with different schemes. In a second step, because of the differences between each feature the feature matrix was standardized. As a third step, a PCA analysis was performed without removing vectors from the transformation matrix. Finally, the obtained PCA matrix was multiplied by the feature matrix and the resulting matrix was used as input to train the SVM classifier.

Steps one and two were repeated for the test data set; later, the feature matrix obtained was multiplied by the PCA matrix and the resulting matrix was used to test the SVM.

3.3. Experimentation

Six classes of foot movement plus rest were considered for the research: lift the toe (LP), lift the heel (LT), move the toe to the right (PD), move the toe to the left (PI), lean on the heel (AT), lean on the toe (AP), and rest foot (RR). In the experiment, the subjects were sitting and started from a relaxation state and then performed the movement and held it for 5 s, and then they returned to the relaxation position. The movements were repeated 20 times with a resting period of 25 s between the movements by each subject. Tests were done in a single session.

The first window size considered was 250 ms, since 300 ms is an acceptable delay from the system in case that the intended use the system is controlling a prosthesis [3]. Also, considering other possible usages, another two window sizes were considered, namely, 500 and 1000 ms.

Finally, the collected data were divided into two groups, the training, and the testing data; ten samples for each group. In other words, the database is composed of 1120 movements, from eight different people (four females and four males) and seven different movements. Of these movements, 560 were used to train the SVM and the other 560 were used to test the classification accuracy.

4. Results

The obtained results are shown in Tables 1 and 2. The first table shows the best results in accuracy for each window size, considering one, two or three channels; with an additional row with the values for four channels. The first column contains the number of channels considered, and the last one has the channels with which the result was obtained.

Table 1. Best accuracy results obtained among the eight subjects.

Number	Samples			Channels
	250	500	1000	
1	90.00%	91.43%	95.71%	VL
2	95.71%	97.14%	97.14%	GM & VL
3	95.71%	100.00%	98.57%	TA, GM & VL
4	95.71%	100.00%	100.00%	TA, GM, BF & VL

Table 2. Results obtained with the lowest accuracy among the eight subjects.

Number	Samples			Channels
	250	500	1000	
1	52.86%	55.71%	64.29%	TA
2	70.00%	72.86%	75.71%	TA & VL
3	78.57%	78.57%	87.14%	TA, GM & BF
4	81.43%	81.43%	87.14%	TA, GM, BF & VL

Data shown in Table 1 indicates that the best muscle to extract movement information is *VL* since it appears with one, two or three channels; and the second-best option is *gastroctemius medialis*, also appearing with two or three channels. Additionally, the difference in the accuracy obtained with 500 and 1000 samples using two, three and four channels is of one sample at the most.

The results in Table 2 are the lowest scores, and these in turn show that *tibialis anterior* has not enough information to make a good classification, even if it is combined with the VL muscle. Also, the combination of three channels without the VL muscle has the worst performance. The accuracy of the classification increases less by increasing the window size than by increasing the number of channels.

Figure 2 shows a graphic with the average of the results obtained with a single channel, where the VL muscle presents the best accuracy classification and the TA muscle the worst. Additionally, the results of varying the sampling window size are not conclusive enough to state the recommended size.

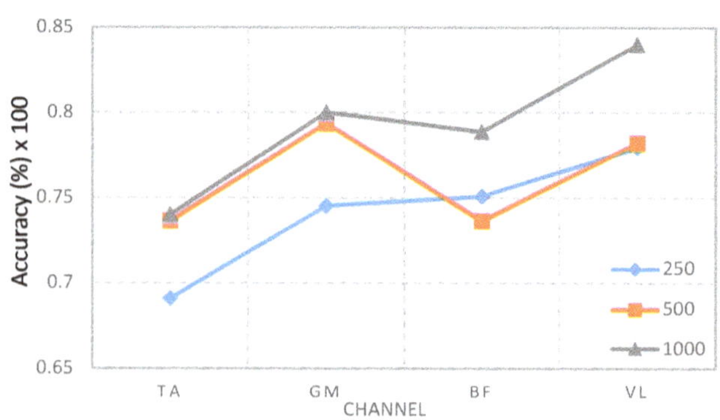

Figure 2. Classification accuracy with a single channel.

As shown in the comparison of two channels in Figure 3 the GM and VL muscles have a better performance than the rest. Furthermore, a more consistent performance can be achieved with a sampling rate of 1000 than with any other number of samples, but the difference with 500 is minimal in most cases. Figure 4 shows that the combination with GM, BF, and VL is better for classification than those that include channel TA; again, the difference between 500 and 1000 samples is minimal.

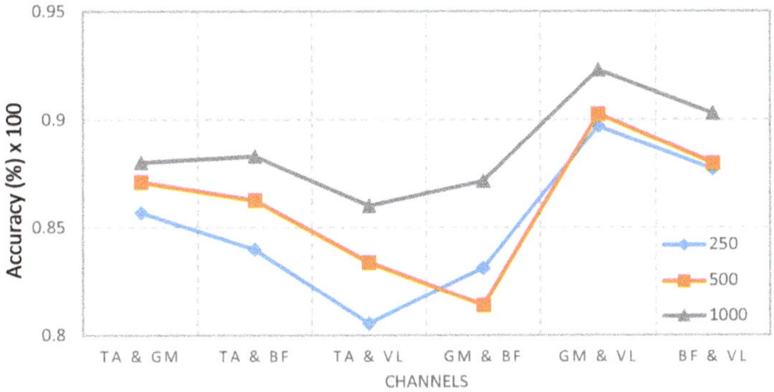

Figure 3. Classification accuracy with two channels.

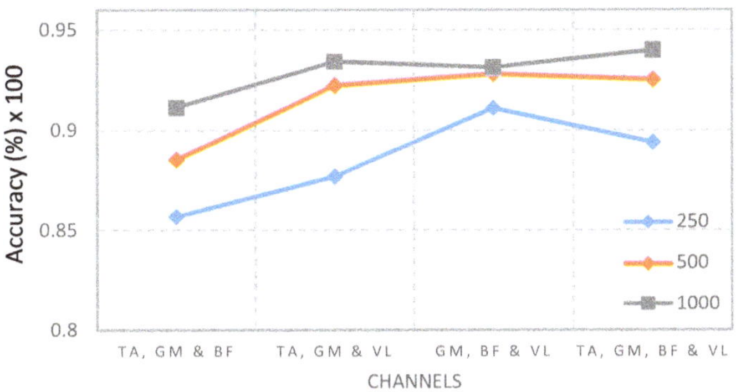

Figure 4. Classification accuracy with three and four channels.

In addition to the tables and graphics with accuracy scores, a channel forward selection of variables was also made based on the area of ROC (Receiver Operating Characteristic) curve multi-class and a classification error rate. The results obtained and their corresponding 95% confidence intervals with a sample size of 24 are shown in Table 3.

Table 3. Results of Channel forward selection of variables based on the area of ROC curve multi-class.

Step	Selection	ROC Area	ROC Area CI	C. E.	C. E. CI
1	Channel VL	0.9397	(0.8770, 1.00)	0.1952	(0.0224, 0.3680)
2	Channel GM & VL	0.9517	(0.8675, 1.00)	0.1000	(0.00, 0.2152)
3	Channel TA, GM & VL	0.9673	(0.9147, 1.00)	0.0839	(0.00, 0.1925)
4	All Channels	0.9866	(0.9426, 1.00)	0.0506	(0.00, 0.1604)

Table 3 shows that there is no statistical evidence to affirm that the increase of channels offers an improvement in the quality indicators of the classification. Similarly, there is also not enough statistical evidence either to assert that a lower quantity channels proves beneficial. Subsequently, the window size effect was analyzed in a fixed channel selection. Evidence that this does impact classifier quality indicators is illustrated in Figure 5.

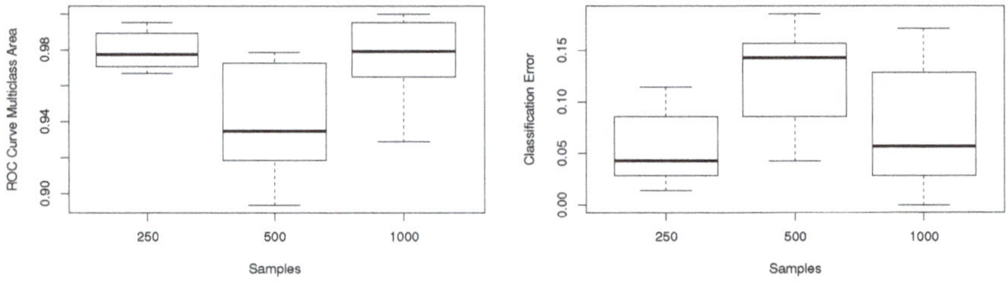

(**a**) (**b**)

Figure 5. (**a**) Area under the curve estimation of ROC curve multi-class for different window sizes using all channels. (**b**) Error classification estimation for different window sizes using all channels.

Moreover, an ANalysis Of VAriance (ANOVA) was carried out to perform the hypothesis testing, and the obtained p-values were 0.0402 and 0.00768 for the effect of the area under the curve for the ROC curve multi-class and the error classification, respectively.

Furthermore, Figure 5 shows that a 250- or 1000-sample window size has a similar accuracy classification, i.e., the percentage of true positives increases in relation to the number of true positives and false positive resulting a positive effect. Furthermore, this same quantity also increases in comparison with the sum of false negative cases with true positive cases.

5. Discussion

In the first stage, the experiments were developed only with four subjects, three women and one man, the other three men and one woman were added in the second stage. We found a trend, that is, in the second stage we also obtained that the muscles individually analyzed, the one that obtained the least accuracy for the classification was the TA and the one with the highest precision was the VL.

The TA muscle (TA) presents the worst results when analyzed individually or jointly; this is probably because this is the muscle responsible for the dorsiflexion and inversion of the ankle, which helps the stabilization of the ankle during gait, so the selected movements do not require much of it. However, 100% of the classification accuracy was only obtained when this muscle was taken into account. However, it is also a muscle with a relatively small volume, compared to the others; this implies that the potential differential generated at the moment of movement is more difficult to measure. The muscle that offered the highest precision was the VL muscle.

On the other hand, it was expected that the difference in the accuracy of the classification, when increasing more channels, was significantly higher; however, the better results with two and three channels were similar to four channels, with the biggest difference being in the number of samples selected. In this sense, it was observed that when duplicating the number of samples, from 500 to 1000, the difference was not higher than 5% in most cases, so it is considered that it is not necessary to have such a large window size.

6. Conclusions

The obtained results with four channels were better than those with one single channel, but the difference with two and three channels is negligible. Even with 250 sample size, the results in three channels were better on average compared with four channels. The muscle with the worst performance was the TA. Additionally, the best results are obtained by taking the signal of opposing muscles. Finally, this work aims to help the researcher decide how necessary it is to increase the resources used in the classification process to obtain the accuracy that is required.

Nevertheless, owing to the observed response variability presents a reduction as the number of channels increases, it is recommended employ a high number of channels to avoid changes in the classification by factors of sample size or subject. However, by considering just two channels, it is possible to achieve the same accuracy by making some adjustments to the classification algorithm.

Author Contributions: Conceptualization, D.C.T.-P. and J.R.-R.; Methodology, D.C.T.-P.; Software, D.C.T.-P.; Validation, D.C.T.-P., W.J.P.-G., and M.A.M.-P.; Formal analysis, D.C.T.-P.; Investigation and Visualization, D.C.T.-P., W.J.P.-G. and J.R.-R.; Data curation, D.C.T.-P., W.J.P.-G., and M.A.M.-P.; Writing—original draft preparation; Writing—original draft, review & editing, all the authors.

Funding: This research was funded by the "Consejo Nacional de Ciencia y Tecnología (CONACYT)" under the scholarship 561144.

Acknowledgments: We would like to thank the Graduate Studies Division from the Faculty of Computing at Universidad Autónoma de Querétaro by allowing me to make Ph.D. studies.

Conflicts of Interest: The authors declare no conflict of interest.

Abbreviations

The following abbreviations are used in this manuscript:

EMG	Electromyography
MAV	Mean Absolute Value
ZC	Zero Crossings
WL	Waveform Length
SSC	Slope Sign Changes
SVM	Support Vector Machine
PCA	Principal Component Analysis
TA	Tibialis Anterioris
GM	Gastroctemius Medials
BF	Biceps Femoris
VL	Vastus Lateralis
QP	Quadratic Programming
ADC	Analog Digital Converter
LP	Lift the toe
LT	Lift the heel
PD	Toe to the right
PI	Toe to the left
AT	Recharge on the heel
AP	Recharge on the toe
RR	Rest of the foot
ROC	Receiver Operating Characteristic
ANOVA	ANalysis Of VAriance

References

1. Oskoei, M.A.; Hu, H. Support Vector Machine-based classification scheme for myoelectric control applied to upper limb. *IEEE Trans. Biomed. Eng.* **2008**, *55*, 1956–1965. [CrossRef] [PubMed]
2. She, Q.; Luo, Z.; Meng, M.; Xu, P. Multiple kernel learning SVM-based EMG pattern classification for lower limb control. In Proceedings of the 11th International Conference on Control Automation Robotics Vision (ICARCV), Singapore, 7–10 December 2010; pp. 2109–2113.
3. Englehart, K.; Hudgins, B. A robust, real-time control scheme for multifunction myoelectric control. *IEEE Trans. Biomed. Eng.* **2003**, *50*, 848–854. [CrossRef] [PubMed]
4. Raj, S.; Ray, K.C. ECG Signal Analysis Using DCT-Based DOST and PSO Optimized SVM. *IEEE Trans. Instrum. Meas.* **2017**, *66*, 470–478. [CrossRef]

5. Sukawattanavijit, C.; Chen, J.; Zhang, H. GA-SVM Algorithm for Improving Land-Cover Classification Using SAR and Optical Remote Sensing Data. *IEEE Geosci. Remote Sens. Lett.* **2017**, *14*, 284–288. [CrossRef]
6. AlOmari, F.; Liu, G. Novel hybrid soft computing pattern recognition system SVM-GAPSO for classification of eight different hand motions. *Optik* **2015**, *126*, 4757–4762. [CrossRef]
7. Meng, M.; Luo, Z.; She, Q.; Ma, Y. Automatic recognition of gait mode from EMG signals of lower limb. In Proceedings of the 2nd International Conference on Industrial Mechatronics and Automation, Wuhan, China, 30–31 May 2010; pp. 282–285.
8. Chan, A.D.C.; Englehart, K.B. Continuous myoelectric control for powered prostheses using hidden Markov models. *IEEE Trans. Biomed. Eng.* **2005**, *52*, 121–124. [CrossRef] [PubMed]
9. Fukuda, O.; Tsuji, T.; Kaneko, M.; Otsuka, A. A human-assisting manipulator teleoperated by EMG signals and arm motions. *IEEE Trans. Robot. Automat.* **2003**, *19*, 210–222. [CrossRef]
10. Vuskovoc, M.; Du, S. Classification of prehensile EMG patterns with simplified fuzzy ARTMAP networks. In Proceedings of the 2002 International Joint Conference on Neural Networks (IJCNN'02), Honolulu, HI, USA, 12–17 May 2002; pp. 2539–2544.
11. Englehart, K.B.; Hudgin, B.; Parker, P.A. A wavelet-based continuous classification scheme for multifunction myoelectric control. *IEEE Trans. Biomed. Eng.* **2001**, *48*, 302–311. [CrossRef] [PubMed]
12. Park, S.H.; Lee, S.P. EMG pattern recognition based on artificial intelligence techniques. *IEEE Trans. Rehabil. Eng.* **1998**, *6*, 400–405. [CrossRef] [PubMed]
13. Hudgins, B.; Parker, P.; Scott, R.N. A new strategy for multifunction myoelectric control. *IEEE Trans. Biomed. Eng.* **1993**, *40*, 82–94. [CrossRef] [PubMed]
14. Purushothaman, G.; Vikas, R. Identification of a feature selection based pattern recognition scheme for finger movement recognition from multichannel EMG signals. *Australas Phys. Eng. Sci. Med.* **2018**, *41*, 549–559. [CrossRef] [PubMed]
15. Li, N.; Zhou, L.; Li, W.; Liu, Y.; Wang, J.; He, P. Protective effects of ginsenosides Rg1 and Rb1 on an Alzheimer's disease mouse model: A metabolomics study. *J. Chromatogr. B Anal. Technol. Biomed. Life Sci.* **2015**, *985*, 54–61. [CrossRef] [PubMed]
16. Miller, J.D.; Beazer, M.S.; Hahn, M.E. Myoelectric Walking Mode Classification for Transtibial Amputees. *IEEE Trans. Biomed. Eng.* **2013**, *60*, 2745–2750. [CrossRef] [PubMed]
17. Lucas, M.F.; Gaufriau, A.; Pascual, S.; Doncarli, C.; Farina, D. Multi-channel surface EMG classification using support vector machines and signal-based wavelet optimization. *Biomed. Signal Process. Control* **2008**, *3*, 169–174. [CrossRef]
18. Ceseracciu, E.; Reggiani, M.; Sawacha, Z.; Sartori, M.; Spolaor, F.; Cobelli, C.; Pagello, E. SVM classification of locomotion modes using surface electromyography for applications in rehabilitation robotics. In Proceedings of the 19th International Symposium in Robot and Human Interactive Communication, Viareggio, Italy, 13–15 September 2010; pp. 165–170.
19. Alizadeh, J.; Vahid, A.; Bahrami, F. Recognizing subjects who are learned how to write with foot from unlearned subjects using EMG signals. In Proceedings of the 23rd Iranian Conference on Biomedical Engineering and 2016 1st International Iranian Conference on Biomedical Engineering (ICBME), Tehran, Iran, 23–25 November 2016; pp. 331–335.
20. Phinyomark, A.; Phukpattaranont, P.; Limsakul, C. Feature reduction and selection for EMG signal classification. *Expert Syst. Appl.* **2012**, *39*, 7420–7431. [CrossRef]
21. Pancholi, S.; Joshi, A.M. Portable EMG Data Acquisition Module for Upper Limb Prosthesis Application. *IEEE Sens. J.* **2018**, *18*, 3436–3443. [CrossRef]
22. Oskoei, M.A.; Hu, H. GA-based Feature Subset Selection for Myoelectric Classification. In Proceedings of the IEEE International Conference on Robotics and Biomimetics (ROBIO 2006), Kunming, China, 17–20 December 2006; pp. 1465–1470.
23. Al-Ani, A.; Koprinska, I.; Naik, G.; Khushaba, R.N. A dynamic channel selection algorithm for the classification of EEG and EMG data. In Proceedings of the International Joint Conference on Neural Networks (IJCNN), Vancouver, BC, Canada, 24–29 July 2016; pp. 4076–4081.
24. Al-Timemy, A.H.; Bugmann, G.; Outram, N.; Escudero, J. Single channel-based myoelectric control of hand movements with Empirical Mode Decomposition. In Proceedings of the Annual International Conference of the IEEE Engineering in Medicine and Biology Society, Boston, MA, USA, 30 August–3 September 2011; pp. 6059–6062.

25. Pulido-Rojas, C.; Solaque-Guzmán, L.; Velasco-Toledo, N. Weed recognition by SVM texture feature classification in outdoor vegetable crops images. *Ing. Investig.* **2017**, *37*, 68–74. [CrossRef]
26. Yang, M.; Zheng, H.; Wang, H.; McClean, S. Feature selection and construction for the discrimination of neurodegenarative diseases based on gait analysis. In Proceedings of the 3rd International Conference on Pervasive Computing Technologies for Healthcare, London, UK, 1–3 April 2009; pp. 1–7.
27. Jolliffe, I.T. *Principal Component Analysis*, 2nd ed.; Springer: Aberdeen, UK, 2002; pp. 21–26.
28. Dunteman, G.H. *Principal Components Analysis*, 1st ed.; SAGE: Des Moines, IA, USA, 1989; pp. 15–46.
29. Sacco, I.C.; Gomes, A.A.; Otuzi, M.E.; Pripas, D.; Onodera, A.N. A method for better positioning bipolar electrodes for lower limb EMG recordings during dynamic contractions. *J. Neurosci. Methods* **2009**, *180*, 133–137. [CrossRef] [PubMed]
30. Pearson, F.R.S.K. LIII on lines and planes of closest fit to systems of points in space. *Philos. Mag. Ser. 6* **1901**, *2*, 559–572. [CrossRef]
31. Hotelling, H. Analysis of a complex of statistical variables into principal components. *J. Educ. Psychol.* **1933**, *24*, 417–441. [CrossRef]
32. Vapnik, V.; Corinna, C. Support-Vector Networks. *Mach. Learn.* **1995**, *20*, 273–297. [CrossRef]
33. Pérez-Hernández, L.P.; Mora-Flórez, J.J.; Bedoya-Cebayos J. A linear approach to determining an SVM-based fault locator's optimal parameters. *Ing. Investig.* **2009**, *29*, 76–81.
34. Afsharipour, B.; Soedirdjo, S.; Merletti, R. Two-dimensional surface EMG: The effects of electrode size, interelectrode distance and image truncation. *Biomed. Signal Process. Control* **2019**, *49*, 298–307. [CrossRef]
35. Chang, C.C.; Lin, C.J. LIBSVM: A Library for Support Vector Machines. *ACM Trans. Intell. Syst. Technol.* **2011**, *2*, 1–27. [CrossRef]

electronics

MDPI

Article

A Countermeasure against DPA on SIMON with an Area-Efficient Structure

Yuanyuan Zhang, Ning Wu *, Fang Zhou, Jinbao Zhang and Muhammad Rehan Yahya

College of Electronic and Information Engineering, Nanjing University of Aeronautics and Astronautics, Nanjing 211106, China; zhangyuanyuan@nuaa.edu.cn (Y.Z.); zfnuaa@nuaa.edu.cn (F.Z.); zjb4050811@126.com (J.Z.); rehanyahya@yahoo.com (M.R.Y.)

* Correspondence: wunee@nuaa.edu.cn; Tel.: +86-139-5189-3307

Received: 8 January 2019; Accepted: 19 February 2019; Published: 20 February 2019

Abstract: Differential power analysis (DPA) is an effective side channel attack method, which poses a critical threat to cryptographic algorithms, especially lightweight ciphers such as SIMON. In this paper, we propose an area-efficient countermeasure against DPA on SIMON based on the power randomization. Firstly, we review and analyze the architecture of SIMON algorithm. Secondly, we prove the threat of DPA attack to SIMON by launching actual DPA attack on SIMON 32/64 circuit. Thirdly, a low-cost power randomization scheme is proposed by combining fault injection with double rate technology, and the corresponding circuit design is implemented. To the best of our knowledge, this is the first scheme that applies the combination of fault injection and double rate technology to the DPA-resistance. Finally, the *t*-test is used to evaluate the security mechanism of the proposed designs with leakage quantification. Our experimental results show that the proposed design implements DPA-resistance of SIMON algorithm at certain overhead the cost of 47.7% LUTs utilization and 39.6% registers consumption. As compared to threshold implementation and bool mask, the proposed scheme has greater advantages in resource consumption.

Keywords: differential power analysis (DPA), SIMON; fault injection; double rate; power randomization

1. Introduction

Differential power analysis (DPA) is a typical side channel attack method that performs a correlation analysis by collecting the dynamic power consumption of the operation. According to the correlation between sensitive information in the operation and the instantaneous power consumption of the CMOS circuit, DPA attack can complete the stealing of the key information of the circuit. Because of its high efficiency and operability, DPA has posed a serious threat to the security of integrated circuits.

SIMON algorithm is a lightweight block cryptographic algorithm proposed by the National Security Bureau in 2013, which is mainly used for resource-constrained encryption applications such as radio frequency identification (RFID) tags, Internet of Things (IoT) sensors referenced in [1–3]. Due to a pursuit of compact structure, SIMON sacrifices part of security which leads to the fact that encryption intensity cannot be matched with advanced encryption standard (AES) algorithm [4]. Reference [5] pointed out that the security of lightweight ciphers can be theoretically guaranteed by increasing the number of encryption rounds of the algorithm, but the round function of the lightweight cryptographic algorithm such as SIMON is too simplified and with no strong security [6]. That leads to security and privacy concerns of IoT devices, especially wearable devices. Accordingly, it is of great significance to carry out study on attacks and countermeasures on lightweight cryptography and seek a strategy to thwart side channel attacks at low resource utilization.

At present, countermeasures against power consumption attack can be divided into circuit level, algorithm level, and transistor level [7]. According to the different application scenarios, conventional countermeasures include power randomization and constant power consumption. The current researches on the power attack resistant measures for the lightweight cipher algorithm are mainly focused on some classic methods, such as the random mask used in Ref. [8], the bool mask in Ref. [2], and the threshold implementation in Refs. [9,10]. These classic countermeasures can indeed provide power attack resilience for lightweight cryptographic algorithms, however the consumption of a large number of resources makes it contrary to the design philosophy of lightweight cipher algorithm.

In this work, we propose a compact countermeasure against DPA attack on SIMON by using power randomization method. In order to reduce the consumption of additional resources, a power randomization design scheme based on fault injection and double rate technology is proposed in this paper. By randomly injecting a 1-bit fault into the plaintext, a random data will be generated according to the fault propagation characteristics of SIMON, which can be used to complete the power consumption randomization. The encrypted operation of fault plaintext is randomly inserted into the first half cycle or the second half cycle of normal encrypted operation by double rate technology so that the attacker cannot accurately locate the position of each round of encryption operations in the power curve. Compared with existing countermeasure based on the threshold implementation and bool mask [2,6,11], our scheme is area-efficient.

The rest of this manuscript is organized as follows: Section 2 introduces SIMON algorithm in detail. Section 3 analyzes the feasibility of DPA attack on SIMON encryption algorithm according to the principle of DPA, and the attack on SIMON 32/64 is carried out on SAKURA side channel attack board, which proves the threat of DPA to SIMON. Section 4 details the compact countermeasure against DPA attack on SIMON through power randomization. In order to reduce the circuit area, we propose a power randomization scheme based on random fault injection and double rate technology. We also detail the design of the fault injection circuit, the double rate circuit, and the random bit generator, and give the resource consumption of the designed anti-DPA SIMON circuit under the Xilinx xc7k160tffg-1 FPGA. In Section 5, we study the practical security of the proposed designs with leakage quantification. Section 6 summarizes the conclusions of this work.

2. Background

2.1. Notation

- m: the keyword size in SIMON algorithm
- n: the word size in SIMON algorithm
- T: the round number of SIMON
- L_i, R_i: the left and right half output of the ith round
- $L_i(j)$, $R_i(j)$: the jth bit of L_i, R_i, $j \in \{1, \ldots, n\}$
- k_i: the ith of the master-key group, $i \in \{1, \ldots, m\}$
- K_i: the ith of round-key, $i \in \{1, \ldots, T\}$
- $K_i(j)$: the jth bit of K_i, $j \in \{1, \ldots, n\}$
- L^*, R^*: the left and right half faulty output of the each round
- $L_i^*(j)$: the jth bit of L^*, $j \in \{1, \ldots, n\}$
- $P_L(i)$, $P_R(i)$: the ith bit of left and right half part of plaintext, $i \in \{1, \ldots, n\}$

2.2. Description of SIMON

SIMON is a typical cryptographic algorithm of Feistel structure. The algorithm has a group size of $2n$ ($n = 16, 24, 32, 48, 64$) and a key size of mn ($m = 2, 3, 4$). The combination of m and n can constitute the SIMON $2n/mn$ algorithm, which is called the SIMON family cryptography algorithm [12].

The SIMON $2n/mn$ consists of two parts: the round operation and the key generation. According to different modes, the algorithm need to perform j ($j = 32, 36, 42, 44, 52, 54, 68, 69, 72$) rounds of encryption operations repeatedly. The ith round encryption operation can be estimated by the Equation (1):

$$\begin{cases} L_i = F(L_{i-1}) \oplus R_{i-1} \oplus k_{i-1} \\ R_i = L_{i-1} \end{cases} \tag{1}$$

where the function F can be manifested as:

$$F(x) = (x <<< 8)\&(x <<< 1) \oplus (x <<< 2) \tag{2}$$

The round key of SIMON algorithm is generated from the master key. The master key is expressed as $\{k_1, k_2, ..., k_m\}$ and the round key is denoted as $\{K_1, K_2, ..., K_{2n}\}$. According to different keys length, the calculation methods of round keys can be described as follow: If $I <= m$, then $K_i = k_i$, otherwise round key generation can be expressed by Equation (3):

$$\begin{cases} m = 2 : K_i = c \oplus (z_j)_{i-m} \oplus K_{i-m} \oplus (K_{i-m+1} >>> 3) \oplus (K_{i-m+1} >>> 4) \\ m = 3 : K_i = c \oplus (z_j)_{i-m} \oplus K_{i-m} \oplus (K_{i-m+2} >>> 3) \oplus (K_{i-m+2} >>> 4) \\ m = 4 : K_i = c \oplus (z_j)_{i-m} \oplus K_{i-m} \oplus K_{i-m+1} \oplus (k_{i-m+1} >>> 1) \oplus (K_{i-m+3} >>> 3) \oplus (K_{i-m+3} >>> 4) \end{cases} \tag{3}$$

where z is a sequence discussed in Ref [1] and c is a constant determined by the algorithm parameter n, which can be described as $c = 2^n - 4$. We used the linear feedback shift register (*LFSR*) circuit with characteristic polynomial as x^4+x^2+x+1 to generate z. Taking computation of round key of SIMON 32/64 as an example, the key generation circuit structure is shown in Figure 1.

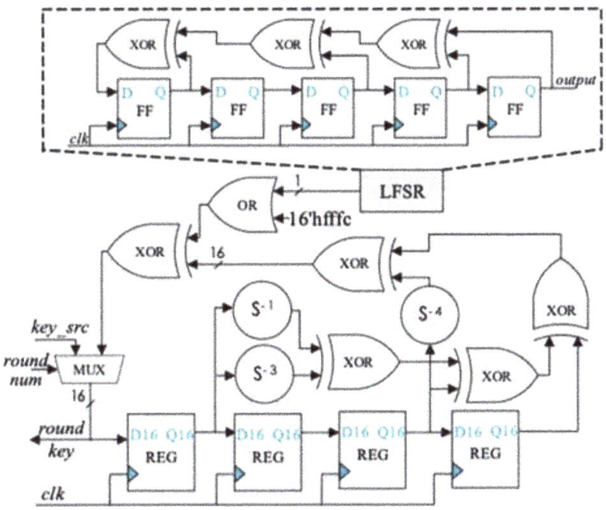

Figure 1. Key generation circuit structure of SIMON 32/64.

3. Differential Power Attack on SIMON

Because of its lightweight structure, SIMON circuit has weak resistance to power analysis attack. Some studies have clearly demonstrated or implemented the crack of SIMON circuits through power analysis such as Refs [2,8,10,11]. The purpose of this section is to show intuitively through analysis and experiment that the SIMON circuit can be cracked by simple differential power analysis.

3.1. Selection of Power Model

The power consumption of CMOS integrated circuits consists of leakage power, short-circuit power and charge–discharge power. In the current process, leakage power and short circuit power consumption are relatively small, and the power consumption of the CMOS circuit mainly comes from the charging and discharging of the capacitive load. The charging and discharging power consumption is formed by the electrical level change of the output logic, accounting for more than 60% of the total power consumption of the chip.

According to the power consumption characteristics of CMOS circuits, Hamming Weight (HW) and Hamming Distance (HD) are the two most popular power consumption models. In addition, some more accurate models such as "switch distance" have been proposed in Ref [13] to improve the compatibility between the model and the actual power consumption. It is well known that the more precise the power model is, the more accurate the power analysis attack results will be. But those complex models make it difficult to implement power analysis attack. Therefore, HW and HD are still the most mainstream power consumption models for side channel analysis.

For power attack on SIMON circuit, Ref [13] uses HD model while Ref [2] uses a modified HD model to improve accuracy. This paper intends to reduce the impact of environmental noise by increasing the number of power traces, and to complete the power attack on SIMON circuit with a simpler and more practical HW model. HW is a power consumption representation method based on statistics, which represents the power consumption of the circuit by the number of high-level nodes in the circuit. It is usually used to simulate the power consumption for side channel analysis such as Refs [14,15]. When using the HW model, the power consumption of circuit can be expressed as:

$$\widetilde{P} \approx kHW(Y) + n \tag{4}$$

where k denoted as the proportional coefficient between HW and power consumption, and n represents the noise in the circuit. Y represents the current state of the circuit.

3.2. Implementation of DPA on SIMON

We take the SIMON 32/64 circuit with a cyclic structure as the target to perform our DPA attack. According to the encryption process of SIMON algorithm, we chose the third round of SIMON algorithm as the attack position and lowest bit of third-round operation can be meant as Equation (5).

$$L_3(1) = K_2(1) + R_2(1) + L_2(15) + (L_2(16)\&L_2(9)) \tag{5}$$

The expression shows that there is a non-linear relationship between $L_3(1)$, $L_2(16)$, and $L_2(9)$. According to the expression of round function of SIMON, Equation (5) can be further expanded into expressions of plaintext and round key as Equation (6):

$$\begin{aligned} L_3(1) = {} & K_2(1) + K_1(15) + L_1(1) + R_1(15) + L_1(13) + [L_1(14)\,\&\,L_1(7)] + \\ & \{[K_1(16) + R_1(16) + L_1(14) + (L_1(15)\,\&\,L_1(8))]\,\&\,[K_1(9) + R_1(9) + L_1(7) + (L_1(8)\,\&\,L_1(1))]\} \end{aligned} \tag{6}$$

The plaintext in Equation (6) can be divided into three parts, marked with different colors. Because the plaintext of each part has a linear relationship with the key bit, so we can select one bit of each part as a representative, and the others can be set to 0. Here, $L_1(7)$ and $L_1(14)$ are selected as the representations; then Equation (6) can be further simplified.

$$L_3(1) = K_2(1) + K_1(15) + [L_1(14)\&L_1(7)] + \{[K_1(16) + L_1(14)]\&[K_1(9) + L_1(7)]\} \tag{7}$$

The constraints condition for the establishment of Equation (7) is $L_1(1)$, $L_1(13)$, $L_1(15)$, and $L_1(8)$ bits are all 0. If we only to deduce the $K_1(16)$ and $K_1(9)$, owing to the $K_2(1)$ and $K_1(15)$ involve only linear operations and have no effect the results of DPA, thus the Equation (7) can be simplified as follows:

$$L_3(1) = [L_1(14)\&L_1(7)] + \{[K_1(16) + L_1(14)]\&[K_1(9) + L_1(7)]\} \tag{8}$$

The derived Equation (8) is the discriminant function at the location. Equation (8) shows that the Hamming Weight at $L_3(1)$ is determined by the plaintext combination $\{L_1(14), L_1(7)\}$ and the key combination $\{K_1(16), K_1(9)\}$. By enumerating the plaintext combination $\{L_1(14), L_1(7)\}$, the $K_1(16)$ and $K_1(9)$ bits can be decoded by DPA attack. According to the deduction method, the discriminant function of the rest of the first round key is shown in Table 1.

Table 1. The discriminant function of first round key.

Attack Position and Distinguishing Function	Attack Bits	Correlation Bits	Constraint Condition
$L_3(1) = [L_1(14)\&L_1(7)]\oplus\{[K_1(16)\oplus L_1(14)]\&[K_1(9)\oplus L_1(7)]\}$	$K_1(16)$ $K_1(9)$	$L_1(14)$ $L_1(7)$	$L_1(1)$ $L_1(13)$ $L_1(15)$ $L_1(8)$
$L_3(3) = [L_1(16)\&L_1(9)]\oplus\{[K_1(2)\oplus L_1(16)\oplus(L_1(1)\&L_1(10))]\&[K_1(11)\oplus L_1(9)]\}$	$K_1(2)$ $K_1(11)$	$L_1(16)$ $L_1(9)$ $L_1(10)$ $L_1(1)$	$L_1(3)$ $L_1(15)$
$L_3(4) = L_1(16)\oplus[L_1(1)\&L_1(10)]\oplus\{[K_1(3)\oplus L_1(1)]\&[K_1(12)\oplus L_1(10)]\}$	$K_1(3)$ $K_1(12)$	$L_1(16)$ $L_1(1)$ $L_1(10)$	$L_1(4)$ $L_1(11)$ $L_1(2)$
$L_3(5) = L_1(5)\oplus[L_1(2)\&L_1(11)]\oplus\{[K_1(4)\oplus(L_1(3)\&L_1(12))]\&[K_1(13)\oplus(L_1(12)\&L_1(5))]\}$	$K_1(4)$ $K_1(13)$	$L_1(3)$ $L_1(12)$ $L_1(5)$ $L_1(2)$ $L_1(11)$	$L_1(1)$
$L_3(6) = L_1(2)\oplus[L_1(3)\&L_1(12)]\oplus\{[K_1(5)\oplus L_1(3)]\&[K_1(14)\oplus L_1(12)]\}$	$K_1(5)$ $K_1(14)$	$L_1(3)$ $L_1(12)$ $L_1(2)$	$L_1(6)$ $L_1(13)$ $L_1(4)$
$L_3(7) = L_1(7)\oplus L_1(3)\oplus\{[K_1(6)\oplus(L_1(5)\&L_1(14)]\&[K_1(15)\oplus(L_1(14)\&L_1(7))]\}$	$K_1(6)$ $K_1(15)$	$L_1(7)$ $L_1(3)$ $L_1(5)$ $L_1(14)$	$L_1(4)$ $L_1(13)$
$L_3(8) = [L_1(5)\&L_1(14)]\oplus\{[K_1(7)\oplus L_1(5)]\&[K_1(16)\oplus L_1(14)]\}$	$K_1(7)$ $K_1(16)$	$L_1(5)$ $L_1(14)$	$L_1(4)$ $L_1(8)$ $L_1(6)$ $L_1(15)$ $L_1(16)$ $L_1(9)$
$L_3(9) = L_1(9)\oplus\{[K_1(8)\oplus(L_1(7)\&L_1(16))]\&[K_1(1)\oplus(L_1(16)\&L_1(9))]\}$	$K_1(8)$ $K_1(1)$	$L_1(9)$ $L_1(7)$ $L_1(16)$	$L_1(5)$ $L_1(6)$ $L_1(15)$
$L_3(10) = L_1(10)\oplus[L_1(7)\&L_1(16)]\oplus\{[K_1(9)\oplus L_1(7)\oplus(L_1(8)\&L_1(1))]\&[K_1(2)\oplus L_1(16)\oplus(L_1(1)\&L_1(10))]\}$	$K_1(9)$ $K_1(2)$	$L_1(10)$ $L_1(7)$ $L_1(16)$ $L_1(8)$ $L_1(1)$	$L_1(6)$
$L_3(11) = L_1(7)\oplus[L_1(8)\&L_1(1)]\oplus\{[K_1(10)\oplus L_1(8)]\&[K_1(3)\oplus L_1(1)]\}$	$K_1(10)$ $K_1(3)$	$L_1(7)$ $L_1(8)$ $L_1(1)$	$L_1(11)$ $L_1(2)$ $L_1(9)$
$L_3(14) = L_1(14)\oplus\{[K_1(13)\oplus(L_1(12)\&L_1(5))]\&[K_1(6)\oplus(L_1(5)\&L_1(14))]\}$	$K_1(13)$ $K_1(6)$	$L_1(14)$ $L_1(12)$ $L_1(5)$	$L_1(4)$ $L_1(10)$ $L_1(11)$
$L_3(15) = [L_1(12)\&L_1(5)]\oplus\{[K_1(14)\oplus L_1(12)]\&[K_1(7)\oplus L_1(5)]\}$	$K_1(14)$ $K_1(7)$	$L_1(12)$ $L_1(5)$	$L_1(15)$ $L_1(13)$ $L_1(6)$ $L_1(11)$

For the reason that DPA is a statistical-based attack method, the power consumption curve collected during the attack must reach a certain threshold to meet the statistical law. Therefore, during the attack process, the groups whose constraint conditions and the correlation bits have no conflicts can be selected to attack at the same time. In this way, not only the number of consumption curve is increased, but also the cracking efficiency is improved. Taking the attack process to key group $\{K_1(16), K_1(14), K_1(13), K_1(9), K_1(7), K_1(5), K_1(4)\}$ as example, the correlation bits that need to be enumerated are $\{L_1(14), L_1(12), L_1(11), L_1(7), L_1(5), L_1(3), L_1(2)\}$, and a total of 128 kinds of plaintext are needed to enumerate the seven plaintext bits. To reduce the error caused by environmental noise, each plaintext is collected 50 times, and only 6400 power consumption curves are needed to complete the decoding of seven key bits in the first round of SIMON 32/64 algorithm.

3.3. DPA Experimental Evaluation

In this section, we present the experimental validation of the DPA attack on SIMON 32/64. In our actual attack, the key value of SIMON 32/64 is randomly set to 0x8522_a01e_83f3_a35e and $\{K_1(16), K_1(14), K_1(13), K_1(9), K_1(7), K_1(5), K_1(4)\}$ is taken as our target of retrieving.

Our DPA attack platform is shown in Figures 2 and 3, including SAKURA-X board, Multi-channel digital storage oscilloscope, and PC. Our DUT (device under test) i.e., SIMON 32/64 circuit implemented on a Xilinx Kintex-7 FPGA mounted on a SAKURA-X board and the Spartan-6 FPGA on SAKURA-X board is used as a control chip to apply the excitation signal to the DUT and transfer the encryption results to the PC via the USB. At the same time, Spartan-6 FPGA also triggers a signal after each new excitation is applied to start the record of power consumption.

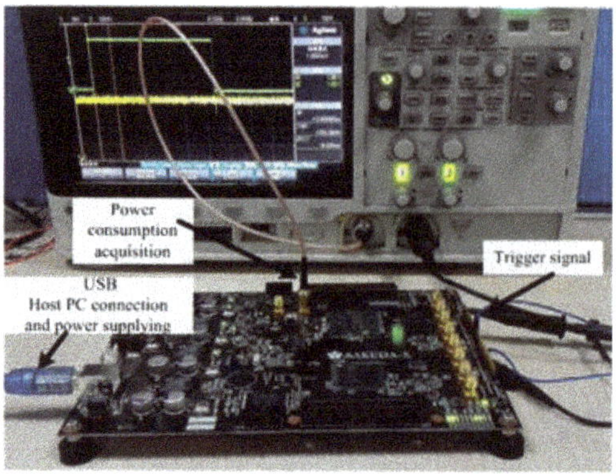

Figure 2. Photo of differential power analysis (DPA) attack platform.

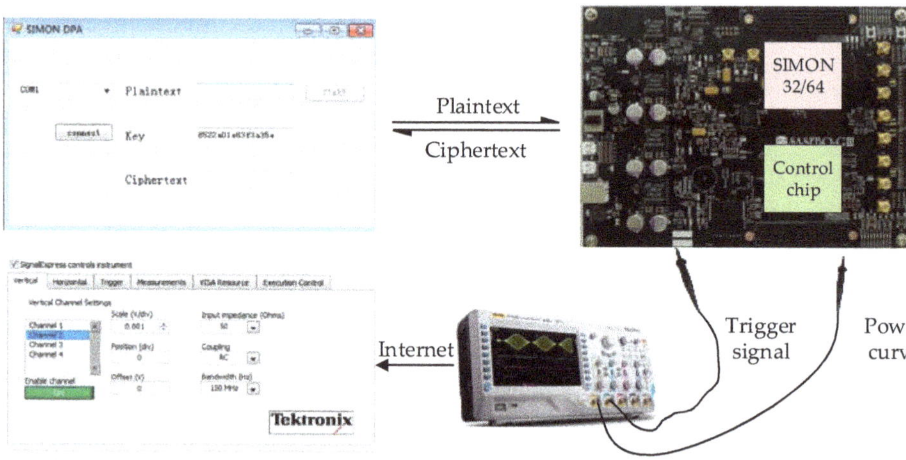

Figure 3. The structure of DPA attack platform.

According to the previous analysis, we collected a total of 600 curves by enumerating the relevant bits of the plaintext multiple times. Through the calculation of the average of 6400 power consumption curves, the simple power analysis is completed to realize the positioning of the power attack point. Figure 4 shows a simple power analysis curve and the position of each encryption process in the power consumption curve.

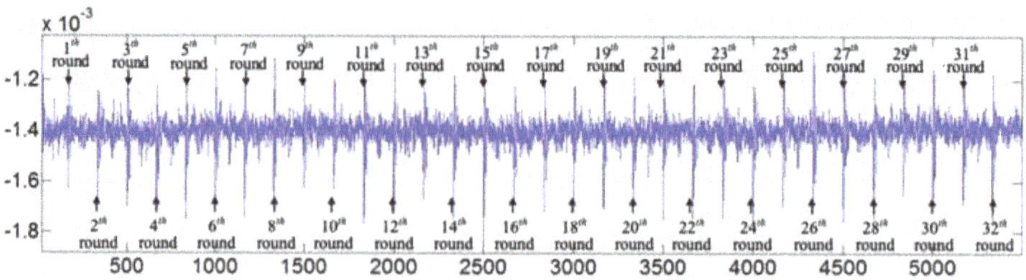

Figure 4. Simple power analysis curve and location of the encryption process.

After locating the attack position, DPA attacks are carried out on the decrypted key groups according to the discriminant function shown in Table 1. This paper develops power analysis software based on Matlab. The execution flow of differential power analysis software is shown in Figure 5.

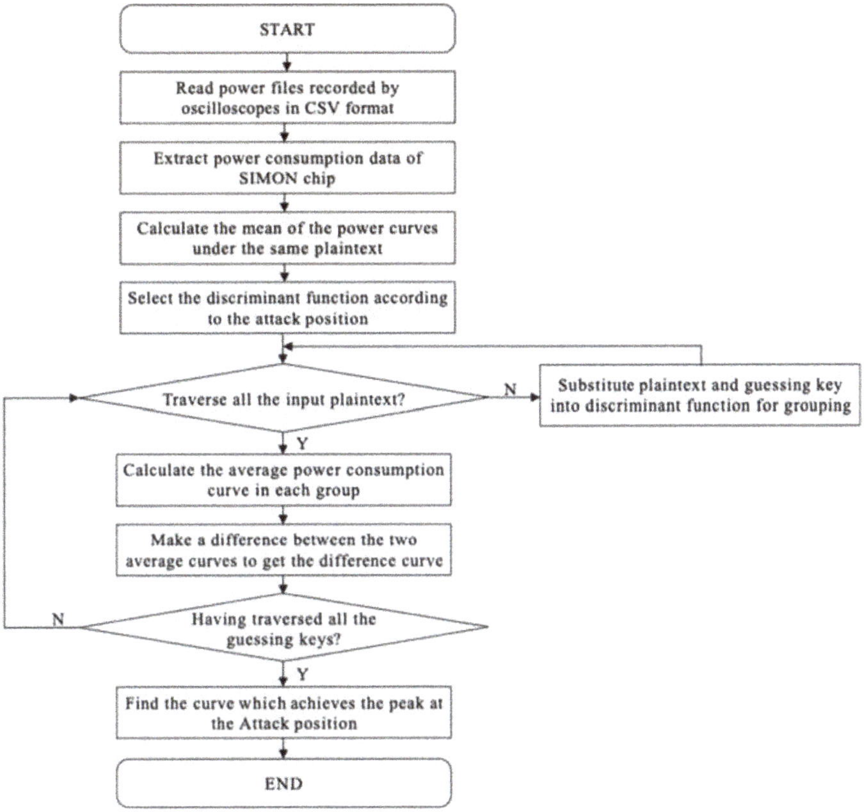

Figure 5. Flow graph of data analysis software.

Firstly, the software reads the power consumption data file of the csv format recorded by the oscilloscope, and extracts the power voltage of the SIMON chip which represents its power consumption. Secondly, it calculates the average value of the power consumption data collected under the same plaintext to reduce the impact of environmental noise on the attack results. Subsequently, a typical differential power analysis calculation is performed according to the discriminant function in Table 1.

The results of DPA attack of $\{K_1(16), K_1(14), K_1(13), K_1(9), K_1(7), K_1(5), K_1(4)\}$ are given in Figure 6.

The guessed key shown in Figure 6 is $\{K_1(16), K_1(14), K_1(13), K_1(9), K_1(7), K_1(5), K_1(4)\}$=7'b1001000, which is consistent with the preset first round key 16'h8522, indicating that the DPA attack successfully cracked the 7 bits of key.

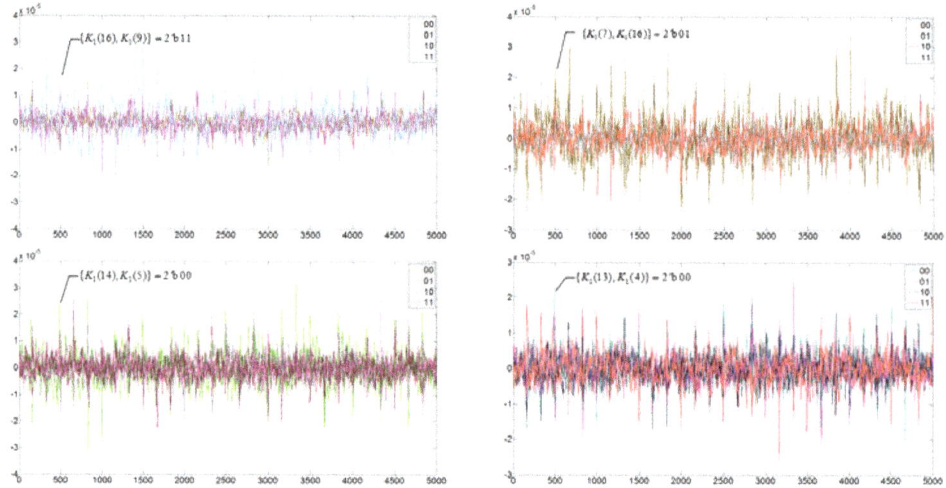

Figure 6. The results of the DPA attack of $\{K_1(16), K_1(14), K_1(13), K_1(9), K_1(7), K_1(5), K_1(4)\}$.

4. DPA-Resistant SIMON Based on Power Randomization

4.1. Design of DPA-Resistant SIMON

The lightweight cryptographic algorithm was originally designed to provide security for resource-constrained scenarios such as the IoT system. Therefore, for SIMON, the resource consumption of encryption circuits and circuit security are almost equally important. In this Section, according to the characteristics of round function on SIMON, a power randomization method for round function is proposed as a compact countermeasure against DPA. Figure 7 shows the circuit structure of the round function circuit of SIMON algorithm.

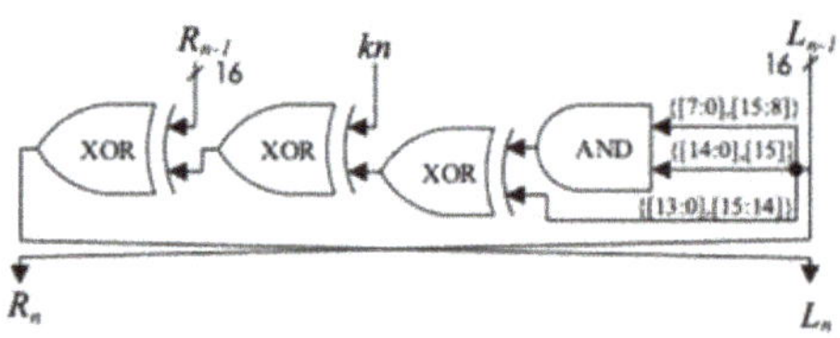

Figure 7. The structure of round function circuit.

According to the principle of DPA attack, it can be known that as long as the power consumption of the SIMON round function circuit is randomized, the DPA cannot get the key information of the circuit through the differential operation. Therefore, we can randomly insert a redundant round operation before or after each encryption round to randomize the power consumption. However, with the insertion of redundant operations, the calculation period of the round function will be doubled, and the data throughput of the whole circuit will become half of the original. In order to solve this problem, we use the double rate technique for the compact structure of SIMON algorithm round function. It can be seen from Figure 7 that the structure of the round function is quite compact, consisting only of one set of AND gates and three sets of XOR gates that means the critical path of the SIMON round circuit is quite short, and it will not become a critical path for a complex system, so the double rate technology is feasible.

In this paper, the SIMON 32/64 circuit is implemented in a cyclic structure, and the structure of a circuit optimized by the anti-power attack is shown in Figure 8.

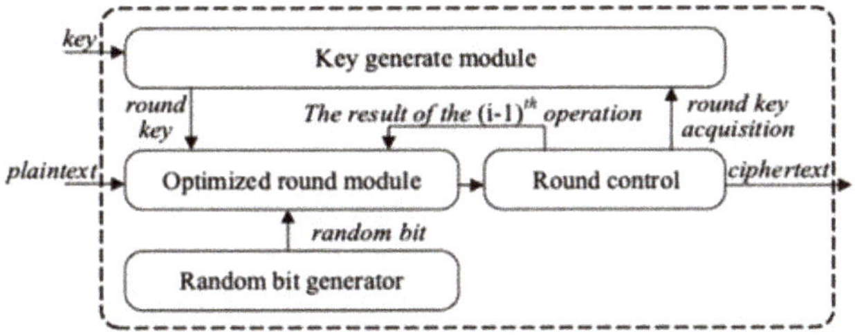

Figure 8. Overall circuit architecture of DPA-resistant SIMON.

The double rate technology was first used in the Ref [16] to resist power attacks. That work proposes to use double rate technology to pre-charge each register in round function of AES so that the Hamming distance in the encryption process will be changed randomly. However, that method needs to input a set of random data before normal encryption to randomize power consumption. The generation and preservation of multi-bit random data require a certain circuit area, which is not advisable for area-sensitive lightweight cipher circuits such as SIMON.

To complete the power randomization with low resource consumption, random data in this paper is generated by injecting a fault into the plaintext and utilizing the fault diffusion effect of SIMON algorithm. We inject a 1-bit fault into the Jth bit of plaintext, according to the operation of the round function of SIMON, the influence of this 1-bit fault on the subsequent rounds is as shown in Table 2.

Table 2. The influence of the 1-bit fault injected in the Jth bit of plaintext.

Round num	L^*	R^*
0	J	NULL
1	$\overline{J+8}$ $\overline{J+1}$ $J+2$	J
2	$\overline{J}$ $\overline{J+9}$ $\overline{J+10}$ $\overline{J+2}$ $\overline{J+3}$ $J+4$	$\overline{J+8}$ $\overline{J+1}$ $J+2$
3	$\overline{J+12}$ $\overline{J+11}$ $\overline{J+10}$ $\overline{J+8}$ $\overline{J+5}$ $\overline{J+4}$ $\overline{J+3}$ $\overline{J+2}$ $\overline{J+1}$ $J+6$	$\overline{J}$ $\overline{J+9}$ $\overline{J+2}$ $\overline{J+3}$ $\overline{J+10}$ $J+4$
...	...	...

In Table 2, we only consider the effect of the fault bit in the left half part on the subsequent encryption round. The overlined bits such as $\overline{J}$ in the table indicate there is a possibility that the location is affected by the fault bit. The reason for this phenomenon is that the round function of SIMON algorithm contains an AND operation. Taking the $(J+8)^{th}$ bit of the left part in the first round as an example, according to the SIMON round function calculation, this bit can be expressed as Equation (9).

$$L_1(J+8) = [P_L(J)\&P_L(J+7)] \oplus P_L(J+6) \oplus K_1(J+8) \oplus P_R(J+8) \quad (9)$$

Whether the $(J+8)^{th}$ bit in the first round will be affected by the fault injected in $P_L(J)$ depends on the value of $P_L(J+7)$, and $L_1(J+8)$ is affected by the fault only if $P_L(J+7)$ is 1.

As shown in Table 2, a 1-bit fault injected in the plaintext has an increasing influence on the output of each round as the number of encryption rounds increases, and the specific diffusion effect is affected by the different plaintext. Thus, it can be assumed that injecting a 1-bit fault into the plaintext will have an extremely complicated effect on the encryption of SIMON algorithm. According to the fault diffusion characteristic of SIMON algorithm, the power consumption of SIMON circuit can be randomized by introducing a 1-bit random fault into the input plaintext, which can be used to replace the random data in Ref [16]. The schematic diagram of fault injection is shown in Figure 9.

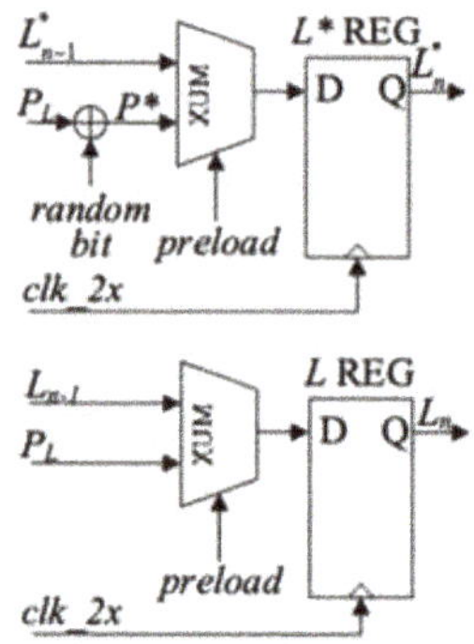

Figure 9. The schematic diagram of fault injection.

The upper part of Figure 9 schematically describes the register and fault inject circuit for the faulty plaintext, and the lower part of Figure 9 is the register to store normal left part data of round function. The specific injection circuit is shown in Figure 10. When the input plaintext preloads the L^* register, the 1-bit fault is introduced in the last two bits of the plaintext in the left half part by a random bit.

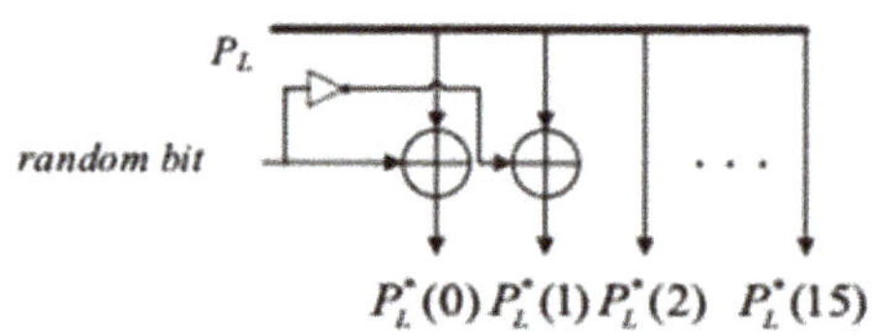

Figure 10. The structure of fault introducing circuit.

Based on this concept, a compact power attack countermeasure is proposed in this paper. Its structure is shown in Figure 11.

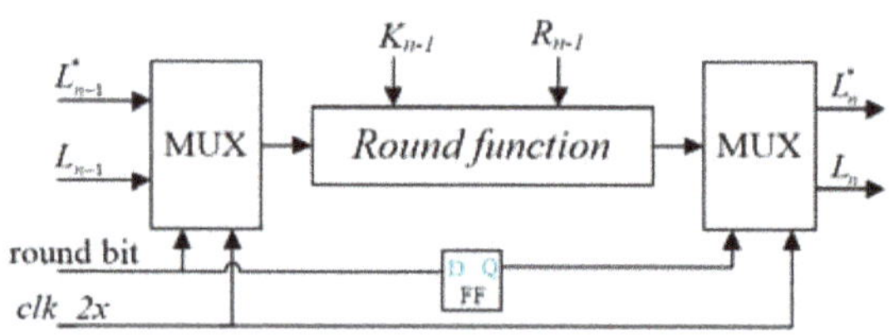

Figure 11. Block diagram of SIMON round circuit by double rate technology.

By using double rate system clock and a random selection bit, the encryption operation and the power hiding operation are carried out randomly in the first half cycle and the second half cycle of the round function circuit, which makes the attacker unable to carry out the differential analysis correctly.

4.2. Random Bit Generation Circuit

The countermeasure proposed in this paper needs to use a random bit. A two-stage ring oscillator (RO) is designed to form a random bit generator, the structure is shown in Figure 12.

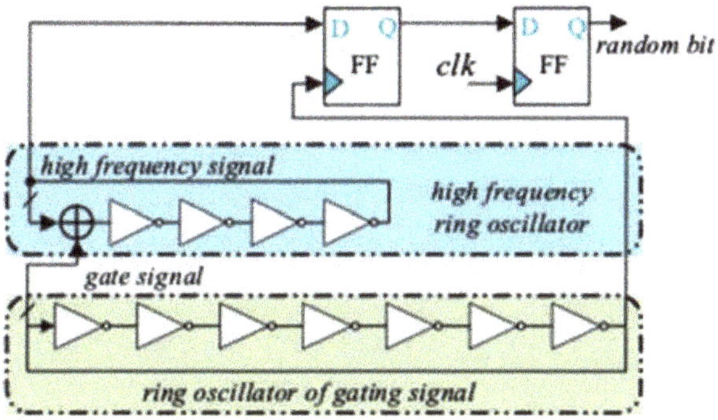

Figure 12. Circuit diagram of 1-bit random generator.

In this circuit, a long RO circuit composed of an odd number of NOT gates generates a low-frequency gating signal, a short RO circuit composed of an even number of NOT gates and an XOR gate generates a high-frequency signal. The short RO will oscillate as long as the gate signal generated by long RO chain is equal to 1. Otherwise, the short RO will stop oscillate and keep current state (Figure 13).

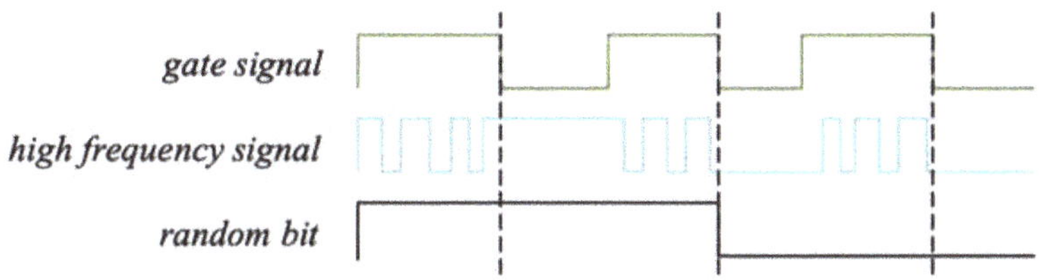

Figure 13. 1-bit random number generator working principle diagram.

The transmission delay of the gate circuit fluctuates with changes in temperature and voltage, so the period of the oscillating chain also exhibits a small random fluctuation. Since the period of the gating signal period and the high-frequency oscillating signal are all randomized when the gating signal is 0, the state of the short RO will be a random value.

The smaller the inverters number of short RO and the greater the difference in the inverters number between the two RO, the randomness of the generated bit will be better. We made the long RO contain 27 inverters, and the short RO consisted of 4 inverters as an example. This random bit generator requires 33 LUTs and 2 registers. We implemented this architecture in a Xilinx Kintex-7 series FPGA with speed grade -1, the long RO has an average oscillation frequency of 77 MHz, and the short RO has an average oscillation frequency of 478 MHz.

4.3. Implementation of Optimized SIMON Circuit

The designed circuit is verified in xc7k160tfbg676-1 FPGA and analysis of resource utilization in comparison to the original design are shown in Table 3.

Table 3. Resource utilization and performance report of original and optimized SIMON 32/64.

	Original Circuit	Optimized Circuit
LUTs	99	146
Registers	116	162
Maximum frequency	312 MHz	277 MHz
Throughput	9.75 Mbps	8.65 Mbps

The optimized circuit increases 47 LUTs and 46 registers as compared to the original circuit. The maximum frequency of the FPGA is reduced by 11.2%, but still can achieve 277 MHz, meeting the needs of most IoT and embedded systems.

Table 4 shows the comparisons with other DPA-resistant SIMON circuits of threshold implementation [11] and bool mask [2]. It can be seen that our countermeasure consume lower resource overhead and keep high performance.

Table 4. Comparisons with other DPA-resistant SIMON circuits.

	This Work	Ref [2]	Ref [11]
LUTs overhead	47.7%	141.1%	66.6%
Registers overhead	39.6%	141.1%	40.0%
Frequency overhead	11.2%	20.6%	13.4%

5. Leakage Quantification

As we all know, the countermeasures based on dual-rate technology have good resistance to power analysis which based on HD model. This paper also randomizes the execution sequence of redundant operations and normal encryption operations in dual-rate technology. The randomization of the execution time of the round operation makes the Hamming weight of any half cycle present a certain degree of randomness, which can also resist the power analysis based on the Hamming weight model.

In order to analyze the countermeasure more objectively, we use *t*-test to evaluate the practical security of the proposed designs with leakage quantification. T-test is a statistical method used to judge whether two sample sets come from the same group. It is used to evaluate the power leakage of circuits in Refs [17–19]. Compared with power analysis attack, *t*-test can quantitatively analyze the DPA-resistant ability of circuits, which is more convincing. The *t*-test is then computed on two sets, one with a fixed plaintext while the other with randomly varying plaintexts, and *t*-test can be expressed as follows:

$$t = \frac{\mu_a - \mu_b}{\sqrt{\sigma_a^2/N_a + \sigma_b^2/N_b}} \tag{10}$$

where μ_a and μ_b are the sample means of two data sets, N denotes the trace number of each set, and σ_a and σ_b refer to the standard deviation. As in Ref [11], we use $|t| > 4.5$ as a threshold to determine whether there is any information disclosure.

We executed 10,000 times of fixed plaintext and random plaintext encryption operations respectively and collect a total of 20,000 power traces. Substituting the collected power consumption values into Equation (10) to complete the *t*-test calculation. The *t*-test result of the optimized circuit is shown in Figure 14.

Figure 14a reports the power trace collected from the optimized SIMON 32/64 circuit and Figure 14b reports the *t*-test result. The original SIMON circuit that has been proven in Section 3 can be cracked by DPA attack, while the protected circuit does pass the *t*-test which again supports our claim of secrecy.

This section proves the resistance of the proposed method to DPA from both quantitative and qualitative analysis, but does not elaborate whether it can resist high-order differential analysis (HO-DPA). Although HO-DPA attacks are more complex to implement, they can crack many circuits that are resistant to common DPA attacks. It is necessary for the chip designer to conduct research on the anti-HO-DPA capabilities of cryptographic chips, which will be our further work.

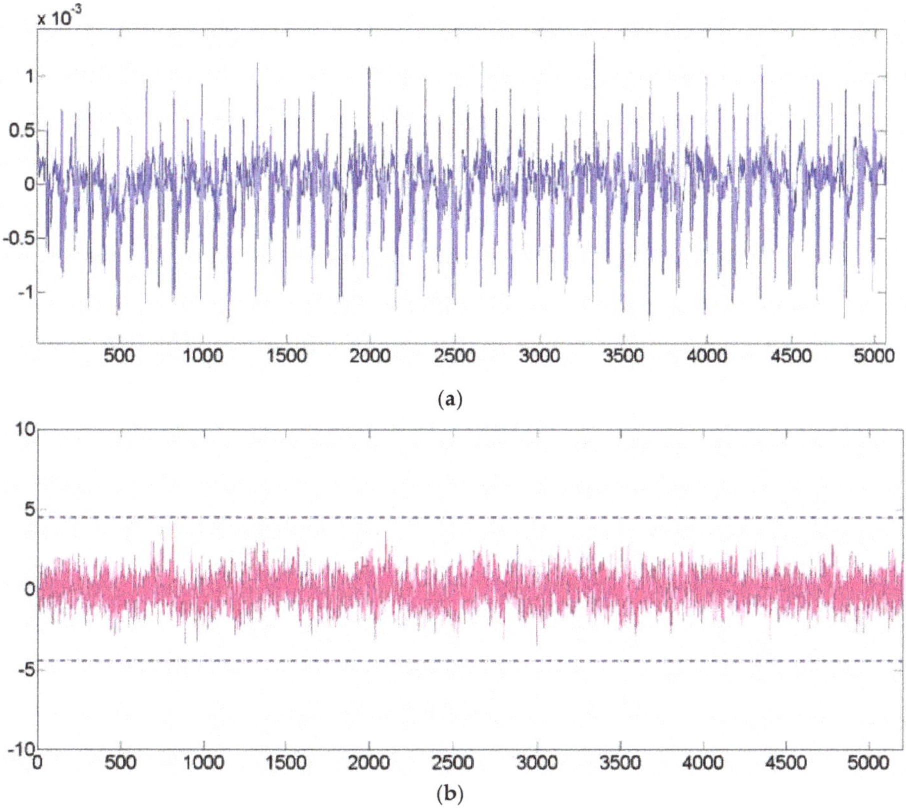

Figure 14. (**a**) The power trace of the optimized SIMON 32/64. (**b**) The *t*-test result of the optimized circuit.

6. Conclusions

This paper proposes a compact countermeasure against DPA on SIMON. Firstly, we present that SIMON algorithm can be threatened by DPA attack, and implement an example of 7-bit key cracking on SAKURA-X board. Subsequently, based on the fault injection technique and the double rate technique, we propose a low-cost DPA-resistant design scheme. By injecting a 1-bit fault into plaintext to form a random data, and uses the double rate technique to insert the encrypting process of random plaintext before or after normal encryption operation randomly to realize the randomization of power consumption. According to the proposed optimal scheme, the circuit structure, random bit generator, and other circuits are implemented. As well as, the evaluation of resources and performance is carried out in Xilinx FPGA. The evaluation results show that the proposed scheme completes the DPA-resistant optimization of SIMON circuit at the cost of 47 LUTs and 46 registers.

Compared with existing works, proposed work is the first one to combine fault injection and double rate technology for DPA attack defense, which makes the SIMON circuit achieve DPA-resistant with 47.7% LUTs and 39.6% registers overhead. Compared with threshold implementation and bool mask, our work has greater advantages in resource consumption, which allows the design philosophy of lightweight cipher algorithm.

Author Contributions: Conceive and structure of the concept of this paper, Y.Z.; Resources, F.Z. and N.W.; Supervision, N.W.; Writing-original draft, Y.Z.; Writing-review and editing, M.R.Y. and J.Z.

Funding: This work was supported by National Natural Science Foundation of China (No. 61774086, 61376025), Natural Science Foundation of Jiangsu Province (No. BK20160806), the Fundamental Research Funds for Central Universities (No. NS2017023, NO. NS2016041).

Acknowledgments: The authors would like to thank Xiaoqiang Zhang for his beneficial suggestions and comments.

Conflicts of Interest: The authors declare no conflict of interest.

References

1. Beaulieu, R.; Shors, D.; Smith, J.; Treatman-Clark, S.; Weeks, B.; Wingers, L. The SIMON and SPECK lightweight block ciphers. In Proceedings of the 52nd ACM/EDAC/IEEE Design Automation Conference (DAC), San Francisco, CA, USA, 8–12 June 2015; pp. 1–6. [CrossRef]
2. Bhasin, S.; Graba, T.; Danger, J.L.; Najm, Z. A look into SIMON from a side-channel perspective. In Proceedings of the 2014 IEEE International Symposium on Hardware-Oriented Security and Trust (HOST), Arlington, VA, USA, 6–7 May 2014; pp. 56–59. [CrossRef]
3. Fu, K.; Sun, L.; Wang, M. New integral attacks on SIMON. *IET Inf. Secur.* **2017**, *11*, 277–286. [CrossRef]
4. McCann, D.; Eder, K.; Oswald, E. Characterising and Comparing the Energy Consumption of Side Channel Attack Countermeasures and Lightweight Cryptography on Embedded Devices. In Proceedings of the 2015 International Workshop on Secure Internet of Things (SIoT), Vienna, Austria, 21–25 September 2015; pp. 65–71.
5. Zhang, F.; Guo, S.; Zhao, X.; Wang, T.; Yang, J.; Standaert, F.X.; Gu, D. A Framework for the Analysis and Evaluation of Algebraic Fault Attacks on Lightweight Block Ciphers. *IEEE Trans. Inf. Forensics Secur.* **2016**, *11*, 1039–1054. [CrossRef]
6. Matsuda, S.; Moriai, S.; Zhang, W.; Bao, Z.; Lin, D.; Rijmen, V.; Yang, B.; Verbauwhede, I. RECTANGLE: a bit-slice lightweight block cipher suitable for multiple platforms. *Sci. China Inf. Sci.* **2015**, *58*, 408–425.
7. Al-Qutayri, M.; Marzouqi, H.; Salah, K. Review of gate-level differential power analysis and fault analysis countermeasures. *IET Inf. Secur.* **2014**, *8*, 51–66.
8. Yoshikawa, M.; Nozaki, Y. Power Analysis Attack and Its Countermeasure for a Lightweight Block Cipher Simon. In *Information Technology: New Generations*; Springer: Cham, Switzerland, 2016; pp. 151–160.
9. Shahverdi, A.; Taha, M.; Eisenbarth, T. Lightweight Side Channel Resistance: Threshold Implementations of Simon. *IEEE Trans. Comput.* **2017**, *66*, 661–671. [CrossRef]
10. Chen, C.; İnci, M.S.; Taha, M.; Eisenbarth, T. SpecTre: A tiny side-channel resistant speck core for FPGAs. In *International Conference on Smart Card Research and Advanced Applications*; Springer: Cham, Switzerland, 2017; pp. 73–88.
11. Shahverdi, A.; Taha, M.; Eisenbarth, T. Silent Simon: A threshold implementation under 100 slices. In Proceedings of the 2015 IEEE International Symposium on Hardware Oriented Security and Trust (HOST), Washington, DC, USA, 5–7 May 2015.
12. Ahir, P.; Mozaffari-Kermani, M.; Azarderakhsh, R. Lightweight Architectures for Reliable and Fault Detection Simon and Speck Cryptographic Algorithms on FPGA. *ACM Trans. Embed. Comput. Syst.* **2017**, *16*, 1–17. [CrossRef]
13. Peeters, E.; Standaert, F.X.; Quisquater, J.J. Power and electromagnetic analysis: Improved model, consequences and comparisons. *Integration* **2007**, *40*, 52–60. [CrossRef]
14. Sasaki, A.; Abe, K. Algorithm-level evaluation of DPA resistance to cryptosystems. *Electr. Eng. Jpn.* **2008**, *165*, 37–45. [CrossRef]
15. Prouff, E.; Rivain, M.; Bévan, R. Statistical analysis of second order differential power analysis. *IEEE Trans. Comput.* **2009**, *58*, 799–811. [CrossRef]
16. Bellizia, D.; Bongiovanni, S.; Monsurrò, P.; Scotti, G.; Trifiletti, A.; Trotta, F.B. Secure Double Rate Registers as an RTL Countermeasure Against Power Analysis Attacks. *IEEE Trans. Very Large Scale Integr. Syst.* **2018**, *26*, 1368–1376. [CrossRef]
17. Goodwill, G.; Jun, B.; Jaffe, J.; Rohatgi, P. A testing methodology for side-channel resistance validation. *NIST Non-Invasive Attack Test. Workshop* **2011**, *7*, 115–136.

18. Leiserson, A.J.; Marson, M.E.; Wachs, M.A. Gate-Level Masking under a Path-Based Leakage Metric. In *International Workshop on Cryptographic Hardware and Embedded Systems*; Springer: Berlin/Heidelberg, Germany, 2014; pp. 580–597.
19. Bilgin, B.; Gierlichs, B.; Nikova, S.; Nikov, V.; Rijmen, V. Higher-Order Threshold Implementations. In *International Conference on the Theory and Application of Cryptology and Information Security*; Springer: Berlin/Heidelberg, Germany, 2014; pp. 326–343.

electronics

MDPI

Article

Performance Analysis of Single-Step Localization Method Based on Matrix Eigen-Perturbation Theory with System Errors

Tianzhu Qin *, Bin Ba and Daming Wang

PLA Strategic Support Force Information Engineering University, Zhengzhou 450001, China; xidianbabin@163.com (B.B.); wdm_wangdaming@163.com (D.W.)

* Correspondence: skypillar@outlook.com; Tel.: + 86-188-3896-7979

Received: 16 January 2019; Accepted: 11 February 2019; Published: 19 February 2019

Abstract: Direct position determination (DPD) is a novel technique in passive localization field recently, receiving superior localization performance compared with the conventional two-step method. The DPD estimator using Doppler shifts is first proposed by Weiss, but it is not suitable for antenna arrays. Additionally, the performance analysis of this method with system errors is absent. This study discusses the single-step localization problem based on moving arrays and exhibits the performance analysis via matrix eigen-perturbation theory with system errors. First, the DPD method using angle of arrival and Doppler shifts is introduced. Then, by adding the eigenvalue perturbations to the estimated Hermitian matrix, the asymptotic linear formulation of localization errors is derived. Consequently, the mean square error of the DPD method is available. Finally, Cramér–Rao bound without system errors is presented, providing a benchmark for the best localization precision and revealing the influence of system errors on the localization precision. Simulation results demonstrate the theoretical analysis in this study.

Keywords: direct position determination; array signal processing; Doppler shifts; matrix eigen-perturbation theory; system errors; Cramér–Rao bound

---

1. Introduction

Transmitter localization has attracted significant attention in wireless communication systems. Generally, the conventional localization approach employs a two-step processing. In the first step, the measurement parameters (e.g., direction of arrival (DOA) [1], time of arrival (TOA) [2], time difference of arrival (TDOA) [3], Doppler shifts [4–6], and frequency difference of arrival (FDOA) [7]) are extracted from the received signal. In the second step, the transmitter position is determined by these estimated parameters via maximum likelihood criterion [8] of subspace data fusion criterion [9]. Although the conventional two-step localization method has been extensively investigated in social location system, it cannot achieve high localization accuracy. Indeed, it is suboptimal, because the intermediate parameters are extracted independently by each receiver station, with ignoring the constraint that all observations must relate to the same geolocation of the emitter. Recently, direct position determination (DPD), which exploits the intrinsic constraint and determines the source position from the received signals in a single step, is regarded as an emerging technology in the field of localization. Compared with the conventional two-step approach, the DPD technique [10–18] receives superior localization performance especially under low signal-to-noise (SNR) conditions.

The DPD algorithms have been intensively investigated in recent years. Weiss first proposes the DPD method for narrowband source based on Doppler shifts in Reference [10]. Then, Tirer and Weiss investigate a high-resolution method based on minimum variance distortonless response (MVDR) without the knowledge of the number of emitters in Reference [11]. To locate the wideband random emitter, the DPD approach based on time delay and Doppler shifts is developed by Weiss [12]. Additionally, a fast DPD method for known wideband signal waveforms is developed by [13]. It should be emphasized that each receiver in above methods is equipped with only one antenna. As a result, the DOA information is not fully utilized. In Reference [14], a DPD approach using antenna arrays is first investigated, demonstrating the effectiveness of the DOA information on improving performance. Based on the work of [14], the DPD methods applied in special conditions are developed. A DPD estimator for a novel localization architecture, called "Multiple Transponders and Multiple Receivers for Multiple Emitters Positioning System" is proposed [15]. DPD approaches are further designed for sources with special properties such as constant modulus [16], orthogonal frequency division multiplexing [17], and cyclostationary [18]. It is easily observed that the DOA information is conducive to the improvement of positioning accuracy. Moreover, note that all results in Reference [10–18] reveal that the direct positioning method has a superior localization performance than that of the conventional two-step method, especially under low SNRs.

In wireless localization scenarios, system errors (i.e., the uncertainty of the receiver position and velocity) often occur. Obviously, the localization performance is deteriorated by environment noise and system errors in this condition. It is well known that the intermediate parameter estimation is sensitive to system errors. When system errors exist, Cherchar [19] and Vincent [20] give statistical performance analyses of the DOA estimation based on SDF and ML criterions, respectively; Hu [21] deduces the localization performance analysis using TDOA and FDOA; and Hari [22] provides an effect of spatial smoothing on the performance analysis of subspace methods. However, the above analyses are only served for the conventional two-step location estimator. It can be predicted that system errors also affect the localization precision of the DPD estimator. The performance of the DPD approach is available in Reference [23–25], but it is only useful for known signal waveforms. Following the work of [14], Wang presents the performance analysis for unknown signal waveforms in presence of array model errors in Reference [26]. Furthermore, Tirer provides the performance analysis of a high-resolution DPD method based on MVDR in Reference [27]. However, the results in Reference [27,28] could not be applied in moving arrays application. Consequently, in the presence of system errors, there is a strong demand for the performance analysis of single-step method for unknown signal waveforms with moving arrays.

Because the single-step approach in Reference [10] plays a fundamental role in this field, we make related improvement and analysis based on it. Following the requirements of current situations, this paper extends the DPD estimator in Reference [10] to moving arrays application, and exhibits the performance analysis via matrix eigen-perturbation theory with system errors. First of all, the signal model is reconstructed by using Doppler and DOA information. Then, since the solution in DPD method is expressed by finding the maximum eigenvalue of the Hermitian matrix in the cost function, system errors can be shown as eigenvalue perturbations on this Hermitian matrix. Based on matrix eigen-perturbation results, which express the perturbations as an additive noise on the Hertmitian matrix, a theoretical analysis is presented. Moreover, the expression of the mean square error (MSE) of direct localization with system errors is provided. Finally, the Cramér–Rao bound (CRB) formulation for the single-step method is also derived, which gives a benchmark for the best localization accuracy for any estimator. Note that the localization errors of the DPD estimator can asymptotically reach the associated CRB in Reference [10]. It is worth mentioning that the CRB is in absence of system errors, which plays a reference to measure the precision loss resulted from system errors.

The rest of this paper is organized as follows. Section 2 lists the notations used in this paper. Section 3 constructs the signal model and formulates the problem. Section 4 discusses the extension of Weiss's method. Section 5 gives the statistically performance analysis of the DPD estimator with system errors. Section 6 presents several numerical simulations to verify the theoretical analysis. Finally, Section 7 draws the conclusions.

2. Notations

In this section, some mathematical notation explanations that will be used through this paper are listed in Table 1.

Table 1. Mathematics notation explanation

Notation	Explanation
$[\cdot]^{\mathrm{T}}$	transpose
$[\cdot]^{\mathrm{H}}$	conjugate transpose
$.^{(R)}$	real part
$.^{(I)}$	imaginary part
diag$\{\cdot\}$	diagonal matrix with diagonal entries
blkdiag$\{\cdot\}$	diagonal matrix with diagonal matrices
$\otimes$	Kronecker product
$\boldsymbol{I}_n$	$n \times n$ identity matrix
0_n	$n \times n$ matrix with zero

3. Signal Model

We consider L moving receivers and a stationary emitter in this scenario. These receivers intercept the received signal at K short intervals along their trajectory. To introduce the DPD signal model, two assumptions are made.

Assumption 1. *Let $\boldsymbol{o}_{l,k}$ and $\boldsymbol{v}_{l,k}$ denote the coordinate and velocity vector of the lth receiver at the kth interception interval. For easy expression, let $\boldsymbol{p}_{l,k} = [\boldsymbol{o}_{l,k}^{\mathrm{T}}, \boldsymbol{v}_{l,k}^{\mathrm{T}}]^{\mathrm{T}}$ denote the system parameter of the lth receiver. The observation is quiye short, thereby these two vectors are unchanged at each interception interval. Furthermore, let $\boldsymbol{z}$ denote the emitter position.*

Assumption 2. *The signal bandwidth is small compared to the inverse of the propagation time among receivers (i.e., $B < 1/\tau_{\max}$, where $\tau_{\max}$ denotes the maximal propagation time among the receivers). Consequently, the observer's spatial separation receives a limitation for a given signal bandwidth.*

After being sampled at $t = nT_s$, the complex signals $\widetilde{y}_{l,k}(n)$ observed by the lth receiver at the kth interception interval is expressed as

$$\widetilde{\boldsymbol{y}}_{l,k}(n) = b_{l,k}\boldsymbol{a}(\boldsymbol{z}, \boldsymbol{o}_{l,k})s_k(n)e^{j2\pi f_{l,k}} + \boldsymbol{n}_{l,k}(n) n = 1, \ldots, N \tag{1}$$

for $l = 1, \ldots, L$ and $k = 1, \ldots, K$, where N denotes the number of sample points at each interval. During the kth interception interval, $b_{l,k}$ and $\boldsymbol{a}(\boldsymbol{z}, \boldsymbol{o}_{l,k})$ are the channel attenuation and the steering vector between the emitter and the lth receiver, $s_k(n)$ is the unknown complex signal envelope of the emitter, $\boldsymbol{n}_{l,k}(n)$ denotes the Gaussian noise vector, and $f_{l,k}$ is the Doppler frequency observed by the lth receiver is expressed by [10]

$$f_{l,k} = \Delta f_k + f_c \mu_{l,k}\left(\boldsymbol{z}, \boldsymbol{p}_{l,k}\right) \tag{2}$$

where Δf_k is the unknown transmitted frequency, f_c is the nominal frequency, and $\mu_{l,k}\left(\boldsymbol{z}, \boldsymbol{p}_{l,k}\right)$ is shown as

$$\mu_{l,k}\left(\boldsymbol{z}, \boldsymbol{p}_{l,k}\right) = \frac{1}{c}\frac{\boldsymbol{v}_{l,k}^{\mathrm{T}}(\boldsymbol{z} - \boldsymbol{o}_{l,k})}{\|\boldsymbol{z} - \boldsymbol{o}_{l,k}\|}. \tag{3}$$

here c is the signal speed. Then, (1) can be expressed by a vector form as

$$\widetilde{\boldsymbol{y}}_{l,k} = b_{l,k}\boldsymbol{C}\left(\boldsymbol{z}, \boldsymbol{p}_{l,k}\right)\boldsymbol{B}_k\boldsymbol{s}_k + \boldsymbol{n}_{l,k} = \boldsymbol{y}_{l,k} + \boldsymbol{n}_{l,k} \tag{4}$$

where

$$\begin{aligned}
&\widetilde{\boldsymbol{y}}_{l,k} = \left[\widetilde{\boldsymbol{y}}_{l,k}^{\mathrm{T}}(1), \widetilde{\boldsymbol{y}}_{l,k}^{\mathrm{T}}(2), \ldots, \widetilde{\boldsymbol{y}}_{l,k}^{\mathrm{T}}(N)\right]^{\mathrm{T}} \\
&\boldsymbol{s}_k = [s_k(1), s_k(2), \ldots, s_k(N)]^{\mathrm{T}} \\
&\boldsymbol{n}_{l,k} = \left[\boldsymbol{n}_{l,k}^{\mathrm{T}}(1), \boldsymbol{n}_{l,k}^{\mathrm{T}}(2), \ldots, \boldsymbol{n}_{l,k}^{\mathrm{T}}(N)\right]^{\mathrm{T}} \\
&\boldsymbol{C}\left(\boldsymbol{z}, \boldsymbol{p}_{l,k}\right) = \boldsymbol{a}(\boldsymbol{z}, \boldsymbol{o}_{l,k}) \otimes \boldsymbol{A}\left(\boldsymbol{z}, \boldsymbol{p}_{l,k}\right) \\
&\boldsymbol{a}(\boldsymbol{z}, \boldsymbol{o}_{l,k}) = \left[1, e^{j2\pi\frac{d}{\lambda}\sin\theta_{l,k}}, \ldots, e^{j2\pi\frac{d}{\lambda}(M-1)\sin\theta_{l,k}}\right]^{\mathrm{T}} \\
&\boldsymbol{A}\left(\boldsymbol{z}, \boldsymbol{p}_{l,k}\right) = \mathrm{diag}\left\{\exp\left(j2\pi f_c \mu\left(\boldsymbol{z}, \boldsymbol{p}_{l,k}\right)\widetilde{\boldsymbol{N}} T_s\right)\right\} \\
&\boldsymbol{B}_k = \mathrm{diag}\left\{\exp\left(j2\pi\Delta f_k \widetilde{\boldsymbol{N}} T_s\right)\right\}
\end{aligned} \tag{5}$$

with $\widetilde{\boldsymbol{N}} = [1, 2, \ldots, N]^{\mathrm{T}}$. Note that λ denotes signal wavelength.

4. Improvement on Previous Work

This section discusses the DPD methods, which locate the emitter directly through the raw data. Weiss [10] first proposed a ML-based DPD method using Doppler shifts. However, when array sensors are adopted in receivers, angle information should be used to enhance localization performance. Hence, we extend Weiss's method through the combination of angle and Doppler. The likelihood function for $\widetilde{\boldsymbol{y}}$ can be formulated by

$$L(\boldsymbol{\zeta}) = \frac{1}{(\pi\sigma^2)^{LKMN}} \exp\left(-\frac{1}{\sigma^2}\sum_{k=1}^{K}\sum_{l=1}^{L}\left\|\widetilde{\boldsymbol{y}}_{l,k} - b_{l,k}\boldsymbol{C}\left(\boldsymbol{z}, \boldsymbol{p}_{l,k}\right)\boldsymbol{B}_k \boldsymbol{s}_k\right\|_2^2\right) \tag{6}$$

$\boldsymbol{\zeta}$ denotes all unknown parameters, where

$$\boldsymbol{\theta} = \left[\boldsymbol{b}^{(R)\mathrm{T}}, \boldsymbol{b}^{(I)\mathrm{T}}, \boldsymbol{s}^{(R)\mathrm{T}}, \boldsymbol{s}^{(I)\mathrm{T}}, \Delta\boldsymbol{f}^{\mathrm{T}}, \boldsymbol{z}^{\mathrm{T}}\right]^{\mathrm{T}} \tag{7}$$

here, $\boldsymbol{b} = \left[\boldsymbol{b}_1^{\mathrm{T}}, \boldsymbol{b}_2^{\mathrm{T}}, \ldots, \boldsymbol{b}_K^{\mathrm{T}}\right]^{\mathrm{T}}$ with $\boldsymbol{b}_k = [b_{1,k}, b_{2,k}, \ldots, b_{L,k}]^{\mathrm{T}}$, $\boldsymbol{s} = \left[\boldsymbol{s}_1^{\mathrm{T}}, \boldsymbol{s}_2^{\mathrm{T}}, \ldots, \boldsymbol{s}_K^{\mathrm{T}}\right]^{\mathrm{T}}$, $\Delta\boldsymbol{f} = [\Delta f_1, \Delta f_2, \ldots, \Delta f_K]^{\mathrm{T}}$. The associated logarithmic likelihood function can be written as

$$L_{Ln}(\boldsymbol{\zeta}) = -LKMN\ln\pi\sigma^2 - \frac{1}{\sigma^2}\sum_{k=1}^{K}\sum_{l=1}^{L}\left\|\widetilde{\boldsymbol{y}}_{l,k} - b_{l,k}\boldsymbol{C}\left(\boldsymbol{z}, \boldsymbol{p}_{l,k}\right)\boldsymbol{B}_k \boldsymbol{s}_k\right\|_2^2 \tag{8}$$

Therefore, the estimation of noise power σ^2 is

$$\hat{\sigma}^2 = \frac{1}{LKMN}\sum_{k=1}^{K}\sum_{l=1}^{L}\left\|\widetilde{\boldsymbol{y}}_{l,k} - b_{l,k}\boldsymbol{C}\left(\boldsymbol{z}, \boldsymbol{p}_{l,k}\right)\boldsymbol{B}_k \boldsymbol{s}_k\right\|_2^2 \tag{9}$$

By substituting (9) into (8), the estimation of parameter $\boldsymbol{\zeta}$ can be determined by

$$\left\{\hat{b}_{l,k}, \hat{\boldsymbol{s}}_k, \Delta\hat{f}_k, \hat{\boldsymbol{z}}\right\} = \mathrm{argmin}\sum_{k=1}^{K}\sum_{l=1}^{L}\left\|\widetilde{\boldsymbol{y}}_{l,k} - b_{l,k}\boldsymbol{C}\left(\boldsymbol{z}, \boldsymbol{p}_{l,k}\right)\boldsymbol{B}_k \boldsymbol{s}_k\right\|_2^2 \tag{10}$$

Next, the optimization solution of (10) is provided. First, by minimizing the above expression, $\hat{b}_{l,k}$ is estimated by [14]

$$\hat{b}_{l,k} = \frac{\left(\boldsymbol{C}\left(\boldsymbol{z}, \boldsymbol{p}_{l,k}\right)\boldsymbol{B}_k \boldsymbol{s}_k\right)^{\mathrm{H}}\widetilde{\boldsymbol{y}}_{l,k}}{\left(\boldsymbol{C}\left(\boldsymbol{z}, \boldsymbol{p}_{l,k}\right)\boldsymbol{B}_k \boldsymbol{s}_k\right)^{\mathrm{H}}\left(\boldsymbol{C}\left(\boldsymbol{z}, \boldsymbol{p}_{l,k}\right)\boldsymbol{B}_k \boldsymbol{s}_k\right)} = \frac{1}{M}\left(\boldsymbol{C}\left(\boldsymbol{z}, \boldsymbol{p}_{l,k}\right)\boldsymbol{B}_k \boldsymbol{s}_k\right)^{\mathrm{H}}\widetilde{\boldsymbol{y}}_{l,k} \tag{11}$$

Then, after applying (11) to (10) and eliminating the constant part, (10) can be written as

$$\left\{\hat{\mathbf{s}}_k, \Delta\hat{f}_k, \hat{\mathbf{z}}\right\} = \operatorname{argmax}\sum_{k=1}^{K} \omega_k^{\mathrm{H}} \mathbf{D}_k(\mathbf{z}, \mathbf{p}_k, \mathbf{n}_k)\mathbf{D}_k^{\mathrm{H}}(\mathbf{z}, \mathbf{p}_k, \mathbf{n}_k)\omega_k \tag{12}$$

where

$$\begin{cases} \omega_k = \mathbf{B}_k \mathbf{s}_k \\ \mathbf{D}_k(\mathbf{z}, \mathbf{p}_k, \mathbf{n}_k) &= \left[\mathbf{c}^{\mathrm{H}}\left(\mathbf{z}, \mathbf{p}_{1,k}\right)\widetilde{\mathbf{y}}_{1,k}, \mathbf{c}^{\mathrm{H}}\left(\mathbf{z}, \mathbf{p}_{2,k}\right)\widetilde{\mathbf{y}}_{2,k}, \ldots, \mathbf{c}^{\mathrm{H}}\left(\mathbf{z}, \mathbf{p}_{L,k}\right)\widetilde{\mathbf{y}}_{L,k}\right] \\ &= \overline{\mathbf{C}}(\mathbf{z}, \mathbf{p}_k)(\mathbf{Y}_k + \mathbf{N}_k) \end{cases} \tag{13}$$

with

$$\begin{cases} \overline{\mathbf{C}}(\mathbf{z}, \mathbf{p}_k) = \left[\mathbf{c}^{\mathrm{H}}\left(\mathbf{z}, \mathbf{p}_{1,k}\right), \mathbf{c}^{\mathrm{H}}\left(\mathbf{z}, \mathbf{p}_{2,k}\right), \ldots, \mathbf{c}^{\mathrm{H}}\left(\mathbf{z}, \mathbf{p}_{L,k}\right)\right] \\ \mathbf{Y}_k = \operatorname{blkdiag}\left\{\mathbf{y}_{1,k}, \mathbf{y}_{2,k}, \ldots, \mathbf{y}_{L,k}\right\} \\ \mathbf{N}_k = \operatorname{blkdiag}\left\{\mathbf{n}_{1,k}, \mathbf{n}_{2,k}, \ldots, \mathbf{n}_{L,k}\right\} \end{cases} \tag{14}$$

Note that ω_k is unknown to receivers. The maximization of (12) is solved by choosing the vector ω_k as the eigenvector associated with the largest eigenvalue of matrix $\mathbf{D}_k(\mathbf{z}, \mathbf{p}_k, \mathbf{n}_k)\mathbf{D}_k^{\mathrm{H}}(\mathbf{z}, \mathbf{p}_k, \mathbf{n}_k)$. Therefore, the optimization problem in (12) respect to $\mathbf{z}$ is expressed by

$$\hat{\mathbf{z}} = \operatorname{argmax}\sum_{k=1}^{K} \lambda_{\max}\left\{\mathbf{D}_k(\mathbf{z}, \mathbf{p}_k, \mathbf{n}_k)\mathbf{D}_k^{\mathrm{H}}(\mathbf{z}, \mathbf{p}_k, \mathbf{n}_k)\right\} \tag{15}$$

where $\lambda_{\max}\{\cdot\}$ denotes the largest eigenvalues of the matrix.

It must be emphasized that the matrices $\mathbf{D}_k(\mathbf{z}, \mathbf{p}_k, \mathbf{n}_k)\mathbf{D}_k^{\mathrm{H}}(\mathbf{z}, \mathbf{p}_k, \mathbf{n}_k)$ and $\mathbf{D}_k^{\mathrm{H}}(\mathbf{z}, \mathbf{p}_k, \mathbf{n}_k)\mathbf{D}_k(\mathbf{z}, \mathbf{p}_k, \mathbf{n}_k)$ share the same nonzero eigenvalues. Generally, the dimension of matrix $\mathbf{D}_k^{\mathrm{H}}(\mathbf{z}, \mathbf{p}_k, \mathbf{n}_k)\mathbf{D}_k(\mathbf{z}, \mathbf{p}_k, \mathbf{n}_k) \in \mathbb{C}^{L\times L}$ is significantly smaller than that of $\mathbf{D}_k(\mathbf{z}, \mathbf{p}_k, \mathbf{n}_k)\mathbf{D}_k^{\mathrm{H}}(\mathbf{z}, \mathbf{p}_k, \mathbf{n}_k) \in \mathbb{C}^{N\times N}$. Hence, to reach for lower computational cost, the estimation of $\mathbf{z}$ can be replaced by

$$\hat{\mathbf{z}} = \operatorname{argmax}\sum_{k=1}^{K} \lambda_{\max}\left\{\mathbf{D}_k^{\mathrm{H}}(\mathbf{z}, \mathbf{p}_k, \mathbf{n}_k)\mathbf{D}_k(\mathbf{z}, \mathbf{p}_k, \mathbf{n}_k)\right\} \tag{16}$$

To fully describe the proposed method, we make a computational complexity analysis. Based on the above derivation, the calculation of $\mathbf{D}_k^{\mathrm{H}}(\mathbf{z}, \mathbf{p}_k, \mathbf{n}_k)\mathbf{D}_k(\mathbf{z}, \mathbf{p}_k, \mathbf{n}_k)$ and grid search in the position set of interest make a major contribution to the computational load. The total number of calculation equals $O(LM^2N^3 + (1+M)N^2 + (1+N)L^3 + M)KN_p$, where N_p is the number of grid search points in terms of emitter position. Since Weiss's method uses only an antenna at each receiver, the value of M should be 1. Therefore, the computational load of Weiss's method is $O(LN^3 + 2N^2 + (1+N)L^3)KN_p$. It is readily observed that the computational complexity of out method is heavier than that of Weiss's method. Even with more computing resources, on the other hand, our method can offer superior performance (see Section 6.2).

5. Statistical Performance Analysis

It is well known that the above DPD method can reach asymptotic optimal with precise system parameters. However, system errors (i.e., the position and velocity uncertainties of airplanes or UAVs) often occur in real life, which deteriorate the localization precision of the above DPD method greatly. For this reason, in this section, the perturbation analysis and the MSE of the DPD method in presence of system errors will be discussed.

We assume that the real parameters are defined as $\mathbf{p}_k$ ($k = 1, \ldots, K$), the observed parameters are written as $\overline{\mathbf{p}}_k$ ($k = 1, \ldots, K$), and the system errors are expressed by $\widetilde{\mathbf{p}}_k$ ($k = 1, \ldots, K$). The relation between these parameters is

$$\widetilde{\mathbf{p}}_k = \overline{\mathbf{p}}_k - \mathbf{p}_k \; k = 1, \ldots, K \tag{17}$$

Therefore, the estimation of $\boldsymbol{z}$ in presence of system errors should be determined by

$$\hat{\boldsymbol{z}} = \operatorname{argmax} \sum_{k=1}^{K} \lambda_{\max} \left\{ \boldsymbol{D}_k^{\mathrm{H}}(\boldsymbol{z}, \overline{\boldsymbol{p}}_k, \boldsymbol{n}_k) \boldsymbol{D}_k(\boldsymbol{z}, \overline{\boldsymbol{p}}_k, \boldsymbol{n}_k) \right\} \tag{18}$$

Obviously, the localization performance analysis related to noise $\boldsymbol{n}_k$ ($k = 1, \ldots, K$) and system errors $\widetilde{\boldsymbol{p}}_k$ ($k = 1, \ldots, K$) should be considered simultaneously. To complete this result analysis, matrix eigen-perturbation theory needs to be applied on (16).

5.1. Basic Theoretical Analysis Tool

Note that the key part of Weiss's method is finding the maximal eigenvalue of Hermitian matrix, which is disturbed by other error matrix. Relevant theory can be expressed by:

Proposition 1. *Assume that $\boldsymbol{Q} \in \mathbb{C}^{N \times N}$ is a positive semidefinite respect to eigenvalues $\lambda_n |_{1 \le n \le N}$ and unit eigenvectors $\boldsymbol{\alpha}_n |_{1 \le n \le N}$. Moreover, assume that $\boldsymbol{Q}$ is disturbed by a matrix $\widetilde{\boldsymbol{Q}} \in \mathbb{C}^{N \times N}$, hence the perturbed matrix can be written as $\overline{\boldsymbol{Q}} = \boldsymbol{Q} + \widetilde{\boldsymbol{Q}}$. Finally, the relation of the eigenvalues ($\overline{\lambda}_n |_{1 \le n \le N}$) of $\overline{\boldsymbol{Q}}$ and $\lambda_n |_{1 \le n \le N}$ is shown as*

$$\overline{\lambda}_n = \lambda_n + \boldsymbol{\alpha}_n^{\mathrm{H}} \widetilde{\boldsymbol{Q}} \boldsymbol{\alpha}_n + \boldsymbol{\alpha}_n^{\mathrm{H}} \widetilde{\boldsymbol{Q}} \boldsymbol{E}_n \widetilde{\boldsymbol{Q}} \boldsymbol{\alpha}_n + o\left(\|\widetilde{\boldsymbol{Q}}\|_2^2 \right) \quad n = 1, \ldots, N \tag{19}$$

where $\boldsymbol{E}_n = \sum_{\substack{i=1 \\ i \neq n}}^{N} (\lambda_n - \lambda_i)^{-1} \boldsymbol{\alpha}_i \boldsymbol{\alpha}_i^{\mathrm{H}}$. The detailed proof of this proposition can be found in [25,26].

5.2. Perturbation Analysis on The Cost Function

As mentioned earlier, our purpose is investigating the relationship between the MSE of the DPD estimator and noise as well as system errors. Herein, we adopt a second-order perturbation analysis to (18), which follows

$$\boldsymbol{D}_k(\hat{\boldsymbol{z}}, \overline{\boldsymbol{p}}_k, \boldsymbol{n}_k) = \overline{\boldsymbol{C}}(\hat{\boldsymbol{z}}, \overline{\boldsymbol{p}}_k)(\boldsymbol{Y}_k + \boldsymbol{N}_k) \approx \boldsymbol{D}_k^{(0)} + \widetilde{\boldsymbol{D}}_k^{(1)} + \widetilde{\boldsymbol{D}}_k^{(2)} \tag{20}$$

It is necessary to emphasize that $\boldsymbol{D}_k^{(0)}$ is the non-perturbation terms, and $\widetilde{\boldsymbol{D}}_k^{(1)}$ as well as $\widetilde{\boldsymbol{D}}_k^{(2)}$ denote the first and second-order perturbation terms, respectively. Their expression is specified by

$$\left\{ \begin{array}{l} \boldsymbol{D}_k^{(0)} = \overline{\boldsymbol{C}}(\boldsymbol{z}, \boldsymbol{p}_k) \boldsymbol{Y}_k \\ \widetilde{\boldsymbol{D}}_k^{(1)} = \sum\limits_{d=1}^{D} \langle \widetilde{\boldsymbol{z}} \rangle_d \dot{\overline{\boldsymbol{C}}}_d^{(a)}(\boldsymbol{z}, \boldsymbol{p}_k) \boldsymbol{Y}_k + \sum\limits_{d=1}^{2DL} \langle \widetilde{\boldsymbol{p}}_k \rangle_d \dot{\overline{\boldsymbol{C}}}_d^{(b)}(\boldsymbol{z}, \boldsymbol{p}_k) \boldsymbol{Y}_k + \overline{\boldsymbol{C}}(\boldsymbol{z}, \boldsymbol{p}_k) \boldsymbol{N}_k \\ \widetilde{\boldsymbol{D}}_k^{(2)} = \frac{1}{2} \sum\limits_{d_1=1}^{D} \sum\limits_{d_2=1}^{D} \langle \widetilde{\boldsymbol{z}} \rangle_{d_1} \langle \widetilde{\boldsymbol{z}} \rangle_{d_2} \ddot{\overline{\boldsymbol{C}}}_{d_1 d_2}^{(aa)}(\boldsymbol{z}, \boldsymbol{p}_k) \boldsymbol{Y}_k + \frac{1}{2} \sum\limits_{d_1=1}^{2DL} \sum\limits_{d_2=1}^{2DL} \langle \widetilde{\boldsymbol{p}}_k \rangle_{d_1} \langle \widetilde{\boldsymbol{p}}_k \rangle_{d_2} \ddot{\overline{\boldsymbol{C}}}_{d_1 d_2}^{(bb)}(\boldsymbol{z}, \boldsymbol{p}_k) \boldsymbol{Y}_k + \\ \sum\limits_{d_1=1}^{D} \sum\limits_{d_2=1}^{2DL} \langle \widetilde{\boldsymbol{z}} \rangle_{d_1} \langle \widetilde{\boldsymbol{p}}_k \rangle_{d_2} \ddot{\overline{\boldsymbol{C}}}_{d_1 d_2}^{(ab)}(\boldsymbol{z}, \boldsymbol{p}_k) \boldsymbol{Y}_k + \sum\limits_{d=1}^{D} \langle \widetilde{\boldsymbol{z}} \rangle_d \dot{\overline{\boldsymbol{C}}}_d^{(a)}(\boldsymbol{z}, \boldsymbol{p}_k) \boldsymbol{N}_k + \sum\limits_{d=1}^{2DL} \langle \widetilde{\boldsymbol{p}}_k \rangle_d \dot{\overline{\boldsymbol{C}}}_d^{(b)}(\boldsymbol{z}, \boldsymbol{p}_k) \boldsymbol{N}_k \end{array} \right. \tag{21}$$

where

$$\left\{ \begin{array}{l} \dot{\overline{\boldsymbol{C}}}_d^{(a)}(\boldsymbol{z}, \boldsymbol{p}_k) = \frac{\partial \overline{\boldsymbol{C}}(\boldsymbol{z}, \boldsymbol{p}_k)}{\partial \langle \boldsymbol{z} \rangle_d}, \dot{\overline{\boldsymbol{C}}}_d^{(b)}(\boldsymbol{z}, \boldsymbol{p}_k) = \frac{\partial \overline{\boldsymbol{C}}(\boldsymbol{z}, \boldsymbol{p}_k)}{\partial \langle \boldsymbol{p}_k \rangle_d}, \ddot{\overline{\boldsymbol{C}}}_{d_1 d_2}^{(aa)}(\boldsymbol{z}, \boldsymbol{p}_k) = \frac{\partial^2 \overline{\boldsymbol{C}}(\boldsymbol{z}, \boldsymbol{p}_k)}{\partial \langle \boldsymbol{z} \rangle_{d_1} \partial \langle \boldsymbol{z} \rangle_{d_2}} \\ \ddot{\overline{\boldsymbol{C}}}_{d_1 d_2}^{(bb)}(\boldsymbol{z}, \boldsymbol{p}_k) = \frac{\partial^2 \overline{\boldsymbol{C}}(\boldsymbol{z}, \boldsymbol{p}_k)}{\partial \langle \boldsymbol{p}_k \rangle_{d_1} \partial \langle \boldsymbol{p}_k \rangle_{d_2}}, \ddot{\overline{\boldsymbol{C}}}_{d_1 d_2}^{(ab)}(\boldsymbol{z}, \boldsymbol{p}_k) = \frac{\partial^2 \overline{\boldsymbol{C}}(\boldsymbol{z}, \boldsymbol{p}_k)}{\partial \langle \boldsymbol{z} \rangle_{d_1} \partial \langle \boldsymbol{p}_k \rangle_{d_2}} \end{array} \right. \tag{22}$$

The derivation of (21) is exhibited in Appendix A, and the matrices in (22) are listed in Appendix B.

For easy derivation, we define $\boldsymbol{Q}_k(\hat{\boldsymbol{z}},\overline{\boldsymbol{p}}_k,\boldsymbol{n}_k)=\boldsymbol{D}_k^{\mathrm{H}}(\hat{\boldsymbol{z}},\overline{\boldsymbol{p}}_k,\boldsymbol{n}_k)\cdot\boldsymbol{D}_k(\hat{\boldsymbol{z}},\overline{\boldsymbol{p}}_k,\boldsymbol{n}_k)$. Following the result in (20), the Hermitian matrix $\boldsymbol{Q}_k(\hat{\boldsymbol{z}},\overline{\boldsymbol{p}}_k,\boldsymbol{n}_k)$ is approximated by

$$\boldsymbol{Q}_k(\hat{\boldsymbol{z}},\overline{\boldsymbol{p}}_k,\boldsymbol{n}_k)\approx\boldsymbol{Q}_k^{(0)}+\widetilde{\boldsymbol{Q}}_k^{(1)}+\widetilde{\boldsymbol{Q}}_k^{(2)} \tag{23}$$

where

$$\begin{cases}\boldsymbol{Q}_k^{(0)}&=\boldsymbol{D}_k^{(0)\mathrm{H}}\boldsymbol{D}_k^{(0)}\\ \widetilde{\boldsymbol{Q}}_k^{(1)}&=\boldsymbol{D}_k^{(0)\mathrm{H}}\widetilde{\boldsymbol{D}}_k^{(1)}+\widetilde{\boldsymbol{D}}_k^{(1)\mathrm{H}}\boldsymbol{D}_k^{(0)}\\ \widetilde{\boldsymbol{Q}}_k^{(2)}&=\boldsymbol{D}_k^{(0)\mathrm{H}}\widetilde{\boldsymbol{D}}_k^{(2)}+\widetilde{\boldsymbol{D}}_k^{(1)\mathrm{H}}\widetilde{\boldsymbol{D}}_k^{(1)}+\widetilde{\boldsymbol{D}}_k^{(2)\mathrm{H}}\boldsymbol{D}_k^{(0)}\end{cases} \tag{24}$$

Note that error matrix is defined as

$$\widetilde{\boldsymbol{Q}}_k=\boldsymbol{Q}_k(\hat{\boldsymbol{z}},\overline{\boldsymbol{p}}_k,\boldsymbol{n}_k)-\boldsymbol{Q}_k^{(0)}\approx\widetilde{\boldsymbol{Q}}_k^{(1)}+\widetilde{\boldsymbol{Q}}_k^{(2)} \tag{25}$$

which is obtained by neglecting the high-order error issues.

Assume that $\boldsymbol{Q}_k^{(0)}$ is related to eigenvalues $\lambda_{k,l}^{(0)}|_{1\le l\le L}$ as well as unit eigenvectors $\boldsymbol{\alpha}_{k,l}^{(0)}|_{1\le l\le L}$, and $\boldsymbol{Q}_k(\hat{\boldsymbol{z}},\overline{\boldsymbol{p}}_k,\boldsymbol{n}_k)$ is associated with eigenvalues $\overline{\lambda}_{k,l}|_{1\le l\le L}$ as well as unit eigenvectors $\overline{\boldsymbol{\alpha}}_{k,l}|_{1\le l\le L}$. By following the result in Proposition 1, we obtain

$$\overline{\lambda}_{k,L}=\lambda_{k,L}^{(0)}+\boldsymbol{\alpha}_{k,L}^{(0)\mathrm{H}}\widetilde{\boldsymbol{Q}}_k\boldsymbol{\alpha}_{k,L}^{(0)}+\boldsymbol{\alpha}_{k,L}^{(0)\mathrm{H}}\widetilde{\boldsymbol{Q}}_k\boldsymbol{E}_{k,L}\widetilde{\boldsymbol{Q}}_k\boldsymbol{\alpha}_{k,L}^{(0)}+o\left(\|\widetilde{\boldsymbol{Q}}_k\|_2^2\right)k=1,\ldots,K \tag{26}$$

where $\boldsymbol{E}_{k,L}=\sum\limits_{i=1}^{L}\left(\lambda_{k,L}^{(0)}-\lambda_{k,i}^{(0)}\right)^{-1}\boldsymbol{\alpha}_{k,i}^{(0)}\boldsymbol{\alpha}_{k,i}^{(0)\mathrm{H}}$.

Inserting (25) into (26) leads to

$$\overline{\lambda}_{k,L}\approx\lambda_{k,L}^{(0)}+\widetilde{\lambda}_{k,L}^{(1)}+\widetilde{\lambda}_{k,L}^{(2)} \tag{27}$$

where $\widetilde{\lambda}_{k,L}^{(1)}$ and $\widetilde{\lambda}_{k,L}^{(2)}$ denote the first- and second-order distributed issues, respectively

$$\begin{cases}\widetilde{\lambda}_{k,L}^{(1)}= & \boldsymbol{\alpha}_{k,L}^{(0)\mathrm{H}}\boldsymbol{D}_k^{(0)\mathrm{H}}\widetilde{\boldsymbol{D}}_k^{(1)}\boldsymbol{\alpha}_{k,L}^{(0)}+\boldsymbol{\alpha}_{k,L}^{(0)\mathrm{H}}\widetilde{\boldsymbol{D}}_k^{(1)\mathrm{H}}\boldsymbol{D}_k^{(0)}\boldsymbol{\alpha}_{k,L}^{(0)}\\ \widetilde{\lambda}_{k,L}^{(2)}= & \boldsymbol{\alpha}_{k,L}^{(0)\mathrm{H}}\boldsymbol{D}_k^{(0)\mathrm{H}}\widetilde{\boldsymbol{D}}_k^{(2)}\boldsymbol{\alpha}_{k,L}^{(0)}+\boldsymbol{\alpha}_{k,L}^{(0)\mathrm{H}}\widetilde{\boldsymbol{D}}_k^{(1)\mathrm{H}}\widetilde{\boldsymbol{D}}_k^{(1)}\boldsymbol{\alpha}_{k,L}^{(0)}+\boldsymbol{\alpha}_{k,L}^{(0)\mathrm{H}}\widetilde{\boldsymbol{D}}_k^{(2)\mathrm{H}}\boldsymbol{D}_k^{(0)}\boldsymbol{\alpha}_{k,L}^{(0)}+\\ & \boldsymbol{\alpha}_{k,L}^{(0)\mathrm{H}}\boldsymbol{D}_k^{(0)\mathrm{H}}\widetilde{\boldsymbol{D}}_k^{(1)}\boldsymbol{E}_{k,L}\boldsymbol{D}_k^{(0)\mathrm{H}}\widetilde{\boldsymbol{D}}_k^{(1)}\boldsymbol{\alpha}_{k,L}^{(0)}+\boldsymbol{\alpha}_{k,L}^{(0)\mathrm{H}}\widetilde{\boldsymbol{D}}_k^{(1)\mathrm{H}}\boldsymbol{D}_k^{(0)}\boldsymbol{E}_{k,L}\boldsymbol{D}_k^{(0)\mathrm{H}}\widetilde{\boldsymbol{D}}_k^{(1)}\boldsymbol{\alpha}_{k,L}^{(0)}+\\ & \boldsymbol{\alpha}_{k,L}^{(0)\mathrm{H}}\boldsymbol{D}_k^{(0)\mathrm{H}}\widetilde{\boldsymbol{D}}_k^{(1)}\boldsymbol{E}_{k,L}\widetilde{\boldsymbol{D}}_k^{(1)\mathrm{H}}\boldsymbol{D}_k^{(0)}\boldsymbol{\alpha}_{k,L}^{(0)}+\boldsymbol{\alpha}_{k,L}^{(0)\mathrm{H}}\widetilde{\boldsymbol{D}}_k^{(1)\mathrm{H}}\boldsymbol{D}_k^{(0)}\boldsymbol{E}_{k,L}\widetilde{\boldsymbol{D}}_k^{(1)\mathrm{H}}\boldsymbol{D}_k^{(0)}\boldsymbol{\alpha}_{k,L}^{(0)}\end{cases} \tag{28}$$

Define $J_{\mathrm{cost}}(\hat{\boldsymbol{z}},\overline{\boldsymbol{p}},\boldsymbol{n})=\sum\limits_{k=1}^{K}\lambda_{\max}\{\boldsymbol{D}_k^{\mathrm{H}}(\boldsymbol{z},\overline{\boldsymbol{p}}_k,\boldsymbol{n}_k)\boldsymbol{D}_k(\boldsymbol{z},\overline{\boldsymbol{p}}_k,\boldsymbol{n}_k)\}$ and apply (27) in (18). Then, $J_{\mathrm{cost}}(\hat{\boldsymbol{z}},\overline{\boldsymbol{p}},\boldsymbol{n})$ can be approximated by

$$J_{\mathrm{cost}}(\hat{\boldsymbol{z}},\overline{\boldsymbol{p}},\boldsymbol{n})\approx J_{\mathrm{cost}}^{(0)}+\widetilde{J}_{\mathrm{cost}}^{(1)}+\widetilde{J}_{\mathrm{cost}}^{(2)} \tag{29}$$

where

$$J_{\cos\mathrm{t}}^{(0)}=\sum_{k=1}^{K}\lambda_{k,L}^{(0)},\widetilde{J}_{\cos\mathrm{t}}^{(1)}=\sum_{k=1}^{K}\widetilde{\lambda}_{k,L}^{(1)},\widetilde{J}_{\cos\mathrm{t}}^{(2)}=\sum_{k=1}^{K}\widetilde{\lambda}_{k,L}^{(2)} \tag{30}$$

more specially, $\widetilde{J}_{\mathrm{cost}}^{(1)}$ is written as

$$\begin{aligned}\widetilde{J}_{\mathrm{cost}}^{(1)}&=\sum_{k=1}^{K}\boldsymbol{\alpha}_{k,L}^{(0)\mathrm{H}}\boldsymbol{D}_k^{(0)\mathrm{H}}\widetilde{D}_k^{(1)}\boldsymbol{\alpha}_{k,L}^{(0)}+\sum_{k=1}^{K}\boldsymbol{\alpha}_{k,L}^{(0)\mathrm{H}}\widetilde{D}_k^{(1)\mathrm{H}}\boldsymbol{D}_k^{(0)}\boldsymbol{\alpha}_{k,L}^{(0)}\\&=\sum_{k=1}^{K}\boldsymbol{f}_{1k}^{\mathrm{H}}\widetilde{z}+\sum_{k=1}^{K}\boldsymbol{f}_{2k}^{\mathrm{H}}\widetilde{p}_k+\sum_{k=1}^{K}\boldsymbol{f}_{3k}^{\mathrm{H}}\widetilde{n}_k\end{aligned} \tag{31}$$

where

$$\begin{cases} \widetilde{\boldsymbol{n}}_k = \left[\boldsymbol{n}_k^{\mathrm{T}}\ \boldsymbol{n}_k^{\mathrm{H}}\right]^{\mathrm{T}} \\ f_{nk} = \boldsymbol{F}_{nk}^{(a)\mathrm{H}}\left(\boldsymbol{\alpha}_{k,L}^{(0)}\right)\boldsymbol{D}_k^{(0)}\boldsymbol{\alpha}_{k,L}^{(0)} + \boldsymbol{F}_{nk}^{(b)\mathrm{H}}\left(\boldsymbol{D}_k^{(0)}\boldsymbol{\alpha}_{k,L}^{(0)}\right)\boldsymbol{\alpha}_{k,L}^{(0)} n = 1,2,3 \end{cases} \tag{32}$$

with

$$\begin{cases} \boldsymbol{F}_{1k}^{(a)}(\boldsymbol{q}) = \frac{\partial\left(\overline{\boldsymbol{C}}(\boldsymbol{z},\boldsymbol{p}_k)\boldsymbol{Y}_k\boldsymbol{q}\right)}{\partial \boldsymbol{z}^{\mathrm{T}}}, \boldsymbol{F}_{2k}^{(a)}(\boldsymbol{q}) = \frac{\partial\left(\overline{\boldsymbol{C}}(\boldsymbol{z},\boldsymbol{p}_k)\boldsymbol{Y}_k\boldsymbol{q}\right)}{\partial \boldsymbol{p}_k^{\mathrm{T}}}, \\ \boldsymbol{F}_{3k}^{(a)}(\boldsymbol{q}) = \overline{\boldsymbol{C}}(\boldsymbol{z},\boldsymbol{p}_k)(\mathrm{diag}\{\boldsymbol{q}\}\otimes \boldsymbol{I}_{MN})\boldsymbol{\Pi}_1 \end{cases}$$
$$\begin{cases} \boldsymbol{F}_{1k}^{(b)}(\boldsymbol{q}) = \frac{\partial\left(\boldsymbol{Y}_k^{\mathrm{H}}\overline{\boldsymbol{C}}^{\mathrm{H}}(\boldsymbol{z},\boldsymbol{p}_k)\boldsymbol{q}\right)}{\partial \boldsymbol{z}^{\mathrm{T}}}, \boldsymbol{F}_{2k}^{(b)}(\boldsymbol{q}) = \frac{\partial\left(\boldsymbol{Y}_k^{\mathrm{H}}\overline{\boldsymbol{C}}^{\mathrm{H}}(\boldsymbol{z},\boldsymbol{p}_k)\boldsymbol{q}\right)}{\partial \boldsymbol{p}_k^{\mathrm{T}}}, \\ \boldsymbol{F}_{3k}^{(b)}(\boldsymbol{q}) = (\boldsymbol{I}_L\otimes \boldsymbol{q}^{\mathrm{T}})\overline{\overline{\boldsymbol{C}}}(\boldsymbol{z},\boldsymbol{p}_k)\boldsymbol{\Pi}_2 \end{cases} \tag{33}$$

where

$$\begin{cases} \boldsymbol{\Pi}_1 = [\boldsymbol{I}_{MNL}\ \boldsymbol{0}_{MNL}] \\ \boldsymbol{\Pi}_2 = [0_{MNL}\ \boldsymbol{I}_{MNL}] \\ \overline{\overline{\boldsymbol{C}}}(\boldsymbol{z},\boldsymbol{p}_k) = \mathrm{blkdiag}\left\{\boldsymbol{C}^{\mathrm{T}}\left(\boldsymbol{z},\boldsymbol{p}_{1,k}\right),\boldsymbol{C}^{\mathrm{T}}\left(\boldsymbol{z},\boldsymbol{p}_{2,k}\right),\ldots,\boldsymbol{C}^{\mathrm{T}}\left(\boldsymbol{z},\boldsymbol{p}_{L,k}\right)\right\} \end{cases} \tag{34}$$

The detailed derivation of (31) to (34) can be seen in Appendix C.

Furthermore, $\widetilde{J}_{\mathrm{cost}}^{(2)}$ can be formulated as

$$\begin{aligned} \widetilde{J}_{\mathrm{cost}}^{(2)} &= \sum_{k=1}^{K}\boldsymbol{\alpha}_{k,L}^{(0)\mathrm{H}}\boldsymbol{D}_k^{(0)\mathrm{H}}\widetilde{D}_k^{(2)}\boldsymbol{\alpha}_{k,L}^{(0)} + \sum_{k=1}^{K}\boldsymbol{\alpha}_{k,L}^{(0)\mathrm{H}}\widetilde{D}_k^{(2)\mathrm{H}}\boldsymbol{D}_k^{(0)}\boldsymbol{\alpha}_{k,L}^{(0)} + \sum_{k=1}^{K}\boldsymbol{\alpha}_{k,L}^{(0)\mathrm{H}}\widetilde{D}_k^{(1)\mathrm{H}}\widetilde{D}_k^{(1)}\boldsymbol{\alpha}_{k,L}^{(0)} + \\ &\sum_{k=1}^{K}\boldsymbol{\alpha}_{k,L}^{(0)\mathrm{H}}\boldsymbol{D}_k^{(0)\mathrm{H}}\widetilde{D}_k^{(1)}\boldsymbol{E}_{k,L}\boldsymbol{D}_k^{(0)\mathrm{H}}\widetilde{D}_k^{(1)}\boldsymbol{\alpha}_{k,L}^{(0)} + \sum_{k=1}^{K}\boldsymbol{\alpha}_{k,L}^{(0)\mathrm{H}}\widetilde{D}_k^{(1)\mathrm{H}}\boldsymbol{D}_k^{(0)}\boldsymbol{E}_{k,L}\boldsymbol{D}_k^{(0)\mathrm{H}}\widetilde{D}_k^{(1)}\boldsymbol{\alpha}_{k,L}^{(0)} + \\ &\sum_{k=1}^{K}\boldsymbol{\alpha}_{k,L}^{(0)\mathrm{H}}\boldsymbol{D}_k^{(0)\mathrm{H}}\widetilde{D}_k^{(1)}\boldsymbol{E}_{k,L}\widetilde{D}_k^{(1)\mathrm{H}}\boldsymbol{D}_k^{(0)}\boldsymbol{\alpha}_{k,L}^{(0)} + \sum_{k=1}^{K}\boldsymbol{\alpha}_{k,L}^{(0)\mathrm{H}}\widetilde{D}_k^{(1)\mathrm{H}}\boldsymbol{D}_k^{(0)}\boldsymbol{E}_{k,L}\widetilde{D}_k^{(1)\mathrm{H}}\boldsymbol{D}_k^{(0)}\boldsymbol{\alpha}_{k,L}^{(0)} \\ &= \sum_{k=1}^{K}\widetilde{z}^{\mathrm{T}}\boldsymbol{\xi}_{1k}\widetilde{z} + \sum_{k=1}^{K}\widetilde{p}_k^{\mathrm{T}}\boldsymbol{\xi}_{2k}\widetilde{p}_k + \sum_{k=1}^{K}\widetilde{z}^{\mathrm{T}}\boldsymbol{\xi}_{3k}\widetilde{p}_k + \sum_{k=1}^{K}\widetilde{z}^{\mathrm{T}}\boldsymbol{\xi}_{4k}\widetilde{n}_k + \sum_{k=1}^{K}\widetilde{p}_k^{\mathrm{T}}\boldsymbol{\xi}_{5k}\widetilde{n}_k + \sum_{k=1}^{K}\widetilde{n}_k^{\mathrm{H}}\boldsymbol{\xi}_{6k}\widetilde{n}_k \end{aligned} \tag{35}$$

where

$$\begin{aligned} \boldsymbol{\xi}_{ik} = &\boldsymbol{\Sigma}_{ik}^{(a)}\left(\boldsymbol{\alpha}_{k,L}^{(0)},\boldsymbol{I}_N,\boldsymbol{\alpha}_{k,L}^{(0)}\right) + \boldsymbol{\Sigma}_{ik}^{(a)}\left(\boldsymbol{\alpha}_{k,L}^{(0)},\boldsymbol{D}_k^{(0)}\boldsymbol{E}_{k,L}\boldsymbol{D}_k^{(0)\mathrm{H}},\boldsymbol{\alpha}_{k,L}^{(0)}\right) + \\ &\boldsymbol{\Sigma}_{ik}^{(b)}\left(\boldsymbol{D}_k^{(0)}\boldsymbol{\alpha}_{k,L}^{(0)},\boldsymbol{E}_{k,L},\boldsymbol{D}_k^{(0)}\boldsymbol{\alpha}_{k,L}^{(0)}\right) + \boldsymbol{\Sigma}_{ik}^{(c)}\left(\boldsymbol{D}_k^{(0)}\boldsymbol{\alpha}_{k,L}^{(0)},\boldsymbol{E}_{k,L}\boldsymbol{D}_k^{(0)\mathrm{H}},\boldsymbol{\alpha}_{k,L}^{(0)}\right) + \\ &\boldsymbol{\Sigma}_{ik}^{(c)*}\left(\boldsymbol{D}_k^{(0)}\boldsymbol{\alpha}_{k,L}^{(0)},\boldsymbol{E}_{k,L}\boldsymbol{D}_k^{(0)\mathrm{H}},\boldsymbol{\alpha}_{k,L}^{(0)}\right) + \boldsymbol{\Sigma}_{ik}^{(d)}\left(\boldsymbol{D}_k^{(0)}\boldsymbol{\alpha}_{k,L}^{(0)},\boldsymbol{\alpha}_{k,L}^{(0)}\right) + \\ &\boldsymbol{\Sigma}_{ik}^{(d)*}\left(\boldsymbol{D}_k^{(0)}\boldsymbol{\alpha}_{k,L}^{(0)},\boldsymbol{\alpha}_{k,L}^{(0)}\right)(1\le i\le 3) \end{aligned} \tag{36}$$

$$\begin{aligned} \boldsymbol{\xi}_{jk} = &\boldsymbol{\Sigma}_{jk}^{(a)}\left(\boldsymbol{\alpha}_{k,L}^{(0)},\boldsymbol{I}_N,\boldsymbol{\alpha}_{k,L}^{(0)}\right) + \boldsymbol{\Sigma}_{jk}^{(a)}\left(\boldsymbol{\alpha}_{k,L}^{(0)},\boldsymbol{D}_k^{(0)}\boldsymbol{E}_{k,L}\boldsymbol{D}_k^{(0)\mathrm{H}},\boldsymbol{\alpha}_{k,L}^{(0)}\right) + \\ &\boldsymbol{\Sigma}_{jk}^{(b)}\left(\boldsymbol{D}_k^{(0)}\boldsymbol{\alpha}_{k,L}^{(0)},\boldsymbol{E}_{k,L},\boldsymbol{D}_k^{(0)}\boldsymbol{\alpha}_{k,L}^{(0)}\right) + \boldsymbol{\Sigma}_{jk}^{(c)}\left(\boldsymbol{D}_k^{(0)}\boldsymbol{\alpha}_{k,L}^{(0)},\boldsymbol{E}_{k,L}\boldsymbol{D}_k^{(0)\mathrm{H}},\boldsymbol{\alpha}_{k,L}^{(0)}\right) + \\ &\boldsymbol{\Sigma}_{jk}^{(c)*}\left(\boldsymbol{D}_k^{(0)}\boldsymbol{\alpha}_{k,L}^{(0)},\boldsymbol{E}_{k,L}^{\mathrm{H}}\boldsymbol{D}_k^{(0)\mathrm{H}},\boldsymbol{\alpha}_{k,L}^{(0)}\right)\boldsymbol{\Pi}_3 + \boldsymbol{\Sigma}_{jk}^{(d)}\left(\boldsymbol{D}_k^{(0)}\boldsymbol{\alpha}_{k,L}^{(0)},\boldsymbol{\alpha}_{k,L}^{(0)}\right) + \\ &\boldsymbol{\Sigma}_{jk}^{(d)*}\left(\boldsymbol{D}_k^{(0)}\boldsymbol{\alpha}_{k,L}^{(0)},\boldsymbol{\alpha}_{k,L}^{(0)}\right)\boldsymbol{\Pi}_3(j=4,5) \end{aligned} \tag{37}$$

$$\begin{aligned} \boldsymbol{\xi}_{6k} = &\boldsymbol{\Sigma}_{6k}^{(a)}\left(\boldsymbol{\alpha}_{k,L}^{(0)},\boldsymbol{I}_N,\boldsymbol{\alpha}_{k,L}^{(0)}\right) + \boldsymbol{\Sigma}_{6k}^{(a)}\left(\boldsymbol{\alpha}_{k,L}^{(0)},\boldsymbol{D}_k^{(0)}\boldsymbol{E}_{k,L}\boldsymbol{D}_k^{(0)\mathrm{H}},\boldsymbol{\alpha}_{k,L}^{(0)}\right) + \\ &\boldsymbol{\Sigma}_{6k}^{(b)}\left(\boldsymbol{D}_k^{(0)}\boldsymbol{\alpha}_{k,L}^{(0)},\boldsymbol{E}_{k,L},\boldsymbol{D}_k^{(0)}\boldsymbol{\alpha}_{k,L}^{(0)}\right) + \boldsymbol{\Sigma}_{6k}^{(c)}\left(\boldsymbol{D}_k^{(0)}\boldsymbol{\alpha}_{k,L}^{(0)},\boldsymbol{E}_{k,L}\boldsymbol{D}_k^{(0)\mathrm{H}},\boldsymbol{\alpha}_{k,L}^{(0)}\right) + \\ &\boldsymbol{\Sigma}_{6k}^{(c)\mathrm{H}}\left(\boldsymbol{D}_k^{(0)}\boldsymbol{\alpha}_{k,L}^{(0)},\boldsymbol{E}_{k,L}^{\mathrm{H}}\boldsymbol{D}_k^{(0)\mathrm{H}},\boldsymbol{\alpha}_{k,L}^{(0)}\right) \end{aligned} \tag{38}$$

with

$$\begin{cases} \boldsymbol{\Sigma}_{1k}^{(a)}(\boldsymbol{q}_1,\boldsymbol{\Phi},\boldsymbol{q}_2) = \boldsymbol{F}_{1k}^{(a)\mathrm{H}}(\boldsymbol{q}_1)\boldsymbol{\Phi}\boldsymbol{F}_{1k}^{(a)}(\boldsymbol{q}_2), \boldsymbol{\Sigma}_{2k}^{(a)}(\boldsymbol{q}_1,\boldsymbol{\Phi},\boldsymbol{q}_2) = \boldsymbol{F}_{2k}^{(a)\mathrm{H}}(\boldsymbol{q}_1)\boldsymbol{\Phi}\boldsymbol{F}_{2k}^{(a)}(\boldsymbol{q}_2) \\ \boldsymbol{\Sigma}_{3k}^{(a)}(\boldsymbol{q}_1,\boldsymbol{\Phi},\boldsymbol{q}_2) = \boldsymbol{F}_{1k}^{(a)\mathrm{H}}(\boldsymbol{q}_1)\boldsymbol{\Phi}\boldsymbol{F}_{2k}^{(a)}(\boldsymbol{q}_2) + \boldsymbol{F}_{1k}^{(a)\mathrm{T}}(\boldsymbol{q}_2)\boldsymbol{\Phi}^{\mathrm{T}}\boldsymbol{F}_{2k}^{(a)*}(\boldsymbol{q}_1) \\ \boldsymbol{\Sigma}_{4k}^{(a)}(\boldsymbol{q}_1,\boldsymbol{\Phi},\boldsymbol{q}_2) = \boldsymbol{F}_{1k}^{(a)\mathrm{H}}(\boldsymbol{q}_1)\boldsymbol{\Phi}\boldsymbol{F}_{3k}^{(a)}(\boldsymbol{q}_2) + \boldsymbol{F}_{1k}^{(a)\mathrm{T}}(\boldsymbol{q}_2)\boldsymbol{\Phi}^{\mathrm{T}}\boldsymbol{F}_{3k}^{(a)*}(\boldsymbol{q}_1)\boldsymbol{\Pi}_3 \\ \boldsymbol{\Sigma}_{5k}^{(a)}(\boldsymbol{q}_1,\boldsymbol{\Phi},\boldsymbol{q}_2) = \boldsymbol{F}_{2k}^{(a)\mathrm{H}}(\boldsymbol{q}_1)\boldsymbol{\Phi}\boldsymbol{F}_{3k}^{(a)}(\boldsymbol{q}_2) + \boldsymbol{F}_{2k}^{(a)\mathrm{T}}(\boldsymbol{q}_2)\boldsymbol{\Phi}^{\mathrm{T}}\boldsymbol{F}_{3k}^{(a)*}(\boldsymbol{q}_1)\boldsymbol{\Pi}_3 \\ \boldsymbol{\Sigma}_{6k}^{(a)}(\boldsymbol{q}_1,\boldsymbol{\Phi},\boldsymbol{q}_2) = \boldsymbol{F}_{3k}^{(a)\mathrm{H}}(\boldsymbol{q}_1)\boldsymbol{\Phi}\boldsymbol{F}_{3k}^{(a)}(\boldsymbol{q}_2) \end{cases} \tag{39}$$

$$\begin{cases} \boldsymbol{\Sigma}_{1k}^{(b)}(\boldsymbol{q}_1,\boldsymbol{\Phi},\boldsymbol{q}_2) = \boldsymbol{F}_{1k}^{(b)\mathrm{H}}(\boldsymbol{q}_1)\boldsymbol{\Phi}\boldsymbol{F}_{1k}^{(b)}(\boldsymbol{q}_2), \boldsymbol{\Sigma}_{2k}^{(b)}(\boldsymbol{q}_1,\boldsymbol{\Phi},\boldsymbol{q}_2) = \boldsymbol{F}_{2k}^{(b)\mathrm{H}}(\boldsymbol{q}_1)\boldsymbol{\Phi}\boldsymbol{F}_{2k}^{(b)}(\boldsymbol{q}_2) \\ \boldsymbol{\Sigma}_{3k}^{(b)}(\boldsymbol{q}_1,\boldsymbol{\Phi},\boldsymbol{q}_2) = \boldsymbol{F}_{1k}^{(b)\mathrm{H}}(\boldsymbol{q}_1)\boldsymbol{\Phi}\boldsymbol{F}_{2k}^{(b)}(\boldsymbol{q}_2) + \boldsymbol{F}_{1k}^{(b)\mathrm{T}}(\boldsymbol{q}_2)\boldsymbol{\Phi}^{\mathrm{T}}\boldsymbol{F}_{2k}^{(b)*}(\boldsymbol{q}_1) \\ \boldsymbol{\Sigma}_{4k}^{(b)}(\boldsymbol{q}_1,\boldsymbol{\Phi},\boldsymbol{q}_2) = \boldsymbol{F}_{1k}^{(b)\mathrm{H}}(\boldsymbol{q}_1)\boldsymbol{\Phi}\boldsymbol{F}_{3k}^{(b)}(\boldsymbol{q}_2) + \boldsymbol{F}_{1k}^{(b)\mathrm{T}}(\boldsymbol{q}_2)\boldsymbol{\Phi}^{\mathrm{T}}\boldsymbol{F}_{3k}^{(b)*}(\boldsymbol{q}_1)\boldsymbol{\Pi}_3 \\ \boldsymbol{\Sigma}_{5k}^{(b)}(\boldsymbol{q}_1,\boldsymbol{\Phi},\boldsymbol{q}_2) = \boldsymbol{F}_{2k}^{(b)\mathrm{H}}(\boldsymbol{q}_1)\boldsymbol{\Phi}\boldsymbol{F}_{3k}^{(b)}(\boldsymbol{q}_2) + \boldsymbol{F}_{2k}^{(b)\mathrm{T}}(\boldsymbol{q}_2)\boldsymbol{\Phi}^{\mathrm{T}}\boldsymbol{F}_{3k}^{(b)*}(\boldsymbol{q}_1)\boldsymbol{\Pi}_3 \\ \boldsymbol{\Sigma}_{6k}^{(b)}(\boldsymbol{q}_1,\boldsymbol{\Phi},\boldsymbol{q}_2) = \boldsymbol{F}_{3k}^{(b)\mathrm{H}}(\boldsymbol{q}_1)\boldsymbol{\Phi}\boldsymbol{F}_{3k}^{(b)}(\boldsymbol{q}_2) \end{cases} \quad (40)$$

$$\begin{cases} \boldsymbol{\Sigma}_{1k}^{(c)}(\boldsymbol{q}_1,\boldsymbol{\Phi},\boldsymbol{q}_2) = \boldsymbol{F}_{1k}^{(b)\mathrm{H}}(\boldsymbol{q}_1)\boldsymbol{\Phi}\boldsymbol{F}_{1k}^{(a)}(\boldsymbol{q}_2), \boldsymbol{\Sigma}_{2k}^{(c)}(\boldsymbol{q}_1,\boldsymbol{\Phi},\boldsymbol{q}_2) = \boldsymbol{F}_{2k}^{(b)\mathrm{H}}(\boldsymbol{q}_1)\boldsymbol{\Phi}\boldsymbol{F}_{2k}^{(a)}(\boldsymbol{q}_2) \\ \boldsymbol{\Sigma}_{3k}^{(c)}(\boldsymbol{q}_1,\boldsymbol{\Phi},\boldsymbol{q}_2) = \boldsymbol{F}_{1k}^{(b)\mathrm{H}}(\boldsymbol{q}_1)\boldsymbol{\Phi}\boldsymbol{F}_{2k}^{(a)}(\boldsymbol{q}_2) + \boldsymbol{F}_{1k}^{(a)\mathrm{T}}(\boldsymbol{q}_2)\boldsymbol{\Phi}^{\mathrm{T}}\boldsymbol{F}_{2k}^{(b)*}(\boldsymbol{q}_1) \\ \boldsymbol{\Sigma}_{4k}^{(c)}(\boldsymbol{q}_1,\boldsymbol{\Phi},\boldsymbol{q}_2) = \boldsymbol{F}_{1k}^{(b)\mathrm{H}}(\boldsymbol{q}_1)\boldsymbol{\Phi}\boldsymbol{F}_{3k}^{(a)}(\boldsymbol{q}_2) + \boldsymbol{F}_{1k}^{(a)\mathrm{T}}(\boldsymbol{q}_2)\boldsymbol{\Phi}^{\mathrm{T}}\boldsymbol{F}_{3k}^{(b)*}(\boldsymbol{q}_1)\boldsymbol{\Pi}_3 \\ \boldsymbol{\Sigma}_{5k}^{(c)}(\boldsymbol{q}_1,\boldsymbol{\Phi},\boldsymbol{q}_2) = \boldsymbol{F}_{2k}^{(b)\mathrm{H}}(\boldsymbol{q}_1)\boldsymbol{\Phi}\boldsymbol{F}_{3k}^{(a)}(\boldsymbol{q}_2) + \boldsymbol{F}_{2k}^{(a)\mathrm{T}}(\boldsymbol{q}_2)\boldsymbol{\Phi}^{\mathrm{T}}\boldsymbol{F}_{3k}^{(b)*}(\boldsymbol{q}_1)\boldsymbol{\Pi}_3 \\ \boldsymbol{\Sigma}_{6k}^{(c)}(\boldsymbol{q}_1,\boldsymbol{\Phi},\boldsymbol{q}_2) = \boldsymbol{F}_{3k}^{(b)\mathrm{H}}(\boldsymbol{q}_1)\boldsymbol{\Phi}\boldsymbol{F}_{3k}^{(a)}(\boldsymbol{q}_2) \end{cases} \quad (41)$$

$$\begin{cases} \boldsymbol{\Sigma}_{1k}^{(d)}(\boldsymbol{q}_1,\boldsymbol{q}_2) = \frac{1}{2}\frac{\partial^2(\boldsymbol{q}_1^{\mathrm{H}}\overline{\boldsymbol{C}}(\boldsymbol{z},\boldsymbol{p}_k)\boldsymbol{Y}_k\boldsymbol{q}_2)}{\partial\boldsymbol{z}\partial\boldsymbol{z}^{\mathrm{T}}}, \boldsymbol{\Sigma}_{2k}^{(d)}(\boldsymbol{q}_1,\boldsymbol{q}_2) = \frac{1}{2}\frac{\partial^2(\boldsymbol{q}_1^{\mathrm{H}}\overline{\boldsymbol{C}}(\boldsymbol{z},\boldsymbol{p}_k)\boldsymbol{Y}_k\boldsymbol{q}_2)}{\partial\boldsymbol{p}_k\partial\boldsymbol{p}_k^{\mathrm{T}}} \\ \boldsymbol{\Sigma}_{3k}^{(d)}(\boldsymbol{q}_1,\boldsymbol{q}_2) = \frac{\partial^2(\boldsymbol{q}_1^{\mathrm{H}}\overline{\boldsymbol{C}}(\boldsymbol{z},\boldsymbol{p}_k)\boldsymbol{Y}_k\boldsymbol{q}_2)}{\partial\boldsymbol{z}\partial\boldsymbol{p}_k^{\mathrm{T}}}, \boldsymbol{\Sigma}_{4k}^{(d)}(\boldsymbol{q}_1,\boldsymbol{q}_2) = \left(\frac{\partial\left(\overline{\boldsymbol{C}}^{\mathrm{H}}(\boldsymbol{z},\boldsymbol{p}_k)\boldsymbol{q}_1\right)}{\partial\boldsymbol{z}^{\mathrm{T}}}\right)^{\mathrm{H}}\cdot(\mathrm{diag}\{\boldsymbol{q}_2\}\otimes\boldsymbol{I}_{MN})\boldsymbol{\Pi}_1 \\ \boldsymbol{\Sigma}_{5k}^{(d)}(\boldsymbol{q}_1,\boldsymbol{q}_2) = \left(\frac{\partial\left(\overline{\boldsymbol{C}}^{\mathrm{H}}(\boldsymbol{z},\boldsymbol{p}_k)\boldsymbol{q}_1\right)}{\partial\boldsymbol{p}_k^{\mathrm{T}}}\right)^{\mathrm{H}}\cdot(\mathrm{diag}\{\boldsymbol{q}_2\}\otimes\boldsymbol{I}_{MN})\boldsymbol{\Pi}_1 \end{cases} \quad (42)$$

The detailed derivation is exhibited in Appendix D.

In sight of the above analysis, as a result, the second-order approximation of $J_{\mathrm{cost}}(\hat{\boldsymbol{z}},\overline{\boldsymbol{p}},\boldsymbol{n})$ can be drawn as

$$\begin{aligned} J_{\mathrm{cost}}(\hat{\boldsymbol{z}},\overline{\boldsymbol{p}},\boldsymbol{n}) \approx\ & J_{\mathrm{cost}}^{(0)} + \sum_{k=1}^{K}\boldsymbol{f}_{1k}^{\mathrm{H}}\widetilde{\boldsymbol{z}} + \sum_{k=1}^{K}\boldsymbol{f}_{2k}^{\mathrm{H}}\widetilde{\boldsymbol{p}}_k + \sum_{k=1}^{K}\boldsymbol{f}_{3k}^{\mathrm{H}}\widetilde{\boldsymbol{n}}_k \\ & + \sum_{k=1}^{K}\widetilde{\boldsymbol{z}}^{\mathrm{T}}\boldsymbol{\xi}_{1k}\widetilde{\boldsymbol{z}} + \sum_{k=1}^{K}\widetilde{\boldsymbol{p}}_k^{\mathrm{T}}\boldsymbol{\xi}_{2k}\widetilde{\boldsymbol{p}}_k + \sum_{k=1}^{K}\widetilde{\boldsymbol{z}}^{\mathrm{T}}\boldsymbol{\xi}_{3k}\widetilde{\boldsymbol{p}}_k \\ & + \sum_{k=1}^{K}\widetilde{\boldsymbol{z}}^{\mathrm{T}}\boldsymbol{\xi}_{4k}\widetilde{\boldsymbol{n}}_k + \sum_{k=1}^{K}\boldsymbol{p}_k^{\mathrm{T}}\boldsymbol{\xi}_{5k}\widetilde{\boldsymbol{n}}_k + \sum_{k=1}^{K}\widetilde{\boldsymbol{n}}_k^{\mathrm{H}}\boldsymbol{\xi}_{6k}\widetilde{\boldsymbol{n}}_k \end{aligned} \quad (43)$$

Note that $\boldsymbol{f}_{nk}|_{1\le n\le 3}$ can act as the gradient vector, and $\boldsymbol{\xi}_{jk}|_{1\le j\le 6}$ can form the Hessian matrix, respectively. It is easily found that the single-step localization errors is linearly associated with the environment noise and system errors. Furthermore, the MSE of DPD estimator is presented in the next subsection.

5.3. MSE of The Single-Step Method with System Errors

Following the analysis presented above, it can be easily obtained that

$$\begin{cases} \frac{\partial J_{\mathrm{cost}}^{(0)}}{\partial\boldsymbol{z}} = 0 \\ \frac{\partial J_{\mathrm{cost}}(\hat{\boldsymbol{z}},\overline{\boldsymbol{p}},\boldsymbol{n})}{\partial\hat{\boldsymbol{z}}} = 0 \end{cases} \quad (44)$$

Then, via combining the first expression in (44) with (43), we have

$$\frac{\partial J_{\mathrm{cost}}^{(0)}}{\partial\boldsymbol{z}} = \sum_{k=1}^{K}\boldsymbol{f}_{1k}^{*} = 0 \quad (45)$$

Through the second equality in (44), we imply

$$\widetilde{\boldsymbol{z}} = \underset{\boldsymbol{q}}{\mathrm{argmax}}\left(\sum_{k=1}^{K}\boldsymbol{f}_{1k}^{\mathrm{H}}\boldsymbol{q} + \sum_{k=1}^{K}\boldsymbol{q}^{\mathrm{T}}\boldsymbol{\xi}_{1k}\boldsymbol{q} + \sum_{k=1}^{K}\boldsymbol{q}^{\mathrm{T}}\boldsymbol{\xi}_{3k}\widetilde{\boldsymbol{p}}_k + \sum_{k=1}^{K}\boldsymbol{q}^{\mathrm{T}}\boldsymbol{\xi}_{4k}\widetilde{\boldsymbol{n}}_k\right) \quad (46)$$

moreover, (46) can be specified by

$$\begin{aligned}\tilde{z} &= -\tfrac{1}{2}\left(\sum_{k=1}^{K}\xi_{1k}\right)^{-1}\left(\sum_{k=1}^{K}\xi_{3k}\tilde{p}_k + \sum_{k=1}^{K}\xi_{4k}\tilde{n}_k + \sum_{k=1}^{K}f_{1k}^{*}\right)\\ &= -\tfrac{1}{2}\left(\sum_{k=1}^{K}\xi_{1k}\right)^{-1}\left(\sum_{k=1}^{K}\xi_{4k}\tilde{n}_k\right) - \tfrac{1}{2}\left(\sum_{k=1}^{K}\xi_{1k}\right)^{-1}\left(\sum_{k=1}^{K}\xi_{3k}\tilde{p}_k\right)\end{aligned} \tag{47}$$

It is readily observed that the localization error parameter $\tilde{z}$ is composed of two terms. The first formulation in (47) is associated with the environment noise, which is shown as

$$\tilde{z}_1 = -\frac{1}{2}\left(\sum_{k=1}^{K}\xi_{1k}\right)^{-1}\left(\sum_{k=1}^{K}\xi_{4k}\tilde{n}_k\right) \tag{48}$$

The second equality in (47) is corresponding to the system errors, which is exhibited as

$$\tilde{z}_2 = -\frac{1}{2}\left(\sum_{k=1}^{K}\xi_{1k}\right)^{-1}\left(\sum_{k=1}^{K}\xi_{3k}\tilde{p}_k\right) \tag{49}$$

To perfect the analysis, we make a statistical assumption that the system error vectors $\tilde{\boldsymbol{p}}_k|_{1\le k\le K}$ obey zero-mean with covariance matrix $\Omega_k|_{1\le k\le K}$. As a result, we have the location error covariance matrices

$$\begin{aligned}\boldsymbol{R} = \mathrm{E}\left[\tilde{z}\tilde{z}^{\mathrm{T}}\right] = \; & \tfrac{\sigma^2}{4}\left(\sum_{k=1}^{K}\xi_{1k}\right)^{-1}\left(\sum_{k=1}^{K}\xi_{4k}\xi_{4k}^{\mathrm{H}}\right)\left(\sum_{k=1}^{K}\xi_{1k}^{\mathrm{H}}\right)^{-1} + \\ & \tfrac{1}{4}\left(\sum_{k=1}^{K}\xi_{1k}\right)^{-1}\left(\sum_{k=1}^{K}\xi_{3k}\Omega_k\xi_{3k}^{\mathrm{H}}\right)\left(\sum_{k=1}^{K}\xi_{1k}^{\mathrm{H}}\right)^{-1}\end{aligned} \tag{50}$$

Note that the first part in (50) is related to environment noise and the second part in (50) is attached by system errors. It should be emphasized that trace$\{\boldsymbol{R}\}$ can represent the MSE of the single-step approach in presence of two kinds of disturbance issues.

To better exhibit the analysis process, we summarize it as Algorithm 1 as follows.

Algorithm 1. The main steps of the analysis process

Input:
The observed data: $\tilde{y}_{l,k}$, the real parameter and the error parameter of the lth receiver: $\boldsymbol{p}_{l,k}$ and $\tilde{\boldsymbol{p}}_{l,k}$, $l = 1,\ldots,L$ $k = 1,\ldots,K$;

1. Calculate a second-order perturbation expression of $\boldsymbol{D}_k(\hat{\boldsymbol{z}},\overline{\boldsymbol{p}}_k,\boldsymbol{n}_k)$ via Equation (20);
2. Substitute $\boldsymbol{D}_k(\hat{\boldsymbol{z}},\overline{\boldsymbol{p}}_k,\boldsymbol{n}_k)$ into (23) to obtain the expression of the estimated Hermitian matrix $\boldsymbol{Q}_k(\hat{\boldsymbol{z}},\overline{\boldsymbol{p}}_k,\boldsymbol{n}_k)$;
3. Based on the matrix-perturbation analysis, calculate $\overline{\lambda}_{k,L}$ through Equation (26);
4. Approximate $J_{\text{cost}}(\hat{\boldsymbol{z}},\overline{\boldsymbol{p}},\boldsymbol{n})$ by (29);
5. Obtain the location error covariance matrices R.

Output: The MSE of the estimated location error trace$\{\boldsymbol{R}\}$.

5.4. CRB under Precise Known Receiver Conditions

For any unbiased estimator, the CRB provides a lower bound on emitter localization variance. This section presents the derivation of the CRB under the precise known positions and velocities of the receivers. It is not difficult to find that although the MSE in Section 5.3 is given with system errors, the CRB is provided without system errors. Therefore, the comparison between this CRB and the MSE can reveal the performance difference caused by system errors.

The unknown parameter vector η can be defined by

$$\eta = \left[z^{\mathrm{T}}, \omega^{\mathrm{T}}\right]^{\mathrm{T}} \tag{51}$$

where ω denotes all real parameters except the target position. The expression of ω is written as

$$\omega = \left[s^{\mathrm{T}}, b^{\mathrm{T}}, \Delta f^{\mathrm{T}}\right]^{\mathrm{T}} \tag{52}$$

here $s = \left[s_1^{(R)\mathrm{T}}, s_2^{(R)\mathrm{T}}, \ldots, s_K^{(R)\mathrm{T}}, s_1^{(I)\mathrm{T}}, s_2^{(I)\mathrm{T}}, \ldots, s_K^{(I)\mathrm{T}}\right]^{\mathrm{T}}$ with $s_k = [s_k(1), s_k(2), \ldots, s_k(N)]^{\mathrm{T}}$, $b = \left[b_1^{(R)\mathrm{T}}, b_2^{(R)\mathrm{T}}, \ldots, b_L^{(R)\mathrm{T}}, b_1^{(I)\mathrm{T}}, b_2^{(I)\mathrm{T}}, \ldots, b_L^{(I)\mathrm{T}}\right]^{\mathrm{T}}$ with $b_l = [b_{l,1}, b_{l,2}, \ldots, b_{l,K}]^{\mathrm{T}}$, and $\Delta f = [\Delta f_1, \Delta f_2, \ldots, \Delta f_K]^{\mathrm{T}}$.

Let $d_{l,k}(\eta) = b_{l,k} C\left(z, p_{l,k}\right) B_k s_k$. According to [28], the fisher information matrix of unknown parameter vector η is shown as

$$J_{\eta\eta} = \frac{2}{\sigma^2} \sum_{k=1}^{K} \sum_{l=1}^{L} Re\left(\frac{\partial d_{l,k}(\eta)}{\partial \eta^{\mathrm{T}}}\right)^{\mathrm{H}} \left(\frac{\partial d_{l,k}(\eta)}{\partial \eta^{\mathrm{T}}}\right) \tag{53}$$

Define

$$\begin{cases} Y_{zz} = \sum_{k=1}^{K} \sum_{l=1}^{L} Re\left\{ \left(\frac{d_{l,k}(\eta)}{z^{\mathrm{T}}}\right)^{\mathrm{H}} \frac{d_{l,k}(\eta)}{z^{\mathrm{T}}} \right\} \\ Y_{z\omega} = \sum_{k=1}^{K} \sum_{l=1}^{L} Re\left\{ \left(\frac{d_{l,k}(\eta)}{z^{\mathrm{T}}}\right)^{\mathrm{H}} \frac{d_{l,k}(\eta)}{w^{\mathrm{T}}} \right\} \\ Y_{\omega\omega} = \sum_{k=1}^{K} \sum_{l=1}^{L} Re\left\{ \left(\frac{d_{l,k}(\eta)}{\omega^{\mathrm{T}}}\right)^{\mathrm{H}} \frac{d_{l,k}(\eta)}{\omega^{\mathrm{T}}} \right\} \end{cases} \tag{54}$$

The expression of $J_{\eta\eta}$ can be rewritten as

$$J_{\eta\eta} = \frac{2}{\sigma_n^2} \begin{bmatrix} Y_{zz} & Y_{z\omega} \\ Y_{z\omega}^{\mathrm{T}} & Y_{\omega\omega} \end{bmatrix} \tag{55}$$

Following the matrix inversion formula in Reference [29], the block matrix form of $J_{\eta\eta}$ is formulated as

$$CRB = \frac{\sigma_n^2}{2} \left(Y_{zz} - Y_{zw} Y_{ww}^{-1} Y_{zw}^{\mathrm{T}}\right)^{-1} \tag{56}$$

Therefore, substituting the sub-blocks into (56), which are shown in Appendix E, will get the CRB value.

6. Simulation Results

This section provides 200 Monte Carlo trials to corroborate the above theoretical analysis based on MATLAB 2015b (MathWorks, Natick, MA, USA), and source data is generated as a Gaussian random signal. Firstly, the localization performance of the proposed method and Weiss's method [10] are performed. Secondly, when system errors exist, the related theoretical values developed in Section 5 are exhibited. Unless otherwise specified, we collect $N = 32$ sample points in each interval at a sampling rate of $f_s = 15$ kHz, use $L = 3$ receivers, perform a total of $K = 8$ observations, set the velocity of receiver as $v = 300$ m/s and select the unknown transmitted frequency from [−100 100] Hz randomly. Additionally, the propagation channel is an additive white Gaussian noise channel, and the channel attenuation is drawn from a normal distribution with mean of 1 and standard deviation of 0.1, as well as the channel phase is selected from a uniform distribution over $[-\pi, \pi]$. The target locates at [1.5 1.5] km, and the receivers move along the trajectories (three scenarios are included) shown in Figure 1. Note that the simulations in Sections 6.2 and 6.3 are based on the scenario (a) in Figure 1. Finally, root mean square error (RMSE) is adopted to evaluate localization accuracy in this paper, which is defined by

$$RMSE = \sqrt{\frac{1}{200}\sum_{j=1}^{200}\left\|z - \hat{z}^{(j)}\right\|^2} \tag{57}$$

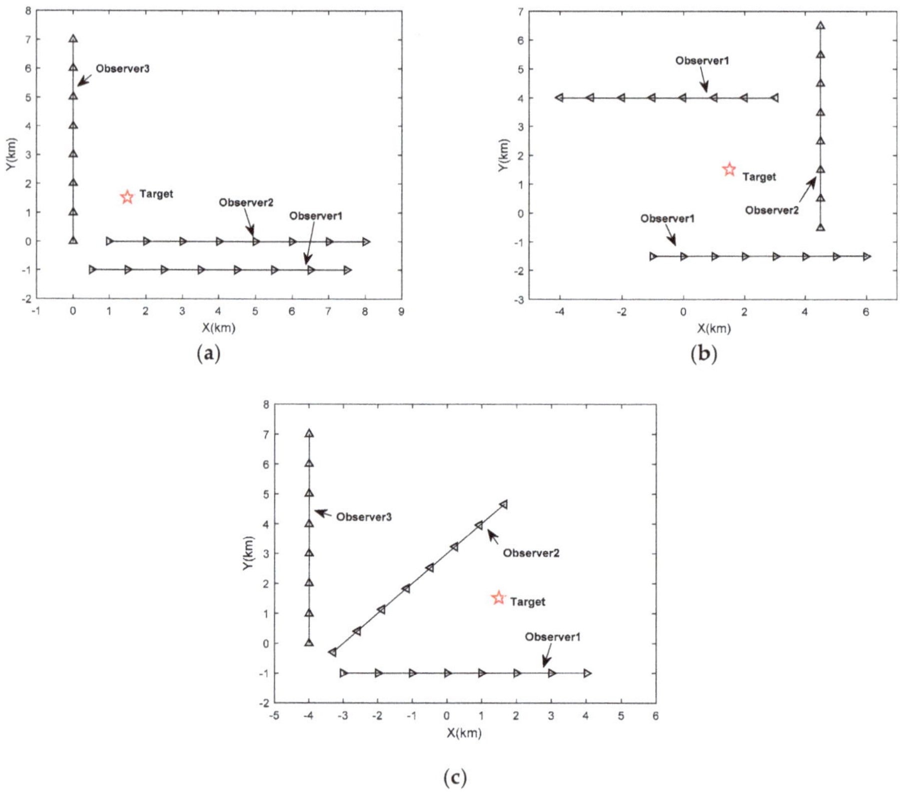

Figure 1. Position of target and the trajectories of receivers. (**a**) Scenario a; (**b**) scenario b; (**c**) scenario c.

6.1. Effect of Reveicer Trajectories

In order to test the test whether our algorithm is sensitive to motion trajectories, we exhibit the localization performance in the different scenarios in Figure 1. Figure 2 indicts that CRB for scenario (a) can generate best localization accuracy, and CRBs for scenario (b) as well as (c) have similar positioning precision. It is easily found that our method has the same trend as with CRB curves. Consequently, the performance of our method is satisfied with theoretical analysis and our method is robust to the receiver trajectories.

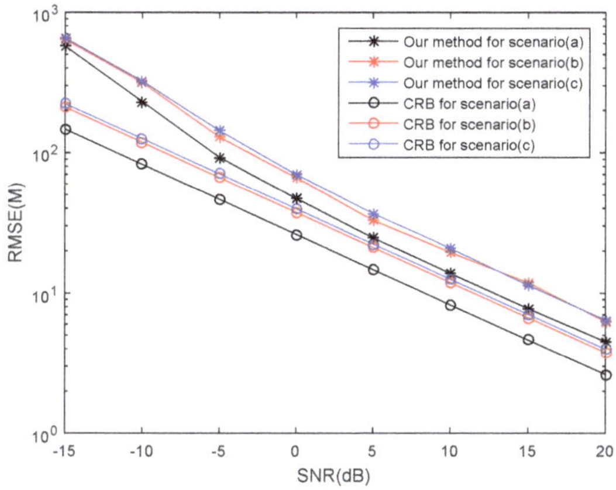

Figure 2. RMSEs versus SNR under different trajectories of receivers.

6.2. *Effect of DOA Information*

To verify the influence of investigating DOA information in signal model on localization performance, we take the following simulations. Firstly, the pseudo spatial spectra of the DPD estimator with different parameter information at SNR = -10 dB are presented in Figure 3. It is easily observed in Figure 3a,b that by using additional DOA information, the true peak of the spectrum is more prominent and the pseudo peaks are significantly reduced. Additionally, the 2D plots in Figure 3c,d indicate that with the utilization of DOA information, the estimated target position is closer to the true target position.

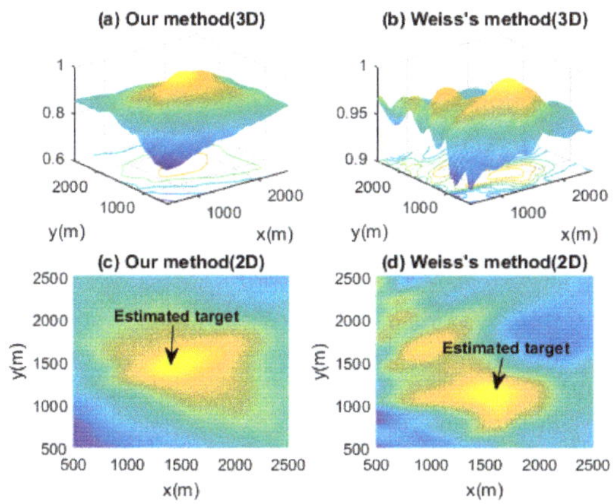

Figure 3. Pseudo spatial spectra of the DPD method using different parameter information. (**a**) Our method (3D); (**b**) Welss's method (3D); (**c**) Our method (2D); (**d**) Welss's method (2D).

Secondly, the performance comparison between the two methods is available in Figure 4. It is straightforward to see that compared with Weiss's method, our method performs superior at each SNR level. More specifically, our method receives higher localization performance at low SNRs, which shows strong robustness to harsh environments. Additionally, our method is closer to the corresponding CRB. Consequently, DOA information gives a significant improvement on positioning accuracy of this single-step approach.

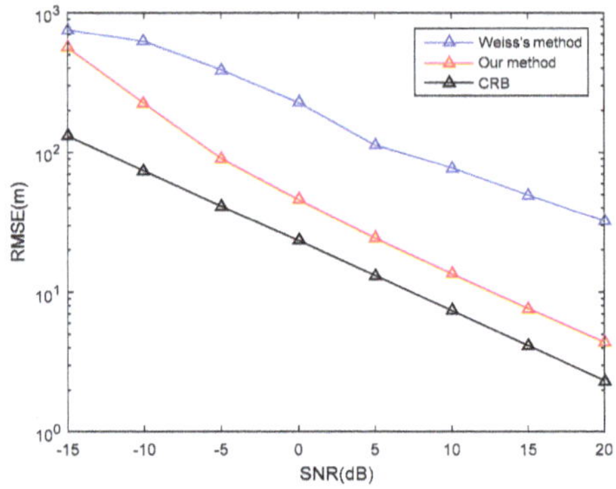

Figure 4. RMSEs versus SNR.

6.3. Effect of System Errors

This subsection mainly reveals the performance loss caused by system errors. The disturbances from the receiver position and velocity are assumed to be a Gaussian distribution with zero-mean and variances of σ_p^2 as well as σ_v^2, respectively. Note that the disturbances from different receivers at different observed interval have the same value in this paper. Additionally, the single-step method is exhibited at two conditions: (1) both system errors and environment noise present; (2) only environment noise attends. Furthermore, the MSE with system errors provided by (50) and the CRB without system errors provided by (56) are also included in the simulations.

Firstly, the localization performance versus SNR are presented and both σ_p^2 and σ_v^2 are set at 15. As shown in Figure 5, whether system errors exist or not, there is no difference of the DPD localization performance at SNR ranging from -5 to 0 dB. This phenomenon indicates that positioning accuracy has not received too much influence on system error and is mainly caused by environment noise at low SNRs. However, as SNR increases, the localization performance in presence of system errors deteriorates. It tells us that the localization errors are affected by environment noise and system errors together at high SNRs. Additionally, when SNR reaches 20 dB, the RMSE of our algorithm is almost constant. The reason is that when SNR is relatively large, the localization precision mainly comes from system errors and cannot be reduced by the increase of SNR. Meanwhile, the localization errors of Weiss's method continue to decline, which implies our method can achieve the best performance faster in presence of system errors at the same SNR condition. Furthermore, when two errors exist, the curve of our method approximates the MSE curve, demonstrating the effectiveness of the theoretical analysis in Section 5.

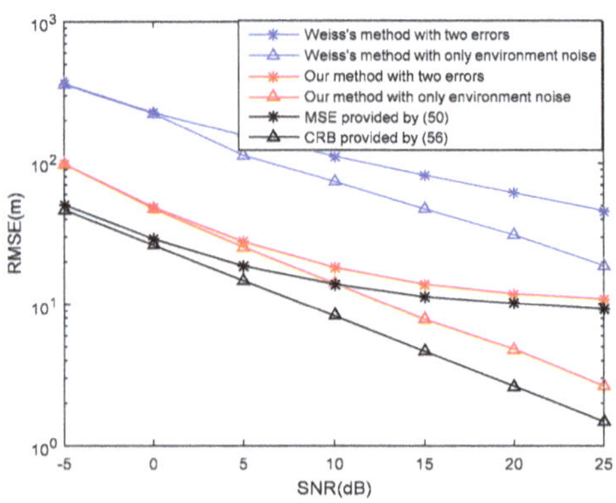

Figure 5. RMSEs versus SNR under different scenarios.

Then, the localization errors versus the perturbation variance of system errors at SNR = 10 dB are plotted in Figure 6. Unsurprisingly, the localization errors of the CRB and our method in absence of system errors have hardly changed. On the other hand, it is evidently seen that the curves of the MSE and our method with two errors are on the rise. The reason is that the DPD estimator could not solve the influence of system errors.

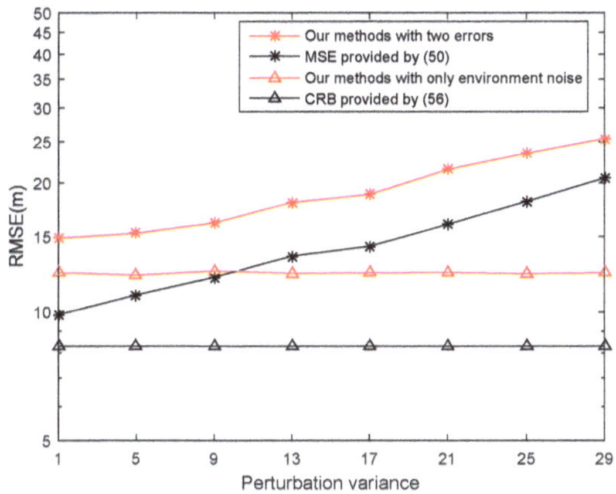

Figure 6. RMSEs versus perturbation variance of system errors.

Finally, in Figure 7, the localization RMSEs versus the number of snapshots is provided, under the scenario that SNR is 10, and both σ_p^2 as well as σ_v^2 are set at 20. As snapshots increase, the localization performance of the CRB and our method without system errors can improve continuously. Unfortunately, the DPD method remains approximately changeless no matter how much snapshots increase. As we mentioned above, this phenomenon can be explained as system errors being the main contributor to positioning precision under this condition, whose affects cannot be erased by the DPD approach.

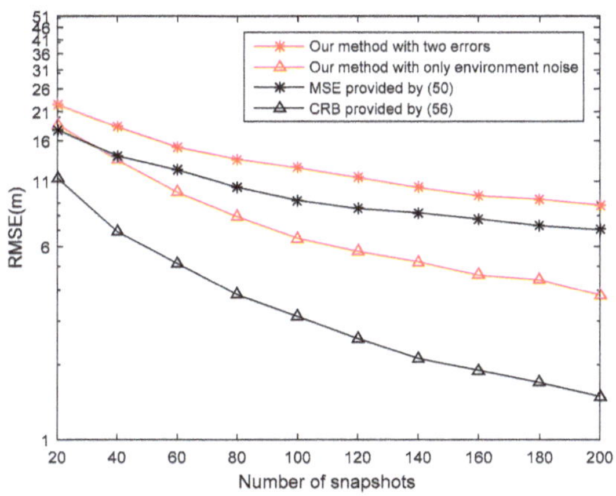

Figure 7. RMSEs versus number of snapshots.

7. Conclusions

In this paper, an improved work to the DPD method proposed by Weiss [10] is studied, and the performance analysis of this method with system errors is provided. We start to reconstruct the signal model by using Doppler and DOA information, which is more suitable for the moving arrays application. Then, the theoretical analysis is presented based on matrix eigen-perturbation results, which express the perturbations as an additive noise on the Hertmitian matrix. Besides, the MSE formulation of direct localization with system errors is provided. Finally, the CRB formulation for the single-step method is also derived, which indicates the localization performance loss caused by system errors. Several simulations demonstrate the analysis that system errors can deteriorate the localization performance of the DPD estimator especially in high SNRs. Consequently, an improved DPD approach considering system errors should be developed in the future work.

Author Contributions: T.Q. derived the proposed method. T.Q. conceived and designed the experiments. B.B. performed the simulations. D.W. analyzed the results. T.Q. wrote the paper. B.B. reviewed the paper.

Funding: This research was funded by the National Natural Science Foundation of China, grant number 61401513 and The APC was funded by 61401513.

Conflicts of Interest: The authors declare no conflict of interest.

Appendix A. Derivation of The Expressions in (21)

A second-order Taylor series expansion of $\overline{C}(\hat{z},\overline{p}_k)$ around (z,p_k) is shown as

$$\begin{aligned}\overline{C}(\hat{z},\overline{p}_k) = \; & \overline{C}(z,p_k) + \sum_{d=1}^{D}\langle\widetilde{z}\rangle_d \overline{C}_d^{\cdot(a)}(z,p_k) + \sum_{d=1}^{2DL}\langle\widetilde{p}_k\rangle_d \overline{C}_d^{\cdot(b)}(z,p_k) + \\ & \frac{1}{2}\sum_{d_1=1}^{D}\sum_{d_2=1}^{D}\langle\widetilde{z}\rangle_{d_1}\langle\widetilde{z}\rangle_{d_2}\overline{C}_{d_1d_2}^{\cdot\cdot(aa)}(z,p_k) + \frac{1}{2}\sum_{d_1=1}^{2DL}\sum_{d_2=1}^{2DL}\langle\widetilde{p}_k\rangle_{d_1}\langle\widetilde{p}_k\rangle_{d_2}\overline{C}_{d_1d_2}^{\cdot\cdot(bb)}(z,p_k) + \\ & \sum_{d_1=1}^{D}\sum_{d_2=1}^{2DL}\langle\widetilde{z}\rangle_{d_1}\langle\widetilde{p}_k\rangle_{d_2}\overline{C}_{d_1d_2}^{\cdot\cdot(ab)}(z,p_k) + o(\varepsilon^2)\end{aligned} \tag{A1}$$

where $\overline{C}_d^{\cdot(a)}(z,p_k)$, $\overline{C}_d^{\cdot(b)}(z,p_k)$, $\overline{C}_{d_1d_2}^{\cdot\cdot(aa)}(z,p_k)$, $\overline{C}_{d_1d_2}^{\cdot\cdot(bb)}(z,p_k)$ and $\overline{C}_{d_1d_2}^{\cdot\cdot(ab)}(z,p_k)$ are exhibited in (22).

Substituting (A1) into (13) leads to

$$\begin{aligned}D_k(\hat{z},\overline{p}_k,n_k) & = \overline{C}(\hat{z},\overline{p}_k)(Y_k+N_k) \approx \overline{C}(z,p_k)Y_k + \\ & \sum_{d=1}^{D}\langle\widetilde{z}\rangle_d \overline{C}_d^{\bullet(a)}(z,p_k)Y_k + \sum_{d=1}^{2DL}\langle\widetilde{p}_k\rangle_d \overline{C}_d^{\bullet(b)}(z,p_k)Y_k + \overline{C}(z,p_k)N_k + \\ & \frac{1}{2}\sum_{d_1=1}^{D}\sum_{d_2=1}^{D}\langle\widetilde{z}\rangle_{d_1}\langle\widetilde{z}\rangle_{d_2}\overline{C}_{d_1d_2}^{\bullet\bullet(aa)}(z,p_k)Y_k + \frac{1}{2}\sum_{d_1=1}^{2DL}\sum_{d_2=1}^{2DL}\langle\widetilde{p}_k\rangle_{d_1}\langle\widetilde{p}_k\rangle_{d_2}\overline{C}_{d_1d_2}^{\bullet\bullet(bb)}(z,p_k)Y_k + \\ & \sum_{d_1=1}^{D}\sum_{d_2=1}^{2DL}\langle\widetilde{z}\rangle_{d_1}\langle\widetilde{p}_k\rangle_{d_2}\overline{C}_{d_1d_2}^{\bullet\bullet(ab)}(z,p_k)Y_k + \sum_{d=1}^{D}\langle\widetilde{z}\rangle_d \overline{C}_d^{\bullet(a)}(z,p_k)N_k + \sum_{d=1}^{2DL}\langle\widetilde{p}_k\rangle_d \overline{C}_d^{\bullet(b)}(z,p_k)N_k \\ & = D_k^{(0)} + \widetilde{D}_k^{(1)} + \widetilde{D}_k^{(2)}\end{aligned} \tag{A2}$$

This ends the derivation.

Appendix B. Derivation of The Expressions in (22)

Firstly, we start with developing the required derivatives of $\dot{\delta}^{(a)}(z,o_{l,k})$ and $\dot{\mu}^{(a)}\left(z,p_{l,k}\right)$ as

$$\dot{\delta}^{(a)}(z,o_{l,k}) = \frac{\partial\delta(z,o_{l,k})}{\partial z} = \frac{\begin{bmatrix}1 & 0\end{bmatrix}^{\mathrm{T}}}{\|z-o_{l,k}\|_2} - \frac{\left(z^{(x)}-o_{l,k}^{(x)}\right)(z-o_{l,k})}{\|z-o_{l,k}\|_2^3} \tag{A3}$$

$$\dot{\delta}^{(b)}(z,o_{l,k}) = \frac{\partial\delta(z,o_{l,k})}{\partial p_{l,k}} = \begin{bmatrix}\frac{\partial\delta(z,o_{l,k})}{\partial o_{l,k}} \\ \frac{\partial\delta(z,o_{l,k})}{\partial v_{l,k}}\end{bmatrix} = \begin{bmatrix}\frac{\left(z^{(x)}-o_{l,k}^{(x)}\right)(z-o_{l,k})}{\|z-o_{l,k}\|_2^3} - \frac{\begin{bmatrix}1 & 0\end{bmatrix}^{\mathrm{T}}}{\|z-o_{l,k}\|_2} \\ \mathbf{0}_D\end{bmatrix} \tag{A4}$$

$$\overset{..(aa)}{\delta}(z, o_{l,k}) = \frac{\partial\delta^2(z,o_{l,k})}{\partial z\partial z^{\mathrm{T}}} = 3\frac{\left(z^{(x)}-o_{l,k}^{(x)}\right)(z-o_{l,k})(z-o_{l,k})^{\mathrm{T}}}{\|z-o_{l,k}\|_2^5} - \frac{\left(z^{(x)}-o_{l,k}^{(x)}\right)I_D}{\|z-o_{l,k}\|_2^3} - \frac{(z-o_{l,k})\begin{bmatrix}1 & 0\end{bmatrix}}{\|z-o_{l,k}\|_2^3} - \frac{\begin{bmatrix}1 & 0\end{bmatrix}^{\mathrm{T}}(z-o_{l,k})^{\mathrm{T}}}{\|z-o_{l,k}\|_2^3} \tag{A5}$$

$$\overset{..(ab)}{\delta}(z, o_{l,k}) = \frac{\partial\delta^2(z,o_{l,k})}{\partial z\partial p_{l,k}^{\mathrm{T}}} = \left[\begin{array}{c|c} \frac{\begin{bmatrix}1 & 0\end{bmatrix}^{\mathrm{T}}\cdot(z-o_{l,k})^{\mathrm{T}}}{\|z-o_{l,k}\|_2^3} + \frac{\left(z^{(x)}-o_{l,k}^{(x)}\right)\cdot I_D}{\|z-o_{l,k}\|_2^3} + \frac{(z-o_{l,k})\cdot\begin{bmatrix}1 & 0\end{bmatrix}}{\|z-o_{l,k}\|_2^3} & \\ -3\frac{\left(z^{(x)}-o_{l,k}^{(x)}\right)(z-o_{l,k})(z-o_{l,k})^{\mathrm{T}}}{\|z-o_{l,k}\|_2^5} & \mathbf{0}_D \end{array}\right] \tag{A6}$$

$$\overset{..(bb)}{\delta}(z, o_{l,k}) = \frac{\partial\delta^2(z,o_{l,k})}{\partial p_{l,k}\partial p_{l,k}^{\mathrm{T}}} = \left[\begin{array}{c|c} 3\frac{\left(z^{(x)}-o_{l,k}^{(x)}\right)(z-o_{l,k})(z-o_{l,k})^{\mathrm{T}}}{\|z-o_{l,k}\|_2^5} - \frac{\left(z^{(x)}-o_{l,k}^{(x)}\right)\cdot I_D}{\|z-o_{l,k}\|_2^3} & \\ -\frac{(z-o_{l,k})\cdot\begin{bmatrix}1 & 0\end{bmatrix}}{\|z-o_{l,k}\|_2^3} - \frac{\begin{bmatrix}1 & 0\end{bmatrix}^{\mathrm{T}}\cdot(z-o_{l,k})^{\mathrm{T}}}{\|z-o_{l,k}\|_2^3} & \mathbf{0}_D \\ \hline \mathbf{0}_D & \mathbf{0}_D \end{array}\right] \tag{A7}$$

$$\dot{\mu}^{(a)}\left(z, p_{l,k}\right) = \frac{\partial\mu\left(z, p_{l,k}\right)}{\partial z} = \frac{1}{c}\left(\frac{v_{l,k}}{\|z-o_{l,k}\|_2} - \frac{(z-o_{l,k})\left(v_{l,k}^{\mathrm{T}}(z-o_{l,k})\right)}{\|z-o_{l,k}\|_2^3}\right) \tag{A8}$$

$$\dot{\mu}^{(b)}\left(z, p_{l,k}\right) = \frac{\partial\mu\left(z, p_{l,k}\right)}{\partial p_{l,k}} = \begin{bmatrix} \frac{\partial\mu(z,p_{l,k})}{\partial o_{l,k}} \\ \frac{\partial\mu(z,p_{l,k})}{\partial v_{l,k}} \end{bmatrix} = \frac{1}{c}\left[\begin{array}{c} \frac{v_{l,k}}{\|z-o_{l,k}\|_2} - \frac{(z-o_{l,k})(v_{l,k}^{\mathrm{T}}(z-o_{l,k}))}{\|z-o_{l,k}\|_2^3} \\ \hline \frac{z-o_{l,k}}{\|z-o_{l,k}\|_2} \end{array}\right] \tag{A9}$$

$$\overset{..(aa)}{\mu}\left(z, p_{l,k}\right) = \frac{\partial\mu^2\left(z, p_{l,k}\right)}{\partial z\partial z^{\mathrm{T}}} = \frac{1}{c}\left(\begin{array}{c} 3\frac{(v_{l,k}^{\mathrm{T}}(z-o_{l,k}))}{\|z-o_{l,k}\|_2^5}(z-o_{l,k})(z-o_{l,k})^{\mathrm{T}} - \frac{v_{l,k}(z-o_{l,k})^{\mathrm{T}}}{\|z-o_{l,k}\|_2^3} \\ -\frac{v_{l,k}^{\mathrm{T}}(z-o_{l,k})}{\|z-o_{l,k}\|_2^3}I_D - \frac{(z-o_{l,k})v_{l,k}^{\mathrm{T}}}{\|z-o_{l,k}\|_2^3} \end{array}\right) \tag{A10}$$

$$\overset{..(ab)}{\mu}\left(z, p_{l,k}\right) = \frac{\partial\mu^2(z,p_{l,k})}{\partial z\partial p_{l,k}^{\mathrm{T}}} = \frac{1}{c}\left[\begin{array}{c|c} \frac{v_{l,k}(z-o_{l,k})^{\mathrm{T}}}{\|z-o_{l,k}\|_2^3} + \frac{v_{l,k}^{\mathrm{T}}(z-o_{l,k})}{\|z-o_{l,k}\|_2^3}I_D + \frac{(z-o_{l,k})v_{l,k}^{\mathrm{T}}}{\|z-o_{l,k}\|_2^3} & \frac{1}{\|z-o_{l,k}\|_2}I_D - \\ -3\frac{(v_{l,k}^{\mathrm{T}}(z-o_{l,k}))}{\|z-o_{l,k}\|_2^5}(z-o_{l,k})(z-o_{l,k})^{\mathrm{T}} & \frac{(z-o_{l,k})(z-o_{l,k})^{\mathrm{T}}}{\|z-o_{l,k}\|_2^3} \end{array}\right] \tag{A11}$$

$$\overset{..(bb)}{\mu}\left(z, p_{l,k}\right) = \frac{\partial\mu^2(z,p_{l,k})}{\partial p_{l,k}\partial p_{l,k}^{\mathrm{T}}} = \frac{1}{c}\left[\begin{array}{c|c} 3\frac{(v_{l,k}^{\mathrm{T}}(z-o_{l,k}))}{\|z-o_{l,k}\|_2^5}(z-o_{l,k})(z-o_{l,k})^{\mathrm{T}} - \frac{v_{l,k}(z-o_{l,k})^{\mathrm{T}}}{\|z-o_{l,k}\|_2^3} & \frac{(z-o_{l,k})(z-o_{l,k})^{\mathrm{T}}}{\|z-o_{l,k}\|_2^3} \\ -\frac{v_{l,k}^{\mathrm{T}}(z-o_{l,k})}{\|z-o_{l,k}\|_2^3}I_D - \frac{(z-o_{l,k})v_{l,k}^{\mathrm{T}}}{\|z-o_{l,k}\|_2^3} & -\frac{1}{\|z-o_{l,k}\|_2}I_D \\ \hline \frac{(z-o_{l,k})(z-o_{l,k})^{\mathrm{T}}}{\|z-o_{l,k}\|_2^3} & 0_D \end{array}\right] \tag{A12}$$

Following the expression of $\overline{C}(z, p_k)$ in (14), it gives

Appendix B.1. Expression of $\dot{\overline{C}}_d^{(a)}(z, p_k)$

$$\dot{\overline{C}}_d^{(a)}(z,p_k) = \frac{\partial \overline{C}(z,p_k)}{\partial \langle z \rangle_d} = \left[\frac{\partial C^{\mathrm{H}}\left(z,p_{1,k}\right)}{\partial \langle z \rangle_d}, \frac{\partial C^{\mathrm{H}}\left(z,p_{2,k}\right)}{\partial \langle z \rangle_d}, \ldots, \frac{\partial C^{\mathrm{H}}\left(z,p_{L,k}\right)}{\partial \langle z \rangle_d} \right] \tag{A13}$$

where

$$\frac{\partial C^{\mathrm{H}}\left(z,p_{l,k}\right)}{\partial \langle z \rangle_d} = \left[\frac{\partial a(z,o_{l,k})}{\partial \langle z \rangle_d} \otimes A\left(z,p_{l,k}\right) + a(z,o_{l,k}) \otimes \frac{\partial A\left(z,p_{l,k}\right)}{\partial \langle z \rangle_d} \right]^{\mathrm{H}} \tag{A14}$$

with

$$\frac{\partial a(z,o_{l,k})}{\partial \langle z \rangle_d} = \dot{\psi}_{1,d}^{(a)}(z,o_{l,k}) \cdot a(z,o_{l,k}) \tag{A15}$$

$$\dot{\psi}_{1,d}^{(a)}(z,o_{l,k}) = \left\langle \dot{\delta}^{(a)}(z,o_{l,k}) \right\rangle_d \cdot \mathrm{diag}\left\{ j2\pi \frac{d}{\lambda} \widetilde{M} \right\} \tag{A16}$$

and

$$\frac{\partial A\left(z,p_{l,k}\right)}{\partial \langle z \rangle_d} = A\left(z,p_{l,k}\right) \cdot \dot{\psi}_{2,d}^{(a)}\left(z,p_{l,k}\right) \tag{A17}$$

$$\dot{\psi}_{2,d}^{(a)}\left(z,p_{l,k}\right) = \left\langle \dot{\mu}^{(a)}\left(z,p_{l,k}\right) \right\rangle_d \cdot \mathrm{diag}\left\{ j2\pi f_c \widetilde{N} T_s \right\} \tag{A18}$$

Appendix B.2. Expression of $\dot{\overline{C}}_d^{(b)}(z, p_k)$

$$\dot{\overline{C}}_d^{(b)}(z,p_k) = \frac{\partial \overline{C}(z,p_k)}{\partial \langle p_k \rangle_d} = \left[\frac{\partial C^{\mathrm{H}}\left(z,p_{1,k}\right)}{\partial \langle p_k \rangle_d}, \frac{\partial C^{\mathrm{H}}\left(z,p_{2,k}\right)}{\partial \langle p_k \rangle_d}, \ldots, \frac{\partial C^{\mathrm{H}}\left(z,p_{L,k}\right)}{\partial \langle p_k \rangle_d} \right] \tag{A19}$$

where

$$\frac{\partial C^{\mathrm{H}}\left(z,p_{l,k}\right)}{\partial \langle p_k \rangle_d} = \left[\frac{\partial a(z,o_{l,k})}{\partial \langle p_k \rangle_d} \otimes A\left(z,p_{l,k}\right) + a(z,o_{l,k}) \otimes \frac{\partial A\left(z,p_{l,k}\right)}{\partial \langle p_k \rangle_d} \right]^{\mathrm{H}} \tag{A20}$$

with

$$\frac{\partial a(z,o_{l,k})}{\partial \langle p_k \rangle_d} = \dot{\psi}_{1,d}^{(b)}(z,o_{l,k}) \cdot a(z,o_{l,k}) \tag{A21}$$

$$\dot{\psi}_{1,d}^{(b)}(z,o_{l,k}) = \left\langle \dot{\delta}^{(b)}(z,o_{l,k}) \right\rangle_d \cdot \mathrm{diag}\left\{ j2\pi \frac{d}{\lambda} \widetilde{M} \right\} \tag{A22}$$

and

$$\frac{\partial A\left(z,p_{l,k}\right)}{\partial \langle p_k \rangle_d} = A\left(z,p_{l,k}\right) \cdot \dot{\psi}_{2,d}^{(b)}\left(z,p_{l,k}\right) \tag{A23}$$

$$\dot{\psi}_{2,d}^{(b)}\left(z,p_{l,k}\right) = \left\langle \dot{\mu}^{(b)}\left(z,p_{l,k}\right) \right\rangle_d \cdot \mathrm{diag}\left\{ j2\pi f_c \widetilde{N} T_s \right\} \tag{A24}$$

therefore $\overset{\cdot(b)}{\overline{C}_d}(z,p_k)$ can be written as

$$\overset{\cdot(b)}{\overline{C}_d}(z,p_k)=\begin{cases}\left[\frac{\partial C^{\mathrm{H}}(z,p_{1,k})}{\partial\langle p_k\rangle_d},0_{N\times MN},\ldots,0_{N\times MN}\right] & 1\le d\le 2D\\ \left[0_{N\times MN},\frac{\partial C^{\mathrm{H}}(z,p_{2,k})}{\partial\langle p_k\rangle_d},\ldots,0_{N\times MN}\right] & 2D+1\le d\le 4D\\ \vdots & \vdots\\ \left[0_{N\times MN},\ldots,0_{N\times MN},\frac{\partial C^{\mathrm{H}}(z,p_{L,k})}{\partial\langle p_k\rangle_d}\right] & 2D(L-1)+1\le d\le 2DL\end{cases}\tag{A25}$$

Appendix B.3. Expression of $\overset{\cdot\cdot(aa)}{\overline{C}_{d_1d_2}}(z,p_k)$

$$\overset{\cdot\cdot(aa)}{\overline{C}_{d_1d_2}}(z,p_k)=\frac{\partial^2\overline{C}(z,p_k)}{\partial\langle z\rangle_{d_1}\partial\langle z\rangle_{d_2}}=\left[\frac{\partial^2C^{\mathrm{H}}\left(z,p_{1,k}\right)}{\partial\langle z\rangle_{d_1}\partial\langle z\rangle_{d_2}},\frac{\partial^2C^{\mathrm{H}}\left(z,p_{2,k}\right)}{\partial\langle z\rangle_{d_1}\partial\langle z\rangle_{d_2}},\ldots,\frac{\partial^2C^{\mathrm{H}}\left(z,p_{L,k}\right)}{\partial\langle z\rangle_{d_1}\partial\langle z\rangle_{d_2}}\right]\tag{A26}$$

where

$$\frac{\partial^2C^{\mathrm{H}}\left(z,p_{l,k}\right)}{\partial\langle z\rangle_{d_1}\partial\langle z\rangle_{d_2}}=\begin{bmatrix}\frac{\partial^2a(z,o_{l,k})}{\partial\langle z\rangle_{d_1}\partial\langle z\rangle_{d_2}}\otimes A\left(z,p_{l,k}\right)+\frac{\partial a(z,o_{l,k})}{\partial\langle z\rangle_{d_1}}\otimes\frac{\partial A(z,p_{l,k})}{\partial\langle z\rangle_{d_2}}\\+\frac{\partial a(z,o_{l,k})}{\partial\langle z\rangle_{d_2}}\otimes\frac{\partial A(z,p_{l,k})}{\partial\langle z\rangle_{d_1}}+a(z,o_{l,k})\otimes\frac{\partial^2A(z,p_{l,k})}{\partial\langle z\rangle_{d_1}\partial\langle z\rangle_{d_2}}\end{bmatrix}^{\mathrm{H}}\tag{A27}$$

with

$$\begin{aligned}\frac{\partial^2a(z,o_{l,k})}{\partial\langle z\rangle_{d_1}\partial\langle z\rangle_{d_2}}&=\dot{\psi}^{(a)}_{1,d_1}(z,o_{l,k})\cdot\frac{\partial a(z,o_{l,k})}{\partial\langle z\rangle_{d_2}}+a(z,o_{l,k})\cdot\frac{\partial\dot{\psi}^{(a)}_{1,d_1}(z,o_{l,k})}{\partial\langle z\rangle_{d_2}}\\&=\dot{\psi}^{(a)}_{1,d_1}(z,o_{l,k})\dot{\psi}^{(a)}_{1,d_2}(z,o_{l,k})a(z,o_{l,k})+a(z,o_{l,k})\overset{\cdot\cdot(aa)}{\psi}_{1,d_1d_2}(z,o_{l,k})\end{aligned}\tag{A28}$$

$$\overset{\cdot\cdot(aa)}{\psi}_{1,d_1d_2}(z,o_{l,k})=\left\langle\overset{\cdot\cdot(aa)}{\delta}(z,o_{l,k})\right\rangle_{d_1d_2}\cdot\mathrm{diag}\left\{j2\pi\frac{d}{\lambda}\widetilde{M}\right\}\tag{A29}$$

and

$$\begin{aligned}\frac{\partial^2A(z,p_{l,k})}{\partial\langle z\rangle_{d_1}\partial\langle z\rangle_{d_2}}&=A\left(z,p_{l,k}\right)\cdot\frac{\partial\dot{\psi}^{(a)}_{2,d_1}(z,p_{l,k})}{\partial\langle z\rangle_{d_2}}+\dot{\psi}^{(a)}_{2,d_1}\left(z,p_{l,k}\right)\cdot\frac{\partial A(z,p_{l,k})}{\partial\langle z\rangle_{d_2}}\\&=A\left(z,p_{l,k}\right)\overset{\cdot\cdot(aa)}{\psi}_{2,d_1d_2}\left(z,p_{l,k}\right)+\dot{\psi}^{(a)}_{2,d_1}\left(z,p_{l,k}\right)A\left(z,p_{l,k}\right)\dot{\psi}^{(a)}_{2,d_2}\left(z,p_{l,k}\right)\end{aligned}\tag{A30}$$

$$\overset{\cdot\cdot(aa)}{\psi}_{2,d_1d_2}\left(z,p_{l,k}\right)=\left\langle\overset{\cdot\cdot(aa)}{\mu}\left(z,p_{l,k}\right)\right\rangle_{d_1d_2}\cdot\mathrm{diag}\left\{j2\pi f_c\widetilde{N}T_s\right\}\tag{A31}$$

Appendix B.4. Expression of $\overset{\cdot\cdot(ab)}{\overline{C}_{d_1d_2}}(z,p_k)$

$$\begin{aligned}\overset{\cdot\cdot(ab)}{\overline{C}_{d_1d_2}}(z,p_k)&=\frac{\partial^2\overline{C}(z,p_k)}{\partial\langle z\rangle_{d_1}\partial\langle p_k\rangle_{d_2}}\\&=\left[\frac{\partial^2C^{\mathrm{H}}(z,p_{1,k})}{\partial\langle z\rangle_{d_1}\partial\langle p_k\rangle_{d_2}},\frac{\partial^2C^{\mathrm{H}}(z,p_{2,k})}{\partial\langle z\rangle_{d_1}\partial\langle p_k\rangle_{d_2}},\ldots,\frac{\partial^2C^{\mathrm{H}}(z,p_{L,k})}{\partial\langle z\rangle_{d_1}\partial\langle p_k\rangle_{d_2}}\right]\end{aligned}\tag{A32}$$

where

$$\frac{\partial^2C^{\mathrm{H}}\left(z,p_{l,k}\right)}{\partial\langle z\rangle_{d_1}\partial\langle p_k\rangle_{d_2}}=\begin{bmatrix}\frac{\partial^2a(z,o_{l,k})}{\partial\langle z\rangle_{d_1}\partial\langle p_k\rangle_{d_2}}\otimes A\left(z,p_{l,k}\right)+\frac{\partial a(z,o_{l,k})}{\partial\langle z\rangle_{d_1}}\otimes\frac{\partial A(z,p_{l,k})}{\partial\langle p_k\rangle_{d_2}}\\+\frac{\partial a(z,o_{l,k})}{\partial\langle p_k\rangle_{d_2}}\otimes\frac{\partial A(z,p_{l,k})}{\partial\langle z\rangle_{d_1}}+a(z,o_{l,k})\otimes\frac{\partial^2A(z,p_{l,k})}{\partial\langle z\rangle_{d_1}\partial\langle p_k\rangle_{d_2}}\end{bmatrix}^{\mathrm{H}}\tag{A33}$$

with

$$\begin{aligned}\frac{\partial^2 \boldsymbol{a}(\boldsymbol{z},\boldsymbol{o}_{l,k})}{\partial\langle \boldsymbol{z}\rangle_{d_1}\partial\langle \boldsymbol{p}_k\rangle_{d_2}} &= \dot{\boldsymbol{\psi}}^{(a)}_{1,d_1}(\boldsymbol{z},\boldsymbol{o}_{l,k})\cdot\frac{\partial \boldsymbol{a}(\boldsymbol{z},\boldsymbol{o}_{l,k})}{\partial\langle \boldsymbol{p}_k\rangle_{d_2}} + \boldsymbol{a}(\boldsymbol{z},\boldsymbol{o}_{l,k})\cdot\frac{\partial \dot{\boldsymbol{\psi}}^{(a)}_{1,d_1}(\boldsymbol{z},\boldsymbol{o}_{l,k})}{\partial\langle \boldsymbol{p}_k\rangle_{d_2}} \\ &= \dot{\boldsymbol{\psi}}^{(a)}_{1,d_1}(\boldsymbol{z},\boldsymbol{o}_{l,k})\dot{\boldsymbol{\psi}}^{(b)}_{1,d_2}(\boldsymbol{z},\boldsymbol{o}_{l,k})\boldsymbol{a}(\boldsymbol{z},\boldsymbol{o}_{l,k}) + \boldsymbol{a}(\boldsymbol{z},\boldsymbol{o}_{l,k})\ddot{\boldsymbol{\psi}}^{(ab)}_{1,d_1d_2}(\boldsymbol{z},\boldsymbol{o}_{l,k})\end{aligned} \tag{A34}$$

$$\ddot{\boldsymbol{\psi}}^{(ab)}_{1,d_1d_2}(\boldsymbol{z},\boldsymbol{o}_{l,k}) = \left\langle \ddot{\boldsymbol{\delta}}^{(ab)}(\boldsymbol{z},\boldsymbol{o}_{l,k})\right\rangle_{d_1d_2}\cdot\mathrm{diag}\left\{j2\pi\frac{d}{\lambda}\widetilde{\boldsymbol{M}}\right\} \tag{A35}$$

and

$$\frac{\partial^2 \boldsymbol{A}\left(\boldsymbol{z},\boldsymbol{p}_{l,k}\right)}{\partial\langle \boldsymbol{z}\rangle_{d_1}\partial\langle \boldsymbol{p}_k\rangle_{d_2}} = \boldsymbol{A}\left(\boldsymbol{z},\boldsymbol{p}_{l,k}\right)\cdot\ddot{\boldsymbol{\psi}}^{(ab)}_{2,d_1d_2}\left(\boldsymbol{z},\boldsymbol{p}_{l,k}\right) + \dot{\boldsymbol{\psi}}^{(a)}_{2,d_1}\left(\boldsymbol{z},\boldsymbol{p}_{l,k}\right)\cdot\boldsymbol{A}\left(\boldsymbol{z},\boldsymbol{p}_{l,k}\right)\cdot\dot{\boldsymbol{\psi}}^{(b)}_{2,d_2}\left(\boldsymbol{z},\boldsymbol{p}_{l,k}\right) \tag{A36}$$

$$\ddot{\boldsymbol{\psi}}^{(ab)}_{2,d_1d_2}(\boldsymbol{z},\boldsymbol{o}_{l,k}) = \left\langle \ddot{\boldsymbol{\mu}}^{(ab)}\left(\boldsymbol{z},\boldsymbol{p}_{l,k}\right)\right\rangle_{d_1d_2}\cdot\mathrm{diag}\left\{j2\pi f_c\widetilde{\boldsymbol{N}}T_s\right\} \tag{A37}$$

Therefore, $\ddot{\overline{\boldsymbol{C}}}^{(ab)}_{d_1d_2}(\boldsymbol{z},\boldsymbol{p}_k)$ can be written as

$$\ddot{\overline{\boldsymbol{C}}}^{(ab)}_{d_1d_2}(\boldsymbol{z},\boldsymbol{p}_k) = \begin{cases} \left[\frac{\partial^2\boldsymbol{C}^{\mathrm{H}}(\boldsymbol{z},\boldsymbol{p}_{1,k})}{\partial\langle\boldsymbol{z}\rangle_{d_1}\partial\langle\boldsymbol{p}_k\rangle_{d_2}},\boldsymbol{0}_{N\times MN},\ldots,\boldsymbol{0}_{N\times MN}\right] & \begin{array}{l}1\le d_1\le 2D\\ 1\le d_2\le 2D\end{array} \\ \left[\boldsymbol{0}_{N\times MN},\frac{\partial^2\boldsymbol{C}^{\mathrm{H}}(\boldsymbol{z},\boldsymbol{p}_{2,k})}{\partial\langle\boldsymbol{z}\rangle_{d_1}\partial\langle\boldsymbol{p}_k\rangle_{d_2}},\ldots,\boldsymbol{0}_{N\times MN}\right] & \begin{array}{l}2D+1\le d_1\le 4D\\ 2D+1\le d_2\le 4D\end{array} \\ \vdots & \vdots \\ \left[\boldsymbol{0}_{N\times MN},\ldots,\boldsymbol{0}_{N\times MN},\frac{\partial^2\boldsymbol{C}^{\mathrm{H}}(\boldsymbol{z},\boldsymbol{p}_{L,k})}{\partial\langle\boldsymbol{z}\rangle_{d_1}\partial\langle\boldsymbol{p}_k\rangle_{d_2}}\right] & \begin{array}{l}2D(L-1)+1\le d_1\le 2DL\\ 2D(L-1)+1\le d_2\le 2DL\end{array} \end{cases} \tag{A38}$$

Appendix B.5. Expression of $\ddot{\overline{\boldsymbol{C}}}^{(bb)}_{d_1d_2}(\boldsymbol{z},\boldsymbol{p}_k)$

$$\begin{aligned}\ddot{\overline{\boldsymbol{C}}}^{(bb)}_{d_1d_2}(\boldsymbol{z},\boldsymbol{p}_k) &= \frac{\partial^2\overline{\boldsymbol{C}}(\boldsymbol{z},\boldsymbol{p}_k)}{\partial\langle\boldsymbol{p}_k\rangle_{d_1}\partial\langle\boldsymbol{p}_k\rangle_{d_2}} \\ &= \left[\frac{\partial^2\boldsymbol{C}^{\mathrm{H}}(\boldsymbol{z},\boldsymbol{p}_{1,k})}{\partial\langle\boldsymbol{p}_k\rangle_{d_1}\partial\langle\boldsymbol{p}_k\rangle_{d_2}},\frac{\partial^2\boldsymbol{C}^{\mathrm{H}}(\boldsymbol{z},\boldsymbol{p}_{2,k})}{\partial\langle\boldsymbol{p}_k\rangle_{d_1}\partial\langle\boldsymbol{p}_k\rangle_{d_2}},\ldots,\frac{\partial^2\boldsymbol{C}^{\mathrm{H}}(\boldsymbol{z},\boldsymbol{p}_{L,k})}{\partial\langle\boldsymbol{p}_k\rangle_{d_1}\partial\langle\boldsymbol{p}_k\rangle_{d_2}}\right]\end{aligned} \tag{A39}$$

where

$$\frac{\partial^2\boldsymbol{C}^{\mathrm{H}}\left(\boldsymbol{z},\boldsymbol{p}_{l,k}\right)}{\partial\langle\boldsymbol{p}_k\rangle_{d_1}\partial\langle\boldsymbol{p}_k\rangle_{d_2}} = \begin{bmatrix} \frac{\partial^2\boldsymbol{a}(\boldsymbol{z},\boldsymbol{o}_{l,k})}{\partial\langle\boldsymbol{p}_k\rangle_{d_1}\partial\langle\boldsymbol{p}_k\rangle_{d_2}}\otimes\boldsymbol{A}\left(\boldsymbol{z},\boldsymbol{p}_{l,k}\right) + \frac{\partial\boldsymbol{a}(\boldsymbol{z},\boldsymbol{o}_{l,k})}{\partial\langle\boldsymbol{p}_k\rangle_{d_1}}\otimes\frac{\partial\boldsymbol{A}(\boldsymbol{z},\boldsymbol{p}_{l,k})}{\partial\langle\boldsymbol{p}_k\rangle_{d_2}} \\ +\frac{\partial\boldsymbol{a}(\boldsymbol{z},\boldsymbol{o}_{l,k})}{\partial\langle\boldsymbol{p}_k\rangle_{d_2}}\otimes\frac{\partial\boldsymbol{A}(\boldsymbol{z},\boldsymbol{p}_{l,k})}{\partial\langle\boldsymbol{p}_k\rangle_{d_1}} + \boldsymbol{a}(\boldsymbol{z},\boldsymbol{o}_{l,k})\otimes\frac{\partial^2\boldsymbol{A}(\boldsymbol{z},\boldsymbol{p}_{l,k})}{\partial\langle\boldsymbol{p}_k\rangle_{d_1}\partial\langle\boldsymbol{p}_k\rangle_{d_2}} \end{bmatrix}^{\mathrm{H}} \tag{A40}$$

with

$$\frac{\partial^2\boldsymbol{a}(\boldsymbol{z},\boldsymbol{o}_{l,k})}{\partial\langle\boldsymbol{p}_k\rangle_{d_1}\partial\langle\boldsymbol{p}_k\rangle_{d_2}} = \dot{\boldsymbol{\psi}}^{(b)}_{1,d_1}(\boldsymbol{z},\boldsymbol{o}_{l,k})\dot{\boldsymbol{\psi}}^{(b)}_{1,d_2}(\boldsymbol{z},\boldsymbol{o}_{l,k})\boldsymbol{a}(\boldsymbol{z},\boldsymbol{o}_{l,k}) + \boldsymbol{a}(\boldsymbol{z},\boldsymbol{o}_{l,k})\ddot{\boldsymbol{\psi}}^{(bb)}_{1,d_1d_2}(\boldsymbol{z},\boldsymbol{o}_{l,k}) \tag{A41}$$

$$\ddot{\boldsymbol{\psi}}^{(bb)}_{1,d_1d_2}(\boldsymbol{z},\boldsymbol{o}_{l,k}) = \left\langle \ddot{\boldsymbol{\delta}}^{(bb)}(\boldsymbol{z},\boldsymbol{o}_{l,k})\right\rangle_{d_1d_2}\cdot\mathrm{diag}\left\{j2\pi\frac{d}{\lambda}\widetilde{\boldsymbol{M}}\right\} \tag{A42}$$

and

$$\frac{\partial^2 \boldsymbol{A}\left(\boldsymbol{z},\boldsymbol{p}_{l,k}\right)}{\partial\langle \boldsymbol{p}_k\rangle_{d_1}\partial\langle \boldsymbol{p}_k\rangle_{d_2}} = \boldsymbol{A}\left(\boldsymbol{z},\boldsymbol{p}_{l,k}\right)\dot{\boldsymbol{\psi}}^{(bb)}_{2,d_1d_2}\left(\boldsymbol{z},\boldsymbol{p}_{l,k}\right) + \dot{\boldsymbol{\psi}}^{(b)}_{2,d_1}\left(\boldsymbol{z},\boldsymbol{p}_{l,k}\right)\boldsymbol{A}\left(\boldsymbol{z},\boldsymbol{p}_{l,k}\right)\dot{\boldsymbol{\psi}}^{(b)}_{2,d_2}\left(\boldsymbol{z},\boldsymbol{p}_{l,k}\right) \tag{A43}$$

$$\ddot{\psi}^{(bb)}_{2,d_1d_2}\left(z,p_{l,k}\right)=\left\langle \ddot{\mu}^{(bb)}\left(z,p_{l,k}\right)\right\rangle_{d_1d_2}\cdot\mathrm{diag}\left\{j2\pi f_c\widetilde{N}T_s\right\} \tag{A44}$$

Therefore, $\ddot{\overline{C}}^{(bb)}_{d_1d_2}(z,p_k)$ can be written as

$$\ddot{\overline{C}}^{(bb)}_{d_1d_2}(z,p_k)=\begin{cases}\left[\frac{\partial^2 C^{\mathrm{H}}(z,p_{1,k})}{\partial\langle p_k\rangle_{d_1}\partial\langle p_k\rangle_{d_2}},O_{N\times MN},\ldots,O_{N\times MN}\right] & \begin{array}{l}1\le d_1\le 2D\\ 1\le d_2\le 2D\end{array}\\ \left[O_{N\times MN},\frac{\partial^2 C^{\mathrm{H}}(z,p_{2,k})}{\partial\langle p_k\rangle_{d_1}\partial\langle p_k\rangle_{d_2}},\ldots,O_{N\times MN}\right] & \begin{array}{l}2D+1\le d_1\le 4D\\ 2D+1\le d_2\le 4D\end{array}\\ \vdots & \vdots\\ \left[O_{N\times MN},\ldots,O_{N\times MN},\frac{\partial^2 C^{\mathrm{H}}(z,p_{L,k})}{\partial\langle p_k\rangle_{d_1}\partial\langle p_k\rangle_{d_2}}\right] & \begin{array}{l}2D(L-1)+1\le d_1\le 2DL\\ 2D(L-1)+1\le d_2\le 2DL\end{array}\end{cases} \tag{A45}$$

This completes the derivation.

Appendix C. Derivation of (31) to (34)

Associated with the second formulation in (21), it follows for any vector $q_1\in\mathbb{C}^{L\times 1}$ and $q_2\in\mathbb{C}^{N\times 1}$ that

$$\begin{aligned}\widetilde{D}^{(1)}_k q_1 &= \sum_{d=1}^{D}\langle\widetilde{z}\rangle_d\dot{\overline{C}}^{(a)}_d(z,p_k)Y_k q_1+\sum_{d=1}^{2DL}\langle\widetilde{p}_k\rangle_d\dot{\overline{C}}^{(b)}_d(z,p_k)Y_k q_1+\overline{C}(z,p_k)N_k q_1\\ &=\frac{\partial(\overline{C}(z,p_k)Y_k q_1)}{\partial z^{\mathrm{T}}}\widetilde{z}+\frac{\partial(\overline{C}(z,p_k)Y_k q_1)}{\partial p_k^{\mathrm{T}}}\widetilde{p}_k+\overline{C}(z,p_k)(\mathrm{diag}\{q_1\}\otimes I_N)\Pi_1\widetilde{n}_k\\ &=F^{(a)}_{1k}(q_1)\widetilde{z}+F^{(a)}_{2k}(q_1)\widetilde{p}_k+F^{(a)}_{3k}(q_1)\widetilde{n}_k\end{aligned} \tag{A46}$$

$$\begin{aligned}\widetilde{D}^{(1)\mathrm{H}}_k q_2 &= \sum_{d=1}^{D}\langle\widetilde{z}\rangle_d Y_k^{\mathrm{H}}\overset{\bullet}{\overline{C}}^{(a)\mathrm{H}}_d(z,p_k)q_2+\sum_{d=1}^{2DL}\langle\widetilde{\mathbf{p}}_k\rangle_d Y_k^{\mathrm{H}}\overset{\bullet}{\overline{C}}^{(b)\mathrm{H}}_d(z,p_k)q_2+N_k^{\mathrm{H}}\overline{C}(z,p_k)q_2\\ &=\frac{\partial\left(Y_k^{\mathrm{H}}\overline{C}^{\mathrm{H}}(z,p_k)q_2\right)}{\partial z^{\mathrm{T}}}\widetilde{z}+\frac{\partial\left(Y_k^{\mathrm{H}}\overline{C}^{\mathrm{H}}(z,p_k)q_2\right)}{\partial p_k^{\mathrm{T}}}\widetilde{p}_k+(I_L\otimes q_2^{\mathrm{T}})\overline{\overline{C}}(z,p_k)\Pi_2\widetilde{n}_k\\ &=F^{(b)}_{1k}(q_2)\widetilde{z}+F^{(b)}_{2k}(q_2)\widetilde{p}_k+F^{(b)}_{3k}(q_2)\widetilde{n}_k\end{aligned} \tag{A47}$$

Consequently, the formulation of $\widetilde{J}^{(1)}_{\mathrm{cost}}$ can be shown as

$$\begin{aligned}\widetilde{J}^{(1)}_{\mathrm{cost}}=&\sum_{k=1}^{K}\alpha^{(0)\mathrm{H}}_{k,L}D^{(0)\mathrm{H}}_k\left(F^{(a)}_{1k}\left(\alpha^{(0)}_{k,L}\right)\widetilde{z}+F^{(a)}_{2k}\left(\alpha^{(0)}_{k,L}\right)\widetilde{p}_k+F^{(a)}_{3k}\left(\alpha^{(0)}_{k,L}\right)\widetilde{n}_k\right)+\\ &\sum_{k=1}^{K}\alpha^{(0)\mathrm{H}}_{k,L}\left(F^{(b)}_{1k}\left(D^{(0)}_k\alpha^{(0)}_{k,L}\right)\widetilde{z}+F^{(b)}_{2k}\left(D^{(0)}_k\alpha^{(0)}_{k,L}\right)\widetilde{p}_k+F^{(b)}_{3k}\left(D^{(0)}_k\alpha^{(0)}_{k,L}\right)\widetilde{n}_k\right)\end{aligned} \tag{A48}$$

This completes the derivation.

Appendix D. Derivation of (35)

Associated with (A46), it follows for any vector q_1 and q_2 that

$$\begin{aligned} q_1^{\mathrm{H}}\widetilde{D}_k^{(1)\mathrm{H}}\boldsymbol{\Phi}\widetilde{D}_k^{(1)}q_2 = & \widetilde{z}^{\mathrm{T}}F_{1k}^{(a)\mathrm{H}}(q_1)\boldsymbol{\Phi}F_{1k}^{(a)}(q_2)\widetilde{z}+\widetilde{p}_k^{\mathrm{T}}F_{2k}^{(a)\mathrm{H}}(q_1)\boldsymbol{\Phi}F_{2k}^{(a)}(q_2)\widetilde{p}_k+ \\ & \widetilde{z}^{\mathrm{T}}\left(F_{1k}^{(a)\mathrm{H}}(q_1)\boldsymbol{\Phi}F_{2k}^{(a)}(q_2)+F_{1k}^{(a)\mathrm{T}}(q_2)\boldsymbol{\Phi}^{\mathrm{T}}F_{2k}^{(a)*}(q_1)\right)\widetilde{p}_k+ \\ & \widetilde{z}^{\mathrm{T}}\left(F_{1k}^{(a)\mathrm{H}}(q_1)\boldsymbol{\Phi}F_{3k}^{(a)}(q_2)+F_{1k}^{(a)\mathrm{T}}(q_2)\boldsymbol{\Phi}^{\mathrm{T}}F_{3k}^{(a)*}(q_1)\Pi_3\right)\widetilde{n}_k+ \\ & \widetilde{p}_k^{\mathrm{T}}\left(F_{2k}^{(a)\mathrm{H}}(q_1)\boldsymbol{\Phi}F_{3k}^{(a)}(q_2)+F_{2k}^{(a)\mathrm{T}}(q_2)\boldsymbol{\Phi}^{\mathrm{T}}F_{3k}^{(a)*}(q_1)\Pi_3\right)\widetilde{n}_k+ \\ & \widetilde{n}_k^{\mathrm{T}}F_{3k}^{(a)\mathrm{H}}(q_1)\boldsymbol{\Phi}F_{3k}^{(a)}(q_2)\widetilde{n}_k \\ = & \widetilde{z}^{\mathrm{T}}\boldsymbol{\Sigma}_{1k}^{(a)}(q_1,\boldsymbol{\Phi},q_2)\widetilde{z}+\widetilde{p}_k^{\mathrm{T}}\boldsymbol{\Sigma}_{2k}^{(a)}(q_1,\boldsymbol{\Phi},q_2)\widetilde{p}_k+ \\ & \widetilde{z}^{\mathrm{T}}\boldsymbol{\Sigma}_{3k}^{(a)}(q_1,\boldsymbol{\Phi},q_2)\widetilde{p}_k+\widetilde{z}^{\mathrm{T}}\boldsymbol{\Sigma}_{4k}^{(a)}(q_1,\boldsymbol{\Phi},q_2)\widetilde{n}_k+ \\ & \widetilde{p}_k^{\mathrm{T}}\boldsymbol{\Sigma}_{5k}^{(a)}(q_1,\boldsymbol{\Phi},q_2)\widetilde{n}_k+\widetilde{n}_k^{\mathrm{T}}\boldsymbol{\Sigma}_{6k}^{(a)}(q_1,\boldsymbol{\Phi},q_2)\widetilde{n}_k \end{aligned} \tag{A49}$$

Meanwhile, the similar result respect to (A47) is drawn as

$$\begin{aligned} q_1^{\mathrm{H}}\widetilde{D}_k^{(1)}\boldsymbol{\Phi}\widetilde{D}_k^{(1)\mathrm{H}}q_2 = & \widetilde{z}^{\mathrm{T}}F_{1k}^{(b)\mathrm{H}}(q_1)\boldsymbol{\Phi}F_{1k}^{(b)}(q_2)\widetilde{z}+\widetilde{p}_k^{\mathrm{T}}F_{2k}^{(b)\mathrm{H}}(q_1)\boldsymbol{\Phi}F_{2k}^{(b)}(q_2)\widetilde{p}_k+ \\ & \widetilde{z}^{\mathrm{T}}\left(F_{1k}^{(b)\mathrm{H}}(q_1)\boldsymbol{\Phi}F_{2k}^{(b)}(q_2)+F_{1k}^{(b)\mathrm{T}}(q_2)\boldsymbol{\Phi}^{\mathrm{T}}F_{2k}^{(b)*}(q_1)\right)\widetilde{p}_k+ \\ & \widetilde{z}^{\mathrm{T}}\left(F_{1k}^{(b)\mathrm{H}}(q_1)\boldsymbol{\Phi}F_{3k}^{(b)}(q_2)+F_{1k}^{(b)\mathrm{T}}(q_2)\boldsymbol{\Phi}^{\mathrm{T}}F_{3k}^{(b)*}(q_1)\Pi_3\right)\widetilde{n}_k+ \\ & \widetilde{p}_k^{\mathrm{T}}\left(F_{2k}^{(b)\mathrm{H}}(q_1)\boldsymbol{\Phi}F_{3k}^{(b)}(q_2)+F_{2k}^{(b)\mathrm{T}}(q_2)\boldsymbol{\Phi}^{\mathrm{T}}F_{3k}^{(b)*}(q_1)\Pi_3\right)\widetilde{n}_k+ \\ & \widetilde{n}_k^{\mathrm{T}}F_{3k}^{(b)\mathrm{H}}(q_1)\boldsymbol{\Phi}F_{3k}^{(b)}(q_2)\widetilde{n}_k \\ = & \widetilde{z}^{\mathrm{T}}\boldsymbol{\Sigma}_{1k}^{(b)}(q_1,\boldsymbol{\Phi},q_2)\widetilde{z}+\widetilde{p}_k^{\mathrm{T}}\boldsymbol{\Sigma}_{2k}^{(b)}(q_1,\boldsymbol{\Phi},q_2)\widetilde{p}_k+ \\ & \widetilde{z}^{\mathrm{T}}\boldsymbol{\Sigma}_{3k}^{(b)}(q_1,\boldsymbol{\Phi},q_2)\widetilde{p}_k+\widetilde{z}^{\mathrm{T}}\boldsymbol{\Sigma}_{4k}^{(b)}(q_1,\boldsymbol{\Phi},q_2)\widetilde{n}_k+ \\ & \widetilde{p}_k^{\mathrm{T}}\boldsymbol{\Sigma}_{5k}^{(b)}(q_1,\boldsymbol{\Phi},q_2)\widetilde{n}_k+\widetilde{n}_k^{\mathrm{T}}\boldsymbol{\Sigma}_{6k}^{(b)}(q_1,\boldsymbol{\Phi},q_2)\widetilde{n}_k \end{aligned} \tag{A50}$$

Additionally, we can obtain the following formulations

$$\begin{aligned} q_1^{\mathrm{H}}\widetilde{D}_k^{(1)}\boldsymbol{\Phi}\widetilde{D}_k^{(1)}q_2 = & \widetilde{z}^{\mathrm{T}}F_{1k}^{(b)\mathrm{H}}(q_1)\boldsymbol{\Phi}F_{1k}^{(a)}(q_2)\widetilde{z}+\widetilde{p}_k^{\mathrm{T}}F_{2k}^{(b)\mathrm{H}}(q_1)\boldsymbol{\Phi}F_{2k}^{(a)}(q_2)\widetilde{p}_k+ \\ & \widetilde{z}^{\mathrm{T}}\left(F_{1k}^{(b)\mathrm{H}}(q_1)\boldsymbol{\Phi}F_{2k}^{(a)}(q_2)+F_{1k}^{(a)\mathrm{T}}(q_2)\boldsymbol{\Phi}^{\mathrm{T}}F_{2k}^{(b)*}(q_1)\right)\widetilde{p}_k+ \\ & \widetilde{z}^{\mathrm{T}}\left(F_{1k}^{(b)\mathrm{H}}(q_1)\boldsymbol{\Phi}F_{3k}^{(a)}(q_2)+F_{1k}^{(a)\mathrm{T}}(q_2)\boldsymbol{\Phi}^{\mathrm{T}}F_{3k}^{(b)*}(q_1)\Pi_3\right)\widetilde{n}_k+ \\ & \widetilde{p}_k^{\mathrm{T}}\left(F_{2k}^{(b)\mathrm{H}}(q_1)\boldsymbol{\Phi}F_{3k}^{(b)}(q_2)+F_{2k}^{(a)\mathrm{T}}(q_2)\boldsymbol{\Phi}^{\mathrm{T}}F_{3k}^{(b)*}(q_1)\Pi_3\right)\widetilde{n}_k+ \\ & \widetilde{n}_k^{\mathrm{T}}F_{3k}^{(b)\mathrm{H}}(q_1)\boldsymbol{\Phi}F_{3k}^{(a)}(q_2)\widetilde{n}_k \\ = & \widetilde{z}^{\mathrm{T}}\boldsymbol{\Sigma}_{1k}^{(c)}(q_1,\boldsymbol{\Phi},q_2)\widetilde{z}+\widetilde{p}_k^{\mathrm{T}}\boldsymbol{\Sigma}_{2k}^{(c)}(q_1,\boldsymbol{\Phi},q_2)\widetilde{p}_k+ \\ & \widetilde{z}^{\mathrm{T}}\boldsymbol{\Sigma}_{3k}^{(c)}(q_1,\boldsymbol{\Phi},q_2)\widetilde{p}_k+\widetilde{z}^{\mathrm{T}}\boldsymbol{\Sigma}_{4k}^{(c)}(q_1,\boldsymbol{\Phi},q_2)\widetilde{n}_k+ \\ & \widetilde{p}_k^{\mathrm{T}}\boldsymbol{\Sigma}_{5k}^{(c)}(q_1,\boldsymbol{\Phi},q_2)\widetilde{n}_k+\widetilde{n}_k^{\mathrm{T}}\boldsymbol{\Sigma}_{6k}^{(c)}(q_1,\boldsymbol{\Phi},q_2)\widetilde{n}_k \end{aligned} \tag{A51}$$

$$\begin{aligned} q_1^{\mathrm{H}}\widetilde{D}_k^{(1)\mathrm{H}}\boldsymbol{\Phi}\widetilde{D}_k^{(1)\mathrm{H}}q_2 = & \left(q_2^{\mathrm{H}}\widetilde{D}_k^{(1)}\boldsymbol{\Phi}^{\mathrm{H}}\widetilde{D}_k^{(1)}q_1\right)^{\mathrm{H}} \\ = & \widetilde{z}^{\mathrm{T}}\boldsymbol{\Sigma}_{1k}^{(c)*}(q_2,\boldsymbol{\Phi}^{\mathrm{H}},q_1)\widetilde{z}+\widetilde{p}_k^{\mathrm{T}}\boldsymbol{\Sigma}_{2k}^{(c)*}(q_2,\boldsymbol{\Phi}^{\mathrm{H}},q_1)\widetilde{p}_k+ \\ & \widetilde{z}^{\mathrm{T}}\boldsymbol{\Sigma}_{3k}^{(c)*}(q_2,\boldsymbol{\Phi}^{\mathrm{H}},q_1)\widetilde{p}_k+\widetilde{z}^{\mathrm{T}}\boldsymbol{\Sigma}_{4k}^{(c)*}(q_2,\boldsymbol{\Phi}^{\mathrm{H}},q_1)\Pi_3\widetilde{n}_k+ \\ & \widetilde{p}_k^{\mathrm{T}}\boldsymbol{\Sigma}_{5k}^{(c)*}(q_2,\boldsymbol{\Phi}^{\mathrm{H}},q_1)\Pi_3\widetilde{n}_k+\widetilde{n}_k^{\mathrm{T}}\boldsymbol{\Sigma}_{6k}^{(c)\mathrm{H}}(q_2,\boldsymbol{\Phi}^{\mathrm{H}},q_1)\widetilde{n}_k \end{aligned} \tag{A52}$$

Finally, associated with the third formulation in (21), we have

$$\begin{aligned} q_1^{\mathrm{H}}\widetilde{D}_k^{(2)}q_2 &= \tfrac{1}{2}\sum_{d_1=1}^{D}\sum_{d_2=1}^{D}\langle\widetilde{z}\rangle_{d_1}\langle\widetilde{z}\rangle_{d_2}q_1^{\mathrm{H}}\overline{\ddot{C}}_{d_1d_2}^{(aa)}(z,p_k)Y_kq_2+ \\ &\quad \tfrac{1}{2}\sum_{d_1=1}^{2DL}\sum_{d_2=1}^{2DL}\langle\widetilde{p}_k\rangle_{d_1}\langle\widetilde{p}_k\rangle_{d_2}q_1^{\mathrm{H}}\overline{\ddot{C}}_{d_1d_2}^{(bb)}(z,p_k)Y_kq_2+ \\ &\quad \sum_{d_1=1}^{D}\sum_{d_2=1}^{2DL}\langle\widetilde{z}\rangle_{d_1}\langle\widetilde{p}_k\rangle_{d_2}q_1^{\mathrm{H}}\overline{\ddot{C}}_{d_1d_2}^{(ab)}(z,p_k)Y_kq_2+ \\ &\quad \sum_{d=1}^{D}\langle\widetilde{z}\rangle_d q_1^{\mathrm{H}}\overline{\dot{C}}_d^{(a)}(z,p_k)N_kq_2+\sum_{d=1}^{2DL}\langle\widetilde{p}_k\rangle_d q_1^{\mathrm{H}}\overline{\dot{C}}_d^{(b)}(z,p_k)N_kq_2 \\ &= \widetilde{z}^{\mathrm{T}}\Sigma_{1k}^{(d)}(q_1,q_2)\widetilde{z}+\widetilde{p}_k^{\mathrm{T}}\Sigma_{2k}^{(d)}(q_1,q_2)\widetilde{p}_k+\widetilde{z}^{\mathrm{T}}\Sigma_{3k}^{(d)}(q_1,q_2)\widetilde{p}_k+ \\ &\quad \widetilde{z}^{\mathrm{T}}\Sigma_{4k}^{(d)}(q_1,q_2)\widetilde{n}_k+\widetilde{p}_k^{\mathrm{T}}\Sigma_{5k}^{(d)}(q_1,q_2)\widetilde{n}_k \end{aligned} \tag{A53}$$

$$\begin{aligned} q_1^{\mathrm{H}}\widetilde{D}_k^{(2)\mathrm{H}}q_2 &= \left(q_2^{\mathrm{H}}\widetilde{D}_k^{(2)}q_1\right)^{\mathrm{H}} \\ &= \widetilde{z}^{\mathrm{T}}\Sigma_{1k}^{(d)*}(q_2,q_1)\widetilde{z}+\widetilde{p}_k^{\mathrm{T}}\Sigma_{2k}^{(d)*}(q_2,q_1)\widetilde{p}_k+\widetilde{z}^{\mathrm{T}}\Sigma_{3k}^{(d)*}(q_2,q_1)\widetilde{p}_k+ \\ &\quad \widetilde{z}^{\mathrm{T}}\Sigma_{4k}^{(d)*}(q_2,q_1)\Pi_3\widetilde{n}_k+\widetilde{p}_k^{\mathrm{T}}\Sigma_{5k}^{(d)*}(q_2,q_1)\Pi_3\widetilde{n}_k \end{aligned} \tag{A54}$$

By using the above results, the expression of $\widetilde{J}_{\mathrm{cost}}^{(2)}$ is written as

$$\begin{aligned} \widetilde{J}_{\mathrm{cost}}^{(2)} = &\sum_{k=1}^{K}\widetilde{z}^{\mathrm{T}}\xi_{1k}\widetilde{z}+\sum_{k=1}^{K}\widetilde{p}_k^{\mathrm{T}}\xi_{2k}\widetilde{p}_k+\sum_{k=1}^{K}\widetilde{z}^{\mathrm{T}}\xi_{3k}\widetilde{p}_k+ \\ &\sum_{k=1}^{K}\widetilde{z}^{\mathrm{T}}\xi_{4k}\widetilde{n}_k+\sum_{k=1}^{K}\widetilde{p}_k^{\mathrm{T}}\xi_{5k}\widetilde{n}_k+\sum_{k=1}^{K}\widetilde{n}_k^{\mathrm{H}}\xi_{6k}\widetilde{n}_k \end{aligned} \tag{A55}$$

This ends the derivation in this part.

Appendix E. Derivation of CRB

Appendix E.1. The Partial of $d_{l,k}(\eta)$ Respect to z

Define $G_{l,k} = b_{l,k}\dot{C}\left(z,p_{l,k}\right)(I_2\otimes B_ks_k)$ with $\dot{C}\left(z,p_{l,k}\right) = \frac{\partial C(z,p_{l,k})}{\partial z^{\mathrm{T}}} = \frac{\partial a(z,p_{l,k})}{\partial z^{\mathrm{T}}}\otimes A\left(z,p_{l,k}\right)+a(z,o_{l,k})\otimes\frac{\partial A(z,p_{l,k})}{\partial z^{\mathrm{T}}}$. The derivation with respect to z can be expressed by

$$\frac{\partial d_{l,k}(\eta)}{\partial z^{\mathrm{T}}} = G_{l,k} \tag{A56}$$

where

$$\frac{\partial a\left(z,p_{l,k}\right)}{\partial z^{\mathrm{T}}} = \mathrm{diag}\{a(z,o_{l,k})\}\cdot\left[0,j2\pi\frac{d}{\lambda},\ldots,j2\pi\frac{d}{\lambda}(M-1)\right]^{\mathrm{T}}\cdot\dot{\delta}^{(a)\mathrm{T}}(z,o_{l,k}) \tag{A57}$$

$$\begin{aligned} \frac{\partial A(z,p_{l,k})}{\partial z^{\mathrm{T}}} = &A\left(z,p_{l,k}\right)\cdot \\ &\mathrm{diag}\left\{0,j2\pi f_cT_s\dot{\mu}^{(a)\mathrm{T}}\left(z,p_{l,k}\right),\ldots,j2\pi f_c(N-1)T_s\dot{\mu}^{(a)\mathrm{T}}\left(z,p_{l,k}\right)\right\} \end{aligned} \tag{A58}$$

Appendix E.2. The Partial of $d_{l,k}(\eta)$ Respect to b

Define $H_{l,k} = C\left(z,p_{l,k}\right)B_ks_k\delta_{l,m}\delta_{k,n}$. Note that b is a complex vector, so we obtain

$$\begin{aligned} \frac{\partial d_{l,k}(\eta)}{\partial b_{m,n}^{(R)\mathrm{T}}} &= H_{l,k} \\ \frac{\partial d_{l,k}(\eta)}{\partial b_{m,n}^{(I)\mathrm{T}}} &= jH_{l,k} \end{aligned} \tag{A59}$$

Appendix E.3. The Partial of $\boldsymbol{d}_{l,k}(\boldsymbol{\eta})$ *Respect to* $\boldsymbol{s}$

Define $\boldsymbol{K}_{l,k} = b_{l,k}\boldsymbol{C}\left(z, \boldsymbol{p}_{l,k}\right)\boldsymbol{B}_k\delta_{k,n}$. Note that $\boldsymbol{s}$ is also a complex vector, so we obtain

$$\begin{aligned} \frac{\partial \boldsymbol{d}_{l,k}(\boldsymbol{\eta})}{\partial \boldsymbol{s}_n^{(R)\mathrm{T}}} &= \boldsymbol{K}_{l,k} \\ \frac{\partial \boldsymbol{d}_{l,k}(\boldsymbol{\eta})}{\partial \boldsymbol{s}_n^{(I)\mathrm{T}}} &= j\boldsymbol{K}_{l,k} \end{aligned} \tag{A60}$$

Appendix E.4. The Partial of $\boldsymbol{d}_{l,k}(\boldsymbol{\eta})$ *Respect to* $\boldsymbol{\Delta f}$

Define $\boldsymbol{M}_{l,k} = \boldsymbol{b}_{l,k}\boldsymbol{C}\left(z, \boldsymbol{p}_{l,k}\right)\dot{\boldsymbol{B}}_k(\boldsymbol{I}_K \otimes \boldsymbol{s}_k)$ with $\dot{\boldsymbol{B}}_k = \boldsymbol{B}_k \cdot \mathrm{diag}\{j2\pi\widetilde{N}T_s\}\delta_{k,n}$. The derivation with respect to Δf can be expressed by

$$\frac{\partial \boldsymbol{d}_{l,k}(\eta)}{\partial \Delta \boldsymbol{f}^{\mathrm{T}}} = \boldsymbol{M}_{l,k} \tag{A61}$$

By substituting (A56), (A59), (A60), (A61) into (53), the sub-blocks of $\boldsymbol{J}_{\eta\eta}$ is formulated as

$$\left\{\begin{array}{l}
\left\{\begin{array}{l}
\boldsymbol{Y}_{zz} = \sum\limits_{k=1}^{K}\sum\limits_{l=1}^{L} Re\left\{\boldsymbol{G}_{l,k}^{\mathrm{H}}\boldsymbol{G}_{l,k}\right\}\ \boldsymbol{Y}_{z\boldsymbol{b}^{(R)}} = \sum\limits_{k=1}^{K}\sum\limits_{l=1}^{L} Re\left\{\boldsymbol{G}_{l,k}^{\mathrm{H}}\boldsymbol{H}_{l,k}\right\}\ \boldsymbol{Y}_{z\boldsymbol{b}^{(I)}} = -\sum\limits_{k=1}^{K}\sum\limits_{l=1}^{L} Im\left\{\boldsymbol{G}_{l,k}^{\mathrm{H}}\boldsymbol{H}_{l,k}\right\} \\
\boldsymbol{Y}_{z\boldsymbol{s}^{(R)}} = \sum\limits_{k=1}^{K}\sum\limits_{l=1}^{L} Re\left\{\boldsymbol{G}_{l,k}^{\mathrm{H}}\boldsymbol{K}_{l,k}\right\}\ \boldsymbol{Y}_{z\boldsymbol{s}^{(I)}} = -\sum\limits_{k=1}^{K}\sum\limits_{l=1}^{L} Im\left\{\boldsymbol{G}_{l,k}^{\mathrm{H}}\boldsymbol{K}_{l,k}\right\}\ \boldsymbol{Y}_{z\Delta f} = \sum\limits_{k=1}^{K}\sum\limits_{l=1}^{L} Re\left\{\boldsymbol{G}_{l,k}^{\mathrm{H}}\boldsymbol{M}_{l,k}\right\}
\end{array}\right. \\
\left\{\begin{array}{l}
\boldsymbol{Y}_{\boldsymbol{b}^{(R)}\boldsymbol{b}^{(R)}} = \sum\limits_{k=1}^{K}\sum\limits_{l=1}^{L} Re\left\{\boldsymbol{H}_{l,k}^{\mathrm{H}}\boldsymbol{H}_{l,k}\right\}\ \boldsymbol{Y}_{\boldsymbol{b}^{(R)}\boldsymbol{b}^{(I)}} = -\sum\limits_{k=1}^{K}\sum\limits_{l=1}^{L} Im\left\{\boldsymbol{H}_{l,k}^{\mathrm{H}}\boldsymbol{H}_{l,k}\right\}\ \boldsymbol{Y}_{\boldsymbol{b}^{(R)}\boldsymbol{s}^{(R)}} = \sum\limits_{k=1}^{K}\sum\limits_{l=1}^{L} Re\left\{\boldsymbol{H}_{l,k}^{\mathrm{H}}\boldsymbol{K}_{l,k}\right\} \\
\boldsymbol{Y}_{\boldsymbol{b}^{(R)}\boldsymbol{s}^{(I)}} = -\sum\limits_{k=1}^{K}\sum\limits_{l=1}^{L} Im\left\{\boldsymbol{H}_{l,k}^{\mathrm{H}}\boldsymbol{K}_{l,k}\right\}\ \boldsymbol{Y}_{\boldsymbol{b}^{(R)}\Delta f} = \sum\limits_{k=1}^{K}\sum\limits_{l=1}^{L} Re\left\{\boldsymbol{H}_{l,k}^{\mathrm{H}}\boldsymbol{M}_{l,k}\right\}
\end{array}\right. \\
\left\{\begin{array}{l}
\boldsymbol{Y}_{\boldsymbol{b}^{(I)}\boldsymbol{b}^{(I)}} = \sum\limits_{k=1}^{K}\sum\limits_{l=1}^{L} Re\left\{\boldsymbol{H}_{l,k}^{\mathrm{H}}\boldsymbol{H}_{l,k}\right\}\ \boldsymbol{Y}_{\boldsymbol{b}^{(I)}\boldsymbol{s}^{(R)}} = \sum\limits_{k=1}^{K}\sum\limits_{l=1}^{L} Im\left\{\boldsymbol{H}_{l,k}^{\mathrm{H}}\boldsymbol{K}_{l,k}\right\} \\
\boldsymbol{Y}_{\boldsymbol{b}^{(I)}\boldsymbol{s}^{(I)}} = \sum\limits_{k=1}^{K}\sum\limits_{l=1}^{L} Re\left\{\boldsymbol{H}_{l,k}^{\mathrm{H}}\boldsymbol{K}_{l,k}\right\}\ \boldsymbol{Y}_{\boldsymbol{b}^{(I)}\Delta f} = \sum\limits_{k=1}^{K}\sum\limits_{l=1}^{L} Im\left\{\boldsymbol{H}_{l,k}^{\mathrm{H}}\boldsymbol{M}_{l,k}\right\}
\end{array}\right. \\
\boldsymbol{Y}_{\boldsymbol{s}^{(R)}\boldsymbol{s}^{(R)}} = \sum\limits_{k=1}^{K}\sum\limits_{l=1}^{L} Re\left\{\boldsymbol{K}_{l,k}^{\mathrm{H}}\boldsymbol{K}_{l,k}\right\}\ \boldsymbol{Y}_{\boldsymbol{s}^{(R)}\boldsymbol{s}^{(I)}} = -\sum\limits_{k=1}^{K}\sum\limits_{l=1}^{L} Im\left\{\boldsymbol{K}_{l,k}^{\mathrm{H}}\boldsymbol{K}_{l,k}\right\}\ \boldsymbol{Y}_{\boldsymbol{s}^{(R)}\Delta f} = \sum\limits_{k=1}^{K}\sum\limits_{l=1}^{L} Re\left\{\boldsymbol{K}_{l,k}^{\mathrm{H}}\boldsymbol{M}_{l,k}\right\} \\
\boldsymbol{Y}_{\boldsymbol{s}^{(I)}\boldsymbol{s}^{(I)}} = \sum\limits_{k=1}^{K}\sum\limits_{l=1}^{L} Re\left\{\boldsymbol{K}_{l,k}^{\mathrm{H}}\boldsymbol{K}_{l,k}\right\}\ \boldsymbol{Y}_{\boldsymbol{s}^{(I)}\Delta f} = \sum\limits_{k=1}^{K}\sum\limits_{l=1}^{L} Im\left\{\boldsymbol{K}_{l,k}^{\mathrm{H}}\boldsymbol{M}_{l,k}\right\} \\
\boldsymbol{Y}_{\Delta f\Delta f} = \sum\limits_{k=1}^{K}\sum\limits_{l=1}^{L} Re\left\{\boldsymbol{M}_{l,k}^{\mathrm{H}}\boldsymbol{M}_{l,k}\right\}
\end{array}\right. \tag{A62}$$

References

1. Zhang, Y.; Xu, X.; Sheikh, Y.A.; Ye, Z. A rank-reduction based 2-D DOA estimation algorithm for three parallel uniform linear arrays. *Signal Process.* **2016**, *120*, 305–310. [CrossRef]
2. Oh, D.; Kim, S.; Yoon, S.H.; Chong, J.W. Two-Dimensional ESPRIT-Like Shift-Invariant TOA Estimation Algorithm Using Multi-Band Chirp Signals Robust to Carrier Frequency Offset. *IEEE Trans. Wirel. Common.* **2013**, *12*, 3130–3139. [CrossRef]
3. Cao, H.; Chan, Y.T.; So, H.C. Maximum likelihood TDOA estimation from compressed sensing samples without reconstruction. *IEEE Signal Process. Lett.* **2017**, *24*, 564–568. [CrossRef]
4. Tahat, A.; Kaddoum, G.; Yousefi, S.; Valaee, S.; Gagnon, F. A look at the recent wireless positioning techniques with a focus on algorithms for moving receivers. *IEEE Access* **2016**, *4*, 6652–6680. [CrossRef]
5. Gajewski, P.; Ziolkowski, C.; Kelner, J.M. Using SDF method for simultaneous location of multiple radio transmitters. In Proceedings of the IEEE 19th International Conference on Microwave Radar and Wireless Communications (MIKON), Warsaw, Poland, 21–23 May 2012; Volume 2, pp. 634–637.
6. Kelner, J.M.; Ziolkowski, C.; Nowosielski, L.; Wnuk, M. Localization of emission source in urban environment based on the Doppler effect. In Proceedings of the 2016 IEEE 83rd Vehicular Technology Conference (VTC Spring), Nanjing, China, 15–18 May 2016; pp. 1–5.
7. Li, J.; Guo, F.; Yang, L.; Jiang, W.; Pang, H. On the use of calibration sensors in source localization using TDOA and FDOA measurements. *Digit. Signal Process.* **2014**, *27*, 33–43. [CrossRef]

8. Luo, X.; Jiu, B.; Chen, S.; Ge, Q. ML estimation of transition probabilities for an unknown maneuvering emitter tracking. *Signal Process.* **2015**, *109*, 248–260. [CrossRef]
9. Viberg, M.; Ottersten, B. Sensor array processing based on subspace fitting. *IEEE Trans. Signal Process.* **1991**, *39*, 1110–1121. [CrossRef]
10. Amar, A.; Weiss, A.J. Localization of narrowband radio emitters based on Doppler frequency shifts. *IEEE Trans. Signal Process.* **2008**, *56*, 5500–5508. [CrossRef]
11. Tier, T.; Weiss, A.J. High resolution localization of narrowband radio emitters based on doppler frequency shifts. *Signal Process.* **2017**, *141*, 288–298. [CrossRef]
12. Weiss, A.J. Direct geolocation of wideband emitters based on delay and Doppler. *IEEE Trans. Signal Process.* **2011**, *59*, 2513–2521. [CrossRef]
13. Qin, T.; Li, L.; Ba, B.; Wang, D. A fast ML-based single-step localization using EM algorithm based on time delay and Doppler shift for a far-field scenario. *Sensors* **2018**, *18*, 4139. [CrossRef] [PubMed]
14. Weiss, A.J. Direct position determination of narrowband radio frequency transmitters. *IEEE Signal Process. Lett.* **2004**, *11*, 513–516. [CrossRef]
15. Du, J.; Wang, D.; Yu, W.; Yu, H.; Du, J.; Wang, D.; Yu, W.; Yu, H. Direct position determination of unknown signals in the presence of multipath propagation. *Sensors* **2018**, *18*, 892.
16. Wang, D.; Zhang, G.; Shen, C.; Zhang, J. A direct position determination algorithm for constant modulus signals with single moving observer. *Acta Aeronaut. Astronaut. Sin.* **2016**, *37*, 1622–1633.
17. Lu, Z.; Wang, J.; Ba, B.; Wang, D. A novel direct position determination algorithm for orthogonal frequency division multiplexing signals based on the time and angle of arrival. *IEEE Access.* **2017**, *5*, 25312–25321. [CrossRef]
18. Reuven, A.M.; Weiss, A.J. Direct position determination of cyclostationary signals. *Signal Process.* **2009**, *89*, 2448–2464. [CrossRef]
19. Cherchar, A.; Thameri, M.; Belouchrani, A. Performance improvement of direction finding algorithms in non-homogeneous environment through data fusion. *Digit. Signal Process.* **2015**, *41*, 41–47. [CrossRef]
20. Vincent, F.; Besson, O.; Chaumette, E. Approximate maximum likelihood estimation of two closely spaced sources. *Signal Process.* **2014**, *97*, 83–90. [CrossRef]
21. Hu, D.; Huang, Z.; Zhang, S.; Lu, J. Joint TDOA, FDOA and differential Doppler rate estimation: Method and its performance analysis. *Chin. J Aeronaut.* **2018**, *31*, 137–147. [CrossRef]
22. Hari, K.V.S.; Gummadavelli, U. Effect of spatial smoothing on the performance of subspace methods in the presence of array model errors. *Automatica* **1994**, *30*, 11–26. [CrossRef]
23. Amar, A.; Weiss, A.J. Direct position determination in the presence of model errors-known waveforms. *Digit. Signal Process.* **2006**, *16*, 52–83. [CrossRef]
24. Amar, A.; Weiss, A.J. Analysis of direct position determination approach in the presence of model errors. In Proceedings of the IEEE Convention on Electrical and Electronics Engineers, Telaviv, Israel, 6–7 September 2004; pp. 408–411.
25. Wang, D.; Yin, J.; Liu, R.; Yu, H.; Wang, Y. Performance analysis and improvement of direct position determination based on Doppler frequency shifts in presence of model errors: Case of known waveforms. *Multidimens. Syst. Sign Process.* **2018**, 1–42. [CrossRef]
26. Wang, D.; Yu, H.; Wu, Z.; Wang, C. Performance Analysis of the Direct Position Determination Method in the Presence of Array Model Errors. *Sensors* **2017**, *17*, 1550. [CrossRef] [PubMed]
27. Tirer, T.; Weiss, A.J. Performance Analysis of a High-Resolution Direct Position Determination Method. *IEEE Trans. Signal Process.* **2016**, *65*, 544–554. [CrossRef]
28. Li, J.; Yang, L.; Guo, F.; Jiang, W. Coherent summation of multiple short-time signals for direct positioning of a wideband source based on delay and Doppler. *Digit. Signal Process.* **2016**, *48*, 58–70. [CrossRef]
29. Wang, S.; Wu, M.; Jia, Z. *Matrix Inequality*; Science Press: Beijing, China, 2006.

electronics

MDPI

Article

Stochastic Gradient Matching Pursuit Algorithm Based on Sparse Estimation

Liquan Zhao [1,*], Yunfeng Hu [1] and Yulong Liu [2]

[1] School of Electrical Engineering, Northeast Electric Power University, Jilin 132012, China; huyunfeng22@163.com

[2] Guangxi Power Grid Corporation, Nanning 530023, China; liuyulong163@126.com

* Correspondence: zhao_liquan@163.com; Tel.: +86-150-432-01901

Received: 20 November 2018; Accepted: 17 January 2019; Published: 1 February 2019

Abstract: The stochastic gradient matching pursuit algorithm requires the sparsity of the signal as prior information. However, this prior information is unknown in practical applications, which restricts the practical applications of the algorithm to some extent. An improved method was proposed to overcome this problem. First, a pre-evaluation strategy was used to evaluate the sparsity of the signal and the estimated sparsity was used as the initial sparsity. Second, if the number of columns of the candidate atomic matrix was smaller than that of the rows, the least square solution of the signal was calculated, otherwise, the least square solution of the signal was set as zero. Finally, if the current residual was greater than the previous residual, the estimated sparsity was adjusted by the fixed step-size and stage index, otherwise we did not need to adjust the estimated sparsity. The simulation results showed that the proposed method was better than other methods in terms of the aspect of reconstruction percentage in the larger sparsity environment.

Keywords: compressed sensing; estimated sparsity; least squares solution; stochastic gradient; reconstruction probability

1. Introduction

Compressed sensing (CS) [1–3] theory has aroused significant concern over the past few years. It asserts that a signal can be conducted using compressive sampling, which has a much lower frequency than that of Nyquist. The signal processing of an electrical circuit includes an analog-to-digital converter (ADC). The ADC receives an analog input signal, samples the analog input signal based on a sampling clock signal and converts the sampled analog input signal into a digital output signal. The compressed sensing method can be used to sample the analog signal with a lower sample rate than the Nyquist sampling rate. CS theory mainly includes three core issues [4]: (1) The signal sparsity representation, which designs the sparsity basis or the over-complete dictionary with the capability of sparse representation; (2) The compressive measurement of the sparse signal or compressive signal for designing the sensing matrix, which satisfies the incoherence of atoms or restricted isometry property (RIP) [5]; and (3) The reconstruction of the sparse signal is to design the efficiency signal recovery algorithm. In terms of the aspects of signal sparse representation and sensing matrix design, there have been several better solutions. However, extending CS theory to practical applications requires a crucial step to implement, which is the design of a signal recovery algorithm. Therefore, the design of a recovery algorithm is still an important topic in the field of CS research.

Currently, several mature signal recovery algorithms have been proposed. Among the existing recovery algorithms, two major approaches are the l_1-norm minimization (or convex optimization) and l_0-norm minimization (or greedy pursuit) methods. Convex optimization methods approach the signal by changing the non-convex problem into convex ones such as the basis pursuit (BP) [6] algorithm, the gradient projection for sparse reconstruction (GPSR) [7] algorithm, the interior-point method Bergman iteration (BT) [8] and total-variation (TV) [9]. While the convex optimization methods work correctly for all sparse signals and provide theoretical performance guarantees, its high computational complexity may prevent it from encountering practical large-scale recovery problems. The other category is the greedy pursuit algorithm, which iteratively identifies the true support of the original signal and constructs an approximation signal based on a set of chosen supports until the halt iteration stop condition is satisfied. This can more efficiently solve large-scale data recovery problems. An example of an earlier typical greedy algorithm is the matching pursuit (MP) [10] algorithm. The orthogonal matching pursuit (OMP) [11] algorithm was developed based on the MP algorithm to optimize MP by orthogonalizing the atoms of the support set. However, the OMP algorithm selects one of the columns of preliminary atoms to add the candidate atoms set, which will increase the number of iterations, thereby reducing the speed of the OMP algorithm. Subsequently, some researchers have proposed several modified methods and as for the shortcoming where OMP places only one atom (or column) onto the support atom set at each round of iteration, the stage-wise OMP (StOMP) [12] algorithm has been proposed. StOMP can select multiple atoms to add to the support atom set by using the thresholds. Regularization is introduced in OMP and can provide a powerful theoretical guarantee. This recovery algorithm is called the regularized OMP (ROMP) [13] algorithm. The computational complexity of these algorithms is significantly lower than that of the convex optimization methods; however, they require more measurement of values for exact recovery and have poor reconstruction performance in a noisy environment. To date, subspace pursuit (SP) [14] and compressive sampling matching pursuit (CoSaMP) [15,16] algorithms have been proposed by incorporating a backtracking strategy. These algorithms offer strong theoretical guarantees and provide robustness to noise. However, both of these algorithms require the sparsity K as priority information, which may not be available in most practical applications. In order to overcome this weakness, the sparsity adaptive matching pursuit (SAMP) [17] algorithm was proposed for blind signal recovery when the sparsity is unknown. The SAMP algorithm divides the recovery process of the algorithm into several stages with a fixed step-size and without the prior information of the sparsity. In the SAMP algorithm, the step-size is fixed at the initial stage of the SAMP algorithm. Additional iterations are required if the step-size is much smaller than the signal's sparsity. This will lead to a long reconstruction time. Furthermore, the fixed step-size cannot estimate the real sparsity precisely because this method can only set the estimated sparsity to a multiple integer of the step-size. Although these traditional greedy pursuit algorithms are widely used due to their simple structure, convenient calculation and better reconstruction effect, they still have many drawbacks. These methods do not directly solve the original optimization problem, which will result in the quality of the signal recovery being of poorer quality than the convex optimization method-based l_1-norm. In addition, these greedy pursuit algorithms have the disadvantage of a high computing complexity and large storage capacity for large-scale date recovery.

Since calculating the orthogonal projection requires a large number of calculations using traditional greedy algorithms, this will result in a decline in the recovery efficiency of the greedy algorithm. Thomas et al. first proposed a gradient pursuit (GP) [18] algorithm for the sake of overcoming this shortcoming. This algorithm uses the update of the gradient direction to replace the calculation of the orthogonal projection, which reduces the computational complexity of the greedy pursuit algorithms. Their successors include the Newton pursuit (NP) [19] algorithm, the conjugate gradient pursuit (CGP) [20] algorithm, the approximate conjugate gradient pursuit (ACGP) [21] algorithm and the variable metric method-based gradient pursuit (VMMGP) [22] algorithm. These methods reduce the computational complexity and storage space of the traditional greedy algorithm in terms of the large-scale recovery problem but the reconstruction performance still requires improvement. Therefore, based on the GP algorithm, the stage-wise weak gradient pursuit (SwGP) [23] algorithm was proposed to improve the reconstruction efficiency and convergence speed of the GP algorithm via the weak selection strategy. Although the SwGP algorithm makes the fashioning of atom selection more flexible and improves the reconstruction precision, the time taken for atom selection is greatly increased. Recently, motivated by the stochastic gradient descent methods, the stochastic gradient matching pursuit (StoGradMP) [24] algorithm was proposed for the optimization problem with sparsity constraints. The StoGradMP algorithm not only improves the reconstruction efficiency of the greedy recovery algorithm for the large-scale data recovery problem but also reduces the computational complexity of the algorithm. However, the StoGradMP algorithm still requires the sparsity of the signal as a priori information, which restricts the capacity of the algorithm's availability in practical situations. This study proposed a sparsity pre-evaluation strategy to estimate the sparsity of the signal and utilized the estimated sparsity as the input parameter of the algorithm. This strategy will make the algorithm eliminate the dependence on signal sparsity and decrease the number of iterations of the algorithm. This algorithm then approaches the real sparsity of the signal by adjusting its initial sparsity estimation, thereby realizing the expansion of the support atoms set and the signal reconstruction.

In recent years, a variety of reconstruction algorithms have been proposed, which have further enhanced the application prospect of CS theory in the field of signal processing such as channel estimation and blind source separation. There is no denying that the application research of reconstruction algorithms will even further highlight the importance of such algorithms. In the literature [25], novel subspace-based blind schemes have been proposed and applied to the sparse channel identification problem. Moreover, the adaptive sparse subspace tracking method was proposed to provide efficient real-time implementations. In Reference [26], a novel unmixing method based on the simultaneously sparse and low-rank constrained non-negative Matrix factorization (NMF) was applied to the remote sensing image analysis.

2. Preliminaries and Problem Statement

In CS theory, for $x \in R^{n\times 1}$, here, n is the length of signal x. If the number of non-zero entries is K in original signal, then we regard the signal x as the K-sparse signal or compressive signal (in noiseless environments). Generally, the signal x can be expressed as follows:

$$x = \sum_{i=1}^{n} \beta_i \psi_i = \Psi\beta \tag{1}$$

$$\| \beta \|_0 = K \tag{2}$$

where $\psi_i (i = 1, 2, \ldots, n)$ are the basis vectors of the sparse basis matrix $\Psi^{n\times n}$, that is, Ψ is the matrix constituted by the $\{\psi_i\}_{i=1}^{n}$. $\beta \in R^n$ is a projection coefficient vector and $K \ll n$. $\| \cdot \|_0$ denotes that the number of non-zero entries in the projection coefficient vector β.

When the sparse representation of the original signal is completed, we need to construct a measurement matrix Φ for the compression measurement of the sparse signal x to obtain the observation values u, this process can be described as follows:

$$u = \Phi x \tag{3}$$

where $\Phi \in R^{m\times n}$, $u \in R^{m\times 1}$ and $m \ll n$. According to Equation (3), the observation vector nearly contains the whole information of the n-dimensional signal x. Furthermore, this process is non-adaptive, which will ensure that the crucial information of the original signal is not lost when the dimensional signal is decreased from n to m. The m is called the number of observation values in the later description.

When the original signal x itself is not sparse, the original signal measurement process cannot be directly utilized in Equation (3). Thus, we need the compressive measurement on the projection coefficient vector β to obtain the measurement value. According to Equations (1) and (3), we can obtain the follow equation:

$$u = \Phi\Psi\beta = \Gamma\beta \tag{4}$$

where $\Gamma = \Phi\Psi \in R^{m \times n}$ is the sensing matrix. According to Equation (4), we know that the dimensional of the observation vector u is much lower than the dimensional of signal x, that is, $m \ll n$. Therefore, Equation (4) is regarded as an under-determined problem and indicates that Equation (4) has an infinite number of solutions. That is to say, it is hard to reconstruct the projection coefficient vector β from observation vector u.

Whereas, according to the literature [27], we know that the sufficient condition for exact sparse signal recovery is that sensing matrix Γ satisfies the RIP condition. Thus, if the sensing matrix satisfies the RIP condition, the reconstruction on signal β is equivalent to the l_0-norm optimization problem [28]:

$$\min_{\beta \in R^{n \times 1}} \| \beta \|_0 \text{ subject to } u = \Gamma\beta \tag{5}$$

where $\| \cdot \|_0$ represents the number of non-zero entries in projection coefficient β. Unfortunately, Equation (5) is a NP-hard optimization problem. When the isometry constant δ_K of the sensing matrix Γ is less than or equal to $\sqrt{2} - 1$, Equation (5) is equivalent to the l_1-norm optimization problem:

$$\min_{\beta \in R^{n \times 1}} \| \beta \|_1 \text{ subject to} u = \Gamma\beta \tag{6}$$

where $\| \cdot \|_1$ denotes that the absolute sum of the non-zero entries in projection coefficient β. Equation (6) is a convex optimization problem. Meanwhile, when the sparse basis is determined, in order to ensure that the sensing matrix Γ also satisfies the RIP condition, the measurement matrix Φ must meet certain conditions. However, in Reference [29,30], the researchers found that when the measurement matrix Φ was a random matrix with a Gaussian distribution, the sensing matrix Γ could satisfy the RIP condition with a large probability. This will greatly reduce the difficultly of the design of the measurement matrix.

However, in most practical applications and conditions, the original signal ordinarily contains noise signals. In this setting, this sensing process can be represented in the following equation:

$$u = \Gamma\beta + \varepsilon \tag{7}$$

where $\varepsilon \in R^{n \times 1}$ is the noise signal. In this study, for simplicity, we supposed that the signal x itself was K-sparse, thus, the original signal x and sensing matrix Γ were equal to the projection coefficient β and measurement matrix Φ, respectively. According to Equation (7), it can be written as $u = \Phi x + \varepsilon$. We minimized the follow equation to reconstruct the original sparse signal x:

$$\min_{x \in R^{n \times 1}} \frac{1}{2m} \| u - \Phi x \|_2^2 \text{ subject to} \| x \|_0 \leq K \tag{8}$$

where $u - \Phi x$ is the residual of the original signal x, which is represented as r_k. That is, $r_k = u - \Phi x$. $\| \cdot \|_2$ represents the square of l_2-norm of the signal residual vector r_k. To analyze Equation (8), we combined Equation (1). In Equation (1), β_i is the projection coefficient of the sparse signal x. This notion is general enough to address many important sparse models such as group sparsity and low rankness (see studies [31,32] for examples). Then, we can express Equation (9) in the form of

$$\min_{x} \underbrace{\frac{1}{M} \sum_{i=1}^{M} f_i(x)}_{F(x)} \text{ subject to} \| x \|_{0,\Psi} \leq K \tag{9}$$

where $f_i(x)$ is a smooth function, that is, it is a non-convex function. $\| x \|_{0,\Psi}$ is defined as the norm that captures the sparsity of signal x.

For a sparse signal recovery problem, the sparse basis Ψ consists of n basic vectors, each of size n in the Euclidean space. This problem can be regarded as a special case of Equation (9) with $f_i(x) =$

$(u_i - <\phi_i, x>)^2$ and $M = m$. The observation vector u is decomposed into the non-overlapping block observation vectors u_{b_i} with a size of b. $\Phi_{b_i \times n}$ denotes the block-matrix of the measurement matrix of size b. According to Equations (8) and (9), the objective function $F(x)$ can be represented as in the following form:

$$F(x) = \frac{1}{M}\sum_{i=1}^{M}\frac{1}{2b}\| u_{b_i} - \Phi_{b_i}x \|_2^2 = \frac{1}{M}\sum_{i=1}^{M} f_i(x) \tag{10}$$

where $M = m/b$, which is a positive integer. According to the equation, each smooth function $f_i(x)$ can be represented as $f_i(x) = \frac{1}{2b}\| u_{b_i} - \Phi_{b_i}x \|_2^2$. Obviously, in this case, each sub-function $f_i(x)$ accounts for a collection (or block) of measurements of size b, rather than only one observation. Here, the smooth function $F(x)$ is divided into multiple smooth sub-functions $f_i(x)$ and the measurement matrix Φ block into multiple block matrix Φ_{b_i}, which will contribute to the computation of the gradient in the stochastic gradient matching pursuit algorithm, thereby improving the reconstruction performance of the algorithm.

3. StoGradMP Algorithm

The CoSaMP algorithm is fast for small-scale signals with a lower dimensional but for large-scale signals with a higher dimensional and noise signal, the reconstruction precise is not very accurate and the robustness of the algorithm itself is poorly. Therefore, in Reference [30], the researchers generalized the idea of the CoSaMP algorithm and proposed the GradMP algorithm for the reconstruction problem of large-scale signals with sparsity constraints and noise signals. Regrettably, the GradMP algorithm needs to calculate the overall gradient of the smooth function $F(x)$, which increases the computational complexity of the GradMP algorithm. After the GradMP algorithm, Needell et al. proposed a stochastic version of the GradMP algorithm called the StoGradMP [24] algorithm. This algorithm only computes the gradient of the sub-function $f_i(x)$ at each round of iterations.

According to the literature [24], the StoGradMP algorithm is described in Algorithm 1, which consists of the following steps at each round of iterations:

Randomize: The measurement matrix Φ is randomly divided into blocks, that is, it searches the row index of the measurement matrix constituting a block matrix Φ_{b_i} of size $b_i \times n$ by the row vector corresponding to those row indexes. Then, according to Equation (10) and the block matrix, execute the calculation operation of sub-function $f_{i_k}(x_k)$.

Proxy: Compute the gradient G_k of $f_{i_k}(x_k)$, where the gradient G_k is a $n \times 1$ column vector.

Identify: The absolute value of the gradient vector is ranked in descending order, the first $2K$ absolute value of the gradient coefficients are selected, the column index (atomic index) of the measurement matrix corresponding to those coefficients is found, then form a preliminary index set P_k.

Merge: Constitute the candidate atomic index set C_k, which is consists of the preliminary index set P_k and the support index set S_{k-1} of the previous iteration.

Estimation: The transition estimation of the signal b_k by the least square method.

Prune: The absolute value of the estimation vector of the signal transition is ranked in descending order, the first K absolute value of signal estimation coefficients is determined, then conduct a search for the atomic index of the measurement matrix corresponding to those coefficients, forming the support atomic index set S_k.

Update: Update the final estimation of signal $x_k = b_{kS}$ at the current iteration, which corresponds to the support atomic index set S_k.

Check: When the l_2-norm of the signal residual is less than the tolerance error of the StoGradMP algorithm, the iteration is halted. Or, if the loop index k is greater than the maximum number of iterations, the proposed method ends and the approximation of signal $\hat{x} = x_k$ is the output. Otherwise, continue the iteration until the halting condition is met.

4. Proposed Algorithm

The StoGradMP algorithm selects $2K$ atoms in each preliminary stage of iteration. Here, K is a fixed number. Therefore, the StoGradMP algorithm requires the sparsity as a priori information, which is not available in practical applications. We first proposed a sparsity pre-evaluation strategy to obtain an estimation of sparsity as a way to overcome this problem. The next step was to put forward a sparsity adjustment strategy to adjust the estimation of sparsity, approaching the real sparsity of the signal.

4.1. Pre-Evaluation Strategy

In this section, we propose a sparsity pre-evaluation strategy to estimate the real sparsity of the original signal. This process is described below.

Firstly, we provided an initial estimation of sparsity, which is $K_0 = 1$. Next, we calculate the atom correlation g, which is expressed as:

$$g = \Phi^T u \tag{11}$$

where Φ, u represents the measurement matrix and observation vector, respectively.

Second, when the calculation of atom correlation is completed, we selected the K_0 atoms from the measurement matrix Φ to expand the support atom set Φ_V, where the support atomic index can be expressed as:

$$V = \max(|g|, K_0) \tag{12}$$

where $|g|$ is the absolute value of the atom correlation coefficients g. $\max(|g|, K_0)$ represents finding the atomic (or column) index of matrix Φ, corresponding to the K_0 maximal value from $|g|$.

Finally, we checked the iterative stopping condition of the sparsity evaluation to determine whether to continue to the next iteration and update the iterative parameters. This condition is expressed as:

$$\| \Phi_V^T u \|_2 \geq \frac{1-\delta_K}{\sqrt{1+\delta_K}} \| u \|_2 \tag{13}$$

where Φ_V represents the support atomic set (or matrix) corresponding to the support atomic index set V. $\| . \|_2$ denotes the l_2-norm of a vector. The element δ_K is the isometry constant and $\delta_K \in (0, 1)$. If the iteration stopping criteria is satisfied, then the output is the estimated sparsity K_0 and the support atomic index set V, otherwise, the iteration is continued and the estimated sparsity $K_0 = K_0 + 1$ is updated to gradually approach the real sparsity of the original signal until the conditions are satisfied. In addition, the set V will be used for the initial support atomic set estimation in the recovery algorithm. This is S_0= V, which will be used to reduce the selection time of the support atoms set in the recovery algorithm and improve the reconstruction precision.

4.2. Adjustment Strategy

In Section 4.1, we utilized the sparsity pre-evaluation strategy to obtain the sparsity estimation K_0 and the support atomic index set V. However, the sparsity estimation level was lower than the real sparsity of the original signal. If we used it as an input for the recovery algorithm, it would have resulted in the lack of sparsity estimation, which would have led to a decline in the proposed method in terms of reconstruction performance.

Therefore, we proposed an adjustment strategy for the sparsity estimation to control the convergence conditions of the recovery algorithm and adjust the estimated sparsity K_0. This strategy is described below.

We started by checking the iterative stopping condition that is expressed as:

$$\| r_{new} \|_2 \leq tol \text{ or} k \geq \text{maxIter} \tag{14}$$

where tol is a threshold and k and maxIter is the number of iterations and the maximum number of iterations, respectively. In addition, r_{new} is the residual at the k-th iteration. It can be expressed as:

$$r_{new} = u - \Phi x_k \tag{15}$$

$$x_k = b_{kS} \tag{16}$$

where x_k is the approximation of the signal x at the k-th iteration. Furthermore, b_{kS} is the estimation vector corresponding to the support atomic index set S. The set S is expressed as:

$$S = \max(|b_k|, K_0) \tag{17}$$

$$b_k = \Phi_{C_k}^{+} u \tag{18}$$

where b_k, K_0 is the least solution of the signal and the estimated sparsity at the k-th iteration. In addition, $\max(|b_k|, K_0)$ represents finding the atomic (or column) index of the measurement matrix Φ corresponding to the largest K_0 value from $|b_k|$ and constitutes the final (or support) atomic set S. Furthermore, $\Phi_{C_K}^{+}$ is the pseudo inverse matrix of the candidate atomic set (or matrix) Φ_{C_k} and its definition is consistent with the definition in the StoGradMP algorithm.

Second, according to Equation (13), we can see that if the iteration stopping condition is not satisfied, we can judge the stage switching condition to complete the goal of adjusting the estimated sparsity. The condition can then be described as:

$$\| r_{new} \|_2 \geq \| r_{k-1} \|_2 \tag{19}$$

then

$$j = j + 1 \text{ and} K_0 = j * s \tag{20}$$

where j, s are the stage index and the iterative step-size, respectively. Among these, s is a fixed number. In this paper, we set the step-size set as $s = 1, 5, 10, 15$, with K_0 as the estimated sparsity at the j-th stage. If

$$\| r_{new} \|_2 \leq \| r_{k-1} \|_2 \tag{21}$$

then continue to iterate and update the parameters:

$$S_k = S \text{ and } r_k = r_{new} \tag{22}$$

where r_k and S_k are the current residual and the support index set at the k-th iteration, respectively.

4.3. Reliability Verification Condition

Finally, according to Equation (18), before obtaining the least square solutions of the signal, we needed to add a reliability verification condition to ensure that the proposed method was correct and effective. This condition was that the number of rows was greater than the number of columns in the candidate atomic matrix Φ_{C_k}, that is, Φ_{C_k} is a full column-rank matrix. This condition can then be described as:

$$length(C_k) \leq m \tag{23}$$

then

$$b_k = \Phi_{C_k}^{+} u \tag{24}$$

where m is the number of the rows in the measurement matrix. The definition of b_k, $\Phi_{C_k}^{+}$ and u keep pace with the definition in Equation (18). If the condition is not met, that is to say, the candidate atomic matrix is not inverse, then we set $b_k = 0$ and the exit loop.

Figure 1 is a block diagram of the proposed algorithm. As can be seen from Figure 1, the algorithm includes sparsity estimation and restoration. In the sparsity estimation part, the real sparsity estimation is obtained by using the sparsity pre-evaluation strategy. In the recovery part, the sparsity adjustment strategy is proposed to approach the real sparse gradually. This improves the reconstruction accuracy and convergence of the proposed algorithm. The key innovation of the algorithm is that the signal can be recovered without prior sparsity information K.

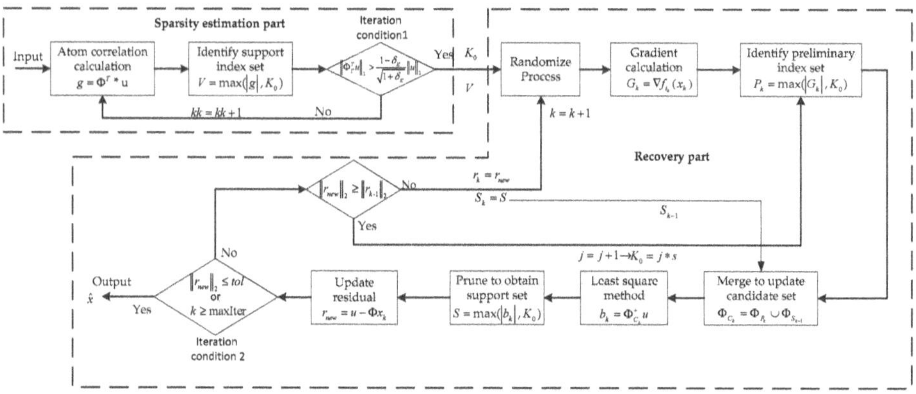

Figure 1. Block diagram of the proposed algorithm.

The entire procedure is shown in Algorithm 1.

Algorithm 1 Proposed algorithm

Input: Measurement matrix $\Phi^{m\times n}$, Observation vector u, Block size b
Step-size s, Isometry constant δ_K, Initial sparsity estimation $K_0 = 1$
Tolerance used to exit loop tol, Maximum number of iterations maxIter
Output1: K_0 sparsity estimation of the original signal
V the support atomic index set
Output2: $\hat{x} = x_k$ K-sparse approximation of signal x
Set parameters:

$\hat{x} = 0$	{initialize signal approximation}
$k = 0$	{loop index used to loop 2}
$kk = 0$	{loop index used to loop 1}
$done1 = 0$	{while loop 1 flag}
$done2 = 0$	{while loop 2 flag}
$r_k = u$	{initialize residual}
$\mathrm{M} = \mathrm{floor(m/b)}$	{number of blocks}
$P_0 = []$	{empty preliminary index set}
$C_0 = []$	{empty candidate index set}
$V = []$	{empty support index set used to loop 1}
$S_0 = []$	{empty support index set used to loop 2}
$j = 0$	{stage index}

Part 1: Sparsity Estimation
While ($\sim done1$)
$kk = kk + 1$

(1) **Compute the atom correlation:** $g = \Phi^T * u$
(2) **Identify the support index set:** $V = \max(|g|, K_0)$
(3) **Check the iteration condition** If $(\| \Phi_\Gamma^T u \|_2 > \frac{1-\delta_K}{\sqrt{1+\delta_K}} \| u \|_2)$ $\quad done1 = 1$ $\quad$ quit iteration
else $\quad K_0 = K_0 + 1$ $\quad$ Sparsity approach $\quad$ end

end

Part 2: Recovery part
$S_0 = V$ Update the support index set
While ($\sim$ *done2*)
$k = k + 1$

(1) **Randomize**

$ii = ceil(rand * M) \rightarrow block = b * (ii - 1) + 1 : b * ii \rightarrow f_{i_k}(x_k) = \frac{1}{2b} \| u_{b_{i_k}} - \Phi_{b_{i_k}} x \|_2^2$

(2) **Computation of gradient:** $G_k = \nabla f_{i_k}(x_k) = -2 * \Phi_{b_{i_k}}^T (u_{b_{i_k}} - \Phi_{b_{i_k}} x_{k-1})$

(3) **Identify the large K_0 components:** $P_k = \max(|G_k|, K_0)$

(4) **Merge to update candidate index set:** $\Phi_{C_k} = \Phi_{P_k} \cup \Phi_{S_{k-1}}$ **Reliability verification condition**

If $length(C_k) \leq m$
$b_k = \Phi_{C_k}^{+} u$ Signal estimation by the least square method
else
$b_k = 0$
break;
end

(5) **Prune to obtain current support index set:** $S = \max(|b_k|, K_0)$
(6) **Signal approximation by the support set:** $x_k = b_{kS}, r_{new} = u - \Phi x_k$
(7) **Check the iteration condition**

If ($\| r_{new} \|_2 \leq tol$ or $k \geq$ maxIter)
$done2 = 1$ quit iteration
else if ($\| r_{new} \|_2 \geq \| r_{k-1} \|_2$) **sparsity adjustment condition**
$j = j + 1$ shift into stage
$K_0 = j * s$ approach the real sparsity
else
$r_k = r_{new}$ update the residual
$S_k = S$ update the support index set
end
end

5. Proof of the Proposed Algorithm

In this section, we prove the correctness of the pre-evaluation strategy. The main idea of this strategy is to carry out the matching test of atoms to obtain the support atomic index set V. The size of the potential of the set V is K_0 and K_0 is smaller than K. Here, K_0, K is the estimated sparsity and the real sparsity of the original signal, respectively. The potential of a set is represented by supp(.). We assumed that the real support of the original signal x could be represented by F and supp(F) $=K$. Φ_F represents a sub-matrix formed by the atoms (or columns) of the measurement matrix Φ, whose indices correspond to the real support index set F. Moreover, $g = \Phi^T u$, g_i is the i-th element of the atomic correlation coefficient g. In addition, the set V consists of indices corresponding to the K_0 largest absolute value of g_i and supp(V) $=K_0$. Finally, the proposition can be explained as follows.

Proposition 5.1. *Assume that measurement matrix Φ satisfies the restricted isometry property with parameters K and δ_K. If $K_0 \geq K$, then we can obtain the formula in the form:*

$$\| \Phi_V^T u \|_2 \geq \frac{1 - \delta_K}{\sqrt{1 + \delta_K}} \| u \|_2 \tag{25}$$

Proof. Select the atomic index of matrix Φ corresponding to the K largest value from $|g|$ and form the real support atomic index set F. When $K_0 \geq K$, $F \subseteq V$. Then, we can obtain

$$\| \Phi_V^T u \|_2 \geq \| \Phi_F^T u \| \tag{26}$$

Furthermore, we have

$$\begin{aligned} \| \Phi_V^T u \|_2 &= \max_{|F|=K} \sqrt{\sum_{i \in F} |\langle \Phi_i, u \rangle|^2} \\ &\geq \| \Phi_F^T u \|_2 = \| \Phi_F^T \Phi_F x \|_2 \end{aligned} \tag{27}$$

According to the definition of RIP, the range of the singular value of Φ_F is $\sqrt{1-\delta_K} \leq \sigma(\Phi_F) \leq \sqrt{1+\delta_K}$. Here, $\sigma(.)$ represents a singular value of the matrix. If we denote $\lambda\left(\Phi_F^T \Phi_F\right)$ as the eigenvalue of matrix $\Phi_F^T \Phi_F$, we have $1-\delta_K \leq \lambda\left(\Phi_F^T \Phi_F\right) \leq 1+\delta_K$. Therefore, we can obtain a formula in the form:

$$\| \Phi_F^T \Phi_F x \|_2 \geq (1-\delta_K) \| x \|_2 \tag{28}$$

On the other hand, according to the definition of RIP properties, we can obtain the following formula:

$$\| x \|_2 \geq \frac{\| u \|_2}{\sqrt{1+\delta_K}} \tag{29}$$

Combining the inequalities of Equations (27)–(29), the following formula can be obtained:

$$\| \Phi_V^T u \|_2 \geq \frac{1-\delta_K}{\sqrt{1+\delta_K}} \| u \|_2 \tag{30}$$

Therefore, the proof is completed. □

In light of the relationship between Proposition 5.1 and its converse-negative propositions, that is to say, if Proposition 1 is true, then its converse-negative proposition is also true. Therefore, for Proposition 1 in this paper, we have

$$\| \Phi_V^T u \|_2 < \frac{1-\delta_K}{\sqrt{1+\delta_K}} \| u \|_2 \tag{31}$$

Then $K_0 < K$.

According to Proposition 1, we can obtain an estimation method of true sparsity K. That is, if we obtain an index set V satisfying the inequality (Equation (31)), then the sparsity estimation K_0 can be obtained. We can describe this as follows: first, we set the initial estimated sparsity as $K_0 = 1$ and if the inequality (Equation (31)) is true, then $K_0 = K_0 + 1$. Exit the loop when inequality (Equation (31) is false. Meanwhile, we can obtain an initial index set V, which is the estimation of the true support index set F.

6. Discussion

In this section, we used the signal with different K-sparsity as the original signal. The measurement matrix was randomly generated with a Gaussian distribution. All performances were an average calculated after running the simulation 100 times using a computer with a 32-core, 64-bit processor, two processors and a 32 G memory. We also set the recovery error of all recovery methods as $1 \times e^{-6}$ and the tolerance error as $1 \times e^{-7}$. The maximum number of iterations of the recovery part of the proposed method was $500 * M$.

In Figure 2, we compared the reconstruction percentage of different step-sizes of the proposed method with different sparsities in different isometry constants. We set the step size set and the range of sparsity as $s \in [1, 5, 10, 15]$ and $K \in [10\ 100]$, respectively. The isometry constant parameter set was $\delta_K \in [0.1, 0.2, 0.3, 0.4, 0.5, 0.6]$. From Figure 2, we can see that the reconstruction percentage was very close, with almost no difference for all isometry constants δ_K. This means that the selection of the isometry constants had almost no effect on the reconstruction percentage of the signal.

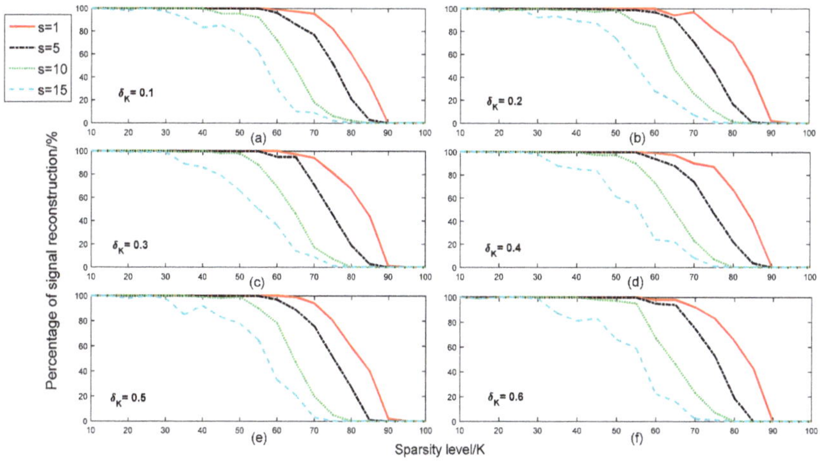

Figure 2. Reconstruction percentage of different step-sizes with different sparsities in different isometry constant conditions ($n = 400$, $s \in [1, 5, 10, 15]$, $\delta_K \in [0.1, 0.2, 0.3, 0.4, 0.5, 0.6]$ and $m = 170$, Gaussian signal).

In Figure 3, we compared the reconstruction percentage of different isometry constants δ_K with different sparsities in different step-size conditions. In order to better analyze the effects of different step-sizes on the reconstruction percentage, the setting of parameters in Figure 3 was consistent with the parameters in Figure 2. From Figure 3, we can see that when the step-size s was 1, the reconstruction performance was the best for different isometry constants. When the step size continued to increase, the reconstruction percentage of the proposed method gradually declined. In particular, when the step-size s was 15, the reconstruction performance was the worst. This shows that a smaller step-size benefits the reconstruction of the signal.

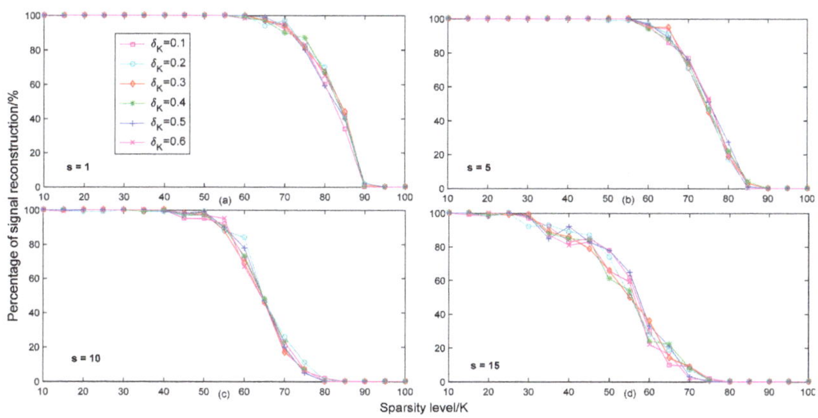

Figure 3. Reconstruction percentage of different isometry constants with different sparsities in different step-size conditions ($n = 400$, $s \in [1, 5, 10, 15]$, $\delta_K \in [0.1, 0.2, 0.3, 0.4, 0.5, 0.6]$ and $m = 170$, Gaussian signal).

In Figure 4, we compared the average estimate of the sparsity of different isometry constants δ_K of the proposed method with different real sparsity K. We set the range of the real sparsity and isometry constant set as $K \in [10\ 60]$ and $\delta_K \in [0.1, 0.2, 0.3, 0.4, 0.5, 0.6]$, respectively. From Figure 4, we can see that when the isometry constant was equal to 0.1, the estimated sparsity K_0 was closer to the real sparsity of the original, rather than the other isometry constant. When the isometry constant was equal to 0.6, the estimated sparsity was much lower than the real sparsity of the signal. Therefore, we can say that a smaller isometry constant may be useful for estimating sparsity. Furthermore, this indicates that a smaller isometry constant can reduce the runtime of sparsity adjustments, making the recovery algorithm able to more quickly approach the real sparsity of the signal, thereby decreasing the overall recovery runtime of the proposed method.

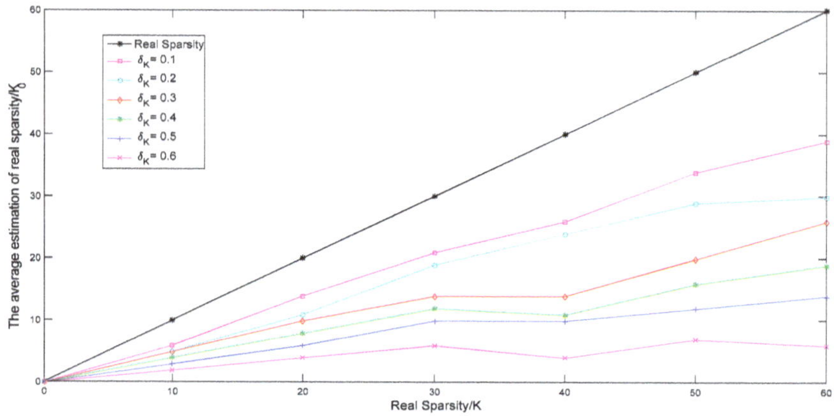

Figure 4. The average estimated sparsity of different isometry constants with different sparsities ($n = 400$, $\delta_K \in [0.1, 0.2, 0.3, 0.4, 0.5, 0.6]$ and $m = 170$, Gaussian signal).

In Figure 5, we compared the reconstruction percentage of different algorithms with different sparsities in different real sparsity conditions. We set the range of the real sparsity of the original signal and the assumed sparsity as $K \in [20, 30, 40, 50]$ and $L \in [10\ 100]$, respectively. From Figure 4, we can see that when the isometry constant was equal to 0.1, the estimated level of sparsity was higher than the other isometry constants. Therefore, we set the isometry constant as 0.1 in the simulation in Figure 5. In Figure 5a,b, we can see that the proposed method had a higher reconstruction percentage than other algorithms when the real sparsity was equal to 20 and 30, almost all of them reached 100%. In Figure 5a, for real sparsity $K = 20$, we can see that when the assumed sparsity $L < 20$, the reconstruction percentage of the StoIHT, GradMP and StoGradMP algorithms was 0%, that is to say, these algorithms could not complete the signal recovery. When $20 \leq L \leq 28$, all recovery methods almost achieved a higher reconstruction percentage. When $28 \leq L \leq 34$, the reconstruction percentage of the StoIHT algorithm began to decline from approximately 100% to 0%, while the other algorithms still had a higher reconstruction percentage. When $34 \leq L$, the reconstruction percentage of the StoIHT algorithm was 0%. For $63 \leq L \leq 72$, the reconstruction percentage of the GradMP and StoGradMP algorithms began to decline from approximately 100% to 0%. Moreover, the reconstruction percentage of the GradMP algorithm was higher than the StoGradMP algorithm in the variation range of this sparsity. In Figure 5b, we can see that the reconstruction percentage of the StoIHT algorithm was still 0% for all assumed sparsity. When $L < 30$, the reconstruction percentage of the GradMP and StoGradMP algorithms was equal to 0%, while the proposed method had a higher reconstruction percentage and was approximately 100%. For $30 \leq L \leq 61$, the reconstruction percentage of all recovery methods was approximately equal to 100%. When $61 \leq L \leq 65$, the reconstruction percentage of the StoGradMP algorithm began to decline from approximately 99% to 1%, while the GradMP algorithm still had a higher reconstruction percentage. In Figure 5c,d, we can see that the reconstruction percentage of the proposed method with $s = 15$ decreased from approximately 99% to 84% and 69%, respectively. Furthermore, from all of the sub-figures in Figure 5, we can see that when the assumed sparsity was close to the real sparsity, the reconstruction percentage of the GradMP and StoGradMP algorithms were very close, with almost no difference. In addition, when the real sparsity of the original signal gradually increased, the range of sparsity that maintained a higher reconstruction percentage became smaller. This means that the GradMP and StoGradMP algorithms were more sensitive to larger real sparsity.

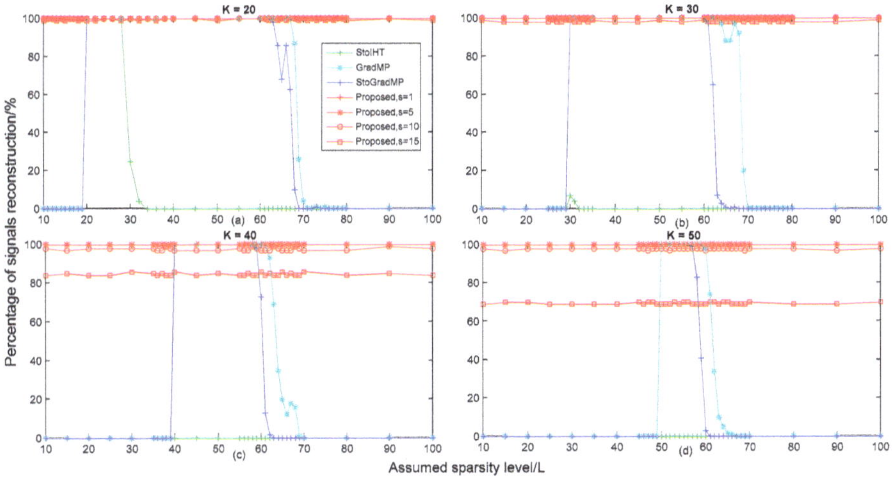

Figure 5. Reconstruction percentage of different algorithms with different sparsities in different real sparsity K conditions ($n = 400$, $s \in [1, 5, 10, 15]$, $\delta_K = 0.1$, and $m = 170, L \in [10\ 100]$, Gaussian signal).

In Figure 6, we compared the reconstruction percentage of different algorithms with different measurements in different real sparsity conditions. We set the range of real sparsity as $K \in [20, 30, 40, 50]$ in the simulation of Figure 6 to keep it consistent with Figure 5. The range of the measurement was $m = 2 * K : 5 : 300$. From Figure 6, we can see that when the real sparsity ranged from 20 to 50, the proposed method was gradually higher than the other algorithms. In Figure 6a, we can see that when $50 \leq m \leq 65$, the reconstruction percentage of the proposed method with $s = 1$ was higher than other methods. For $65 \leq m \leq 115$, the reconstruction percentage of the proposed method was lower than the StoGradMP and GradMP algorithms, except for the StoIHT algorithm. When $65 \leq m \leq 145$, the reconstruction percentage that the StoIHT algorithm was superior to the proposed method was $s = 15$. When $150 \leq m$, all of the recovery methods almost achieved higher reconstruction probabilities. In Figure 6b, we can see that when $65 \leq m \leq 92$, the reconstruction percentage of the proposed method with $s = 1$ and $s = 5$ was higher than the StoGradMP and StoIHT algorithms. When $92 \leq m \leq 165$, the recovery percentage of the proposed method with $s = 5, 10, 15$ was higher than the StoIHT algorithm, except for the StoGradMP and GradMP algorithms. For $95 \leq m \leq 145$, the reconstruction percentage of the proposed method with $s = 5$ was higher than the proposed method with $s = 10$ and $s = 15$, while the StoIHT algorithm still could not complete a recovery of the signal. When $145 \leq m \leq 165$, the reconstruction percentage of the SoIHT algorithm began to dramatically increase from approximately 0% to 100%, while the other algorithms still had a higher recovery percentage. When $165 \leq m$, all of the methods almost achieved higher reconstruction probabilities. In Figure 6c, we can see that when $90 \leq m \leq 127$, the reconstruction percentage of the proposed method with $s = 1$ and $s = 5$ was superior to the StoGradMP and StoIHT algorithms. For $130 \leq m \leq 185$, the recovery percentage of the proposed method with $s = 5$ was higher than the proposed method with $s = 10$ and $s = 15$ and the StoIHT algorithm, except for the StoGradMP algorithm. In Figure 6d, we can see that the reconstruction percentage of the proposed method with $s = 1$ still had a higher recovery percentage than the other methods. When $105 \leq m \leq 153$, the reconstruction percentage of the proposed method with a random step-size was higher than the StoGradMP and StoIHT algorithms. For $110 \le m \le 150$, the reconstruction percentage of the proposed method with $s = 1$ and $s = 5$ was higher than the other methods. When $155 \le m \le 215$, the reconstruction percentage of the proposed method with $s = 5$ and $s = 10$ was superior to the StoIHT algorithm. When $245 \le m \le 270$, the reconstruction percentage of the StoIHT algorithm ranged from approximately 0% to 100%. When $m \ge 270$, all of the methods could achieve complete recovery. Overall, based on all of the sub-figures in Figure 6, we can see that the reconstruction performance of the proposed method with $s = 1$ was the best and the proposed method was more suitable for signal recovery under larger sparsity conditions.

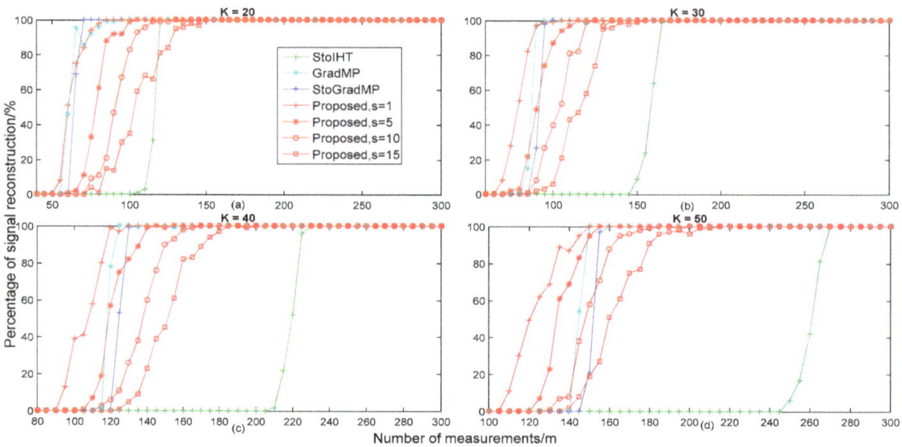

Figure 6. Reconstruction percentage of different algorithms with different measurements in different real sparsity K conditions ($n = 400$, $s \in [1, 5, 10, 15]$, $\delta_K = 0.1$ and $m = 2 * K : 5 : 300$, Gaussian signal).

Based on the above analysis, in a noise-free signal interference environment, the proposed method with $s = 1$ and $\delta_K = 0.1$ has a better recovery performance for different sparsity and measurements in comparison to other methods. Furthermore, the proposed method is more sensitive to larger sparsity signals. In other words, signals are more easily recovered in large sparsity environments.

In Figure 7, we compared the average runtime of different algorithms with different sparsities. From Figure 5a, we can see that the reconstruction percentage was 100% for the StoIHT algorithm with sparsity $L \in [20\ 28]$ and the real sparsity of $K = 20$ and for the GradMP, StoGradMP and the proposed method with $s = 1, 5, 10$ with $L \in [20\ 60]$. Therefore, we set the range of the assumed sparsity as $L \in [20\ 28]$ and $L \in [20\ 60]$ in Figure 7a, respectively. From Figure 7a, we can see that the average runtime of the proposed algorithm with $s = 5, 10$ was less than the StoGradMP algorithm, except for the proposed method $s = 1$.

From Figure 5b, we can see that the reconstruction percentage of all algorithms was 100% when the range of the assumed sparsity was $L \in [30\ 60]$ and the real sparsity was $K = 30$, except for the StoIHT and the proposed method with $s = 1, 5$. Therefore, the range of the assumed sparsity was set as $L \in [30\ 60]$ in the simulation of Figure 7b. From Figure 7b, we can see that the average runtime of the proposed method with $s = 5, 10$ was still lower than the StoGradMP algorithm.

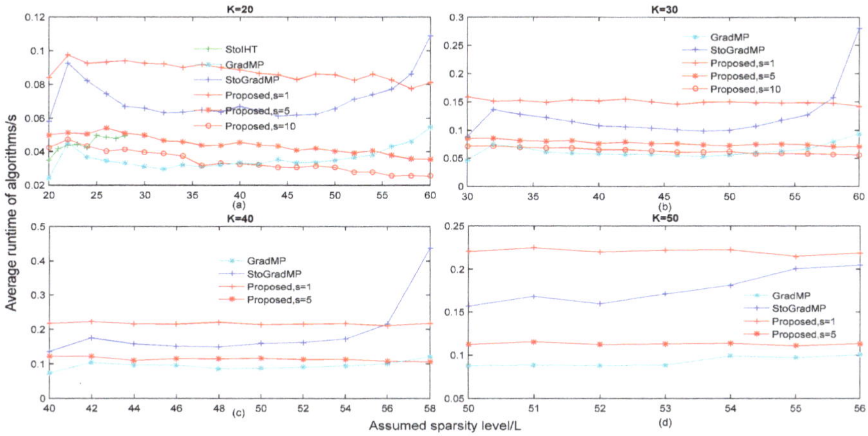

Figure 7. The average runtime of different algorithms with different sparsities in different sparsity conditions. ($n = 400$, $s \in [1, 5, 10, 15]$, $\delta_K = 0.1$ and $m = 170$, Gaussian signal).

From Figure 5c,d, we can see that the reconstruction percentage of all reconstruction methods was 100% when the assumed sparsity level was $L = \in [40\ 58]$ and $L = \in [50\ 56]$, respectively, except for the StoIHT and the proposed method with $s = 10$ and $s = 15$. Therefore, we set the range of the assumed sparsity as $L \in [40\ 58]$ and $L = \in [50\ 56]$ in the simulation of Figure 7c,d, respectively. From Figure 7c,d, we can see that the proposed algorithm with $s = 5$ had a shorter runtime than the StoGradMP algorithm. Although the proposed method with $s = 1$ had a longer runtime than the other method, it required less measurements to achieve the same reconstruction percentage as the others shown in Figure 6. Furthermore, from all sub-figures in Figure 7, we discovered that the average runtime of all algorithms increased when the assumed sparsity was gradually greater than the real sparsity, except for in the proposed method. This means that the inaccuracy of the sparsity estimation will increase the computational complexity of these algorithms. Meanwhile, it is indicated that the proposed method removes the dependence of the state-of-the-art algorithms on real sparsity and enhances the practical application capacity of the proposed algorithm.

In Figure 8, we compared the average runtime of different algorithms with different measurements in different real sparsity conditions. From Figure 6, for the different sparsity levels, we can see that all algorithms could achieve 100% reconstruction when the number of measurements was greater than 180, 200, 220 and 240, respectively, except for the StoIHT algorithm. Therefore, we set the range of measurements as $m \in [180\ 300]$, $m \in [200\ 300]$,$m \in [220\ 300]$, and $m \in [240\ 300]$ in Figure 8a–d, respectively. In particular, in Figure 6c,d, we can see that the reconstruction percentage was 100% when the number of measurements of the StoIHT algorithm was greater than 230 and 270, respectively. Therefore, we set the range of measurements as $m \in [230\ 300]$ and $m \in [270\ 300]$ in the simulation of Figure 8c,d, respectively.

From Figure 8, we can see that the GradMP algorithm had the lowest runtime, the next lowest were the StoIHT algorithm, the proposed algorithm with $s = 5, 10, 15$ and the StoGradMP algorithm. This means that the proposed method with $s = 5, 10, 15$ had a lower computational complexity than the StoGradMP algorithm, except for the GradMP and StoIHT algorithms. Meanwhile, in terms of the proposed algorithm, we can see that when the size of the step-size was $s = 15$, the average runtime was the shortest, the next shortest were the proposed method with $s = 10$, the proposed method with $s = 5$ and the proposed method with $s = 1$, respectively. This shows that a larger step-size will be beneficial to approach the real sparsity K of the original signal, thereby reducing the computational complexity of the proposed method. Furthermore, from Figures 6 and 8, although the proposed method with

$s = 1$ had the highest runtime, it could achieve reconstruction with fewer measurements than the other algorithms.

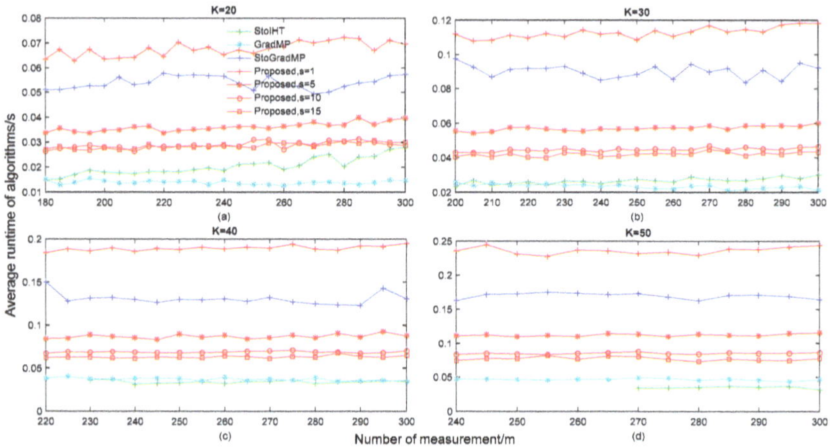

Figure 8. The average runtime of different algorithm with different measurements in different sparsity conditions ($n = 400$, $s \in [15, 10, 15]$, $\delta_K = 0.1$ and $m = 2 * K : 5 : 300$, Gaussian signal).

Based on the above analysis, in a noise-free signal interference environment, the proposed algorithm had a lower computational complexity with a larger step-size than a smaller step-size. Although, the proposed method had a higher computational complexity than some existing algorithms in some conditions, it is more suitable for applications without knowing the sparsity information.

In Figure 9, we compared the average mean square error of different algorithms with different SNR levels in different real sparsity conditions to better analyze the reconstruction performance of the different algorithms when the original sparse signal was corrupted with different levels of noise. We set the range of the noise signal level as $SNR = 10 : 5 : 50$ in simulation of Figure 9. Furthermore, to better analyze the reconstruction performance of all reconstruction algorithms in different real sparsity levels conditions, we set the real sparsity level K as 20, 30, 40 and 50, respectively. Here, the noise signal was a Gaussian white noise signal. In particular, all of the experimental parameters were consistent with Figure 5. In Figure 5a,b, the reconstruction percentage of the proposed method was 100% with a step-size of $s = 1, 5, 10$. Therefore, we set the size of s as 1, 5 and 10 in the simulation of Figure 9a,b, respectively. In Figure 5c,d, the reconstruction percentage of the proposed method was 100% with a step-size of $s = 1, 5$. Thus, we set the size of the step-size of the proposed method as 1 and 5 in Figure 9c,d, respectively.

From Figure 9, we can see that the proposed methods with different step-sizes had a higher error than other algorithms for different SNR levels. This is because the proposed methods supposed that the sparsity prior information of the source signal was unknown, while the other methods used the real sparsity as prior information. The estimated sparsity by our proposed method was still different to the real sparsity. This made the proposed method have a higher error than the others. In particular, the error was very small for all algorithms with a larger SNR, which had little effect on the reconstruction signal. Although the proposed method was inferior to other algorithms in terms of reconstruction performance when the original sparse signal was corrupted by different levels of noise, it provides a reconstruction scheme that is more suitable for practical applications. In this paper, we mainly focused on the no noise environments. Recently, in Reference [33–36], the researchers focused on the reconstruction solutions for the original signal in the presence of noise corruption and several algorithms were proposed. In the future, we can use their ideas to improve our proposed method in anti-noise interference performance.

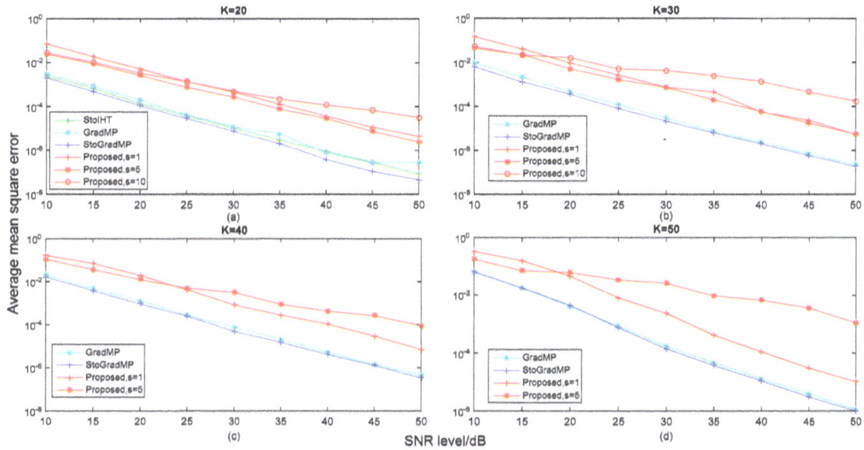

Figure 9. The average mean square error of different algorithms with different SNR levels in different real sparsity conditions ($n = 400$, $s \in [1, 5, 10]$, $\delta_K = 0.1$ and $m = 170$, $SNR = 10 : 5 : 50$, Gaussian signal).

In Figure 10, we test the application efficiency of our proposed method in remote sensing image compressing and reconstructing. The Figure 10a–d show the original remote sensing image, its sparse coefficient, compressed image(observation signal) and reconstructed image by our proposed method. By comparing the Figure 11a with Figure 10d, we can see that our proposed method reconstructs the compressed remote sensing image successfully.

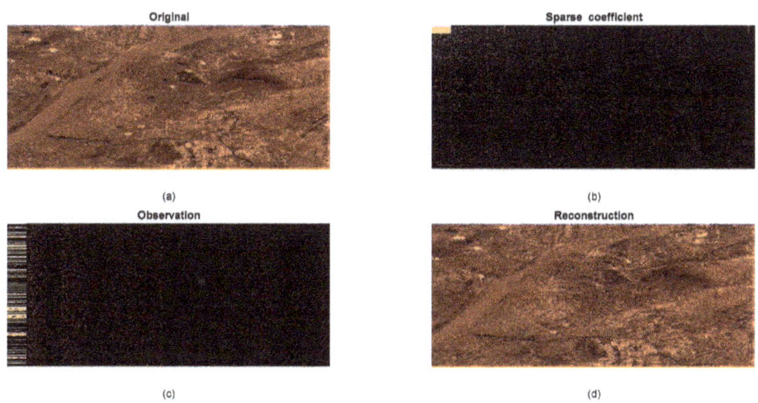

Figure 10. Application in remote sensing image compressing and reconstructing with our proposed method.

In Figure 11, we test the efficiency of our proposed method in application of power quality signal compressing and reconstruction. The Figure 11a–c show the inter-harmonic signal, compressed signal (observation signal) and reconstructed inter-harmonic signal by our proposed method respectively. It can be seen from Figure 11a,c that the waveforms of two figures are basically the same. This proves that our proposed method is efficiency for inter-harmonic reconstruction.

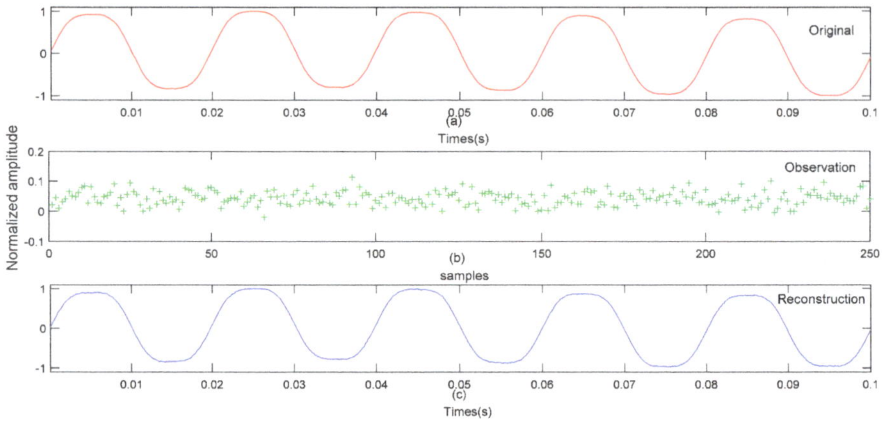

Figure 11. Application in power quality signal compressing and reconstructing with our proposed method.

We also used the National Instruments PXI (peripheral component interconnect extensions for instrumentation) system to test the efficiency of our proposed method in application. The hardware of the PXI system includes an arbitrary waveform and signal generator and oscilloscopes. The hardware architecture of arbitrary waveform and signal generator is shown in Figure 12. The hardware architecture of oscilloscopes is shown in Figure 13. Figure 14 shows the PXI chassis and controller, which are used to control the arbitrary waveform and signal generator and oscilloscopes. We insert the arbitrary waveform and signal generator and oscilloscopes into PXI chassis to construct the complete measurement device. As is shown in Figure 15. Mixed programming of Labview and MATLAB were used to realize the compressed and reconstructed algorithm. From the experimental results, it can be seen that the proposed method successfully reconstructed the source signal from the compressed signal.

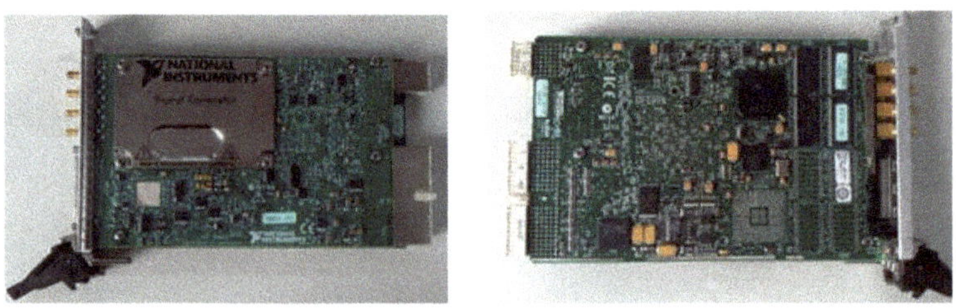

Figure 12. Hardware architecture of arbitrary waveform and signal generator.

Figure 13. Hardware architecture of oscilloscopes.

Figure 14. Peripheral component interconnect extensions for instrumentation chassis and controller.

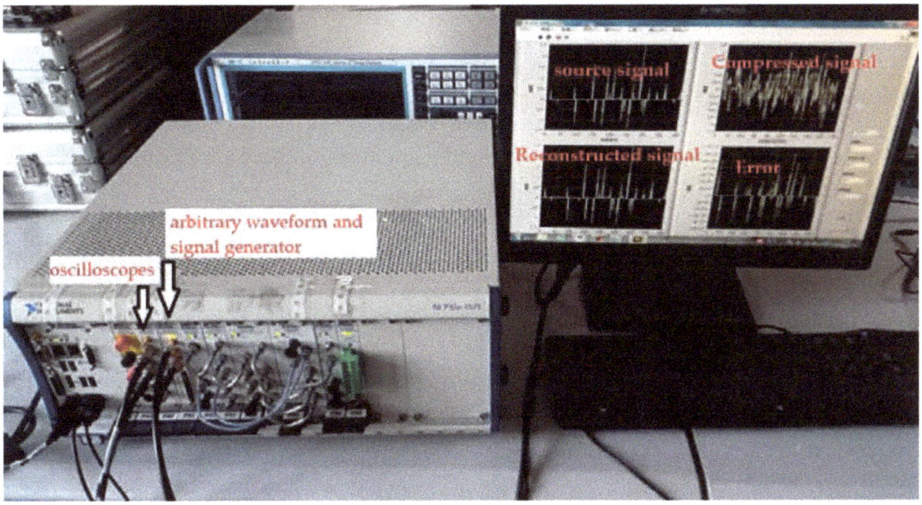

Figure 15. Measurement device for real applications.

7. Conclusions

This paper proposed a new recovery method. This method first utilized the sparsity pre-evaluation strategy to estimate the real sparsity of the original signal and used the estimated sparsity as the length of the support set in the initial stage, which allows the proposed method to eliminate the dependency of sparsity, thereby reducing the computational complexity of the proposed method. The proposed algorithm then adopts the adjustment strategy of sparsity estimation to control the convergence of the proposed method and adjust the estimated sparsity, which makes the proposed method more accurately approach the real sparsity of the original signal. Furthermore, a reliability verification condition was added to ensure the correctness and effectiveness of the proposed method. The proposed method not only solved the problem of the sparsity estimation of the original signal but also improved the recovery performance of the practical applications of the proposed method. The simulation results proved that the proposed method performed better than other stochastic greedy pursuit methods in larger sparsity environments and smaller step-sizes.

Author Contributions: L.Z. and Y.H. were responsible for the overall work and proposed the idea and experiments of the proposed algorithm in the paper, and the paper was written mainly by the two authors. Y.H. built the simulation program and performed part of the simulation experiments and contributed to many effective discussions in both ideas and simulation design. Y.L. performed part of the simulation experiments and provided some positive technical suggestions for the paper.

Acknowledgments: This work was supported by the National Natural Science Foundation of China (61271115) and the Foundation of Jilin Educational Committee (2015235).

Conflicts of Interest: The authors declare no conflicts of interest.

References

1. Vehkaperä, M.; Kabashima, Y.; Chatterjee, S. Analysis of Regularized LS Reconstruction and Random Matrix Ensembles in Compressed Sensing. *IEEE Trans. Inf. Theory* **2016**, *62*, 2100–2124. [CrossRef]
2. Laue, H.E.A. Demystifying Compressive Sensing [Lecture Notes]. *IEEE Signal Process. Mag.* **2017**, *34*, 171–176. [CrossRef]
3. Arjoune, Y.; Kaabouch, N.; El Ghazi, H.; Tamtaoui, A. Compressive sensing: Performance comparison of sparse recovery algorithms. In Proceedings of the 2017 IEEE 7th Annual Computing and Communication Workshop and Conference (CCWC), Las Vegas, NV, USA, 9–11 January 2017; pp. 1–7.
4. Liu, J.K.; Du, X.L. A gradient projection method for the sparse signal reconstruction in compressive sensing. *Appl. Anal.* **2018**, *97*, 2122–2131. [CrossRef]
5. Wang, Q.; Qu, G. Restricted isometry constant improvement based on a singular value decomposition-weighted measurement matrix for compressed sensing. *IET Commun.* **2017**, *11*, 1706–1718. [CrossRef]
6. Lopes, M.E. Unknown Sparsity in Compressed Sensing: Denoising and Inference. *IEEE Trans. Inf. Theory* **2016**, *62*, 5145–5166. [CrossRef]
7. Guo, J.; Song, B.; He, Y.; Yu, F.R.; Sookhak, M. A Survey on Compressed Sensing in Vehicular Infotainment Systems. *IEEE Commun. Surv. Tutor.* **2017**, *19*, 2662–2680. [CrossRef]
8. Chen, W.; You, J.; Chen, B.; Pan, B.; Li, L.; Pomeroy, M.; Liang, Z. A sparse representation and dictionary learning based algorithm for image restoration in the presence of Rician noise. *Neurocomputing.* **2018**, *286*, 130–140. [CrossRef]
9. Li, K.; Chandrasekera, T.C.; Li, Y.; Holland, D.J. A nonlinear reweighted total variation image reconstruction algorithm for electrical capacitance tomography. *IEEE Sens. J.* **2018**, *18*, 5049–5057. [CrossRef]
10. He, Q.; Song, H.; Ding, X. Sparse signal reconstruction based on time–frequency manifold for rolling element bearing fault signature enhancement. *IEEE Trans. Instrum. Meas.* **2016**, *65*, 482–491. [CrossRef]
11. Schnas, K. Average performance of Orthogonal Matching Pursuit (OMP) for sparse approximation. *IEEE Signal Process. Lett.* **2018**, *25*, 1865–1869. [CrossRef]
12. Meena, V.; Abhilash, G. Robust recovery algorithm for compressed sensing in the presence of noise. *IET Signal Process.* **2016**, *10*, 227–236. [CrossRef]
13. Pei, L.; Jiang, H.; Li, M. Weighted double-backtracking matching pursuit for block-sparse reconstruction. *IET Signal Process.* **2016**, *10*, 930–935. [CrossRef]
14. Fu, W.; Chen, J.; Yang, B. Source recovery of underdetermined blind source separation based on SCMP algorithm. *IET Signal Process.* **2017**, *11*, 877–883. [CrossRef]
15. Satpathi, S.; Chakraborty, M. On the number of iterations for convergence of CoSaMP and Subspace Pursuit algorithms. *Appl. Comput. Harmon. Anal.* **2017**, *43*, 568–576. [CrossRef]
16. Golbabaee, M.; Davies, M.E. Inexact gradient projection and fast data driven compressed sensing. *IEEE Trans. Inf. Theory* **2018**, *64*, 6707–6721. [CrossRef]
17. Gao, Y.; Chen, Y.; Ma, Y. Sparse-bayesian-learning-based wideband spectrum sensing with simplified modulated eideband converter. *IEEE Access* **2018**, *6*, 6058–6070. [CrossRef]
18. Lin, Y.; Chen, Y.; Huang, N.; Wu, A. Low-complexity stochastic gradient pursuit algorithm and architecture for robust compressive sensing reconstruction. *IEEE Trans. Signal Process.* **2017**, *65*, 638–650. [CrossRef]
19. Mamandipoor, B.; Ramasamy, D.; Madhow, U. Newtonized orthogonal matching pursuit: Frequency estimation over the continuum. *IEEE Trans. Signal Process.* **2016**, *64*, 5066–5081. [CrossRef]
20. Rakotomamonjy, A.; Flamary, R.; Gasso, G. DC proximal Newton for Non-convex optimization problems. *IEEE Trans. Neural Netw. Learn. Syst.* **2016**, *27*, 636–647. [CrossRef]

21. Chou, C.; Chang, E.; Li, H.; Wu, A. Low-Complexity Privacy-Preserving Compressive Analysis Using Subspace-Based Dictionary for ECG Telemonitoring System. *IEEE Trans. Biomed. Circuits Syst.* **2018**, *12*, 801–811.
22. Bonettini, S.; Prato, M.; Rebegoldi, S. A block coordinate variable metric linesearch based proximal gradient method. *Comput. Optim. Appl.* **2018**, *71*, 5–52. [CrossRef]
23. Rani, M.; Dhok, S.B.; Deshmukh, R.B. A systematic review of compressive sensing: Concepts, implementations and applications. *IEEE Access* **2018**, *6*, 4875–4894. [CrossRef]
24. Nguyen, N.; Needell, D.; Woolf, T. Linear convergence of stochastic iterative greedy algorithms with sparse constraints. *IEEE Trans. Inf. Theory* **2017**, *63*, 6869–6895. [CrossRef]
25. Tsinos, C.G.; Berberidis, K. Spectrum Sensing in Multi-antenna Cognitive Radio Systems via Distributed Subspace Tracking Techniques. In *Handbook of Cognitive Radio*; Springer: Singapore, 2017; pp. 1–32.
26. Tsinos, C.G.; Rontogiannis, A.A.; Berberidis, K. Distributed Blind Hyperspectral Unmixing via Joint Sparsity and Low-Rank Constrained Non-Negative Matrix Factorization. *IEEE Trans. Comput. Imaging* **2017**, *3*, 160–174. [CrossRef]
27. Li, H.; Zhang, J.; Zou, J. Improving the bound on the restricted isometry property constant in multiple orthogonal least squares. *IET Signal Process.* **2018**, *12*, 666–671. [CrossRef]
28. Wang, J.; Li, P. Recovery of Sparse Signals Using Multiple Orthogonal Least Squares. *IEEE Trans. Signal Process.* **2017**, *65*, 2049–2062. [CrossRef]
29. Wang, J.; Kwon, S.; Li, P.; Shim, B. Recovery of sparse signals via generalized orthogonal matching pursuit: A new analysis. *IEEE Trans. Signal Process.* **2016**, *64*, 1076–1089. [CrossRef]
30. Soltani, M.; Hegde, C. Fast algorithms for de-mixing sparse signals from nonlinear observations. *IEEE Trans. Signal Process.* **2017**, *65*, 4209–4222. [CrossRef]
31. Li, H.; Liu, G. Perturbation analysis of signal space fast iterative hard thresholding with redundant dictionaries. *IET Signal Process.* **2017**, *11*, 462–468. [CrossRef]
32. Rakotomamonj, A.; Koço, S.; Ralaivola, L. Greedy Methods, Randomization Approaches and Multiarm Bandit Algorithms for Efficient Sparsity-Constrained Optimization. *IEEE Trans. Neural Netw. Learn. Syst.* **2017**, *28*, 2789–2802. [CrossRef]
33. Srimanta, M.; Bhavsar, A.; Sao, A.K. Noise Adaptive Super-Resolution from Single Image via Non-Local Mean and Sparse Representation. *Signal Process.* **2017**, *132*, 134–149.
34. Dziwoki, G. Averaged properties of the residual error in sparse signal reconstruction. *IEEE Signal Process. Lett.* **2016**, *23*, 1170–1173. [CrossRef]
35. Stanković, L.; Daković, M.; Vujović, S. Reconstruction of sparse signals in impulsive disturbance environments. *Circuits Syst. Signal. Process.* **2016**, *36*, 767–794. [CrossRef]
36. Metzler, C.A.; Maleki, A.; Baraniuk, R.G. From denoising to compressed sensing. *IEEE Trans. Inf. Theory* **2016**, *62*, 5117–5144. [CrossRef]

electronics

MDPI

Article

Development of a Miniaturized Frequency Standard Comparator Based on FPGA

Sheng Tang [1], Jing Ke [2,*], Tianxiang Wang [1] and Zhouhu Deng [1]

1 School of Information Science and Technology, Northwest University, Xi'an 710127, China; tangsheng@nwu.edu.cn (S.T.); wangtianxiangnwu@hotmail.com (T.W.); dengzh@nwu.edu.cn (Z.D.)

2 National Time Service Center, Chinese Academy of Sciences, Xi'an 710600, China

* Correspondence: kejing@ntsc.ac.cn; Tel.: +86-29-8389-4565

Received: 30 December 2018; Accepted: 19 January 2019; Published: 23 January 2019

Abstract: Frequency standard comparison measurement has important practical significance for the rational use of frequency standard in engineering. This paper was devoted to the study of frequency standard comparison measurement based on classical dual mixing time difference method. However, in the actual system design and implementation, the commonly used counter was discarded and the phase difference was measured by a digital signal processing method based on Field Programmable Gate Array (FPGA). A miniaturized 10 MHz frequency standard comparator with good noise floor was successfully developed. The size of the prototype circuit board is only about 292.1 cm^2. The experimental results showed that the noise floor of the frequency standard comparator was typically better than $7.50 \times 10^{-12}/s$, and its relative error of phase difference measurement was less than 1.70×10^{-5}.

Keywords: frequency standard comparator; dual Mixing Time difference; phase difference; correlation function; chebyshev polynomial

1. Introduction

A frequency standard is a device that can provide sinusoidal signal with high accuracy and stability, and its frequency value is usually 10 MHz, although in some cases it can be 5 MHz or 1 MHz [1–4]. It is undeniable that any specific device that generates standard values is not absolutely stable, and the frequency standard is no exception. Under the influences of internal and external factors, the frequency standard output will change slowly and eventually lead to the failure of its standard reference function [5–10]. In this case, it will need to be calibrated. The calibration of frequency standards is performed by high precision frequency standard comparison measurement. A frequency standard comparison measurement is conducted to measure and evaluate the accuracy and stability of frequency standard, which has important practical significance for the rational use of frequency standard in engineering [11–17].

Nowadays, commonly used frequency standard comparison methods include oscilloscope method, time interval counting method, beat frequency method, etc. Among these methods, the oscilloscope method is the simplest one. The oscilloscope can graphically display the frequency difference relationship between two sinusoidal signals. When the frequencies of the two standard frequencies in the comparison are strictly equal, a fixed Lissajous-Figure will be displayed on the screen of the oscilloscope. If there is a difference between the two frequencies, the Lissajous-Figure will move relatively on the oscilloscope display screen. The length of time consumed by the period of Lissajous-Figure's movement will reflect the frequency difference between two frequency standard signals. A stopwatch is usually used to measure the accuracy and stability of the frequency standard indirectly in this measurement method. In order to reduce the human-controlled error of stopwatch, the measuring time can be prolonged appropriately. Many Lissajous-Figure movement cycles can be included in one measurement. Consequently, the human-controlled error can be weakened relatively, and the ultimate measurement error can be reduced. Usually, it takes a long time to realize high precision frequency standard comparison measurement, thus, this method is not suitable for short-term stability measurement of frequency standard. The basic principle of the time interval counting method is that the frequency to be measured and the reference frequency are both shaped into square waves by a voltage comparator, and then, the time difference between them is measured by the time interval counter. The measurement accuracy of this method is determined by the measurement ability of the time interval counter. The internal time scale error and trigger error of the counter itself will directly reflect the measurement error of this method. The beat frequency method is a classical method that can obtain a high measurement resolution by using a common counter. Its core technology is down-conversion, that is to say, mixing the frequency to be measured and the reference frequency to obtain the frequency difference signal (also known as the beat signal) of the frequency to be measured relative to the reference frequency. Because the frequency of the beat signal after mixing is relatively low, the cycle of the beat signal can be counted and measured by common counter. Because the frequency value of the beat signal is much less than the nominal value of the frequency to be measured, compared with the direct measurement of the frequency to be measured, this method can greatly improve the measurement resolution. To be exact, this method can improve the resolution of frequency measurement system by a multiple of beat factor. However, the beat frequency measurement method also has its drawbacks. For example, this method requires that the reference frequency stability be higher than the frequency to be measured. In addition, the internal time scale error and trigger error of the counter used for measurement can also directly reflect the measurement error of this measurement method.

Another frequency standard comparison method that has to be mentioned is the DMTD (Dual Mixing Time Difference) method [18,19]. This method combines the advantages of the time interval counting method and the beat frequency method. It down-converts the frequency to be measured and the reference frequency to two low frequency beat signals at the same time. Then, the time difference of the two low frequency beat signals is measured by a time interval counter. The frequency to be measured and the reference frequency in the system have the same nominal value, and there is a frequency deviation between the common oscillator and the frequency to be measured. The measurement resolution of the DMTD system is usually determined by the resolution of the time interval counter and the size of the beat factor. The DMTD method is one of the most accurate methods to realize the comparison measurement between two frequency standards. The implementation of this method requires that the parameters of dual-channel devices should be as similar as possible, so that the common errors of the system can be well offset. Additionally, this method does not require a high stability of the common oscillator, because the error effect of the common oscillator will be mostly offset in the double balanced measurement.

Due to its higher measurement resolution, the DMTD method is adopted by many excellent commercial frequency standard comparison products on sale, such as Timetech's Phase-comp, Symmetricom's MMS (Multi-channel Measurement System) and so on. These instruments or systems based on the classical DMTD method have achieved high measurement accuracy. However, they are generally large in size and lack of portability. Even some instruments or systems must use computers with pre-installed high-precision data acquisition cards. High prices are also a common feature of them. Meanwhile, the counter is still used in some DMTD measurement system. The counter itself has some measurement errors, which can directly affect the measurement errors of this method, for example, the ± 1 counting error.

This paper focused on the study of frequency standard comparison measurement method based on DMTD, tried to displace the counter and introduce digital signal processing into the classical DMTD method and developed a high precision and miniaturized frequency standard comparator.

2. Frequency Standard Comparator Using Modified DMTD

The structural principle of a frequency standard comparator based on traditional DMTD method is shown in Figure 1. It mixes the frequency to be measured and the reference frequency with the common oscillator respectively at the same time. Then the beat signals were filtered, amplified and reshaped to form two square wave signals. Finally, the time interval counter was used to measure the last phase difference.

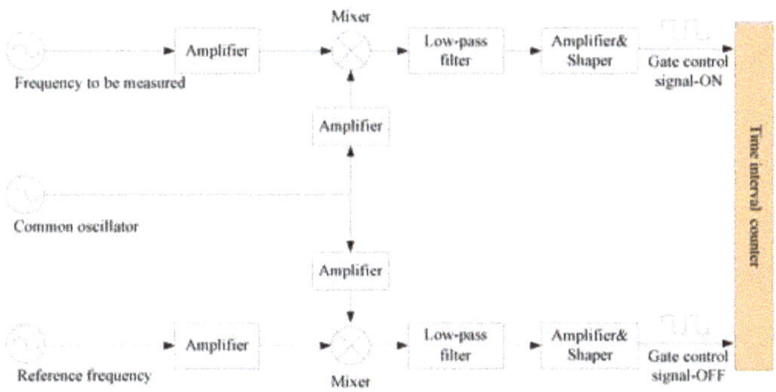

Figure 1. Block diagram of frequency standard comparator using traditional Dual Mixing Time Difference (DMTD).

However, the result of the phase difference measurement by a counter is not very accurate. For example, the ±1 counting error is inevitable in the process of phase difference counting measurement. The reason for the ±1 counting error is the uncertainty of the relative displacement between the counting pulse signal and the gate signal of the counter. As shown in the Figure 2, there are two gate signals with the same length of time t_m, A and B. The counting with gate switching B can get a count of two, and the counting with gate switching A can get a count of only one.

Figure 2. Counter's ±1 counting error.

In this paper, a modified method of the frequency standard comparison measurement based on classical DMTD method was proposed, and a 10 MHz frequency standard comparator was developed. The comparator mainly consisted of a 9.9999 MHz common oscillator, a frequency down-beater that output 100 Hz sinusoidal signals, a dual-channel data simultaneous acquisition module and a digital signal processing module. Unlike the traditional DMTD measurement system, the last unit of the comparator abandoned the counter and replaced it with a digital signal processing module. Two sinusoidal beat signals were sampled at the same time, and then, the phase difference between the frequency to be measured and the reference frequency was calculated by digital signal processing. After down-conversion of the frequency beater, the resolution of phase difference measurement was increased by a multiple of the beat factor. The frequency to be measured and the reference frequency were mixed with the same 9.9999 MHz common oscillator. The symmetrical circuit structure determined that the two sinusoidal beat signals were disturbed by roughly the same noise. The effect of noise from electronic components in two channels on measurement results can be largely offset by subsequent digital correlation processing, thus, it is expected to achieve a good measurement performance.

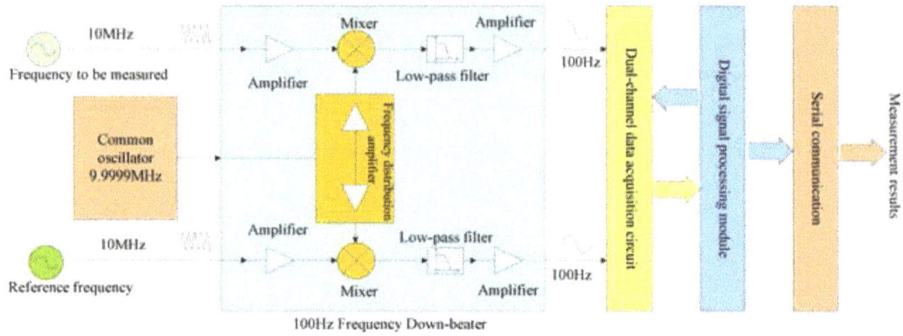

Figure 3. Structural block diagram of frequency standard comparator using modified DMTD.

Just as shown in Figure 3, the digital signal processing module of the frequency standard comparator was designed based on Field Programmable Gate Array (FPGA). With the development of microprocessors and large scale integrated circuits, digital measurement methods show more advantages, such as higher accuracy, smaller volume, lower cost, better flexibility, etc. [20–25]. The realization of signal processing algorithms in digital measurement mostly depends on the platform of computer, MCU (Microcontroller Unit), DSP (Digital Signal Processor) or FPGA (Field Programmable Gate Array). At present, the computer platforms with multi-core processors usually do not have the problem of insufficient computing speed when implementing large data volume algorithms, however, it is difficult to meet the needs of portability and miniaturization of measurement system. The comparatively smaller measurement systems usually use MCU or DSP to implement data processing, where algorithm instructions are usually executed sequentially within their CPU (Central Processing Unit), and the consequent speed bottleneck is inevitable. The application of modern high-speed and large-capacity FPGA is expected to overcome the shortcomings of the above technical solutions. In our design, besides the task of digital signal processing, the FPGA also took into account the functions of controlling data acquisition and output measurement results [26–30].

2.1. Common Oscillator

The photograph of the common oscillator is shown in Figure 4. The core device was an oven controlled crystal oscillator MV200, of which the nominal frequency was 10 MHz. The operating voltage of the common oscillator module is +12 V. Three resistors and one precision potentiometer VR (the blue block devices on the right side of circuit board in Figure 4b) provided an accurate bias for MV200. By adjusting the potentiometer, the output of the common oscillator SMA-0 (SMA-1 as a backup output, which can be used as a test point) could be stable at 9.999999 MHz. Figure 4a is the PCB (Printed Circuit Board) photograph of the common oscillator designed in this paper, and Figure 4b is its physical photograph.

(a) (b)

Figure 4. Photograph of the common oscillator. (**a**) PCB photograph; (**b**) Physical photograph.

2.2. Frequency Down-Beater

The frequency down-beater mainly consisted of one frequency distribution amplifier and two mixers. Its function was to down-convert the nominal 10 MHz frequencies to 100 Hz low frequency sinusoidal waves. In this paper, two OPA (Operational Amplifier) chips LMH6609 were selected to build an active frequency distribution amplifier with "one in two out", which realized the function of dividing one 9.9999 MHz signal into two without loss and sending them to the mixers, respectively. Additionally, the four quadrant multiplier AD835 was used in the mixers design. Figure 5a is the PCB photograph of the frequency down-beater designed in this paper, and Figure 5b is its physical photograph.

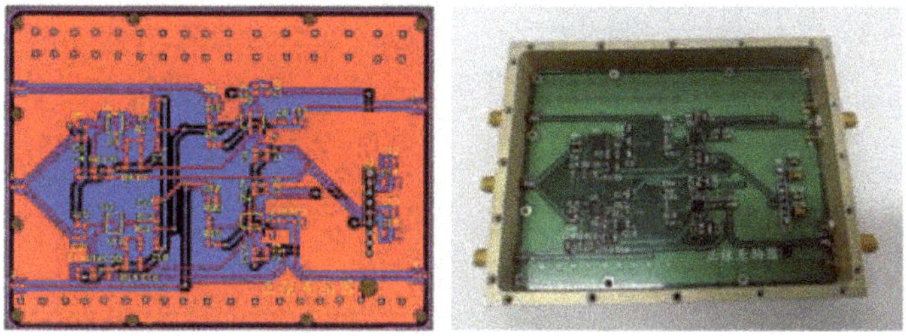

(a) (b)

Figure 5. Photograph of the frequency down-beater. (**a**) PCB photograph; (**b**) Physical photograph.

When the output of the common oscillator is 9.9999 MHz and 2.88 Vpp, and the signals to be measured and the reference signal are10 MHz and 3.20 Vpp, the actual outputs of the frequency down-beater were as shown in Figure 6. The frequency of the two sinusoidal beat signals was 100 Hz, and the peak to peak voltage was about 2.30 Vpp.

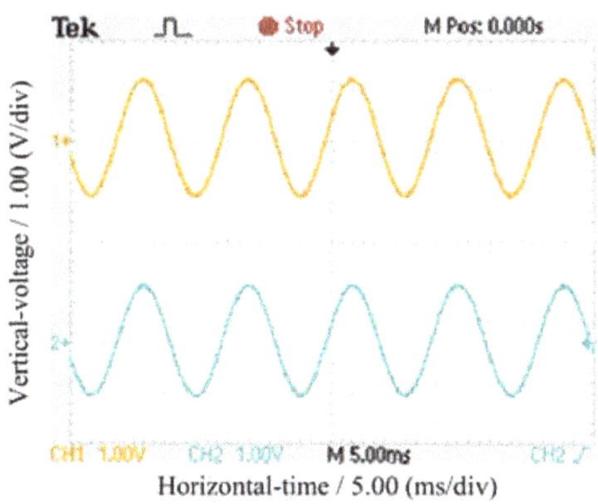

Figure 6. Oscilloscope measurement of sinusoidal beat signal (by Tektronix TBS1102, Beaverton, Oregon, USA).

2.3. Circuit Module of Signal Sampling, Processing and Transmission

The dual-channel signal sampling circuit proposed in this paper was implemented based on ADS8364 (Texas Instruments Incorporated, Dallas, USA). The non-simultaneous sampling error of ADS8364 was mainly determined by the aperture jitter of the device itself. Aperture jitter was the sampling signal phase error caused by the delay uncertainty of sampling and holding switch in AD converter. Referring to the official data sheet, the typical value of the aperture jitter of ADS8364 was 50 ps. That is to say, the phase error caused by the aperture jitter of ADS8364 was about

1.8×10^{-6} degrees, which can be neglected in the digital measurement of the phase difference between 100 Hz DMTD beat signals.

In our design, the digital signal processing module was based on the Hurricane Series FPGA (EP1C12Q240, Altera Company, San Jose, CA, USA). Additionally, the frequency standard comparator used asynchronous serial communication to output the measurement results. The integrated layout of dual-channel data simultaneous sampling circuit, FPGA circuit and serial communication circuit is shown in Figure 7a, and the corresponding circuit is shown in Figure 7b.

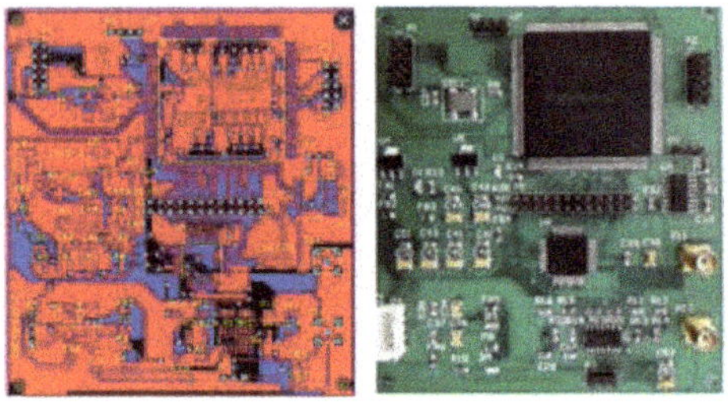

(a) (b)

Figure 7. Photograph of the circuit module of data acquisition, processing and transmission. (**a**) PCB photograph; (**b**) Physical photograph.

3. Phase Difference Measurement by Correlation Method Based on FPGA

The phase difference measurement method based on correlation operation was a digital phase difference measurement method, which is widely used in telecommunication, geological exploration, power distribution, aerospace and many other fields. Suppose there are two sinusoidal signals:

$$x(t) = A\sin 2\pi ft + N_x(t) \tag{1}$$

$$y(t) = B\sin(2\pi ft + \Delta\varphi) + N_y(t) \tag{2}$$

where A and B represent the amplitudes of the two sinusoidal signals $x(t)$ and $y(t)$, respectively, $N_x(t)$ and $N_y(t)$ are the noise signals superimposed on $x(t)$ and $y(t)$, respectively, and $\triangle\varphi$ is the phase difference between two sinusoidal signals.

In the time T (integer multiple of the signal period), the correlation operation on $x(t)$ and $y(t)$ is:

$$\begin{aligned} R_{xy}(\tau) &= \tfrac{1}{T}\int_0^T x(t)y(t+\tau)dt \\ &= \tfrac{1}{T}\int_0^T [A\sin 2\pi ft + N_x(t)] \times \left[B\sin(2\pi f(t+\tau) + \Delta\varphi) + N_y(t+\tau)\right]dt \end{aligned} \tag{3}$$

Equation (3) shows that the delay amount τ affects the cross-correlation function value of the signals $x(t)$ and $y(t)$. When $\tau = 0$, the phase difference $\triangle\varphi$ between $x(t)$ and $y(t)$ is related to the value of the cross-correlation function $R_{xy}(0)$. Since there is usually no correlation between noise and noise, and there is usually no correlation between signal and noise too, if $\tau = 0$, Equation (3) can be simplified as:

$$R_{xy}(0) = \frac{1}{T}\int_0^T [A\sin 2\pi ft \times B\sin(2\pi ft + \Delta\varphi)]dt = \frac{A \cdot B \cdot \cos\Delta\varphi}{2} \tag{4}$$

Then, the phase difference between $x(t)$ and $y(t)$:

$$\Delta\varphi = 2k\pi + \arccos\left(\frac{2R_{xy}(0)}{A \cdot B}\right) \tag{5}$$

where, k = 0, 1, 2 ... If two sinusoidal signals have the same nominal frequency value, then k = 0. Moreover, because the relationships between the two sinusoidal autocorrelation functions and their phases are $A = \sqrt{2R_{xx}(0)}$ and $B = \sqrt{2R_{yy}(0)}$, then:

$$\Delta\phi = \arccos\left(\frac{2R_{xy}(0)}{\sqrt{2R_{xx}(0)} \cdot \sqrt{2R_{yy}(0)}}\right) = \arccos\left(\frac{R_{xy}(0)}{\sqrt{R_{xx}(0)} \cdot \sqrt{R_{yy}(0)}}\right) \tag{6}$$

It can be seen from Equation (6) that the phase difference between two sinusoidal signals can be solved by calculating their autocorrelation values and cross-correlation values from the sampled values of them.

In recent years, many scholars have done research on the theory of phase difference measurement based on digital correlation method, and proposed various improved algorithms. In this paper, the phase difference measurement method based on correlation operation was also improved to make it suitable for FPGA implementation. Although FPGA has the advantages of high-speed parallel processing compared with MCU and DSP, it is undeniable that it is less flexible in numerical calculations, especially for arithmetic processing with signed numbers. In order to make our calculation method universally applicable, that is to say, the algorithm could be easily programmed and implemented in both high and low versions of Verilog language, in the system design, the sampling data of the two sinusoidal signals to be measured were simultaneously shifted up by half a peak-to-peak value a, thereby changing the operation of measuring the phase difference of the entire correlation method into an unsigned operation. Therefore, the two sinusoidal signals to be measured (corresponding to the two 100 Hz signals output by the sinusoidal frequency beater) become:

$$x(t) = A\sin 2\pi ft + N_x(t) + a \tag{7}$$

$$y(t) = B\sin(2\pi ft + \Delta\varphi) + N_y(t) + a \tag{8}$$

The autocorrelation coefficients of the above two signals can be calculated as:

$$R_{xx}(0) = \frac{A^2}{2} + a^2 \tag{9}$$

$$R_{yy}(0) = \frac{B^2}{2} + a^2 \tag{10}$$

Correspondingly, the relationship between the amplitudes of the two new sinusoidal beat signals and their autocorrelation coefficients are:

$$A = \sqrt{2R_{xx}(0) - a^2},\ B = \sqrt{2R_{yy}(0) - a^2} \tag{11}$$

The correlation coefficients between two signals can be derived:

$$R_{xy}(0) = \frac{A \cdot B \cdot \cos\Delta\varphi}{2} + a^2 \tag{12}$$

Then, the phase difference between $x(t)$ and $y(t)$ is:

$$\Delta\varphi = \arccos\left(\frac{R_{xy}(0) - a^2}{\sqrt{R_{xx}(0) - a^2} \cdot \sqrt{R_{yy}(0) - a^2}}\right) \tag{13}$$

Because:

$$\tan\varphi = \frac{\sqrt{1 - \cos^2\varphi}}{\cos\varphi} \tag{14}$$

Substituting the Equation (14) into (13), the second expression of the phase difference formula can be obtained, just as shown as:

$$\Delta\varphi = \arctan\left(\frac{\sqrt{(R_{xx}(0)-a^2)(R_{yy}(0)-a^2)-(R_{xy}(0)-a^2)^2}}{R_{xy}(0)-a^2}\right) \tag{15}$$

It can be known from Equations (13) and (15) that the core task of measuring phase difference based on the FPGA correlation method is to implement inverse trigonometric function converter based on Verilog hardware description language. In this paper, Chebyshev polynomial was used to approximate the arctangent function shown in Equation (15), and the calculation of the function was reduced to the form of accumulating polynomial with coefficients, which is implemented in FPGA by iterative algorithm.

Chebyshev polynomial $\{Tn(x) = \cos(n\arccos x) = \cos n\theta\}_{n=0}^{\infty}$ is an orthogonal polynomial group with a weight function on [-1,1], which can be expressed as:

$$\begin{aligned}\left(Tn, Tm\right) &= \int_{-1}^{1} \frac{1}{\sqrt{1-x^2}} Tn(x)Tm(x)dx \overset{x=\cos\theta}{=} \cos(n\theta)\cos(m\theta)d\theta \\ &= \begin{cases} 0, m \neq n \\ \pi/2, m = n \neq 0 \\ \pi, m = n = 0 \end{cases}\end{aligned} \tag{16}$$

The recursive formula of Chebyshev is:

$$\begin{cases} T0(x) = 1 \\ T1(x) = x \\ \ldots \\ Tk+1(x) = 2xTk(x) - Tk-1(x), k = 1,2,\ldots x \in [-1,1] \end{cases} \tag{17}$$

The expression that approximates the function to be implemented using the Chebyshev polynomial is:

$$f(x) \approx \left[\sum_{n=0}^{N-1} cnTn(x)\right] - \frac{c0}{2} \tag{18}$$

where the Chebyshev coefficient is:

$$cn = \frac{2}{N}\sum_{n=0}^{N-1} f\left[\cos\left(\frac{\pi(k+1/2)}{N}\right)\right]\cos\left(\frac{\pi n(k+1/2)}{N}\right) \tag{19}$$

Then:

$$f(x) = \arctan x = \left[\sum_{n=0}^{N-1} cnTn(x)\right] - \frac{c0}{2} \tag{20}$$

Since the domain of the Chebyshev function is [-1, 1], the calculation formula outside this range is calculated as:

$$\arctan x = \pi/2 - \arctan(1/x), x < 1 or x > 1 \tag{21}$$

It can be seen from Chebyshev's recursive Equation (17) that as the number of iterations increases, it will cause too many multiplications and addition, which leads to the algorithm being too complicated. According to the Chebyshev recursion Equation (17), the function can be implemented in a recursive manner, thereby avoiding the problem that the hardware is difficult to implement as the approximation precision increases. The specific operation process can be described as:

$$\begin{cases} dm+1 = dm = 0 \\ di = 2xdi+1-di+2+ci \\ f(x) = d0 = xd1-d2+c0/2 \end{cases} \tag{22}$$

where $i = m-1, m-2, ..., 1$ is the number of Chebyshev estimation coefficients, d_i is the iterative estimation process value, and *f(x)* is the estimation result.

The circuit structure of Chebyshev-based arctangent algorithm generated by Quartus II compilation is shown in Figure 8. Where 'clk' is the clock signal of Chebyshev-based arctangent algorithm. 'x_in [16..0]' corresponds to the x in Equation (22) and 'fout [16..0]' is the radians accumulator, corresponding to the *f(x)* in Equation (22).

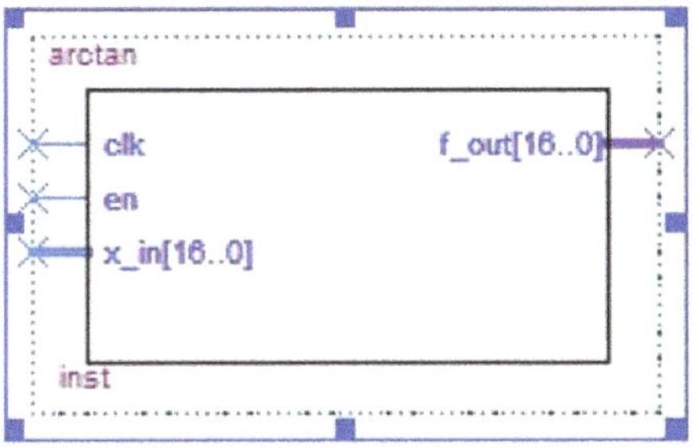

Figure 8. Chebyshev-based arctangent converter structure diagram. In this figure, 'clk' is the clock signal of Chebyshev-based arctangent algorithm. 'x_in [16..0]' corresponds to the x in Equation (22) and 'fout [16..0]' is the radians accumulator, corresponding to the f(x) in Equation (22).

It can be seen from the recurrence process described in Equation (22) that the number of estimation coefficient *M* is the main factor affecting the accuracy of phase difference estimation. Additionally, in order to evaluate the accuracy of Chebyshev estimation mentioned above, we designed the following experiment. Two 100 Hz sinusoidal signals (with a presupposed phase difference, e.g. 5) were generated by the function signal generator SDG5162. After data acquisition and processing by the circuit board shown in Figure 7, the accuracy of phase difference calculation by Chebyshev estimation method can be observed with the on-line debugging tool SignalTap II. Figure 9 shows the effect of the number of estimation coefficients *M* on the phase difference estimation accuracy. As can be seen from Figure 9, with the increase of the number of estimation coefficient, the accuracy of phase difference estimation gradually improved, and the relative error of phase difference estimation was close to 0.005 when *M* = 5. However, this improvement becomes insignificant when the coefficient is greater than 6. In our design, the number of estimation coefficient was set to 6 on a trade-off between estimation accuracy and logical resource consumption of system.

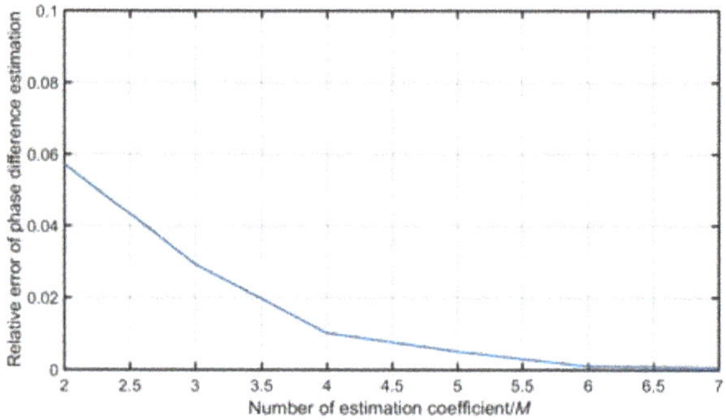

Figure 9. Effect of M on the Chebyshev estimation accuracy.

4. Experimental Evaluation

After single board debugging and hardware joint debugging of the common oscillator, sinusoidal frequency beater, dual-channel data simultaneous sampling circuit, FPGA digital signal processing circuit and serial communication circuit, it was necessary to test the whole machine of the frequency standard comparator to determine its indicators, such as relative channel delay, noise floor, etc. In order to obtain more credible experimental results, all of the experimental instruments and equipment were preheated six hours before each measurement.

4.1. Relative Channel Delay

In order to measure the delay difference between the two channels of the comparator, an experimental platform was built as shown in Figure 10. First of all, the 10 MHz sine wave signal output by the Agilent 58503 GPS (Santa Rosa, CA, USA) time-frequency reference receiver was divided into two identical signals by a frequency distribution amplifier Agilent 5087A (Santa Rosa, CA, USA). Subsequently, they were sent to the comparator's input channel (CH A) and reference frequency channel (CH B).

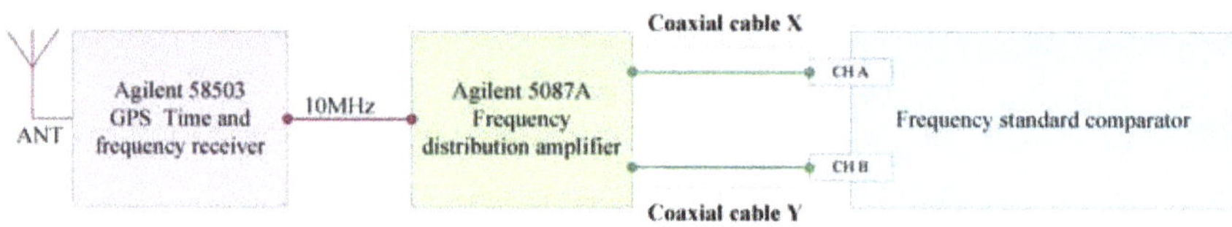

(a)

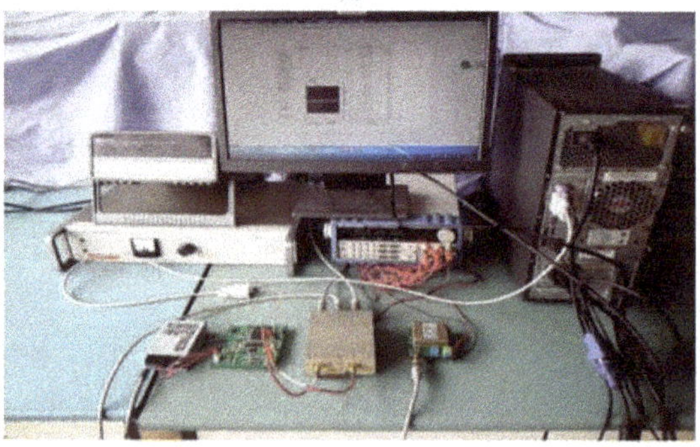

(b)

Figure 10. Experimental platform for testing the characteristics of the comparator. (**a**) Block diagram of the experimental platform; (**b**) photograph of the experimental platform.

In Figure 10a, the phase differences of the two 10 MHz sinusoidal signals output by the frequency distribution amplifier Agilent 5087A relative to the 10 MHz output signal of the Agilent 58503 were defined as T_{in1} and T_{in2}, respectively. Moreover, the time delays caused by the CHA and CHB channels of the frequency standard comparator to the two 10 MHz signals output by the Agilent 5087A were defined as T_{CHA} and T_{CHB}. The relative channel delay between CHA and CHB measured by the comparator at this time can be expressed as:

$$\Delta T_1 = (T_{in1} + T_{CHA}) - (T_{in2} + T_{CHB}) \tag{23}$$

Disconnect the comparator from the coaxial cable X and Y, and the other devices remain the same. Next, the coaxial cable X was connected to the CHB of the comparator, and correspondingly, Y was connected to the CHA. The relative channel delay between CHA and CHB measured by the comparator at this time can be expressed as:

$$\Delta T_2 = (T_{in2} + T_{CHA}) - (T_{in1} + T_{CHB}) \tag{24}$$

The channel delay of the frequency standard comparator can be obtained by Equations (23) and (24), which can be expressed as:

$$T_{CHA} - T_{CHB} = \frac{\Delta T_1 + \Delta T_2}{2} \tag{25}$$

According to the above method, the relative channel delay of the comparator developed in this paper was 199.60 ps, which is an averaged value of 10 measurements. In order to ensure the authenticity of the measurement results, this delay difference of the channels should be deducted when the FPGA solves the phase difference of the frequency to be measured and the reference frequency.

4.2. Noise Floor

The noise floor is usually used to represent the measurement capability of the frequency standard comparator, which is typically characterized by ADEV (Allan Deviation). The experimental platform used to test the noise floor performance of the comparator is also shown in Figure 10. The Allan deviation stability of the comparator noise floor that we achieved was less than 7.50×10^{-12}/s, which was obtained by statistics of more than 10000 phase difference measurements (i.e., a measurement result was obtained every second for three consecutive hours), as shown in Figure 11.

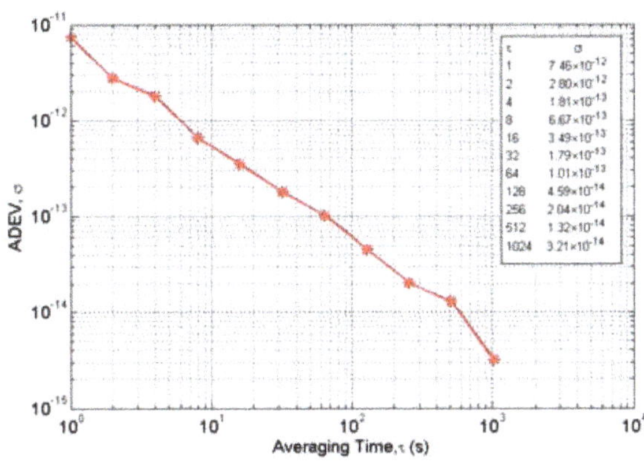

Figure 11. Noise floor performance of the frequency standard comparator.

4.3. Measurement Accuracy

In order to test the phase difference measurement accuracy of the comparator developed in this paper, an experimental platform was built; the structure is shown in Figure 12. Two sets of experiments were designed.

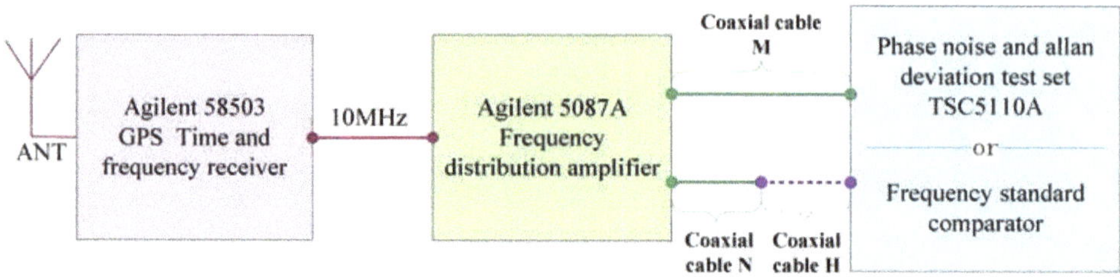

Figure 12. Experimental platform for testing the measurement accuracy.

Experiment 1: The 10 MHz output of Agilent 58503 GPS time-frequency reference receiver was divided into two identical signals by the Agilent 5087A frequency distribution amplifier. Subsequently, they were sent to a phase noise and Allan Deviation test set TSC5110A's input channel and its reference frequency channel via coaxial cable M and N. Subsequently, a phase difference measurement was carried out, and the measurement result was recorded as $\triangle\varphi_1$. In order to add a small amount of phase delay to one of the signals, the coaxial cable H was continued at the rear end of the coaxial cable N. At this time, the two signals of the Agilent 5087A output were input to the TSC5110A by the coaxial cable M and the coaxial cable N+H to perform phase difference measurement, and the result was recorded as $\triangle\varphi_2$. Record the amount of phase delay introduced by coaxial cable H as $\triangle\varphi_{\text{C-5110A}}$. Then, there is:

$$\Delta\varphi_{C-5110A} = \Delta\varphi_2 - \Delta\varphi_1 \tag{26}$$

The measurement results of this experiment were $\triangle\varphi_1 = 6.876548 \times 10^{-10}$ s and $\triangle\varphi_2 = 1.261997 \times 10^{-9}$ s. Substituting $\triangle\varphi_1$ and $\triangle\varphi_2$ into Equation (26), the phase delay of the coaxial cable H for a 10 MHz signal can be calculated, i.e.$\triangle\varphi_{\text{C-5110A}} = 5.743422 \times 10^{-10}$ s.

Experiment 2: The above experiment was repeated, however, the phase difference measuring instrument TSC5110(Symmetricom, San Jose, CA, USA) was replaced with the frequency standard comparator proposed in this paper. The two phase difference measurements were defined as $\triangle\varphi'_1$ and $\triangle\varphi'_2$, respectively. Moreover the amount of phase delay introduced by coaxial cable H was recorded as $\triangle\varphi_{\text{C-FPGA}}$. Then, there was:

$$\Delta\varphi_{C-FPGA} = \Delta\varphi\prime_2 - \Delta\varphi\prime_1 \tag{27}$$

The measurement results of this experiment were $\triangle\varphi'_1 = 7.394913 \times 10^{-10}$ s and $\triangle\varphi'_2 = 1.313824 \times 10^{-9}$ s. Substituting $\triangle\varphi'_1$ and $\triangle\varphi'_2$ into Equation (27), the phase delay of the coaxial cable H for a 10 MHz signal can be calculated, i.e. $\triangle\varphi_{\text{C-FPGA}} = 5.743327 \times 10^{-10}$ s.

In view of the excellent measurement accuracy of the TSC5110A, it can be approximated that the $\triangle\varphi_{\text{C-5110A}}$ measured by the TSC5110A is the true value of the phase delay of the coaxial cable H for the 10 MHz signal. Then, the relative error of the phase difference measurement of the frequency standard comparator developed in this paper can be considered to be about 1.654066×10^{-5}, which was an averaged value of 10 measurements.

5. Conclusions

This paper designed and produced a 9.9999 MHz common oscillator, 100 Hz frequency down-beater, dual-channel data simultaneous sampling circuit, FPGA circuit and serial communication circuit, and performed functional tests on each of the above circuit modules. Finally, these circuit modules were connected together to form a frequency standard comparator. The characteristic work of this paper can be summarized as follows:

(1) An improvement was made to the traditional DMTD method, the counter was discarded, and the phase difference was measured by a digital signal processing method.

(2) The classical phase difference measurement method based on correlation operation was improved to make it suitable for FPGA implementation. Furthermore, the logic operation unit such as inverse trigonometric function converter, digital correlator, multiplier and divider was designed on the FPGA using hardware description language.

(3) A miniaturized 10 MHz frequency standard comparator with good noise floor was successfully developed. The size of the prototype circuit board of the system was only about 292.1 cm^2. The noise floor of the comparator was typically better than $7.50 \times 10^{-12}/s$, and its relative error of phase difference measurement was less than 1.70×10^{-5}.

The related methods and technical schemes proposed in this paper are expected to provide references for miniaturized frequency standard comparator engineering. However, it should be admitted that there is still a gap between the phase difference measurement resolution of our comparator and some excellent frequency standard comparison instruments on the market. For example, the noise floor of the TSC5110 was only $2.50 \times 10^{-14}/s$. Therefore, our future work will continue to pursue miniaturization design, while at the same time strive to reduce the noise floor of the frequency standard comparator. Several preliminary research plans can be described briefly as follows:

(1) Seeking or designing better anti-tangent solution than Chebyshev algorithm.

(2) Developing a floating-point unit in FPGA implementation in order to achieve higher phase difference measurement accuracy.

(3) Developing a sinusoidal beater with a higher beat factor. Our next goal is to make the frequency down-beater output a 10 Hz sinusoidal wave with a corresponding beat factor of 10^{-6}, which can theoretically improve the current measurement resolution by an order of magnitude.

(4) Improving circuit manufacture and packaging skill. Miniaturization of the frequency standard comparator is still our unremitting pursuit.

Author Contributions: All of the authors were working together very tightly. The theoretical works on the design of the miniaturized frequency standard comparator were mainly driven by S.T. and J.K. The hardware and software implementation of the comparator was done by T.W. Z.D. and T.W. were responsible for performing the experiments. S.T. and J.K. analyzed the experimental results and wrote the paper together.

Funding: The Project supported by the Natural Science Basic Research Plan in Shaanxi Province of China (Program NO. 2018JM1025), the Open Fund of Key Laboratory of Precision Navigation and Timing Technology of National Time Service Center (No.2014PNTT04) and the National Natural Science Foundation of China (Program NO. 11403018).

Acknowledgments: The authors would like to thank Yuqing Hou and Xiaowei He of Northwest University, for partially supporting this work.

Conflicts of Interest: The authors declare no conflict of interest.

References

1. Fujieda, M.; Yang, S.; Gotoh, T.; Hwang, S.; Hachisu, H.; Kim, H.; Lee, Y.; Tabuchi, R.; Ido, T.; Lee, W.; et al. Advanced satellite-based frequency transfer at the 10(-16) level. *IEEE Trans. Ultrason. Ferroelectr. Freq. Control* **2018**, *65*, 973–978. [CrossRef] [PubMed]
2. Heo, M.; Park, S.; Lee, W.; Lee, S.; Hong, H.; Kwon, T.; Park, C.; Yu, D.; Santarelli, G.; Hilton, A.; et al. Drift-compensated low noise frequency synthesis based on a cryoCSO for the KRISS-F1. *IEEE Trans. Instrum. Meas.* **2017**, *66*, 1343–1348. [CrossRef]
3. Zhuang, Y.; Shi, D.; Li, D.; Wang, Y.; Zhao, X.; Zhao, J.; Wang, Z. An accurate frequency control method and atomic clock based on coherent population beating phenomenon. *Chin. Phys. Lett.* **2016**, *33*, 040601. [CrossRef]
4. Wang, X.; Meng, Y.; Wang, Y.; Wan, J.; Yu, M.; Wang, X.; Xiao, L.; Li, T.; Cheng, H.; Liu, L. Dick effect in the integrating sphere cold atom clock. *Chin. Phys. Lett.* **2017**, *34*, 063702. [CrossRef]

5. Formichella, V.; Camparo, J.; Sesia, I.; Signorile, G.; Galleani, L.; Huang, M.; Tavella, P. The ac stark shift and space-borne rubidium atomic clocks. *J. Appl. Phys.* **2016**, *120*, 194501. [CrossRef]
6. Galleani, L.; Tavella, P. Robust detection of fast and slow frequency jumps of atomic clocks. *IEEE Trans. Ultrason. Ferroelectr. Freq. Control* **2017**, *64*, 475–485. [CrossRef] [PubMed]
7. Zucca, C.; Tavella, P.; Peskir, G. Detecting atomic clock frequency trends using an optimal stopping method. *Metrologia* **2016**, *53*, S89. [CrossRef]
8. Khare, A.; Arora, R.; Banik, A.; Mehta, S. Autonomous rubidium clock weak frequency jump detector for onboard navigation satellite system. *IEEE Trans. Ultrason. Ferroelectr. Freq. Control* **2016**, *63*, 326–335. [CrossRef]
9. Feng, L.; Li, G. Research on self-monitoring method for anomalies of satellite atomic clock. *Int. J. Aerosp. Eng.* **2016**, *2016*, 1759512. [CrossRef]
10. Formichella, V.; Camparo, J.; Tavella, P. Atomic clocks and the continuous-time random-walk. *Eur. Phys. J. B* **2017**, *90*, 206. [CrossRef]
11. Kudeyarov, K.; Vishnyakova, G.; Khabarova, K.; Kolachevsky, N. 2.8 km fiber link with phase noise compensation for transportable Yb+ optical clock characterization. *Laser Phys.* **2018**, *28*, 105103. [CrossRef]
12. Wang, Q.; Wei, R.; Wang, Y. Atomic fountain frequency standard: Principle and development. *Acta Phys. Sin.* **2018**, *67*, 163202.
13. Mehlstaubler, T.; Grosche, G.; Lisdat, C.; Schmidt, P.; Denker, H. Atomic clocks for geodesy. *Rep. Prog. Phys.* **2018**, *81*, 064401. [CrossRef] [PubMed]
14. Guena, J.; Weyers, S.; Abgrall, M.; Grebing, C.; Gerginov, V. First international comparison of fountain primary frequency standards via a long distance optical fiber link. *Metrologia* **2017**, *54*, 348–354. [CrossRef]
15. Abgrall, M.; Guena, J.; Lours, M.; Santarelli, G.; Tobar, M.; Bize, S.; Grop, S.; Dubois, B.; Fluhr, C.; Giordano, V. High-stability comparison of atomic fountains using two different cryogenic oscillators. *IEEE Trans. Ultrason. Ferroelectr. Freq. Control* **2016**, *63*, 1198–1203. [CrossRef] [PubMed]
16. Calosso, C.; Clivati, C.; Micalizio, S. Avoiding Aliasing in Allan Variance: An Application to fiber link data analysis. *IEEE Trans. Ultrason. Ferroelectr. Freq. Control* **2016**, *63*, 646–655. [CrossRef] [PubMed]
17. Huang, Y.; Guan, H.; Liu, P.; Bian, W.; Ma, L.; Liang, K.; Li, T.; Gao, K. Frequency comparison of two 40Ca+ optical clocks with an uncertainty at the 10(-17) level. *Phys. Rev. Lett.* **2016**, *116*, 013001. [CrossRef]
18. Carsten, A.; Alexander, I.; Julia, B.; Niklas, B.; Giovanni, D. High-precision measurement of sine and pulse reference signals using software-defined radio. *IEEE Trans. Instrum. Meas.* **2018**, *67*, 1132–1141.
19. Wang, S.; Cao, P.; Shang, L.; An, Q. A precise clock distribution network for MRPC-based experiments. *J. Instrum.* **2016**, *11*, C06006. [CrossRef]
20. Sun, S.; Xu, L.; Cao, Z.; Sun, J.; Yang, W. A recursive demodulator for real-time measurement of multiple sinusoids. *IEEE Sens. J.* **2018**, *18*, 6281–6289. [CrossRef]
21. Lv, Y. The influence of digital signal processing on electronic measurement and instrument. *Agro Food Ind. Hi-Tech* **2017**, *28*, 1861–1865.
22. Najafi, E.; Yatim, A.; Mirzaei, A. An improved sag detection approach based on modified Goertzel algorithm. *Int. J. Electron.* **2019**, *106*, 36–47. [CrossRef]
23. Wu, J.; Xu, K.; Xu, W.; Yu, X. Transient process based electromagnetic flow measurement methods and implementation. *Rev. Sci. Instrum.* **2018**, *89*, 095108. [CrossRef] [PubMed]
24. Liu, B.; Xu, K.; Mu, L.; Tian, L. Echo energy integral based signal processing method for ultrasonic gas flow meter. *Sens. Actuators A* **2018**, *277*, 181–189.
25. Hace, A.; Curkovic, M. Accurate FPGA-based velocity measurement with an incremental encoder by a fast generalized divisionless MT-type algorithm. *Sensors* **2018**, *18*, 3250. [CrossRef] [PubMed]
26. Saha, A.; Das, S.; Suresh, M.; Kiran, V.; Dey, N. FPGA based self-vibration compensated two dimensional non-contact vibration measurement using 2D position sensitive detector with remote monitoring. *Measurement* **2017**, *111*, 271–278. [CrossRef]
27. Yang, D.; Wang, J.; Feng, Y.; Tang, Q.; Zhang, H.; Chen, T. A FPGA-based energy measurement approach for high-repetition rate narrow laser pulses. *IEEE Trans. Nucl. Sci.* **2018**, *65*, 2665–2669. [CrossRef]
28. Dam, M.; Nguyen, V.; Lee, J. A carry chain-based ADMFC design on an FPGA for EMI reduction and noise compensation. *J. Circuits Syst. Comput.* **2019**, *28*, 1950018. [CrossRef]

29. Sarnago, H.; Lucia, O.; Burdio, J. FPGA-based resonant load identification technique for flexible induction heating appliances. *IEEE Trans. Ind. Electron.* **2018**, *65*, 9421–9428. [CrossRef]
30. Dantism, S.; Rohlen, D.; Wagner, T.; Wagner, P.; Schoning, M. Optimization of cell-based multi-chamber LAPS measurements utilizing FPGA-controlled laser-diode modules. *Phys. Status Solidi A* **2018**, *215*, 1800058. [CrossRef]

electronics

Article

Fully Integrated Low-Ripple Switched-Capacitor DC–DC Converter with Parallel Low-Dropout Regulator

Jeong-Yun Lee, Gwang-Sub Kim, Kwang-Il Oh and Donghyun Baek *

Microwave Embedded Circuit and System Laboratory, School of Electrical Engineering, Chung-Ang University, Seoul 06974, Korea; lostria1985@gmail.com (J.-Y.L.); gsubkim@naver.com (G.-S.K.); dhrhkddlf6763@hanmail.net (K.-I.O.)

* Correspondence: dhbaek@cau.ac.kr; Tel.: +82-02-820-5828

Received: 20 November 2018; Accepted: 11 January 2019; Published: 16 January 2019

Abstract: In this paper, we propose a fully integrated switched-capacitor DC–DC converter with low ripple and fast transient response for portable low-power electronic devices. The proposed converter reduces the output ripple by filtering the control ripple via combining a low-dropout regulator with a main switched-capacitor DC–DC converter with a four-bit digital capacitance modulation control. In addition, the four-phase interleaved technique applied to the main converter reduces the switching ripple. The proposed converter provides an output voltage ranging from 1.2 to 1.5 V from a 3.3 V supply. Its peak efficiency reaches 73% with ripple voltages below 55 mV over the entire output power range. The transient response time for a load current variation from 100 μA to 50 mA is measured to be 800 ns. Importantly, the converter chip, which is fabricated using 0.13 μm complementary metal–oxide–semiconductor (CMOS) technology, has a size of 2.04 mm^2. We believe that our approach can contribute to advancements in power sources for applications such as wearable electronics and the Internet of Things.

Keywords: DC–DC converter; switched capacitor; power management integrated circuit; CMOS technology

1. Introduction

Rapid advances in the Internet of Things and wearable electronic devices have led to an increasing demand for various types of sensors [1]. For portability, such devices/applications are usually powered by small batteries, which limit the operating time of sensor-based devices. Therefore, in order to increase the battery efficiency to provide longer operating times, power management units such as power management integrated circuits (ICs) are used to control power consumption [2,3]. The power management IC can be mounted on the same printed circuit board as the sensor IC, as shown in Figure 1a. Meanwhile, certain off-chip passive components such as inductors and capacitors are additionally required for external support of the power management IC because they cannot be integrated into the chip. In this regard, although multichip configurations are convenient for a sensor module design, the cost and size of the resulting modules increase. Thus, integration of the power management unit and passive components into a single sensor chip (Figure 1b) is being actively pursued to reduce the module size and to increase market competitiveness [4–9]. The power management unit normally comprises a high-efficiency switching DC–DC converter and a linear low-dropout regulator. The switching DC–DC converter adjusts the battery voltage that drops over time to a fixed supply voltage, and its output drives the low dropout regulator to provide a voltage with low ripple and low noise to supply-sensitive analog components on the sensors [10,11].

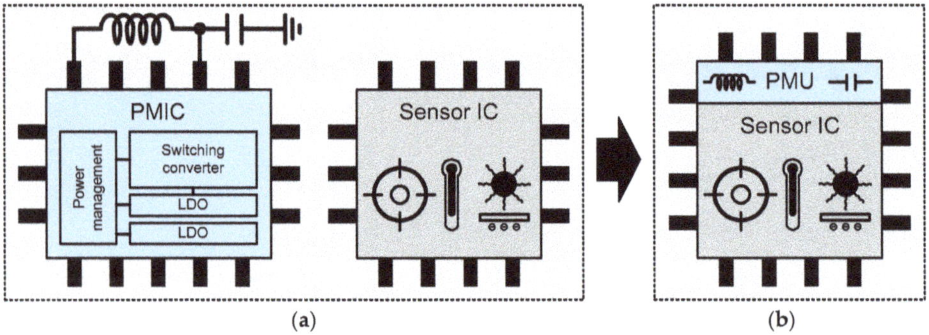

Figure 1. Types of sensor modules: (**a**) Sensor module with multichip architecture and (**b**) fully integrated sensor module (PMIC, power management IC; PMU, power management unit; and LDO, linear low dropout regulator).

Figure 2 shows the two types of currently available switching DC–DC converters according to the choice of passive components utilized for energy storage. The first type is the inductor-based converter, which employs inductor L_S and a push–pull stage as shown in Figure 2a. The output voltage is regulated by controlling the switches (Φ_1 and Φ_2) with either a pulse–width modulator or a pulse–frequency modulator [12,13]. The second type is the capacitor-based converter, which employs a flying capacitor C_F and four switches (Φ_1 and Φ_2). The output voltage is regulated by controlling the switches with a pulse–width modulator, pulse–frequency modulator, or digital capacitor modulator (DCpM) [14–17]. The converting power depends on the storage capacity of the passive components as per the relation $P_L = L{\cdot}I^2/2$ for the inductor-based converter and $P_C = C{\cdot}V^2/2$ for the capacitor-based converter. Here, we note that inductor-based converters can deliver more power than capacitor-based converters via increasing the current at a fixed battery voltage, V_{BAT}. Moreover, high power efficiency can be achieved by use of an off-chip inductor with high inductance and high Q values while maintaining a low ripple voltage. Thus, the traditional inductor-based buck converter has been widely adopted for moderate- to high-power applications.

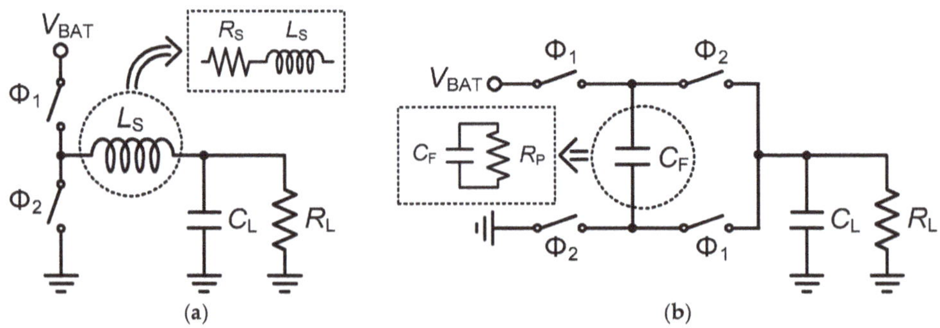

Figure 2. Configuration of the step-down switching DC–DC converters: (**a**) Inductor-based converter and (**b**) capacitor-based converter.

On the other hand, integrated inductors based on complementary metal–oxide–semiconductor (CMOS) technology present many limitations. First, the feasible inductance L_S on a chip is limited from a few to some tens of nanohenries due to the planar layout structure and fabrication cost. Thus, integrated inductor-based converters should increase the modulation frequency to maintain ripple levels; however, this also increases the switching loss. Second, series resistance R_S is very high, which leads to an increased inductor loss over the switching loss, regardless of the use of expensive additional manufacturing processes involving thick metals or integrated magnetic materials. Finally, the integrated inductor exhibits power loss due to the large parasitic capacitance in relation to the substrate. In contrast, integrated capacitors afford either high parallel resistance R_F or high Q factor via the metal–insulator–metal (MIM) structure. Therefore, when passive components are realized with CMOS technology, capacitors afford better energy density per chip area relative to inductors, as explained in References [18–20]. Consequently, capacitor-based converters exhibit better power and cost efficiency than inductor-based converters in low-power applications, such as sensors and Internet of Things devices.

Figure 3 shows the block diagram of a commonly used switched-capacitor (SC) DC–DC converter utilizing one-boundary hysteresis feedback for output voltage regulation and its output ripple voltage. The controller provides switching control signals to the converter in phase with the input clock CLK. The one-boundary hysteresis configuration employs only one comparator for the feedback control to compare the output voltage with the reference voltage V_{REF} [21]. In the steady state, this feedback causes a low-frequency control ripple. In addition, the SC DC–DC converter "dumps" the charge from the input to the flying capacitor and from the capacitor to the output at discrete time intervals according to the clock frequency. This discrete charge transfer causes an unavoidable switching ripple. The switching ripple is usually lower than the control ripple because the switching frequency is higher than the control frequency.

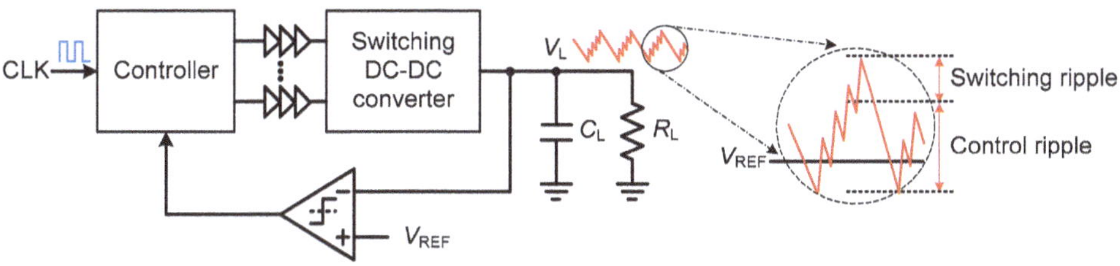

Figure 3. Block diagram of the switched-capacitor (SC) DC–DC converter using one-boundary hysteresis feedback and its output ripple voltage.

Fully integrated SC DC–DC converters require additional techniques to suppress the ripple due to the size limitations of the load and flying capacitors. Figure 4 shows three representative ripple mitigation techniques applied to the SC DC–DC converters. The capacitance modulation technique regulates the capacity of the flying capacitor, which transfers the charge to the load, to suppress the ripple. Flying capacitors are divided into several capacitors controlled by digital codes [15]. The capacitance modulation operates as a low-capacity flying capacitor in the light-load state and is controlled to operate as a high-capacity flying capacitor in the heavy-load state. Further, pulse–width modulation controls the time for which the flying capacitor is connected to the load. This method reduces the ripple by regulating the amount of charge delivered to the load per clock cycle [22].

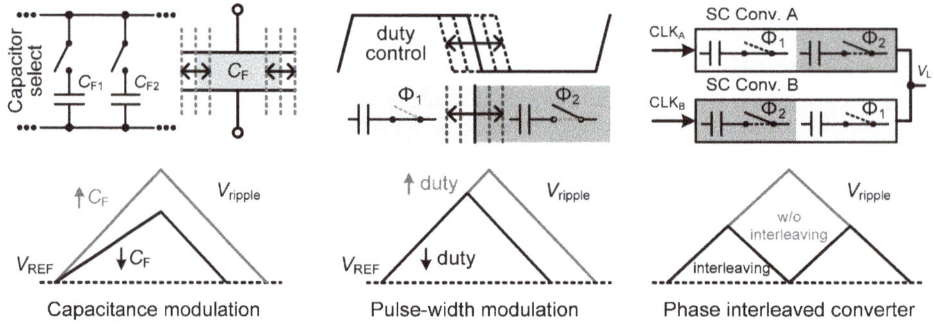

Figure 4. Ripple mitigation techniques.

The multiphase interleaving technique divides a converter into multiple units and drives each unit in a different clock phase [19]. Because each converter operates in different phases, it appears that the ripple waveform is operating at a frequency that is equal to the number of interleaved phases. The ripple is reduced by the number of interleaved phases.

Against this backdrop, here, we propose a low-ripple fast-transient SC DC–DC converter operating over the output current range, which integrates all the active and passive components on a single chip. The converter employs a two-boundary hysteresis control with interleaving through a four-bit DCpM to reduce the switching ripple and a parallel low-dropout regulator (LDR) to considerably mitigate the ripple.

2. Principles of SC DC–DC Converters

2.1. Operation of the 2:1 Step-Down SC DC–DC Converter

The 2:1 step-down SC DC–DC converter operates in the two phases, as shown in Figure 5. The output voltage is half the input voltage under ideal operation. Hence, maximum efficiency can only be achieved if each phase operates at 50% duty cycle. During phase 1 (Φ_1), the flying capacitor is connected between the input node V_{BAT} and output node V_L, as shown in Figure 5b. In this phase, the flying capacitor is charged up to the voltage difference between V_{BAT} and V_L. During phase 2 (Φ_2), the flying capacitor is connected to V_L and the ground, as shown in Figure 5c. The charge acquired by the flying capacitor during phase 1 is supplied to the output node. The repeated charging and discharging during these phases produce output voltage ripple ΔV_L, as illustrated in Figure 5d.

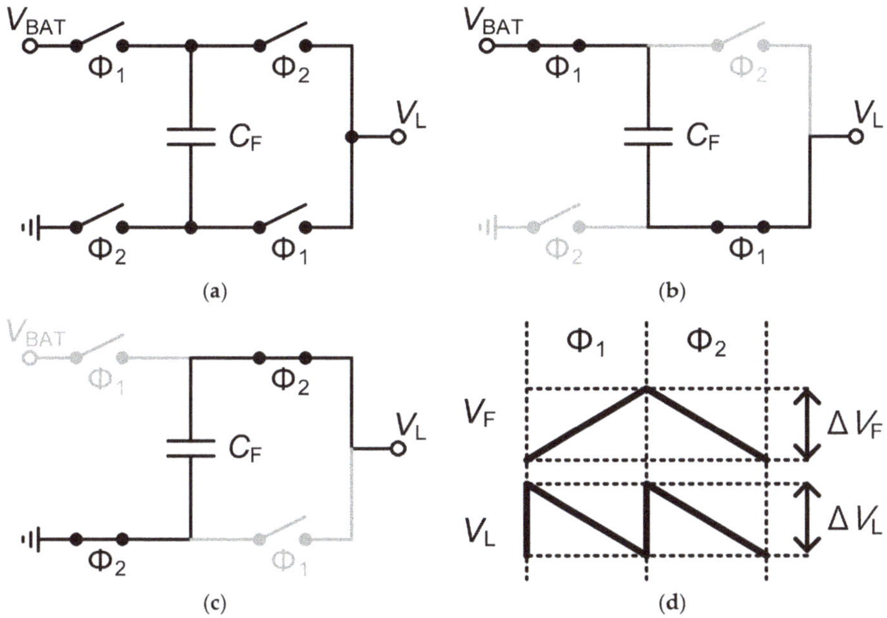

Figure 5. Operation of a step-down SC DC–DC converter. (**a**) Block diagram of a 2:1 step-down SC DC–DC converter; (**b**) operation during phase 1; (**c**) operation during phase 2; and (**d**) the output voltage ripple.

Figure 6 shows a simplified model of the 2:1 step-down SC DC–DC converter. The parallel resistor R_P represents the shunt loss due to parasitic capacitances in the switches and flying capacitors. We note here that R_P is independent of the output current. The output impedance R_O is connected in series with the load resistor R_L. R_O changes the load voltage, and its power loss, called series loss, is the sum of the switch conductance loss and the intrinsic SC loss. The switch conductance loss is caused by the resistance in the on state of the switch. Increasing the size of the switch reduces the conductance loss but increases the shunt loss via the parasitic capacitance of the switch [23]. The intrinsic SC loss is caused by voltage ripple ΔV_F due to the charge and discharge of the capacitor, as shown in Figure 5d. The intrinsic SC loss of a 2:1 step-down SC DC–DC converter can be expressed as [24,25]

$$P_{C_F} = I_L \cdot \frac{\Delta V_F}{2} = \frac{{I_L}^2}{4 \cdot C_F \cdot f_{SW}} \tag{1}$$

where f_{SW} denotes the switching frequency related to the two-phase operation. A fully-integrated SC DC–DC converter provides a relatively large load current with a small flying capacitance due to chip size limitations. Therefore, the intrinsic SC loss is larger than the switch conductance loss. In this paper, assuming an ideal switch, only the intrinsic SC loss is expressed as the series loss. Upon applying Equation (1) to this simplified model, the load current can be approximated as

$$I_L \approx \frac{(V_{BAT}/2 - V_L)}{R_O} = 4 \cdot C_F \cdot f_{SW} \cdot (V_{BAT}/2 - V_L) \tag{2}$$

The SC DC–DC converters regulate the output voltage via changing the value of Ro, which is adjusted through either frequency or pulse–width modulation of the switching clock.

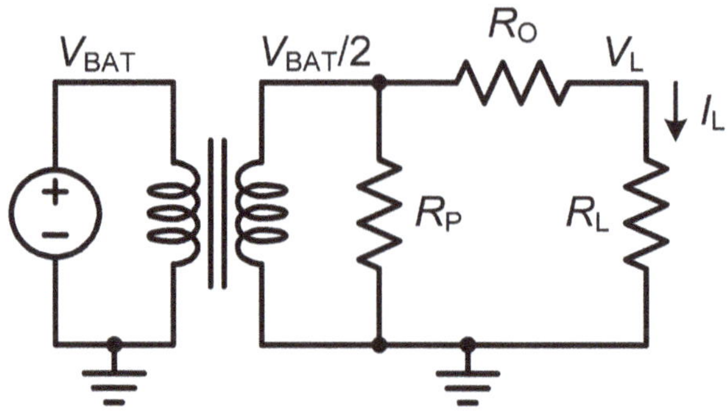

Figure 6. Simplified model of a 2:1 step-down switched-capacitor (SC) DC–DC converter.

2.2. Multiphase Interleaved SC DC–DC Converter for Low Switching Ripple

As the SC DC–DC converter performs repeated charging and discharging, the output voltage exhibits an inherent switching ripple. Multiphase interleaving aims to mitigate this ripple via dividing the converter into multiple units and driving each unit with different clock phases. Figure 7 illustrates a four-phase interleaved converter, with each unit utilizing a quarter of the total capacitance and operating at a 45° phase shift relative to the clocks of the neighboring nodes. The flying capacitances of all units are equal, and hence, the output charge per cycle is also identical. The output current of each unit is the same as that of the converter without interleaving. Thus, the charge flowing through each unit of the flying capacitor in multiphase interleaving is the same as that in the case of the original converter.

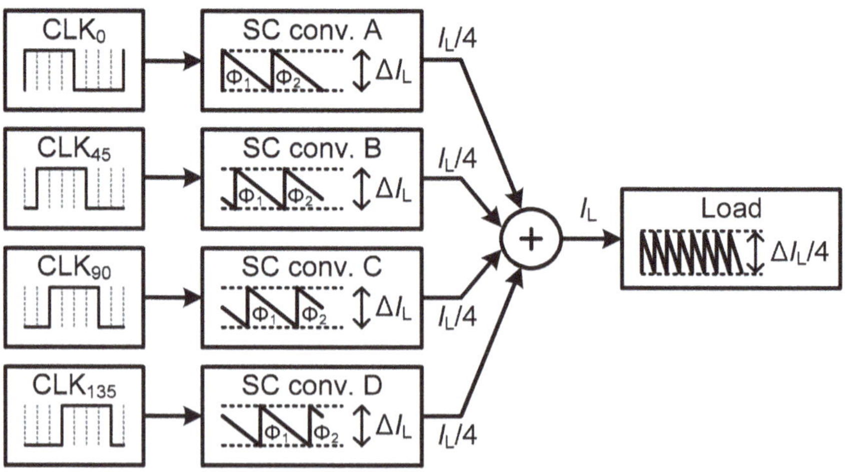

Figure 7. Block diagram of s four-phase interleaved SC DC–DC converter.

Figure 8 shows the operation of a four-phase interleaved SC DC–DC converter including the output voltage ripple with and without phase interleaving. In Figure 8a, each SC DC–DC converter without phase interleaving operates at the same clock phase (Φ_1 and Φ_2), producing output ripple ΔV_L. In Figure 8b, each converter of the interleaving configuration operates with 45° phase-shifted clocks (Φ_{A_1}, Φ_{B_1}, Φ_{C_1}, and Φ_{D_1}). Therefore, the effective switching frequency f_{ripple} in the converter increases by a factor of four relative to the case with no interleaving, thereby reducing the output ripple to 25% of the original ΔV_L. Multiphase interleaving reduces the voltage ripple by increasing the effective switching frequency but maintains switching losses. To mitigate the output voltage ripple, a load capacitor is generally used. Multiphase interleaving also decreases this load capacitor value by a factor of four due to the increased ripple frequency.

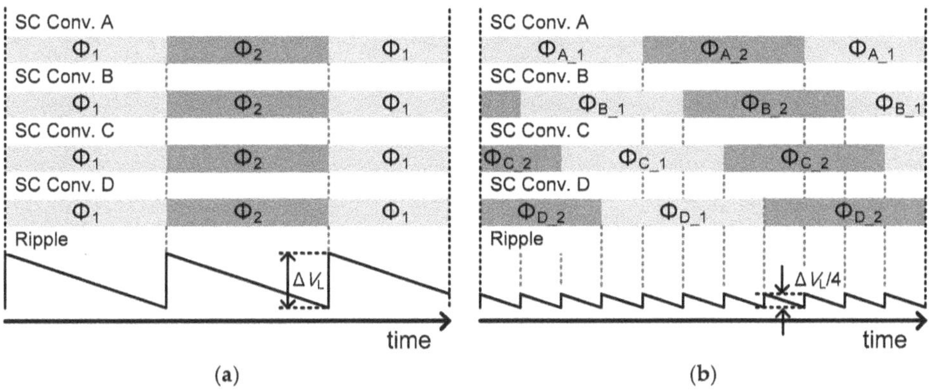

Figure 8. Operation of an interleaved switched-capacitor (SC) DC–DC converter: (**a**) Without interleaving and (**b**) upon applying interleaving.

2.3. Output Voltage Regulation

The output voltage of the SC DC–DC converter can be modulated by the three methods depicted in Figure 9. First, frequency modulation enables the adjustment of the operating frequency of switching according to the load impedance, with the duty cycle usually set to 50%. This method changes the output impedance of the converter by varying the charge transferred from the flying capacitor to the load. However, it requires an additional voltage-controlled oscillator for frequency modulation. Second, time modulation enables the adjustment of the pulse width of the switching signal, which allows for control of the output current for the flying capacitors to charge or discharge. This method modulates the output current by varying the connection time to the output node. However, efficiency is low due to switching losses under light loads, given the low output current of the converter; nevertheless, the switching loss is maintained constant under this condition. Third, capacitance modulation of the charge transfer can be achieved by dividing the SC DC–DC converter into multiple converter cells in parallel and utilizing some cells to provide the required current to the load, thereby establishing "digital" operation. In this method, only the flying capacitors and switches of the converter cells involved in the output current circuit operate, thus improving the efficiency with respect to switching loss. However, the main limitation of this method is the required division of the SC DC–DC converter into cells for accurate output current control. This division increases the complexity of both the chip layout and the state machine to select the appropriate number of cells, thus imposing a tradeoff between efficiency and complexity.

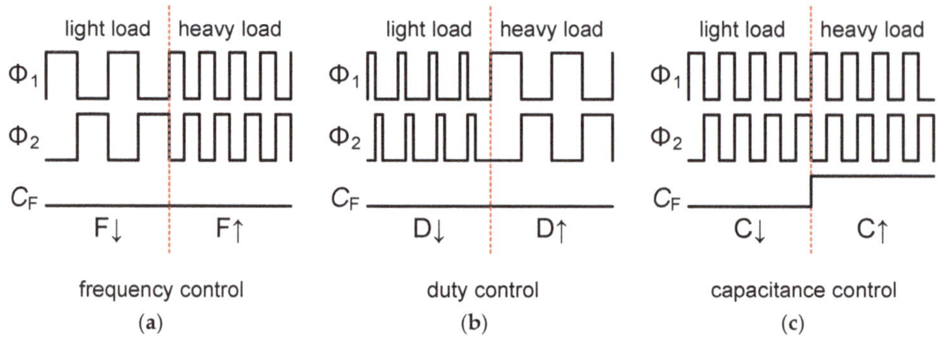

Figure 9. Methods for output voltage regulation: (**a**) Pulse–frequency modulation; (**b**) pulse–width modulation; and (**c**) capacitance modulation.

2.4. DCpM Control

The DCpM approach allows control of the amount of flying capacitance associated with the charge transfer in the converter, thereby enabling load current regulation given that the amount of charge transferred in one clock cycle is proportional to this capacitance. With this method, the total switch size involved in the output current of the SC DC–DC converter can be adjusted according to the size of the flying capacitance. Thus, the shunt loss originating from parasitic capacitances of the flying capacitors and switches and the conduction loss due to the switch resistance are reduced when the load current is low, thereby maintaining high efficiency under light load.

In the implementation of the SC DC–DC converter with DCpM control, the flying capacitor is divided into a binary-weight bank. Figure 10 shows the structure of the SC DC–DC converter with a four-bit DCpM control. The flying capacitance is divided into four different banks of size $x1$, $x2$, $x4$, and $x8$. These four converter cells form a single matrix, and the charge transfer operation is enabled by control code C[3:0]. Figure 11 shows a model of the proposed SC DC–DC converter based on a four-bit DCpM. The 2:1 transformer represents the required voltage step-down process. The output impedance R_O and the shunt impedance R_P are binary-weighted according to the DCpM control signal. The output impedance is determined as $1/(4{\cdot}C_F{\cdot}f_{SW})$, where f_{SW} and C_F denote the switching frequency and the unit flying capacitance, respectively. The load current I_L of the converter can be expressed as

$$I_L = 4 \cdot (0.5 \cdot V_{BAT} - V_L) \cdot f_{SW} \cdot \sum_{n=0}^{3} C[n] \cdot 2^n \cdot C_F \tag{3}$$

where V_I and V_L represent the input and output voltages, respectively, and DCpM code $C[n]$ determines the output current.

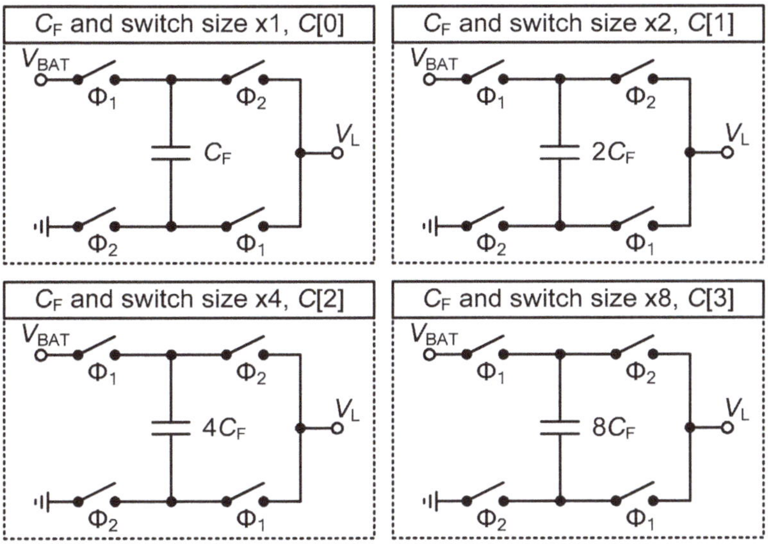

Figure 10. Binary-weighted switched-capacitor (SC) DC–DC converter cells for DCpM.

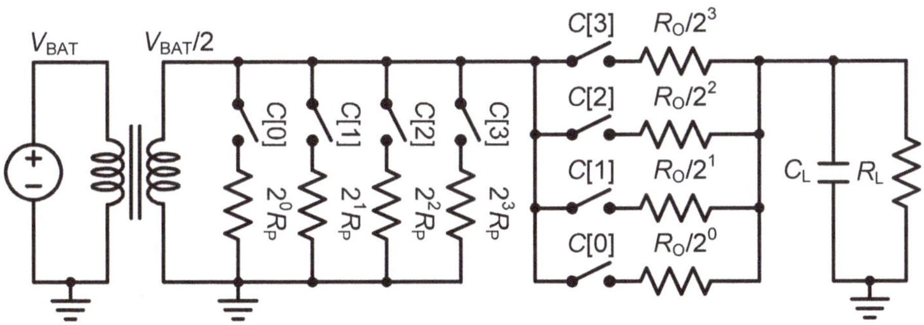

Figure 11. Model of the proposed switched-capacitor (SC) DC–DC converter using a four-bit DCpM.

3. Proposed Low-Ripple SC DC–DC Converter

Figure 12 shows the block diagram of the proposed SC DC–DC converter, which is composed of a main converter, an auxiliary LDR, and a DCpM controller. The main converter provides most of the current to the load, whereas the LDR assists the main converter to provide an accurate output current. The LDR is powered by a small four-phase interleaved SC converter to improve efficiency. To reduce the switching ripple, four interleaved phases (0°, 45°, 90°, and 135°) are adopted for the SC DC–DC converter cells. The current of the main converter is controlled by the DCpM, which compares the output voltage with two reference voltages using two clocked comparators. If output voltage $V_O > V_{REF} + \Delta V$ or $< V_{REF} - \Delta V$, the binary code decreases or increases, respectively. If V_L lies between $V_{REF} + \Delta V$ and $V_{REF} - \Delta V$, the binary code remains unchanged.

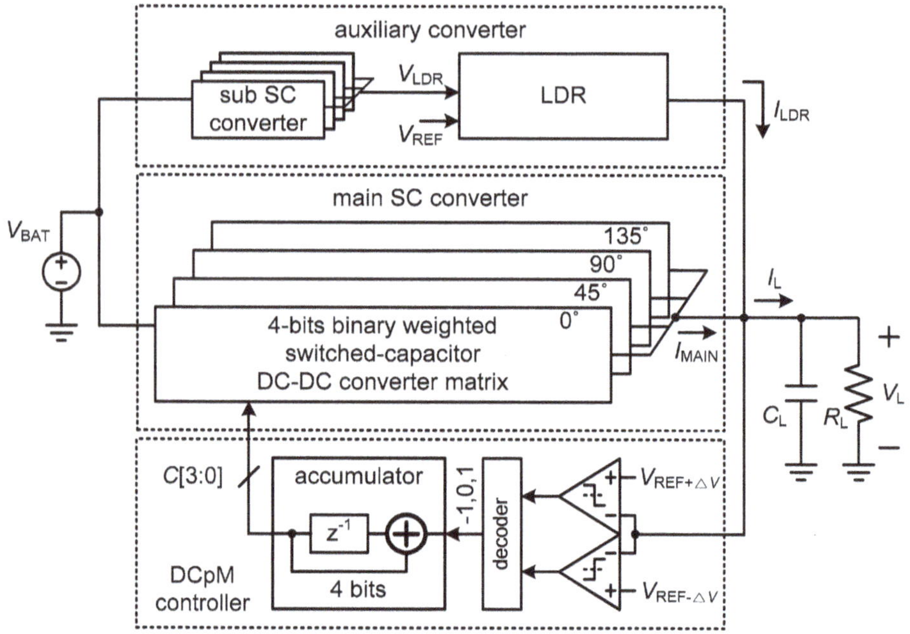

Figure 12. Block diagram of the proposed switched-capacitor (SC) DC–DC converter.

Figure 13a shows one of the four-phase interleaved SC DC–DC converter matrices used in the main converter, which is composed of four converter cells. Each cell employs a 2:1 step-down topology and operates in a bi-phase mode (Φ_1 and Φ_2) with 50% duty cycle. The magnitudes of the flying capacitors C_F and switches are four-bit binary-weighted. Binary code C[3:0] of the DCpM controller either enables or disables the operation of each converter cell to adjust the output current. As shown in Figure 13b, the auxiliary LDR powered by the small four-phase SC converter employs a p–channel metal–oxide–semiconductor (PMOS) pass transistor and a two-stage operational amplifier. Figure 14 shows the block diagram of the proposed LDR-assisted SC DC–DC converter with a low output ripple. The proposed converter exhibits only a switching ripple, and the main converter is controlled by the DCpM via two-boundary hysteresis feedback, which also produces a low-frequency control ripple. Nevertheless, the two-boundary controller can limit the control ripple between $V_{REF} - \Delta V$ and $V_{REF} + \Delta V$. Therefore, the LDR with a low output current capability can compensate for the output current fluctuation due to the feedback control ripple by providing an opposite-phase accurate current to the load. This approach ensures that the DCpM control bits performing coarse tuning are fixed at every output current range, and hence, the output voltage ripple of the proposed converter presents no control ripple due to hysteresis feedback but only switching ripple.

Figure 15a shows a simplified model of the proposed SC DC–DC converter, where the 2:1 transformer represents the 2:1 voltage step-down process. The main converter is described using a binary-weighed unit-resistance R_O, which equals $1/(4 \cdot C_{FLY} \cdot f_{SW})$, where f_{SW} and C_{FLY} represent the switching frequency and unit flying capacitance, respectively. Current I_{MAIN} of the main SC DC–DC converter can be expressed as

$$I_{MAIN} = 4 \cdot k_i \cdot (0.5 \cdot V_{BAT} - V_L) \cdot f_{SW} \cdot \sum_{n=0}^{3} C[n] \cdot 2^n \cdot C_F \tag{4}$$

where V_I, V_O, and k_i denote the input voltage, output voltage, and number of interleaved phases, respectively. The auxiliary LDR is modeled as a fixed resistance R_{SUB} for each SC DC–DC converter cell and a variable resistance R_{LDR} for the LDR. Consequently, output current I_{LDR} of the auxiliary LDR can be expressed as

$$I_{LDR} = 4 \cdot k_i \cdot (0.5 \cdot V_{BAT} - V_L - V_{DO}) \cdot f_{SW} \cdot 2 \cdot C_F \tag{5}$$

where V_{DO} represents the dropout voltage of the pass transistor in the LDR. Hence, the auxiliary LDR can finely adjust the output current. From Figure 15b, we note that the main converter provides a discrete coarse current that is determined by the DCpM code, whereas the auxiliary converter "fills" the discrete steps using the linear LDR. Thus, the proposed SC DC–DC converter can provide any output current in its operating range without requiring a complex pulse–width modulated or pulse–frequency modulated controller.

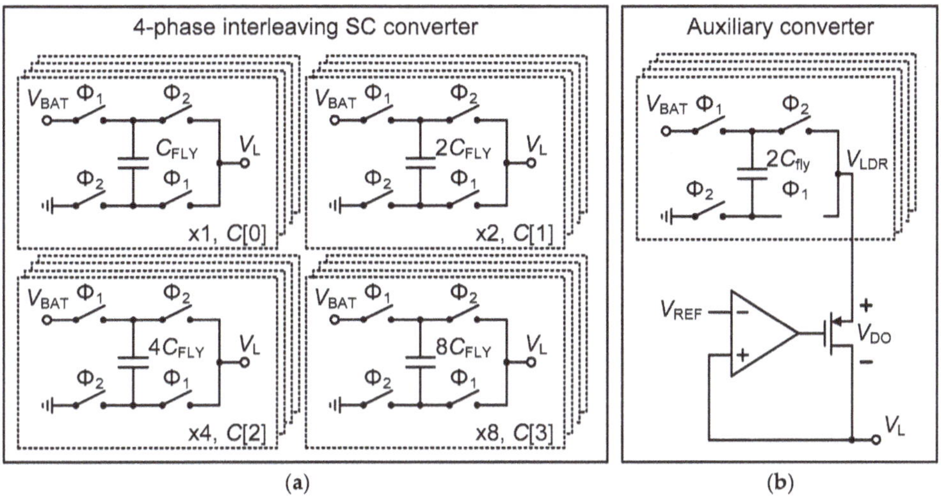

Figure 13. Schematic of (**a**) the main switched-capacitor (SC) DC–DC converter and (**b**) the auxiliary SC DC–DC converter.

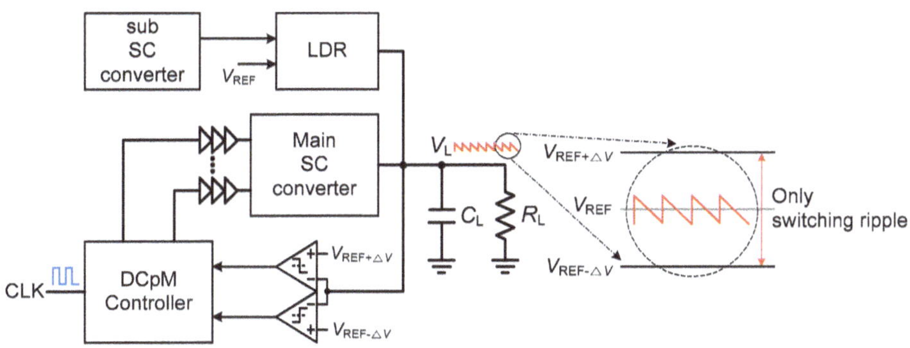

Figure 14. Block diagram of the proposed LDR-assisted SC DC–DC converter and its output ripple voltage.

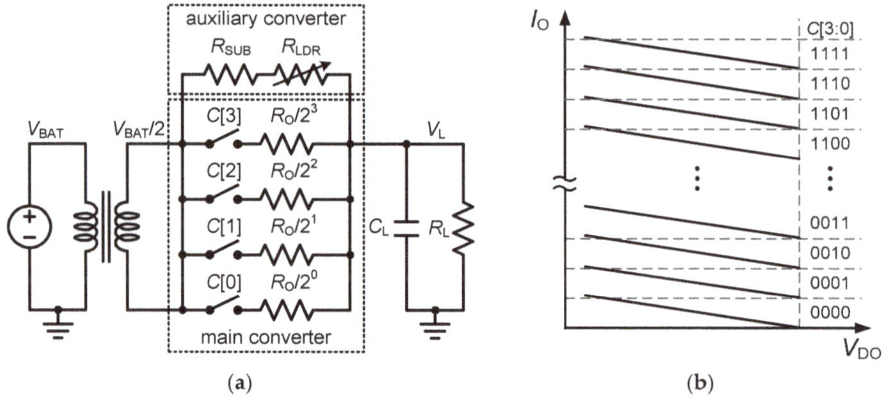

Figure 15. (**a**) The simplified model of the proposed switched-capacitor (SC) DC–DC converter and (**b**) output current versus dropout voltage of the LDR pass transistor.

4. Results and Discussion

The proposed SC DC–DC converter was implemented using a 0.13 µm CMOS process (Dongbu HiTek, Seoul, Korea), which provides triple-well CMOS devices and MIM capacitors with eight metal layers and one poly layer. Figure 16 shows the microphotograph of the fabricated SC DC–DC converter. The core chip has an area of 2.04 mm^2. Several pads are allocated to the input and output ports to reduce interconnection loss during measurement. The area of the capacitors is the major contributor to the size of the main SC DC–DC converter, converter cells, and load capacitor. Stacked capacitors utilizing the MIM and metal–oxide–semiconductor (MOS) capacitors are used to increase the capacitance per unit area, which are 1 fF/µm^2 and 2.5 fF/µm^2 for the MIM and MOS capacitors, respectively. Figure 17 shows the measured output voltage and current. The proposed converter has an output voltage range of 1.2 to 1.5 V from a 3.3 V supply. The output voltage waveforms were measured with the use of an MSO7104B oscilloscope (Keysight Technologies, Santa Rosa, CA, USA). The output voltage and LDR control signal are shown for the LDR in the on and off states in Figure 17a. The output voltage and current were set to 1.2 V and 100 µA, respectively. When the LDR was deactivated, a high ripple of approximately 380 mV was obtained. This is because the DCpM control code does not converge to one value at light loads, and the variation in the control code generates a large control ripple. However, the ripple drops below 10 mV upon activation of the LDR, which fine-tunes the output current and limits the DCpM control code to one value in the main SC DC–DC converter. Thus, the control ripple disappears due to the bounded DCpM control code, and only the switching ripple appears in the output voltage waveform. Figure 17b shows the load transient performance when the current suddenly changes from 120 µA to 50 mA. The output current and output voltage are restored to their regulated values in less than 800 ns.

Figure 18a shows the measured efficiency according to the output current at the input voltage of 3.3 V. The efficiency depends on the output voltage, with the output voltage of 1.5 V corresponding to the highest efficiency and lowest output current. This is because the voltage ratio of the input to output is the closest to the transformer ratio of the 2:1 step-down topology in this case. The peak efficiency is 73, 70, and 65% at output voltages of 1.5, 1.35, and 1.2 V, respectively. Figure 18b shows the measured output voltage ripple according to the output current. The maximum ripple values remain below 26, 36, and 55 mV at output voltages of 1.5, 1.35, and 1.2 V, respectively. Figure 19 shows the loss contributions and their ratio according to the output current. At the very low output current, the DCpM loss and the LDR quiescent loss decrease the power efficiency, but as the current increases, the switching loss and conduction loss dominates. Figure 19a shows an efficiency reduction of 2.3% due to the LDR loss at the output current of 5 mA but only a 0.23% reduction at the output current of 53 mA.

Figure 16. Microphotograph of the proposed switched-capacitor (SC) converter.

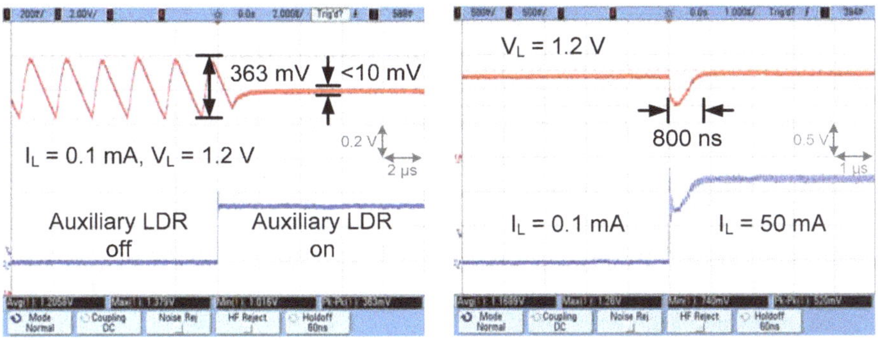

(a) (b)

Figure 17. Measured output voltage and current waveforms: (**a**) the ripple voltage at a low output current with and without the low dropout regulator (LDR) in operation and (**b**) the load transient responses to a sudden current variation.

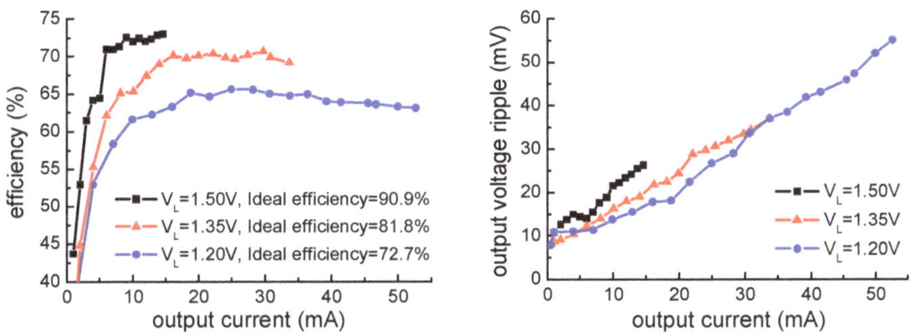

(a) (b)

Figure 18. (**a**) Measured efficiency and (**b**) voltage ripple according to output current.

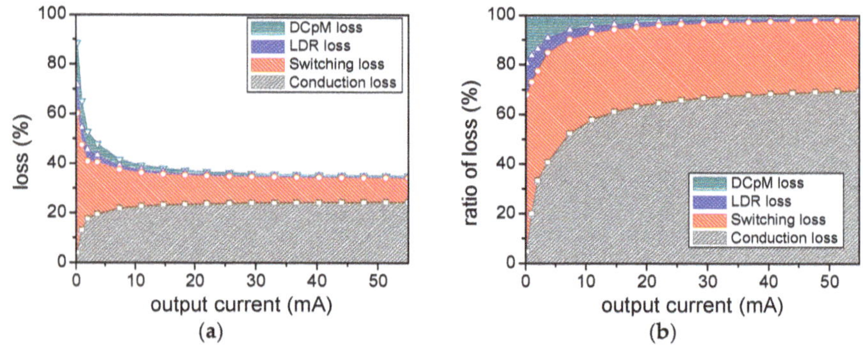

Figure 19. Loss contributions from DCpM, LDR, switching, and conduction losses versus output current for $V_L = 1.2$ V: (**a**) the loss contributions and (**b**) the ratio of loss contribution.

Table 1 compares the performance of the proposed SC DC–DC converter with similar low-ripple converters. As the ripple depends on the output current, load capacitance, and switching frequency, we used the following figure of merit for a fair ripple comparison [26,27]:

$$FoM_{\text{ripple}} = I_L / \left(C_L \cdot f_{SW} \cdot V_{\text{ripple}}\right) \tag{6}$$

As can be observed from the table, our approach affords the highest figure of merit.

Table 1. Comparison of results of previously reported studies and current study.

Characteristic	[15]	[22]	[26]	[28]	This Work
Technology (nm)	45	130	130	130	130
Input voltage (V)	1.8	1.8	1–1.2	1.2	3.3
Output voltage (V)	0.8–1	0.3–0.55	1.8–2.1	0.2–1.1	1.2–1.5
Maximum load current (mA)	10	55	2.61	2.53	53
Power density (mW/mm^2)	50	24.5	0.67	7.56	31.2
Flying capacitance (pF)	534	-	400	840	2176
Load capacitance (pF)	700	5000	400	764	1000
Ripple (mV)	<50	<50	<10	30 @I_L = 30 μA	8–55
Peak efficiency (%)	69	70	82	80.6	73
Switching frequency (MHz)	30	100	20	5	40
Figure of merit, Equation (6)	9.5	2.2	14	3.25	24.1
Active area (mm^2)	0.16	0.97	2.25	0.291	2.04

5. Conclusions

We proposed a fully-integrated SC DC–DC converter with low ripple and high efficiency. The proposed converter employs a four-bit DCpM control and includes an auxiliary LDR in parallel. The DCpM efficiently optimizes the gate switching losses according to the output current, and a two-boundary hysteresis method supports the ripple control. In the proposed structure, the feedback ripple is removed by means of an auxiliary LDR connected in parallel. Moreover, four-phase interleaving is employed to reduce the switching ripple in the main converter. The proposed SC DC–DC converter affords an output voltage range of 1.2–1.5 V from a 3.3 V supply and achieves a peak efficiency of 73% and ripple below 55 mV over the entire output current range. We believe that our approach will find utility in power sources for sensor-based devices of the future.

Author Contributions: J.-Y.L., G.-S.K., and K.-I.O. carried out the experiments and computer simulation and wrote the paper under the supervision of D.B.

Funding: This research received no external funding.

Acknowledgments: This work was supported by the Institute for Information and Communications Technology Promotion (IITP) grant funded by the Korea government (MSIT) (No. 2018-0-01663, Development of Communication-Sensing Converged B5G Millimeter Wave System) and by the Chung-Ang University Graduate Research Scholarship in 2018.

Conflicts of Interest: The authors declare no conflict of interest.

References

1. Su, W.; Wu, Z.; Fang, Y.; Bahr, R.; Raj, P.M.; Tummala, R.; Tentzeris, M.M. 3d printed wearable flexible SIW and microfluidics sensors for internet of things and smart health applications. In Proceedings of the 2017 IEEE MTT-S International Microwave Symposium (IMS), Honololu, HI, USA, 4–9 June 2017; pp. 544–547.
2. Bang, S.; Blaauw, D.; Sylvester, D. A Successive-Approximation Switched-Capacitor DC–DC Converter With Resolution of $V_{IN}/2^N$ for a Wide Range of Input and Output Voltages. *IEEE J. Solid-State Circuits* **2016**, *51*, 543–556.
3. Saif, H.; Lee, Y.; Lee, H.; Kim, M.; Khan, M.; Chun, J.H.; Lee, Y. A Wide Load Current and Voltage Range Switched Capacitor DC–DC Converter with Load Dependent Configurability for Dynamic Voltage Implementation in Miniature Sensors. *Energies* **2018**, *11*, 3092. [CrossRef]
4. Pei, C.; Booth, R.; Ho, H.; Kusaba, N.; Li, X.; Brodsky, M.; Iyer, S. A novel, low-cost deep trench decoupling capacitor for high-performance, low-power bulk CMOS applications. In Proceedings of the 2008 9th International Conference on Solid-State and Integrated-Circuit Technology, Beijing, China, 20–23 October 2008; pp. 1146–1149.
5. Kudva, S.S.; Harjani, R. Fully-integrated on-chip DC-DC converter with a 450X output range. *IEEE J. Solid-State Circuits* **2011**, *46*, 1940–1951. [CrossRef]
6. Ahsanuzzaman, S.M.; Prodić, A.; Johns, D.A. An integrated high-density power management solution for portable applications based on a multioutput switched-capacitor circuit. *IEEE Trans. Power Electron.* **2016**, *31*, 4305–4323. [CrossRef]
7. Zimmer, B.; Lee, Y.; Puggelli, A.; Kwak, J.; Jevtić, R.; Keller, B.; Bailey, S.; Blagojević, M.; Chiu, P.F.; Le, H.P.; et al. A RISC-V vector processor with simultaneous-switching switched-capacitor DC–DC converters in 28 nm FDSOI. *IEEE J. Solid-State Circuits* **2016**, *51*, 930–942.
8. Sung, E.T.; Park, S.; Baek, D. A Fast-Transient Output Capacitor-Less Low-Dropout Regulator Using Active-Feedback and Current-Reuse Feedforward Compensation. *Energies* **2018**, *11*, 688. [CrossRef]
9. Lu, Y.; Jiang, J.; Ki, W.H. Design Considerations of Distributed and Centralized Switched-Capacitor Converters for Power Supply On-Chip. *IEEE J. Emerg. Sel. Top. Power Electron.* **2018**, *6*, 515–525. [CrossRef]
10. Dini, M.; Romani, A.; Filippi, M.; Tartagni, M. A nanocurrent power management IC for low-voltage energy harvesting sources. *IEEE Trans. Power Electron.* **2016**, *31*, 4292–4304. [CrossRef]
11. Avalur, K.K.G.; Azeemuddin, S. A 6–18 V hybrid power management IC with adaptive dropout for improved system efficiency up to 150 °C. *IEEE J. Emerg. Sel. Top. Power Electron.* **2018**, *6*, 477–484. [CrossRef]
12. Wibben, J.; Harjani, R. A high-efficiency DC–DC converter using 2 nH integrated inductors. *IEEE J. Solid-State Circuits* **2008**, *43*, 844–854. [CrossRef]
13. Li, Q.; Dong, Y.; Lee, F.C.; Gilham, D.J. High-density low-profile coupled inductor design for integrated point-of-load converters. *IEEE Trans. Power Electron.* **2013**, *28*, 547–554. [CrossRef]
14. Chang, L.; Montoye, R.K.; Ji, B.L.; Weger, A.J.; Stawiasz, K.G.; Dennard, R.H. A fully-integrated switched-capacitor 2:1 voltage converter with regulation capability and 90% efficiency at 2.3 A/mm^2. In Proceedings of the 2010 Symposium on VLSI Circuits, Honolulu, HI, USA, 16–18 June 2010; pp. 55–56.
15. Ramadass, Y.K.; Fayed, A.A.; Chandrakasan, A.P. A fully-integrated switched-capacitor step-down DC-DC converter with digital capacitance modulation in 45 nm CMOS. *IEEE J. Solid-State Circuits* **2010**, *45*, 2557–2565. [CrossRef]
16. Lee, H.; Mok, P.K. An SC voltage doubler with pseudo-continuous output regulation using a three-stage switchable opamp. *IEEE J. Solid-State Circuits* **2007**, *42*, 1216–1229. [CrossRef]
17. Calhoun, B.H.; Chandrakasan, A.P. Ultra-dynamic voltage scaling (UDVS) using sub-threshold operation and local voltage dithering. *IEEE J. Solid-State Circuits* **2006**, *41*, 238–245. [CrossRef]

18. Sanders, S.R.; Alon, E.; Le, H.P.; Seeman, M.D.; John, M.; Ng, V.W. The road to fully integrated DC–DC conversion via the switched-capacitor approach. *IEEE Trans. Power Electron.* **2013**, *28*, 4146–4155. [CrossRef]
19. Le, H.P.; Sanders, S.R.; Alon, E. Design techniques for fully integrated switched-capacitor DC-DC converters. *IEEE J. Solid-State Circuits* **2011**, *46*, 2120–2131. [CrossRef]
20. Shenoy, P.S.; Amaro, M.; Morroni, J.; Freeman, D. Comparison of a buck converter and a series capacitor buck converter for high-frequency, high-conversion-ratio voltage regulators. *IEEE Trans. Power Electron.* **2016**, *31*, 7006–7015. [CrossRef]
21. Jain, R.; Kim, S.T.; Vaidya, V.; Ravichandran, K.; Tschanz, J.W.; De, V. Conductance modulation techniques in switched-capacitor DC-DC converter for maximum-efficiency tracking and ripple mitigation in 22 nm tri-gate CMOS. *IEEE J. Solid-State Circuits* **2015**, *50*, 1809–1819. [CrossRef]
22. Kudva, S.S.; Harjani, R. Fully Integrated Capacitive DC–DC Converter with All-Digital Ripple Mitigation Technique. *IEEE J. Solid-State Circuits* **2013**, *48*, 1910–1920. [CrossRef]
23. Jeon, H.; Kim, K.K.; Kim, Y.B. Fully Integrated on-Chip Switched DC–DC Converter for Battery-Powered Mixed-Signal SoCs. *Symmetry* **2017**, *9*, 18. [CrossRef]
24. Seeman, M.D.; Sanders, S.R. Analysis and optimization of switched-capacitor DC–DC converters. *IEEE Trans. Power Electron.* **2008**, *23*, 841–851. [CrossRef]
25. Butzen, N.; Steyaert, M.S. Scalable parasitic charge redistribution: Design of high-efficiency fully integrated switched-capacitor DC–DC converters. *IEEE J. Solid-State Circuits* **2016**, *51*, 2843–2853. [CrossRef]
26. Van Breussegem, T.; Steyaert, M. A 82% efficiency 0.5% ripple 16-phase fully integrated capacitive voltage doubler. In Proceedings of the 2009 Symposium on VLSI Circuits, Kyoto, Japan, 16–18 June 2009; pp. 198–199.
27. Yoo, A.; Chang, M.; Trescases, O.; Wang, H.; Ng, W.T. FOM (Figure of Merit) Analysis for Low Voltage Power MOSFETs in DC-DC Converter. *Proc. IEEE Conf. Electron Devices Solid-State Circuits* **2007**, 1039–1042.
28. Lisboa, P.C.; Pérez-Nicoli, P.; Veirano, F.; Silveira, F. General top/bottom-plate charge recycling technique for integrated switched capacitor DC-DC converters. *IEEE Trans. Circuits Syst. I* **2016**, *63*, 470–481. [CrossRef]

electronics

MDPI

Article

Design of Voltage Mode Electronically Tunable First Order All Pass Filter in ±0.7 V 16 nm CNFET Technology

Muhammad Masud [1,*], Abu A'ain [1,2], Iqbal Khan [3] and Nasir Husin [1]

1 School of Electrical Engineering, Universiti Teknologi Malaysia, Johor Bahru 81310, Malaysia; abukhari@uthm.edu.my (A.A.); nasirsh@fke.utm.my (N.H.)
2 Institute for Integrated Engineering, Universiti Tun Hussein Onn, Johor 86400, Malaysia
3 Department of Electrical Engineering, Umm Al Qura University, Makkah 21955, Saudi Arabia; iqbalakhan19@rediffmail.com
* Correspondence: idreesmasud@yahoo.com; Tel.: +6-0166-128-954

Received: 11 November 2018; Accepted: 9 January 2019; Published: 15 January 2019

Abstract: A novel voltage mode first order active only tuneable all pass filter (AOTAPF) circuit configuration is presented. The AOTAPF has been designed using ±0.7 V, 16 nm carbon nanotube field effect transistor (CNFET) Technology. The circuit uses CNFET based varactor and unity gain inverting amplifier (UGIA). The presented AOTAPF is realized with three N-type CNFETs and without any external passive components. It is to be noted that the realized circuit uses only two CNFETs between its supply-rails and thus, suitable for low-voltage operation. The electronic tunability is achieved by varying the voltage controlled capacitance of the employed CNFET varactor. By altering the varactor tuning voltage, a wide tunable range of pole frequency between 34.2 GHz to 56.9 GHz is achieved. The proposed circuit does not need any matching constraint and is suitable for multi-GHz frequency applications. The presented AOTAPF performance is substantiated with HSPICE simulation program for 16 nm technology-node, using the well-known Stanford CNFET model. AOTAPF simulation results verify the theory for a wide frequency-range.

Keywords: APF; CNFET; pole-frequency; chirality; phase angle; tuning

1. Introduction

First order active all pass filter (APF) is an important analog signal processing (ASP) module. It is used for design of multiphase oscillators, phase-equalizers and high-quality-factor frequency-selective circuits. Several first order voltage mode (VM) single-ended (SE) APF circuit realizations have been reported in technical literature [1–8]. These APF circuits use a variety of efficient active-building-blocks (ABBs). However, these realized APFs are based on passive elements and large number of transistors count, which result in larger chip-area, lower slew rate, higher power dissipation and limitations to higher frequency operations. Few such APF configurations with low active and passive component counts are also available in the technical literature [9–18]. Some of these APF circuits also employ one or more ideal DC current-sources for biasing, which further increase the transistor count [15,17,18].

From the integrated circuit point of view, the active only filters (AOFs) provide several attractive advantages like capability of operating at much higher frequencies, lesser chip area, low power dissipations and electronic tunabilty. As a result, few first order AOTAPFs are reported in the technical literature [19–21]. These AOTAPFs use the MOSFETs transconductance and intrinsic gate to source parasitic capacitance as filter design components; still, the frequency of operations falls with in MHz range. Moreover, these reported AOTAPF circuits also contain a large number of transistors.

The APF circuits proposed in the technical literature [1–21] are based on bulk-semiconductor-technology. This technology faces serious challenges due to the persistent focus on transistor-scaling in nano meter regime for further continuation of Moore's law. These obstacles contain scattering-effect, decreased gate control, parasitic-capacitance, drain to source tunneling, channel mobility, threshold-voltage-variability [22,23]. It has been proven experimentally that below 65 nm-node, high-frequency analog circuit performance of silicon based semiconductor CMOS-technology is seriously degraded [24,25]. These emerging difficulties led the integrated-circuits industry to explore alternative materials and devices for below 65 nm-node that work equally well for future high-frequency ASP applications and more than Moore's technologies devised by ITRS recently [26]. These include double gate FETs, FinFETs and carbon nanotube field effect transistors (CNFETs) etc. [22].

Among these new devices, CNFETs are self-evident frontrunners for future continuation of downscaling the feature length to further extend the saturated Moore's law in nanometer-regime in the case of CMOS-technology [23,25,27]. CNFET has potential to minimize the serious emerging problems of current CMOS-based technology due to its near ballistic-charge-conduction, smaller feature size, fast switching-speed, lower power-dissipation, higher cutoff frequency, lower parasitic capacitances and larger drive-current [26,27]. These outstanding features make the CNFET a potential candidate for future high-frequency analog circuit applications. Since CNFET introduction as an alternative to MOSFET, limited studies on CNFET-based analog filter design have been carried out [28–36].

In this paper, a new VM SE CNFET-only APF is proposed. The realized APF has a very compact circuit structure and it is free from external passive capacitors and resistors. The proposed AOTAPF employs only three N-type CNFETs instead of massive ABBs. Moreover, the proposed topology is tuneable over a wide frequency range. In addition, the proposed circuit is free from any matching constraint and it is a potential candidate for low power, low voltage and high-frequency applications. The AOTAPF circuit is substantiated with HSPICE-simulation using the Stanford-CNFET-model.

The rest of this paper is organized as follows. Section 2 describes a brief overview of CNFET. The unity gain inverting amplifier (UGIA) with its equivalent parasitic model is discussed in Section 3. Section 4, illustrates the proposed CNFET-based VM AOTAPF. The performance and simulation results of the proposed AOTAPF are given in Section 5. Comparison of the proposed circuit with other compact topologies of APFs in the technical literature is presented in Section 6. Finally, Section 7 concludes the work.

2. Carbon Nano-Tube Field Effect Transistor

Carbon nanotubes (CNTs) are graphite-cylindrical-sheets (GCSs), which are considered as the most promising material for future nano-electronic devices and applications, due to their exceptional electronic, mechanical, chemical and thermal properties. CNTs are classified as single-wall CNTs and multi-wall CNTs. Single-wall CNT is based on single GCS while multi-wall CNT consists of more than one GCS. The properties of single-wall CNT are dependent on the chirality-vector (C_h) [28,29]. The C_h is defined by vector indices n_1 and n_2, which are positive-integers. The arrangement-angle of carbon atoms along the CNT is determined by C_h. The single-wall CNT can be of metallic or semiconducting behavior depending on the vector indices n_1 and n_2. If $\mid n_1 - n_2 \mid$ is an integer-multiple of three or $n_1 = n_2$, the single-wall CNT behaves as metallic, otherwise it behaves as semiconductor. The C_h, diameter (D_T) and threshold-voltage (V_{th}) of a CNT are related by the following equations.

$$C_h = a\sqrt{n_1^2 + n_2^2 + n_1 n_2} \tag{1}$$

$$D_T = C_h / \pi \tag{2}$$

$$V_{th} = aV_\pi / \sqrt{3}eD_T \tag{3}$$

where, e is the unit-electron-charge and a is the graphene-lattice-constant with a value of $2.49A^\circ$. V_π is the π to π bond-energy in tight-bonding-model with a value of 3.033 eV [30]. CNFET is one of the most attractive applications of CNT, which is obtained by replacing the MOSFET channel with one or more single-wall semiconducting CNTs as a channel material, as shown in Figure 1. Like conventional-MOSFET, CNFET is also a voltage-controlled-device and the current through the CNT based channel is controlled via gate voltage. CNFET gate is coupled capacitively with the beneath channel that consists of one or more narrow CNT. A single-CNT provides a limited amount of current. To enhance the channel current significantly, multiple parallel CNTs are incorporated in the channel. As compared to CMOS, where the design is dependent on the aspect ratio of transistors, a CNFET is usually optimized in terms of D_T, number of CNTs (N_T) and inter-CNT pitch (S_T). The S_T is basically the distance between the centers of two adjacent CNTs under the same gate. The width of the CNFET gate (W_g) is determined by the following equation [35]:

$$W_g = min(W_{min}, (N_T - 1)S_T + D_T) \tag{4}$$

where, W_{min} is the minimum gate width. The CNFET gate capacitance (C_g), is one of the key device features and it significantly affects the performance, especially at high-frequencies. The C_g is composed of three different capacitive components; coupling capacitance among the gate and adjacent contacts (C_{gtg-t}), gate outer-fringe capacitance (C_{fr-t}) and gate to channel capacitance (C_{gc-t}). The C_{gc-t} is further composed of two capacitances C_{gc-m} and C_{gc-e}, which are capacitances of single-wall CNTs located in the middle and edge of CNFET respectively. The C_{gc-t} components are shown in Figure 2. The CNFET gate capacitance parameters are:

$$C_{gtg-t} = W_g C_{gtg} \tag{5}$$

$$C_{fr-t} = L_s C_{fr} \tag{6}$$

$$C_{gc-t} = L_g C_{gc} \tag{7}$$

where, W_g and L_g are the CNFET channel-width and channel-length respectively. L_s is the length of doped source-side extension region. C_{gtg} is the gate-coupling capacitance per unit gate-width, C_{gc} is the gate to channel capacitance per unit channel-length and C_{fr} is the gate outer-fringe capacitance per unit CNT length. Comparatively to C_{gc} and C_{gtg}, the C_{fr} capacitance magnitude is quite smaller and thus its effect can be ignored [35]. The C_g thus can be approximated as:

$$C_g \approx (C_{gtg} * W_g) + (C_{gc} * L_g) \tag{8}$$

The drain/source capacitance ($C_{d/s}$) can be determined by following relation.

$$C_{d/s} \approx (C_{sub}/C_{ox} + 1) + (C_{gd/gs}) \tag{9}$$

where, C_{sub} is the capacitance between the CNFET channel and substrate, C_{ox} is the capacitance between the CNFET gate and channel and $C_{gd/gs}$ is the capacitance of the CNFET gate to the drain/source contact. The ratio C_{sub}/C_{ox} is only important when CNFET substrate drive (switches) the gate. To assess the potential performance of CNFET, an accurate and efficient device-model is required, which incorporates typical non-idealities of CNFET device. Most available models of CNFETs in recent literature bear the drawback of ideal modeling, resulting in ignoring numerous important effects [36,37]. The Stanford CNFET model overcomes shortcomings of previous models by including several non-idealities like drain to source series resistance, interconnect wiring capacitance, finite scattering mean free path, inter CNT charge-screening-effect, effect of drain-source extension region and many more [38]. The Stanford CNFET model has been experimentally validated and it efficiently predicts the dynamic and transient performance with more than 90% accuracy [34].

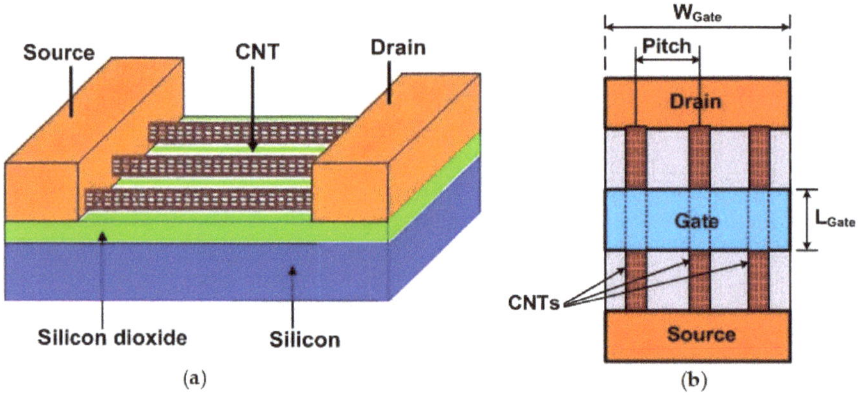

Figure 1. Carbon nanotube field effect transistor (CNFET) (**a**) Schematic; (**b**) Top-View.

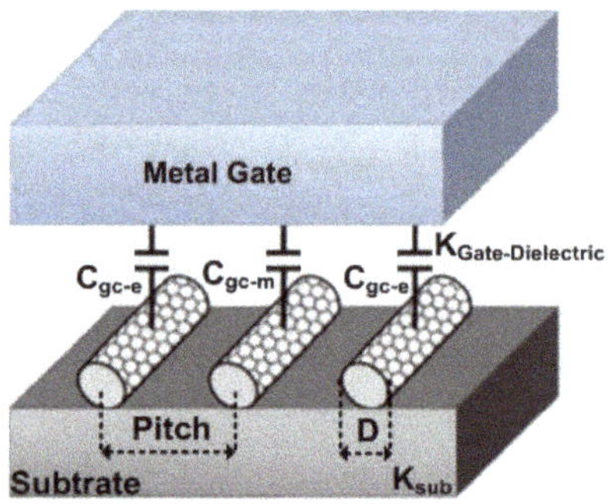

Figure 2. CNFET gate to channel capacitance.

Some important Stanford CNFET-model parameters are shown in Table 1.

Table 1. The Stanford CNFET model parameters.

Parameter	Description	Value
V	Power-supply	0.7 V
L_g	Physical-channel-length	16 nm
S_T	CNTs-pitch	10 nm
(n_1, n_2)	CNTs-chirality	(19, 0)
Lceff	Mean free-path in intrinsic-CNT	200 nm
V_{fbn}	N-type CNFET flatband-voltage	0
High-K_{ox}	Dielectric material of top-gate	HfO_2 (16)
L_s	Source-side length of doped-CNT	16 nm
L_d	Drain-side length of doped-CNT	16 nm
T_{ox}	Oxide-thickness	4 nm
K_{sub}	Dielectric constant	SiO_2 (4)
Leff	Mean free-path in doped-CNT	15 nm
E_{fo}	Fermi-level of n+ doped drain/source CNT-region	0.6 eV
N_T	Total number of CNT used per CNFET	~

~: Variable parameter.

3. CNFET Based UGIA

The UGIA is one of the simplest types of ABB, which employs two N-type CNFETs as shown in Figure 3a [16] and its symbol is shown in Figure 3b. Its transfer gain can be expressed as follows.

$$\frac{V_o}{V_i} = -\frac{g_{m2}}{g_{m1}} \tag{10}$$

where, g_{m1} and g_{m2} are the transconductance gains of the CNFETs T_1 and T_2 respectively. With symmetrical T_1 and T_2 on the same die, the $g_{m1} = g_{m2}$, the Equation (10) reduces to

$$\frac{V_o}{V_i} = -1 \tag{11}$$

Thus, the circuit of Figure 3a, realizes an unity gain inverting amplifier (UGIA). The UGIA equivalent parasitic-model is shown in Figure 3c. Its input and output port resistances can be expressed as

$$r_i = r_g \tag{12}$$

$$r_o = r_{ds1} || r_{ds2} \tag{13}$$

where, r_g represents the gate-resistance of transistor T_2 and r_{ds1}, r_{ds2} are the channel-resistances of transistors T_1 and T_2 respectively. The UGIA input port has very high-resistance. The UGIA output port, being the voltage source, exhibits small resistance.

The impact of increasing CNTs (N_T) of both the CNFETs of the UGIA on its performance is studied using HSPICE simulation tool. In the simulations, the Stanford CNFET model is used for the CNFETs with transistor parameters of Table 1. Figure 4 demonstrates the impact of N_T on r_o, C_i, C_o, power dissipation and -3 dB bandwidth of the UGIA. A single CNT carries approximately a constant current of 20 μA [28]. The increase of N_T of transistors increase the overall current drive capability and hence the transconductance [29]. The impact of increasing N_T on UGIA r_o is shown in Figure 4a. It is seen that by increasing N_T, the output resistance r_o decreases. Since an increase in N_T is equivalent to an increase in channel width of the CNFETs, r_o decreases with the increase of N_T. The effects of increasing N_T on input and output parasitic capacitances C_i and C_o of the UGIA, are shown in Figure 4b. It is observed that by increasing N_T, both the parasitic capacitances C_i and C_o increase. Figure 4c demonstrates the effect of increasing N_T on the UGIA power dissipation. The power dissipation of UGIA increases as N_T increases. The current drive capability of employed CNFETs increases with an increase of N_T, which leads to an increase in power dissipation. Figure 4d demonstrates the effect of increasing N_T, on UGIA -3 dB bandwidth. It is observed that by increasing N_T, the -3 dB bandwidth of UGIA increases.

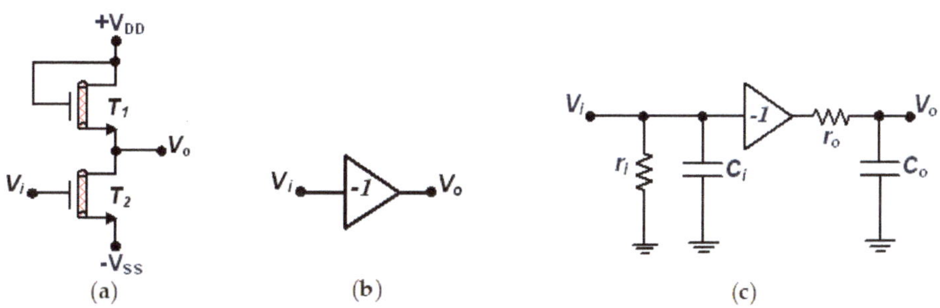

Figure 3. CNFET based UGIA: (**a**) Transistor-level realization; (**b**) Symbol; (**c**) Parasitic model.

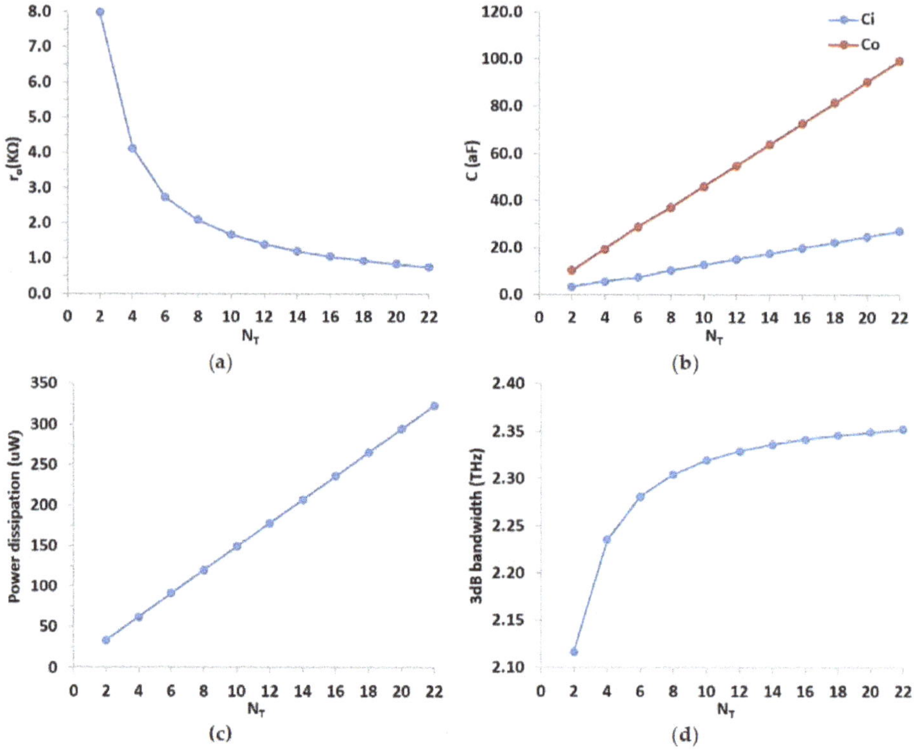

Figure 4. Effect of variation of N_T on unity gain inverting amplifier (UGIA): (**a**) r_o; (**b**) C_i and C_o; (**c**) Power dissipation; (**d**) -3 dB bandwidth.

The transient and AC-analysis were performed with CNFET parameters of Table 1, with $N_T = 2$. Figure 5a shows the transient-response of UGIA input and output voltage at 50 GHz. Figure 5b displays the UGIA AC simulation results of voltage-gain (V_o/V_i). It is seen that the obtained voltage-gain magnitude is unity over a wide range of frequency. The -3 dB frequency of employed UGIA voltage-gain is 2.1172 THz. This massive value of -3 dB cutoff frequency makes the UGIA a potential candidate for the design of high frequency ASP modules.

The UGIA parasitic capacitance C_i and resistance r_i are found as 3.54 aF and 1 TΩ respectively. Figure 5c displays the frequency response of UGIA output port resistance (r_o), which is constant at r_o = 7.9921 kΩ over wide frequency-range. The -3 dB cutoff frequency of the UGIA output-impedance is obtained as 1.9017 THz. The UGIA output parasitic-capacitance, C_o is found as 10.472 aF, which is nearly insignificant for frequency-range up to several GHz. The total-harmonic-distortion (THD) of UGIA is determined by applying a 50 GHz sinusoidal-signal to input with different voltage amplitudes. Simulation results are presented in Figure 6a. It can be seen that THD is less than 1% for sinsoidal signal with amplitude of 200 mV. Monte Carlo simulation results of the UGIA were performed for 30-trials with transient and AC-sweep environment to see the influence of process-variations. Figure 6b,c illustrate the results of Monte Carlo analysis for UGIA transient and AC-sweep respectively.

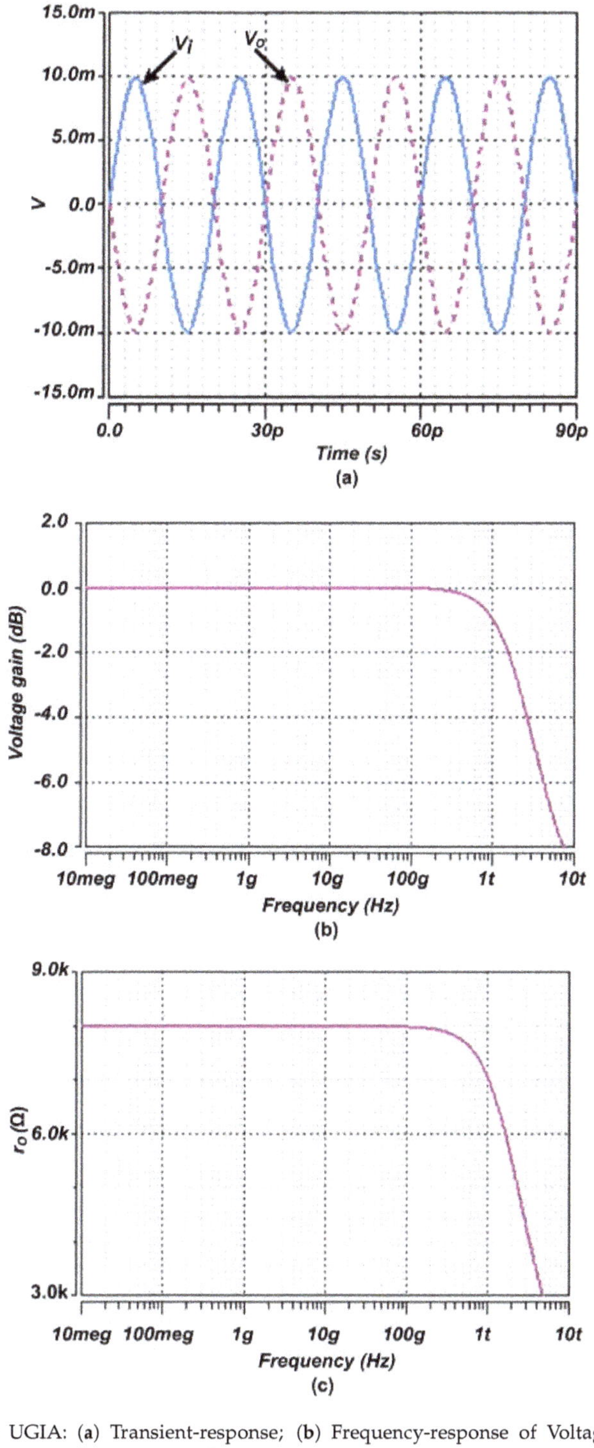

Figure 5. The UGIA: (**a**) Transient-response; (**b**) Frequency-response of Voltage-gain (V_o/V_i); (**c**) Frequency-response of Output-impedance.

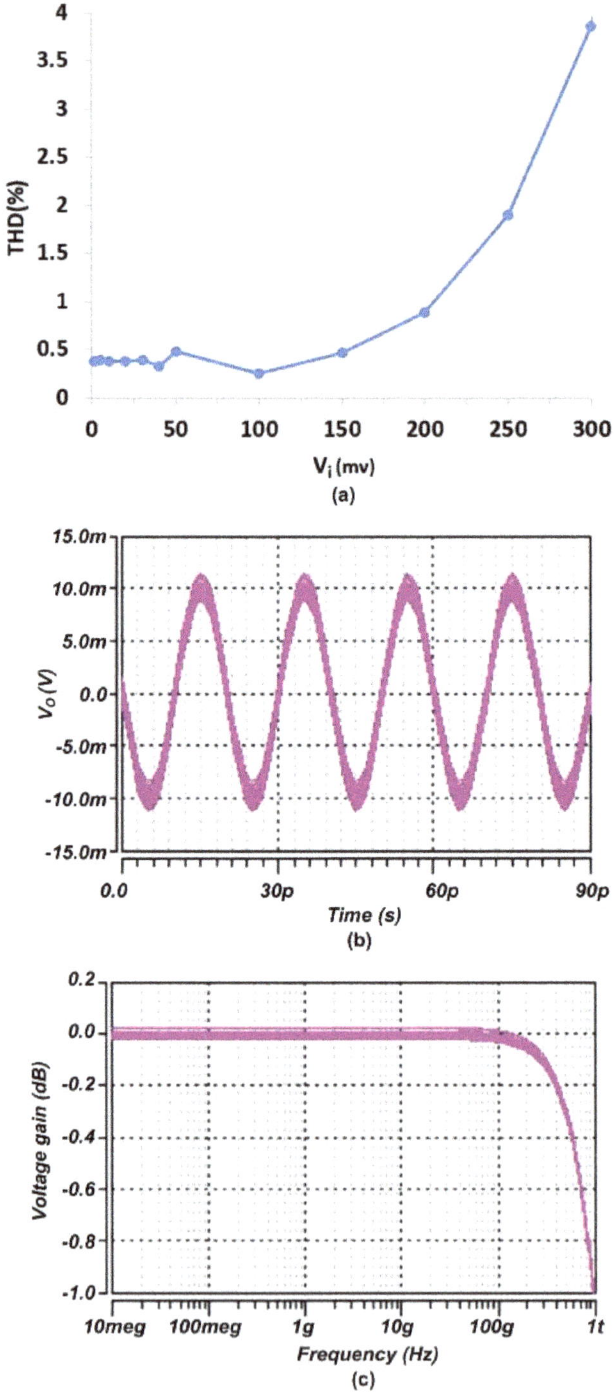

Figure 6. The UGIA: (**a**) THD Vs input voltage at 50 GHz; (**b**) Monte Carlo simulations for V_o in time domain; (**c**) Monte Carlo simulations for voltage gain in frequency domain.

4. AOTAPF Circuit Description

The basic scheme for first order APF section is given in Figure 7a. Its transfer function can be expressed as follows.

$$\frac{V_o}{V_i} = \frac{s-a}{s+a} \tag{14}$$

Its equivalent RC circuit along with a unity gain inverting amplifier is shown in Figure 7b, where pole frequency $\omega_o = a = 1/RC$. The CNFET version of Figure 7b is given in Figure 7c, where the unity gain inverting amplifier is replaced with UGIA of Figure 3a and the capacitor C is replaced with a CNFET based varactor capacitance C_{var} between input and output. The N-Type CNFET based varactor used in Figure 7c is given in Figure 8a. Its equivalent symbol is shown in Figure 8b. The varactor CNFET source and drain are tied together and connected to the tuning control voltage (V_{tune}) to form one capacitor terminal (x), while the gate form the other terminal (y). The varactor capacitance (C_{var}) can be controlled by varying V_{tune}. The output resistance r_o of UGIA is utilized to the benefit, to replace resistor R of Figure 7b. The circuit of Figure 7c results in an active only tunable all pass filter (AOTAPF).

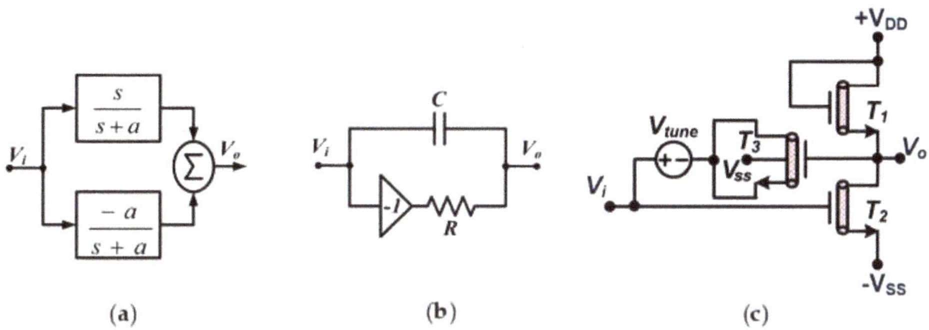

Figure 7. First order APF: (**a**) Basic scheme; (**b**) Equivalent circuit; (**c**) CNFET based implementation.

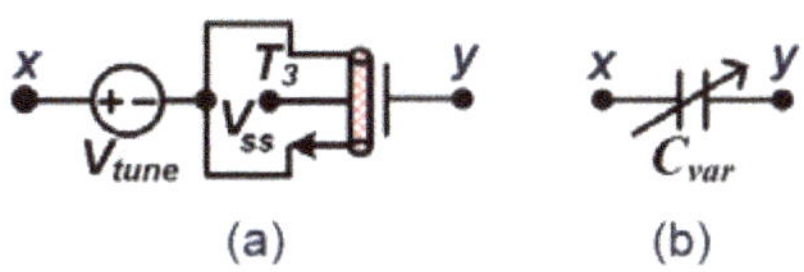

Figure 8. CNFET based varactor: (**a**) Transistor-level realization; (**b**) Symbol.

Ignoring the effect of extremely low valued output capacitance C_o of UGIA, the proposed VM SE AOTAPF circuit shown in Figure 7c results in the following voltage transfer function (VTF).

$$\frac{V_o}{V_i} = \frac{(s - \frac{1}{r_o C_{var}})}{(s + \frac{1}{r_o C_{var}})} \tag{15}$$

From Equation (15), the pole-frequency ($\omega_o = \omega_z = \omega_p$) and the phase-angle ($\phi$), can be expressed respectively as:

$$\omega_o = \frac{1}{r_o C_{var}} \tag{16}$$

$$\phi = \pi - 2tan^{-1}(\omega r_o C_{var}) \tag{17}$$

The Proposed AOTAPF pole-frequency incremental sensitivity with respect to the components C_{var} and r_o can be expressed as:

$$S^{\omega_o}_{C_{var}} = S^{\omega_o}_{r_o} = -1 \tag{18}$$

From Equation (18), it is observed that the incremental sensitivities of the pole-frequency (ω_o) with respect to C_{var} and r_o are within unity in magnitude. By considering the UGIA non-ideal voltage-gain (α) and parasitic resistance (r_s) of tuning control voltage (V_{tune}) into consideration, the VTF of Equation (15) can be expressed as follows.

$$\frac{V_o}{V_i} = \frac{(s(1-\frac{r_s}{r_o}) - \frac{\alpha}{r_o C_{var}})}{(s(1+\frac{r_s}{r_o}) + \frac{1}{r_o C_{var}})} \tag{19}$$

From Equation (19) it is seen that the AOTAPF gain and pole-frequency (ω_p) is insensitive to α. However, the zero-frequency (ω_z) is affected slightly due to α. Moreover, the impact of source resistance (r_s) on the performance of APF is negligible due to the presence of high valued output resistance (r_o) of the UGIA ($r_o \gg r_s$). Thus, the effect of r_s can be ignored. By considering α into account, the non-ideal phase-angle for the realized filter can be expressed as follows.

$$\phi = \pi - tan^{-1}(\frac{\omega r_o C_{var}}{\alpha}) - tan^{-1}(\omega r_o C_{var}) \tag{20}$$

Thus, it is seen from Equation (20) that the phase-angle is slightly affected by α. To examine the high-frequency performance of the realized AOTAPF, the UGIA parasitic impedances must be evaluated. By considering the α and non-ideal parasitic impedances of UGIA, the ideal VTF of the realized AOTAPF as illustrated by Equation (15) turns out to be

$$\frac{V_o}{V_i} = \frac{C_{var}}{C_{var}+C_o} * \frac{(s - \frac{\alpha}{r_o C_{var}})}{(s + \frac{1}{r_o (C_{var}+C_o)})} \tag{21}$$

From Equation (21), the ω_z and ω_p can be written as

$$\omega_z = \frac{\alpha}{r_o C_{var}} \tag{22}$$

$$\omega_p = \frac{1}{r_o (C_{var}+C_o)} \tag{23}$$

It is evident from Equation (22) that the non-ideal factor α sightly affects the zero-frequency. In addition, it can be noticed from Equation (23) that UGIA parasitic capacitance C_o affects the pole-frequency. The influence of the C_o on the performance of the AOTAPF can be minimized by making $C_{var} \gg C_o$.

5. Design and Verification

To verify the proposed AOTAPF circuit, it is designed and simulated using HSPICE simulation tool with the Stanford CNFET model parameters of Table 1. Based on Equation (21), the value of the varactor capacitance (C_{var}) is to be set sufficiently higher than the parasitic capacitance (C_o), to evade the mismatch between zero and pole frequencies as well as non-unity gain for higher frequencies design. Figure 9 shows the capacitance tuning characteristics (C-V curves) of the realized CNFET varactor of Figure 8a, with different values of N_T. It has been observed that by increasing N_T, the capacitance spread ($C_{max} - C_{min}$), increases, which ultimately determines the frequency tuning range of AOTAPF. The C-V relationship approximated by polynomial curve fitting is given in Appendix A. For instance, with $N_T = 100$ and by setting $V_{tune} = -0.32$ V for CNFET T_3, the observed C_{var} is 0.40423 fF. Thus, with r_o = 7.9921 kΩ, Equation (16) yields the pole frequency f_o = 49.26 GHz.

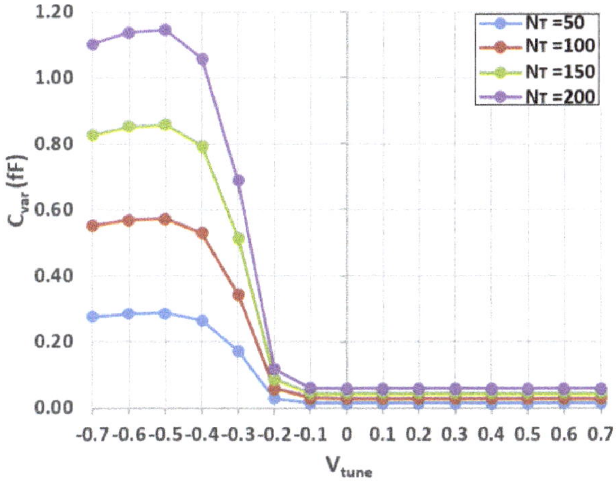

Figure 9. CV characteristics of varactor with different N_T.

The designed circuit was simulated with a sinusoidal input signal of 10 mV peak. The results thus obtained are given in Figures 10–12. The transient response at the designed frequency of f_o = 49.26 GHz is shown in Figure 10, where a phase shift of 90° is evident. Figure 11a,b show the ideal and simulated magnitude and phase responses respectively. The proposed AOTAPF power dissipation is found to be 33.76 μW. It is noticed that the realized APF dissipates very small power, even at very high frequency of operation. Figure 12 shows the equivalent input and output noises against the frequency. It is noticed that the equivalent input noise and output noise for the realized AOTAPF at a designed pole-frequency of 49.26 GHz are found as 6.822 nv/Hz and 6.761 nv/Hz respectively, which are satisfactorily low values. Monte Carlo simulation results of AOTAPF were performed for 30-trials with transient and AC-sweep environment to see the influence of process-variations. Figure 13a–c illustrate the results of the AOTAPF Monte Carlo analysis for transient, voltage gain and phase responses respectively. Here, it is observed from Figure 13 that there are no considerable variations of the filter performance characteristics.

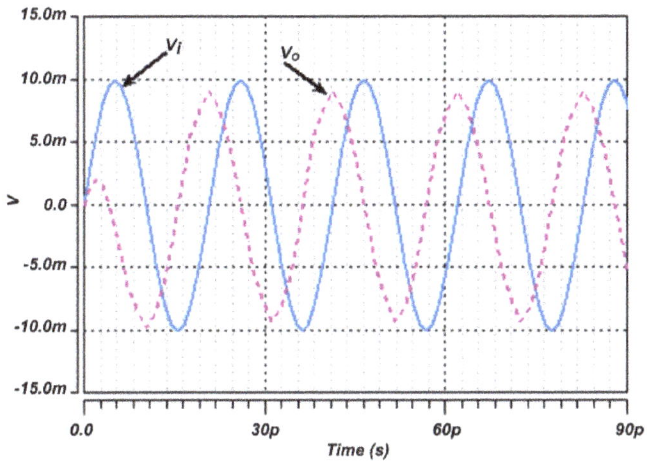

Figure 10. Transient-response of AOTAPF at pole-f_o = 49.26 GHz and $V_{tune} = -0.32$ V.

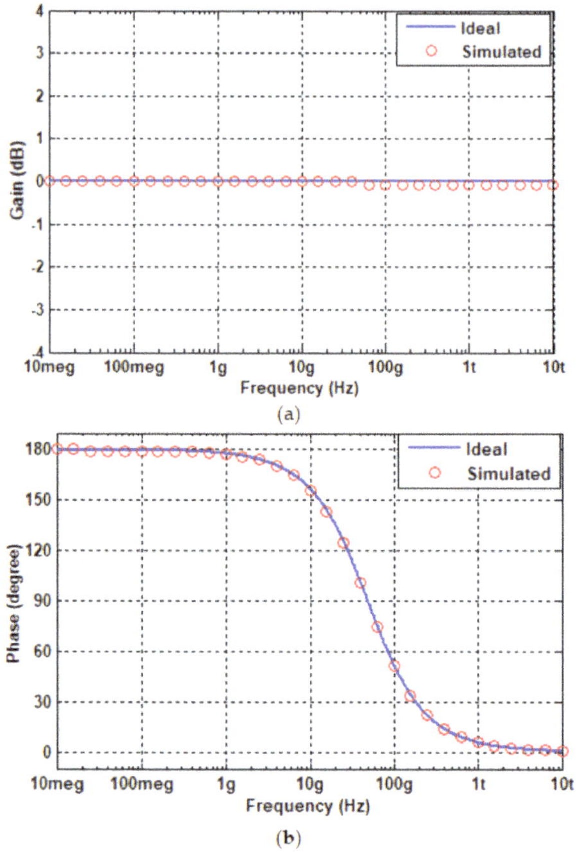

Figure 11. Ideal and simulated frequency-response of AOTAPF at V_{tune} = −0.32 V: (**a**) Voltage gain; (**b**) Phase.

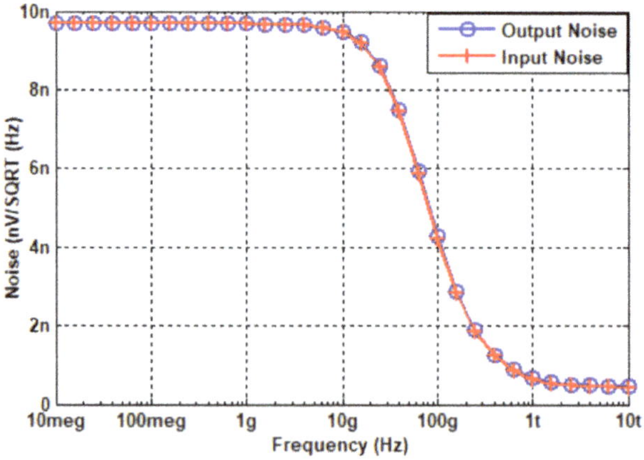

Figure 12. Frequency-response of input and output noise of AOTAPF at $V_{tune} = -0.32$ V.

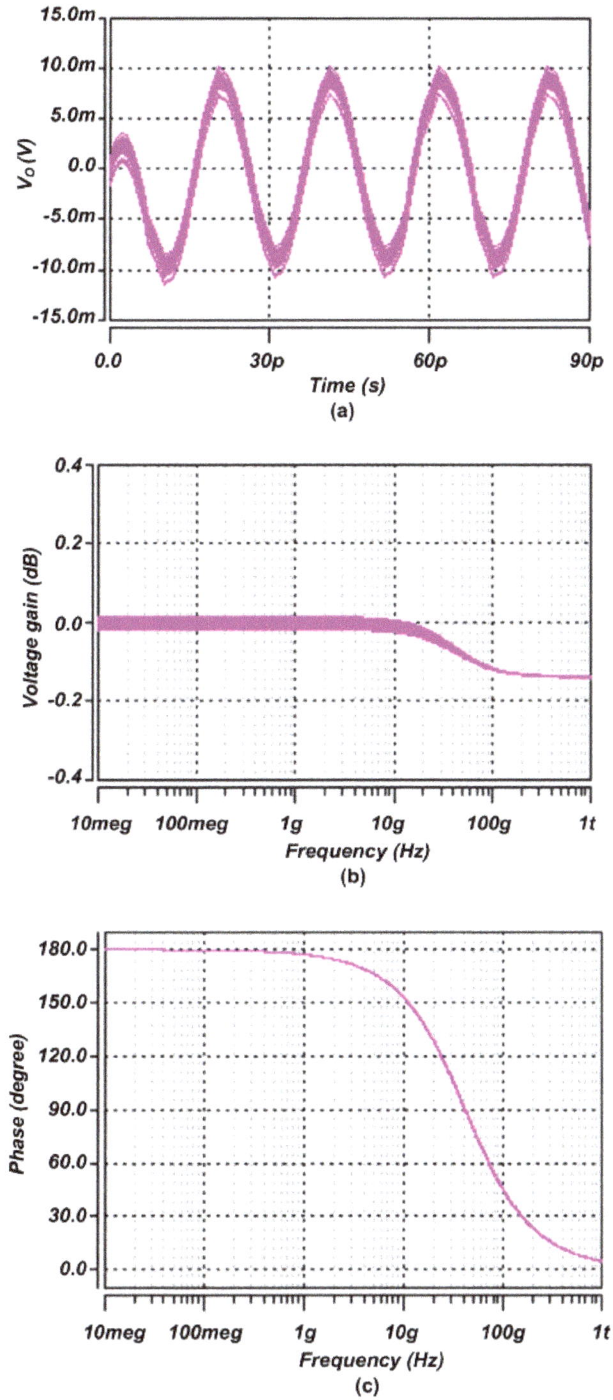

Figure 13. Monte Carlo simulations of AOTAPF for: (**a**) Time domain; (**b**) Voltage-gain (V_o/V_i); (**c**) Phase.

Next, to demonstrate the proposed circuit tunabilty feature, different tuning voltages (V_{tune}) are applied to the varactor. By varying the V_{tune} from −0.5 V to −0.3 V the varactor capacitance (C_{var}) varies in the range of 0.574 fF to 0.346 fF respectively. Figure 14a,b demonstrate the magnitude and phase responses respectively of the realized AOTAPF, at different values of V_{tune}. It is noticed from Figure 14b that by varying the V_{tune} from −0.5 V to −0.3 V, the pole frequency of the proposed filter varies in the range of 34.2 GHz to 56.9 GHz. This wide range of pole frequency by adjusting V_{tune} makes the proposed circuit as a potential candidate for multi GHz applications. The transient responses of the proposed filter for different tune voltages are shown in Figure 15. A phase shift of 90° is noticed for each pole frequency. The THD variations are found as 3.81%, 2.6% and 1.72% for V_{tune} equal to −0.30 V, −0.33 V and −0.50 V respectively. Thus, all the simulation results on the proposed AOTAPF support the theory.

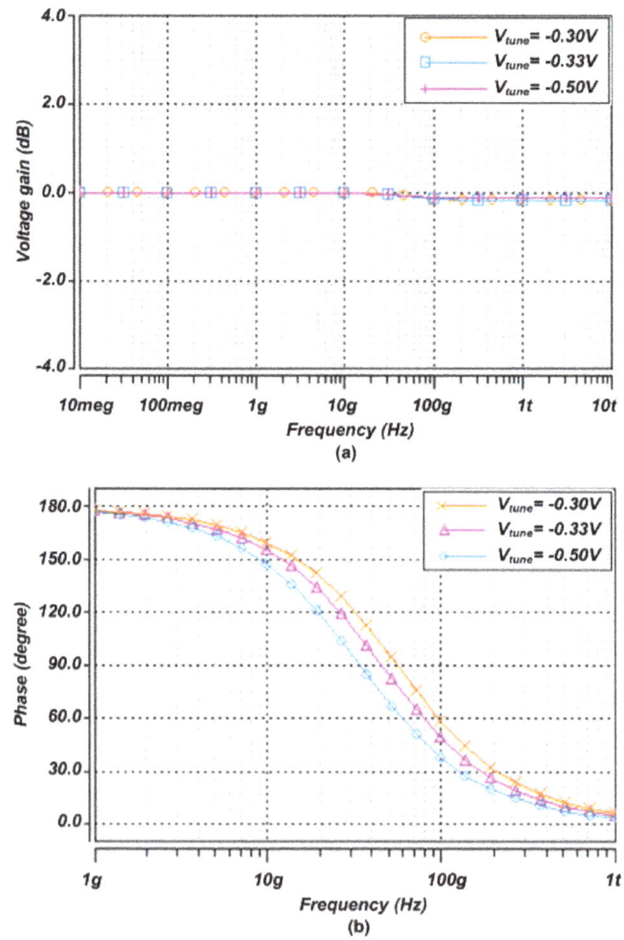

Figure 14. Frequency-response of AOTAPF at different values of V_{tune}: (**a**) Voltage gain; (**b**) Phase.

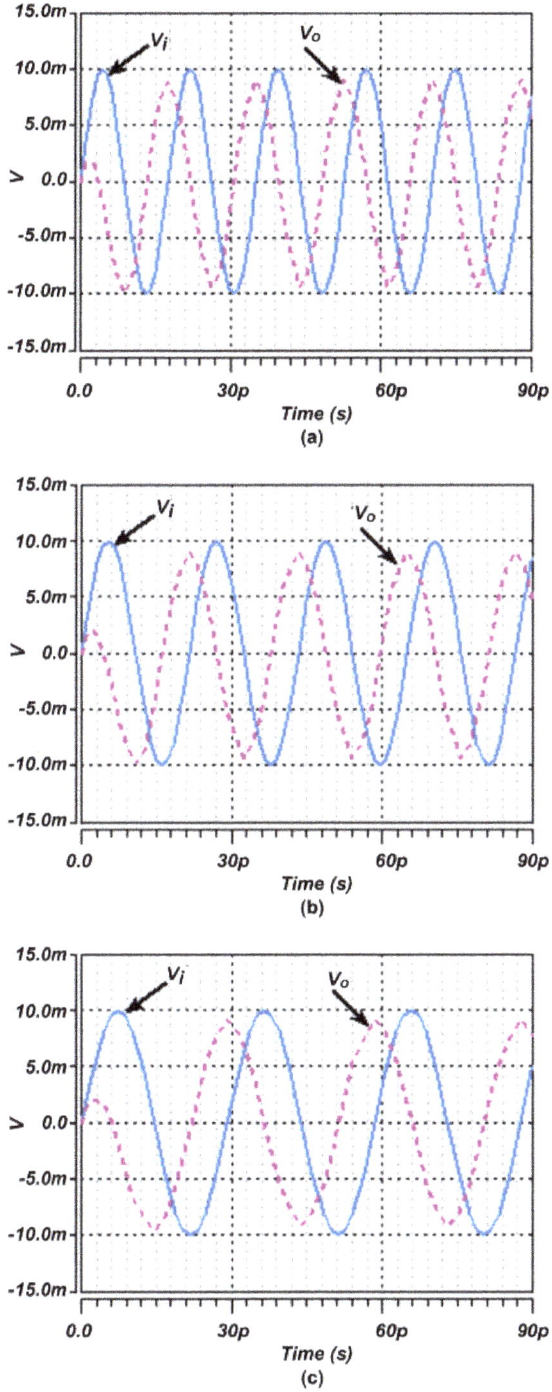

Figure 15. Transient-response of AOTAPF at different values of V_{tune}: (**a**) −0.30 V; (**b**) −0.33 V; (**c**) −0.50 V.

6. Performance Comparison of the Proposed APF

A brief comparison of the proposed AOTAPF with other available VM SE tunable APF circuit configurations is given in the Table 2. For comparison, only APFs realized with not more than 10 transistors are chosen. The APFs of [12–18] employ one or more external passive components, which result in occupying larger chip area and also suffer from slew rate limitations as well as wide tolerance. However, the proposed AOTAPF is free from any external passive component. The APFs presented in [15,17–21] utilize one or more DC current sources for tunability of pole frequency via altering biasing current. However, additional transistors need to be employed for realization of these DC current supplies and thus the transistor count will further increase. It is to be noted like previously presented APFs of [14,15], that the proposed AOTAPF is also suitable for low voltage operation as it employs only two active devices between its supply rails. The proposed AOTAPF circuit configuration is based on only three transistors, while the realized AOTAPF circuits of [19–21] use several transistors as they utilize ideal current sources. Although, the previously presented APF circuits of [14,18] are also based on three transistors like the proposed AOTAPF, but they use one or more external passive components. In addition, the reported APF of [18] uses an ideal DC current source which will ultimately increase the transistor count. Table 2 shows that the CMOS-based APF circuit configurations are limited to MHz range while the proposed circuit operates in several GHz ranges.

Table 2. Comparison of AOTAPF with other reported APFs.

Ref	Ideal Current Source Used	Number of Transistors	Number of External R/C	Technology (nm)	Supply Voltage	Tuning Range (Hz)	Power Dissipation (mW)
[12]	No	4	2/1	180	±0.9	544.8 K to 2.9 M	20.4
[13]	No	9	2/1	350	±1.5	10 K to 56 K	-
[14]	No	3	2/1	90	±0.45	103 K to 18.3 M	0.418
[15]	Yes	5	0/1	350	±1.5	-	-
[16]	No	5	0/1	180	±0.9	3.48 M to 26.1 M	-
[17]	Yes	6	0/1	180	±0.9	1.07 M to 9.44 M	10.5
[18]	Yes	3	0/1	130	±0.75	-	20.6
[19]	Yes	4	0/0	350	±1.65	105 M to 205 M	-
[20]	Yes	8	0/0	350	-	-	-
[21]	Yes	4	0/0	350	±1.5	-	-
Proposed	No	3	0/0	16	±0.7	34.2 G to 56.9 G	0.0337

-: Not Available.

7. Conclusions

In this paper, a new single ended voltage mode first order all pass filter using CNFET based unity gain inverting amplifier and a varactor is presented. The proposed circuit is constructed with only three N-type CNFETs and thus it consumes very little area on chip. Since there are only two CNFETs stacked between the power-supply rails, it is able to work equally well at low voltages. The realized all pass filter circuit is free from external passive components and thus it is suitable for integrated circuit implementation. The proposed AOTAPF circuit non-ideal performance is also evaluated. The filter circuit is designed and verified with HSPICE, using the well-known Stanford CNFET model.

Initially, the CNFET-based unity gain inverting amplifier is studied for different numbers of CNTs. It was observed that with only two CNTs, the unity gain inverting amplifier yields optimal performance. Afterward, the CNFET-based varactor is simulated for different CNTs and variable DC voltages. This study enables the designer to choose the number of CNTs for the desired frequency range of operation. Then the realized AOTAPF circuit is studied in detail including gain, phase, and transient performance. The Monte Carlo analysis for process variations as well as THD simulation studies were also performed. The simulation results show a very good gain and phase characteristics at high frequencies with tunable pole-frequency range from 34.2 GHz to 56.9 GHz. This makes the proposed topology a potential contestant for high frequency applications. It will be interesting to substantiate the all pass filter simulation results experimentally; however, due to the current non-availability of needed resources, experimental authentication is not performed. Physical realization of the presented AOTAPF may be a vital-direction for future extension of the proposed work.

Author Contributions: Conceptualization, M.M. and I.K.; methodology, M.M. and A.A.; software, M.M. and N.H.; validation, M.M., A.A. and I.K.; formal analysis, M.M. and N.H.; investigation, N.H.; data curation, M.M., A.A.; writing—original draft preparation, M.M., A.A., I.K. and N.H.; writing—review and editing, M.M. and I.K.; supervision, A.A., I.K. and N.H.

Funding: This research received no external funding.

Conflicts of Interest: The authors declare no conflict of interest.

Abbreviations

The following abbreviations are used in this manuscript:

APF	All pass filter
AOTAPF	Active only tuneable all pass filter
ASP	Analog signal processing
CNFET	Carbon Nanotube Field Effect Transistor
GCSs	Graphite-cylindrical-sheets
SE	Single-ended
THD	Total-harmonic-distortion
UGIA	Unity gain inverting amplifier
VTF	Voltage transfer function
ω_o	Pole-frequency
ω_z	Zero-frequency
C_{var}	varactor capacitance

Appendix A

The capacitance tuning characteristics (C-V curves) of the realized CNFET varactor of Figure 8a, are obtained by sweeping the V_{tune} from -0.7 V to +0.7 V for different N_{Ts} using HSPICE simualtion tool. The analytical relationship between the capacitance C_{var} and the control voltage V_{tune} can be obtained by polynomial curve fitting for fixed N_T. For $N_T = 100$, the C-V relationship is approximated by following 2nd order polynomial expression:

$$C_{var} = (-6.0601 V_{tune}^2 - 5.9693 V_{tune} - 0.8934)\ fF \tag{A1}$$

References

1. Anju, U.; Kirat, P. First Order All Pass, Low Pass and High Pass Filters Using Differential Voltage Current Conveyors. *J. Act. Passive Electron. Devices* **2017**, *12*, 275–284.
2. Nandi, R.; Koushick, M.; Sandhya, P. Single-CFA first-order allpass filter. *IEICE Electron. Express* **2016**, *13*, 1–8. [CrossRef]
3. Maheshwari, S. Some analog filters of reduced complexity with shelving and multifunctional characteristics. *J. Circuits Syst. Comput.* **2018**, *27*, 1850150. [CrossRef]
4. Kumar, A.; Ajay, K.; Sajal, K. DXCCII-Based First Order Voltage-Mode All-Pass Filter. In *Advances in Power Systems and Energy Management*; Springer: Singapore, 2018; pp. 709–717.
5. Channumsin, O.; Worapong, T. Single VDBA-based phase shifter with low output impedance. In Proceedings of the 14th International Conference on Electrical Engineering/Electronics, Computer, Telecommunications and Information Technology (ECTI-CON), Phuket, Thailand, 27–30 June 2017; pp; 427–430.
6. Iqbal, S.; Psychalinos, C.; Parveen, N. First-order allpass filter using multi-input OTA. *Int. J. Electron.* **2013**, *100*, 1373–1382. [CrossRef]
7. Minaei, S.; Yuce, E. Novel voltage-mode all-pass filter based on using DVCCs. *Circuits Syst. Signal Process.* **2010**, *29*, 391–402. [CrossRef]
8. Herencsar, N.; Koton, J.; Hanak, P. Universal Voltage Conveyor and its Novel Dual-Output Fully-Cascadable VM APF Application. *Appl. Sci.* **2017**, *3*, 307. [CrossRef]

9. Maundy, B.J.; Aronhime, P. A novel CMOS first-order all-pass filter. *Int. J. Electron.* **2002**, *89*, 739–743. [CrossRef]
10. Yuce, E. A novel CMOS-based voltage-mode first-order phase shifter employing a grounded capacitor. *Circuits Syst. Signal Process.* **2010**, *29*, 235–245. [CrossRef]
11. Herencsar, N.; Minaei, S.; Koton, J.; Vrba, K. In Voltage-mode all-pass filter design using simple CMOS transconductor: Non-ideal case study. In Proceedings of the 2015 38th International Conference on Telecommunications and Signal Processing (TSP), Prague, Czech Republic, 9–11 July 2015; pp. 677–681.
12. Yuce, E.; Minaei, S. A novel phase shifter using two NMOS transistors and passive elements. *Analog Integr. Circuits Signal Process.* **2010**, *62*, 77. [CrossRef]
13. Metin, B.; Cicekoglu, O. Tunable all-pass filter with a single inverting voltage buffer. In Proceedings of the 2008 Ph.D. Research in Microelectronics and Electronics, Istanbul, Turkey, 22 June–25 April 2008; pp. 261–263.
14. Metin, B.; Herencsar, N.; Cicekoglu, O. A low-voltage electronically tunable MOSFET-C voltage-mode first-order all-pass filter design. *Radioengineering* **2013**, *22*, 985–994.
15. Toker, A.; Özoğuz, S. Tunable allpass filter for low voltage operation. *Electron. Lett.* **2003**, *39*, 175–176. [CrossRef]
16. Minaei, S.; Yuce, E. High input impedance NMOS-based phase shifter with minimum number of passive elements. *Circuits Syst. Signal Process.* **2012**, *31*, 51–60. [CrossRef]
17. Herencsar, N.; Minaei, S.; Koton, J.; Yuce, E.; Vrba, K. New resistorless and electronically tunable realization of dual-output VM all-pass filter using VDIBA. *Analog Integr. Circuits Signal Process.* **2013**, *74*, 141–154. [CrossRef]
18. Yucel, F.; Yuce, E. A new electronically tunable first-order all-pass filter using only three NMOS transistors and a capacitor. *Turk. J. Electr. Eng. Comput. Sci.* **2016**, *24*, 3286–3292. [CrossRef]
19. Yildiz, H.A.; Ozoguz, S.; Toker, A.; Cicekoglu, O. On the realization of MOS-only allpass filters. *Circuits Syst. Signal Process.* **2013**, *32*, 1455–1465. [CrossRef]
20. Yıldız, H.A.; Toker, A.; Elwakil, A.S.; Ozoguz, S. MOS-only allpass filters with extended operating frequency range. *Analog Integr. Circuits Signal Process.* **2014**, *81*, 17–22. [CrossRef]
21. Metin, B.; Arslan, E.; Herencsar, N.; Cicekoglu, O. Voltage-mode MOS-only all-pass filter. In Proceedings of the 2011 34th International Conference on Telecommunications and Signal Processing (TSP), Budapest, Hungary, 18–20 August 2011; pp. 317–318.
22. Kuhn, K.J. Considerations for ultimate CMOS scaling. *IEEE Trans. Electron Devices* **2012**, *59*, 1813–1828. [CrossRef]
23. Frank, D.J.; Dennard, R.H.; Nowak, E.; Solomon, P.M.; Taur, Y.; Wong, H.-S.P. Device scaling limits of Si MOSFETs and their application dependencies. *Proc. IEEE* **2001**, *89*, 259–288. [CrossRef]
24. Voinigescu, S.P.; Tomkins, A.; Dacquay, E.; Chevalier, P.; Hasch, J.; Chantre, A.; Sautreuil, B. A study of SiGe HBT signal sources in the 220–330-GHz range. *IEEE J. Solid-State Circuits* **2013**, *48*, 2011–2021. [CrossRef]
25. Schröter, M.; Claus, M.; Hermann, S.; Tittman-Otto, J.; Haferlach, M.; Mothes, S.; Schulz, S. In CNTFET-based RF electronics—State-of-the-art and future prospects. In Proceedings of the 2016 IEEE 16th Topical Meeting on Silicon Monolithic Integrated Circuits in RF Systems (SiRF), Austin, TX, USA, 24–27 January 2016; pp. 97–100.
26. Hayat, K.; Cheema, H.; Shamim, A. Potential of carbon nanotube field effect transistors for analogue circuits. *J. Eng.* **2013**, *2013*, 70–76. [CrossRef]
27. Prakash, P.; Sundaram, K.M.; Bennet, M.A. A review on carbon nanotube field effect transistors (CNTFETs) for ultra-low power applications. *Renew. Sustain. Energy Rev.* **2018**, *89*, 194–203. [CrossRef]
28. Nizamuddin, M.; Loan, S.A.; Alamoud, A.R.; Abbassi, S.A. Design, simulation and comparative analysis of CNT based cascode operational transconductance amplifiers. *Nanotechnology* **2015**, *26*, 395201. [CrossRef] [PubMed]
29. Loan, S.A.; Nizamuddin, M.; Alamoud, A.R.; Abbasi, S.A. Design and comparative analysis of high performance carbon nanotube-based operational transconductance amplifiers. *NANO* **2015**, *10*, 1550039. [CrossRef]
30. Masud, M.; A'ain, A.K.B.; Khan, I.A. In Reconfigurable CNTFET based fully differential first order multifunctional filter, Multimedia. In Proceedings of the 2017 International Conference on Signal Processing and Communication Technologies (IMPACT), Aligarh, India, 24–26 November 2017; pp. 55–59.

31. Tripathi, S.; Ansari, M. S.; Joshi, A. M. In Low-Noise Tunable Band-Pass Filter for ISM 2.4 GHz Bluetooth Transceiver in ±0.7 V 32 nm CNFET Technology. In Proceedings of the International Conference on Data Engineering and Communication Technology, Maharashtra, India, 15–16 December 2017; pp. 435–443.
32. Sharma, J.; Ansari, M.S.; Sharma, J. In Current-Mode Electronically Tunable Resistor-less Universal Filter in ±0.5 V 32 Nm CNFET, Devices. In Proceedings of the 2014 International Conference on Circuits and Communications (ICDCCom), Ranchi, India, 12–13 September 2014; pp. 1–6.
33. Masud, M.; A'ain, A.K.B.; Khan, I.A. In CNFET Based Reconfigurable First Order Filter. In Proceedings of the 2017 9th IEEE-GCC Conference and Exhibition (GCCCE), Manama, Bahrain, 8–11 May 2017; pp. 1–9.
34. Sharma, J.; Ansari, M.S.; Sharma, J. In Electronically Tunable Resistor-less Universal Filter in ±0.5 V 32 nm CNFET. In Proceedings of the 2014 Fifth International Symposium on Electronic System Design (ISED), Surathkal, Mangalore, India, 15–17 December 2014; pp. 206–207.
35. Moaiyeri, M.H.; Jahaniana, A.; Navia, K. Comparative performance evaluation of large FPGAs with CNFET and CMOS-based switches in nanoscale. *Nano-Micro Lett.* **2011**, *3*, 178–188. [CrossRef]
36. Guo, J.; Lundstrom, M.; Datta, S. Performance projections for ballistic carbon nanotube field-effect transistors. *Appl. Phys. Lett.* **2002**, *80*, 3192–3194. [CrossRef]
37. Natori, K.; Kimura, Y.; Shimizu, T. Characteristics of a carbon nanotube field-effect transistor analyzed as a ballistic nanowire field-effect transistor. *J. Appl. Phys.* **2005**, *97*, 034306. [CrossRef]
38. Deng, J.; Wong, H.-S.P. A compact SPICE model for carbon-nanotube field-effect transistors including nonidealities and its application—Part II: Full device model and circuit performance benchmarking. *IEEE Trans. Electron Devices* **2007**, *54*, 3195–3205. [CrossRef]

electronics

MDPI

Article

Shannon Entropy Index and a Fuzzy Logic System for the Assessment of Stator Winding Short-Circuit Faults in Induction Motors

Arturo Mejia-Barron [1], J. Jesus de Santiago-Perez [1], David Granados-Lieberman [2], Juan P. Amezquita-Sanchez [1] and Martin Valtierra-Rodriguez [1,*]

[1] ENAP-Research Group, CA-Sistemas Dinámicos, Facultad de Ingeniería, Universidad Autónoma de Querétaro (UAQ), Campus San Juan del Río, Río Moctezuma 249, Col. San Cayetano, San Juan del Río, Qro., C. P. 76807, Mexico; arturo.mejia@enap-rg.org (A.M.-B.); jjdesantiago@hspdigital.org (J.J.d.S.-P.); juan.amezquita@enap-rg.org (J.P.A.-S.)

[2] ENAP-Research Group, CA-Fuentes Alternas y Calidad de la Energía Eléctrica, Departamento de Ingeniería Electromecánica, Instituto Tecnológico Superior de Irapuato (ITESI), Carr. Irapuato-Silao km 12.5, Colonia El Copal, Irapuato, Guanajuato C. P. 36821, Mexico; david.granados@enap-rg.org

* Correspondence: martin.valtierra@enap-rg.org

Received: 28 November 2018; Accepted: 9 January 2019; Published: 15 January 2019

Abstract: The induction motor (IM) is one of the most important elements in industry. Although IMs are robust machines, they are susceptible to faults, where the stator winding short-circuit fault is one of the most common ones. In this work, the Shannon entropy (SE) index and a fuzzy logic (FL) system are proposed to diagnose short-circuit faults, considering both different severity levels and different load conditions. In the proposed methodology, a filtering stage based on brick-wall band-pass filters is firstly carried out. After this stage, the SE index is computed to quantify the fault severity and a FL system is applied to diagnose the IM condition in an automatic way. Unlike other works that propose some types of space transformations, the proposal is only based on a filtering stage and a time domain index, requiring low computational resources. The obtained results demonstrate the effectiveness of the proposal, i.e., the SE index quantifies the fault severity, regardless of the mechanical load, and the proposed FL system achieves a positive classification rate of 98%.

Keywords: brick-wall filter; fuzzy logic; induction motor; Shannon entropy; short-circuit fault

1. Introduction

In recent years, the development of monitoring systems to assess the physical condition of rotatory machinery has been vital to guaranteeing the reliability of industrial processes [1–3]. Among the rotatory machinery, the three-phase induction motor (IM), representing ~85% of the consumed power in the industry, is a default implementation in industrial processes [4] because it offers great benefits, such as low maintenance, low cost, high robustness to aggressive environments and easy control under different load conditions [5,6]. Despite these great benefits, IMs are susceptible to present electrical and mechanical faults during their service-life, which are produced mainly by power quality problems, prolonged activity times and harsh operating conditions, among other factors [1–3,5,6]. Regarding electrical faults, stator winding faults (SWFs) are one of the most dangerous and common faults in IMs [7], representing about a 36–38% of faults that can take place [8,9]. This fault, even in its incipient/early state, can produce alterations and increments in current consumption, temperature and vibrations, putting at risk the personnel, the production, the machine itself and other machines in the same line of production.

During the last 15 years, an important number of techniques and methodologies for SWF detection using the analysis of acoustic, current and vibration signals have been proposed [10–14]. Motor current signature analysis (MCSA) is one of the most used methods because of its advantages, such as possessing a non-invasive capacity, possible remote sensing, easy implementation and low implementation costs [2,9]. MCSA is mainly used to identify faults in the IM according to the analysis of frequency components found in the measured signal. Particularly, MCSA for SWF detection is employed to identify frequencies around the fundamental frequency or harmonic components [15]. In the literature, diverse signal processing algorithms for stator winding short-circuit (SWSC) fault detection using MCSA have been introduced; for instance, fast Fourier transform (FFT) [14,16,17], wavelet transform (WT) [18–20], empirical mode decomposition-based methods (EMD) [21,22], Wigner-Ville distribution (WVD) [22], Hilbert transform (HT) [23], statistical time series model (STSM) [24], and statistical analysis (SA) [25]. Despite obtaining promising results, diverse limitations still remain. For instance, the FFT is a proficient tool to analyze time signals with stationary properties; yet, current applications in industry require continuous changes of the load applied to IMs, which can generate fluctuations in the voltage and current signals, producing non-stationary properties, therefore making the FFT method unsuitable [26]. WT is a suitable tool for analyzing signals of non-stationary nature; regrettably, it requires a fine election of the decomposition level and the wavelet mother in order to estimate adequate features that allow for correct evaluation of the IM's condition [27]. In this sense, EMD-based algorithms are used to analyze or decompose time signals of non-stationary nature according to their frequency components; yet, they are susceptible to present a phenomenon called mode mixing, which produces waves with different frequency components that are assigned to the same frequency band, complicating the identification of frequencies associated to the SWSC fault. Furthermore, the computational resources can increase depending on the EMD-method used, e.g., when the ensemble-EMD method is used [28]. HT is employed for obtaining the instantaneous frequency and the instantaneous amplitude of a time signal; but its results can be affected by the noise and the number of frequency components found in the analyzed signal [29]. WVD is a method capable of providing a time-frequency representation of time signals; yet, its results can be contaminated with spurious frequencies, frequency components that do not exist in the measured signal due to a problem called cross-term [30], compromising the ability for adequate location of the frequencies associated to the SWSC fault. STSMs are employed for modelling signals with a linear or time-invariant behavior; but, they can present problems for modelling nonlinear behaviors [31], which are greatly produced in an IM because of the dynamic loading. Further, their results are susceptible to errors due to the quantity of noise contained in the measured signal. The SA methods are employed for calculating statistical parameters of the time-domain signals, such as median, variance, standard deviation and among others, but their results can fail due to the noise and nonlinearities found in the time signal [32].

Although diverse methods for SWSC fault detection have been introduced in the literature, most of them are negatively affected by the non-stationary properties of the measured signal. These properties are generated by different factors, e.g., the variations in current consumption associated to changes in the mechanical load. In this regard, the proposal and development of efficient and reliable methodologies in terms of processing and performance are still required, mainly if they are not susceptible to the motor load, e.g., they have to be independent of the motor mechanical load in order to provide a consistent diagnosis for a large variety of industrial processes where the mechanical load can be different and time-variant.

In this paper, a new methodology to diagnose and quantify the severity of SWSC faults, where an independent fault indicator of the mechanical load is presented. The proposed methodology is based on MCSA, using the monitored current during the IM steady-state as input. It is based on three steps. Firstly, a filtering stage based on brick-wall band-pass filters is carried out. This type of filter is used as it presents great advantages, such as a rectangular frequency response and an abrupt transition between the pass and stop bands. Secondly, the Shannon entropy (SE) index is applied to the filtered signal in order to identify the short-circuit faults, considering both different severity levels and different load conditions. Other indices, such as the signal energy and the root mean square (RMS) value are tested and compared with the results obtained by the SE index. Finally, a fuzzy logic (FL) system is developed in order to classify the IM condition in an automatic way. The usefulness and effectiveness of the proposal is validated through experimentation, where a healthy (HLT) IM and an IM with short-circuited turns using four different levels of load are considered. The obtained results show that the proposal is an effective and consistent tool for diagnosing SWSC faults independently of load conditions, making it a promising solution for a large variety of industrial applications.

2. Theoretical Background

2.1. Motor Current Signal Analysis (MCSA)

MCSA is a widely used method for online condition monitoring in IMs, where the current spectra is used to obtain information associated to the motor fault. This fault information is obtained through abnormal harmonics in the stator current produced by the magnetomotive force distribution and the permeance-wave representation of the air gap [15,33].

Regarding the SWSC fault, signature patterns in different frequency components have been associated to the following equation [9,15,33]:

$$f_{st} = f_1\left\{\frac{n}{p}(1-s) \pm k\right\} \qquad k = 1,3,5,\ldots \quad n = 1,2,3,\ldots \tag{1}$$

where the values for f_{st} are the frequency components due to the SWSC fault, f_1 is the supply frequency, p is the pole-pairs and s is the slip. Different values for k and n can be tested in order to obtain the frequencies of interest, where promising results have been obtained for k = 1 with n = 3 and n = 5 [15].

2.2. Brick-Wall Filters

Brick-wall filters or sinc filters are idealized digital FIR (finite impulse response) filters with a rectangular frequency response, which provides an ideally flat amplitude response in the passband and an abrupt transition in the cutoff frequency [34]. Besides, a FIR filter is featured by its stability and linear phase. Then, an ideal brick-wall low-pass filter with bandwidth ω_p and zero phase provides the impulse response, as per Reference [35]:

$$g(t) = \frac{\omega_p}{2}\frac{\sin \omega_p t}{\omega_p t} \tag{2}$$

As the filter impulse response has an infinite length, making the structure implementation impossible [35], a window function $w(t)$ of length τ is applied to $g(t)$ to obtain a practical filter, which can be expressed as per Reference [35]:

$$h(t) = w(t)\frac{\omega_p}{2}\frac{\sin \omega_p t}{\omega_p t} \tag{3}$$

By using two brick-wall low-pass filters, a brick-wall band-pass filter is obtained as follows:

$$h(t)_{BP} = h(t)_u - h(t)_l \tag{4}$$

where $h(t)_u$ and $h(t)_l$ are the upper and lower band edges, respectively.

2.3. Fault Indices

In the literature, several indices have been presented for fault diagnosis. In this work, the SE, energy and RMS indices, which have proven to be efficient in other electric applications related to the diagnosis of faults in induction motors and transformers [36–39], are analyzed as potential indicators to diagnose SWSC faults.

In the information theory, SE is used to describe the uncertainty of information content provided by an event or a signal [36]; as the SWSC fault generates different frequency components, the amount of information can change, making the SE index a promising fault indicator to quantify this change. It is given by:

$$\mathrm{SE}(X) = -\sum_{i=1}^{n} p(x_i)\log_2[p(x_i)] \tag{5}$$

where $x_1, x_2, x_3, \ldots, x_n$ are the possible outcomes of an event or signal given by X, where $p(x_i)$ is the corresponding probability vector.

On the other hand, the energy and RMS indices are obtained by means of the following equations [37]:

$$\mathrm{Energy} = \sum_{i=1}^{N}[x(i)]^2 \tag{6}$$

$$\mathrm{RMS} = \sqrt{\frac{1}{N}\sum_{i=1}^{N}[x(i)]^2} \tag{7}$$

where $x(i)$ is the signal value at the sample i and N is the total number of samples. As can be noticed, these indices somehow increase their value according to the increments in the signal amplitude $x(i)$; therefore, they can be sensitive to SWSC faults, considering that the presence of fault and its severity increase the signal amplitude from the increment of different frequency components.

2.4. Fuzzy Logic Systems

In general, FL systems can be used as control strategies based exclusively on FL or in combination with other methods, such as neural networks (neuro-fuzzy systems) [40–44] or classification algorithms [5,26], where features such as simplicity and flexibility of design, handling of imprecise data and the capability to model nonlinear systems, among others, can be exploited. In particular, a classification task can be carried out if the information behavior is described using 'if-then' classification rules for when information about the input data is known. These rules describe the class of an object according to its features; for instance, if an object is high then its class is big.

FL systems consist of four stages: Fuzzification, rules, inference mechanism and defuzzification [5,44], as shown in Figure 1. In the fuzzification, the inputs are mapped into linguistic variables and quantified through membership functions. These functions can have Gaussian, triangular, trapezoidal or other shapes. As mentioned previously, the 'if-then' rules describe linguistically how an object has to be assigned to a specific class according to its features. These rules are set by an expert that knows the features and classes. In the inference mechanism, the decision-making process is carried out, giving a conclusion for a specific set of inputs. Finally, the conclusion is converted to understandable information for the user using the defuzzification stage. In this stage, there are several defuzzification methods, where the center-of-gravity method is one of the most popular.

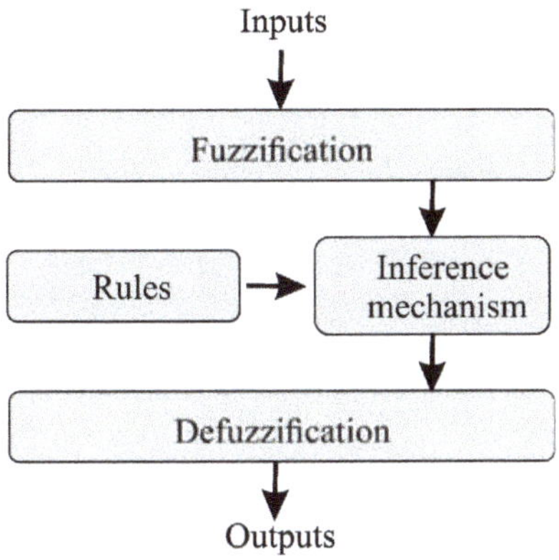

Figure 1. General diagram of a fuzzy logic (FL) system.

3. Proposed Methodology

The proposed methodology to detect SWSC faults in IMs is shown in Figure 2. In general, the methodology is divided into two stages: Design and implementation. In the design stage, the current signals of an IM, ranging from steady state to different fault severities, along with different load conditions are firstly acquired. Then, from the frequency domain analysis, and by following Equation (1), information related to the SWSC condition can be found in f_L (k = 1 and n = 3) and f_R (k = 1 and n = 5). In particular, an IM with two pairs of poles (p = 2) operating in an electric system where f = 60 Hz at no load condition (slip of $s \approx 0$) presents information related to the SWSC fault, where f_L = 150 and f_R = 210. In this regard, two brick-wall band-pass filters (using Equation (4)) are constructed to extract that information. Figure 3 shows their design. In Figure 3a, the region of interest in the frequency domain (f_L and f_R) can be observed. As s in the IM can shift the f_L and f_R components, the brick-wall band-pass filters consider the bandwidths denoted by (f_{L_1},f_{L_2}) and (f_{R_1},f_{R_2}), respectively. By considering a wide range for s according to the nominal motor speed, the values of f_{L_1} = 160, f_{L_2} = 170, f_{R_1} = 200, and f_{R_2} = 210 are used. Figure 3b shows that the band-pass filter, Ff_L, is designed using the difference of two brick-wall low-pass filters with f_{L_1} and f_{L_2} as cutoff frequencies. In a similar way, the band-pass filter, Ff_R, is designed using f_{R_1} and f_{R_2}. The order of the filters is set to 1024 in order to achieve a high attenuation in the stop band. Figure 3c shows the frequency responses for the two brick-wall band-pass filters. Ff_L is the filter that is constructed to extract the f_L component and Ff_R is the filter that is constructed to extract the f_R component. Once the filters are designed, the analysis of fault indices is carried out (see Figure 2); in order to do so, the SE, energy and RMS indices (using Equations (5)–(7) are applied to the filtered signals to determine which index presents the most discriminant information in terms of the fault severity and its susceptibility to the mechanical load. When the most appropriate index to diagnose the SWSC has been obtained, a FL system is designed to automatically determine the IM condition from the information provided by the selected index. The designed FL system consists of the stages presented in Figure 1, i.e., fuzzification, rules, inference mechanism and defuzzification. As the elements that compound each stage depend on the experimental results, they are described in detail in Section 4.3.

In the implementation stage, the designed filters, Ff_L and Ff_R, are applied to the input current signal (see Figure 2). Then, the selected index is computed for each frequency component, namely SE_L for f_L and SE_R for f_R. Finally, on the one hand, the indices are averaged to provide an indicator that quantifies the fault severity, and on the other hand, the indices are analyzed by the FL system to determine the IM condition in an automatic way.

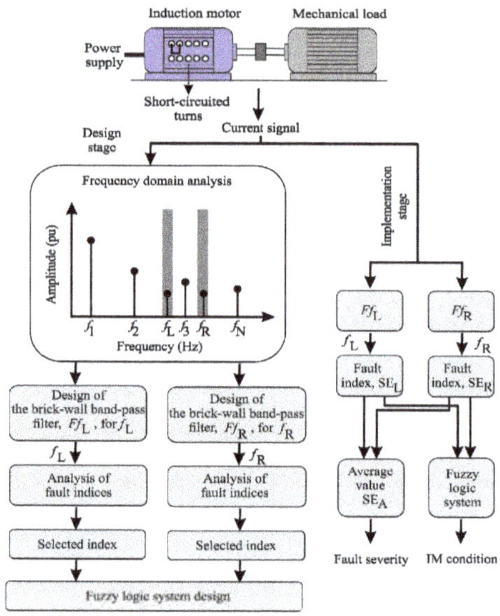

Figure 2. Proposed methodology.

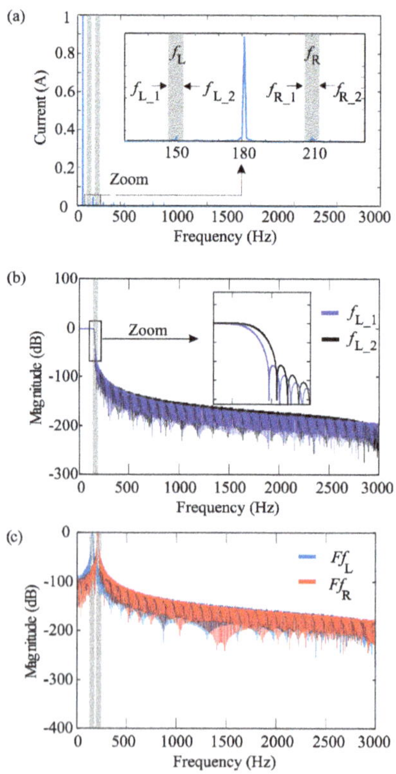

Figure 3. Filter design: (**a**) Frequencies of interest, (**b**) brick-wall low-pass filters and (**c**) brick-wall band-pass filters.

4. Experimentation and Results

4.1. Experimental Setup

In Figure 4, the experimental setup used to test and validate the proposal is depicted. In general, it consists of: A personal computer (PC) to implement the analysis using MATLAB software, an IM in which the stator-winding has been modified with several taps, a motor starter, a data acquisition system (DAS) to acquire the current signals and a dynamometer to generate the mechanical load in a controlled way. The model of the used 3-phase IM was WEG 218ET3EM145TW, featuring 2 poles, 2 hp, 220 VAC and 60 Hz. The SWSC conditions were artificially produced with the insertion of taps in phase A. The analyzed taps correspond to 10, 20, 30 and 40 short-circuited turns (SCTs). The current signal was acquired using a model i200 current clamp from Fluke, a 16-bit analog-to-digital converter model which was incorporated in the NI-USB 6211 board from National Instruments, and a sampling frequency of 6000 samples/s during a time window of 1 s. For the analysis, twenty tests for each motor condition (0, 10, 20, 30 and 40 SCTs) were carried out; therefore, 100 tests were analyzed. Regarding the mechanical load, it was provided by a four-quadrant model 8540 dynamometer from Lab-Volt, where 0.00, 2.04, 4.09 and 6.13 Nm were used as the load torques. These values ranged from no-load to nominal load.

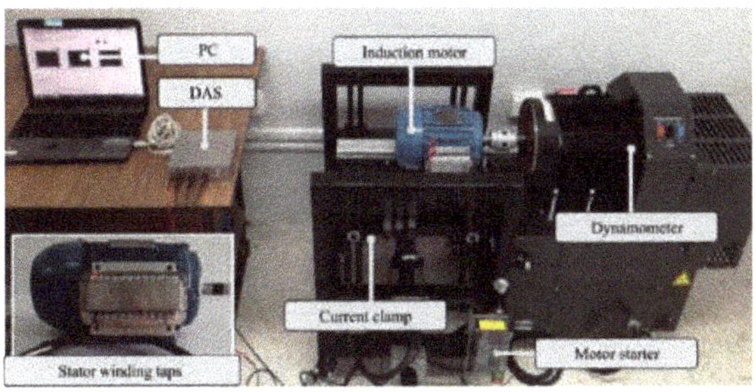

Figure 4. Experimental setup.

Figure 5 shows an example of the acquired current signals, where it was observed that the magnitude of the current signal increased with both the mechanical load and the fault severity. This is very important, as the proposed methodology has to be capable of detecting the SCTs regardless of the mechanical load. For instance, a methodology based on the magnitude of the current signal is inappropriate as the fault can be confused with an increment in the load.

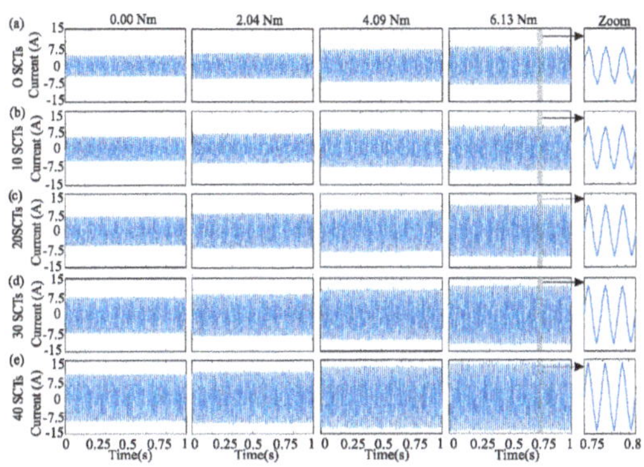

Figure 5. Current signals for (**a**) 0 SCTs, (**b**) 10 SCTs, (**c**) 20 SCTs, (**d**) 30 SCTs and (**e**) 40 SCTs at different loads (0.00, 2.04, 4.09 and 6.13 Nm).

4.2. Results for Real Signals

Following the proposed methodology, the current signals in steady state were filtered using Ff_{L} and Ff_{R}. After the filtering stage, the SE, RMS and energy indices were computed for the output signals given by f_{L} and f_{R} (see Figure 2). In order to have a common reference to quantify the fault severity, the results of the indices were normalized using the numerical value as a normalization factor for the healthy condition (0 SCTs); thus, these indices will have a value of 1 for 0 SCTs, indicating a healthy condition. Figure 6 depicts the obtained results for the analyzed indices. At the left side of this figure, the results for f_{L} under both different fault severities and different load conditions are shown, whereas the results for f_{R} are shown at the right side.

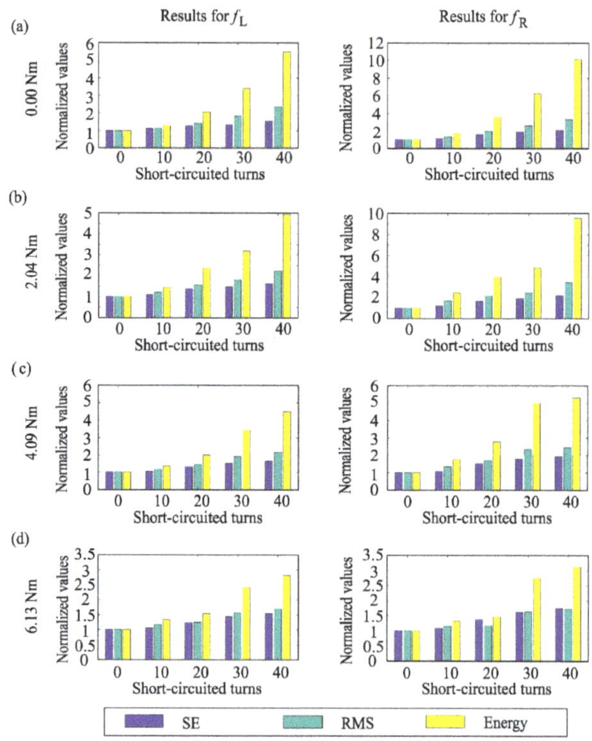

Figure 6. Results for the Shannon entropy (SE), root mean square (RMS), and energy indices at (**a**) 0.00 Nm, (**b**) 2.04 Nm, (**c**) 4.09 Nm and (**d**) 6.13 Nm (left side for f_{L} and right side for f_{R}).

The results presented in Figure 6 show that the values of the indices increased with the fault severity, which was useful for quantification purposes; however, the change rate in some indices was different for different load conditions, which can compromise the diagnosis. For instance, the energy in f_{L} for 30 SCTs under a load of 2.04 Nm was approximately 3, which can be confused with the energy in f_{L} for 40 SCTs under a load of 6.13 Nm, since it was also approximately 3. In the RMS index, a similar behavior was observed; for instance, the RMS in f_{R} for 20 SCTs under a load of 0.00 Nm was approximately 2, which can be confused with the RMS in f_{R} for 30 SCTs under a load of 4.09 Nm, since it was also approximately 2. From these observations and by analyzing the SE behavior, it was found that the SE index provides the most uniform rate of change regardless of the load conditions, making it the most appropriate index to diagnose and quantify the severity of the SWSC fault. For clarity purposes, Figure 7 shows a three-dimensional bar chart of the SE values ($\mathrm{SE_L}$ value for f_{L} and $\mathrm{SE_R}$

value for f_R), where a behavior almost constant for different loads levels and a constant increment according to the fault severity are both observed. This behavior demonstrates that the SE index can diagnose the fault severity in a proper way, regardless of the mechanical load. In order to provide a single fault index, SE_L and SE_R were averaged, where the result, SE_A, was used as indicator for quantifying the fault severity (see Figure 7c).

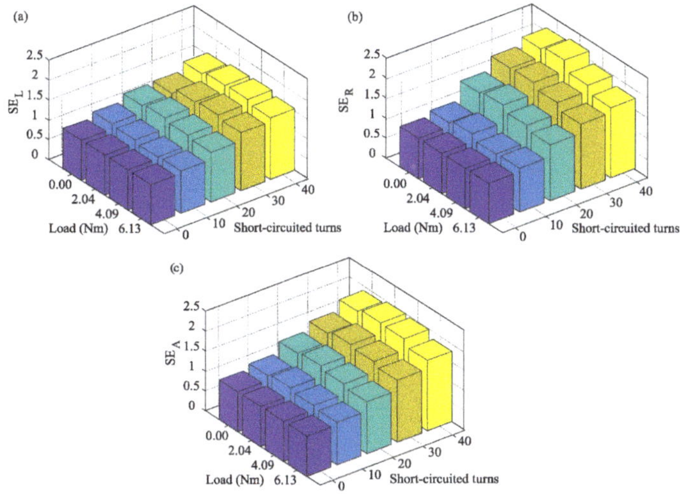

Figure 7. SE values for (**a**) f_L, (**b**) f_R, and (**c**) SE_A at both different loads and different fault severities.

Table 1 presents the mean (μ) and the standard deviation (σ) for the SE values of the twenty tests of each IM condition (0, 10, 20, 30 and 40 SCTs). Figure 8 shows the results of Table 1 as Gaussian distribution functions, where μ and σ are considered. From this figure, it is evident that, in all the cases, the higher the fault severity, the higher the index value, which applies to both SE_L and SE_R. Although the SE index allows for quantification of the fault severity, the classification of the IM condition (0, 10, 20, 30 and 40 SCTs) cannot be directly achieved, since there are small overlaps between some conditions; for instance, there is an overlap between the 0 SCTs condition (dark blue) and the 10 SCTs condition (light blue) in Figure 8a at the different loads. In this regard, a FL system with SE_L and SE_R as inputs was used to provide the automatic classification. It is important to mention that a FL system was used as classifier in this work, since the information presented in Figure 8 (Gaussian distribution functions) can be seized to generate the Gaussian membership functions.

Table 1. μ and σ for SE values.

	f_L Number of Short-Circuited Turns (μ and σ for SE Values)				
Load	**0**	**10**	**20**	**30**	**40**
0.00	1, 0.1229	1.1326, 0.1578	1.3051, 0.1720	1.3297, 0.1160	1.5324, 0.0699
2.04	1, 0.1097	1.0916, 0.0890	1.3673, 0.0483	1.4781, 0.0413	1.6144, 0.0533
4.09	1, 0.0516	1.0358, 0.0530	1.3025, 0.0395	1.5279, 0.0335	1.6354, 0.0359
6.13	1, 0.0558	1.0559, 0.0771	1.2410, 0.0613	1.4507, 0.0477	1.5444, 0.0591
	f_R Number of Short-Circuited Turns (μ and σ for SE Values)				
Load	**0**	**10**	**20**	**30**	**40**
0.00	1, 0.1082	1.1082, 0.1121	1.5639, 0.0951	1.8904, 0.0696	2.0860, 0.0830
2.04	1, 0.1614	1.2044, 0.1315	1.6318, 0.1035	1.9104, 0.0647	2.1743, 0.0580
4.09	1, 0.1204	1.0683, 0.1135	1.5052, 0.1166	1.7788, 0.0960	1.9281, 0.0727
6.13	0.073833	1.0878, 0.0720	1.3794, 0.0838	1.6181, 0.0786	1.7425, 0.0932

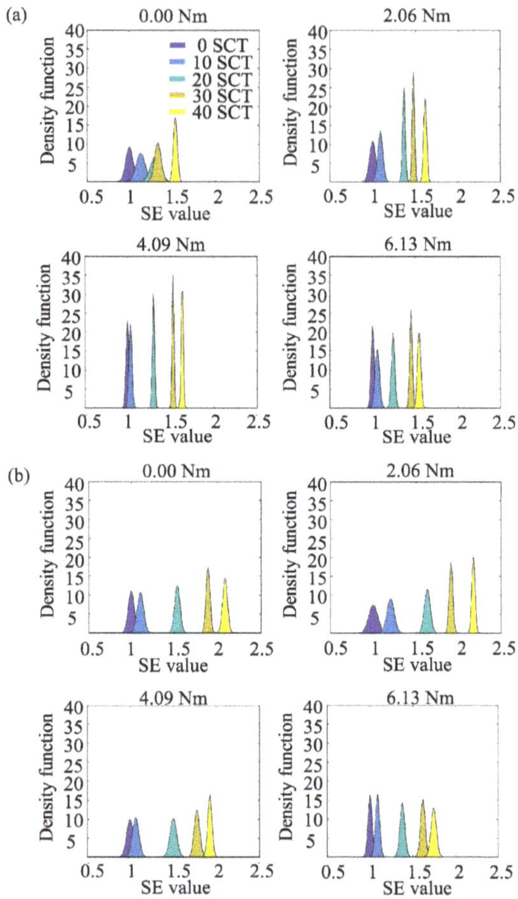

Figure 8. Gaussian distribution functions for (**a**) SE_L and (**b**) SE_R at both different loads and different fault severities.

4.3. Fuzzy Logic System Results

The proposed FL system is a Mamdani-type fuzzy inference system with two inputs, one output and 25 rules. As mentioned previously, the inputs were SE_L and SE_R, while the output was the IM condition. For the fuzzification stage, both inputs were portioned into five Gaussian membership functions, as shown in Figure 9a. These functions were labeled as follows: Very small value (VSV), small value (SV), normal value (NV), high value (HV) and very high value (VHV). The crisp output of the proposed FL system assumes values between 0.5 and 5.5, as shown in Figure 9b; in this figure, 0 SCTs are 1, 10 SCTs are 2, 20 SCTs are 3, 30 SCTs are 4 and 40 SCTs are 5. On the other hand, the 25 functions are presented in Table 2, where one rule can be read as follows: If SE_L is VSV and SE_R is VSV, then the IM condition is 0 SCTs. The minimum composition was used for quantifying the output of the rules and the center-of-gravity method was used for defuzzification [44]. Table 3 shows the classification results for the performed tests. As can be observed, most cases present an effectiveness of 100%; however, two cases present an effectiveness of 95%, implying a general effectiveness of 98%. These cases correspond to 0 SCTs and 10 SCTs. This result can be somehow expected, since the existing overlaps in the Gaussian distribution functions shown in Figure 8 indicate that, in probabilistic terms, there is not a complete separation between cases.

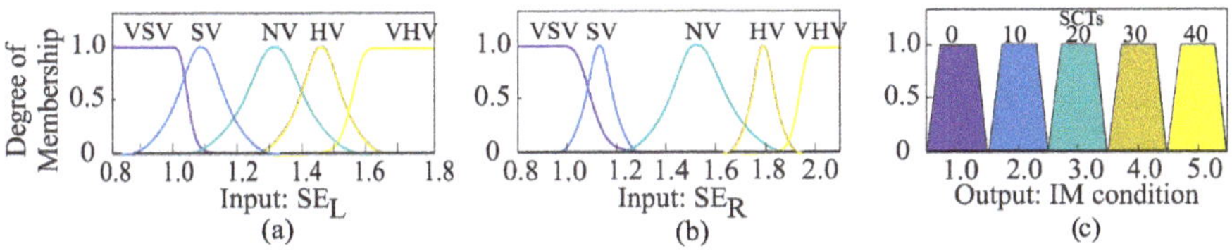

Figure 9. Membership functions for (**a**) SE_L and (**b**) SE_R and (**c**) FL outputs.

Table 2. Rules for the proposed FL system.

Inputs			SE_R		
SE_L	**VSV**	**SV**	**NV**	**HV**	**VHV**
VSV	0 SCTs	0 SCTs	10 SCTs	20 SCTs	20 SCTs
SV	0 SCTs	10 SCTs	20 SCTs	20 SCTs	20 SCTs
NV	10 SCTs	20 SCTs	20 SCTs	20 SCTs	30 SCTs
HV	20 SCTs	20 SCTs	20 SCTs	30 SCTs	40 SCTs
VHV	20 SCTs	20 SCTs	30 SCTs	40 SCTs	40 SCTs

Table 3. Classification results (confusion matrix).

IM Condition	0 SCTs	10 SCTs	20 SCTs	30 SCTs	40 SCTs	EP (%)
0 SCTs	19	1	0	0	0	95
10 SCTs	1	19	0	0	0	95
20 SCTs	0	0	20	0	0	100
30 SCTs	0	0	0	20	0	100
40 SCTs	0	0	0	0	20	100
					Effectiveness	98%

EP: Effectiveness percentage.

4.4. Discussion

Table 4 summarizes a comparison between the proposal and other recent methodologies presented in the literature, where the methods or algorithms applied to diagnose the SWSC fault in the IM and the features or operating conditions that are considered in the experimentation are shown.

From Table 4, it can be observed that the proposed methodology presents an effectiveness percentage of 98% for detecting the SWSC fault, considering both different severity levels (10, 20, 30 and 40 short-circuited turns) and different mechanical load levels (0%, 33%, 66% and 100%), unlike other works reviewed in the literature [12,19,24], which present mainly the analysis of either a level of damage and different operating conditions or different levels of damage and a constant load operating condition.

In the proposal, the obtained effectiveness (98%) is mainly due to the SE index, which allows for both quantifying the severity of damage regardless of the torque load applied to the IM and classifying the SWSC fault using the proposed FL system for an automatic diagnosis. In qualitative terms, it is important to mention that a low computational burden is achieved by the proposal, since a space transformation of the measured signal is not required, allowing for a low complexity implementation, unlike the other introduced proposals, where a signal transformation and several nonlinear indices is required, along with an expert to interpret the obtained results [12,19,23]. It should be pointed out that the expert role is to interpret the results obtained by the analysis of several characteristics, such as: The location of peaks, the spectrum, among other characteristics; in this regard, the aforementioned analyses are performed qualitatively. Yet, the automatic detection of the motor condition can drastically reduce time taken and allow for continuous and online monitoring. In Reference [8], similar features and operating conditions with the proposal can be observed; however, results about the fault indictor as an independent parameter of the mechanical load are not presented. On the contrary, the proposed

SE index demonstrates to be an efficient and insensitive fault indicator to the mechanical load, allowing for consistent diagnosis in different industry applications.

Table 4. Comparison summary between the proposed methodology and works reporting stator winding short-circuit (SWSC) fault diagnosis.

Work	Applied Methods	Domain	Accuracy	Variable Load	Different Fault Severities
[8]	1. Compute the mutual information among current signals. 2. Normalize data. 3. Pattern recognition by means of artificial neural networks (ANN).	Time	>93%	Yes	Yes
[12]	1. Estimate Zero crossing time (ZCT). 2. Compute frequency spectrum of ZCT signal by means of discrete Fourier transform. 3. Locate peaks related to inter-turn fault.	Frequency	NR	Yes	No
[19]	1. Decompose current signal using stationary Wavelet transform (SWT). 2. Obtain fault residues using reconstructed currents. 3. Obtain coefficients by decomposing the residues with discrete Wavelet transform (DWT). 4. Estimate the fault index and compare with an adaptive threshold.	Time-Frequency	NR	No	Yes
[23]	1. Obtain an analytical signal by means of extended Park's vector approach and Hilbert transform (P-H). 2. Estimate frequency domain of the analytical signal via fast Fourier transform (FFT). 3. Calculate the amplitudes and frequencies corresponding to harmonics associated with the fault. 4. Compute the partial relative indexes (PRI) for fault detection.	Frequency	NR	Yes	Yes
[24]	1. Map into the α-β stator-fixed reference frame the stator currents. 2. Compute the instantaneous space phasor (ISP) module. 3. Evaluate the final prediction criterion (FPE) for the proposed ISP autoregressive model by the different operation condition.	Time	95%	No	Yes
This work	1. Brick-wall band-pass FIR filters for extraction of frequency components. 2. Compute the SE index as fault indicator. 3. FL system for automatic classification	Time	98%	Yes	Yes

5. Conclusions

Winding faults are one of the most common faults in IM. In this work, a new method based on filters, fault indices and a FL system for the assessment of SWSC faults in IMs was presented. The SE, RMS and energy indices were tested. These indices evaluated the information that was extracted by the brick-wall band-pass filters from the steady-state current signal. Our results indicated that the SE was the most suitable index for the assessment of SWSC faults. For the analyzed cases, i.e., 10, 20, 30 and 40 SCTs under different load torque conditions (0, 2.04, 4.09 and 6.13 Nm), this index has been demonstrated to be sensitive to fault severity and insensitive to mechanical load, i.e., the SE index can properly assess the fault severity regardless of the mechanical load, which is very important, as the mechanical load can change or be different for different industrial applications. On the other hand, the proposed FL system uses the SE values to classify the IM condition in an automatic way. The obtained results indicate that the proposed FL system provides a general effectiveness of 98%.

In a future work, the proposal will be tested under an unbalanced power supply voltage condition (a common electrical condition in industry) in order to increase its robustness and applicability. Furthermore, as the proposal is based on low complexity algorithms (filters and indices based on time-domain formulas), it may be implemented into an embedded system in order to provide an online condition monitoring system. On the other hand, it is important to mention that at this stage of research, the proposal is focused on the diagnosis of SWSC faults in steady state conditions; however, adaptive filters and time-frequency techniques will be also explored in order to provide a solution for transient operating conditions.

Author Contributions: Conceptualization, A.M.-B. and M.V.-R.; Data curation and formal analysis, A.M.-B; Funding acquisition, J.J.d.S.-P., D.G.-L. and J.P.A.-S.; Methodology, A.M.-B. and M.V.-R.; Investigation, Resources and Visualization, J.J.d.S.-P., D.G.-L. and J.P.A.-S.; Writing—original draft, review & editing, all the Authors.

Funding: This research was funded by the "Consejo Nacional de Ciencia y Tecnología (CONACYT)" under the scholarship 296868.

Conflicts of Interest: The authors declare no conflict of interest.

References

1. Glowacz, A. Acoustic-based fault diagnosis of commutator motor. *Electronics* **2018**, *7*, 299. [CrossRef]
2. Singh, A.; Grant, B.; DeFour, R.; Sharma, C.; Bahadoorsingh, S. A review of induction motor fault modeling. *Electr. Power Syst. Res.* **2016**, *133*, 191–197. [CrossRef]
3. Mrizalde, Y.; Hernandez-Callejo, L.; Duque-Perez, O. State of the art and trends in the monitoring, detection and diagnosis of failures in electric induction motors. *Energies* **2017**, *10*, 1056. [CrossRef]
4. Rangari, S.; Suryawanshi, H.; Renge, M. New fault-tolerant control strategy of five-phase induction motor with four-phase and three-phase modes of operation. *Electronics* **2018**, *7*, 159. [CrossRef]
5. Amezquita-Sanchez, J.P.; Valtierra-Rodriguez, M.; Perez-Ramirez, C.A.; Camarena-Martinez, D.; Garcia-Perez, A.; Romero-Troncoso, R.J. Fractal dimension and fuzzy logic systems for broken rotor bar detection in induction motors at start-up and steady-state regimes. *Meas. Sci. Technol.* **2017**, *28*, 075001. [CrossRef]
6. Gyftakis, K.N.; Spyropoulos, D.V.; Kappatou, J.C.; Mitronikas, E.D. A novel approach for broken bar fault diagnosis in induction motors through torque monitoring. *IEEE Trans. Energy Convers.* **2013**, *28*, 267–277. [CrossRef]
7. Florkowski, M.; Furgał, J. Detection of winding faults in electrical machines using the frequency response analysis method. *Meas. Sci. Technol.* **2004**, *15*, 2067. [CrossRef]
8. Bazan, G.H.; Scalassara, P.R.; Endo, W.; Goedtel, A.; Godoy, W.F.; Palácios, R.H.C. Stator fault analysis of three-phase induction motors using information measures and artificial neural networks. *Electr. Power Syst. Res.* **2017**, *143*, 347–356. [CrossRef]
9. Thomson, W.T.; Fenger, M. Current signature analysis to detect induction motor faults. *IEEE Ind. Appl. Mag.* **2001**, *7*, 26–34. [CrossRef]
10. Arthur, N.; Penman, J. Induction machine condition monitoring with higher order spectra. *IEEE Trans. Ind. Electron.* **2000**, *47*, 1031–1041. [CrossRef]
11. Ballal, M.S.; Khan, Z.J.; Suryawanshi, H.M.; Sonolikar, R.L. Adaptive neural fuzzy inference system for the detection of inter-turn insulation and bearing wear faults in induction motor. *IEEE Trans. Ind. Electron.* **2007**, *54*, 250–258. [CrossRef]
12. Ukil, A.; Chen, S.; Andenna, A. Detection of stator short circuit faults in three-phase induction motors using motor current zero crossing instants. *Electr. Power Syst. Res.* **2011**, *81*, 1036–1044. [CrossRef]
13. Drif, M.H.; Cardoso, A.J.M. Stator fault diagnostics in squirrel cage three-phase induction motor drives using the instantaneous active and reactive power signature analyses. *IEEE Trans. Ind. Inform.* **2014**, *10*, 1348–1360. [CrossRef]
14. Glowacz, A.; Glowacz, W.; Glowacz, Z.; Kozik, J. Early fault diagnosis of bearing and stator faults of the single-phase induction motor using acoustic signals. *Measurement* **2018**, *113*, 1–9. [CrossRef]

15. Thomson, W.T. On-line MCSA to diagnose shorted turns in low voltage stator windings of 3-phase induction motors prior to failure. In Proceedings of the Electric Machines and Drives Conference (IEMDC 2001), Cambridge, MA, USA, 17–20 June 2001; pp. 891–898.
16. Bouzid, M.B.K.; Champenois, G. New expressions of symmetrical components of the induction motor under stator faults. *IEEE Trans. Ind. Electron.* **2013**, *60*, 4093–4102. [CrossRef]
17. Surya, G.N.; Khan, Z.J.; Ballal, M.S.; Suryawanshi, H.M. A simplified frequency-domain detection of stator turn fault in squirrel-cage induction motors using an observer coil technique. *IEEE Trans. Ind. Electron.* **2017**, *64*, 1495–1506. [CrossRef]
18. Asfani, D.A.; Muhammad, A.K.; Purnomo, M.H.; Hiyama, T. Temporary short circuit detection in induction motor winding using combination of wavelet transform and neural network. *Expert Syst. Appl.* **2012**, *39*, 5367–5375. [CrossRef]
19. Devi, N.R.; Sarma, D.V.S.; Rao, P.V.R. Detection of stator incipient faults and identification of faulty phase in three-phase induction motor–simulation and experimental verification. *IET Electr. Power Appl.* **2015**, *9*, 540–548. [CrossRef]
20. Lee, S.H.; Kim, S.; Kim, J.M.; Choi, C.; Kim, J.; Lee, S.; Oh, Y. Extraction of induction motor fault characteristics in frequency domain and fuzzy entropy. In Proceedings of the IEEE International Conference on Electric Machines and Drives, San Antonio, TX, USA, 15 May 2005; pp. 35–40.
21. Li, J.; Yu, H.; Zhang, L. Application of ensemble empirical mode decomposition on stator inter-turn short-circuit fault in doubly fed induction generators. In Proceedings of the Second International Conference on Mechatronics and Automatic Control, Beijing, China, 20–21 September 2014; pp. 73–83.
22. Rosero, J.A.; Romeral, L.; Ortega, J.A.; Rosero, E. Short-circuit detection by means of empirical mode decomposition and Wigner–Ville distribution for PMSM running under dynamic condition. *IEEE Trans. Ind. Electron.* **2015**, *56*, 4534–4547. [CrossRef]
23. Sahraoui, M.; Ghoggal, A.; Guedidi, S.; Zouzou, S.E. Detection of inter-turn short-circuit in induction motors using Park–Hilbert method. *Int. J. Syst. Assur. Eng. Manag.* **2014**, *5*, 337–351. [CrossRef]
24. Garcia-Guevara, F.M.; Villalobos-Piña, F.J.; Alvarez-Salas, R.; Cabal-Yepez, E.; Gonzalez-Garcia, M.A. Stator fault detection in induction motors by autoregressive modeling. *Math. Probl. Eng.* **2016**, *2016*, 1–7. [CrossRef]
25. Ghate, V.N.; Dudul, S.V. Optimal MLP neural network classifier for fault detection of three phase induction motor. *Expert Syst. Appl.* **2010**, *37*, 3468–3481. [CrossRef]
26. Valtierra-Rodriguez, M.; Granados-Lieberman, D.; Torres-Fernandez, J.E.; Rodríguez-Rodríguez, J.R.; Gómez-Aguilar, J.F. A new methodology for tracking and instantaneous characterization of voltage variations. *IEEE Trans. Instrum. Meas.* **2016**, *65*, 1596–1604. [CrossRef]
27. Antonino-Daviu, J.A.; Riera-Guasp, M.; Pineda-Sanchez, M.; Pérez, R.B. A critical comparison between DWT and Hilbert–Huang-based methods for the diagnosis of rotor bar failures in induction machines. *IEEE Trans. Ind. Appl.* **2009**, *45*, 1794–1803. [CrossRef]
28. Torres, M.E.; Colominas, M.A.; Schlotthauer, G.; Flandrin, P. A complete ensemble empirical mode decomposition with adaptive noise. In Proceedings of the IEEE International Conference on Acoustics, Speech and Signal Processing (ICASSP 2011), Prague, Czech Republic, 22–27 May 2011; pp. 4144–4147.
29. Elbouchikhi, E.; Choqueuse, V.; Amirat, Y.; Benbouzid, M.E.H.; Turri, S. An efficient Hilbert–Huang transform-based bearing faults detection in induction machines. *IEEE Trans. Energy Convers.* **2017**, *32*, 401–413. [CrossRef]
30. Boashash, B.; Khan, N.A.; Ben-Jabeur, T. Time–frequency features for pattern recognition using high-resolution TFDs: A tutorial review. *Digit. Signal Process.* **2015**, *40*, 1–30. [CrossRef]
31. Adeli, H.; Jiang, X. Dynamic fuzzy wavelet neural network model for structural system identification. *J. Struct. Eng.* **2006**, *132*, 102–111. [CrossRef]
32. Amezquita-Sanchez, J.P.; Adeli, H. Synchrosqueezed wavelet transform-fractality model for locating, detecting, and quantifying damage in smart highrise building structures. *Smart Mater. Struct.* **2015**, *24*, 065034. [CrossRef]
33. Jung, J.H.; Lee, J.J.; Kwon, B.H. Online diagnosis of induction motors using MCSA. *IEEE Trans. Ind. Electron.* **2006**, *53*, 1842–1852. [CrossRef]
34. Roscoe, A.J.; Abdulhadi, I.F.; Burt, G.M. Filters for M class phasor measurement units. In Proceedings of the IEEE International Workshop on Applied Measurements for Power Systems (AMPS 2012), Aachen, Germany, 26–28 September 2012; pp. 1–6.

35. Ramstad, T. Digital methods for conversion between arbitrary sampling frequencies. *IEEE Trans. Acoust. Speech Signal Process.* **1984**, *32*, 577–591. [CrossRef]
36. Cabal-Yepez, E.; Valtierra-Rodriguez, M.; Romero-Troncoso, R.J.; Garcia-Perez, A.; Osornio-Rios, R.A.; Miranda-Vidales, H.; Alvarez-Salas, R. FPGA-based entropy neural processor for online detection of multiple combined faults on induction motors. *Mech. Syst. Signal Process.* **2012**, *30*, 123–130. [CrossRef]
37. Mejia-Barron, A.; Valtierra-Rodriguez, M.; Granados-Lieberman, D.; Olivares-Galvan, J.C.; Escarela-Perez, R. The application of EMD-based methods for diagnosis of winding faults in a transformer using transient and steady state currents. *Measurement* **2018**, *117*, 371–379. [CrossRef]
38. Bafroui, H.H.; Ohadi, A. Application of wavelet energy and Shannon entropy for feature extraction in gearbox fault detection under varying speed conditions. *Neurocomputing* **2014**, *133*, 437–445. [CrossRef]
39. Seryasat, O.R.; Honarvar, F.; Rahmani, A. Multi-fault diagnosis of ball bearing using FFT, wavelet energy entropy mean and root mean square (RMS). In Proceedings of the IEEE International Conference on Systems Man and Cybernetics (SMC 2010), Istanbul, Turkey, 10–13 October 2010; pp. 4295–4299.
40. Kamalapathi, K.; Priyadarshi, N.; Padmanaban, S.; Holm-Nielsen, J.; Azam, F.; Umayal, C.; Ramachandaramurthy, V. A hybrid moth-flame fuzzy logic controller based integrated cuk converter fed brushless DC motor for power factor correction. *Electronics* **2018**, *7*, 288. [CrossRef]
41. Fink, A.; Töpfer, S.; Isermann, R. Nonlinear model-based control with local linear neuro-fuzzy models. *Arch. Appl. Mech.* **2003**, *72*, 911–922. [CrossRef]
42. Nentwig, M.; Mercorelli, P. Throttle valve control using an inverse local linear model tree based on a Fuzzy neural network. In Proceedings of the 7th IEEE International Conference on Cybernetic Intelligent Systems, London, UK, 9–10 September 2008. [CrossRef]
43. De Silva, C.W. *Intelligent Control: Fuzzy Logic Applications*; CRC Press: Boca Raton, FL, USA, 2018.
44. Passino, K.M.; Yurkovich, S.; Reinfrank, M. *Fuzzy Control*; Addison-Wesley: Menlo Park, CA, USA, 1998; ISBN 0-201-18074-X.

electronics

Article

A Capacitance-to-Time Converter-Based Electronic Interface for Differential Capacitive Sensors

Andrea De Marcellis [1], **Càndid Reig** [2,*] and **María-Dolores Cubells-Beltrán** [2]

1 Department of Industrial and Information Engineering and Economics, University of L'Aquila, 67100 L'Aquila, Italy; andrea.demarcellis@univaq.it

2 Department of Electronic Engineering, University of Valencia, 46100 Burjassot, Spain; m.dolores.cubells@uv.es

* Correspondence: candid.reig@uv.es; Tel.: +34-9635-44038

Received: 13 November 2018; Accepted: 4 January 2019; Published: 10 January 2019

Abstract: In this paper we present an oscillating conditioning circuit, operating a capacitance-to-time conversion, which is suitable for the readout of differential capacitive sensors. The simple architecture, based on a multiple-feedbacks structure that avoids ground noise disturbs and system calibrations, employs only three Operational Amplifiers (OAs) and a mixer implementing a square wave oscillator that provides an AC sensor excitation voltage. It performs a Period Modulation (PM) and a Pulse Width Modulation (PWM) of the output signal proportionally to the sensor differential capacitance values. The sensor variation range and the detection sensitivity can be easily set through the additional resistors. Preliminary PSpice simulation results have shown a good agreement with theoretical calculations as well as a linear response with a high detection sensitivity of differential capacitive sensors having a baseline in the range [2.2 ÷ 180 pF]. Moreover, different experimental measurements have been also performed by implementing the circuit on a laboratory breadboard using commercial discrete components so validating the idea and providing the circuit performances with different kind of differential capacitive sensors achieving detection resolutions of about 0.1 fF in an overall differential capacitive variation range that is equal to ±15.8 pF. The achieved results demonstrate that the proposed interface solution is suitable for on-chip integration with different kinds of differential capacitive sensing devices, such as Micro-Electro-Mechanical-System (MEMS), force/position, and humidity sensors in biomedical and robotics applications.

Keywords: PM/PWM; capacitance-to-time conversion; differential capacitive sensor

1. Introduction

Recent developments on integration techniques and circuit miniaturizations, together with advances on capacitive sensing technologies, have led to the design of high-sensitivity and small-size devices, like Micro-Electro-Mechanical-System (MEMS), gyroscopes, accelerometers, position/displacement, pressure/force, flow, and humidity sensors having very high detection capabilities that are widely used in robotics/biomedical sensor applications as well as in bioengineering microsystems [1–3]. Basically, their behavior can be simply described as a planar capacitor (i.e., $C = \varepsilon \cdot A/d$, being ε the relative dielectric constant, A the active surface, and d the distance between capacitor metal plates) whose mechanical features (i.e., A and/or d) are temporarily changed by the physical phenomena to be detected. Furthermore, in several sensory systems, differential capacitive sensing configurations also provide a suitable reduction of the common-mode noise and the parasitic component effects [4–15].

In the literature, capacitive sensor interfaces mainly concern Capacitance-to-Voltage (*C-V*) and Capacitance-to-Time (*C-T*) analogue conversion techniques [15–24]. In particular, the first topology commonly employs voltage/transimpedance amplifiers, charge/chopper amplifiers, and switched capacitors showing limited/reduced detection ranges, sensitivities, and resolutions mainly due to noise issues. On the contrary, the latter approach is typically based on square wave relaxation oscillators in which the sensing operation is performed by the readout of the output signal period (i.e., a Period Modulation, PM) and/or its duty-cycle (i.e., a Pulse Width Modulation, PWM) as a function of the single/differential sensor capacitance. PM-based interfaces are asynchronous and have a measurement time and a resolution generally dependent from the sensor capacitance. On the other hand, PWM-based solutions are synchronous circuit needing a clock line to synchronize the interfacing operation, while their measurement time and resolution are typically independent from the sensor capacitance. These kinds of interface solutions, typically showing straightforward architectures with a high tolerance to common-mode noise/disturbs and to parasitic components as well as to supply voltage drifts, allows for covering wide capacitive variation ranges and can also be combined with a digital system to easily measure the time intervals (e.g., through counters) [25–34]. Recently, also mixed-signal and digital sensor systems are becoming prevalent, so that new topologies of sensor conditioning circuits have been also introduced. By performing a Capacitance-to-Digital (*C-D*) conversion (sometimes also combined with *C-V* or *C-T* conversions), these architectures can be directly interfaced with a microcontroller even if, sometimes, requiring high frequency clock signals to achieve suitable sensitivities and resolutions [35–42].

However, most of the developed solutions are mainly suitable to only single element capacitive sensors. On the contrary, often direct differential measurements with high acquisition rate, accuracy, precision, sensitivity, and resolution are required, since two single-element measurements could provide errors if not performed simultaneously due to the time variations of the capacitances owed to the occurring dynamic physical phenomenon.

In this regard, here we propose a new low-cost portable solution of analogue electronic interface circuit performing a *C-T* conversion that is suitable for differential capacitive sensors with high detection sensitivity and resolution. The developed architecture is based on a relaxation oscillator whose generated square wave signal period and duty-cycle are linearly dependent from the differential capacitance variations combining both the PM and the PWM modulations. Through the setting of few resistor values, it is possible to regulate dynamic range, sensitivity, and resolution of the differential capacitance variation detection. Moreover, it shows a very low sensitivity to common-mode noise and disturbances, as well as to the effects due to the presence of parasitic elements at the circuit sensing nodes. The interface circuit has been designed and preliminary simulated in the OrCAD PSpice environment. Afterwards, after its implementation on a laboratory breadboard employing commercial discrete components, the proposed solution has been tested through experimental measurements confirming the theoretical calculations. In particular, sample components and commercial capacitive sensors have been employed, as well as an ad-hoc liquid level detection system has been fabricated and characterized validating the developed solution and its performances. The block diagram of the conducted research is depicted in Figure 1.

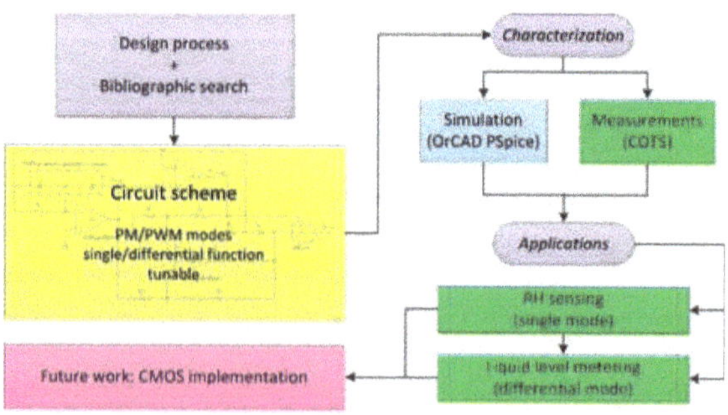

Figure 1. Block diagram of the conducted research.

2. Materials and Methods

The proposed schematic circuit for differential capacitive sensors interfacing is reported in Figure 2, while Figure 3 shows an example of the time response of the circuit reporting the voltage signals at its main output nodes. This solution, designed with a reduced number of active (three Operational Amplifiers, OAs, and one Mixer) and passive (seven resistors) components, is based on a relaxation oscillator performing a *C-T* conversion through a closed multiple-feedback loop architecture combining both the PM and the PWM modulations. This avoids any calibration procedure and it reduces ground noise/disturbs (i.e., a very low sensitivity to common-mode noise/disturbs and to parasitic elements), while it allows for an auto-excitation of the differential capacitive sensor through the output AC square waveform. The designed architecture, in fact, intrinsically provides a square wave output (i.e., a "digitalized" output signal), which is independent from the supply voltage, so offering further benefits, such as immunity to voltage offsets and easiness in digital multiplexing and signal processing (e.g., its period and duty-cycle can be easily read by means of digital counters).

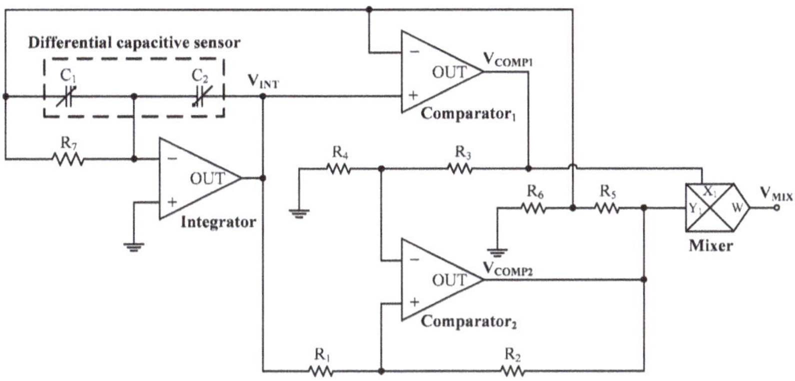

Figure 2. Schematic circuit of the proposed electronic interface for differential capacitive sensors.

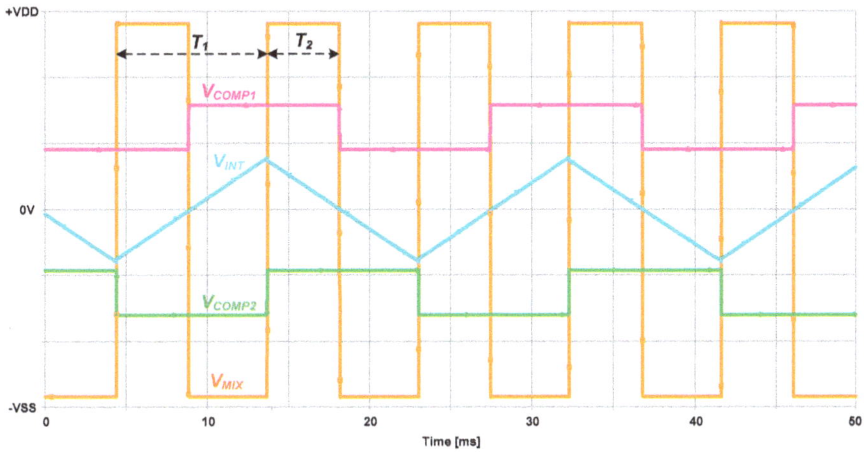

Figure 3. Example of the time response of the proposed interface evaluating the voltage signals at its main nodes.

More in detail, the circuit is composed of a voltage integrator where the differential capacitive sensor is connected, two hysteresis voltage comparators that allow for regulating the dynamic range and the detection sensitivity and resolution through the use of seven resistors and a mixer that combine all the information providing an output pulsed signal whose period and duty-cycle depend on the two capacitances C_1 and C_2 of the differential capacitive sensor. In particular, the mixer also allows for reducing the measurement time, since the period of the output pulsed signal V_{MIX} is equal to a semi-period (i.e., a double frequency) of the internal square wave signal V_{COMP2} generated by the closed loop oscillator composed by the integrator and the two comparators. The mixer, in fact, receives at its input terminals the two square wave signals, V_{COMP1} and V_{COMP2}, generating a further square waveform V_{MIX} whose period T_1 (that is half period of V_{COMP1} and V_{COMP2}) and pulse width T_2 (that is the overlapping time between V_{COMP1} and V_{COMP2}, when they have both positive or negative values) have to be measured so as to estimate and calculate the capacitance values C_1 and C_2. Moreover, it is possible to easily set the interface working range (i.e., the sensor variation range) through the employed resistors, which also allow for fixing the desired detection sensitivity of the overall conditioning circuit.

Through a straightforward circuit node analysis, when considering ideal components, the time period T_1 and the pulse width T_2 of the generated output square waveform V_{MIX} can be expressed, as follows:

$$T_1 = 2R_7\left(C_2\frac{\left(\frac{R_1}{R_1+R_2}+\frac{R_4}{R_3+R_4}\right)}{\left(1-\frac{R_1}{R_1+R_2}\right)\left(\frac{R_6}{R_5+R_6}\right)} - C_1\right) \tag{1}$$

$$T_2 = R_7\left(C_2\frac{\left(\frac{R_1}{R_1+R_2}\frac{R_6}{R_5+R_6}\right)+\frac{R_1}{R_1+R_2}+\frac{R_4}{R_3+R_4}-\frac{R_6}{R_5+R_6}}{\left(1-\frac{R_1}{R_1+R_2}\right)\left(\frac{R_6}{R_5+R_6}\right)} - 2C_1\right) \tag{2}$$

from which (i.e., by inverting them) it is possible to achieve the following relationships to estimate/calculate the two components (i.e., C_1 and C_2) of the differential capacitive sensor as a function of the other circuit parameters and the time values T_1 and T_2:

$$C_1 = \frac{1}{R_7}(T_2 - T_1)\left(\frac{\left(\frac{R_1}{R_1+R_2}+\frac{R_4}{R_3+R_4}\right)}{\left(\frac{R_1}{R_1+R_2}\frac{R_6}{R_5+R_6}\right)-\frac{R_1}{R_1+R_2}-\frac{R_4}{R_3+R_4}-\frac{R_6}{R_5+R_6}}\right) - \frac{1}{R_7}\left(\frac{T_1}{2}\right) \tag{3}$$

$$C_2 = \frac{1}{R_7}(T_2 - T_1)\left(\frac{\left(1-\frac{R_1}{R_1+R_2}\right)\left(\frac{R_6}{R_5+R_6}\right)}{\left(\frac{R_1}{R_1+R_2}\frac{R_6}{R_5+R_6}\right)-\frac{R_1}{R_1+R_2}-\frac{R_4}{R_3+R_4}-\frac{R_6}{R_5+R_6}}\right) \tag{4}$$

It is worth noting that, since the circuit converts a differential capacitance into a pulsed signal, the initial values of C_1 and C_2 impose the starting oscillating period T_1 and duty-cycle T_2 of the output signal. Nevertheless, through the seven resistors it is possible to change these initial values, even if it acts also on the circuit detection range, sensitivity, and resolution. On the other hand, according to Equations (1) and (2), T_1 and T_2 are mostly/directly conditioned by resistor R_7, which mainly regulates the charge/discharge of C_1 and C_2. Therefore, when dealing with small sensor capacitances (in the range of few pF), R_7 is required to be high in order to set T_1 and T_2 in a range (e.g., in the order of µs or ms), which is more suitable for subsequent signal conditioning/processing stages as well as to optimize/maximize the circuit response/performance (i.e., the detectable capacitive variation range and the detection sensitivity/resolution of the interface circuit). For moderate detection of sensitivities/resolutions and/or high capacitive ranges (in the range of hundreds pF), the value of R_7 can be reduced. Finally, if $C_1 = C_2$, the relationship between T_1 and T_2 can be simply expressed, as follows:

$$T_2 = T_1\left(1+\frac{\left(\frac{R_1}{R_1+R_2}\frac{R_6}{R_5+R_6}\right)-\frac{R_1}{R_1+R_2}-\frac{R_4}{R_3+R_4}-\frac{R_6}{R_5+R_6}}{2\left(\left(\frac{R_1}{R_1+R_2}\frac{R_6}{R_5+R_6}\right)+\frac{R_1}{R_1+R_2}+\frac{R_4}{R_3+R_4}-\frac{R_6}{R_5+R_6}\right)}\right) \tag{5}$$

Moreover, we highlight that the output PWM signal can also be easily converted into a DC voltage signal through a low-pass filtering operation. In this way, the information on the duty-cycle of the output square waveform, also evaluated as the ratio T_2/T_1 taking into account the effects due to the variation of both the sensor capacitive elements C_1 and C_2 (i.e., the differential variation) is evaluated by extracting the DC level of the output pulsed signal V_{MIX} (i.e., its mean value that is proportional to the differential capacitive sensor variation $(C_1 - C_2)/(C_1 + C_2)$) whose value can be ideally calculated, as follows:

$$\int V_{MIX}dt = (V_{SAT+} - V_{SAT-})\frac{T_2}{T_1} + V_{SAT-} \tag{6}$$

being V_{SAT+} and V_{SAT-} the output saturation levels of the mixer reached by V_{MIX} signal. In this last case, the overall circuit can be consequently classified as a *C-V* converter.

3. Results

3.1. Simulations

OrCAD PSpice simulations have been preliminary conducted employing low-noise JFET-input TL071 by Texas Instruments as OAs and AD633 by Analog Devices as analog multiplier (i.e., mixer) all being supplied at ±15 V. Different values of the baseline of the differential capacitive sensor have been considered (i.e., $C_1 = C_2 = C_0$ = 2.2 pF, 10 pF, 100 pF, and 180 pF), so to demonstrate the circuit suitability with different kind of commercial and ad-hoc integrated sensors (with responses linear, hyperbolic, etc.). In Figure 4, the simulation results are reported and compared with the related theoretical values (from Equations (1) and (2)) of the period T_1 and the pulse width T_2 of output square waveform V_{MIX} as a function of the relative variation the differential capacitive sensor (i.e., $100 \times (C_1 - C_2)/(C_1 + C_2)$) showing high linearity (i.e., R^2 = 0.9997) and high sensitivity S (i.e., S_{T1} = 0.145 ms/pF; S_{T2} = 0.071 ms/pF). In particular, Figure 4a shows an example of the time response of the circuit when considering the voltage signals at its main nodes for C_1 = 5 pF and C_2 = 15 pF (i.e., considering C_0 = 10 pF and a differential capacitance variation equal to −50%). In addition, Figure 4b reports the T_1 and T_2 time values that were achieved when considering a sensor baseline $C_1 = C_2 = C_0$ = 10 pF (i.e., the central/initial value) and its relative variation of ±50%, so that the differential capacitance (i.e., $C_1 - C_2$) is changed from −10 pF to +10 pF. In particular, C_1 changes from 5 pF to 15 pF, while C_2 varies from 15 pF to 5 pF with a differential capacitance variation step equal to 2 pF (i.e., each single capacitive element varies in opposite way with a step of 1 pF). Moreover, the reported results have been achieved by setting the circuit resistors, as follows: R_1 = 1 kΩ, R_2 = 3 kΩ, R_3 = 20 kΩ, R_4 = 1 kΩ, R_5 = 15 kΩ, R_6 = 1 kΩ and R_7 = 10 MΩ.

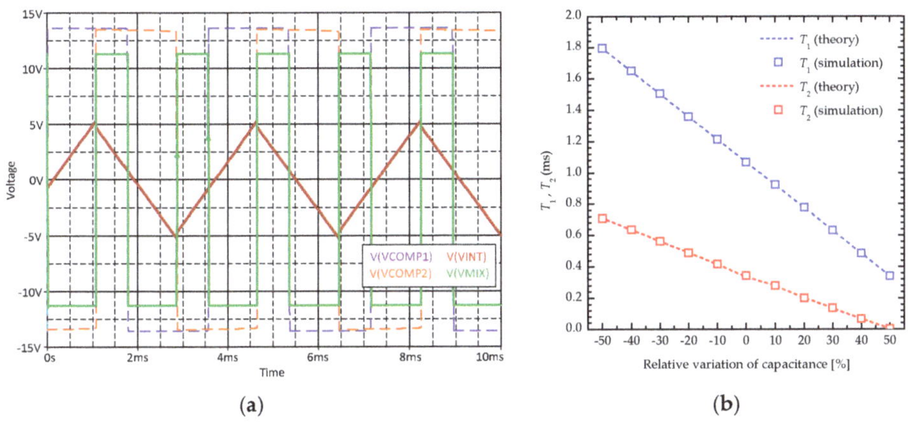

Figure 4. (**a**) Example of the time response of the circuit; and, (**b**) simulation results together with the corresponding theoretical data of the period T_1 and the pulse width T_2 of output square waveform V_{MIX} as a function of the relative variation the differential capacitive sensor for $C_1 = C_2 = C_0$ = 10 pF.

Moreover, further simulations have been conducted in order to evaluate the effects on the circuit of operating temperature variations. More in detail, we have considered/referred to commercial and industrial applications, so simulating the circuit from −20 °C up to 85 °C. In particular, referring to the circuit set-up considered for the capacitive variation range of 5–15 pF for C_1 and C_2, the resulting maximum relative variations of T_1 and T_2 at −20 °C is lower than 0.6%, while at +85 °C it is lower than 9%. These values correspond to maximum relative errors that are lower than 10% at −20 °C and lower than 7% at +85 °C in the estimation of C_1 and C_2 values.

3.2. Preliminary Experimental Measurements

Basic experimental measurements have been performed implementing the circuit on a laboratory breadboard employing commercial discrete components as well as capacitive sensors. In this case, the differential capacitance (i.e., $C_1 - C_2$) is varied from −15.8 pF to +15.8 pF using commercial high-precision high-accuracy discrete capacitors, calibrated/measured by using an ISO-TECH LCR821 high-precision high-accuracy LCR-meter (accuracy better than 0.5%) verifying the maximum deviation from the capacitance nominal value lower than 1%. In particular, C_1 has been changed from 2.2 pF to 18 pF (i.e., 2.2 pF, 4.7 pF, 8.2 pF, 10 pF, 12 pF, 15 pF, 18 pF) and C_2 from 18 pF to 2.2 pF (i.e., 18 pF, 15 pF, 12 pF, 10 pF, 8.2 pF, 4.7 pF, 2.2 pF), while keeping constant the total capacitance value $C_1 + C_2$ at about 20 pF (C_0 = 10 pF). In order to get oscillating periods of few milliseconds and to achieve better results, the following resistance values have been chosen: R_1 = 1 kΩ, R_2 = 1.2 kΩ, R_3 = 15 kΩ, R_4 = 1 kΩ, R_5 = 47 kΩ, R_6 = 1 kΩ, R_7 = 10 MΩ. The parasitic capacitance of the resistors was measured, giving values below 0.5 pF. The resulting measurements of the period T_1 and the pulse width T_2 have been performed while employing a GPIB-based experimental setup and a National Instruments LABVIEW-based automatic acquisition system, including conventional instrumentations, such as a frequency-meter Agilent 34970A (accuracy better than 0.01%), a Data Acquisition/Switch Unit, and digital multimeter Agilent 34401A (accuracy better than 0.01%), as well as an oscilloscope Tektronix TPS2024R.

More in detail, firstly we proved the period and pulse width modulations as a function of the variation of the differential capacitance confirming their proportional linearly dependence. The oscillograms are reported in Figure 5a–c that demonstrate the proper functionality of the proposed interface circuit showing its main signals (in particular, the generated output square waveform V_{MIX}) and the corresponding measured values of T_1 (i.e., CH1 Period in the right part of each picture) and T_2 (i.e., CH1 Pos Width in the right part of each picture) for three different sets of C_1 and C_2 values: Figure 5a, C_1 = 2.2 pF and C_2 = 18 pF; Figure 5b, $C_1 = C_2$ = 10 pF; Figure 5c, C_1 = 18 pF and C_2 = 2.2 pF. The overall measurement results are reported in Figure 5d showing the oscillation period T_1 and the pulse width T_2 as function of the differential capacitance variation (i.e., $100 \times (C_1 - C_2)/(C_1 + C_2)$), which are in a good agreement with the theoretical calculations according Equations (1) and (2) and with the related simulation results achieved with the same operating conditions. In this case, the achieved sensitivities with respect to T_1 and T_2 are S_{T1} = 0.982 ms/pF and S_{T2} = 0.491 ms/pF, respectively (linearity correlation coefficient R^2 of about 0.999). On the other hand, taking into account the measured values of T_1 and T_2, Figure 5e reports on the estimated values of C_1 and C_2 capacitances calculated through the Equations (3) and (4). The corresponding relative error, evaluated between the measured/estimated capacitance values and its nominal/real values, is lower than 3%. Finally, the maximum averaged RMS jitter level, measured on the rising/falling edges of the output square wave signal V_{MIX}, results to be always lower than 50 ns, so that the resulting estimated minimum detectable differential capacitance variation (i.e., the best theoretical detection resolution of the circuit) is about 0.1 fF. The average power consumption of the overall electronic interface circuit is about 68 mW.

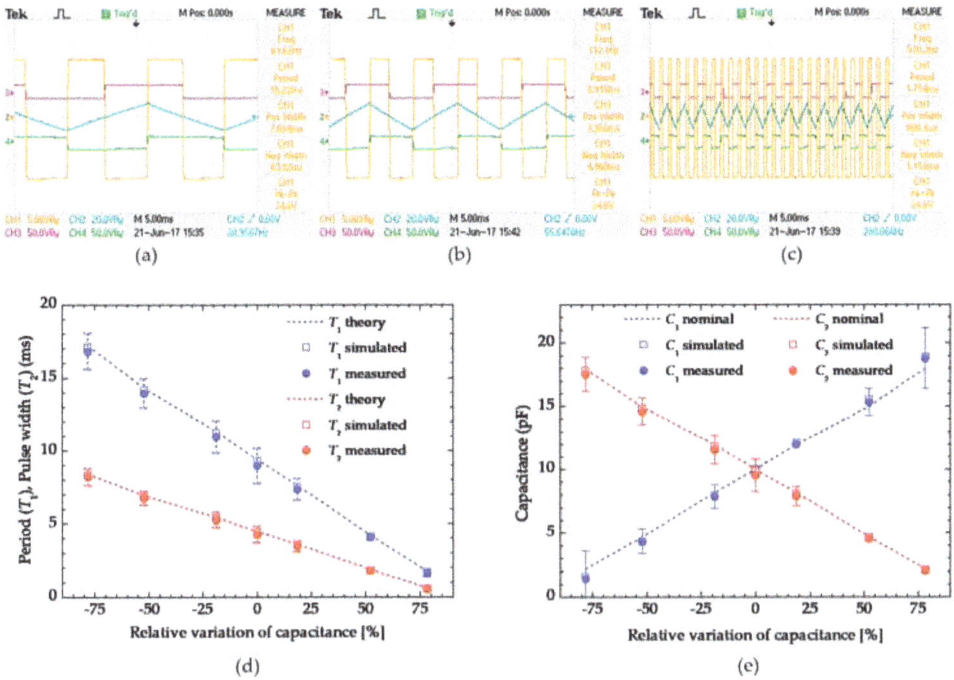

Figure 5. Measured main output signals of the circuit for (**a**) C_1 = 2.2 pF and C_2 = 18 pF, (**b**) $C_1 = C_2$ = 10 pF, and (**c**) C_1 = 18 pF and C_2 = 2.2 pF; (**d**) measured, simulated and theoretical oscillation periods T_1 and pulse widths T_2 as a function of the relative differential capacitance variation; (**e**) calculated/estimated capacitance values.

3.3. Relative Humidity (RH) Sensor

In order to evaluate the potentiality of the proposed circuit as electronic interface in real capacitive sensing applications, we arranged an experimental set-up for the measurement of the Relative Humidity (RH) making use of the commercial capacitive sensor HS1101LF by Sensor Solutions—Measurement Specialties and a climatic chamber Challenge CH600 by ACS—Angelantoni Industries. The employed capacitive sensor provides a nominal capacitance value of about 180 pF @ 55% RH and it can be biased/excited in the operating range [5 kHz–300 kHz] with AC voltage signals up to 10 V.

In this regard, for preliminary setting/testing and optimization of the circuit, we made use of commercial high-precision high-accuracy discrete capacitors, calibrated/measured by using the ISO-TECH LCR821 LCR-meter showing a maximum deviation of lower than 1%. In particular, the capacitor C_1 has been fixed to a value equal to 180 pF (i.e., C_0 = 180 pF), while performing a sweep of the C_2 capacitance in the variation range [160 pF–197 pF] by means of a fixed 150 pF capacitor in parallel with others having smaller different values (i.e., C_2 = 150 pF + [10 pF; 12 pF; 15 pF; 18 pF; 22 pF; 27 pF; 30 pF; 33 pF; 39 pF; 47 pF]). In this case, the following resistance values have been set so to optimize the interface circuit response: $R_1 = 470\ \Omega$, $R_2 = 560\ \Omega$, $R_3 = 2.2\ \text{k}\Omega$, $R_4 = 470\ \Omega$, $R_5 = 4.7\ \text{k}\Omega$, $R_6 = 470\ \Omega$, $R_7 = 47\ \text{k}\Omega$. The employed experimental set-up and measurement instrumentations are those ones already previously reported and described in Section 3.1.

The overall obtained results are shown in Figure 6. In particular, Figure 6a–c report the oscillograms showing the circuit main output square waveform V_{MIX}, including the corresponding measured values of the period T_1 (i.e., CH1 Period in the right part of each picture) and the pulse width T_2 (i.e., CH1 Pos Width in the right part of each picture). They demonstrate the proper performances of the circuit for an operating configuration with a fixed value of C_1 = 180 pF and three different values of C_2: Figure 6a, C_2 = 160 pF; Figure 6b, C_2 = 180 pF; Figure 6c, C_2 = 197 pF.

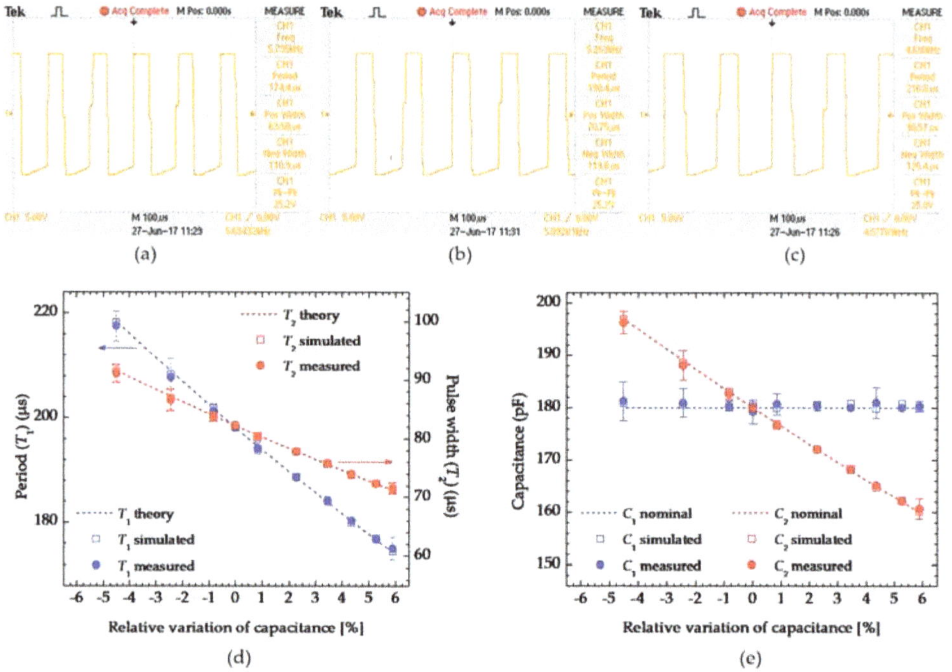

Figure 6. Oscillograms showing the output voltage signal V_{MIX} that demonstrates the proper performances of the circuit for a configuration employing a fixed C_1 = 180 pF and (**a**) C_2 = 160 pF, (**b**) C_2 = 180 pF, and (**c**) C_2 = 197 pF; (**d**) measured, simulated, and theoretical oscillation periods T_1 and pulse widths T_2 as a function of the relative differential capacitance variation; (**e**) calculated/estimated capacitance values.

Furthermore, Figure 6d shows the overall measurement results reporting the oscillation period T_1 and the pulse width T_2 as function of the relative capacitance variation (i.e., $100 \times (C_1 - C_2)/(C_1 + C_2)$) achieved by changing C_2 in the range [160 pF–197 pF]. The collected data confirm the correct functionalities of the interface circuit agreeing with the theoretical calculations, from Equations (1) and (2), and also with the corresponding simulations performed when considering the same circuit parameters. From these results, the two detection sensitivities S_{T1} = 0.001 ms/pF and S_{T2} = 0.0006 ms/pF with respect to T_1 and T_2, respectively, have been calculated (linearity correlation coefficient R^2 of about 0.999) so as to evaluate/estimate the performances of the circuit, especially in terms of the minimum detection resolution of differential capacitance variations that, in this case, is about 83 fF (considering a maximum averaged RMS jitter level, measured on the rising/falling edges of V_{MIX}, lower than 50 ns). Finally, starting from these results and by employing Equations (3) and (4), the values of C_1 and C_2 capacitances have been estimated/calculated, as reported in Figure 6d showing a relative error, evaluated between the measured/estimated capacitance values and its nominal/real values, lower than 0.5 %.

Successively, the C_2 capacitor has been replaced by the commercial HS1101LF RH capacitive sensor, so performing through the controlled climatic chamber a sweep in the RH from 35% to 75%, with steps of 5% and room temperature set to 25 °C. The achieved results are reported in Figure 7 comparing the calculated/estimated RH% with the fixed nominal values as well as with the values that were achieved from direct measurement of the sensor by using the ISO-TECH LCR821 high-precision high-accuracy LCR-meter and extracting the data from sensor datasheet. The reported data have been extracted starting from the measurement of the time period T_1 and the pulse width T_2 as a function of the RH% variation with detection sensitivities of about $S_{T1} = 0.0004$ ms/RH% and $S_{T2} = 0.0002$ ms/RH% with respect to T_1 and T_2, respectively. Subsequently, the values of the capacitances C_1 and C_2 have been estimated/calculated by employing Equations (3) and (4). Finally, from the information reported in the datasheet of the used commercial capacitive sensor, the RH% values have been extracted employing the reverse polynomial response equation reported in the same device datasheet. In this case, the relative error, evaluated between the measured/estimated values and nominal/real values, is lower than 3% and the minimum estimated detection resolution in terms of the RH% variation is about 0.25% (considering a maximum averaged RMS jitter level, measured on the rising/falling edges of V_{MIX}, lower than 50 ns).

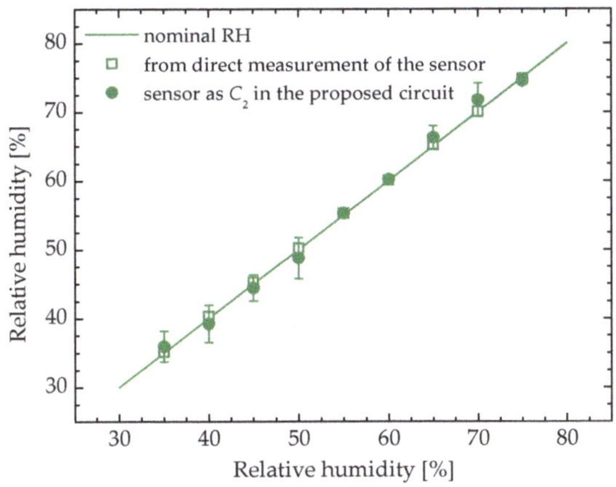

Figure 7. Experimental results concerning the measurement of RH% performed by employing a commercial capacitive sensor.

3.4. *Liquid Level*

Lastly, we developed a liquid level meter, with the aim of evaluating the performance of the proposed electronic interface in real differential capacitance measurement applications. The considered experimental apparatus is depicted in Figure 8. A plexiglas-based cube with a volume lower than 1 ℓ is provided with three Cu-based conductive plates. Two of them are fixed at the top and bottom faces, while the other floats onto the top liquid surface keeping the horizontality through the use of a polystyrene plate fixed under the Cu plate (i.e., the combined plates work like a float). In this way, two complementary (differential) capacitors having a common element (i.e., common plate) are formed: one with liquid (i.e., distilled water) as the dielectric forming the capacitor C_1 and the other with the air providing the capacitor C_2.

This apparatus has been employed as differential capacitive sensor measured through the developed interface circuit. More in detail, the distilled water level has been changed from 1 cm to 7 cm, with steps of 1 cm, so moving the common central floating plate that provides a simultaneous variation of C_1 and C_2 capacitance values. Moreover, in order to optimize the interface circuit response, the following resistance values have been considered: R_1 = 1 kΩ, R_2 = 1.2 kΩ, R_3 = 15 kΩ, R_4 = 1 kΩ, R_5 = 47 kΩ, R_6 = 1 kΩ, R_7 = 10 MΩ. The overall experimental measurement results are reported in Figure 8, when considering that the two capacitors C_1 and C_2 have been connected to the circuit both, as depicted in Figure 8 (see results of Figure 9a) and by interchanging their positions/connections (i.e., C_1 used as C_2, and vice versa; see results of Figure 9b). In particular, by measuring the time period T_1 and the pulse width T_2 values of the circuit main output signal V_{MIX} as a function of the liquid level, the C_1 and C_2 capacitance values have been calculated/estimated through the Equations (3) and (4) and compared with the capacitive values of the same elements achieved from direct measurements through the ISO-TECH LCR821 high-precision high-accuracy LCR-meter. In this last case, the resulting relative error, as calculated among the obtained experimental data, is always lower than 5%, both for the capacitor with higher values and for the capacitor with lower values. The minimum estimated detection resolution in terms of the liquid level variation is about 0.01 mm (considering a maximum averaged RMS jitter level, measured on the rising/falling edges of V_{MIX}, lower than 50 ns). Additional simulation analyses have been also performed, so demonstrating a low sensitivity to common-mode noise and disturbances as well as, in particular, an excellent immunity to additional parasitic capacitances at the main circuit sensing nodes. More in detail, at each terminal node of the differential capacitive sensor connected to the circuit (i.e., the three main input nodes of the interface), a 10 pF grounded capacitor has been considered and added as an external parasitic component (i.e., as a parasitic capacitance provided by a differential capacitive sensor, as those ones related to the plates of C_1 and C_2 of the box shown in Figure 8). In this sense, referring to the circuit set-up considered for the simulation results reported in Section 3.1, and thus considering the capacitive variation range of 5–15 pF for C_1 and C_2, the resulting maximum relative variation of T_1 and T_2 is lower than 0.25% that, on the other hand, corresponds to a maximum relative error lower than 8% in the estimation of C_1 and C_2 values.

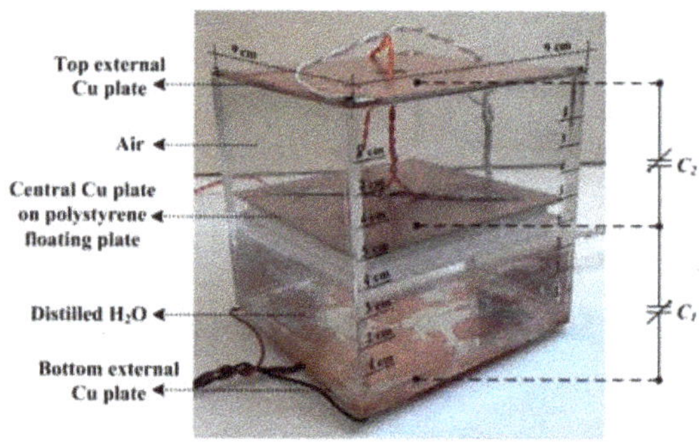

Figure 8. Developed liquid level meter employed as differential capacitive sensor.

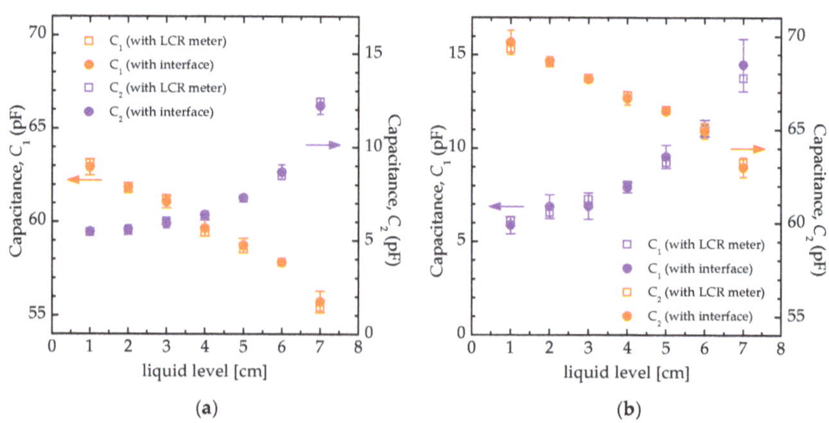

Figure 9. (**a**) Experimental results achieved by using the developed liquid level meter as differential capacitive sensor reported in Figure 8, (**b**) after interchanging C_1 and C_2.

4. Discussion

As a final remark, Table 1 summarizes the main performances and the experimental characteristics of the proposed circuit compared with other similar solutions presented in the literature having linear responses and based on *C-T* conversion architectures. As it can be seen, the presented circuit is an analogue solution that manages differential capacitive sensors combining PM and PWM techniques and, even if only implemented on breadboard with discrete commercial components, shows very satisfactory characteristics, especially in terms of high detection sensitivity, good detection range, and minimum detection resolution.

Table 1. Main performance parameters of the proposed circuit as compared to other similar solutions based on *C-T* conversion.

Ref.	Sensor Topology	Circuit Typology	Circuit Realization	Output Format	Detection Range	Sensitivity	Resolution
[25]	Single element	A/D mixed signal	On chip integration	PWM	0.8–1.2 pF	47 μs/pF @ 20 kHz 15 μs/pF @ 50 kHz	0.9 fF
[26]	Single element	A/D mixed signal	On chip integration	PM	1.8–6.8 pF	1.12 ms/pF	0.2 fF
[27]	Single element	A/D mixed signal	On chip integration	PM	0–8 pF	n.a.	1.4 fF
[28]	Single element	Analogue	On chip integration	PWM	16–256 fF	32 μs/pF	0.8 fF
[29]	Single element	Analogue	On chip integration	PWM	0.013 fF–9 pF	1.82 μs/pF	0.013 fF
[30]	Single element	Analogue	On chip integration	PWM	2.5–2.8255 pF	3.88 μs/pF	2.8 fF
[31]	Single element	Analogue	On chip integration	PWM	1–22 pF	3.62 μs/pF	0.011 fF
[32]	Differential element	Analogue	Discrete components	PM	0–19.8 pF (single) −19.8 ÷ +19.8 pF (differential)	0.49 μs/pF (differential)	2 fF (differential)
[33]	Differential element	A/D mixed signal	On chip integration	PWM	40–60 fF (single) −20 ÷ +20 fF (differential)	127 μs/pF (differential)	0.16 fF (differential)
[35]	Differential element	A/D mixed signal	Discrete components	PM	400 pF (±50%)	n.a.	10 pF
This work	Single or Differential element	Analogue	Discrete components	PWM PM	2.2–197 pF (single) [−15.8 ÷ +15.8 pF] (differential)	1 μs/pF (single) 982 μs/pF (differential)	83 fF (single) 0.1 fF (differential)

5. Conclusions

A simple interface circuit, operating a capacitance-to-time conversion by means of an oscillator-based topology, suitable for the readout of differential capacitive sensors has been presented. The multiple-feedback circuit architecture, employing only four active components (three OAs and one mixer) and seven resistors, provides an AC sensor excitation voltage, reduces ground noise disturbs, and avoids system calibrations. By performing a PM and a PWM modulations, the provided output square wave signal results in being linearly proportional to the sensor differential capacitance values. The sensor variation range and the detection sensitivity can be easily set through the employed resistors. PSpice simulations, together with all the experimental measurements, performed by implementing the circuit on a laboratory breadboard and using commercial discrete components, have validated the proposed idea showing a good agreement with theoretical calculations within an overall capacitive variation range that is equal to [2.2 pF–197 pF] and achieving a minimum capacitance detection resolutions as low as 0.1 fF for a differential capacitive variation range equal to ±15.8 pF. Moreover, the simple schematic of the proposed circuit makes it suitable to be designed at the transistor level in a Si-based standard CMOS integrated technology for integrated portable sensor applications, also employing different kinds of differential capacitive sensing devices, such as MEMS, force/position, and humidity sensors.

Author Contributions: Conceptualization, A.D.M.; methodology, A.D.M., M.-D.C.-B. and C.R.; formal analysis, A.D.M., M.-D.C.-B. and C.R.; investigation, A.D.M.; resources, C. Reig.; writing—original draft preparation, A.D.M.—review and editing, A.D.M., M.-D.C.-B. and C.R.; visualization, C.R.; project administration, A.D.M. and C.R.; funding acquisition, A.D.M. and C.R.

Funding: This research was partially funded by University of Valencia, with the grant program "Attracting Talent"—VLC CAMPUS 2015.

Acknowledgments: The co-author Andrea De Marcellis acknowledges the Research Fellowship from the University of Valencia (Sub-Program "Attracting Talent"—VLC CAMPUS 2015).

Conflicts of Interest: The authors declare no conflict of interest.

References

1. Yuan, J.S.; Bi, Y. Process and temperature robust voltage multiplier design for RF energy harvesting. *Micorelectron. Reliab.* **2015**, *55*, 107–113. [CrossRef]
2. Bi, Y.; Gaillardon, P.E.; Hu, X.S.; Niemier, M.; Yuan, J.S.; Jin, Y. Leveraging Emerging Technology for Hardware Security—Case Study on Silicon Nanowire FETs and Graphene SymFETs. In Proceedings of the 2014 IEEE 23rd Asian Test Symposium, Hangzhou, China, 16–19 November 2014; pp. 342–347. [CrossRef]
3. Bi, Y.; Shamsi, K.; Yuan, J.S.; Jin, Y.; Niemier, M.; Hu, X.S. Tunnel FET Current Mode Logic for DPA-Resilient Circuit Designs. *IEEE Trans. Emerg. Top. Comput.* **2017**, *5*, 340–352. [CrossRef]
4. Chen, X.; Brox, D.; Assadsangabi, B.; Ali, M.S.M.; Takahata, K. A stainless-steel-based implantable pressure sensor chip and its integration by microwelding. *Sens. Actuators A Phys.* **2017**, *257*, 134–144. [CrossRef]
5. Apigo, D.J.; Bartholomew, P.L.; Russell, T.; Kanwal, A.; Farrow, R.C.; Thomas, G.A. An Angstrom-sensitive, differential MEMS capacitor for monitoring the milliliter dynamics of fluids. *Sens. Actuators A Phys.* **2016**, *251*, 234–240. [CrossRef] [PubMed]
6. Rivadeneyra, A.; Fernández-Salmerón, J.; Agudo-Acemel, M.; López-Villanueva, J.A.; Capitan-Vallvey, L.F.; Palma, A.J. Printed electrodes structures as capacitive humidity sensors: A comparison. *Sens. Actuators A Phys.* **2016**, *244*, 56–65. [CrossRef]
7. Aydemir, A.; Terzioglu, Y.; Akin, T. A new design and a fabrication approach to realize a high performance three axes capacitive MEMS accelerometer. *Sens. Actuators A Phys.* **2016**, *244*, 324–333. [CrossRef]
8. Rahman, M.D.T.; Rahimi, A.; Gupta, S.; Panat, R. Microscale additive manufacturing and modeling of interdigitated capacitive touch sensors. *Sens. Actuators A Phys.* **2016**, *246*, 94–103. [CrossRef]
9. Liu, Y.T.; Kuo, Y.L.; Yan, D.W. System integration for on-machine measurement using a capacitive LVDT-like contact sensor. *Adv. Manuf.* **2017**, *5*, 50–58. [CrossRef]
10. Bai, Y.; Lu, Y.; Hu, P.; Wang, G.; Xu, J.; Zeng, T.; Li, Z.; Zhang, Z.; Tan, J. Absolute position sensing based on a robust differential capacitive sensor with a grounded shield window. *Sensors* **2016**, *16*, 680. [CrossRef]
11. Park, S.H.; Kim, H.S.; Bang, J.S.; Cho, G.H.; Cho, G.H. A 0.26-nJ/node, 400-kHz Tx driving, filtered fully differential readout IC with parasitic RC time delay reduction technique for 65-in 169 × 97 capacitive-type touch screen panel. *IEEE J. Solid-State Circuits* **2017**, *52*, 528–542. [CrossRef]
12. Liu, X.; Peng, K.; Chen, Z.; Pu, H.; Yu, Z. A new capacitive displacement sensor with nanometer accuracy and long range. *IEEE Sens. J.* **2016**, *16*, 2306–2316. [CrossRef]
13. Ciccarella, P.; Carminati, M.; Sampietro, M.; Ferrari, M. Multichannel 65zF RMS resolution CMOS monolithic capacitive sensor for counting single micrometer-sized airborne particles on chip. *IEEE J. Solid-State Circuits* **2016**, *51*, 2545–2553. [CrossRef]
14. Wu, X.; Deng, F.; Hao, Y.; Fu, Z.; Zhang, L. Design of a humidity sensor tag for passive wireless applications. *Sensors* **2015**, *15*, 25564–25576. [CrossRef] [PubMed]
15. Arefin, M.S.; Redouté, J.M.; Yuce, M.R. A MEMS interface IC with low-power and wide-range frequency-to-voltage converter for biomedical applications. *IEEE Trans. Biomed. Circuits Syst.* **2016**, *10*, 455–466. [CrossRef] [PubMed]
16. Constandinou, T.G.; Georgiou, J.; Toumazou, C. A micropower front-end interface for differential-capacitive sensor systems. In Proceedings of the 2008 IEEE International Symposium on Circuits and Systems, Seattle, WA, USA, 18–21 May 2008; pp. 2474–2477. [CrossRef]

17. Mochizuki, K.; Watanabe, K.; Masuda, K. A high-accuracy high-speed signal processing circuit of differential-capacitance transducers. *IEEE Trans. Instrum. Meas.* **1998**, *47*, 1244–1247. [CrossRef]
18. Singh, T.; Saether, T.; Ytterdal, T. Current-mode capacitive sensor interface circuit with single-ended to differential output capability. *IEEE Trans. Instrum. Meas.* **2009**, *58*, 3914–3920. [CrossRef]
19. Royo, G.; Sánchez-Azqueta, C.; Gimeno, C.; Aldea, C.; Celma, S. Programmable low-power low-noise capacitance to voltage converter for MEMS accelerometers. *Sensors* **2017**, *17*, 67. [CrossRef] [PubMed]
20. Scotti, G.; Pennisi, S.; Monsurro, P.; Trifiletti, A. 88-μA 1-MHz stray-insensitive CMOS current-mode interface IC for differential capacitive sensors. *IEEE Trans. Circuits Syst.* **2014**, *61*, 1905–1916. [CrossRef]
21. Kyriakis-Bitzaros, E.D.; Stathopoulos, N.A.; Pavlos, S.; Goustouridis, D.; Chatzandroulis, S. A reconfigurable multichannel capacitive sensor array interface. *IEEE Trans. Instrum. Meas.* **2011**, *60*, 3214–3221. [CrossRef]
22. Wang, S.; Koickal, T.J.; Hamilton, A.; Mastropaolo, E.; Cheung, R.; Abel, A.; Smith, L.S.; Wang, L. A power-efficient capacitive read-out circuit with parasitic-cancellation for MEMS cochlea sensors. *IEEE Trans. Biomed. Circuits Syst.* **2016**, *10*, 25–37. [CrossRef]
23. Ignjatovic, Z.; Bocko, M.F. An interface circuit for measuring capacitance changes based upon capacitance-to-duty cycle (CDC) converter. *IEEE Sens. J.* **2005**, *5*, 403–410. [CrossRef]
24. De Marcellis, A.; Ferri, G.; Mantenuto, P. A CCII-based non-inverting Schmitt trigger and its application as astable multivibrator for capacitive sensor interfacing. *Int. J. Circuit Theory Appl.* **2016**, *45*, 1060–1076. [CrossRef]
25. Bruschi, P.; Nizza, N.; Piotto, M. A current-mode, dual slope, integrated capacitance-to-pulse duration converter. *IEEE J. Solid-State Circuits* **2007**, *42*, 1884–1891. [CrossRef]
26. Tan, Z.; Shalmany, S.H.; Meijer, G.C.M.; Pertijs, M.A.P. An energy-efficient 15-bit capacitive-sensor interface based on period modulation. *IEEE J. Solid-State Circuits* **2012**, *47*, 1703–1711. [CrossRef]
27. He, Y.; Chang, Z.Y.; Pakula, L.; Shalmany, S.H.; Pertijs, M.A.P. A 0.05 mm^2 1 V capacitance-to-digital converter based on period modulation. In Proceedings of the IEEE International Solid-State Circuits Conference (ISSCC), San Francisco, CA, USA, 22–26 February 2015; pp. 1–3. [CrossRef]
28. Nizza, N.; Dei, M.; Butti, F.; Bruschi, P. A low-power interface for capacitive sensors with PWM output and intrinsic low pass characteristic. *IEEE Trans. Circuits Syst.* **2013**, *60*, 1419–1431. [CrossRef]
29. Lu, J.H.L.; Inerowicz, M.; Joo, S.; Kwon, J.K.; Jung, B. A low-power wide-dynamic-range semi-digital universal sensor readout circuit using pulsewidth modulation. *IEEE Sens. J.* **2011**, *11*, 1134–1144. [CrossRef]
30. Sheu, M.L.; Hsu, W.H.; Tsao, L.J. A capacitance-ratio-modulated current front-end circuit with pulsewidth modulation output for a capacitive sensor interface. *IEEE Trans. Instrum. Meas.* **2012**, *61*, 447–455. [CrossRef]
31. Arefin, M.S.; Redouté, J.M.; Yuce, M.R. A low-power and wide-range MEMS capacitive sensors interface IC using pulse-width modulation for biomedical applications. *IEEE Sens. J.* **2016**, *16*, 6745–6754. [CrossRef]
32. Brookhuis, R.A.; Lammerink, T.S.J.; Wiegerink, R.J. Differential capacitive sensing circuit for a multi-electrode capacitive force sensor. *Sens. Actuators A Phys.* **2015**, *234*, 168–179. [CrossRef]
33. Aezinia, F.; Bahreyni, B. Low-power parasitic-insensitive interface circuit for capacitive microsensors. *IET Circuits Device Syst.* **2016**, *10*, 104–110. [CrossRef]
34. De Marcellis, A.; Cubells-Beltrán, M.D.; Reig, C.; Madrenas, J.; Zadov, B.; Paperno, E.; Cardoso, S.; Freitas, P.P. Quasi-digital front-ends for current measurement in integrated circuits with GMR technology. *IET Circuits Device Syst.* **2014**, *8*, 291–300. [CrossRef]
35. Mohan, N.M.; Shet, A.R.; Kedarnath, S.; Kumar, V.J. Digital converter for differential capacitive sensors. *IEEE Trans. Instrum. Meas.* **2008**, *57*, 2576–2581. [CrossRef]
36. Reverter, F.; Casas, O. Interfacing differential capacitive sensors to microcontrollers: A direct approach. *IEEE Trans. Instrum. Meas.* **2010**, *59*, 2763–2769. [CrossRef]
37. Nabovati, G.; Ghafar-Zadeh, E.; Mirzaei, M.; Ayala-Charca, G.; Awwad, F.; Sawan, M. A new fully differential CMOS capacitance to digital converter for lab-on-chip applications. *IEEE Trans. Biomed. Circuits Syst.* **2015**, *9*, 353–361. [CrossRef] [PubMed]
38. Omran, H.; Arsalan, M.; Salama, K.N. An integrated energy-efficient capacitive sensor digital interface circuit. *Sens. Actuators A Phys.* **2014**, *216*, 43–51. [CrossRef]
39. Shin, D.Y.; Lee, H.; Kim, S. A delta–sigma interface circuit for capacitive sensors with an automatically calibrated zero point. *IEEE Trans. Circuits Syst. II Express Briefs* **2011**, *58*, 90–94. [CrossRef]

40. Alhoshany, A.; Omran, H.; Salama, K.N. A 45.8 fJ/step, energy-efficient, differential SAR capacitance-to-digital converter for capacitive pressure sensing. *Sens. Actuators A Phys.* **2016**, *245*, 10–18. [CrossRef]
41. Tan, Z.; Daamen, R.; Humbert, A.; Ponomarev, Y.V.; Chae, Y.; Pertijs, M.A.P. A 1.2-V 8.3-nJ CMOS humidity sensor for RFID applications. *IEEE J. Solid-State Circuits* **2013**, *48*, 2469–2477. [CrossRef]
42. Oh, S.; Lee, Y.; Wang, J.; Foo, Z.; Kim, Y.; Blaauw, D. Dual-slope capacitance to digital converter integrated in an implantable pressure sensing system. In Proceedings of the 40th European Solid State Circuits Conference (ESSCIRC), Venice, Italy, 22–26 September 2014; pp. 295–298. [CrossRef]

electronics

MDPI

Article

A Fusion Frequency Feature Extraction Method for Underwater Acoustic Signal Based on Variational Mode Decomposition, Duffing Chaotic Oscillator and a Kind of Permutation Entropy

Yuxing Li [1,*], Xiao Chen [2], Jing Yu [3] and Xiaohui Yang [4,*]

1 Faculty of Information Technology and Equipment Engineering, Xi'an University of Technology, Xi'an 710048, China
2 College of Electrical & Information Engineering, ShaanXi University of Science & Technology, Xi'an 710021, China; chenxiao@sust.edu.cn
3 School of Marine Science and Technology, Northwestern Polytechnical University, Xi'an 710072, China; yujing@nwpu.edu.cn
4 School of Art and Design, Inner Mongolia University of Science & Technology, Baotou 014010, China
* Correspondence: liyuxing@xaut.edu.cn (Y.L.); yxh198715@163.com (X.Y.)

Received: 5 December 2018; Accepted: 26 December 2018; Published: 5 January 2019

Abstract: In order to effectively extract the frequency characteristics of an underwater acoustic signal under sensor measurement, a fusion frequency feature extraction method for an underwater acoustic signal is presented based on variational mode decomposition (VMD), duffing chaotic oscillator (DCO) and a kind of permutation entropy (PE). Firstly, VMD decomposes the complex multi-component underwater acoustic signal into a set of intrinsic mode functions (IMFs), so as to extract the estimated center frequency of each IMF. Secondly, the frequency of the line spectrum can be obtained by using DCO and a kind of PE (KPE). DCO is used to detect the actual frequency of the line spectrum for each IMF and KPE can determine the accurate frequency when the phase space track is in the great periodic state. Finally, the frequency characteristic parameters acted as the input of the support vector machine (SVM) to distinguish different types of underwater acoustic signals. By comparing with the other three traditional methods for simulation signal and different kinds of underwater acoustic signals, the results show that the proposed method can accurately extract the frequency characteristics and effectively realize the classification and recognition for the underwater acoustic signal.

Keywords: variational mode decomposition (VMD); duffing chaotic oscillator (DCO); permutation entropy (PE); feature extraction; frequency characteristic; underwater acoustic signal; ship-radiated noise

1. Introduction

Underwater acoustic signal processing is one of the hot topics in the field of marine science. The denoising, feature extraction and classification of underwater acoustic signals are of great significance to the research of underwater acoustic signals, which can provide convenience and basis for the detection and tracking of underwater acoustic signals [1–4]. Feature extraction methods for underwater acoustic signals mainly include frequency feature extraction, energy feature extraction and complexity feature extraction [5]. The frequency feature extraction method usually consists of three steps: (1) signal processing, (2) feature extraction and (3) classification, among which the first two steps have a great impact on feature extraction. Therefore, we face two challenges: how to select the right signal processing method and how to extract features accurately [6,7].

Traditional time-frequency analysis methods include short-time Fourier transform (STFT), Wigner-Ville distribution (WVD), wavelet transform (WT) and empirical mode decomposition (EMD) [8,9]. However, these methods have certain limitations. For example, the WT need to select wavelet basis functions and wavelet decomposition levels, and EMD has the problem of mode mixing [10]. In recent years, several improved EMD methods have been proposed to suppress mode mixing, which are ensemble EMD (EEMD) [11–14] and complete EEMD with adaptive noise (CEEMDAN) [15–17]. However, EMD and improved EMD methods are all empirical decomposition algorithms, which lack the strict mathematical theory for support [18,19].

Variational mode decomposition (VMD) is a time-frequency analysis method after EMD and improved EMD methods [20]. Compared with EMD and improved EMD methods, VMD decomposes a complex signal into a set of intrinsic mode functions (IMFs) based on a foundation of mature mathematical theories and methods, which are wiener filtering, Hilbert transform, analytic signal and heterodyne demodulation. The sensitivity of VMD to noise is lower than that of EMD and improved EMD methods [21]. VMD has been used in many fields, such as fault diagnosis, clinical medicine and underwater acoustics. In Reference [22], a hybrid fault feature extraction method using VMD combined with multipoint kurtosis was proposed. In Reference [23], focusing on high voltage circuit breakers, a fault diagnosis method using VMD and multi-layer classifier was proposed to improve the accuracy of fault diagnosis. In Reference [24], a new detection method for atrial fibrillation using electrocardiogram signal was proposed, sample entropy and center frequency were extracted from IMFs by VMD, which can effectively distinguish the normal sinus rhythm and atrial fibrillation.

Because the chaotic system is sensitive to a weak signal and immune to noise, it has a wide range of applications in weak signal detection. Duffing system is a kind of common nonlinear system which produces chaotic phenomena. Therefore, duffing chaotic oscillator (DCO) can detect weak signals of low-frequency components. We can detect weak signals in strong background noise by changing the phase space tracks of DCO [25]. Detection of weak signals using DCO has been implemented [26]. In Reference [27], a new weak signal detection method was proposed based on the scale transformation of DCO, which can detect any harmonic signal using a set of determined parameters. In Reference [28], an effective weak signal detection method for underwater acoustic signal was put forward based on DCO and Hilbert transform, which can improve the signal-to-noise ratio greater than the traditional DCO method. The above methods have proved the validity of DCO for weak signal detection.

Permutation entropy (PE) is a method to measure the complexity of time series, which is used in many fields [29,30]. A kind of PE (KPE) was proposed by Bandt in April 2017. KPE, as a novel PE, has better performance than PE in terms of the stability of time series with different lengths [31,32]. Many studies have shown that KPE is superior to PE in medical diagnosis and underwater acoustics.

There are many methods for feature extraction of underwater acoustic signals [33]. Among these methods, the feature extraction methods based on mode decomposition is one of the hot issues of research for underwater acoustic signals. In terms of energy feature extraction of underwater acoustic signals, two methods were put forward using EMD combined with energy entropy and energy spectrum [34,35]. In terms of the complexity of feature extraction of underwater acoustic signals, PE and multi-scale PE (MPE) of IMFs were extracted as new features, where IMFs were obtained by EMD and VMD respectively. Focusing on the frequency feature extraction of underwater acoustic signals, center frequency feature extraction methods were presented using EEMD and VMD. However, these frequency characteristics were not accurate enough.

In the paper, we proposed a new frequency feature extraction method for underwater acoustic signals to effectively extract the frequency characteristics. The proposed method is based on VMD, DCO and KPE. We use VMD to decompose underwater acoustic signals into IMFs. According to the estimated frequency, DCO can detect the frequency of each IMF. When the phase space track is in the great periodic state, we can use KPE to determine the accurate frequency. DCO and KPE are first used to the frequency of IMF for underwater acoustic signals.

The next section is the theory of VMD, DCO and KPE; the novel frequency feature extraction method for underwater acoustic signals is presented in Section 3; the proposed frequency feature extraction method is used to simulate signals and underwater acoustic signals in Sections 4 and 5; and, finally, the concluding remarks are made in the last section.

2. Theory

2.1. VMD

The VMD theory consists of two parts: the constrained variational problem and specific steps to solve. VMD defines the amplitude-modulated-frequency-modulated signal as IMF, which is shown in Equation (1).

$$u_k(t) = A_k(t)\cos(\phi_k(t)) \tag{1}$$

where $u_k(t)$ is the k-th IMF by VMD, $A_k(t)$ and $\phi_k(t)$ are the envelope and phase of the k-th IMF. Each IMF has estimated frequency and limited bandwidth. The constrained variational problem is shown in Equation (2).

$$\begin{cases} \min\limits_{\{u_k\},\{w_k\}} \left\{ \sum\limits_{k=1}^{K} \left\| \partial t[(\delta(t) + \frac{j}{\pi t}) * u_k(t)]e^{-jwkt} \right\|_2^2 \right\} \\ s.t. \sum\limits_{k=1}^{K} u_k = x(t) \end{cases} \tag{2}$$

where $x(t)$ represents the un-decomposed complex signal, K and w_k represent the number of $u_k(t)$ and estimated frequency for the k-th IMF. The solved non-constrained variational problem is shown in Equation (3).

$$L(\{u_k\},\{w_k\},\lambda) = \alpha \sum_{k=1}^{K} \left\| \partial t[(\delta(t) + \tfrac{j}{\pi t}) * u_k(t)]e^{-jwkt} \right\|_2^2 + \left\| x(t) - \sum_{k=1}^{K} u_k(t) \right\|_2^2 + \left\langle \lambda(t), x(t) - \sum_{k=1}^{K} u_k(t) \right\rangle \tag{3}$$

where L is the augmented Lagrangian method, α and λ are the penalty factor and Lagrange multiplier. We use the alternating direction multiplier method to get saddle points and update $\hat{u}_k^{n+1}$, w_k^{n+1} and $\hat{\lambda}^{n+1}$. These updated formulas are shown in Equation (4).

$$\begin{cases} \hat{u}_k^{n+1}(w) = \dfrac{\hat{x}(w) - \sum\limits_{i<k} \hat{u}i^n(w) - \sum\limits_{i>k} \hat{u}i^n(w) + \frac{\hat{\lambda}^n(w)}{2}}{1 + 2\alpha(w - w_k^n)^2} \\ w_k^{n+1} = \dfrac{\int_0^\infty w|\hat{u}_k^{n+1}|^2 dw}{\int_0^\infty |\hat{u}_k^{n+1}|^2 dw} \\ \hat{\lambda}^{n+1}(w) = \hat{\lambda}^n(w) + \tau\left(\hat{x}(w) - \sum\limits_k \hat{u}_n^{n+1}(w)\right) \end{cases} \tag{4}$$

where w represents the frequency domain. The flow diagram of VMD is given in Figure 1. More detailed explanations about VMD can be found [20,21].

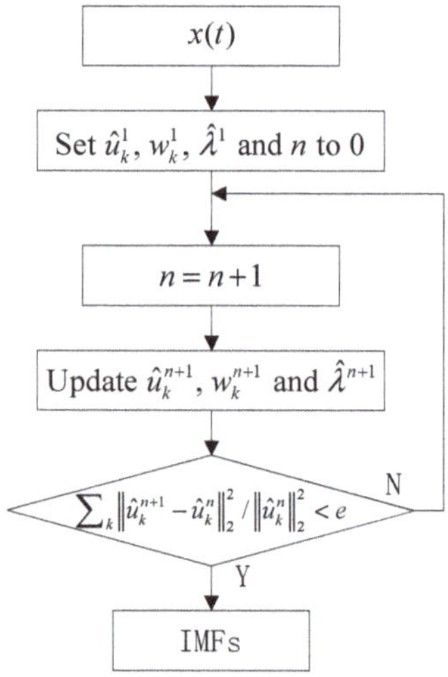

Figure 1. The flow diagram of VMD.

2.2. DCO

The normal form of a duffing chaotic oscillator (DCO) equation is shown in Equation (5).

$$\frac{d^2x}{dt^2} + k\frac{dx}{dt} - x(t) + x^3(t) = F(t) \tag{5}$$

where k is the damping ratio, $-x(t) + x^3(t)$ and $F(t)$ represent the nonlinear resilience item and the driving force. When $F(t)$ equals $\gamma \cos(\omega t)$, DCO equation can be expressed in Equation (6).

$$\frac{d^2x}{dt^2} + k\frac{dx}{dt} - x(t) + x^3(t) = \gamma \cos(\omega t) \tag{6}$$

γ and ω represent the angular frequency and amplitude of the driving force. Due to the existence of the nonlinear resilience item, the DCO equation is rich in nonlinear dynamic characteristics. We make γ increase from 0 while fixing k, the system state changes from homoclinic orbits state to bifurcation state, and then when the threshold γ_d is exceeded, the system state changes from chaos state to the great periodic motion. The steps of periodic signal detection by DCO are as follows:

(1) Put periodic signal $s(t)$ and noise signal $n(t)$ into the system, DCO equation can be expressed in Equation (7).

$$\frac{d^2x}{dt^2} + k\frac{dx}{dt} - x(t) + x^3(t) = \gamma_d \cos(\omega t) + s(t) + n(t) \tag{7}$$

(2) Set k, $x(0)$ and $x'(0)$ to 0.5, 0 and 0. The Runge-Kutta of the fourth order is used for a solution of DCO equation.

(3) We can determine whether the angular frequency of the periodic signal $s(t)$ is close to ω according to the system state. When the system state is the great periodic state, this means that the angular frequency of the periodic signal $s(t)$ is approximated as ω, and vice versa. More detailed explanations about DCO can be found elsewhere [27,28].

2.3. KPE

In order to better understand KPE, we learn KPE by comparing with PE. Both PE and KPE can represent the complexity of time series. However, they have the following differences:

(1) KPE, as an improved PE, is defined as the distance between the time series and white Gaussian noise. Therefore, KPE and PE have a totally opposite trend. For example, when the time series is white Gaussian noise, PE and KPE are close to 1 and 0 respectively.
(2) The equations of KPE and PE are different. KPE and PE can be expressed as

$$\begin{cases} H_{PE} = -\sum_{j=1}^{K} P_j \ln P_j / \ln(m!) \\ H_{KPE} == \sum_{j=1}^{K} P_j^2 - \frac{1}{m!} \end{cases} \tag{8}$$

where H_{KPE} and H_{PE} represent KPE and PE, K and m are the number of reconstructed vectors and the embedded dimension, P_j represents j-th probability of symbol sequence.
(3) Compared with PE, KPE has better robustness for time series of different lengths.

More details of PE and KPE can be found elsewhere [29–32].

3. Frequency Feature Extraction Method for Underwater Acoustic Signal

This paper presents a fusion frequency feature extraction method for underwater acoustic signal based on VMD, DCO and KPE. The flow chart of the feature extraction method is shown in Figure 2. The experimental steps of this frequency feature extraction method are as follows:

Step 1: Signal decomposition.

(1) Collect underwater acoustic signals by sensors;
(2) Decompose underwater acoustic signals by EMD, M IMFs can be obtained;
(3) Set the decomposition layers of VMD to M;
(4) Decompose underwater acoustic signals by VMD.

Step 2: Feature extraction.

(1) Select the low-frequency IMF for the research, such as the last IMF;
(2) Obtain estimated frequency of selected IMF by VMD;
(3) Detect periodic signal of selected IMF using DCO;
(4) When the phase track of selected IMF is in great periodic, and the KPE of DCO system output reaches the maximum, we can determine the accurate frequency of selected IMF.

Step 3: Classification recognition.

(1) Input frequency characteristics of different kinds of underwater acoustic signals into SVM;
(2) Obtain classification results of different kinds of underwater acoustic signals.

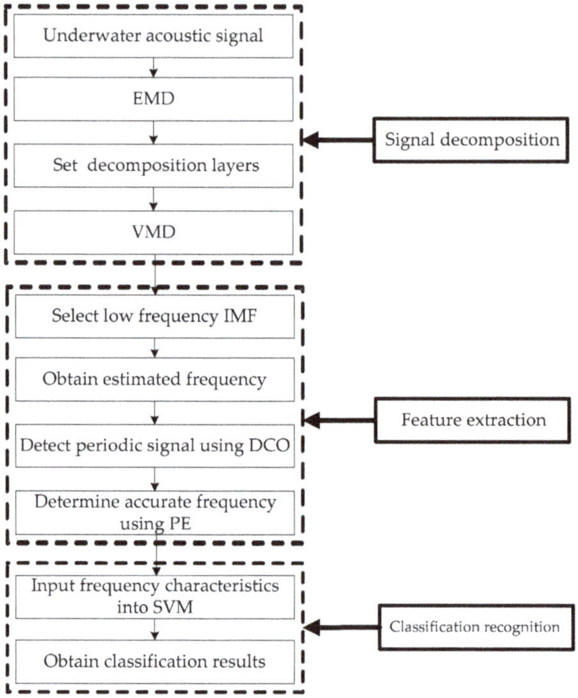

Figure 2. The flow chart of a fusion frequency feature extraction method.

4. Frequency Feature Extraction for Simulation Signal

To prove the reliability of this fusion frequency feature extraction method, we extract the frequency feature for the simulation signal. First, the simulation signal is decomposed by VMD. Secondly, the periodic signal of IMFs can be detected by DCO, frequency characteristics of IMFs can be obtained by KPE. Finally, we compared with three frequency feature extraction methods presented recently.

4.1. VMD of Simulation Signal

Line spectrums of ship-radiated noise can reflect an important frequency feature, and the line spectrum corresponds to the periodic signal in the time domain. Therefore, the clear signal S consists of three cosine signals with different amplitudes and frequencies, and the noisy signal Y consists of both the clear signal and the standard Gaussian white noise N. The specific simulation signals are as follows:

$$\begin{cases} S = 0.4\cos(20\pi t) + 0.5\cos(100\pi t) + 0.3\cos(200\pi t) \\ N = randn(t) \\ Y = S + N \end{cases} \tag{9}$$

The frequencies of three cosine signals are 10 Hz, 50 Hz and 100 Hz, respectively. The sampling frequency is 1 kHz. The time-domain waveforms of both clear and noisy signal are shown in Figure 3. According to the EMD result for noisy signals, we set the decomposition layers of VMD to 9, the VMD result for noisy signals is shown in Figure 4. As seen in Figure 3, the clear signal is submerged in noise. As seen in Figure 4, the order of IMFs by VMD is from high frequency to low frequency. Each IMF has an estimated frequency, the frequency distribution of IMFs by VMD is listed in Table 1. As can be seen in Table 1, IMF9, IMF8 and IMF7 correspond to the cosine signal with the frequency of 10 Hz, 50 Hz and 100 Hz, respectively.

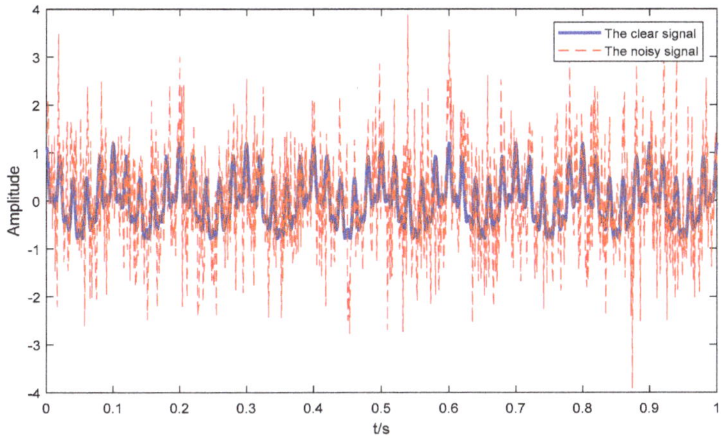

Figure 3. The time-domain waveforms of both the clear and noisy signals.

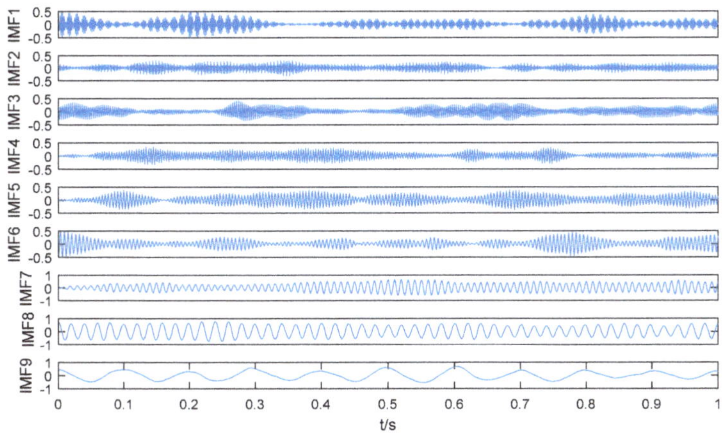

Figure 4. The VMD result for the noisy signal.

Table 1. The frequency distribution of IMFs by VMD.

IMF1	IMF2	IMF3	IMF4	IMF5	IMF6	IMF7	IMF8	IMF9
442.23 Hz	392.59 Hz	321.89 Hz	264.65 Hz	227.73 Hz	168.37 Hz	99.47 Hz	50.12 Hz	10.14 Hz

4.2. Frequency Feature Extraction of IMF Using DCO and KPE

According to the estimated frequency of IMFs, the three periodic signals are in the last three IMFs, and the other IMFs are noise IMFs without periodic signals. Therefore, we extract the frequency features of the last three IMFs using DCO and KPE, respectively.

4.2.1. Frequency Feature Extraction of IMF9

The estimated frequency of IMF9 is 10.14 Hz. A DCO column is used to sweep through the true frequency, and the driving force frequency is close to 10.14 Hz. A DCO column consists of 10 DCOs, frequency interval of each DCO is 0.01 KHz. The phase space tracks of different driving force frequencies are shown in Figure 5. As seen in Figure 5, when the driving force frequencies are

9.74 Hz and 10.24 Hz, the phase space tracks are in chaos state; when the driving force frequencies are 9.94 Hz and 10.04 Hz, the phase space tracks are in the great periodic state.

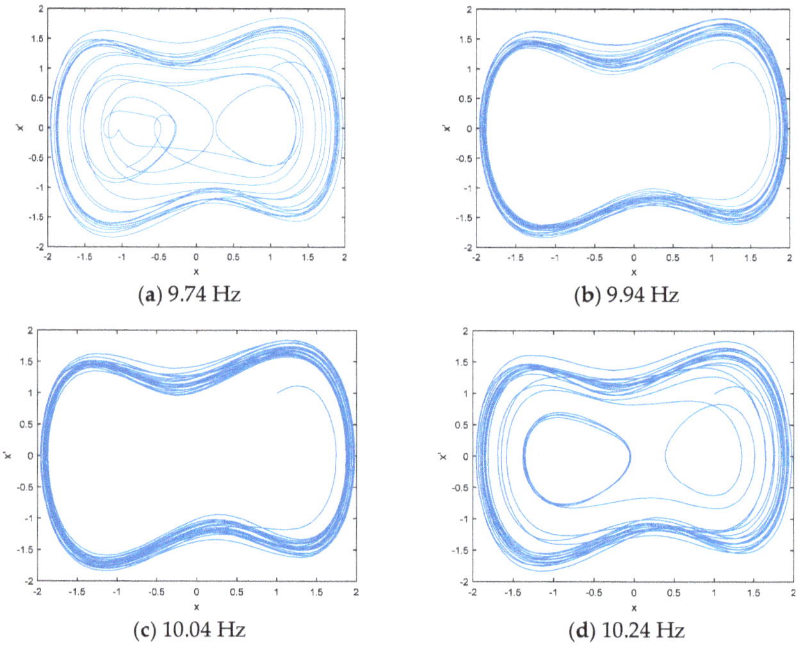

Figure 5. The phase space tracks of different driving force frequencies for IMF9.

When the phase space track is in the great periodic state, we calculated the KPE of the DCO system output under different driving force frequencies. The KPE distribution of IMF9 under different driving force frequencies is listed in Table 2. As can be seen in Table 2, when the driving force frequency is 9.98 Hz, the KPE reaches the maximum. Therefore, the frequency feature of IMF9 is 9.98 Hz using the proposed frequency feature extraction method.

Table 2. The KPE distribution of IMF9 under different driving force frequencies.

9.95 Hz	9.96 Hz	9.97 Hz	9.98 Hz	9.99 Hz	10.00 Hz	10.01 Hz
0.313618	0.313618	0.313622	0.313630	0.313622	0.313618	0.313616

4.2.2. Frequency Feature Extraction of IMF8 and IMF7

The estimated frequencies of IMF8 and IMF7 are 50.12 Hz and 99.47 Hz. Two DCO columns were used to sweep through the true frequency of IMF8 and IMF7 according to the estimated frequencies. The phase space tracks of different driving force frequencies for IMF8 and IMF7 are shown in Figures 6 and 7. As seen in Figure 6, when the driving force frequencies are 49.82 Hz and 50.22 Hz, the phase space tracks are in the chaos state, and when the driving force frequencies are 49.92 Hz and 50.12 Hz, the phase space tracks are in the great periodic state. As seen in Figure 7, when the driving force frequencies are 99.47 Hz and 100.17 Hz, the phase space tracks are in the chaos state, and when the driving force frequencies are 99.97 Hz and 100.07 Hz, the phase space tracks are in the great periodic state.

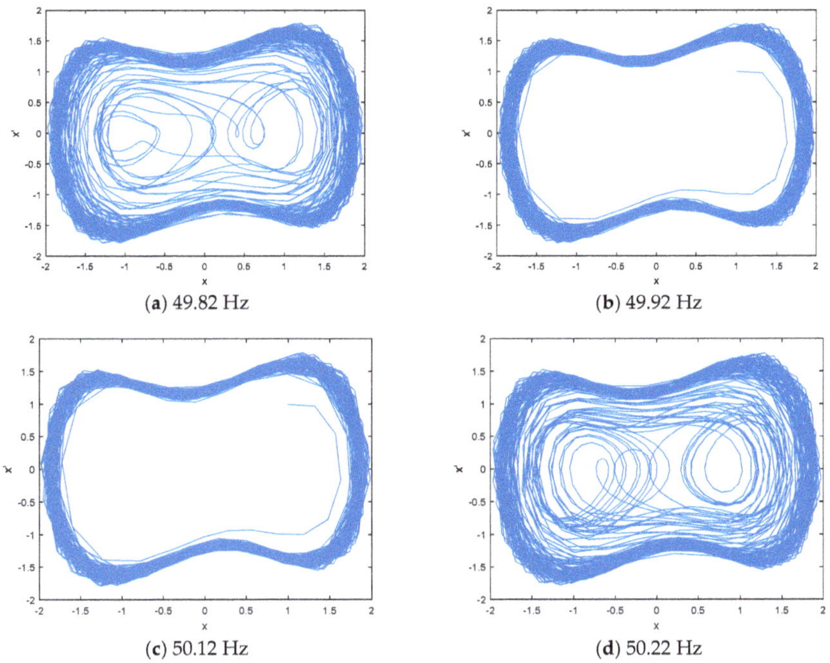

(a) 49.82 Hz (b) 49.92 Hz

(c) 50.12 Hz (d) 50.22 Hz

Figure 6. The phase space tracks of different driving force frequencies for IMF8.

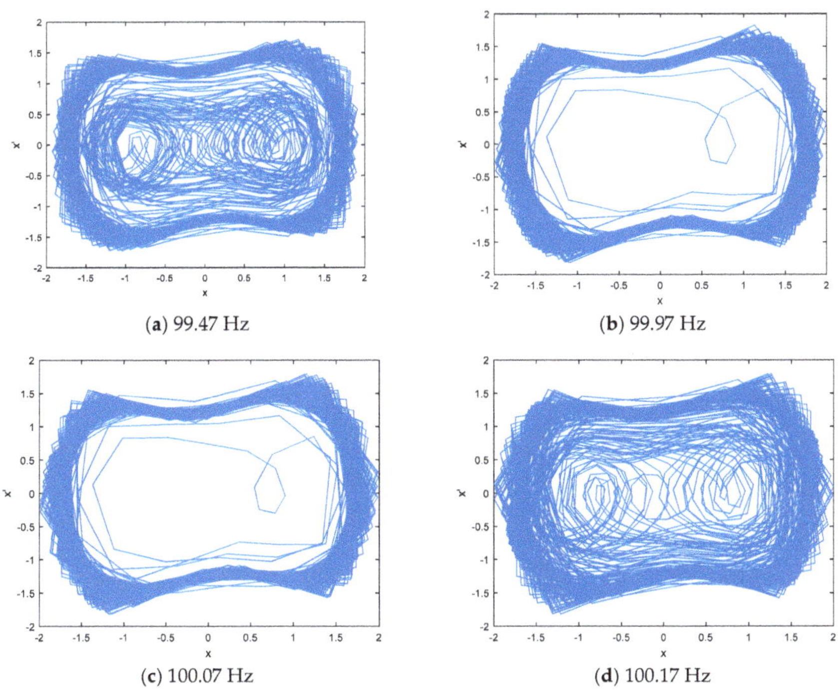

(a) 99.47 Hz (b) 99.97 Hz

(c) 100.07 Hz (d) 100.17 Hz

Figure 7. The phase space tracks of different driving force frequencies for IMF7.

When the phase space tracks of IMF8 and IMF7 were in the great periodic state, we calculated the KPE of the DCO system output under different driving force frequencies. The KPE distributions of IMF8 and IMF7 under different driving force frequencies are listed in Tables 3 and 4. As can be seen in Tables 3 and 4, when the driving force frequency of IMF8 and IMF7 are 49.99 Hz and 100.03 Hz, the KPEs reach the maximum. Therefore, the frequency features of IMF8 and IMF7 are 49.99 Hz and 100.03 Hz.

Table 3. The KPE distribution of IMF8 under different driving force frequencies.

49.96 Hz	49.97 Hz	49.98 Hz	49.99 Hz	50.00 Hz	50.01 Hz	50.02 Hz
0.240762	0.240779	0.240813	0.240961	0.240884	0.240761	0.240753

Table 4. The KPE distribution of IMF7 under different driving force frequencies.

100.00 Hz	100.01 Hz	100.02 Hz	100.03 Hz	100.04 Hz	100.05 Hz	100.06 Hz
0.163238	0.163396	0.164079	0.164159	0.164076	0.164027	0.163390

4.3. Comparison of Different Frequency Feature Extraction Methods

In order to further prove the reliability of this fusion frequency feature extraction method, we compare the results of four different frequency feature extraction methods. The frequency feature extraction methods using different mode decomposition and center frequency are named as EMD-CF, EEMD-CF and VMD-CF, and the proposed frequency feature extraction method is called VMD-DCO-KPE. Frequency features are statistical center frequencies in EMD-CF, EEMD-CF and VMD-CF, and frequency feature is line spectrum frequencies in VMD-DCO-KPE. The EMD and EEMD results for noisy signals are shown in Figure 8. As seen in Figure 8, the number of IMFs are different between EMD and EEMD. The frequency distributions of IMFs by EMD and EEMD are listed in Tables 5 and 6. As can be seen in Tables 5 and 6, IMF6, IMF4 and IMF3 correspond to cosine signal with the frequency of 10 Hz, 50 Hz and 100 Hz, respectively. Frequency features by different frequency feature extraction methods are listed in Table 7. As can be seen in Table 7, the proposed VMD-DCO-KPE method is the closest to the true frequency.

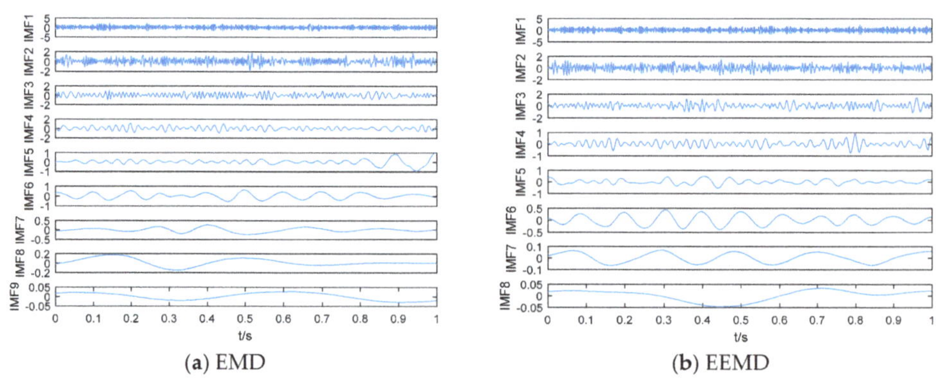

Figure 8. The EMD and EEMD results for noisy signal.

Table 5. The frequency distribution of IMFs by EMD.

IMF1	IMF2	IMF3	IMF4	IMF5	IMF6	IMF7	IMF8	IMF9
319.2 Hz	147.13 Hz	68.94 Hz	43.66 Hz	19.54 Hz	9.68 Hz	6.32 Hz	3.02 Hz	1.97 Hz

Table 6. The frequency distribution of IMFs by EEMD.

IMF1	IMF2	IMF3	IMF4	IMF5	IMF6	IMF7	IMF8
338.07 Hz	149.78 Hz	73.74 Hz	44.07 Hz	16.26 Hz	9.72 Hz	4.41 Hz	2.37 Hz

Table 7. Frequency features by different frequency feature extraction methods.

Methods	10 Hz	50 Hz	100 Hz
EMD-CF	9.68 Hz	43.66 Hz	68.94 Hz
EEMD-CF	9.72 Hz	44.07 Hz	73.74 Hz
VMD-CF	10.14 Hz	50.12 Hz	99.47 Hz
VMD-DCO-KPE	9.98 Hz	49.99 Hz	100.03 Hz

5. Application in Underwater Acoustic Signals

Firstly, three kinds of underwater acoustic signals were decomposed by VMD; then, the frequency features were extracted using the VMD-DCO-KPE method; finally, the frequency feature and classification results of the different methods were compared.

5.1. VMD of Ship-Radiated Noise Signal

Ship-radiated noise is an important part of underwater acoustic signals. In this paper, three kinds of ship-radiated noise samples were selected for frequency feature extraction, namely ship 1, ship 2 and ship 3. Their sampling frequency and sampling points were 44.1 kHz and 2000. Figure 9 depicts a 3D underwater acoustic signal measurement. The depth of the measurement area was about 4 km, and the topography of the seabed was fairly flat. In order to degrade the influence of ocean environmental noise, we measured data at the level 1 sea state by using omnidirectional hydrophones. The research ship carried hydrophones with a depth of 30 m and was not in service during the whole measurement process. The distance between the research ship and the target ship (Ship 1, Ship 2 and Ship 3) was about 2.5 km. When one of the target ships was running, the other ships remained out of work.

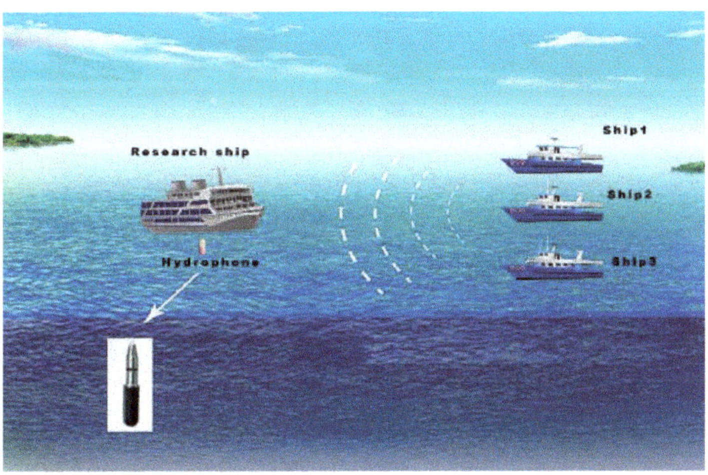

Figure 9. 3D underwater acoustic signal measurement.

The normalized time-domain waveform for three kinds of ship-radiated noise samples is shown in Figure 10. The VMD results of ship-radiated noise samples are shown in Figure 11. As seen in Figure 11, the IMFs of three kinds of ship-radiated noise samples were recorded in descending order of frequency, and the IMF of the lowest frequency is IMF8.

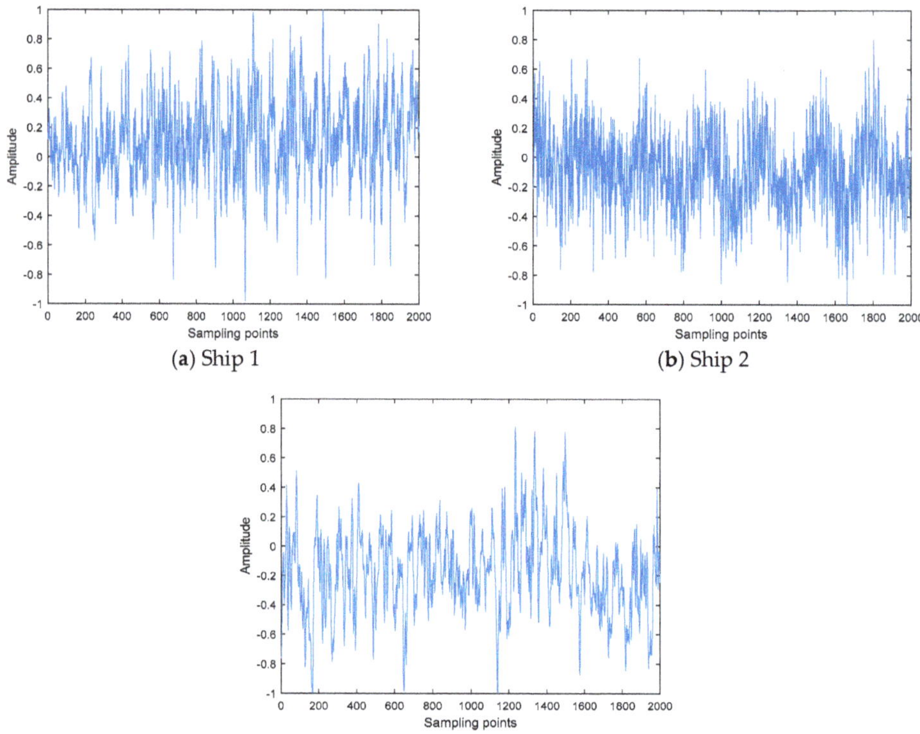

(a) Ship 1 (b) Ship 2

(c) Ship 3

Figure 10. The normalized time-domain waveform for three kinds of ship-radiated noise samples.

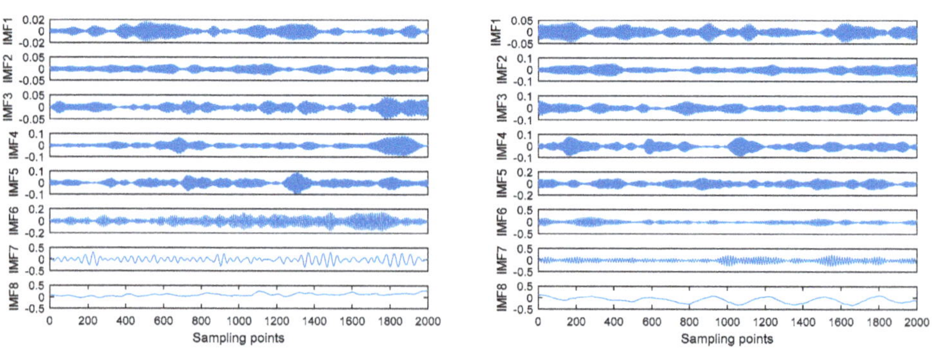

(a) Ship 1 (b) Ship 2

Figure 11. *Cont.*

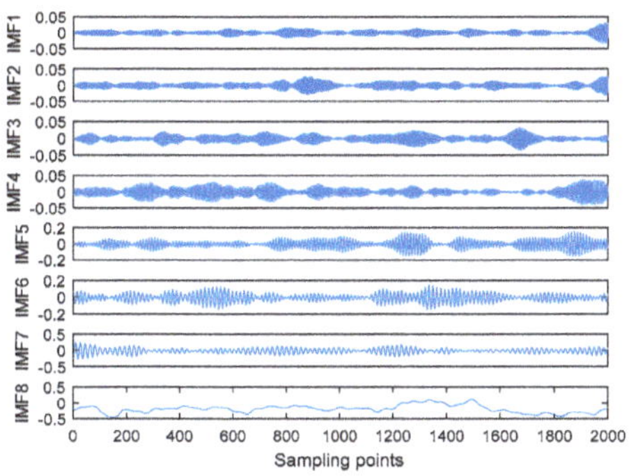

(c) Ship 3

Figure 11. The VMD results of ship-radiated noise samples.

5.2. Frequency Feature Extraction of Line Spectrum

The line spectrums of ship-radiated noise can reflect the important physical characteristics of ships, and line spectrums usually exist in the low frequency of ships. In this paper, we selected the line spectrum of IMF8 as the frequency feature of ship-radiated noise. The frequency distribution of IMF8 by VMD for three kinds of ship-radiated noise samples are listed in Table 8. According to the estimated frequency by VMD, a DCO column is used to sweep through the true frequency. The great periodic states of IMF8 for three kinds of ship-radiated noise samples are shown in Figure 12. When the phase space track is in the great periodic state, we calculated the KPE and obtained the true frequency by using the maximum value of KPE. The frequency distribution of IMF8 by VMD-DCO-KPE for three kinds of ship-radiated noise samples is listed in Table 9.

Table 8. The frequency distribution of IMF8 by VMD for three kinds of ship-radiated noise samples.

Ship 1	Ship 2	Ship 3
15.59 Hz	66.18 Hz	26.11 Hz

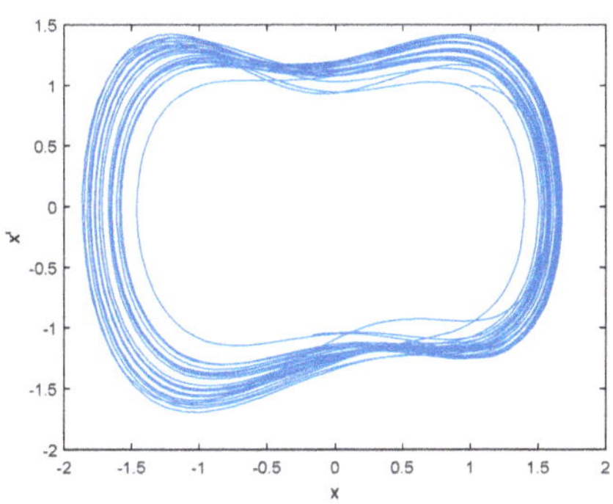

(a) Ship 1

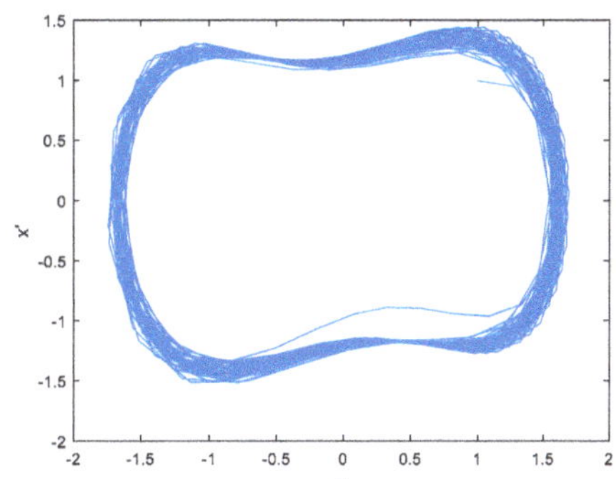

(b) Ship 2

Figure 12. *Cont.*

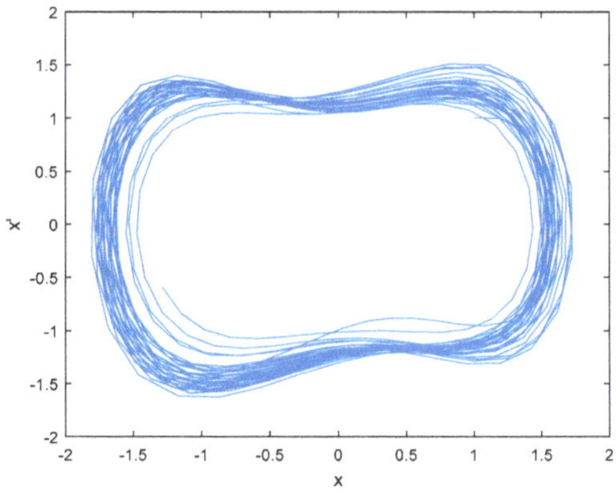

(**c**) Ship 3

Figure 12. The great periodic states of IMF8 for three kinds of ship-radiated noise samples.

Table 9. The frequency distribution of IMF8 by VMD-DCO-KPE for three kinds of ship-radiated noise samples.

Ship 1	Ship 2	Ship 3
11.82 Hz	44.29 Hz	29.85 Hz

5.3. Comparison of Different Frequency Feature Extraction Methods

We extracted the frequency features of 20 samples for each kind of ship. The frequency distributions of VMD-DCO-KPE and VMD-CF are shown in Figure 13. In order to prove the effectiveness of VMD-DCO-KPE, SVM with polynomial kernel function was used for the classification of three kinds of ships. The number of training samples and test samples were 20 and 30, and the classification results of different frequency feature extraction methods are listed in Table 10. As shown in Table 10, the classification result of VMD-DCO-KPE was 100%, which is better than EMD-CF, EEMD-CF and VMD-CF.

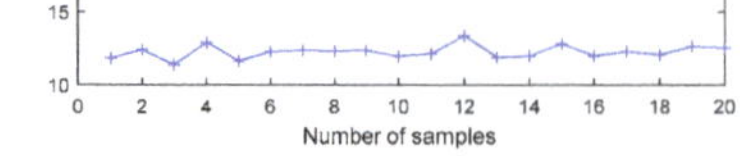

(**a**) VMD-DCO-KPE

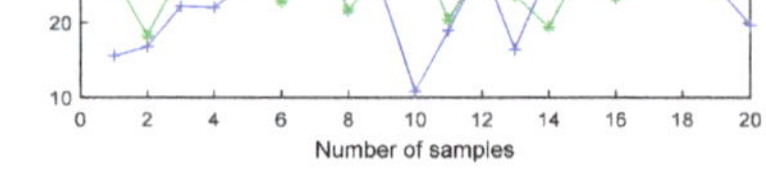

(**b**) VMD-CF

Figure 13. The frequency distributions of VMD-DCO-KPE and VMD-CF.

Table 10. The classification results of different frequency feature extraction methods.

EMD-CF	EEMD-CF	VMD-CF	VMD-DCO-KPE
67.33%	74.67%	80.67%	100%

6. Conclusions

A novel frequency feature extraction method for underwater acoustic signal is proposed in this paper based on VMD, DCO and KPE. The main contributions of this work are as follows:

(1) DCO is first used to detect the frequency of IMF by VMD for underwater acoustic signals in this paper.
(2) KPE is first used to determine the frequency of IMF combined with DCO for underwater acoustic signals in this paper.
(3) VMD-DCO-PE is successfully applied to extract the frequency feature of a simulation signal. Compared with EMD-CF, EEMD-CF and VMD-CF, VMD-DCO-KPE can be more accurate and efficient to extract the frequency feature of a simulation signal.
(4) VMD-DCO-KPE is also applied to extract the frequency feature extraction of line spectrum for underwater acoustic signal. VMD-DCO-KPE has better classification performance than EMD-CF, EEMD-CF and VMD-CF.

Author Contributions: Y.L. contributed the frequency feature extraction method, all authors designed and analyzed the experiments.

Funding: Authors gratefully acknowledge the support by National Natural Science Foundation of China (No. 51409214 and No. 11574250).

Conflicts of Interest: The authors declare no conflicts of interest.

References

1. Urick, R.J. *Principles of Underwater Sound*, 3rd ed.; McGraw-Hill: New York, NY, USA, 1983.
2. Li, Y.; Li, Y.; Chen, X.; Yu, J. A Novel Feature Extraction Method for Ship-Radiated Noise Based on Variational Mode Decomposition and Multi-Scale Permutation Entropy. *Entropy* **2017**, *19*, 342. [CrossRef]
3. Li, Y.; Li, Y.; Chen, X.; Yu, J. Research on Ship-Radiated Noise Denoising Using Secondary Variational Mode Decomposition and Correlation Coefficient. *Sensors* **2018**, *18*, 48. [CrossRef]
4. Villecco, F. On the Evaluation of Errors in the Virtual Design of Mechanical Systems. *Machines* **2018**, *6*, 36. [CrossRef]
5. Wang, S.; Zeng, X. Robust underwater noise targets classification using auditory inspired time-frequencyanalysis. *Appl. Acoust.* **2014**, *78*, 68–76. [CrossRef]
6. Xu, L.; Yang, K.; Yang, Q. Joint time-frequency inversion for seabed properties of ship noise on a vertical line array in South China Sea. *IEEE Access* **2018**, *6*, 62856–62864. [CrossRef]
7. Gassmann, M.; Wiggins, S.M.; Hildebrand, J.A. Deep-water measurements of container ship radiated noise signatures and directionality. *J. Acoust. Soc. Am.* **2017**, *142*, 1563. [CrossRef] [PubMed]
8. Wang, Y.H.; Hu, K.; Lo, M.T. Uniform phase empirical mode decomposition: An optimal hybridization of masking signal and ensemble approaches. *IEEE Access* **2018**, *6*, 34819–34833. [CrossRef]
9. Wang, J.L.; Wei, Q.X.; Zhao, L.Q.; Yu, T.; Han, R. An improved empirical mode decomposition method using second generation wavelets interpolation. *Digit. Signal Process.* **2018**, *79*, 164–174. [CrossRef]
10. Chen, T.; Ju, S.; Yuan, X.; Elhoseny, M.; Ren, F.; Fan, M.; Chen, Z. Emotion recognition using empirical mode decomposition and approximation entropy. *Comput. Electr. Eng.* **2018**, *72*, 383–392. [CrossRef]
11. Zhang, X.; Liang, Y.; Zhou, J.; Zhou, J.; Zang, Y. A novel bearing fault diagnosis model integrated permutation entropy, ensemble empirical mode decomposition and optimized SVM. *Measurement* **2015**, *69*, 164–179. [CrossRef]
12. Chu, H.; Wei, J.; Qiu, J. Monthly Streamflow Forecasting Using EEMD-Lasso-DBN Method Based on Multi-Scale Predictors Selection. *Water* **2018**, *10*, 1486. [CrossRef]
13. Huang, Y.; Liu, S.; Yang, L. Wind Speed Forecasting Method Using EEMD and the Combination Forecasting Method Based on GPR and LSTM. *Sustainability* **2018**, *10*, 3693. [CrossRef]
14. Singh, J.; Darpe, A.K.; Singh, S.P. Bearing damage assessment using Jensen-Rényi Divergence based on EEMD. *Mech. Syst. Signal Process.* **2017**, *87*, 307–339. [CrossRef]

15. Yeh, J.R.; Shieh, J.S.; Huang, N.E. Complementary ensemble empirical mode decomposition: A novel noise enhanced data analysis method. *Adv. Adapt. Data Anal.* **2010**, *2*, 135–156. [CrossRef]
16. Liu, H.; Mi, X.W.; Li, Y.F. Comparison of two new intelligent wind speed forecasting approaches based on Wavelet packet decomposition, complete ensemble empirical mode decomposition with adaptive noise and artificial neural networks. *Energ. Conv. Manag.* **2018**, *155*, 188. [CrossRef]
17. Lv, Y.; Yuan, R.; Wang, T.; Li, H.; Song, G. Health Degradation Monitoring and Early Fault Diagnosis of a Rolling Bearing Based on CEEMDAN and Improved MMSE. *Materials* **2018**, *11*, 1009. [CrossRef]
18. Dai, S.; Niu, D.; Li, Y. Daily Peak Load Forecasting Based on Complete Ensemble Empirical Mode Decomposition with Adaptive Noise and Support Vector Machine Optimized by Modified Grey Wolf Optimization Algorithm. *Energies* **2018**, *11*, 163. [CrossRef]
19. Bin Queyam, A.; Kumar Pahuja, S.; Singh, D. Quantification of Feto-Maternal Heart Rate from Abdominal ECG Signal Using Empirical Mode Decomposition for Heart Rate Variability Analysis. *Technologies* **2017**, *5*, 68. [CrossRef]
20. Dragomiretskiy, K.; Zosso, D. Variational mode decomposition. *IEEE Trans. Signal Process.* **2014**, *62*, 531–544. [CrossRef]
21. Li, Y.; Li, Y.; Chen, X.; Yu, J. Denoising and Feature Extraction Algorithms Using NPE Combined with VMD and Their Applications in Ship-Radiated Noise. *Symmetry* **2017**, *9*, 256. [CrossRef]
22. Cai, W.; Yang, Z.; Wang, Z.; Wang, Y. A New Compound Fault Feature Extraction Method Based on Multipoint Kurtosis and Variational Mode Decomposition. *Entropy* **2018**, *20*, 521. [CrossRef]
23. Wan, S.; Chen, L.; Dou, L.; Zhou, J. Mechanical Fault Diagnosis of HVCBs Based on Multi-Feature Entropy Fusion and Hybrid Classifier. *Entropy* **2018**, *20*, 847. [CrossRef]
24. Tripathy, R.K.; Paternina, M.R.A.; Arrieta, J.G.; Pattanaik, P. Automateddetection of atrialfibrillation ECG signalsusing twostage VMD and atrialfibrillation diagnosis index. *J. Mech. Med. Biol.* **2017**, *17*, 840–844. [CrossRef]
25. Wang, G.; Chen, D.; Lin, J.; Chen, X. The application of chaotic oscillators to weak signal detection. *IEEE Trans. Ind. Electron.* **1999**, *46*, 440–444. [CrossRef]
26. Li, Y.; Shi, Y.; Ma, H.; Yang, B. Chaotic detection method for weak square wave signal submerged in colored noise. *Chin. J. Electron.* **2004**, *32*, 87–90.
27. Lai, Z.; Leng, Y.; Sun, J.; Fan, B. Weak characteristic signal detection based on scale transformation of Duffing oscillator. *Acta Phys. Sin.* **2012**, *61*, 050503.
28. Chen, Z.; Li, Y.; Chen, X. Underwater acoustic weak signal detection based on Hilbert transform and intermittent chaos. *Acta Phys. Sin.* **2015**, *64*, 200502.
29. Bandt, C.; Pompe, B. Permutation entropy: A natural complexity measure for time series. *Phys. Rev. Lett.* **2002**, *88*, 174102. [CrossRef]
30. Zhang, J.; Hou, G.; Cao, K.; Ma, B. Operation conditions monitoring of flood discharge structure based on variance dedication rate and permutation entropy. *Nonlinear Dyn.* **2018**, *93*, 1–15. [CrossRef]
31. Bandt, C. A new kind of permutation entropy used to classify sleep stages from invisible EEG microstructure. *Entropy* **2017**, *19*, 197. [CrossRef]
32. Haruna, T. Partially ordered permutation complexity of coupled time series. *Phys. D Nonlinear Phenom.* **2018**, *388*, 40–44. [CrossRef]
33. Hu, Q.; Hao, B.; Lv, L. Feature extraction model for underwater target radiated noise. *Torpedo Technol.* **2008**, *16*, 38.
34. Liu, S.; Zhang, X.; Niu, Y. Feature extraction and classification experiment of underwater acoustic signals based on energy spectrum of IMF's. *Comput. Eng. Appl.* **2014**, *50*, 203–206.
35. Yang, H.; Li, Y.; Li, G. Energy analysis of ship-radiated noise based on ensemble empirical mode decomposition. *J. Vib. Shock* **2015**, *34*, 55–59.

electronics

MDPI

Article

A Tilt Sensor Node Embedding a Data-Fusion Algorithm for Vibration-Based SHM

Nicola Testoni [1,*], Federica Zonzini [2], Alessandro Marzani [3], Valentina Scarponi [2] and Luca De Marchi [2]

1 Advanced Research Center on Electronic Systems "Ercole De Castro" (ARCES), University of Bologna, Via V. Toffano 2/2, 40125 Bologna, Italy

2 Department of Electrical, Electronic and Information Engineering, University of Bologna, 40123 Bologna, Italy; federica.zonzini@unibo.it (F.Z.); valentina.scarponi3@studio.unibo.it (V.S.); l.demarchi@unibo.it (L.D.M.)

3 Department of Civil, Chemical, Environmental and Materials Engineering, University of Bologna, 40123 Bologna, Italy; alessandro.marzani@unibo.it

* Correspondence: nicola.testoni@unibo.it; Tel.: +39-051-209-3268

Received: 30 November 2018; Accepted: 29 December 2018; Published: 1 January 2019

Abstract: This work describes a miniaturized sensor network based on low-power, light-weight and small footprint microelectromechanical (MEMS) sensor nodes capable to simultaneously measure tri-axial accelerations and tri-axial angular velocities. A real-time data fusion algorithm based on complementary filters is applied to extract tilt angles. The resulting device is designed to show competitive performance over the whole frequency range of the inertial units. Besides the capability to provide accurate measurements both in static and dynamic conditions, an optimization process has been designed to efficiently make the fusion procedure running on-sensor. An experimental campaign conducted on a pinned-pinned steel beam equipped with a network comprising several sensor nodes was used to evaluate the reliability of the developed architecture. Performance metrics revealed a satisfactory agreement to the physical model, thus making the network suitable for real-time tilt monitoring scenarios.

Keywords: tilt sensor; sensor data fusion; complementary filters; overlap-add processing; spectral analysis

1. Introduction

The deployment of vibration-based Structural Health Monitoring (SHM) systems involves a plurality of requirements to be satisfied. Technology non-invasiveness, real-time analysis capability, and compatibility with long-term installation can be listed among them [1–3]. The ability to continuously provide up-to-date information about current structural health conditions requires dedicated hardware and software resources. In time, these can be combined to obtain wide-area sensor networks embedding local data processing functionality.

The goal of providing early anomaly detection and damage localization is pivotal in SHM [4–9]. Compactness and reduced power consumption make microelectromechanical (MEMS) sensors suitable for structural monitoring; also, they can be directly deployed on-structure, all the while allowing for low-cost frameworks and extending electronics life cycle. Accelerometers are particularly suited to capture linear accelerations: despite this, damage metrics applied to data acquired in the proximity of unfavorable locations fail in properly detecting anomalies, primarily due to a reduced Signal to Noise Ratio (SNR) [10].

Recent trends in electronics highlighted the possibility to combine MEMS technology with multi-degree-of-freedom measurement units. As such, monitoring schemes are moving towards redundant but more accurate and reliable configurations, capable to gather both static and dynamic features. Tailoring this necessity to civil and industrial applications, the concurrent usage of accelerometers and gyroscopes provides a set of complementary quantities which can compensate for each other. Experimental validations of this integrated strategy have been conducted for high-rise buildings [11], showing that the joined exploitation of acceleration and tilt sensors yields a more precise understanding of the structural deformation at higher frequencies. Similarly, coupled linear and rotational measurements have shown to have superior performance in monitoring wind induced vibrations in tall infrastructures [12–15]. Furthermore, diagnosis systems for bridge monitoring purposes have been implemented through sensor networks comprising gyroscopes and PCB accelerometers [10,16] showing superior performance in damage localization.

At the same time, current monitoring solutions embrace the idea to provide devices with embedded data fusion algorithms, that is, each sensor node combines multiple sensor signals to reduce the uncertainty of single-source sensing architectures. Among the possible techniques, which may include Kalman [17–20] or particle filtering [21,22], Complementary Filters (CF) based on the simultaneous adoption of low-pass and high-pass filtering demonstrated to be extremely effective. In fact, being the CF overall transfer function constant over the whole spectrum [23–26], their design combines well with wide-band sensing strategies. Similar studies are usually based on processing procedures computed off the sensor node, however, to provide real-time embedded signal processing capabilities, especially while chasing rapid phenomena, fully tunable filter chains which do not strongly impinge on the computational effort should be considered.

Consistently with the aforementioned scenarios, the presented work describes a monitoring system based on small footprint, low-power and light-weight MEMS sensors which can be interchangeably used as accelerometers, gyroscopes, or tilt nodes thanks to an embedded data fusion algorithm. Specific attention was given to the software implementation of the CF technique, essential to extract tilt information directly on the node by means of a low-complexity algorithmic scheme. As a result, the sensor-near electronic design strategy could be adopted. The deriving versatility, scalability and computational efficiency allow to optimally shape the network in relation to each specific monitoring application.

The paper is organized as follows. Section 2 is firstly dedicated to the architectural description of the monitoring system, in which the sensor nodes represent the hardware core blocks dedicated to acquisition tasks. The sensor node schematic diagram and the relative prototype are detailed, highlighting the properties of the digital gyroscope, whose exploitation represents the core of the work. The data fusion algorithm is then introduced, including the calibration steps chosen to set the parameters of the digital filters. Section 3 shows that an Overlap-Add (OLA) processing method provides highly accurate measurements in static conditions. Moreover, in the same Section, a test-bench is introduced to evaluate the validity of the proposed system for modal analysis purposes, comparing the extracted frequencies of vibration to the theoretical predictions. Finally, conclusions will be drawn in Section 4.

2. Materials and Methods

2.1. Sensor Node

The developed sensor node characteristics are: (a) 30 mm × 23 mm stamp-size, (b) 3.5 g light-weight and (c) reduced power consumption, which allow to gradually distribute processing power, hence striving to realize a decentralized monitoring platform. This device represents an improved version of a previous and already validated circuitry customized to acquire acceleration data only [27]. Devices are connected by a Sensor Area Network (SAN) bus exploiting data-over-power (DoP) communication, whereas meaningful information is pre-processed by means of a lossless encoding technique; finally, data is transmitted to a connected PC via a purposely designed companion gateway device. When SAN is powered at 5.0 V, 44.8 mW are drained.

From an architectural point of view, all the building blocks sketched in Figure 1a are controlled by a Micro-Controller Unit (MCU) equipped with Digital Signal Processing (DSP) instructions, Floating Point Unit (FPU) and limited FLASH memory. A serial RAM is integrated for temporary storage, while network connectivity to the bus is accomplished through a low-power transceiver (XCVR). Serial Peripheral Interface (SPI) and I2C serial protocol guarantee internal and external communication between components and connected peripherals.

Angular velocities and acceleration signals are collected using an LSM6DSL iNEMO Inertial Measurement Unit (IMU), an ST Microelectronics system-in-package [28] featuring a 3D digital accelerometer and a 3D digital gyroscope, accessed by means of a dedicated SPI interface. It exhibits full-scale acceleration ranges of ± 2 g, ± 4 g, ± 8 g and ± 16 g and angular rate ranges of ± 125 dps, ± 250 dps, ± 500 dps, ± 1000 dps and ± 2000 dps. In shutdown mode, 3 µA are absorbed from the 3.3 V power supply fixed by a Low-Drop-Out regulator (LDO). It consumes 0.65 mA in the most-demanding configuration, thus enabling always-on low power measurements. Power-down, low-power, normal-mode, and high-performance mode are the four different operating modes available for the sensing elements, whose Output Data Rate spans from 12.5 Hz up to 6.664 kHz and is real-time programmable by means of a digital low-pass filter.

The integrated tri-axial gyroscope belongs to a category of devices producing a positive digital output for counterclockwise rotation around a predefined axis. Its sensitivity of 4.375 mg/LSB (Least Significant Bit) for the chosen output range of ± 125 dps is subjected to minimal drifts over time, also withstanding a thermal excursion between -40 °C and 85 °C. Furthermore, this inertial component features an ultra-low noise density of about 4 mdps/$\sqrt{\text{Hz}}$ in high-performance mode, exhibiting a competitive resolution within its class. In order to reinforce the placement, facilitate the installation step and protect circuitry against electromagnetic coupling or atmospheric-driven failures, common in harsh environments, each node is lodged in a dedicated case weighing less than 6 g on the whole.

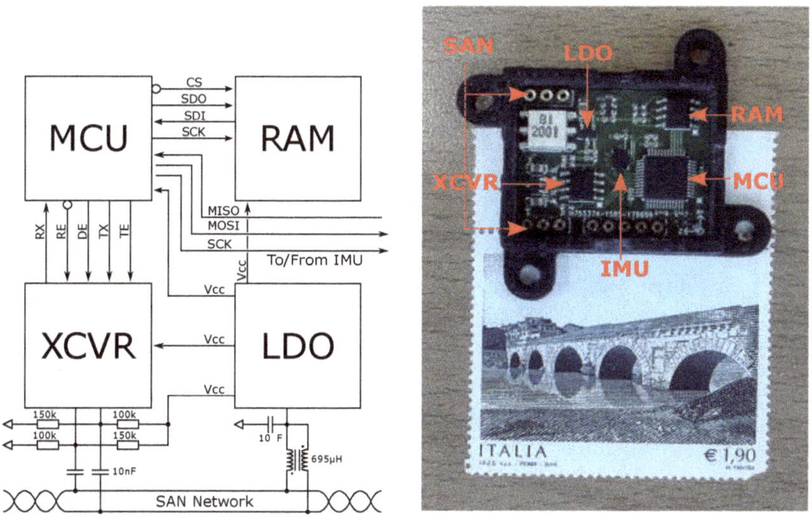

(a) (b)

Figure 1. Hardware instrumentation: (a) Schematic diagram of the sensor node and (b) its relative prototype inside an ad-hoc case.

The resulting sensor node, depicted in Figure 1b, can be permanently installed on the structures to be monitored as its physical and electrical properties do not interfere with their behavior.

2.2. Sensor Data Fusion

Data-level fusion algorithms are encouraged by the widely shared opinion that reliability, resolution, availability, and accuracy are primary issues in every SHM process [29,30]. Compensation and auto-calibration obtained by a complementary and integrated approach strengthen the inspection phase in seizing multiple aspects of the same phenomena [31]. In structural applications, the synchronized measurement of angular and linear displacements enables to estimate inclination with finer precision.

On one hand, accelerometers perform well at low frequencies: in fact, even if dynamic features suffer from crosstalk, this undesired effect is filtered out by the acceleration transfer function. On the other hand, gyroscopes work optimally in the superior spectral band; they, however, suffer from the integration procedure mandatory to transform angular velocities into tilt values. Among the variety of methods theorized for tilt estimation, the strategy here proposed and embedded in the sensor node is based on high-pass filtered angular velocities and low-pass filtered accelerations. The sensor data fusion mechanism, which is consistent with FIR Complementary Filters suggested in [25], is chosen to minimize phase and magnitude distortion around the cutoff frequency.

2.3. Algorithm Definition

The time-dependent acceleration-based and angular-based tilt values, addressed in the following as θ_a and θ_g, characterize the modal behavior of structures undergoing vibrations. For the sake of clarity, in case of devices installed on the top surface of a structure, Figure 2 schematically depicts the problem from a geometric point of view. In detail, the sensor node laying on the xy-plane is programmed to estimate inclinations of the vertical plane, consequently, the tilt is intended as a positive value around the z axis.

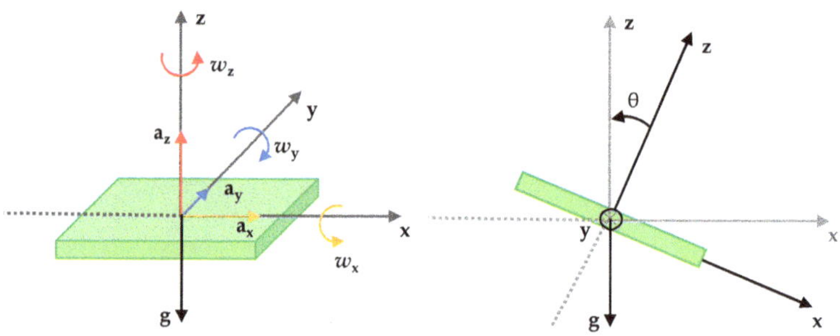

Figure 2. Geometric relation between tilt angles and acceleration referred to z-direction for a device installed on the top of a structure.

The acceleration vector constitutes of three components a_x, a_y, a_z recorded along the three directions, whereas angular rates w_x, w_y, w_z correspond to rotational spins projected on the same axes. Radial acceleration $a_r = a_z$ and tangential acceleration $a_t = \sqrt{a_x^2 + a_y^2}$ are fused together to extract the tilt values $\hat{\theta}_a$ defined as

$$\tan \hat{\theta}_a = \frac{a_t}{a_r} = \frac{\sqrt{a_x^2 + a_y^2} + \xi_c + \xi_a}{a_z + \xi_c + \xi_a} \tag{1}$$

Crosstalk noise ξ_c and accelerometer intrinsic noise ξ_a usually affects the collected data, their contribute becoming evident at higher frequencies. Such disturbances must be filtered out by an appropriate low-pass transfer function, thus providing an accurate estimation only for pseudo-static behavior.

Angles described by rotation around predefined directions can be numerically computed by integrating the absolute angular velocity components

$$\hat{\omega}_g = \sqrt{\omega_x^2 + \omega_y^2 + \omega_z^2} + \xi_b + \xi_g \quad (2)$$

in which drift errors caused by inherently biased and noisy measurements, respectively indicated as ξ_b and ξ_g, typically impact on pseudo-static measurements. The robustness of integration with respect to high-pass filtering leads to precise gyroscope-driven tilt estimations only in the dynamic regime.

According to the aforementioned CF technique, by taking the Fourier Transform (FT) of Equations (1) and (2), the estimated $\hat{\theta}_a(f)$ and $\omega_g(f)$ enter the fundamental fusion step to obtain a unique fused value $\hat{\theta}(f)$ defined as

$$\hat{\theta}(f) = H_L(f)\,\hat{\theta}_a(f) - j\frac{H_H(f)}{2\pi f}\,\hat{\omega}_g(f) \quad (3)$$

This is accomplished by applying in parallel two second order filters: $H_L(f)\hat{\theta}_a(f)$ is the low-pass filtered version of data coming from accelerometer, whereas angular rate signals undergo an high-pass filtering elaboration. The two quantities $H_L(f)$ and $H_H(f)$ designate the following low-pass and high-pass filtering envelopes

$$H_L(f) = \frac{1}{1 + \left(\frac{f}{f_\beta}\right)^{2n}} \qquad H_H(f) = \frac{1}{1 + \left(\frac{f_\beta}{f}\right)^{2n}} \qquad H_L(f) + H_H(f) = 1 \quad (4)$$

where f_β indicates the cut-off frequency of the filters. Fused inclination values at every instant in time are finally computed by applying the Inverse Fourier Transform (IFT) of the output provided by expression (3).

2.4. Embedded Processing

The processing flow implemented on the sensor node is schematically depicted in Figure 3. The sampling frequency F_s is chosen on the basis of the spectral content, following the Nyquist theory. The number of samples N_{tot} is related to the maximum available storage capability of each sensor. The CF data fusion was performed in the Fourier domain by adopting the Overlap-Add (OLA) method [32].

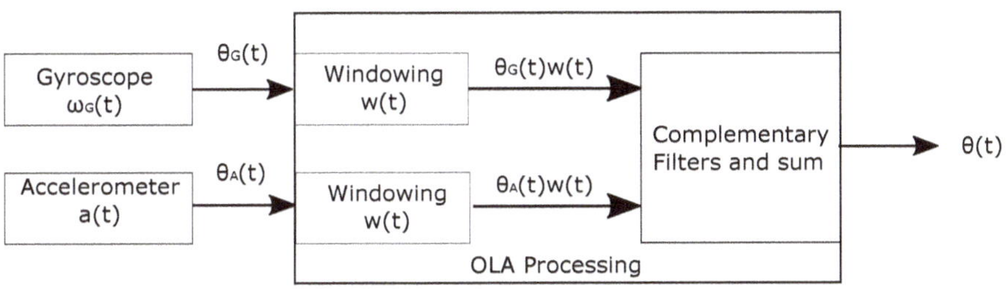

Figure 3. Schematic representation of signal processing method programmed onboard.

According to the OLA paradigm, data must undergo a windowing phase. Window size in the time-domain N_s is at least one order of magnitude smaller than the entire time-series N_{tot}. The window adopted in this study is displayed in Figure 4a, and can be mathematically described as:

$$w(t) = \begin{cases} \sin^2\left(\frac{\pi}{2}\frac{t}{T_{ov}}\right) & 0 \le t < T_{ov} \\ 1 & T_{ov} \le t < T_{hop} \\ \cos^2\left(\frac{\pi}{2}\frac{t-T_{hop}}{T_{ov}}\right) & T_{hop} \le t < T_{frame} \end{cases} \tag{5}$$

where T_{frame} is $\frac{N_s}{F_s}$, T_{ov} is the time interval in which consecutive windows are overlapped, and $T_{hop} = T_{frame} - T_{ov}$ is hop size. Windowed data are then Fourier transformed and filtered, so that the finally derived tilt values can be concatenated.

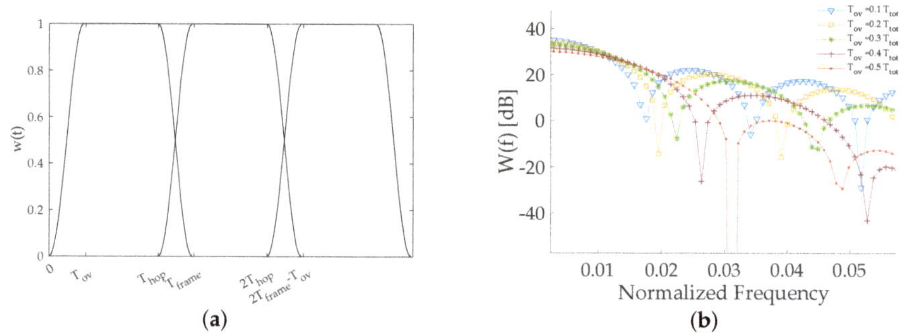

Figure 4. Window function implemented to diminish the computational burden of the data fusion algorithm working on sensor: (**a**) Time-domain working principle of the OLA mechanism and (**b**) window spectral properties for overlapping fraction spanning in the interval [0.1; 0.5].

The optimal selection of the cutoff frequency of the complementary filters is highly dependent on the sensor technology, as well as on the specific application case. In the experiments related to static tilt conditions, such parameter has been selected on the basis of a calibration step which led to the minimization of the mean square error, as shown in Figure 5a.

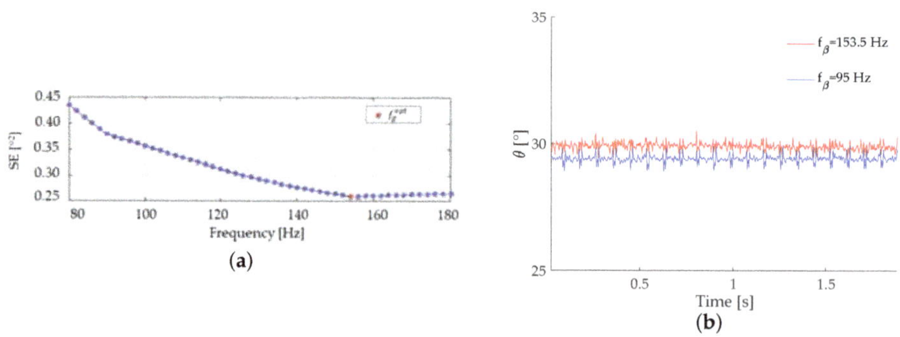

Figure 5. (**a**) Optimal cutoff frequency estimation and (**b**) effect of the cutoff frequency selection on OLA sensor data fusion (actual tilt value: 30°).

For the global accuracy, the selection of the window length N_s is also very relevant, since this parameter directly affects the quality of the filter approximation based on the discrete FT. This implies that the ratio between the frequency resolution of the windowed and non-windowed processing ($\Delta_{ft} = \frac{F_s}{N_{tot}}$ and $\Delta_{fw} = \frac{F_s}{N_s}$, respectively) should be lower bounded:

$$\frac{\Delta_{ft}}{\Delta_{fw}} = \frac{N_s}{N_{tot}} \geq \alpha \quad \rightarrow \quad N_s \geq \alpha\, N_{tot} \tag{6}$$

where α is a predefined accuracy threshold.

It is worth noting that, since the rising and falling edge respectively obey to a $\sin^2(t)$ and a $\cos^2(t)$ trend, the mask of this window is shaped to satisfy the Constant-Overlap-Add (COLA) constraint stated in Equation (7):

$$\sum_{k=0}^{N_w-1} w(t - k\,T_{hop}) = 1, \quad \forall t \tag{7}$$

being $N_w = \frac{T_{tot}}{T_{hop}}$ the total number of iterations. This necessary and sufficient condition allows to correctly reconstruct signals split into successive windowed frames. The COLA constraint implies that the spectral values of the window functions must be zero at all harmonics of the hop rate $F_{hop} = \frac{1}{T_{hop}}$, consequently, it must ensure that

$$W(k\,F_{hop}) = 0, \quad \forall k = 1, \ldots, N_w - 1 \tag{8}$$

Taking the Fourier Transform of the window described in (5) and introducing the overlap fraction $o = \frac{T_{ov}}{T_{frame}}$, it follows that

$$W(f) = -\frac{(2f\,o\,T_{frame})^2}{1-(2f\,o\,T_{frame})^2} T_{hop} \cos(\pi\, f\, o\, T_{frame}) \mathrm{sinc}(f\,T_{hop}) e^{-j\pi\, f\, T_{frame}} \tag{9}$$

clearly showing zero values for $f = k\,F_{hop}$ and then compliant to (8), independently either from the duration of the window and the number of samples to be overlapped.

A narrow amplitude of the first lobe of the window spectrum, together with a highly attenuated second lobe, would be desirable. However, the spectrum obtained by processing windows with increasing values of T_{ov} (see Figure 4b) clearly demonstrates that a wider first lobe corresponds to a deeper attenuation of secondary lobes. As a result, the final choice must be properly balanced among these two opposite behaviors, in order to reach the best performance.

Besides accuracy, also the computational cost to perform the OLA processing is strongly affected by the selection of T_{ov} and N_s. As well known, the FFT has $O(N_s \log_2 N_s)$ complexity, implying a logarithmic decrease when N_s is reduced. Therefore, the computational effort C paid to process a generic sequence of N_{tot} elements divided into N_w frames results in

$$C = N_w N_s \log_2(N_s) \tag{10}$$

Specifically, the contribution associated with the number of overlapped samples is upper bounded to $2N_{tot}$ whenever the maximum allowable T_{ov} is chosen. On the contrary, the logarithmic relation connected to the dimension of the window leads to a consistent reduction of C as N_s downsizes. Consequently, following what was anticipated in (6), there is a clear trade-off between the computational cost and the filter approximation accuracy.

To fully exploit the versatility of the circuitry, all the parameters necessary to perform the processing were stored in registers programmable at run-time: the gyroscope full-scale, the sampling frequency, the total number of samples to be acquired, the overlap fraction, and the output data rate.

3. System test and Discussion

To evaluate the reliability of the developed hardware and software architecture, after a validation phase in which window parameters have been quantified, the accuracy of tilt estimation has been examined experimentally in almost static conditions, and successively in dynamic regimes.

3.1. System Validation in Static Condition

A Newport IG Breadboard anti-vibration table shown in Figure 6 was used to filter out unwanted surrounding vibrations, while a sensor was statically tilted to a fixed angle during an initial trial necessary to extract the proper complementary filters cutoff frequency.

The sampling frequency was set to 1250 Hz. Since the maximum available temporary storage capability is 30 kBytes and each sensor concurrently acquires two data bytes for each one of the six inertial degrees of freedom, the available number of samples on each channel cannot exceed 2500. Obeying to internal DSP functionalities, which impose window length to be a power of two, and assuming a resolution ratio $\alpha = 0.02$, 64 samples shifted with an overlapping ratio equal to 0.25 were selected, ensuring an optimal trade-off among the spectral design of the corresponding window frame and the computational complexity.

Experimental data were processed with f_β values varying from 80 Hz to 180 Hz with an increasing step of 1.5 Hz. The optimal cutoff frequency was obtained by minimizing the square error

$$SE = (\theta_{TUV} - \bar{\theta}_S)^2 \tag{11}$$

between the constant θ_{TUV} reference angle provided by a TUV GS level included in the same setup and the mean value $\bar{\theta}_S$ extracted from collected samples. The global minimum displayed in Figure 5a corresponding to a cutoff frequency of 153.5 Hz was finally set as the optimal cutoff frequency. Figure 5b shows that the selection of the most appropriate cutoff frequency effectively captures the actual tilt value, while a wrong selection may cause periodic artifacts, the periodicity of them being related to the window dimension.

In the following, first and second order statistics have been used to establish the accuracy of the measured inclinations in stationary conditions, with a sensor node fixed at three different inclinations: 30°, 45°, 60°. Table 1 points out the distribution of mean value and standard deviation for each configuration: relative error E_r lower than 0.7% and σ always less than two-tenths of a degree prove that results are highly precise.

It is worth pointing out how variance slightly arises for increasing inclination values, showing an almost linear trend. This evidence paves the way to an auto-calibration procedure transferable onto the node itself: once a finer-scale training would be executed, biased measurements could be internally corrected after inferring the proper compensation curve.

Table 1. Statistics obtained from measurements in different pseudo-static configurations: mean value, relative error and standard deviation.

Reference Tilt [°]	Measured Tilt [°]	E_r [%]	σ [°]
30	30.1832	0.611	0.1399
45	45.0024	0.005	0.1523
60	60.3116	0.519	0.1985

Figure 6. Experimental setup in pseudo-static conditions: anti-vibration table equipped with TUV level.

3.2. Vibration Analysis

The study of the dynamic properties of the system in the frequency domain is typically carried out estimating the most energetic natural frequencies, firstly for the extraction of the instantaneous rate of vibration and subsequently to assess the integrity.

As displayed in Figure 7, a network comprising seven sensor nodes and one interface connected in a daisy-chain fashion was mounted on a pinned-pinned $L = 2140\,\text{mm}$ steel beam with cross-section base $b = 60\,\text{mm}$ and height $h = 10\,\text{mm}$, thus corresponding to a moment of inertia $I = bh^3/12$. The material density is $\rho = 7880\,\text{kg}/\text{m}^3$ and the Young's modulus $E = 195\,\text{GPa}$ can be used to predict the first natural frequencies through the closed formula

$$f_n = \frac{(\pi n)^2}{2\pi L}\sqrt{\frac{EI}{\rho b h}} \tag{12}$$

Sensors were placed at a step of 220 mm starting from the first node, whose distance from the fixed left edge of the beam is 135 mm.

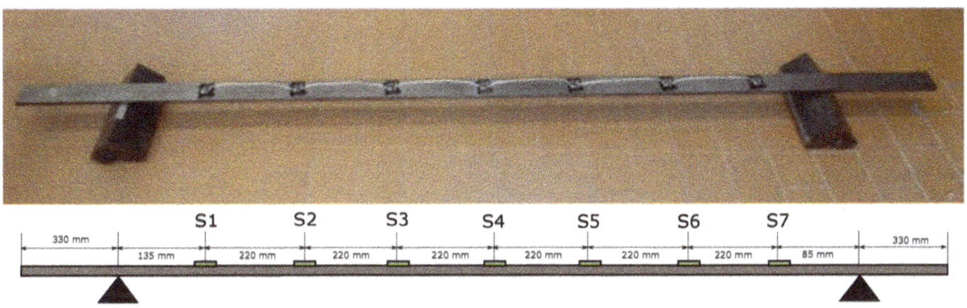

Figure 7. Experimental setup in vibrating condition.

An explicit relationship exists between acceleration and inclination observed with respect to a common direction. Resorting to trigonometric relationships for the scheme introduced in Figure 2, the time-spatial dependent angle θ, described by rotations of the sensor, can be geometrically interpreted as the derivative of vertical position displacements along the longitudinal direction.

Mathematically speaking, the governing equation of a thin rod undergoing transverse motion is defined as [33]

$$z(x,t) = \sum_{n=1}^{\infty} (A_n \cos(\omega_n t) + B_n \sin(\omega_n t)) \sin(\beta_n x) \tag{13}$$

where appearing quantities A_n, B_n are constants deriving from boundary conditions and $\omega_n = 2\pi f_n$ correspond to the nth-cyclic pulsation. Algebraic manipulation of (13) yields to the more compact form

$$z(x,t) = \sum_{n=1}^{\infty} R_n \sin(\omega_n t + \alpha_n) \sin(\beta_n x), \quad R_n = \sqrt{A_n^2 + B_n^2}; \quad \alpha_n = \arctan \frac{A_n}{B_n} \tag{14}$$

on which a derivative operation can be performed providing the final result stated in (15).

$$\theta(x,t) = \frac{\partial z(x,t)}{\partial x} = \sum_{n=1}^{\infty} \beta_n R_n \sin(\omega_n t + \alpha_n) \cos(\beta_n x) \tag{15}$$

Comparing (13) to (15), it can be inferred that the spectral content of $z(x,t)$ and $\theta(x,t)$ is localized at the same angular frequencies ω_n. As a consequence, frequency analysis accomplished on tilt angles or acceleration signals allow identifying the same modes of vibration, predictable through Equation (12). For this reason, in this experiment, we have evaluated how similar is the frequency spectrum computed on the acceleration data with respect to the one computed on the result of the CF data fusion procedure.

It is important to highlight that, besides data and power communication, the bus connecting the sensor nodes also natively allows for time base synchronization in the acquisition. As a consequence, an output-only estimation of vibrating components can be put in place. Gathering data at a sampling frequency $F_s = 1250\,\text{Hz}$ ensures a Nyquists' bandwidth compliant with the theoretical estimation up to a satisfying accuracy. The beam was excited at the two-thirds of the span by means of an impact hammer, thus allowing it to oscillate in a condition of free vibrations. Since the dynamic operating conditions substantially differ from static measurements, a new calibration phase is necessary to be executed.

The most appropriate cutoff frequency was selected according to the spectral range of interest. More explicitly, supposing that the energy of a structure is mainly distributed among the lower spectral components, the analysis here conducted included the characterization of the first and second harmonic, corresponding to $f_1 = 6.195\,\text{Hz}$ and $f_2 = 24.778\,\text{Hz}$ for the setup under test. Thereafter, a value of $f_\beta = 27\,\text{Hz}$ was adopted to properly balance accelerometers and gyroscopes performance. Window size equal to 128 samples was chosen in order to provide high-resolution data, and an overlap factor $o = 0.4$ enabled to smooth the envelope of the window transition bands.

The joint hardware-software optimization of the circuitry allowed us to compute Power Spectral Densities (PSD) of tilt signals, processed exploiting FFT and Welch estimation method. In order to assess the reliability of the proposed multi-type sensor framework, results obtained for windowed and non-windowed processing were compared to the performance obtained applying the aforementioned techniques to the same radial acceleration dataset used for tilt angles estimation.

Improvements in the quality of the vibration analysis can be inferred from Figure 8. Basing on data extracted from a single sensor node installed on the top surface of the beam, the introduction of the right cutoff frequency intensely attenuates spurious peaks. Furthermore, the spectral trend estimated through the windowing strategy is almost perfectly superimposed to the one extracted processing the whole dataset at one time and it is also coherent to numerical predictions.

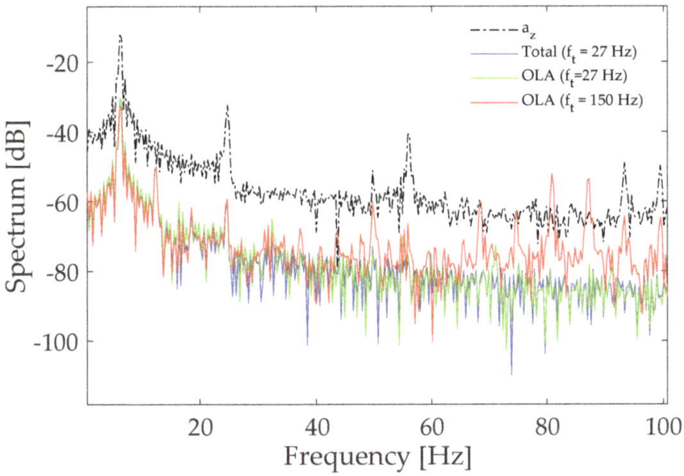

Figure 8. Comparison of spectra resulting from the windowed and non-windowed approach with respect to the spectral content of radial acceleration.

An important observation comes from the analysis of Figure 9: a deep correlation is present between the experimental and theoretical modes of vibration, being the relative error always below 1% under all the investigated methods. Moreover, the FFT estimator not only accomplishes the highest level of accuracy but also implies a lower computational cost if executed by local on-sensor processing units.

Another significant evidence comes from the analysis of Figure 10, were the seven spectra obtained from the sensors placed in the positions described by Figure 7 are superimposed; not only the peaks corresponding to the different vibrating harmonics are distinctly resolved over the whole band and are characterized by a satisfactory peak-to-noise ratio of about 15 dB, but also a high degree of coherence between them is evident.

Finally, it can be observed that the obtained spectra allow detecting both the pinned-pinned frequencies (triangular marked peaks) and the free-free flexural behavior of the beam (red circles), which may arise because the hinging supports do not perfectly anchor the structure. As a result, improved real-time algorithms embedded into electronic equipment permit to capture detailed snapshots of the rotational properties characterizing vibrating structures.

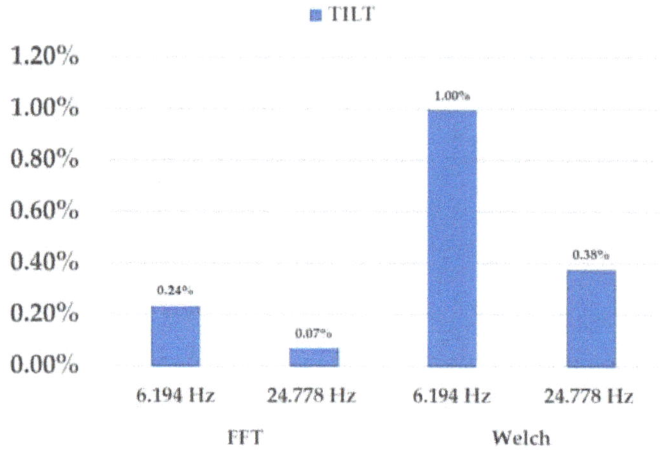

Figure 9. Spectral analysis on tilt values extracted by sensor node S2: comparison of the error distribution in vibrating modes extraction from acceleration and tilt signals via FFT and Welch strategy.

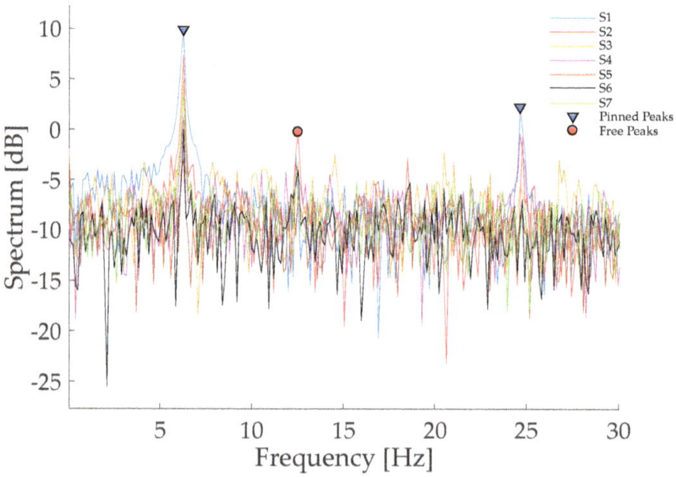

Figure 10. Comparison of spectral density characteristics of tilt signals estimated via FFT elaboration by nodes located at different positions.

4. Conclusions

This work describes a tilt sensor node, along with all the implemented procedures adopted to acquire and process high-quality signals in real-time. Inclination values are extracted with a simple but robust sensor data fusion algorithm supported by an optimized onboard signal processing scheme, accomplishing high accuracy both in pseudo-static and dynamic conditions. Reduced computational complexity, combined with scalability, versatility, and non-invasiveness, perfectly cope with long-term monitoring instances. The suitability of the designed framework to vibration-based SHM instances makes it possible to integrate such a network for modal analysis purposes, comprehending the development of new modal shapes reconstruction strategies which represent a big concern in modal analysis scenarios.

Author Contributions: Conceptualization, N.T., F.Z., A.M., V.S. and L.D.M.; Data curation, F.Z. and V.S.; Methodology, N.T., F.Z. and V.S.; Supervision, N.T., A.M. and L.D.M.; Validation, N.T. and F.Z.; Visualization, F.Z. and V.S.; Writing—original draft, F.Z. and V.S.; Writing—review & editing, N.T., F.Z., A.M. and L.D.M.

Funding: This work has been funded by INAIL within the BRIC/2016, ID = 15 framework, project SMARTBENCH.

Conflicts of Interest: The authors declare no conflict of interest.

References

1. Farrar, C.R.; Worden, K. An introduction to structural health monitoring. *Philos. Trans. R. Soc. Lond. A Math. Phys. Eng. Sci.* **2007**, *365*, 303–315. [CrossRef] [PubMed]
2. Carden, E.P.; Fanning, P. Vibration based condition monitoring: A review. *Struct. Health Monit.* **2004**, 3, 355–377. [CrossRef]
3. Lynch, J.P.; Loh, K.J. A summary review of wireless sensors and sensor networks for structural health monitoring. *Shock Vib. Dig.* **2006**, *38*, 91–130. [CrossRef]
4. Poncelet, F.; Kerschen, G.; Golinval, J.C.; Verhelst, D. Output-only modal analysis using blind source separation techniques. *Mech. Syst. Signal Process.* **2007**, *21*, 2335–2358. [CrossRef]
5. Esfandabadi, Y.K.; De Marchi, L.; Testoni, N.; Marzani, A.; Masetti, G. Full wavefield analysis and Damage imaging through compressive sensing in Lamb wave inspections. *IEEE Trans. Ultrason. Ferroelectr. Freq. Control* **2018**, *65*, 269–280. [CrossRef] [PubMed]
6. Hu, X.; Wang, B.; Ji, H. A wireless sensor network-based structural health monitoring system for highway bridges. *Comput.-Aid. Civ. Infrastruct. Eng.* **2013**, *28*, 193–209. [CrossRef]
7. Manthei, G.; Plenkers, K. Review on In Situ Acoustic Emission Monitoring in the Context of Structural Health Monitoring in Mines. *Appl. Sci.* **2018**, *8*, 1595. [CrossRef]
8. Ngabo, C.I.; El Beqqali, O. 3D tilt sensing by using accelerometer-based wireless sensor networks: Real case study: Application in the smart cities. In Proceedings of the 2018 International Conference on Intelligent Systems and Computer Vision (ISCV), Fez, Morocco, 2–4 April 2018; pp. 1–8.
9. Giri, P.; Ng, K.; Phillips, W. Laboratory simulation to understand translational soil slides and establish movement criteria using wireless IMU sensors. *Landslides* **2018**, *15*, 2437–2447. [CrossRef]
10. Sung, S.H.; Park, J.W.; Nagayama, T.; Jung, H.J. A multi-scale sensing and diagnosis system combining accelerometers and gyroscopes for bridge health monitoring. *Smart Mater. Struct.* **2013**, *23*, 015005. [CrossRef]
11. Hu, R.; Xu, Y.; Lu, X.; Zhang, C.; Zhang, Q.; Ding, J. Integrated multi-type sensor placement and response reconstruction method for high-rise buildings under unknown seismic loading. *Struct. Des. Tall Spec. Build.* **2018**, *27*, e1453. [CrossRef]
12. Li, X.; Rizos, C.; Tamura, Y.; Ge, L.; Yoshida, A.; Cranenbroeck, J. Fundamental bending mode and vibration monitoring with inclinometer and accelerometer on high-rise buildings subject to wind loads. In Proceedings of the 5th World Conference Structural Control and Monitoring, Tokyo, Japan, 12–14 July 2010; pp. 1–15.
13. Yigit, C.O.; Li, X.; Inal, C.; Ge, L.; Yetkin, M. Preliminary evaluation of precise inclination sensor and GPS for monitoring full-scale dynamic response of a tall reinforced concrete building. *J. Appl. Geod.* **2010**, *4*, 103–113. [CrossRef]
14. Su, J.Z.; Xia, Y.; Chen, L.; Zhao, X.; Zhang, Q.L.; Xu, Y.L.; Ding, J.M.; Xiong, H.B.; Ma, R.J.; Lv, X.L.; et al. Long-term structural performance monitoring system for the Shanghai Tower. *J. Civ. Struct. Health Monit.* **2013**, *3*, 49–61. [CrossRef]
15. Dong, L.; Wang, H.; Wang, G.; Qiu, W. A wireless multifunctional monitoring system of tower body running state based on MEMS acceleration sensor. In Proceedings of the 2018 19th International Symposium on Quality Electronic Design (ISQED), Santa Clara, CA, USA, 13–14 March 2018; pp. 357–363.
16. König, S.; Leinfelder, P. First results with MEMS tilt sensors on bridges. In Proceedings of the Intertial Sensors and Systems (ISS), Karlsruhe, Germany, 20–21 Septenber 2016; pp. 1–15.
17. Leavitt, J.; Sideris, A.; Bobrow, J.E. High bandwidth tilt measurement using low-cost sensors. *IEEE/ASME Trans. Mech.* **2006**, *11*, 320–327. [CrossRef]
18. Liu, Y.; Noguchi, N.; Ishii, K. Development of a low-cost IMU by using sensor fusion for attitude angle estimation. *IFAC Proc. Vol.* **2014**, *47*, 4435–4440. [CrossRef]
19. Li, C.; Azzam, R.; Fernández-Steeger, T.M. Kalman Filters in Geotechnical Monitoring of Ground Subsidence Using Data from MEMS Sensors. *Sensors* **2016**, *16*, 1109. [CrossRef] [PubMed]

20. Yean, S.; Lee, B.S.; Yeo, C.K.; Vun, C.H. Algorithm for 3D orientation estimation based on Kalman filter and gradient descent. In Proceedings of the 2016 IEEE 7th Annual Information Technology, Electronics and Mobile Communication Conference , Vancouver, BC, Canada, 13–15 October 2016; pp. 1–6.
21. Ren, Y.; Ke, X. Particle filter data fusion enhancements for MEMS-IMU/GPS. *Intell. Inf. Manag.* **2010**, *2*, 417. [CrossRef]
22. Lim, J.; Hong, D. Cost reference particle filtering approach to high-bandwidth tilt estimation. *IEEE Trans. Ind. Electron.* **2010**, *57*, 3830–3839. [CrossRef]
23. Lee, H.S.; Hong, Y.H.; Park, H.W. Design of an FIR filter for the displacement reconstruction using measured acceleration in low-frequency dominant structures. *Int. J. Numer. Methods Eng.* **2010**, *82*, 403–434. [CrossRef]
24. Park, J.W.; Sim, S.H.; Jung, H.J. Displacement estimation using multimetric data fusion. *IEEE/ASME Trans. Mech.* **2013**, *18*, 1675–1682. [CrossRef]
25. Liu, C.; Park, J.W.; Spencer, B., Jr.; Moon, D.S.; Fan, J. Sensor fusion for structural tilt estimation using an acceleration-based tilt sensor and a gyroscope. *Smart Mater. Struct.* **2017**, *26*, 105005. [CrossRef]
26. Shi, L.; He, Y.; Luo, Q.; He, W.; Li, B. Tilt Angle On-Line Prognosis by Using Improved Sparse LSSVR and Dynamic Sliding Window. *IEEE Trans. Instrum. Meas. IM* **2018**, *67*, 296–306. [CrossRef]
27. Testoni, N.; Aguzzi, C.; Arditi, V.; Zonzini, F.; De Marchi, L.; Marzani, A.; Cinotti, T.S. A Sensor Network with Embedded Data Processing and Data-to-Cloud Capabilities for Vibration-Based Real-Time SHM. *J. Sens.* **2018**, *2018*, 2107679. [CrossRef]
28. STMicroelectronics. *iNEMO Inertial Module: Always-On 3D Accelerometer and 3D Gyroscope*; Technical Report; STMicroelectronics: Geneva, Switzerland, 2017.
29. Khaleghi, B.; Khamis, A.; Karray, F.O.; Razavi, S.N. Multisensor data fusion: A review of the state-of-the-art. *Inf. Fusion* **2013**, *14*, 28–44. [CrossRef]
30. Lahat, D.; Adali, T.; Jutten, C. Multimodal data fusion: an overview of methods, challenges, and prospects. *Proc. IEEE* **2015**, *103*, 1449–1477. [CrossRef]
31. Kok, M.; Hol, J.D.; Schön, T.B. Using inertial sensors for position and orientation estimation. *arXiv* **2017**, arXiv:1704.06053.
32. Smith, J.O. *Spectral Audio Signal Processing*; W3K Publishing: Palo Alto, CA, USA, 2011.
33. Graff, K.F. *Wave Motion in Elastic Solids*; Courier Corporation: Chelmsford, MA, USA, 2012.

electronics

MDPI

Article

A Robust Registration Method for Autonomous Driving Pose Estimation in Urban Dynamic Environment Using LiDAR

Rendong Wang [1], Youchun Xu [1,*], Miguel Angel Sotelo [2], Yulin Ma [3], Thompson Sarkodie-Gyan [4], Zhixiong Li [5,*] and Weihua Li [5]

1. Department of Military Vehicles, Military Transportation University, Tianjin 300161, China; rendongwang.army@gmail.com
2. Department of Computer Engineering, University of Alcalá, 28801 Alcalá de Henares (Madrid), Spain; miguel.sotelo@uah.es
3. Suzhou Automotive Research Institute, Tsinghua University, Suzhou 215134, China; myl@itsc.cn
4. Laboratory for Industrial Metrology and Automation, College of Engineering, University of Texas, El Paso, TX 79968, USA; tsarkodi@utep.edu
5. School of Mechanical, Materials, Mechatronic and Biomedical Engineering, University of Wollongong, Wollongong, NSW 2522, Australia; weihuali@uow.edu.au

* Correspondence: youchunxu.army@gmail.com (Y.X.); zhixiong_li@uow.edu.au (Z.L.); Tel.: +1-515-203-5578 (Z.L.)

Received: 8 November 2018; Accepted: 18 December 2018; Published: 1 January 2019

Abstract: The registration of point clouds in urban environments faces problems such as dynamic vehicles and pedestrians, changeable road environments, and GPS inaccuracies. The state-of-the-art methodologies have usually combined the dynamic object tracking and/or static feature extraction data into a point cloud towards the solution of these problems. However, there is the occurrence of minor initial position errors due to these methodologies. In this paper, the authors propose a fast and robust registration method that exhibits no need for the detection of any dynamic and/or static objects. This proposed methodology may be able to adapt to higher initial errors. The initial steps of this methodology involved the optimization of the object segmentation under the application of a series of constraints. Based on this algorithm, a novel multi-layer nested RANSAC algorithmic framework is proposed to iteratively update the registration results. The robustness and efficiency of this algorithm is demonstrated on several high dynamic scenes of both short and long time intervals with varying initial offsets. A LiDAR odometry experiment was performed on the KITTI data set and our extracted urban data-set with a high dynamic urban road, and the average of the horizontal position errors was compared to the distance traveled that resulted in 0.45% and 0.55% respectively.

Keywords: intelligent vehicles; LiDAR odometry; range sensing; simultaneous localization and mapping (SLAM)

1. Introduction

Pose estimation is one of the key technologies for autonomous driving in the urban environment. Considering the fact that the GPS is unable to keep high accuracy in urban environments, a very common solution to this problem has emerged in recent years. Currently, the most common solution is based on the pose provided by GPS/INS, where real time data acquired from LiDAR or camera are registered with a priori map to achieve accurate localization [1,2]. In fact, in an environment that does not have a map, pose transform can be estimated by registration between the front and back frames in which the vehicle's global pose is calculated by the accumulation of the previous transforms, that is odometry or the front-end of SLAM (simultaneous localization and mapping) [3,4].

The key challenges of point cloud registration in the urban environment involve two aspects: In the first instance, there are large numbers of moving objects, such as vehicles and pedestrians, and various static surroundings (e.g., the number and position of vehicles parking on the roadside vary at different times, and trees or green belts will vary greatly in different seasons) whose changes and shadows greatly increase the difficulty of registration. The second instance involves tall buildings, overpasses, and trees that may lead to multipathing and shadowing. These attributes characterize the most important factors of GPS error, whereby should the precision of INS be not high enough, will lead to the easy generation of large pose error in order to make registration hard.

ICP and its improved algorithms have very high registration accuracy, but they can easily fall into a local optimum with large noise and high initial error. Therefore, it is necessary to use coarse registration to eliminate the outliers and provide a better initial pose for accurate registration. A common coarse registration method is the RANSAC [5], which is mainly used to solve model estimation problems of large amounts of outliers in data sets. There have been many contributions towards its improvement from feature extraction and the RANSAC algorithm itself [5–7].

In this paper, the authors propose a robust solution with the generation of more efficient sample sets and employing a novel framework of the RANSAC. In Section 2 of this paper, we discuss related work and how our work is unique compared to the state of the art. In Section 3, our segmentation strategy and its merits are introduced and the principle of our registration method is analyzed and a novel framework of the RANSAC is proposed. Experimental results are shown in Section 4, and the conclusion and future work are discussed in Section 5.

2. Related Work

Up to today, there has been tremendous contribution towards the further development of pose registration based on LiDAR registration. A state-of-the-art methodology of the LiDAR odometry and SLAM has emerged as the 'LiDAR odometry and mapping in real-time' (LOAM) method [3]. This method is capable of achieving fast and accurate 3D motion estimation and works well in static and low dynamic urban scenes, but exhibits some difficulty in obtaining good results in highly dynamic scenes. In order to overcome the interference of dynamic objects, the registration is combined with detection and tracking of moving object (DATMO). Moosmann and Stiller [8] used an object tracking framework, while Yang et al. [9] used a registration framework (multiple-model RANSAC) as contribution towards this methodology. Although their methods are different, both of them integrated dynamic and static objects into one framework, and obtained good results in sequential frames. However, as time intervals between frames increase, the match of the corresponding objects becomes more difficult [10]. Furthermore, satellite positioning results are usually used as the initial pose of point cloud registration. However, satellite positioning will drift and fluctuate due to the occlusion of GPS signals from buildings and trees in cities, resulting in larger initial position and attitude errors [11]. This will make it more difficult to search and register the corresponding points in the point cloud. All of these factors will make the frame registration and object tracking worse. Particularly, this method cannot be employed in localization using priori maps.

In aspects of map-based localization, Levinson et al. [1] used the LiDAR intensity information on the road to construct high precision map in order to realize the vehicle's localization. This method can effectively overcome dynamic interference, but it is hard to adapt to rainy or snowy weather. Wolcott et al. [2] proposed a multiresolution Gaussian mixture map based on z-height information to realize the accurate localization in different weather conditions. The method may also be used in low dynamic scenes but the a priori map is required to be static. Therefore, the feature extraction of static objects is usually needed during the process of map construction and real-time registration. There are two main methods that may be considered; one method involves the application of machine learning methods for the recognition of moving and static objects [12], whilst the other involves the static object detection by extracting features such as the vertical corner features of buildings [13], and line features of curbs [14] and road lanes [15], respectively. However, in urban areas where moving objects are crowded and without a prior map, large amounts of occlusions between objects and more sparse point clouds (e.g., using LiDAR with fewer beams or detecting objects at longer distances) will make both recognition and feature detection difficult. Although deep learning can achieve good results, it needs a lot of labeled data to train and consumes a lot of computing resources at runtime.

The above research contributions have mainly focused on the problem of dynamic interference. However, in order to achieve robust pose estimation of autonomous driving, both odometry and map-based localization may be required, and the interference of both dynamic objects and high initial pose error may be considered.

Although 4PCS [16,17] and its improved methods can overcome larger initial error, in the environment where large numbers of dynamic objects exist, it is difficult to find four pairs of corresponding static objects from two frame point clouds by random sampling, because of the fact that occlusion makes static and dynamic objects hard to distinguish, and the fact that the point clouds are sparse and have large density differences.

One method to overcome high initial pose error is usually dealt with in most localization systems by keeping the vehicle immobile (no movement) and running a particle filter to launch several potential solutions around the initial one. The algorithm is run until there is convergence. Since it uses hundreds or thousands of particles, it usually converges to a very accurate result even if the initial error is high. Upon localization convergence, the vehicle can start to move.

However, two problems exist in particle filtering. First, when the initial error is high, the convergence time to the right pose is long. In practical applications, it becomes impossible to tolerate long waiting times. Second, in changing dynamic scenes, the surroundings change all the time, and are different from the historical map data. When the particle filter is applied, there arises the risk of converging to the wrong pose. This phenomenon in urban roads is particularly prominent, such as matching to the opposite lane.

In this paper, both the dynamic interference and high initial pose error are considered. We improve the traditional methods from the object segmentation and registration. In object segmentation, three constraint conditions are adopted to optimize region growing method to get more stable object centroid. In registration, all objects are processed in a same framework irrespective as to whether they are dynamic or not. A novel framework with "RANSAC-iteration-RANSAC" three nested layers is proposed as an improvement of the RANSAC algorithm. The proposed registration method can provide higher quality solutions to any localization system, even in cases of high initial errors. Thus, particle filter-based localization methods will greatly benefit from this method both in terms of localization accuracy and computation time (i.e., less particles will be needed and convergence time will consequently decrease).

3. Proposed Method

Currently, the point cloud scanned by a LiDAR on a self-driving car is usually a sparse point cloud with large scale, and point density of each beam decreases with the increase of distance to the sensor. In addition, the distribution and density of point clouds in the same region will be different due to the change of observation pose, which further increases the difficulty of registration. Compared with local features of each scanning point, registration based on object features may be more robust. Because the number of objects is far less than the number of scanning points, the amount of calculation will be much smaller. Therefore, first of all, a complete frame of point cloud data needs to be divided into independent object point clouds.

In order to overcome the moving objects, occlusions and large initial pose errors, we improve the point cloud segmentation method to obtain more stable observation centroid of the object. A more robust coarse registration based on these centroids is achieved by improving the RANSAC algorithm. Through rough registration, the outliers can be eliminated and the pose error can be reduced greatly. On this basis, we use ICP to accurately register the point cloud data corresponding to the interior centroids.

3.1. Object Segmentation

A point cloud scanned by a multi-beam LiDAR usually contains ground and obstacle data. In this paper, the method expressed in [18] is firstly used to remove the ground. Then, the obstacle point cloud is clustered and segmented. A simple, yet effective method is region growing. Its basic idea is to merge similar areas together, likely to DBSCAN, which can cluster point clouds of arbitrary shape. However, compared with DBSCAN, the advantage of region growing is that it uses neighborhood search does not need to build a search tree, and it does not need to search for the nearest point, so it is efficient, and is less affected by the change of point cloud density. The simplest region growing is to vote three-dimensional point clouds into two-dimensional grids containing only x and y information, and cluster them according to whether there are obstacle point clouds in the grids. The result is shown in Figure 1b, where different objects are distinguished with different colors. Obviously, when two objects close to each other, they are likely to be clustered into one object.

In this paper, we still use two-dimensional grids. The height of point cloud is regarded as an attribute of two-dimensional grid [19], and the region growing algorithm is optimized in two aspects based on this attribute:

- Unstable objects, such as low curbs and branches of trees, are eliminated directly by the threshold of both grid height and object height. The gird height is calculated by maximum height and minimum height difference of points in one grid, and the object height is calculated by maximum height and minimum height difference of grids in one object.
- Objects such as the trees, street lamps, and road signs in the environment are stable, feature-obvious, and not easily shadowed. They are important features in the registration. The height of such objects is usually very different from that of the surrounding objects, so they can be separated from other objects according to the height difference between adjacent grids.

There may be a lot of feature information about the object, but we hope to be able to complete the matching through as simple information as possible. Therefore, in this paper we use the observation centroid of the object.

The observation centroid of ith object (x_i, y_i) in a frame is computed by Equation (1), where n_i represents the number of points contained in the object.

$$x_i = \frac{1}{n_i}\sum_{j=1}^{n_i} x_j \text{ and } y_i = \frac{1}{n_i}\sum_{j=1}^{n_i} y_j, \quad (1)$$

Because the point cloud obtained by LiDAR through one scan is only a part of the surface of the object, to be exact, the part facing the sensor, which has an inconsistent spatial relation to the actual centroid of the object. This will lead to some new problems, for example, the observation centroid of an object may change because of the occlusion of other objects or different observation positions. In this paper, the large objects in the environment are compulsively split into multiple small objects, such as buildings, road barriers, green belts, etc. Here are three advantages: (1) The centroid of a large object is more prone to fluctuate because of occlusion generated by dynamic object, and with the movement of observation position, its centroid may also change greatly, whilst the centroid of the small object is more stable, and has less sensitivity to occlusion. (2) When a large object is divided into several small objects, it could correspond to more centroid. This will increase its proportion in the sample set. Thus, large objects will have greater weight in the registration. (3) In scenes where obvious features are deficient, such as expressways and highways, this will increase the static object ratio, which is helpful to improve the success rate and the reliability of the registration.

The improved segmentation results are shown in Figure 1c. After object segmentation and centroid extraction, the data quantity of the point cloud is significantly decreased. In most urban scenes, the quantity of the centroid in a frame segmented by our method is usually less than 200.

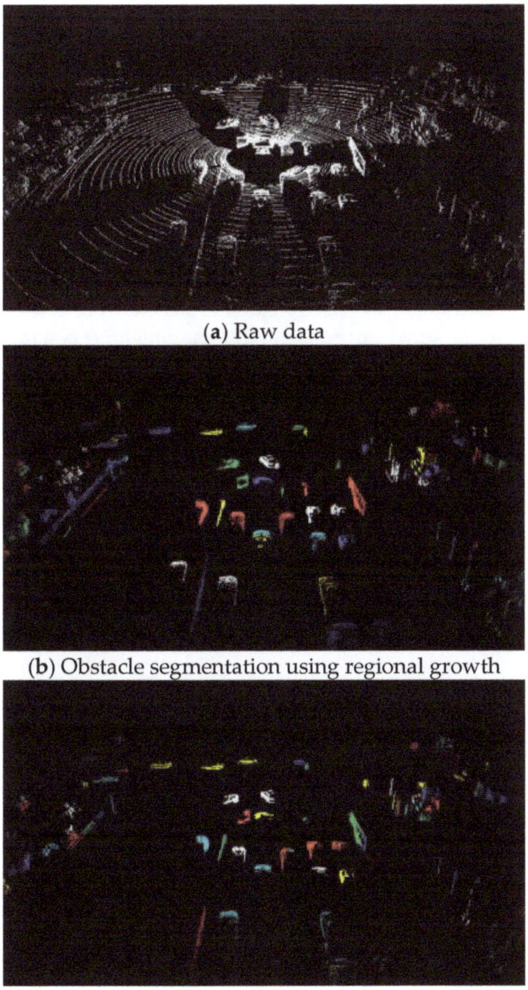

(a) Raw data

(b) Obstacle segmentation using regional growth

(c) Improved segmentation

Figure 1. Object segmentation. (**a**) Raw data. (**b**) Obstacle segmentation using regional growth. (**c**) Improved segmentation.

In summary, in order to obtain more stable observation centroids, we optimize the traditional two-dimensional region growing algorithm in three aspects:

- height information is introduced to denoise the point cloud according to the object height
- more reasonable point cloud segmentation using height difference between grids
- compulsive segmentation of large objects

3.2. Registration

In the process of registration-based localization and odometry, the vehicle's two-dimensional horizon position and heading angle are the most important, yet the most varying elements amongst the six degrees of freedom pose information. In this paper, the coarse registration of these three elements in the horizontal plane is first carried out, and then, accurate registration based on ICP is employed in inliers of 3D point clouds [20].

3.2.1. Overview

A 2D transform needs a minimum of two pairs of corresponding points. This implies that we should find two centroid points of static objects p_1, p_2 in the frame, A, and their corresponding points q_1, q_2 in the frame, B, respectively. However, it is difficult to distinguish between which objects are moving and which objects are static when the sequence of frames is weak and the initial pose is unreliable.

We use RANSAC to solve this problem, as shown in Figure 2. At first, we regarded all objects as moving. Under the consideration that both their moving speed and direction are different, the objects may be divided into several categories. (1) For static objects, the consensus of their moving status makes them classified into the same category. (2) The moving objects with different velocities and directions may be classified into different categories. Therefore, the main hypothesis of this manuscript is that even if the total number of moving objects is greater than the total number of static objects (in most situations the number of objects in the static category is still greater than the number of objects in the other categories), the transformation matrix that has the largest number of inliers found by RANSAC is the transformation matrix of the static class.

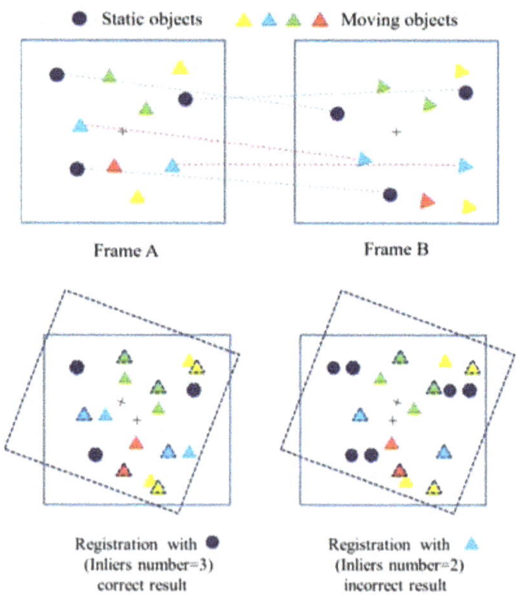

Figure 2. Schematic diagram of RANSAC registration in dynamic environment.

In this scenario, there are three static and seven dynamic objects, and the different colors represent different moving status. Although the total number of moving objects is more than that of static objects, the maximum inliers of the transformation with a set of moving objects which have the same moving status is still less than that with static objects.

3.2.2. Data Association

According to [16], the distance between two points in the same frame should be neither too close nor too far away. If the distance is too close, the error of the transformation matrix will probably increase. If the distance is too far, the corresponding points in the two frames with smaller overlapping may not be found. So, first of all, we choose the moderate two points p_1, p_2 in the frame A. In this paper, the distance between p_1 and p_2 is limited between 10 m and 40 m. Then, the K nearest neighbor points of p_1 in frame B are searched, forming a candidate corresponding points set Q_1 of p_1. Points satisfied with Equation (2) are searched to form candidate corresponding points set Q_2 of p_2.

$$Q_2 = \{q_2 | | \|q_1 - q_2\| - \|p_1 - p_2\| | < \varepsilon, \\ q_1 \in Q_1 \subset \mathrm{B}, q_2 \in \mathrm{B}, p_1 \in \mathrm{A}, p_2 \in \mathrm{A}\} \tag{2}$$

For each candidate pair, the transformation matrix is solved by the least squares method, and B is rotated and translated according to that. Then, the distance between each point in the transformed frame B and its nearest point in frame A is calculated, and the number of inliers whose distance is less than the threshold is counted. The candidate pair which has the greatest number of inliers is chosen as the corresponding pair.

3.2.3. Multi-Layer RANSAC

Assuming that p_A is the ratio of points that meet the transformation of frame A and B in frame A, then, a set of points (n points) is randomly selected in A, and the probability that the transformation between these points and their corresponding points is the real transformation is expressed as

$$p = {p_A}^n \tag{3}$$

After m times of the above sampling process, the probability of obtaining at least one correct result is

$$P_m = 1 - (1 - p)^m = 1 - (1 - {p_A}^n)^m \tag{4}$$

To make this probability greater than p_0, namely

$$1 - (1 - p)^m > p_0 \tag{5}$$

The number of sampling times required is

$$m > \log(1 - p_0) / \log(1 - p_A^n) \tag{6}$$

From Equation (6), it can be seen that the more the number of points needed to extract in a single random sample, the smaller the probability of getting the correct registration relationship, and the more the number of samples required. Therefore, on the premise that the model can be solved, the number of sampling points should be reduced as much as possible.

When initial pose error is small, using method in Part B is easy to get the true corresponding point, and the traditional RANSAC algorithm can register well. However, when the initial error—especially rotation error—increases, the ratio of the static points is constant; but the probability that the corresponding points exist in the set of K nearest neighbor points will significantly reduce, so the success rate of the registration using the traditional RANSAC will decrease. In order to obtain reliable registration results in the presence of high initial errors, a multi-layer nested framework of RANSAC is proposed. The framework consists of three nested layers, as shown in Algorithm 1.

The inner layer is the traditional RANSAC algorithm, where M is the sampling time in frame B. The middle layer of the algorithm is an iterative layer. The best result is selected to update the position of B, and then the RANSAC process is repeated in a new pose. It should be noticed that the aim of N iterations is to reduce the pose error rather than to filter the outliers, and instead of using the inliers, the sample process in each iteration is performed on the whole point set of B. Therefore, the iteration only changes the pose of B. This is to avoid unsuitable filter eliminating useful points. The outer layer is a RANSAC process that nests the above two layers. It is repeated L times until the result satisfies the termination condition.

This framework involves two significant parameters, one is the termination condition ξ and the other parameter is the search range K. ξ is the threshold of the inliers proportion which is used to determine the registration. If ξ is too large, the algorithm becomes time-consuming, while with too small ξ the algorithm may not obtain good results. In this paper, we use the registration error and the algorithm running time to construct loss function, and take large numbers of the urban high dynamic scenes and low dynamic scenes as samples to train this parameter. The quasi Newton method is used to minimize the sample mean of the loss function, and the optimal solution $\xi = 0.3$ is obtained.

Algorithm 1 Multi-layer RANSAC

Step Description of the implementation of the Multi-layer RANSAC algorithm

Start:
K: KNN search parameter
N_{cp}: Number of correspondences in set A and B_j
δ: Threshold of inliers ratio
For $i = 0$ **to** L **do**
    $K = K(i)$
    For $j = 0$ **to** N **do**
        $M = M(j)$
        For $k = 0$ **to** M **do**
            RANSAC(A, B_j)
          End for
        Calculating T_j with the largest number of inliers
        $T = T \cdot T_j$
    $R_j = \max(n_k)/N_{cp}$
        If $R_j > \delta$ then
          Return T
        Else
          $B_{j+1} = T_j\,(B_j)$
        End if
      End for
End for
RANSAC(A, B_j):
      Random sample p_i in A
      Search correspondences of p_i in B_j by KNN
      Pick the points that satisfy conditions
      Calculate matrix of each corresponding pairs
      Transform B_j and count the number of inliers
End.

K is related to the initial pose error of the registration and should be increased as the initial error increases. However, it is difficult to estimate the range of the initial error in practical applications just as the value of K. If K is too big, both the time consumption of the algorithm and the risk of the registration failure will increase. If K is too small, the correct registration results cannot be obtained when the initial error is high. In order to solve this problem, we dynamically adjust K by increasing it in the outer layer and decreasing it in the middle layer, respectively. The initial value of the middle layer is provided by the outer layer. Using this strategy, our algorithm can automatically adapt to various scenes and various initial errors in less time than the method of changing K artificially.

The variables in the algorithm are defined as follows.

S_A is the set of the static objects in set A; S_B is the set of the static objects in set B; a_i is any point in set A; b_i is any point in set B; B_k represents the set B after k-th iterations; $b_i^{(k)}$ is any point in B_k; in set B, q_i is the actual corresponding point of p_i in set A; $q_i^{(k)}$ is the candidate corresponding point in set B_k, Q

is the set of candidates corresponding points; $\widetilde{q}_i^{(k)}$ is the optimum corresponding point that computed by Equations (7) and (8),

$$n_j^{(k)} = \begin{cases} 1, & \|a_j - b_j^{(k)}\| < \varepsilon \\ 0, & else \end{cases} \tag{7}$$

$$\widetilde{q}_i^{(k)} = \underset{q_i^{(k)}}{\operatorname{argmax}} \left\{ \sum_{a_j \in A} n_j^{(k)} \right\} \tag{8}$$

where $e^{(0)}$ is the initial pose error and $e^{(k)}$ is the registration error after kth iteration of the middle layer.

Compared with the traditional RANSAC algorithm, with the same number of samples, the probability of the improved algorithm to achieve the correct registration results is significantly improved, and the proof process is as follows:

According to Bayes formula,

$$\begin{array}{l} P(\widetilde{q}_i^{(k)} = q_i | e^{(k)}) \\ = \frac{P(q_i \in Q | e^{(k)}) \cdot P(\widetilde{q}_i^{(k)} = q_i | q_i \in Q, e^{(k)})}{P(q_i \in Q | \widetilde{q}_i^{(k)} = q_i, e^{(k)})} \end{array} \tag{9}$$

Because all the points in set Q will be iterated through to search the optimum corresponding point,

$$\begin{array}{l} P(\widetilde{q}_i^{(k)} = q_i | q_i \in Q, e^{(k)}) \\ = P(q_i \in Q | \widetilde{q}_i^{(k)} = q_i, e^{(k)}) = 1 \end{array} \tag{10}$$

When the error $e^{(k)}$ increases, especially when the rotation error is higher, the probability of the existence of the corresponding points q_i in the KNN set Q will greatly decrease, so will the value of $P(\widetilde{q}_i^{(k)} = q_i | e^{(k)})$. Therefore, $e^{(k)}$ should be reduced as much as possible to increase the likelihood of finding a corresponding point, which can be achieved through iteration.

In the traditional RANSAC algorithm, the correct registration results can be obtained only when p_i is the static point and the optimal corresponding point $\widetilde{q}_i^{(k)}$ is real corresponding point q_i. Its probability is represented by P_0. However, in the iterative process of this method, we just need the transformation matrix calculated by p_i and $\widetilde{q}_i^{(k)}$ to reduce the initial error, i.e., $e^{(k)} < e^{(0)}$. In the first iteration,

$$\begin{array}{l} P_1 = P(e^{(1)} < e^{(0)} | e^{(0)}) \geq P(\widetilde{q}_i^{(k)} = q_i | e^{(0)}) \\ > P(\widetilde{q}_i^{(k)} = q_i | e^{(0)}) \cdot P(p_i \in S_A | A) = P_0 \end{array} \tag{11}$$

From Equation (11), we can see that P_1 is bigger than P_0, especially in the environment with more moving objects. The advantage of this method is obvious because the value of $P(p_i \in S_A | A)$ is small.

According to Equation (6), the sampling times m_k in the kth iteration should be satisfied with Equation (12).

$$P(e^{(k)} < e^{(0)} | m_k) = 1 - (1 - P_k)^{m_k} > P(p_i \in S_A | A) \tag{12}$$

And in the next iteration,

$$\begin{array}{l} P_k = P(e^{(k)} < e^{(k-1)} | e^{(k-1)}) \\ > P(e^{(k)} < e^{(k-1)}, e^{(k-1)} < e^{(0)} | e^{(k-1)}) \\ = P(e^{(k)} < e^{(k-1)} | e^{(k-1)}, e^{(k-1)} < e^{(0)}) \cdot P(e^{(k-1)} < e^{(0)} | m_{k-1}) \\ \geq P(\widetilde{q}_i^{(k)} = q_i | e^{(k-1)}, e^{(k-1)} < e^{(0)}) \cdot P(e^{(k-1)} < e^{(0)} | m_{k-1}) \\ > P(\widetilde{q}_i^{(k)} = q_i | e^{(0)}) \cdot P(e^{(k-1)} < e^{(0)} | m_{k-1}) \\ > P(\widetilde{q}_i^{(k)} = q_i | e^{(0)}) \cdot P(p_i \in S_A | A) = P_0 \end{array} \tag{13}$$

Therefore, after kth iterations, the probability of registering success is

$$P_s = 1 - \prod_{i=1}^{k} (1 - P_i)^{m_i} > 1 - (1 - P_0)^{\sum_{i=1}^{k} m_i} \tag{14}$$

From the above deduction, it may be observed that within a certain range, the more moving objects, the greater the initial error; and the more iterations, the more obvious the advantages of this method than traditional methods.

The outer layer of our algorithm is a RANSAC process regarded as the middle layer as a unit. This layer can effectively improve the registration success rate when the initial error is high by dynamically adjusting the parameters. Although many iterations may be required during this process, the algorithm is still efficient in most cases due to the small amount of points used for registration.

3.2.4. Accurate Registration

The set B is transformed according to the coarse registration matrix $\boldsymbol{T}$, and the data in the centroid set A and B are divided into inliers and outliers according to the threshold distance of the corresponding points. Then, the segmented point cloud in both P_A and P_B corresponding to the inliers of the centroid set is extracted to consist of P'_A and P'_B, respectively. Finally, weighted ICP is implemented for accurate 3D registration of point cloud in P'_A and P'_B with the initial transformation $\boldsymbol{T}$. Different from the coarse registration, the small and high objects, which are always pole objects such as trees, street lamps, and road signs, are greatly weighted in accurate registrations. Because they are less changeable and the distribution of their point clouds is more concentrated, weighted ICP is more accurate than standard ICP, especially in the registration between the current frame and the historical frame for localization.

4. Experiments

The experiment includes two parts. Part A is the registration on typical urban dynamic scenes with various initial pose errors. It includes registration between sequence frames which is the basis of LiDAR odometry, and the registration between the current frame and the historical frame which is the basis of localization with an a priori map. Part B is the LiDAR odometry experiment in the public dataset provided by KITTI [21] and a data set of urban highly dynamic environments collected by ourselves. For more experimental result, please see the Supplementary video.

Our LiDAR data is collected by velodyne HDL-64E. During the experiments, the algorithms processing the LiDAR data run on a computer with 2.4 GHz quad cores and 8 G memory, on top of the ubuntu14.04 in Linux. The algorithm consumes only one thread.

4.1. Registration

This experiment consists of two groups: registration between sequence frames (short time interval, 0.5 s) and registration between current frame and historical frame (long time interval, three months, winter and spring). Experiments were carried out in large numbers of urban scenes. For each scene, the actual distance between two frames is about 5 m.

The position and heading of the vehicle provided by the RTK-GPS are regarded as the real coordinates, expressed in the form (x, y, θ). An offset $(\Delta x, \Delta y, \Delta\theta)$ is added to the coordinates of one frame as the initial pose of registration. By adjusting the offset distance $|\Delta d|$ ($\sqrt{\Delta x^2 + \Delta y^2}$, units: meter) and offset angles $|\Delta\theta|$ (units: degree), we set four sets of initial errors: The ranges of distance offset are [0, 4], [6, 10], [14, 18], [24, 28]; and the ranges of the angle offsets are [0, 5], [5, 10], [0, 15], [15, 20]. They are denoted as (4 m, 5°), (10 m, 10°), (18 m, 15°), and (28 m, 20°), respectively.

To demonstrate the effectiveness of our work, both segmentation and registration algorithms are compared before and after our improvement. The segmentation is represented as seg0 and seg1, and the registration is represented as RANSAC0 and RANSAC1. They can form four methods, i.e., seg0 + RANSAC0, seg0 + RANSAC1, seg1 + RANSAC0, seg1 + RANSAC1 (proposed). For each scene and each set of initial error, each method is implemented 100 experiments. The initial offset in each experiment is randomly generated within the range of initial error. All the experiments in this paper use the same parameters.

Here, we selected four typical highly urban dynamic scenes in each group for the analysis. Figures 3–10 show the scene registration results at the initial error (28 m, 20°) using the proposed method. In Figures 3–10, (a) is the top view of two obstacle point clouds with an initial offset (they are respectively shown as red and green); (b) is the top view after segmentation and registration (points belong to inliers are shown as blue and most of them are points of overlapping static objects, with moving objects and non-overlapping static objects eliminated effectively); (c) is a local enlarged 3D view of (b), which shows that the static objects, such as buildings, trees and parked vehicles, are registered well.

The four scenes for registration between sequence frames are denoted as scene A1, A2, A3, and A4. A1 is a crossroad with the experimental car turning right; A2 is a crossroad with the experimental car turning left; A3 is a straight narrow road with numbers of static objects occluded by a bus; and A4 is a traffic congestion road with the experimental car changing lane.

The four scenes for registration between current frame and historical frame are denoted as scene B1, B2, B3, and B4. B1 is a straight urban road, with many parked cars whose position has changed greatly; B2 is a crossroad, and our experimental car collected data of these two frames in opposite lanes and directions; B3 is a ramp entrance, in which there are amounts of vegetation changing greatly from winter to spring, and fences existing in winter but not in spring; and B4 is an elevated road lacking geometric features.

Both groups of scenes have numbers of moving objects, such as vehicles and pedestrians.

The registration effectiveness is determined by the deviation between the registration pose and its true pose. However, in practical applications, the true pose of a frame is unknown. In this paper, the ratio of the inliers obtained by the RANSAC is used as an evaluating indicator for the success of the registration. When the ratio of the inliers is greater than the threshold ξ, the coarse registration is considered to be successful.

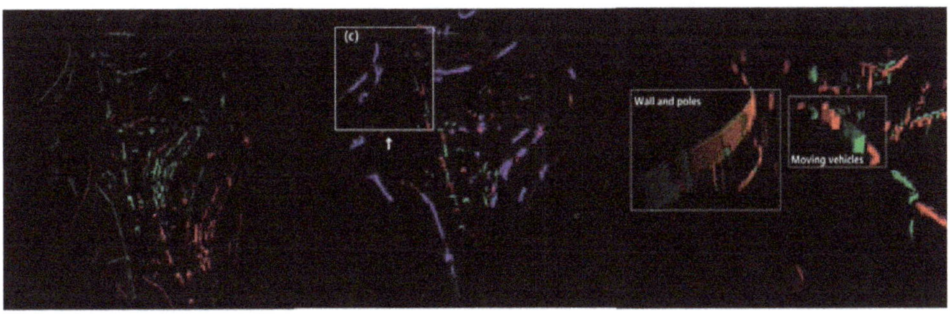

(**a**) Initial pose (**b**) Registration result (**c**) Local 3D display

Figure 3. Scene A1.

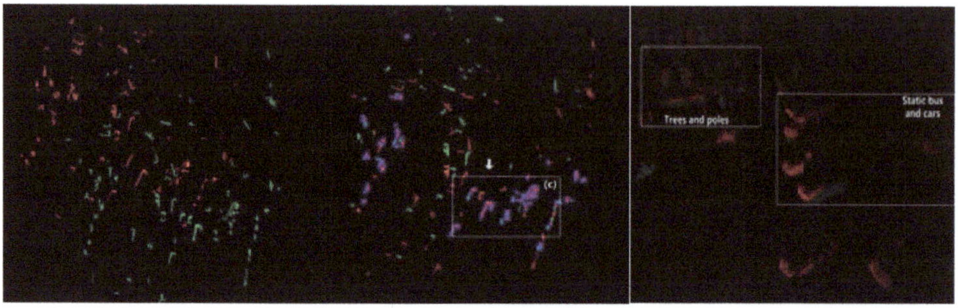

(**a**) Initial pose (**b**) Registration result (**c**) Local 3D display

Figure 4. Scene A2.

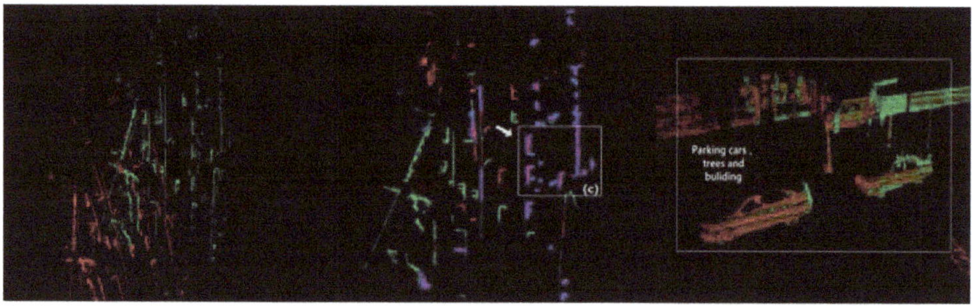

(**a**) Initial pose (**b**) Registration result (**c**) Local 3D display

Figure 5. Scene A3.

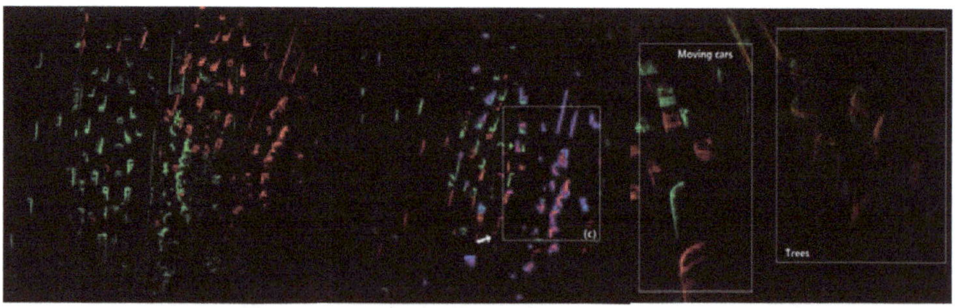

(**a**) Initial pose (**b**) Registration result (**c**) Local 3D display

Figure 6. Scene A4.

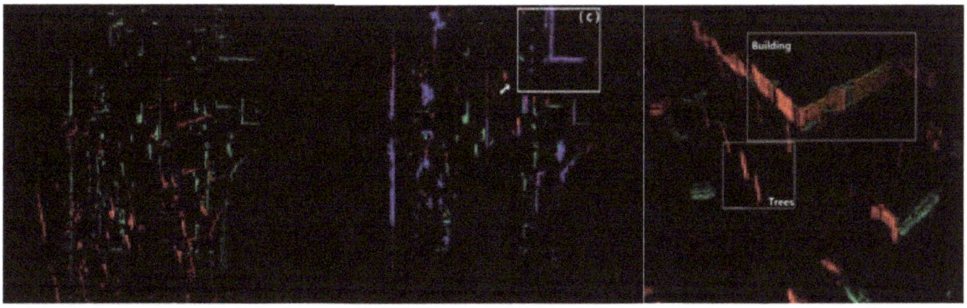

(**a**) Initial pose (**b**) Registration result (**c**) Local 3D display

Figure 7. Scene B1.

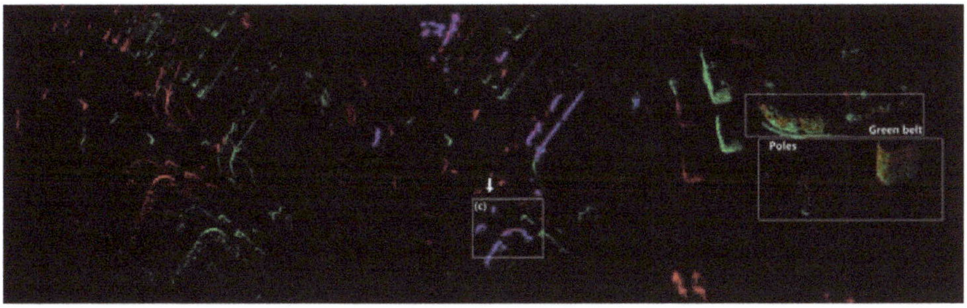

(**a**) Initial pose (**b**) Registration result (**c**) Local 3D display

Figure 8. Scene B2.

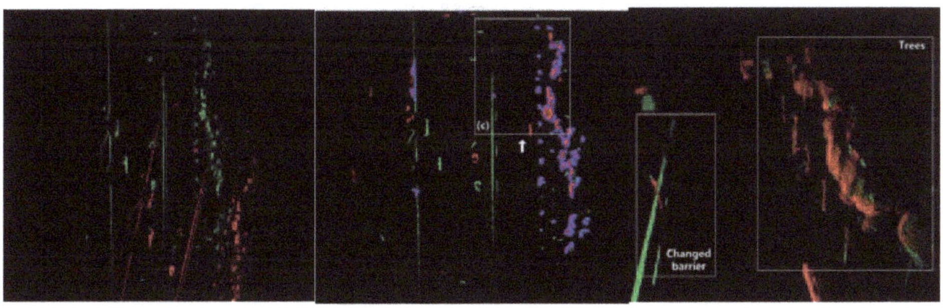

(**a**) Initial pose (**b**) Registration result (**c**) Local 3D display

Figure 9. Scene B3.

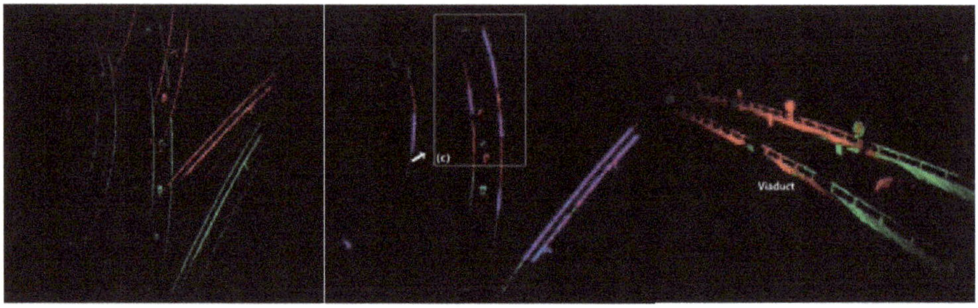

(**a**) Initial pose (**b**) Registration result (**c**) Local 3D display

Figure 10. Scene B4.

In order to verify the rationale of the evaluation method, the 12,800 experimental results of both groups are statistically analyzed, and their error distributions are demonstrated in Figure 11. The x-axis is the distance error between the registration result and the real position, whereas the y-axis is the heading error. In order to represent the distribution of the small error data more clearly, a logarithmic coordinate system is applied. Even scenes, initial offsets and registration algorithms are all different, the error of the registrations recognized as success are obviously smaller than the errors of registrations recognized as failure. Table 1 indicates the relationship between the error and the classification of the registration result in all tests. It may be observed that the majority of the registrations judged as success have errors less than (0.2 m, 0.2 m, 0.5°), and the majority of registrations judged as failures have errors more than (0.2 m, 0.2 m, 0.5°), respectively. There are only a few results that are misjudged. This includes results whose registration error is high whilst the algorithm considers it as a successful registration. In fact, most of these misjudgments occur in the experiments of the high initial error of the scene B4. This is because the static features in this scene are single, and the number of features is very few, which results in the algorithm considering the wrong matching as the correct result, which leads to the big error of our pose estimation. This is also a shortcoming of our method.

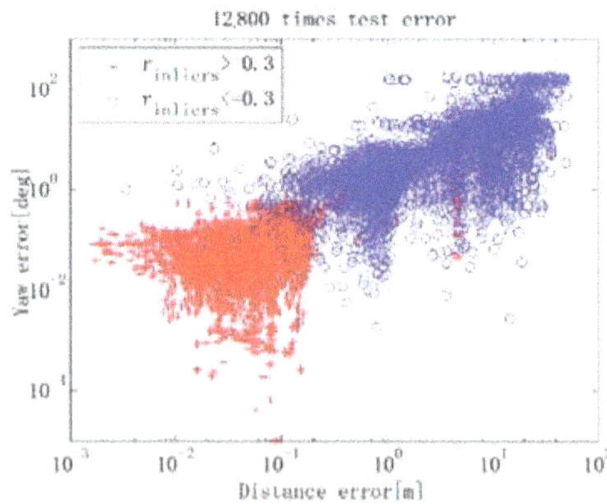

Figure 11. Error distribution and classification of registration results.

Table 1. Statistics of experiment times with result in different error and classification.

	Test A (Short-Term Registration)		**Test B (Long-Term Registration)**	
	Success	**Fail**	**Success**	**Fail**
Error less than (0.2 m, 0.2 m, 0.5°)	4126	40	3561	17
Error more than (0.2 m, 0.2 m, 0.5°)	29	2205	451	2371

Therefore, we make a statistical comparison of the success rate of the algorithm before and after improvement, as shown in Figure 12. The improved algorithm has a very high success rate in different scenes and different initial errors, which is greatly better than the traditional method. Even in the case of initial offset (28 m, 20°), the success rate of our method in the two sets of experiments can achieve 94% and 98%.

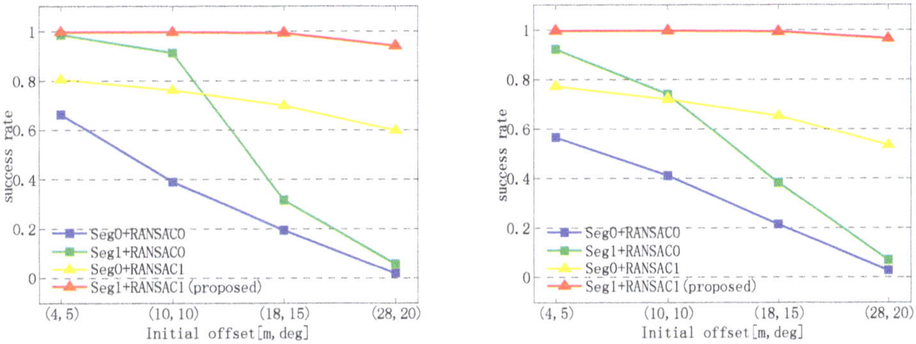

(**a**) Test A (short-term registration) (**b**) Test B (long-term registration)

Figure 12. Comparison of algorithm registration success rates in different scenes.

In addition, the registration errors of each method are also analyzed. Figure 13 shows the proportion of the experiments whose registration errors (lateral, longitudinal, yaw) are less than (0.1 m, 0.1 m, 0.25°) and (0.2 m, 0.2 m, 0.5°) in the two set of experiments with different initial offsets. It can be seen that the proportion of the precise results obtained by the proposed method is higher than that of the traditional methods. In both groups A and B, the proportion of the number of registration errors within (0.2 m, 0.2 m, 0.5°) is kept above 95% by using the proposed method. In the A group, the ratio of the resulting error within (0.1 m, 0.1 m, 0.25°) remained at about 90%. In the B group, due to some changes in the static scene, the proportion of registration errors within (0.1 m, 0.1 m, 0.25°) is much lower, but still more than 70%.

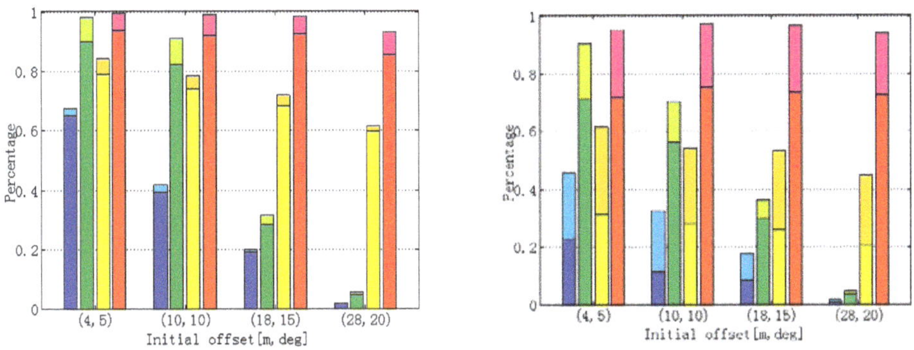

(**a**) Test A (short-term registration) (**b**) Test B (long-term registration)

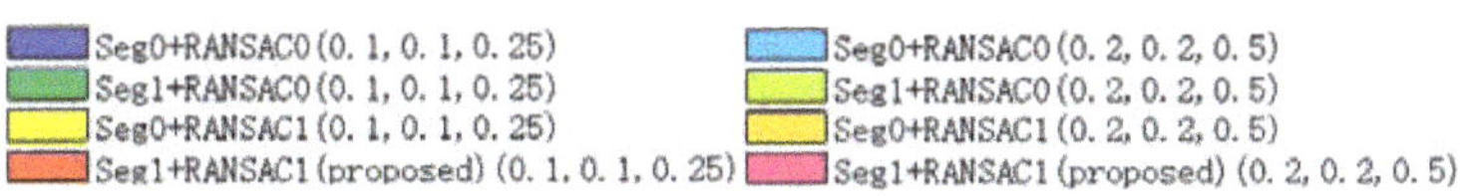

Figure 13. The proportion of high-precision registration results.

According to both Figures 12 and 13, comparing method 1(seg0 + ransac0) and method 2(seg1 + ransac0), it may be observed that the new segmentation method can effectively improve the registration quality of the complex dynamic scene with small initial error. However, as initial error increases, the registration effects of the two methods decline greatly. The comparison between method 1 (seg0 + ransac0) and method 3 (seg0 + ransac1) shows that the improved RANSAC method can work well with greater initial error. As initial error increases, the registration effect of method 3 has only decreased slightly while method 1 decreased greatly. Therefore, using both the improved segmentation and the improved RANSAC, the robustness of the registration is greatly improved in high dynamic environments with high initial error, which can be seen from the contrast between method 1 (seg0 + ransac0) and method 4 (seg1 + ransac1).

Two sets of experiments show that the algorithm can achieve accurate and stable registration results, whether it is short interval registration or long interval registration. The RMSE of the successful registration in different scenes are shown in Table 2. It may be seen that the location accuracy of our algorithm has reached the centimeter level, and the RMSE of the heading angle is within 0.3°. In the scenes with rich features such as crossroads, the angular error of the algorithm is smaller, and it can almost be controlled within 0.1°.

Figure 14 shows the comparison of time consumption of coarse registration between the proposed method and the traditional RANSAC method with different initial offsets. When the offset is small, the average time consumption of our method can be kept at about 80 ms, which is obviously superior to the traditional RANSAC method. With the increase of the initial offset, the average time-consumption of the proposed method increases, but it is still less than 150 ms. In the experiment of the traditional RANSAC method, with more sampling process becoming unavailable due to the failure of corresponding search points, more registration steps are skipped, which results in the decrease of time-consuming as the initial offset increasing.

Table 2. Registration error of our method in different scenes

Scene	RMSE		
	Lateral (m)	**Longitudinal (m)**	**Yaw(°)**
A1 turning right	0.0148	0.0354	0.0954
A2 turning left	0.0298	0.0348	0.1113
A3 bus occlusion	0.0193	0.0753	0.2322
A4 traffic jam	0.0291	0.0249	0.0419
B1 straight road	0.076	0.1561	0.2751
B2 cross road	0.0342	0.044	0.0409
B3 entrance ramp	0.0546	0.0642	0.2328
B4 elevated road	0.0221	0.0373	0.047

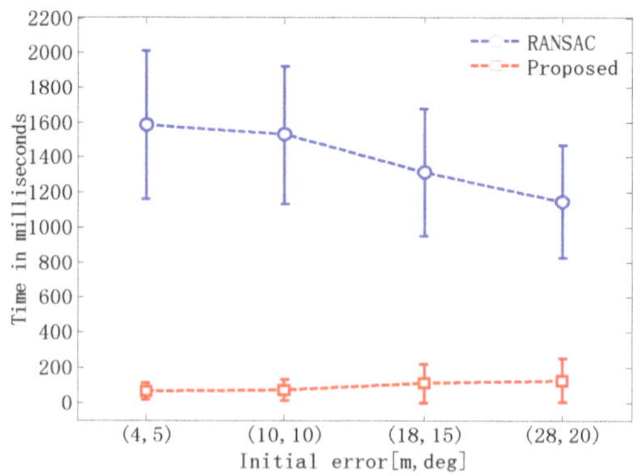

Figure 14. Time consumption of coarse registration.

4.2. LiDAR Odometry

LiDAR odometry based on point cloud registration is one of the key technologies of autonomous driving. It is used to further demonstrate the accuracy and stability of our method.

In this experiment, the starting point is given by GPS, and then the following poses are all calculated by point cloud registration, without any assistance from other sensors and algorithm of smoothing and optimization. To achieve good ego-motion results, each frame for registration is extracted from every five frames on both KITTI dataset and our own dataset. Because the acquisition frequency of LiDAR is 10 Hz, the data acquisition time of the two frames for the registration is 500 ms, which is much larger than the time consumption of the registration calculation, so that the real-time performance of the algorithm can be ensured.

KITTI provides urban public data sets and error evaluation methods that are used specifically for LiDAR odometry testing. The advantages of this method are mainly in the complex urban roads that exist in large numbers of moving objects, but unfortunately, the urban scene for the odometry test in this dataset is mainly static, with only a few moving cars in it. However, to verify the adaptability of the algorithm to different environments, we still do experiments in this data set. Figure 15 is the experimental result of the data set numbered 00, with a total length 3.7 km. Figure 15a shows the location error of the odometer based on the registration method, and the Figure 15b shows the error of the heading angle. Some large fluctuations in Figure 15b are caused by the critical value of the angle just at 360 and 0 degrees.

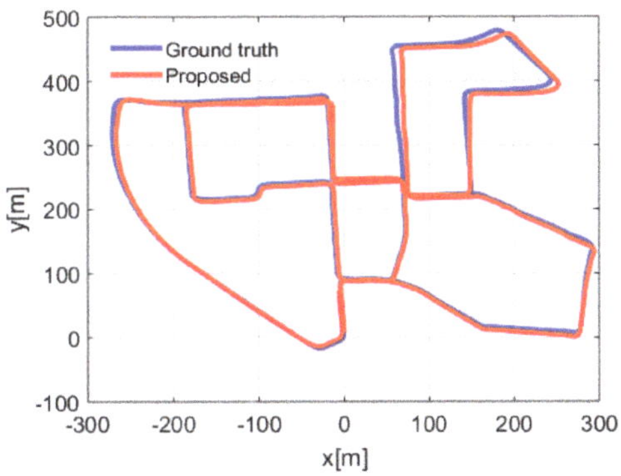

(**a**) Localization result.

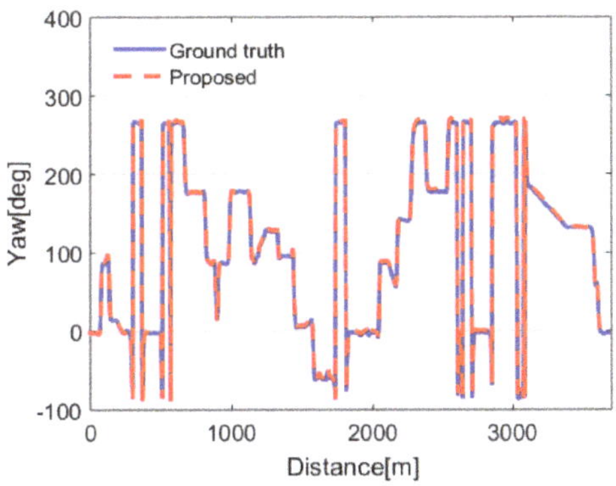

(**b**) Yaw result.

Figure 15. LiDAR odometry result in KITTI dataset.

In addition, experiments were conducted on our urban dataset with large numbers of moving objects in Tianjin. It is shown in Figure 16a, with the full length of 1.6 km. We captured several typical pictures taken by vehicle mounted cameras, including a crossroad, a S curved road, a road with numbers of buses (the bus runs in the opposite lane, causing almost all the static data on this side to be occluded) and a very urgent U-turn road. There are a large number of vehicles and pedestrians in the whole course of the experiment. In this experiment, our method is compared with the MM-RANSAC [9], and the experimental results are shown in Figure 16b,c. Because the frame sequence is weak, the MM-RANSAC algorithm has high errors at many positions. Especially at turning points, such as the S turn and U turn in our experimental dataset, great changes of the heading angle in a short time result in registration failures. Thus, the error of the heading angle calculated by MM-RANSAC based registration is high, causing serious errors in the odometry. However, because the proposed method does not rely on the sequence between the frames and can work well in the case of high initial error, the odometry using our registration method remained stable and accurate, whether on straight road or turning road.

According to the evaluation method provided by KITTI, the average horizontal position error of the LiDAR odometry based on our registration method is 0.45% on the KITTI training dataset sequence 00 with low dynamic scene and 0.55% on our high dynamic scene dataset. We have submitted our horizontal results to KITTI for test.

At present, high precision map has been rapidly developed and has become a necessary module for automatic driving, but it is still difficult to ensure real-time updating. In some sections with no map or map updates, we can use the LiDAR odometry based on the proposed method to achieve reliable localization, even if there are lots of dynamic obstacles in the scene.

It should be mentioned that although our experimental data are all from the LiDAR with 64 beams, our algorithm is not only limited to this LiDAR, but also applies to the LiDAR with 16 beams, 32 beams, and so on.

The experiment shows that our method has strong robustness and can be applied to many complex urban road scenes. The LiDAR odometry based on our method can help to get rid of the dependence on high precision inertial navigation system to reduce cost. In addition, as we can register real-time data with the historical data long before, the requirement for the real-time performance of high precision maps can be reduced.

However, in some cases, the LiDAR localization based on this method is still likely to fail. For example, in a tunnel with single surrounding features, the LiDAR localization is easy to produce a high longitudinal error. Because our method can work in the case of high error, when a vehicle travels from a feature deficient scene to a feature of rich scene, although the pose error may accumulate due to localization failure, the proposed method can still quickly converge to the correct location.

More results can be found in the Supplementary video attached to this paper in the web-site of journal of *Electronics*.

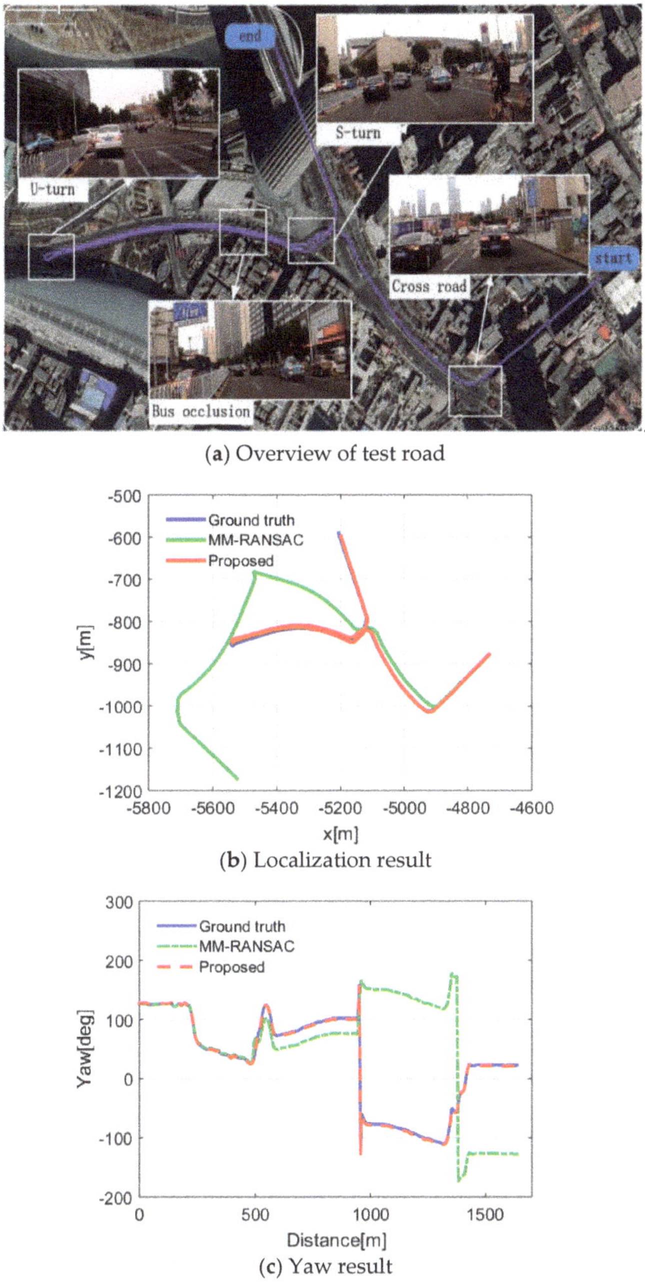

(**a**) Overview of test road

(**b**) Localization result

(**c**) Yaw result

Figure 16. LiDAR odometry result in highly dynamic environment.

5. Conclusions

Aiming at point cloud registration in urban complex dynamic environment for autonomous driving, this paper proposed a more robust registration method by optimizing the segmentation and improving the RANSAC algorithm with a novel framework, named 'multi-layer RANSAC'.

Experimental results demonstrate the robustness of our algorithm. It can achieve fast and accurate registrations in high dynamic scenes with large numbers of moving objects, serious occlusions and static environmental changes; even in case of high initial pose error. The key contributions of our paper are:

(1) The algorithm can solve the urban scene registration with numbers of moving objects without the aid of any other techniques such as object tracking and detection. Therefore, it is fit for both long and short time interval registrations. It can be used not only for LiDAR odometry, but also for precise localization with a priori map.

(2) The algorithm can adapt to higher initial pose error, so it can solve some difficult problems in localization, such as the poor initial localization accuracy caused by poor GPS signal at the vehicle start position and the high accumulated error due to long distance ego-motion estimation.

Our method still has a small amount of over-segmentation and under-segmentation, and our registration is more likely to be wrong due to misjudgment in scenes with sparse features. Next, we will try to extract semantic features from scenes for better segmentation, and set various object weights for better registration. In addition, in view of the fact that our method can provide the initial segmentation of dynamic and static objects, we will try to apply our method to the static map construction in dynamic environments and dynamic object detection and tracking

Supplementary Materials: The research results presented in this paper can be found in the supplementary video at http://www.mdpi.com/2079-9292/8/1/43/s1, Video S1: results.

Author Contributions: R.W. and Y.X. presented the algorithm; R.W., Y.M., and Z.L. wrote the paper; M.A.S. and R.W. provided analysis tool; R.W. and Y.X. analyzed the data; T.S.-G. and W.L. reviewed and revised the paper.

Funding: This work was supported by the National Key Research and Development Program (2016YFB0100903), the National Science Foundation of China (61503284 and 51408417), and Australia ARC DECRA (No. DE190100931).

Acknowledgments: We express our sincere thanks to A. Prof. Reza Malekian at Malmö University, Sweden for his suggestions in writing this paper. We also would like to give our gratitude to the reviewers for their valuable and constructive comments and suggestions.

Conflicts of Interest: The authors declare no conflict of interest.

References

1. Levinson, J.S. Automatic Laser Calibration, Mapping, and Localization for Autonomous Vehicles. Ph.D. Thesis, Stanford University, Stanford, CA, USA, 2011.
2. Wolcott, R.; Eustice, R. Fast LIDAR Localization using Multiresolution Gaussian Mixture Maps. In Proceedings of the 2015 IEEE International Conference on Robotics and Automation (ICRA), Seattle, WA, USA, 26–30 May 2015; IEEE: New York, NY, USA, 2015; pp. 2814–2821.
3. Zhang, J.; Singh, S. Low-drift and real-time lidar odometry and mapping. *Auton. Robots* **2017**, *41*, 401–416. [CrossRef]
4. Cadena, C.; Carlone, L.; Carrillo, H.; Latif, Y.; Scaramuzza, D.; Neira, J.; Reid, I.; Leonard, J. Past, present, and future of simultaneous localization and mapping: Towards the robust-perception age. *IEEE Trans. Robot.* **2016**, *32*, 1309–1332. [CrossRef]
5. Chum, O.; Matas, J.; Kittler, J. Locally Optimized RANSAC. In *Proceedings of Joint Pattern Recognition Symposium*; Springer: Berlin/Heidelberg, Germany, 2003; pp. 236–243.
6. Sattler, T.; Leibe, B.; Kobbelt, L. SCRAMSAC: Improving RANSAC's Efficiency with a Spatial Consistency Filter. In Proceedings of the 2009 IEEE International Conference on Computer vision, Kyoto, Japan, 29 September–2 October 2009; IEEE: New York, NY, USA, 2009; pp. 2090–2097.
7. Pankaj, D.; Nidamanuri, R. A robust estimation technique for 3D point cloud registration. *Image Anal. Stereol.* **2016**, *35*, 15–28. [CrossRef]
8. Moosmann, F.; Stiller, C. Joint Self-Localization and Tracking of Generic Objects in 3D Range Data. In Proceedings of the 2013 IEEE International Conference on Robotics and Automation, Karlsruhe, Germany, 6–10 May 2013; IEEE: New York, NY, USA, 2013; pp. 1146–1152.

9. Yang, S.; Wang, C. Multiple-model RANSAC for Ego-Motion Estimation in Highly Dynamic Environments. In Proceedings of the 2009 IEEE International Conference on Robotics and Automation, Kobe, Japan, 18 August 2009; IEEE: New York, NY, USA, 2009; pp. 3531–3538.
10. Wang, C.; Thorpe, C.; Thrun, S.; Hebert, M.; Durrant-Whyte, H. Simultaneous localization, mapping and moving object tracking. *Int. J. Robot. Res.* **2007**, *26*, 889–916. [CrossRef]
11. Fischler, M.A.; Bolles, R.C. Random sample consensus: A paradigm for model fitting with applications to image analysis and automated cartography. *Commun. ACM* **1981**, *24*, 381–395. [CrossRef]
12. Schlichting, A.; Brenner, C. Vehicle localization by lidar point correlation improved by change detection. *Int. Arch. Photogramm. Remote Sens. Spat. Inf. Sci.* **2016**, *41*, 703–710. [CrossRef]
13. Im, J.; Im, S.; Jee, G. Vertical corner feature based precise vehicle localization using 3D LIDAR in urban area. *Sensors* **2016**, *16*, 1268. [CrossRef] [PubMed]
14. Hata, A.; Wolf, D. Road Marking Detection using LIDAR Reflective Intensity Data and its Application to Vehicle Localization. In Proceedings of the 2014 IEEE International Conference on Intelligent Transportation Systems, Qingdao, China, 8–11 October 2014; IEEE: New York, NY, USA, 2014; pp. 584–589.
15. Hata, A.; Osorio, F.; Wolf, D. Robust Curb Detection and Vehicle Localization in Urban Environments. In Proceedings of the 2014 IEEE Intelligent Vehicles Symposium, Dearborn, MI, USA, 8–11 June 2014; IEEE: New York, NY, USA, 2014; pp. 1257–1262.
16. Aiger, D.; Mitra, N.; Cohen-Or, D. 4-points congruent sets for robust pairwise surface registration. *ACM Trans. Gr.* **2008**, *27*, 85. [CrossRef]
17. Mellado, N.; Aiger, D.; Mitra, N. Super 4pcs fast global point cloud registration via smart indexing. *Comput. Gr. Forum* **2014**, *33*, 205–215. [CrossRef]
18. Su, Z.; Xu, Y.; Peng, Y. Enhanced detection method for structured road edge based on point clouds density. *Automot. Eng.* **2017**, *39*, 833–838.
19. Dimitrievski, M.; Van Hamme, D.; Veelaert, P.; Philips, W. Robust matching of occupancy maps for odometry in autonomous vehicles. In Proceedings of the 11th Joint Conference on Computer Vision, Imaging and Computer Graphics Theory and Applications (VISIGRAPP 2016), Rome, Italy, 27–29 February 2016; Volume 3, pp. 626–633.
20. Anderson, S.; Barfoot, T.D. RANSAC for Motion-Distorted 3D Visual Sensors. In Proceedings of the 2013 IEEE/RSJ International Conference on Intelligent Robots and Systems (IROS), Tokyo, Japan, 3–7 November 2013; IEEE: New York, NY, USA, 2013; pp. 2093–2099.
21. Geiger, A.; Lenz, P.; Stiller, C.; Urtasun, R. Vision meets robotics: The KITTI dataset. *Int. J. Robot. Res.* **2013**, *32*, 1231–1237. [CrossRef]

electronics

MDPI

Article

A Multichannel FRA-Based Impedance Spectrometry Analyzer Based on a Low-Cost Multicore Microcontroller

Arturo Sanchez-Gonzalez, Nicolas Medrano *, Belen Calvo and Pedro A. Martinez

Group of Electronic Design, Aragon Institute for Engineering Research, University of Zaragoza, 50009 Zaragoza, Spain; asgonzalez@unizar.es (A.S.-G.); becalvo@unizar.es (B.C.); pemar2@unizar.es (P.A.M.)

* Correspondence: nmedrano@unizar.es; Tel.: +34-876-553-358

Received: 9 November 2018; Accepted: 21 December 2018; Published: 1 January 2019

Abstract: Impedance spectrometry (IS) is a characterization technique in which a voltage or current signal is applied to a sample under test to measure its electrical behavior over a determined frequency range, obtaining its complex characteristic impedance. Frequency Response Analyzer (FRA) is an IS technique based on Phase Sensitive Detection (PSD) to extract the real and imaginary response of the sample at each input signal, which presents advantages compared to FFT-based (Fast Fourier Transform) algorithms in terms of complexity and speed. Parallelization of this technique has proven pivotal in multi-sample characterization, reducing the instrumentation size and speeding up analysis processes in, e.g., biotechnological or chemical applications. This work presents a multichannel FRA-based IS system developed on a low-cost multicore microcontroller platform which both generates the required excitation signals and acquires and processes the output sensor data with a minimum number of external passive components, providing accurate impedance measurements. With a suitable configuration, the use of this multicore solution allows characterizing several impedance samples in parallel, reducing the measurement time. In addition, the proposed architecture is easily scalable.

Keywords: electrochemical impedance spectroscopy; FRA; multichannel acquisition; impedance spectrometry; microcontroller

1. Introduction

Electrochemical impedance spectroscopy (EIS), in which a sinusoidal signal is applied to a sample under test to evaluate in a determined frequency range its complex impedance—typically modeled as a Randles cell [1]—is a powerful sensing technique that has experienced significant development over the last years due to its broad range of applications. These span from the biotechnological field (rapid detection of foodborne pathogenic bacteria, detection and enumeration of E. coli bacteria in milk samples, real-time detection of milk adulteration, food control, antibiotic susceptibility testing of E. coli and characterization of cellular dielectric properties for cell health evaluation [2–7]), to the characterization of materials (microstructures, dielectric materials, corrosion evaluation, [8–11]), as well as electrical circuit testing and characterization of electrical systems, batteries, and photovoltaic cells [8,12,13]. Unlike other electrochemical techniques, such as potentiometric and amperometric sensing (based on DC voltage and current excitation, respectively [14]), the characterization of a sample by its impedance changes over frequency requires a small stimulus signal, reducing the risk of sample damaging, which is a key point in biological measurement and characterization applications. Information can be then recovered using different readout techniques, being the Fast Fourier Transform (FFT) and the Frequency Response Analyzer (FRA) the two most commonly used. The latter one is based on synchronous demodulation, that is, it relies on phase sensitive detection (PSD) or quadrature modulators to extract the real and imaginary response of the sensor at each input signal f_{in} while noise signals at other frequencies are rejected (Figure 1), presenting advantages compared to FFT algorithms in terms of complexity and speed [15,16]. Thus, the FRA-EIS is a more suitable choice to accomplish an autonomous low-cost real-time multichannel impedance spectroscopy analyzer.

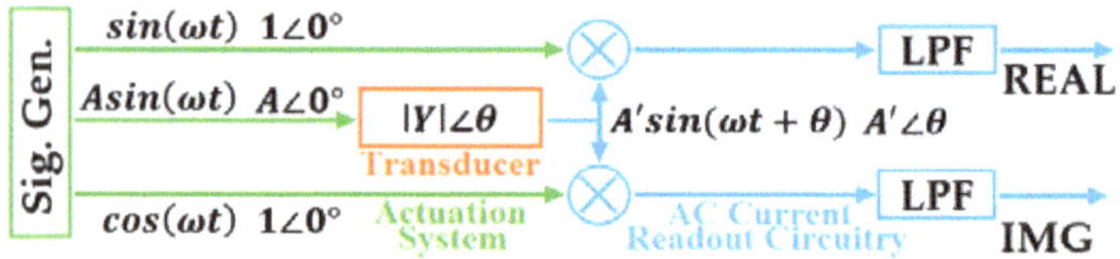

Figure 1. Single channel (Frequency Response Analyzer) FRA-based impedance spectroscopy (IS) system block diagram.

In this attempt, while electrochemical transducers take advantage of Complementary Metal-Oxide-Semiconductor (CMOS) processes to implement the required Micro-Electro-Mechanical Systems (MEMS [17]), the rest of components that conform the data acquisition chain (signal stimuli generators, conditioning, pre-processing and digitization electronics) are still bulky benchtop instruments, making EIS almost exclusively a measurement technique for biochemical, biological, or quality control laboratories, but hindering its use closer to the sampling sources, as milk farms, in food production chains or portable laboratories for on-site tests.

Recent publications in the scientific and technical literature are reporting EIS systems partially implemented using CMOS technologies by addressing specific low-power low-size design techniques to take advantage of the features that miniaturization can provide to the system in terms of portability and high parallelism in the measurement processes [18–20]. These cases succeed in the integration of competitive read-out channels, but the generated real/imaginary analog data must be finally digitized to be processed by a digital processing unit (a microcontroller, a digital signal processor or an external computer), while the generation of the required excitation and control signals are usually assigned either to external resources (commercial waveform generators that provide flexibility in exchange of large size and high power consumption, not being compatible with portability), or small size custom integrated oscillators [21], with exhibit frequency tuning and linearity limitations, especially at high frequencies. Hence, although being fundamental components, both the generation and digitization blocks are not usually considered in the power consumption estimation of the EIS system, thus giving partial information of the real energy required by a complete measurement unit.

In order to reduce electronics complexity, alternative EIS approaches are based on the direct transform of impedance to digital values using impedance-to-digital or dual-slope multiplying ADC (DS-MADC) techniques, achieving accuracies below 10 bits [22,23]. With the goal of further reduce electronics complexity, in order to attain a self-contained low-cost measurement system that renders a true portability while preserving high recovery performance, this paper proposes the complete digitalization of the EIS system through a microcontroller-based FRA implementation, applied to impedance spectroscopy for frequencies in the range of cellular characterization, from 1.1 mHz to 10 kHz. The proposed system uses the internal resources of a Propeller processor core from Parallax [24], to both generate the excitation and control signals required in the process, and to map the read-out and recovery algorithms for the sensed analog signals, so that with minimum additional passive components, it can recover an impedance value with 12-bit accuracy. In addition, the use of this low-cost multicore processor allows for parallelization of the actuation and signal recovery, implementing a multi-channel compact IS instrument on a single microcontroller, able to perform up to 7 real-time in-situ parallel impedance measurements if driven by the same signal generator, or up 4 completely independent parallel impedance measurements; these values can be further extended by accordingly extending the number of cores to allow further parallelization.

This paper is structured as follows: Section 2 describes the proposed FRA-based impedance analyzer, detailing the implementation and the experimental characterization of both the actuation and the signal acquisition blocks. Section 3 validates the recovery performance of the proposed IES system applied to an impedance modeling a bilayer lipid membrane. Finally, Section 4 discusses the proposed approach.

2. Proposed EIS System

The proposed EIS system relies on the use of a single Parallax Propeller microcontroller, characterized by working at up to 80 MHz clock frequency. It presents 8 independent cores plus an additional hub, in charge of controlling the access of each core to the common microcontroller resources (32 kB Main RAM or 32 kB Main ROM), applying a Round Robin schedule. Each core features a video generator, a local 2 kB RAM, and two Counter Modules with Phase-Locked Loops (PLLs) and 32 operation modes (Figure 2). From a software point of view, the microcontroller can be programmed in C, in its own high-level programming language SPIN, or in low-level Propeller Assembly Language (PASM). In order to achieve a suitable implementation of a FRA-based EIS system, optimizing the hardware resources and their access at the rate needed to generate and recover signals of a reasonable frequency, an integral programming in PASM has been adopted.

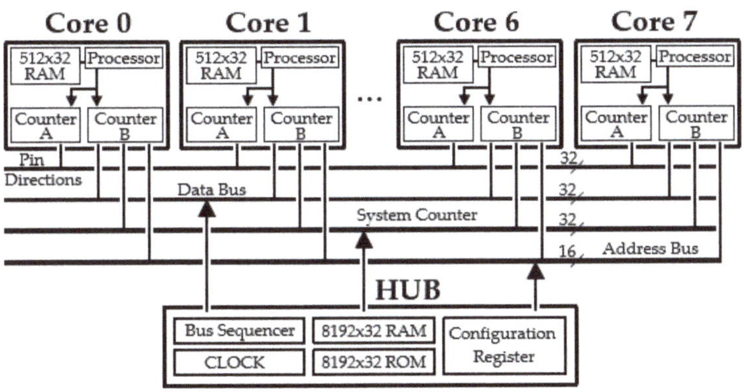

Figure 2. Propeller microcontroller block diagram.

2.1. *Signal Generation: Hardware Implementation and Control*

Figure 3 shows the block diagram of the signal generator hardware implementation. It is based on the D modulation technique to achieve an accurate full range (0 V to 3.3 V) quadrature signal generation needing the minimum number of external passive elements. The first quarter cycle of the two signals to be generated (sine and cosine, from 0 degree to 89.98 degrees) is stored in the shared main RAM memory of the processor. Signal points are stored using a 16-bit representation, with a maximum resolution of 4096 points per quarter. Because the main RAM memory in the processor is composed of 32-bits length registers, each memory position stores the corresponding sine (16 most significant bits—MSB of the memory position) and cosine (16 less significant bits—LSB) values, saving with this choice access time to the global memory and therefore speeding the quadrature signal generation task.

Each core sequentially accesses these data in the main memory at the microcontroller HUB by a Round Robin process schedule, which respectively feeds its two independent hardware Counter Modules, consisting of configurable state machines [25] working up to the maximum 80 MHz clock frequency. Each counter has an adder and an accumulator, which will be employed to implement the D modulator based on pulse density modulation (PDM) using the carry bit of the adder as modulated output. By properly configuring the operation of both counters in the core, the two sinusoidal signals with 90° phase shift required for an EIS channel can be generated with a single core. An external passive integrator converts the resultant modulated pulses into a sinusoidal signal.

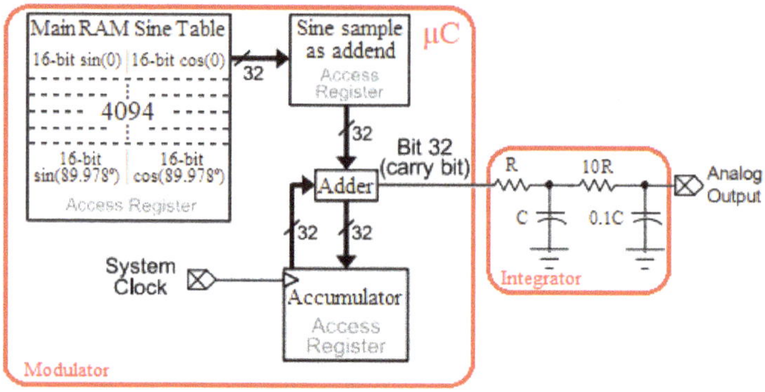

Figure 3. Signal generator hardware implementation based on D modulation.

More in detail (Figure 3), the PDM sine signal generation is performed by a core in the following manner: First, the 32-bit values representing the value of the two quadrature signals at each moment are read consecutively from the main RAM and transferred to the local memory in the selected kernel. Then, the 16 MSB bits (representing the sine value) are extracted and stored in the access register, which is in the adder, and then are added to the accumulator. When the adder overflows, the carry bit changes to 1. The density of 1's in this output depends on the values that are being added: the higher the values accumulated, the faster the carry overflows. Thus, the density increases in the range of the maximum values of the sine function, while it decreases in the minimum values. Finally, the resulting modulated pulse train is converted into an analog signal by means of a passive second order low-pass filter (LPF) (Figure 3) consisting of two cascaded RC circuits (R = 2.2 kW, C = 330 pF) with the same constant time and a factor of 10 in the consecutive R's and C's values to reduce the loading effect. The integration time is selected to keep distortion bounded below 0.75% as design specification, as will be shown next.

This process is iterated until all the values in the table are traversed. Next, the process is repeated using the LSB values. These data represent the cosine values in the first quarter of the cycle and, therefore, the sine values in the second, giving therefore signal continuity. To conclude a complete sinusoidal cycle, this process is repeated but changing the sign of the data in RAM to represent the negative half cycle. The cosine generation is performed in a similar manner.

The frequency of the sine/cosine output signal is determined by two different working frequencies: (i) The frequency of the modulated signal, which corresponds to the frequency of the square signal whose density varies. In this work, this signal matches the microcontroller clock frequency, which is set to its maximum value, 80 MHz; (ii) The data generation sampling frequency, that is, the frequency at which the system picks a sine/cosine value from the Main RAM to be sent to the Adder to provide a new output signal value. This frequency, in turn, mainly depends on the microcontroller clock frequency and the number of clock cycles required to load a value into the Adder register from the RAM, being thus more restrictive. An additional control of the output signal frequency can be achieved by selecting not all the samples, but one of every n values at the RAM memory, thus reducing the number of signal points for the generation and therefore reducing the time to generate a signal period.

Taking into account all the aforementioned, the frequency of the sine/cosine output signals f_{out} can be expressed as

$$f_{out} = f_{clk}\frac{(A+1)}{2^{14}(S+D)} = f_{sampling}\frac{(A+1)}{2^{14}} \quad (1)$$

being the maximum number of samples per signal period equal to 2^{14} (2^{12} samples/quarter × 4), *A* is the number of samples not read between two consecutive readings from the memory, *S* is the number of clock cycles required to load a sample to the Adder register, and *D* is an additional delay that can be added in each reading memory cycle and that serves to achieve a fine tuning in the value of the output signal frequency.

2.2. Signal Generation: Experimental Characterization

By using the degrees of freedom shown in Equation (1), with $f_{clk} = 80$ MHz, S = 65, Arranging from 1 to 2000 and D ranging from 0 to 2^{32}, the frequency can be adjusted to range from 1.1 mHz to 150 kHz in 75 Hz coarse steps (given by parameter *A*), and fine steps given by *D*.

Figure 4 shows the PDM signal (carry bit adder pin) provided by the system, when configured to generate f_{out} = 5 kHz and f_{out} = 150 kHz sinusoidal output, measured by using a Tektronix® DPO4104 oscilloscope. Figure 5 shows the frequency spectrum for the 150 kHz PDM signal (S = 65, *D* = 0). The detail box shows the performance closer to the signal of interest, being the 1.080 MHz peak, corresponding to the $(f_{sampling} - f_{out})$ frequency, ($f_{sampling} = f_{clk}/(S+D)$ = 1.23 MHz, Equation (1)), the most relevant interference source. For signals generated at frequencies below 150 kHz, this $\left(f_{sampling} - f_{out}\right)$ interference peak is kept far away enough to be irrelevant. Therefore, by properly selecting the cutoff frequency of the output LPF (Figure 3), the following sinusoidal signal is set to comply with a maximum distortion (THD and SFDR) below 0.75% over all the frequency range (Figure 6).

Figure 7 shows the generated sine signals within the operating frequency range, for 1 mHz (Figure 7a) and 5 kHz (Figure 7b). Figure 8 shows their corresponding spectra.

Figure 9 shows the two quadrature signals (sine—yellow and cosine—magenta) after the LPF for the upper and lower limit frequencies of the proposed generator (150 kHz, Figure 9a and 100 mHz, Figure 9b). Note that since the counters provide both the carry signal and its negated value, a single core will give two pairs of quadrature signals, that can be used to characterize in parallel two different impedance systems, at the same frequency (signals blue and green in Figure 9a). Figure 10 shows the frequency spectrum for the 150 kHz sine; in this case $THD = -42.8$ dB, $SFDR = 44.6$ dB, constituting the worst case distortion.

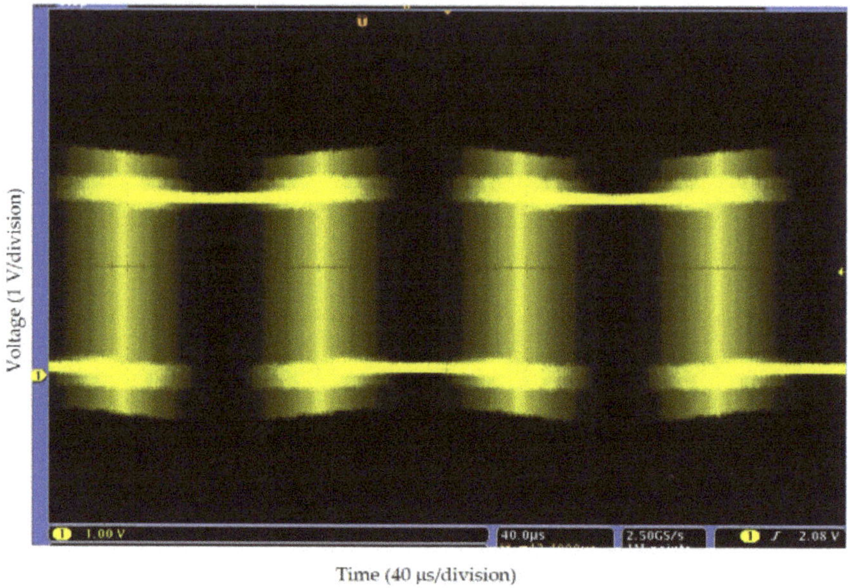

(a)

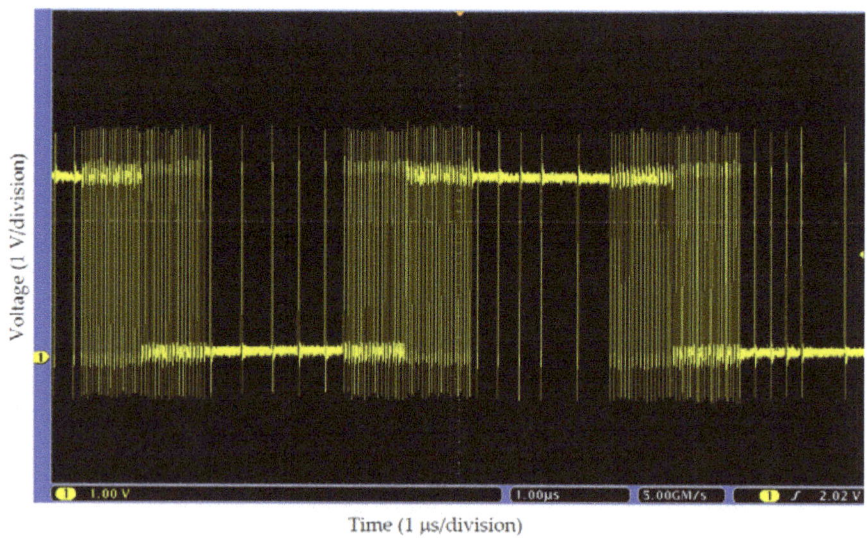

(b)

Figure 4. Pulse density modulation (PDM) signal (voltage versus time) corresponding to (**a**) a 5 kHz and (**b**) 150 kHz sinusoidal outputs. The 80 MHz pulse density is maximum for the rising and falling slopes of the sinusoidal signal and it is minimum for the maximum and minimum sinusoidal signal amplitude.

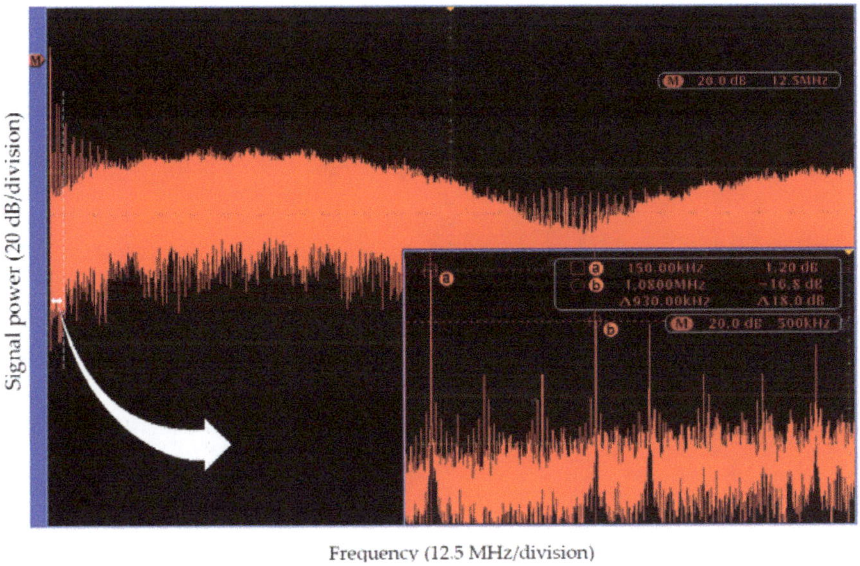

Figure 5. Spectrum (signal power in dB versus frequency) of the PDM signal at the carry bit output (Figure 3), before the low-pass filter (LPF) (output frequency signal of 150 kHz). Peak (**a**) corresponds to the frequency of the sine signal; peak (**b**) is the $\left(f_{sampling} - f_{out}\right)$ frequency. Spectrum peaks between (**a**) and (**b**) corresponds to the 3rd and 5th harmonics of the signal. The rest of the spectrum peaks are due to the intermodal distortion.

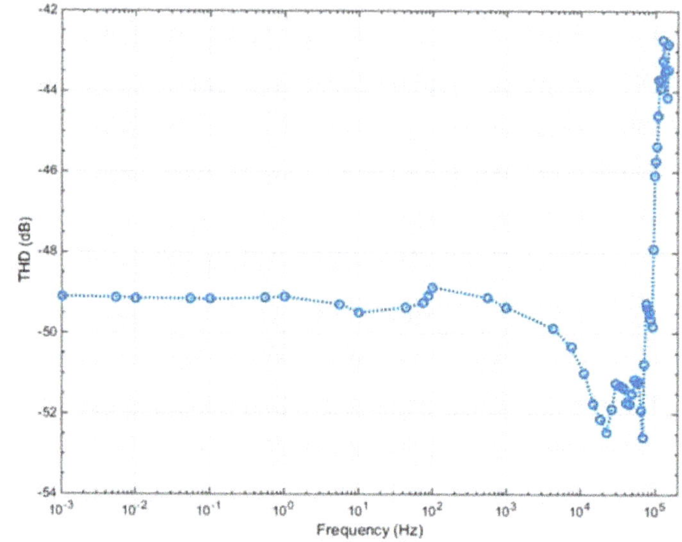

(a)

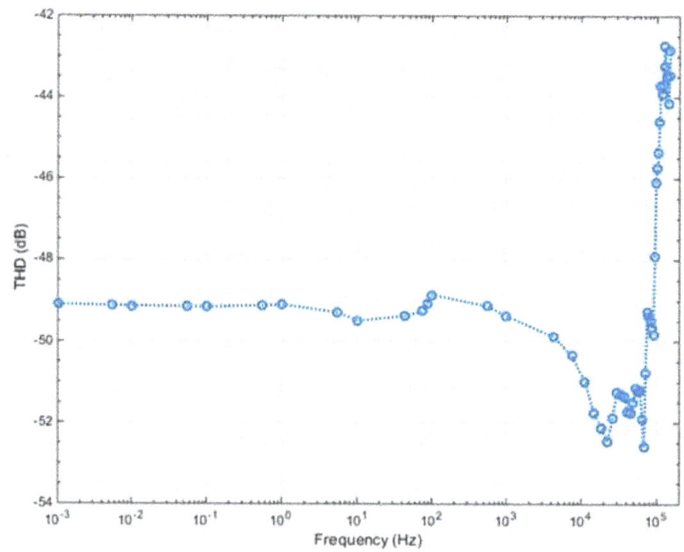

(b)

Figure 6. (**a**) THD and (**b**) SFDR experimental values for the quadrature sinusoidal signals. THD is defined as $\sqrt{\sum_{i=2}^{7} H_i^2 / H_1}$, where H_1 is the contribution of the fundamental frequency of the sinusoidal signal, and H_i are the successive harmonic contributions. SFDR is defined as H_1/M, where M is the tone with the highest contribution which differs from H_1. Both estimators are represented in decibels.

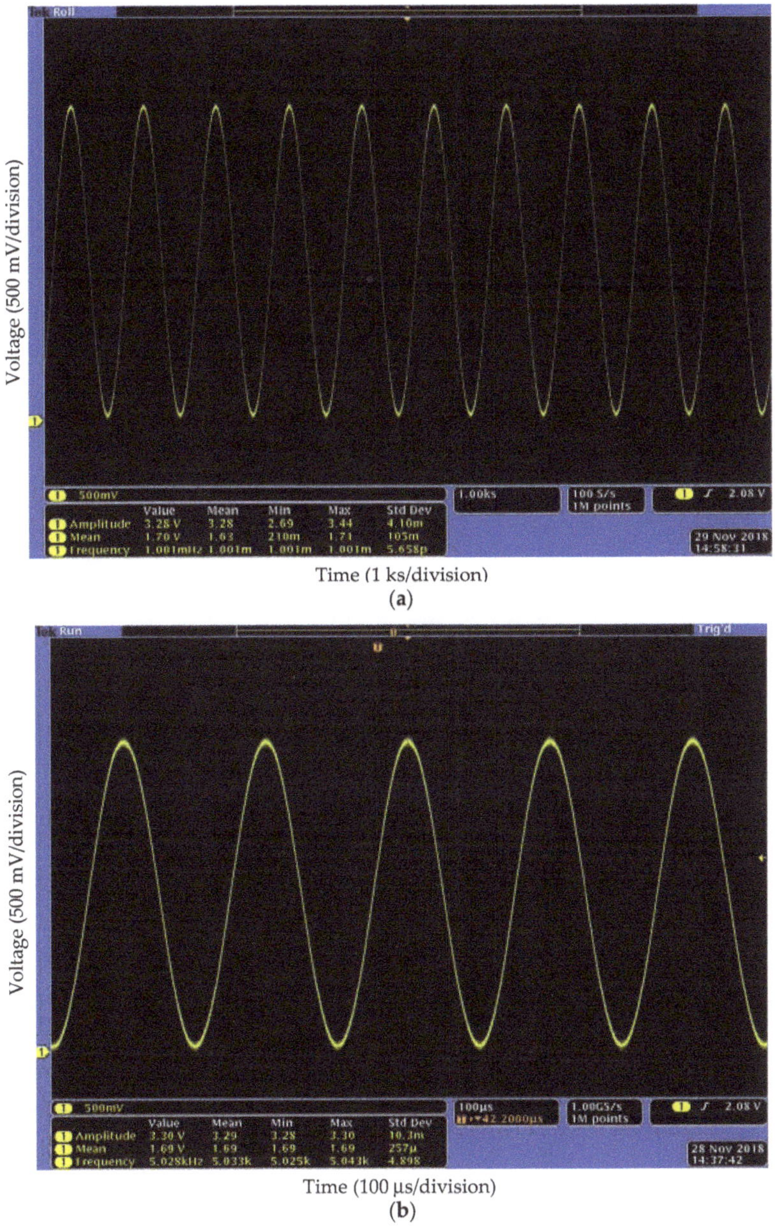

Figure 7. Sine signals (voltage versus time) generated after the corresponding PDM is low-pass filtered for the frequencies in the range of interest: (**a**) 1 mHz and (**b**) 5 kHz.

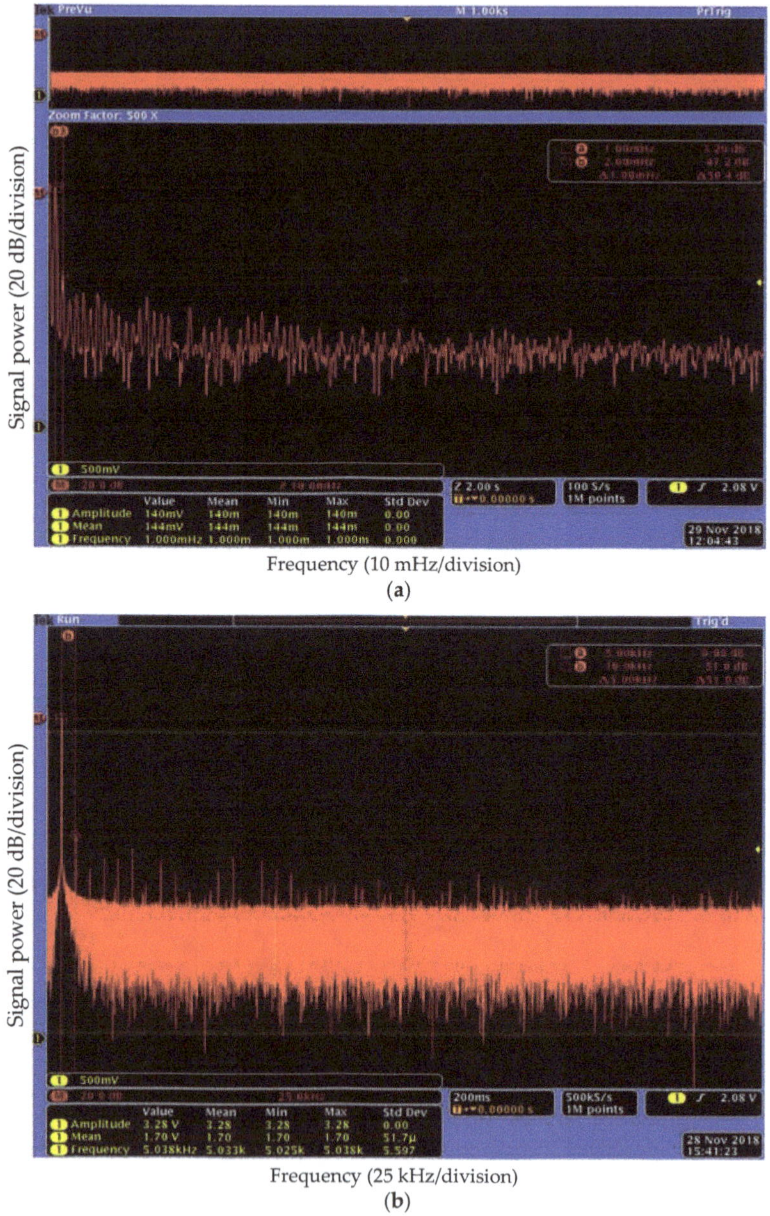

Figure 8. Spectrum (signal power in dB versus frequency) of the sine signals shown in Figure 7: (**a**) 1 mHz and (**b**) 5 kHz.

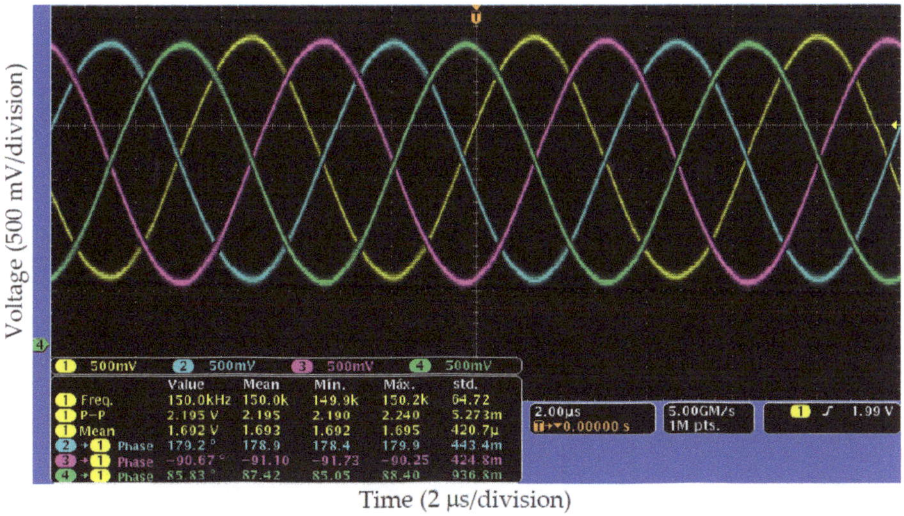

(a)

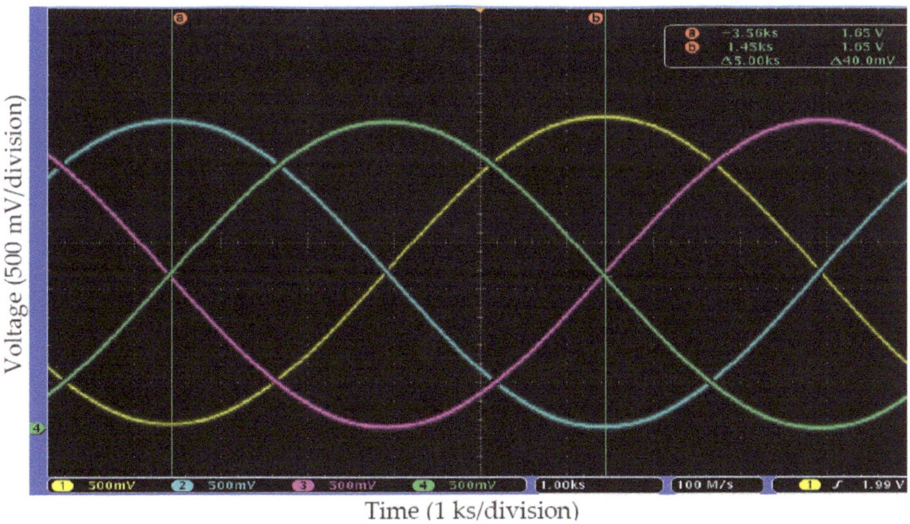

(b)

Figure 9. Quadrature sinusoidal signals (voltage versus time) provided by the proposed Δ modulation Digital-to-Analog Converter (DAC) using the same LPF, for (**a**) 150 kHz, and (**b**) 100 mHz. Because the adder provides both the carry and its negated value, a core can provide two quadrature signal pairs in parallel.

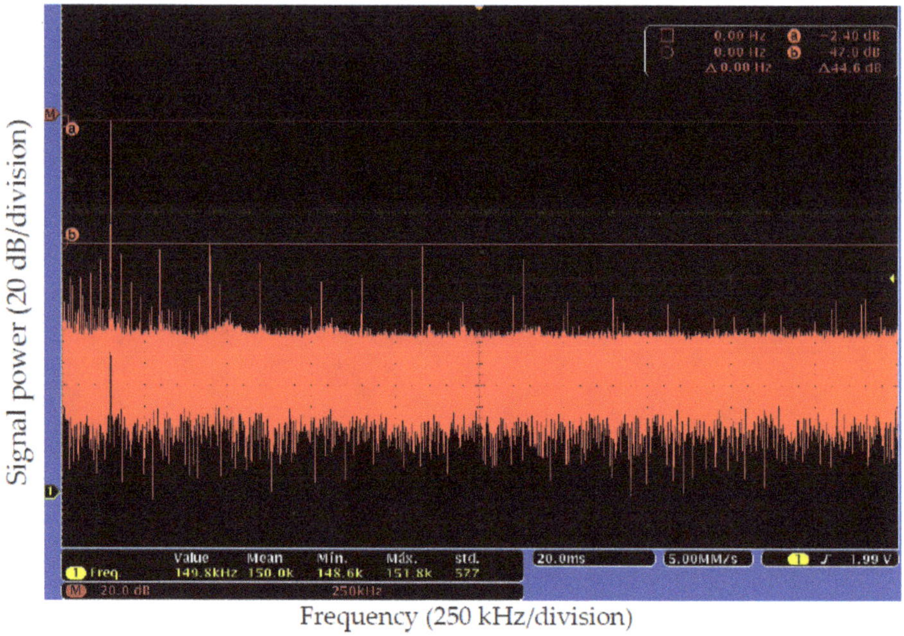

Figure 10. Spectrum (signal power in dB versus frequency) of the generated sine signal at 150 kHz frequency.

2.3. *Signal Acquisition: Hardware Implementation and Control*

The purpose of the signal acquisition stage is to recover the sensor signal to next perform the synchronous mixer operation rendering the corresponding quadrature outputs. The average of these values corresponds to the real and imaginary components or, equivalently, the magnitude and phase of the impedance under test. Note that since the average value of a signal is independent of its frequency, in all this process the signal sampling rate can be relaxed without loss of information. That is why a sigma–delta analog to digital conversion (ΣΔ-ADC) algorithm has been selected in spite of its low conversion rate to accomplish a more accurate conversion. In addition, the ΣΔ-ADC main building blocks can be implemented using the internal resources of a single core, requiring minimum additional external components.

2.3.1. Digitization

Figure 9 shows the block diagram of the ΣΔ-ADC. It consists on a ΣΔ modulator (composed by the integrator, the quantizer and 1-bit Digital-to-Analog Converter DAC blocks) plus a counter module working as a digital decimation filter [26]. Both the quantizer and the decimation filter have been implemented using the registers of the two counters available in a core. The qualitative operation of this system is as follows [27]: the analog input (Figure 11, sensor output signal), through an RC circuit formed by resistor R_1 and capacitor C, provides a voltage value which drives the input of a D flip-flop. While the voltage level in the capacitor is higher than the bi-stable threshold (assuming the threshold voltage in the bi-stable is half the digital bias voltage, $V_{DD}/2$), its output Q remains '1' (and $\overline{Q} = '1'$). This Output Q enables the accumulator operation, increasing the value in this register for each new clock cycle. On the other hand, the output $\overline{Q}$ conforms a negative feedback loop to the input capacitor through resistor R_f that reduces the voltage at the integrating capacitor. Once the voltage at input D gets under the threshold value, outputs Q and $\overline{Q}$ flip their values at the next clock cycle. The accumulator stops increasing its value, providing a binary value related to the number of cycles that the voltage value at the analog input remains greater than the threshold value. The resolution in which the analog value is represented by the Serial Digital Out depends on the selected integration time.

In fact, the application of the circuit shown in Figure 11 to time-dependent signals presents some constrains related to the cutoff frequency of the input low-pass filter, the frequency of the input signal and the conversion frequency to guarantee a suitable estimation of the input sample. For the sake of simplicity, let us suppose the sigma-delta modulator works in linear mode (that is, the clock frequency is much higher than input signal frequency, therefore considering its operation mode as continuous). Then, the block diagram of the ΣΔ modulator in the S domain is given by the scheme in Figure 12 [26].

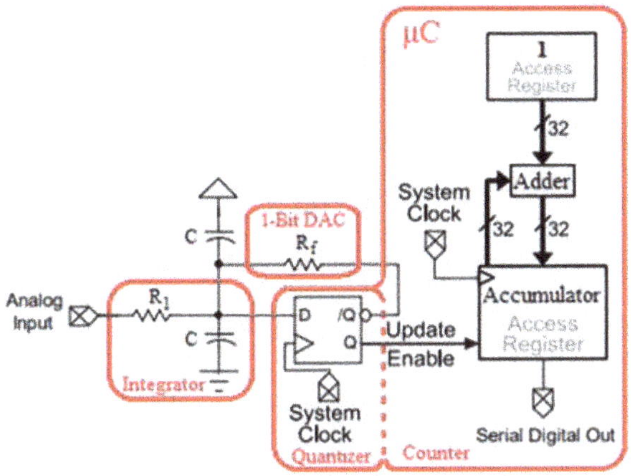

Figure 11. ΣΔ-ADC hardware implementation.

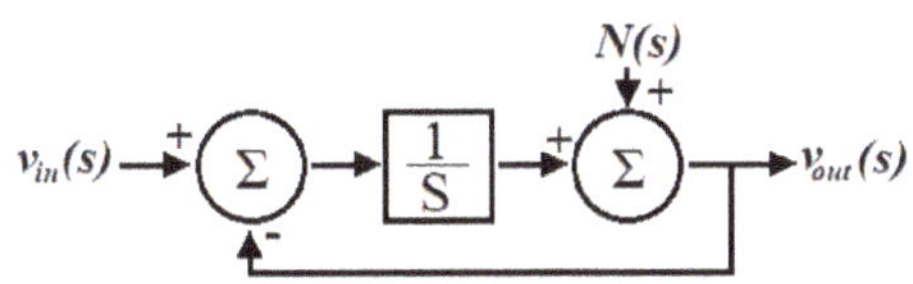

Figure 12. ΣΔ-ADC representation in the S domain.

Where $v_{in}(s)$ is the input voltage and $v_{out}(s)$ is the output Q in Figure 11, while *N(s)* represents the effect of quantization in the transfer function, which is negligible if linear operation is assumed. The ΣΔ modulator transfer function is

$$\frac{v_{out}(s)}{v_{in}(s)} = \frac{1}{1+s} \tag{2}$$

and accordingly, the output voltage can be expressed as

$$v_{out} = \frac{R_f}{R_1} \frac{1}{1+2sR_1C} v_{in} + K \tag{3}$$

where K is proportional to the D flip-flop threshold voltage, $V_{DD}/2$. Therefore, the output voltage depends on the f_{R_1C} passive filter cutoff frequency and the input signal frequency f_{in}. Thus, an input signal with a frequency higher than f_{R_1C} may result in a loss of accuracy. Besides, the conversion frequency f_{conv} must be fast enough to avoid that the capacitor discharge process affects the digitized value, which would reduce the resolution in bits of the ADC. That is, on the overall it must be satisfied

$$f_{in} \leq f_{R_1C} \leq f_{conv} \approx \frac{f_{clk}}{2^N} \quad (4)$$

being necessary to appropriately select the passive components in the modulator stage as well as the conversion rate in order to perform a suitable signal digitization in N bits.

To manage the hardware resources to reliably perform the required operations while minimizing the execution time, a specific code using the microcontroller assembler has been developed. Figure 13 shows the simplified control flowchart describing the signal digitization and data acquisition. After configuring the corresponding input and output pins and the counter register, first a calibration process is carried out. This task, which is performed at the system start up, allows determine both the offset at 0 V in the accumulator and the integration time required to properly acquire the maximum allowable input voltage, therefore maximizing the dynamic range. For completing the calibration step, the Analog Input (Figure 11) is connected to 0 V. After the integration time, the value stored in the accumulator, which ideally should be equal to zero, is the excess offset reading that must be subtracted from the system readings in normal operation mode. The integration time is adjusted by applying the maximum voltage to be digitized in the Analog Input. After the integration time, the accumulator should be filled to the maximum value (2^N), keeping the overflow flag equal to zero. Otherwise, the integration time must be increased/decreased up to reach this condition for a given number N of bits.

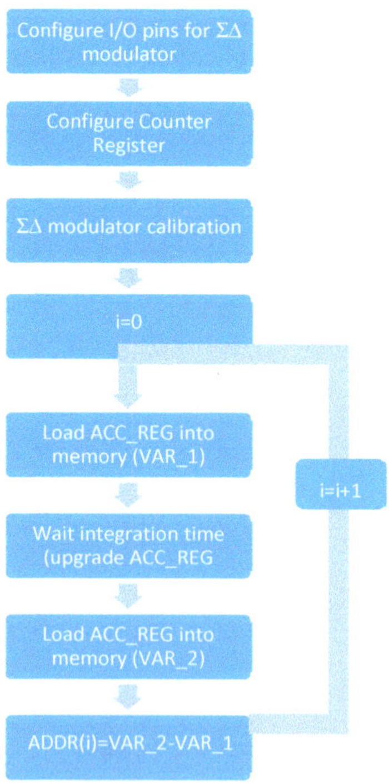

Figure 13. ΣΔ-ADC control flowchart.

Once calibrated, the system temporally saves the value stored in the counter accumulator in a variable, waits the integration time and reads again the value in the register. The difference between both data represents the value of the sensor output signal at this time, which is stored in a memory address so that data from consecutive instants of time use consecutive memory positions.

2.3.2. Mixing and Averaging

In a Propeller microcontroller, the hardware resources allow digitizing up to two different signals in parallel per core. To accurately synchronize the mixers operation, the system makes use of a dedicated core for each impedance measurement according to the following process: one of the counters is dedicated to digitize the signal arriving from the impedance under study, while the other counter synchronously digitizes the original sinusoidal excitation signal. The product of these two data corresponds to the real mixer output. The real component of the impedance under test is then calculated by averaging the products provided by this branch over a minimum $n = 5$ periods of the excitation signal to obtain reliable results over all the operating frequency range. In a single-channel measurement approach, the quadrature signal, which does not excite the impedance under study (cosine) is directly read from the hub memory, feeding the corresponding mixer (Imaginary) before its averaging, thus making unnecessary its digitization, which saves both hardware as computing resources.

2.4. Signal Acquisition: Experimental Characterization

The structure shown in Figure 11 has been implemented for a 12-bit approach. First, the linearity of the analog-to-digital conversion is verified by applying an incremental DC voltage in the biasing range of the microcontroller (from 0 to 3.3 V) to the Analog Input of the ΣΔ-ADC (Figure 11), and recovering the output digital values. Figure 14 shows the results, where y axis corresponds to the analog values represented by the digital words obtained in the conversion, assuming a full scale digitization (that is, 000h represents 0 V and FFFh represents 3.3 V in hexadecimal). It can be seen in this figure that the ADC conversion presents high linearity, resulting in a gain or slope of 0.65, an offset below 18 mV, and with a coefficient of determination $\overline{R}^2 = 1.0000$. The conversion slope can be modified by the feedback loop through $\frac{R_f}{R_1}$ (Equation (3)), to adjust its value according to the conversion requirements. In this case, to allow a full sweep in the supply voltage range avoiding saturation in the digitization module (Figure 11), we kept the output gain < 1 by selecting $R_f = 100\ \text{k}\Omega$ and $R_1 = 155\ \text{k}\Omega$. This choice results in a conservative 0.65 gain, exactly as obtained from the linear fit.

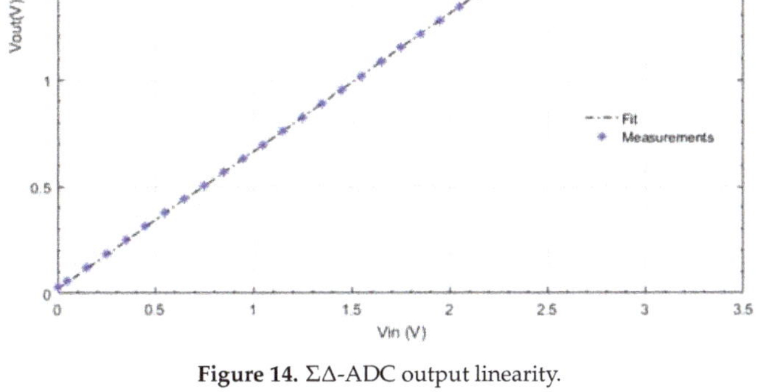

Figure 14. ΣΔ-ADC output linearity.

Figure 15 shows the gain versus frequency characteristic of the proposed ADC configuration, for 10-bit (red dots) and 12-bit (green dots) output resolution, which presents the typical *sinc* digital filter shape, matching with the previous DC characterization. According to this figure, a 12-bit ADC is selected, to enhance resolution and preserving the frequency of operation up to the 10 kHz range.

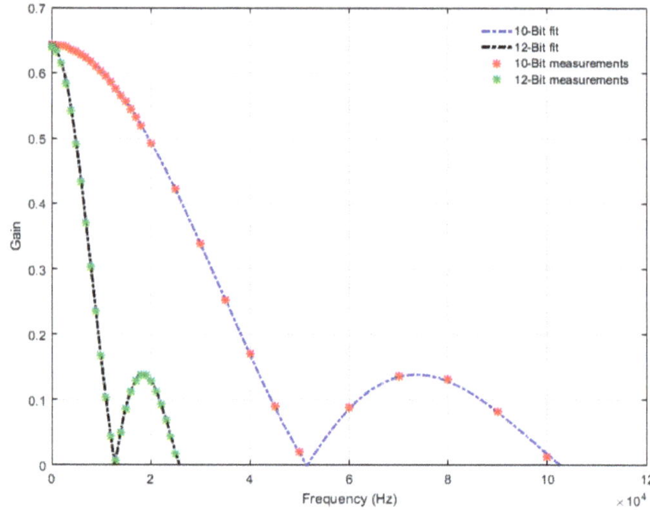

Figure 15. ΣΔ-ADC gain response.

Finally, Figure 16 presents the voltage values acquired using the ADC (red dots) from a 5 kHz sine signal, and the corresponding cosine values (recovered from the RAM memory using the instant sine values acquired). Both show a good matching when represented over their corresponding original full signals. These values are the inputs for the mixing operation.

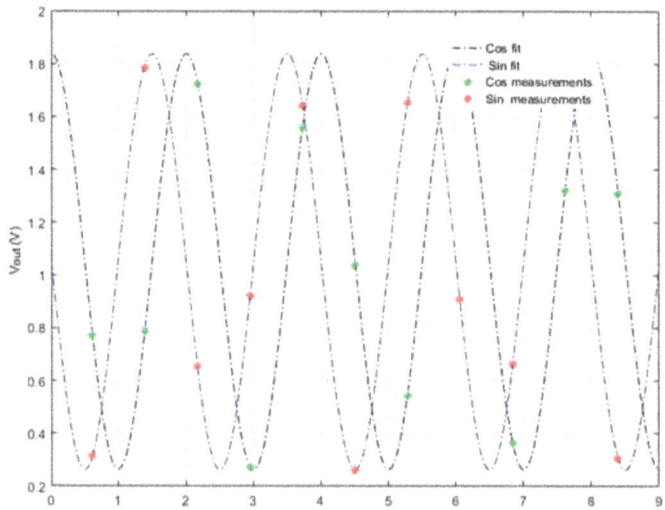

Figure 16. Signal acquisition using the proposed ADC for a sine voltage (red dots). Cosine signal is digitally recovered from the Main Memory using the sine values.

3. Results

The EIS system schematic, considering a two-channel measurement approach, is shown in Figure 17, and the prototype photograph is shown in Figure 18. In Figure 17, Block (a) corresponds to the signal generation module, including the digital Counter Module in the corresponding core and the R-C low-pass filter ($R_\Delta = 2.2\ \text{k}\Omega; C_\Delta = 330\ \text{pF}$). It is followed by an impedance adapter element (a simple voltage follower), which allows transferring the generated signal to the cell under test. For real biological applications, this module can be replaced by voltage or current reduction modules suited for the target application. The Randles cell representing the impedance sample corresponds to Block (b). Note that each Randles cell is followed by a transimpedance amplifier (TIA) consisting on an Operational Amplifier with a feedback resistor R_{f2}, that converts the current I_Z provided by the biocell into a voltage value $V_Z = -R_{f2}I_Z$ for its digitization. Block (c) represents the implemented ΣΔ-ADC -based system that conforms the synchronously digitized impedance extraction; the passive components values are ($R_{f1} = 100\ \text{k}\Omega; R_{\Sigma\Delta} = 150\ \text{k}\Omega; R_C = 4.7\ \text{k}\Omega; C_{\Sigma\Delta} = 250\ \text{pF}$), where $R_C + R_{SD}$ corresponds to R_1 in Figure 11. This configuration allows calibrate the operation of the ADC (Figure 17) by setting the voltage value in (A) at the required values through the ΣΔ calibration pin without the need of deactivating the operation of the cell, thus with minimum waste of time.

Application to Impedance Spectroscopy

The system operation as a frequency response analyzer applied to impedance spectroscopy has been tested using the characteristic impedance of a biological model based on the bilayer lipid membrane presented in [28,29] (Figure 19). The associated Randles cell is modeled using three impedances whose values are: R_m = 434 kW, C_m = 580 nF and C_{dl} = 340 nF (Figure 17). The resistor R_s represents the impedance associated to the sensing electrodes, which can vary from negligible values up to a few MW. In this work, an intermediate value of 500 kW has been selected. The TIA active block in Figure 17, Block (c) is a MAX4231, and resistor R_{f2} = 500 kW to accommodate a full analog voltage input range of 0 to V_{DD}.

For a normalized amplitude excitation signal with operating frequency f_{in}, the corresponding output biosensor signal V_Z is given by

$$V_Z = -\frac{R_{f_2}}{|Z|}\sin(\omega_{in}t + \theta) \tag{5}$$

where Z represents the cell impedance.

The digitized values of the biosensor signal are multiplied in the corresponding microcontroller core by the respective digital sine and cosine values, and the results are averaged over an integer number n of signal periods (with n minimum = 5 as pointed in Section 2.3.2), so that:

$$Re = -\frac{R_{f_2}}{2|Z|}\cos\theta\,\hat{}\,Im = -\frac{R_{f_2}}{2|Z|}\sin\theta \tag{6}$$

Note that since each digitization at the ΣΔ-ADC requires a minimum of 2^{12} clock cycles (for a 12-bit resolution), the two mixing and accumulation cycles are performed in real time by computational resources in the same core. Thus, impedance magnitude and phase shift can be recovered as

$$|Z| = -\frac{R_{f_2}}{2\sqrt{Re^2 + Im^2}}\,\hat{}\,\theta = \tan^{-1}\left(\frac{Im}{Re}\right) \tag{7}$$

The impedance characterization has been performed for 18 frequency values in the 100 mHz to 10 kHz range at 12-bit resolution. Since for each of these measurements an acquisition time of at least 5 signal cycles has been guaranteed, a total time of 100 s is required for the complete characterization over frequency. Figure 20a shows the impedance magnitude recovery performance compared to the ideal values, while Figure 20b presents the phase evolution. Figure 21 shows the impedance magnitude and phase relative error achieved estimating both values.

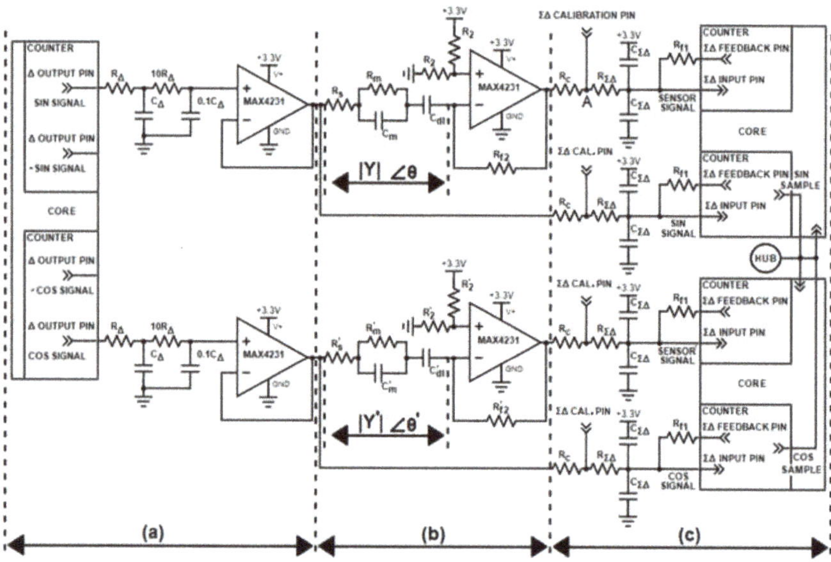

Figure 17. Block diagram for a two-channel Frequency Response Analyzer (FRA) using a single core as signal generator.

Figure 18. Two-channel prototype photograph.

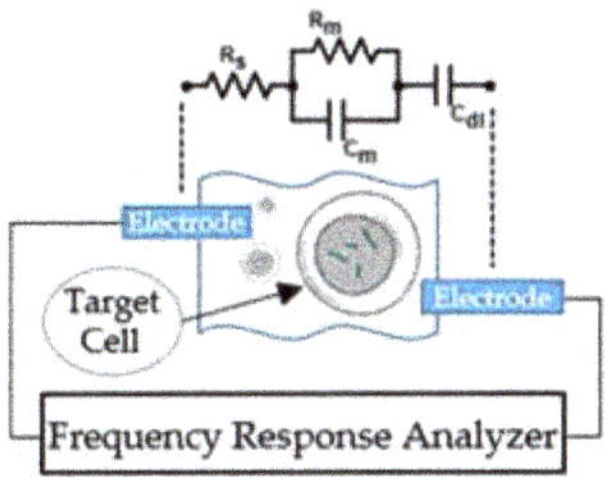

Figure 19. Application example of the proposed FRA system. The bilayer lipid membrane is characterized by applying a frequency-variable signal and measuring its response, using the proposed microcontroller device.

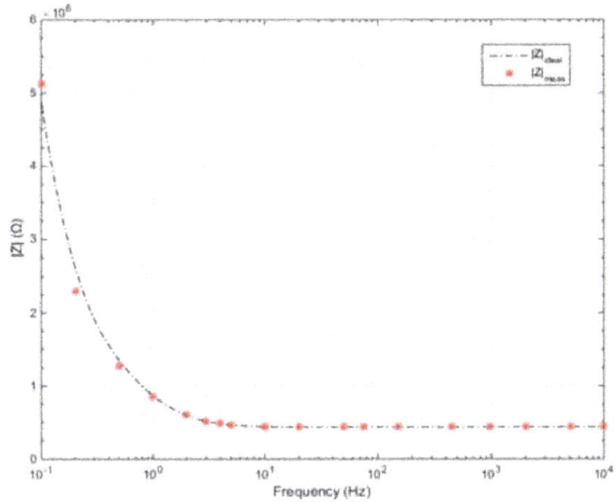

(a)

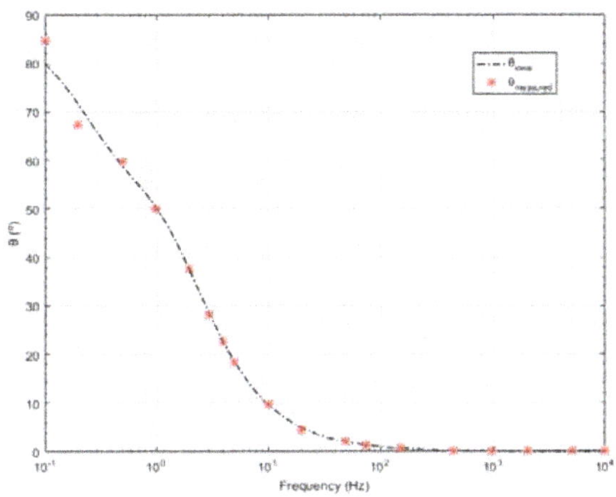

(b)

Figure 20. (**a**) Impedance magnitude recovery for 18 different frequencies in the 100 mHz to 10 kHz range. Red dots represent the experimental values recovered, dashed line is the ideal impedance value. (**b**) Recovered (red dots) phase values and ideal behavior.

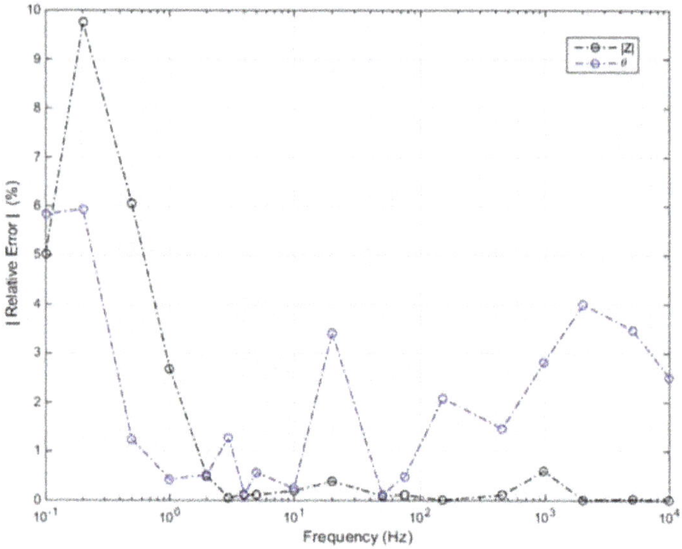

Figure 21. Relative error for estimation of impedance magnitude (black line) and phase (blue line).

4. Discussion

This paper has presented a compact multi-channel FRA-IS instrument that fully relies on a low cost Propeller multicore microcontroller, accomplishing a complete actuation-detection solution needing minimum additional external components. The excitation signal for impedance characterization is generated by a PDM-based generator running on a single core. This module generates up to two pairs of quadrature signals at a single frequency, so that the number of cells to be characterized at the same time can be highly extended by using adaptation modules (voltage followers in Figure 14) connected to the different signal generation ports. In this way, the (bio)impedance characterization processes can be highly parallelized, as it is demanded by current array-based applications. Signal recovery for impedance characterization is performed using the rest of available cores in the microcontroller, being possible to simultaneously acquire up to 7 impedance measurements, one per core. This number can be proportionally widened by extending the number of microcontrollers where the $\Sigma\Delta$ algorithm is implemented in the reading process, provided they receive the excitation and sensor output signals.

On the other hand, the proposed architecture allows impedance characterization using different excitation frequencies for several Randles cells in parallel, just assigning a different generation core per frequency. In fact, a more general solution could consist on assigning the cores of a Propeller microcontroller to generate the different frequencies (implementing the corresponding PDM and LPF per core), while using additional Propeller microcontrollers dedicated to the acquisition and impedance measurement tasks, implementing the $\Sigma\Delta$-ADC in each of the processor cores.

Therefore, the proposed EIS system constitutes a fully operative flexible and modular solution, suitable for multi-channel acquisition while complying the features of portability, and with an enhanced trade-off between low cost and measurement performance compared to similar devices in the literature. Reviewing the state-of-art, a direct approach relies on the use of the component AD5933 or the newer ADuCM350, which shown satisfactory results in different applications [4,11], but performing one measurement process at a time, and at the cost of the high computing power required to implement the Discrete Fourier Transform (DFT) compared to the FRA technique. Similarly, comparable alternative low-cost microcontroller-based EIS architectures [30,31] need more complex external hardware, providing typically worst resolution while covering a similar frequency range and for a single channel impedance measurement. Finally, Table 1 compares the implemented analyzer performances with those of previous multichannel implementations operating at a similar frequency range. It can be seen that our proposal achieves better resolution over a wider frequency range.

Table 1. Comparative analysis. EIS: Electrochemical impedance spectroscopy.

Characteristic	[28]	[23]	This Work
Technology	0.5 mm	0.13 mm	COTS
Supply voltage	3.0 V	1.2 V	3.3 V
Signal Bandwidth	10 mHz to 100 Hz	100 mHz to 10 kHz	1.1 μHz to 10 kHz
Channels	100	16	7/microcontroller
Waveform Generation	External	R-2R DAC	2nd Order Δ-DAC by PDM
Generator Resolution	N/A	8-bit tuning	14-bit coarse tuning + 32-bit fine tuning
Readout Structure	Lock-in IDC	DS-MADC	1st order ΣΔADC
Conversion rate	10 kHz	10 kHz	20 kHz
Effective number of bits (ENOB)	8 bits	9.3 bits	12 bits
THD worst case	N/A	−44 dB	−48.5 dB
EIS max. relative error	N/A	8.4%	10%

Therefore, the proposed approach succeeds in reducing instrument dimensions to allow automatic and in-situ multichannel impedance measurements, while improving the measurement performance using low-cost commercial components off the shelf (COTS).

Author Contributions: Conceptualization, A.S.-G., N.M. and B.C.; Methodology, N.M. and B.C.; hardware and software design and implementation, A.S.-G.; formal analysis, A.S.-G. and P.A.M.; experimental analysis, A.S.-G., N.M. and B.C.; writing—review and editing, N.M., B.C. and A.S.-G.; supervision, N.M. and B.C.

Funding: This research was funded by MINECO-FEDER, UE, grant number TEC2015-65750-R.

Acknowledgments: Authors would like to acknowledge the use of Servicio General de Apoyo a la Investigación-SAI, University of Zaragoza.

Conflicts of Interest: The authors declare no conflict of interest. The funders had no role in the design of the study; in the collection, analyses, or interpretation of data; in the writing of the manuscript, or in the decision to publish the results.

References

1. Randles, J.E.B. Kinetics of rapid electrode reactions. *Discuss. Faraday Soc.* **1947**, *1*, 1–11. [CrossRef]
2. Yang, L.; Bashir, R. Electrical/electrochemical impedance for rapid detection of foodborne pathogenic bacteria. *Biotechnol. Adv.* **2008**, *26*, 135–150. [CrossRef] [PubMed]
3. Liu, J.T.; Settu, K.; Tsai, J.Z.; Chen, C.J. Impedance sensor for rapid enumeration of *E. coli* in milk samples. *Electrochim. Acta* **2015**, *182*, 89–95. [CrossRef]
4. Durante, G.; Becari, W.; Lima, F.; Peres, H. Electrical Impedance Sensor for Real-Time Detection of Bovine Milk Adulteration. *IEEE Sens. J.* **2016**, *16*, 861–865. [CrossRef]
5. Sitkov, N.; Zimina, T.; Soloviev, A. Development of Impedimetric Sensor for *E. coli* M-17 Antibiotic Susceptibility Testing. In Proceedings of the IEEE Conference of Russian Young Researchers in Electrical and Electronic Engineering (EIConRus), Moscow, Russia, 29 January–1 February 2018.
6. Conesa, C.; Ibáñez Civera, J.; Seguí, L.; Fito, P.; Laguarda-Miró, N. An Electrochemical Impedance Spectroscopy System for Monitoring Pineapple Waste Saccharification. *Sensors* **2016**, *16*, 188. [CrossRef] [PubMed]
7. Mansoorifar, A.; Kokñu, A.; Ma, S.; Raj, G.; Beskok, A. Electrical Impedance Measurements of Biological Cells in Response to External Stimuli. *Anal. Chem.* **2018**, *90*, 4320–4327. [CrossRef]
8. Barsoukov, E.; Macdonald, J.R. *Impedance Spectroscopy, Theory, Experiment and Applications*, 3rd ed.; Wiley: Hoboken, NJ, USA, 2018; ISBN 9781119074083.

9. Pérez-Navarrete, J.B. Establishment of Electrical Equivalent Circuits form Electrochemical Impedance Spectroscopy Study of Corrosion Inhibition of Steel by Imidazolium Derived Ionic Liquids in Sulphuric Acid Solution. In Proceedings of the 7th International Conference on Electrical Engineering, Computing Science and Automatic Control (CCE), Tuxtla Gutiérrez, México, 8–10 September 2010.
10. Alchaddoud, A.; Ibrahem, G.; Canale, L.; Zissis, G. Impedance spectroscopy and evolution of the equivalent circuit model for large area organic light emitting diodes aged under stress. In Proceedings of the 18th International Conference on Environmental and Electrical Engineering, Palermo, Italy, 12–15 June 2018.
11. Kaur, N.; Bhalla, S.; Shanker, R.; Panigrahi, R. Experimental Evaluation of Miniature Impedance Chip for Structural Health Monitoring of Prototype Steel/RC Structures. *Exp. Tech.* **2016**, *40*, 981–992. [CrossRef]
12. Deng, Z.; Zhang, Z.; Lai, Y.; Liu, J.; Li, J.; Liu, Y. Electrochemical Impedance Spectroscopy Study of a Lithium/Sulfur Battery: Modeling and Analysis of Capacity Fading. *J. Electrochem. Soc.* **2013**, *160*, A553–A558. [CrossRef]
13. Oprea, M.I.; Spataru, S.V.; Sera, D.; Poulsen, P.B.; Thorsteinsson, S.; Basu, R.; Andersen, A.R.; Frederiksen, K. Detection of Potential Induced Degradation in c-Si PV Panels Using Electrical Impedance Spectroscopy. In Proceedings of the 43rd IEEE Photovoltaic Specialists Conference (PVSC), Portland, OR, USA, 5–10 June 2016.
14. Li, H.; Liu, X.; Li, L.; Mu, X.; Genov, R.; Mason, A.J. CMOS Electrochemical Instrumentation for Biosensor Microsystems: A Review. *Sensors* **2017**, *17*, 74. [CrossRef] [PubMed]
15. Analog Devices. *AD5933 Datasheet*; Analog Devices: Norwood, MA, USA, 2017.
16. Rairigh, D.; Mason, A.; Chao, Y. Analysis of On-Chip Impedance Spectroscopy Methodologies for Sensor Arrays. *Sens. Lett.* **2006**, *4*, 398–402. [CrossRef]
17. Temiz, Y.; Lovchik, R.D.; Kaigala, G.V.; Delamarche, E. Lab-on-a-chip devices: How to close and plug the lab? *Microelectron. Eng.* **2015**, *132*, 156–175. [CrossRef]
18. Kusche, R.; Klimach, P.; Ryschka, M. A Multichannel Real-Time Bioimpedance Measurement Device for Pulse Wave Analysis. *IEEE TBioCAS* **2018**, *12*, 614–622. [CrossRef] [PubMed]
19. Valente, V.; Demosthenous, A. Wideband Fully-Programmable Dual-Mode CMOS Analogue Front-End for Electrical Impedance Spectroscopy. *Sensors* **2016**, *16*, 1159. [CrossRef] [PubMed]
20. Yúfera, A.; Rueda, A. A CMOS Bio-Impedance Measurement System. In Proceedings of the 12th International Symposium on Design and Diagnostics of Electronic Circuits & Systems, Liberec, Czech Republic, 15–17 April 2009.
21. Onet, R.; Rusu, A.; Rodriguez, S. High-Purity and Wide-Range Signal Generator for Bioimpedance Spectroscopy. *IEEE Trans. Circuits-II* **2017**, 1–5. [CrossRef]
22. Chen, T.; Wu, W.; Wei, C.; Darling, R.; Liu, B. Novel 10-bit Impedance-to-Digital Converter for Electrochemical Impedance Spectroscopy Measurements. *IEEE TBioCAS* **2017**, *11*, 370–379. [CrossRef] [PubMed]
23. Mazhab-Jafari, H.; Soleymani, L.; Genov, R. 16-Channel Impedance Spectroscopy DNA Analyzer With Dual-Slope Multiplying ADCs. *IEEE TBioCAS* **2012**, *6*, 468–478. [CrossRef]
24. Parallax Semiconductor. *Propeller P8X32A Datasheet, Rev. 1.4. 6/14/2011*; Parallax Semiconductor: Rocklin, CA, USA, 2011.
25. Parallax Semiconductor. *Propeller P8X32A Counters Application Note AN001*; Parallax Semiconductor: Rocklin, CA, USA, 2011.
26. Park, S. *Principles of Sigma-Delta Modulation for Analog-to-Digital Converters*; Motorola: Chicago, IL, USA, 2008.
27. Parallax Semiconductor. *Sigma-Delta Analog to Digital Conversion Application Note AN008*; Parallax Semiconductor: Rocklin, CA, USA, 2011.
28. Yang, C.; Jadhav, S.R.; Worden, R.M.; Mason, A.J. Compact Low-Power Impedance-to-Digital Converter for Sensor Array Microsystems. *IEEE JSSC* **2009**, *44*, 2844–2855. [CrossRef]
29. Burkhard, R.; Braach-Maksvytis, V.; Cornell, B.A.; King, L.G.; Osman, P.D.; Pace, R.J.; Wieczorek, L. Tethered Lipid Bilayer Membranes: Formation and Ionic Reservoir Characterization. *Langmuir* **1998**, *14*, 648–659.

30. Corbellini, S.; Vallan, A. Arduino-based portable system for bioelectrical impedance measurement. In Proceedings of the IEEE International Symposium on Medical Measurements and Applications (IEEE MeMeA), Lisbon, Portugal, 11–12 June 2014.
31. Grassini, S.; Corbellini, S.; Angelini, E.; Ferraris, E.; Parvis, M. Low-cost impedance spectroscopy system based on logarithmic amplifier. *IEEE Trans. Instrum. Meas.* **2015**, *64*, 1110–1117. [CrossRef]

electronics

MDPI

Article

5GHz CMOS All-Pass Filter-Based True Time Delay Cell

Seyed Rasoul Aghazadeh [1,*], Herminio Martinez [1] and Alireza Saberkari [2]

[1] Department of Electronics Engineering, Technical University of Catalonia (UPC)-BarcelonaTech, 08034 Barcelona, Spain; herminio.martinez@upc.edu

[2] Department of Electronics Engineering, University of Guilan, 4199613776 Rasht, Iran; a_saberkari@guilan.ac.ir

* Correspondence: rasoul.aghazadeh@upc.edu; Tel.: +34-93-413-7290

Received: 9 November 2018; Accepted: 21 December 2018; Published: 22 December 2018

Abstract: Analog CMOS time-delay cells realized by passive components, e.g., lumped LC delay lines, are inefficient in terms of area for multi-GHz frequencies. All-pass filters considered as active circuits can, therefore, be the best candidates to approximate time delays. This paper proposes a broadband first-order voltage-mode all-pass filter as a true-time-delay cell. The proposed true-time-delay cell is capable of tuning delay, demonstrating its potential capability to be used in different systems, e.g., RF beam-formers. The proposed filter achieves a flat group delay of over 60 ps with a pole/zero pair located at 5 GHz. This proposed circuit consumes only 10 mW power from a 1.8-V supply. To demonstrate the performance of the proposed all-pass filter, simulation results are conducted by using Virtuoso Cadence in a standard TSMC 180-nm CMOS process.

Keywords: all-pass filter; CMOS; time delay; broadband; true-time-delay

1. Introduction

All-pass filters as delay cells have a variety of applications in signal processing and communication systems, like equalizers and analog/RF beam-formers [1–6]. In these circuits, the amplitude of the input signal is constant over the desired frequency band, while creating a frequency-dependent delay. There are several reported approaches to approximately realize delay, such as transmission lines and lumped LC delay lines [7,8], which are passive components and, thus, are area inefficient, and also phase shifters for narrow-band frequencies [9–14]. Apart from these circuits, an active RF all-pass filter can be the best option to approximate delay due to its size and delay to area ratio [15,16].

There are many voltage-mode all-pass filters reported over the last one decade, which operate in broadband frequencies and have different applications [15–21]. In some applications, e.g., RF beam-forming, delay stages as delay cells are normally realized by cascading first-order all-pass filters in order to achieve a desired delay [15–17]. However, there are just a few first-order voltage-mode all-pass filters for wide frequency ranges in the literature [15–17,22]. This is because these analog circuits should possess important specifications like wide bandwidth, efficient area, low cost, and power consumption, and high delay amount to be considered as practical and efficient systems. Furthermore, recent circuits have been taking advantage of tunability, since it is one of the key features of signal processing and communication systems [15,16,18,23].

A broadband first-order voltage-mode all-pass filter as a true-time-delay cell is introduced in this paper. The proposed all-pass filter is comprised of two transistors, two resistors, and one grounded inductor. This circuit demonstrates a large amount of delay in a single delay cell through a wide frequency band. The amount of delay can be controlled within the frequency range of interest. Moreover, circuit optimization is carried out to increase the operating frequency and improve the performance of the filter, in particular, in high frequencies.

The structure of this paper is as follows: Section 2 describes the structure of proposed all-pass filter and provides theoretical analyses. In Section 3, circuit optimization technique and tunability are presented, and also the parasitic effects of the proposed filter are evaluated. Section 4 provides results and ultimately a discussion is provided in Section 5.

2. Proposed First-Order All-Pass Filter

Figure 1 shows the block level of the first-order voltage-mode all-pass filter. As shown, a first-order all-pass filter can be approximated by the combination of two sections: a low-pass section with a DC gain of 2 and a unity gain section [24]. Therefore, its ideal transfer function is given as:

$$H(s) = e^{-s\tau} \approx \frac{-2}{1+s(\tau/2)} + 1 = -\frac{1-s(\tau/2)}{1+s(\tau/2)}, \tag{1}$$

where τ is the time delay. Ideally, the gain of the transfer function is 1 and its phase is linear versus the frequency.

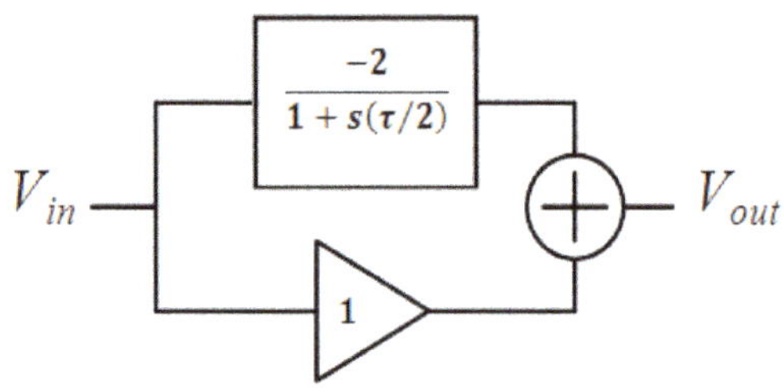

Figure 1. Block diagram of the first-order all-pass filter.

Figure 2 illustrates the block diagram and schematic of the proposed broadband first-order voltage-mode all-pass filter. In this filter, transistor M_1, inductor L, and resistor R_L form the low-pass part, while transistor M_2 and resistor R_L comprise the unity-gain part. In other words, M_1 and M_2 are, respectively, common-source (CS) and common-gate (CG) configurations to convert the input voltage signal into current. At the output node, the drain currents of M_1 and M_2 are subtracted to realize an all-pass function. Then, the output signal will be converted back to voltage by the load resistor R_L.

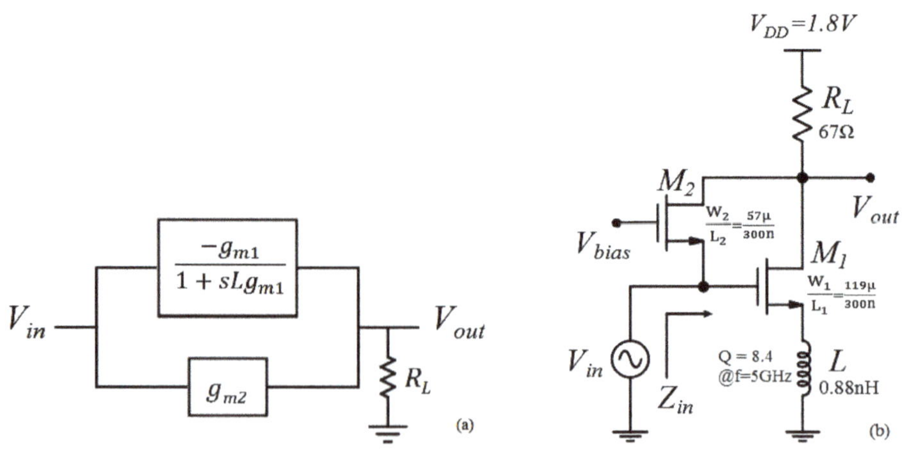

Figure 2. (**a**) Block diagram and (**b**) schematic of the proposed first-order all-pass filter.

Ignoring the parasitics of the transistors (the parasitic effects will be assessed in Section 3) for simplicity, the transfer function of the proposed first-order all-pass filter can be determined by:

$$\frac{V_{out}}{V_{in}}(s) = -\frac{g_{m1}R_L}{1+sLg_{m1}} + g_{m2}R_L = -R_L(g_{m1}-g_{m2})\cdot\frac{1-sL\frac{g_{m1}g_{m2}}{g_{m1}-g_{m2}}}{1+sLg_{m1}}, \tag{2}$$

where g_{m1} and g_{m2} are the transconductances of M_1 and M_2, respectively. If $g_{m1} = 2g_{m2}$ and $g_{m2}R_L = 1$, an all-pass structure will be realized with the same frequency of the left-plane pole and right-plane zero, resulting in twice the phase and group delay responses of an all-pass circuit. As a consequence, the transfer function in (2) can be simplified as:

$$\frac{V_{out}}{V_{in}}(s) = -\frac{1-sLg_{m1}}{1+sLg_{m1}}. \tag{3}$$

The pole/zero frequency and phase response of the first-order all-pass filter can be given as:

$$|\omega_{p,z}| = \frac{1}{Lg_{m1}}, \tag{4}$$

$$\phi(\omega) = -2tan^{-1}(\omega L g_{m1}), \tag{5}$$

respectively, and, thus, group delay response is expressed by:

$$D(\omega) = -\frac{\partial\phi(\omega)}{\partial\omega} = 2Lg_{m1}\cdot\frac{1}{1+(\omega L g_{m1})^2}, \tag{6}$$

where ω is the angular frequency related to the frequency f through $\omega = 2\pi f$. The group delay is approximately equal to $2Lg_{m1}$ at low frequencies. However, this group delay is practically affected by parasitic inductances stemmed from, e.g., bonding wire and PCB and, thus, its value will be increased. The input impedance of the proposed all-pass filter can be simply approximated by considering the Miller effect on the parasitic capacitances of the transistor M_1 plus C_{gs2} given as:

$$C_{in} \approx \frac{\left(C_{gs1}+C_{gd1}\right)(3+sLg_{m1})}{1+sLg_{m1}} + C_{gs2}, \tag{7}$$

which its value affects the next delay stage for cascading purposes.

3. Circuit Optimization and Tunability

In order to contribute to the linearity and increase the operating frequency of the proposed all-pass filter, a variable resistor (R_d) is added to the unity-gain path as shown in Figure 3. In this case, a discrete tuning of delay can be carried out by changing the value of R_d and the bias voltage of M_2 as well, which adjusts g_{m2}. The R_d can be implemented by a switched resistors bank which can be implemented by CMOS transistors, with great ease.

The transfer function of the CG transistor of M_2 (the part inside the dotted box) is, therefore, given as:

$$H_{CG}(s) = \frac{g_{m2}R_L}{1+sC_{gd2}(R_L+R_d)}. \tag{8}$$

Its value for low and high frequencies is $H_{CG,LF} \approx g_{m2}R_L$ and $H_{CG,HF} \approx g_{m2}R_L/C_{gd2}(R_L+R_d)$, respectively. Hence, the R_d will affect the frequency response of the proposed filter at higher frequencies. Note that $g_{m2}R_L$ (i.e., the unity gain section) is no longer equal to 1 at high frequencies, but via varying the bias voltage of M_2, g_{m2} changes and, therefore, the two conditions $g_{m1} = 2g_{m2}$ and $g_{m2}R_L = 1$ will be satisfied.

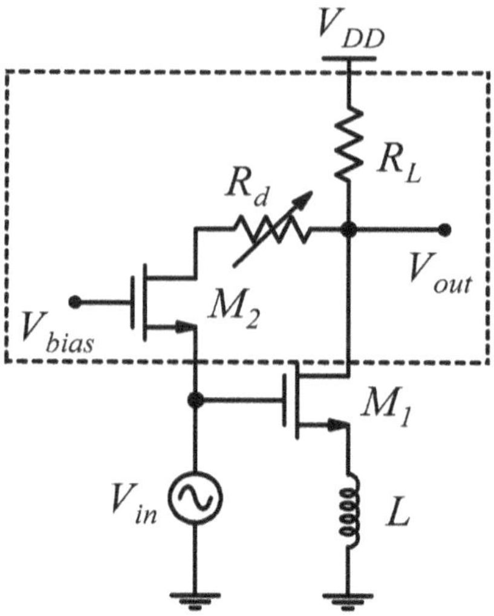

Figure 3. The optimization and tunability technique.

Non-Ideality Analysis

To analyze accurately the performance of the proposed all-pass filter in Figure 3 at high frequencies, the finite output impedances (g_{ds}) and parasitic capacitances (C_{gs} and C_{gd}) of the transistors M_1 and M_2 should be considered. Therefore, the transfer function in (2) can be rewritten as:

$$\frac{V_{out}}{V_{in}}(s) \approx -\frac{R_L(g_{m1}-g_{ds1})-s\left(Lg_{m1}g_{ds1}R_L+C_{gd1}R_L\right)}{sL(g_{m1}+g_{ds1})\left[1+g_{ds1}R_L+R_L\left(g_{ds1}+sC_{gd1}\right)\right]+1+R_L\left(g_{ds1}+sC_{gd1}\right)} + \frac{R_L(g_{m2}+g_{ds2})}{sC_{gd2}(R_L+R_d)+1+g_{ds2}(R_L+R_d)}. \tag{9}$$

If $g_{m1,2} \gg g_{ds1,2}$, $g_{ds1,2}R_L \ll 1$, and $g_{ds2}R_d \ll 1$, the transfer function in Equation (9) can be simplified as:

$$\frac{V_{out}}{V_{in}}(s) = -\frac{g_{m1}R_L\left(1-s\frac{Lg_{m1}g_{ds1}+C_{gd1}}{g_{m1}}\right)}{\left(1+sC_{gd1}R_L\right)(1+sLg_{m1})} + \frac{g_{m2}R_L}{1+sC_{gd2}(R_L+R_d)}, \tag{10}$$

which includes additional parasitic poles and zero. These parasitic high-frequency poles stemmed from C_{gd1} and C_{gd2}, which are located at $1/C_{gd1}R_L$ and $1/C_{gd2}(R_L+R_d)$ respectively, are far beyond the dominant pole in Equation (4) since the values of R_L and R_d are small. Moreover, the additional right-plane zero $(g_{m1}/Lg_{m1}g_{ds1}+C_{gd1})$ is located at considerably higher frequencies, as well.

Additionally, small-signal analysis conducted on the proposed all-pass circuit indicates that the third parasitic pole stemmed from C_{gs1} will be located at:

$$\omega_{p3} = -\frac{\left[g_{m1}\left(1+\sqrt{1-\frac{4C_{gs1}}{Lg_{m1}^2}}\right)\right]}{2C_{gs1}} \approx -\frac{g_{m1}}{C_{gs1}}, \tag{11}$$

which is far beyond the dominant pole in Equation (4). It can be noted that the order of the proposed circuit will increase and convert to the second one if the absolute value of C_{gs1}, which is process-dependent, is large enough. Consequently, choosing an appropriate CMOS process can reduce the effect of the C_{gs1} on the frequency response of the circuit.

4. Results

The proposed first-order all-pass filter is designed in a standard 180-nm TSMC CMOS process and results are obtained using Virtuoso Cadence. The proposed all-pass filter is simulated without and with the R_d. The power consumption of the proposed broadband true-time-delay cell is only 10 mW from a 1.8-V supply voltage.

Figure 4 shows the gain and phase responses of the proposed filter under different values of the R_d. As it can be observed, the gain of the proposed filter without the R_d (i.e., $R_d = 0\ \Omega$) is almost -0.5 dB due to the existence of the parasitic capacitors and finite output impedances of the transistors. Furthermore, the proposed filter does not achieve desired (flat) gain responses at higher frequencies, whereas by varying the value of the R_d, better gain responses are proved at these frequencies. As seen, the pole/zero frequency of the proposed circuit with $R_d = 120\ \Omega$ is 5 GHz (i.e., the point where phase is 90°), indicating a 14% bandwidth improvement compared to once $R_d = 0\ \Omega$ (i.e., 4.4 GHz).

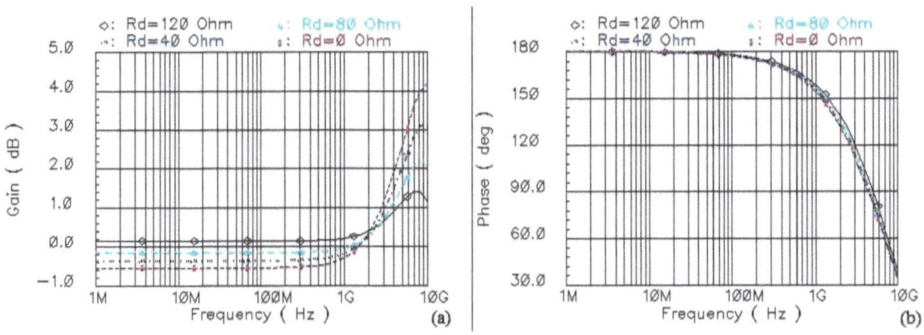

Figure 4. Simulated results for (**a**) gain response and (**b**) phase response of the proposed first-order all-pass filter under different values of the R_d.

The group delay responses of the proposed all-pass filter for different values of the R_d are shown in Figure 5. As it can be seen, the delay can be controlled by varying the R_d. The group delay is equal to about 59 ps, when $R_d = 120\ \Omega$. This group delay value is very close to the theoretical one in Equation (6), with an error of around 11%.

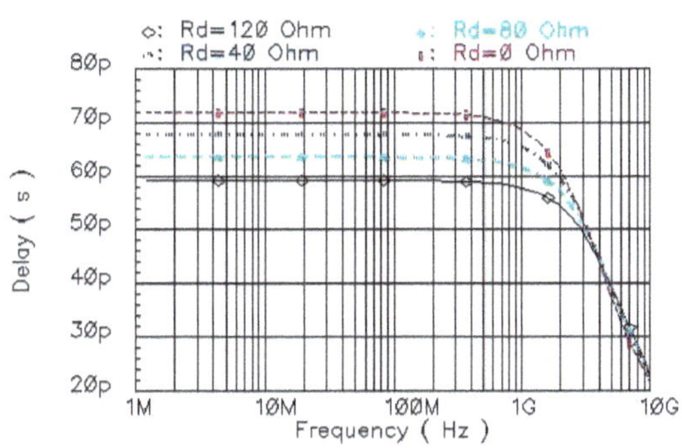

Figure 5. Simulated group delay responses of the proposed first-order all-pass filter under different values of the R_d.

In Figure 6, the input-referred noise response of the all-pass filter is shown when $R_d = 120\ \Omega$. The input-referred noise value is approximately 2.36 nV/sqrt (Hz) by the frequency of 1 GHz. Figure 7 shows the noise figure of the proposed all-pass filter with $R_d = 120\ \Omega$, which is <15 dB over the frequency band. The input-referred 1-dB compression point (P_{1dB}) and input-referred third-order intercept point (IIP3) responses of the first-order all-pass filter with $R_d = 120\ \Omega$ are shown in Figure 8. The input-referred P_{1dB} and IIP3 are -1.9 dBm and 16.6 dBm at 2.5 GHz, respectively.

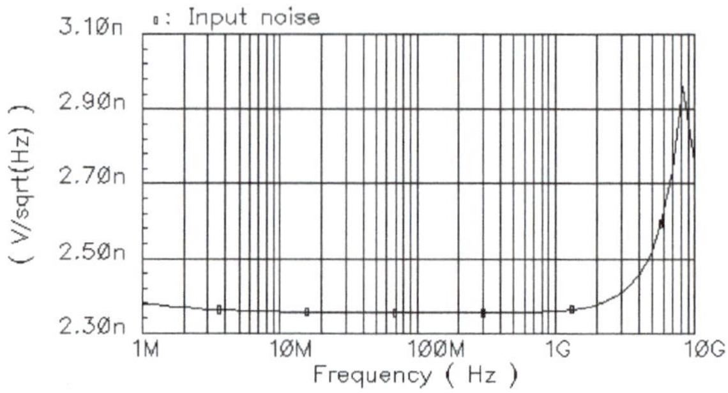

Figure 6. Simulated input-referred noise response of the proposed first-order all-pass filter.

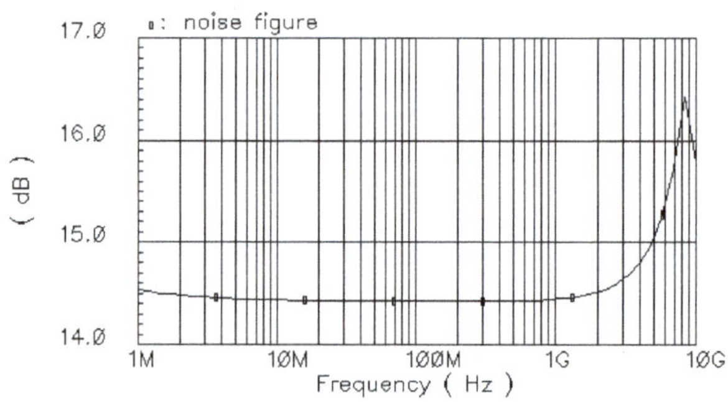

Figure 7. Simulated noise figure response of the proposed first-order all-pass filter.

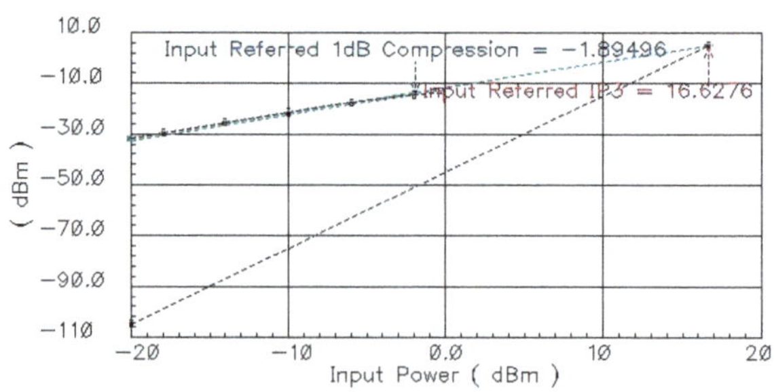

Figure 8. Simulated input-referred P_{1dB} and input-referred IIP3 responses of the proposed first-order all-pass filter.

Since the amount of group delay is affected by the mismatch and is basically process, voltage, and temperature (PVT) dependent, we should therefore consider the effect of these variations on the proposed true-time-delay cell. Figure 9 illustrates Monte Carlo simulation results, which are performed with a Gaussian distribution and 100 iterations, when $R_d = 120\ \Omega$. As it can be seen, the difference between group delay responses due to the mismatch is very small. Although the gain, P_{1dB}, and IIP3 will be affected by the mismatch, these variations can be minimized by changing the bias voltage of M_2. The group delay responses of the proposed filter with $R_d = 120\ \Omega$ for different supply voltages and temperatures are shown in Figure 10. The delay degrades by 15% because of the temperature variations.

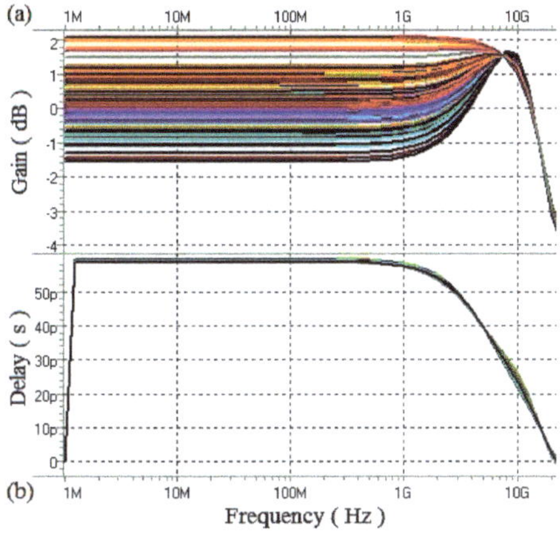

Figure 9. Monte Carlo simulation results for (**a**) gain response and (**b**) group delay response of the proposed first-order all-pass filter.

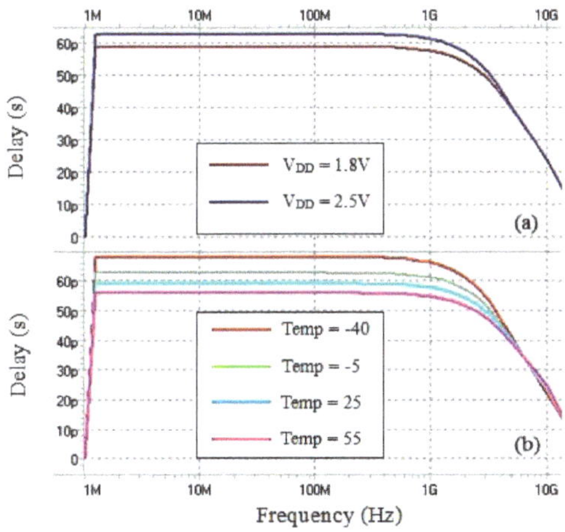

Figure 10. Simulated group delay responses of the proposed first-order all-pass filter for (**a**) different supply voltages and (**b**) different temperatures.

A comparison between recently reported voltage/current all-pass filters and the proposed true-time-delay cell is presented in Table 1. Comparing the results of the first-order voltage-mode all-pass filters, the proposed filter has improved the frequency range compared to the filter in [15]. Moreover, the power consumption and delay tuning can be highlighted and compared with the filter in [22], in which the delay could not be tuned.

Table 1. Performance summary and comparison between broadband all-pass filters.

Reference	Technology	Mode	Order	Frequency (GHz)	Max. Delay (ps)	P_{1dB} (dBm)	IIP3 (dBm)	Power (mW/V)
[15]	140-nm CMOS	Voltage	1st	1–2.5	61 [1]	N/A	N/A	10 [2]/1.5
[19]	SiGe2RF HBT	Voltage	2nd	3–10	75	−1	N/A	38.8/2.5
[20]	130-nm CMOS	Voltage	2nd	6	55	−5.5	2	18.5/1.5
[22]	130-nm CMOS	Voltage	1st	9	49 [3]	−2	8.5	20.4/1.5
[25]	130-nm CMOS	Current	1st	0.3–5.1	82	N/A	N/A	6.15/1.5
[26]	180-nm CMOS	Voltage	2nd	3–12	8.5	14.6	22.6	12/1.8
This work	180-nm CMOS	Voltage	1st	5	59 [4]	−1.9	16.6	10/1.8

[1] A maximum delay of 550 ps was achieved by three fine and six coarse delays. [2] A maximum power of 90 mW was consumed by three fine and six coarse delays. [3] Pre-layout group delay of 33 ps expected for the filter. [4] Simulated group delay value can be increased by varying the value of variable resistor in the proposed filter.

5. Discussion

Compared to the bulky LC delay lines, active filters can be good alternatives to approximate delays as these filters occupy smaller area. This paper presents a broadband first-order voltage-mode all-pass filter as an active circuit. Via an optimization technique, 14% bandwidth extension is achieved. The proposed first-order all-pass filter demonstrates a flat group delay of approximately 60ps through a bandwidth of 5 GHz, while consuming merely 10 mW power. Unlike the active all-pass filter in [22], the proposed filter has a DC-gain of 1 in its voltage transfer function and consequently there is no need for the gain adjustment via additional circuits or components. Furthermore, the proposed circuit proves a frequency range wider than that of the reported active filter in [15] (pre-layout pole frequency of 2.63 GHz), however at a larger area. The proposed all-pass filter is almost linear and achieves the input-referred P_{1dB} of −1.9 dBm and the input-referred IIP3 of 16.6 dBm. We will employ the proposed all-pass filter-based true-time-delay cell in analog RF beam-forming antennas for communication applications in our future work (see Figure 11). In timed-array receivers, tunable true-time-delay cells are exploited to align broadband signals received from a particular direction (θ).

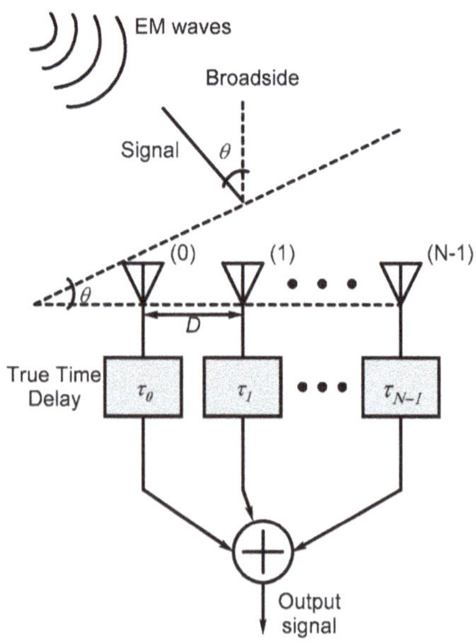

Figure 11. Block diagram of an N-element timed-array receiver.

Author Contributions: S.R.A. designed the circuit, and performed the theoretical analyses and simulation; and all authors wrote the paper.

Funding: This research received no external funding.

Conflicts of Interest: The authors declare no conflict of interest.

References

1. Buckwalter, J.; Hajimiri, A. An active analog delay and the delay reference loop. In Proceedings of the IEEE Radio Frequency Integrated Circuits (RFIC) Systems, Fort Worth, TX, USA, 6–8 June 2004.
2. Wang, Z. A Fully integrated W-band beamformer in 0.13 µm SiGe BiCMOS technology based on distributed true-time-delay architecture. In Proceedings of the IEEE International Nanoelectronics Conference (INEC), Chengdu, China, 9–11 May 2016.
3. Perera, S.M.; Ariyarathna, V.; Udayanga, N.; Madanayake, A.; Wu, G.; Belostotski, L.; Wang, Y.; Mandal, S.; Cintra, R.J.; Rappaport, T.S. Wideband N-beam arrays using low-complexity algorithms and mixed-signal integrated circuits. *IEEE J. Sel. Top. Signal Process.* **2018**, *12*, 368–382. [CrossRef]
4. Mailloux, R.J. *Phased Array Antenna Handbook*, 2nd ed.; Artech House: Norwood, MA, USA, 2005; ISBN 1-58053-689-1.
5. Van Trees, H.L. *Optimum Array Processing: Detection, Estimation, and Modulation Theory*; Wiley: New York, NY, USA, 2002; ISBN 0-471-09390-4.
6. Madanayake, A.; Ariyarathna, V.; Udayanga, N.; Belostotski, L.; Perera, S.K.; Cintra, R.J. Design of a low-complexity wideband analog true-time-delay 5 Beam array in 65 nm CMOS. In Proceedings of the IEEE International Midwest Symposium on Circuits and Systems (MWSCAS), Boston, MA, USA, 6–9 August 2017.
7. Schwartz, J.; Arnedo, I.; Laso, M.A.G.; Lopetegi, T.; Azana, J.; Plant, D. An electronic uwb continuously tunable time-delay system with nanosecond delays. *IEEE Microw. Wirel. Compon. Lett.* **2008**, *18*, 103–105. [CrossRef]
8. Chu, T.; Roderick, J.; Hashemi, H. An integrated ultra-wideband timed array receiver in 0.13 µm cmos using a path-sharing true time delay architecture. *IEEE J. Solid-State Circuits* **2007**, *42*, 2834–2850. [CrossRef]
9. Soer, M.; Klumperink, E.; Nauta, B.; van Vliet, F. A 1.5-to-5.0 GHz input-matched +2 dBm P_{1dB} all-passive switched-capacitor beamforming receiver front-end in 65 nm CMOS. In Proceedings of the IEEE International Solid-State Circuits Conference (ISSCC), San Francisco, CA, USA, 19–23 February 2012.
10. Ghaffari, A.; Klumperink, E.; Soer, M.; Nauta, B. Tunable high-q n-path band-path filters: Modeling and verification. *IEEE J. Solid-State Circuits* **2011**, *46*, 998–1010. [CrossRef]
11. Lien, Y.; Klumperink, E.; Tenbroek, B.; Strange, J.; Nauta, B. A high-linearity CMOS receiver achieving +44 dBm IIP3 and +13 dBm B_{1dB} for SAW-Less LTE radio. In Proceedings of the IEEE International Solid-State Circuits Conference (ISSCC), San Francisco, CA, USA, 5–9 February 2017.
12. Li, W.; Wang, W.; Chen, Y. A 0.5–3 GHz true-time-delay phase shifter for multi-antenna systems. In Proceedings of the IEEE 2nd Advanced Information Technology, Electronic and Automation Control Conference (IAEAC), Chongqing, China, 25–26 March 2017.
13. Elkind, J.; Goldberger, E.; Socher, E. 57–67 GHz highly compact bidirectional 3-bit phase shifter in 28 nm cmos. *IEEE Microw. Wirel. Compon. Lett.* **2018**, *28*, 1017–1019. [CrossRef]
14. Zhang, Y.; Huang, F.; Li, T.; Tang, X.; Jiang, N. A 1 V 2.4–6 GHz 6 bit vector-sum phase shifter with very low rms phase error and gain error. *Microw. Opt. Technol. Lett.* **2018**, *60*, 2467–2471. [CrossRef]
15. Garakoui, S.K.; Klumperink, E.; Nauta, B.; van Vliet, F. Compact cascadable gm-c all-pass true time delay cell with reduced delay variation over frequency. *IEEE J. Solid-State Circuits* **2015**, *50*, 693–703. [CrossRef]
16. Mondal, I.; Krishnapura, N. A 2 GHz bandwidth, 0.25–1.7 ns true-time-delay element using a variable-order all-pass filter architecture in 0.13 µm cmos. *IEEE J. Solid-State Circuits* **2017**, *52*, 2180–2193. [CrossRef]
17. Wijenayake, C.; Madanayake, A.; Belostotski, L.; Xu, Y.; Bruton, L. All-pass filter-based 2-D IIR filter-enhanced beamformers for AESA receivers. *IEEE Trans. Circuits Syst. I Regul. Pap.* **2014**, *61*, 1331–1342. [CrossRef]
18. Chen, Y.; Li, W. Campact and broadband variable true-time delay line with DLL-based delay-time control. *Circuits Syst. Signal Process.* **2018**, *37*, 1007–1027. [CrossRef]
19. Ulusoy, A.; Schleicher, B.; Schumacher, H. A tunable differential all-pass filter for uwb true time delay and phase shift applications. *IEEE Microw. Wirel. Compon. Lett.* **2011**, *21*, 462–464. [CrossRef]

20. Ahmadi, P.; Maundy, B.; Elwakil, A.S.; Belostotski, L.; Madanayake, A. A new 2nd-order all-pass filter in 130 nm cmos. *IEEE Trans. Circuits Syst. II Express Br.* **2016**, *63*, 249–253. [CrossRef]
21. Wijenayake, C.; Xu, Y.; Madanayake, A.; Belostotski, L.; Bruton, L. RF analog beamforming fan filters using cmos all-pass time delay approximations. *IEEE Trans. Circuits Syst. I Regul. Pap.* **2012**, *59*, 1061–1073. [CrossRef]
22. Ahmadi, P.; Taghavi, M.H.; Belostotski, L.; Madanayake, A. 6-GHz all-pass-filter-based delay-and-sum beamformer in 130 nm CMOS. In Proceedings of the IEEE International Midwest Symposium on Circuits and Systems (MWSCAS), College Station, TX, USA, 3–6 August 2014.
23. Maheshwari, S. Tuning approach for first-order filters and new current-mode circuit example. *IET Circuits Devices Syst.* **2018**, *12*, 478–485. [CrossRef]
24. Bult, K.; Wallinga, H. A cmos analog continuous-time delay line with adaptive delay-time control. *IEEE J. Solid-State Circuits* **1988**, *23*, 759–766. [CrossRef]
25. Ahmadi, P.; Belostotski, L.; Madanayake, A.; Haslett, J.W. 0.96-to-5.1 GHz 4-element spatially analog IIR-enhanced delay-and-sum beamformer. In Proceedings of the IEEE International Microwave Symposium (IMS), Honolulu, HI, USA, 4–9 June 2017.
26. Chen, Y.; Li, W. An ultra-wideband pico-second true-time-delay circuit with differential tunable active inductor. *Analog Integr. Circuits Signal Process.* **2017**, *91*, 9–19. [CrossRef]

electronics

MDPI

Article

An 11 GHz Dual-Sided Self-Calibrating Dynamic Comparator in 28 nm CMOS

Athanasios Ramkaj [1,*], Maarten Strackx [2], Michiel Steyaert [1] and Filip Tavernier [1]

1 ESAT-MICAS—KU Leuven, Kasteelpark Arenberg 10, B-3001 Leuven, Belgium; Michiel.Steyaert@esat.kuleuven.be (M.S.); Filip.Tavernier@esat.kuleuven.be (F.T.)

2 Nokia, Bell Labs, Copernicuslaan 50, B-2018 Antwerp, Belgium; maarten.strackx@nokia-bell-labs.com

* Correspondence: Athanasios.Ramkaj@esat.kuleuven.be; Tel.: +32-(0)16-374-727

Received: 28 October 2018; Accepted: 18 December 2018; Published: 21 December 2018

Abstract: This paper demonstrates a high-speed, low-noise dynamic comparator, employing self-calibration. The proposed dual-sided, fully-dynamic offset calibration is able to reduce the input-referred offset voltage by a factor of ten compared to the uncalibrated value without any speed or noise penalty and with less than 5% power overhead. Moreover, the implemented multi-stage topology significantly advances the state-of-the-art comparator performance, achieving the highest reported operating frequency, as well as the lowest delay slope and sensitivity to supply and common mode variations compared to existing works, with similar energy/comparison. This makes the proposed self-calibrating comparator an ideal candidate for high resolution (>10 b) multi-GHz Analog-to-Digital Converters (ADCs). The 28 nm bulk CMOS prototype measures an input-referred noise and calibrated offset of 0.82 mV and 0.99 mV, respectively clocked at 11 GHz, consuming only 0.89 mW from a 1 V supply, for an area of 0.00054 mm^2, including calibration.

Keywords: CMOS; dynamic comparator; offset calibration; high speed; low noise; low power; ADC

1. Introduction

Comparators are omnipresent building blocks in mixed-signal systems. Applications such as memories [1–3], data receivers [4–6], and Analog-to-Digital Converters (ADCs) [7–9] necessitate high speed, low noise/offset, yet power- and area-efficient designs. Their role in ADCs (Successive Approximation Register (SAR), flash, pipeline) (Figure 1) is of special importance, since they need to accurately translate small analog signals into digital information. Therefore, their noise, offset, and speed dictate the overall ADC performance.

Dynamic latch-type comparators [10–14] have become very attractive due to their fast regeneration time, enabled by strong positive feedback, and their zero static current consumption. Owing to their highly digital nature, these comparators are able to scale excellently into deep-submicron nodes. To maximize speed for minimal power, small transistor sizes with minimum parasitic loading on the critical nodes are preferred, but they come at the cost of a significantly increased offset [15,16].

One straightforward approach is to add amplification stages prior to the latch to suppress the offset voltage referred at the input [17]. This comes at the expense of increased power consumption due to the high gain and wide bandwidth requirements of these amplifiers. Alternatively, offset compensation schemes have been presented, in the form of adding digitally-controllable capacitors at the comparator outputs [18–20]. Further, several charge-pump implementations with extra logic and biasing voltages [21–23] have been proposed. However, all these approaches degrade the comparator and its calibration loop speed, increase design complexity and area, and limit robustness.

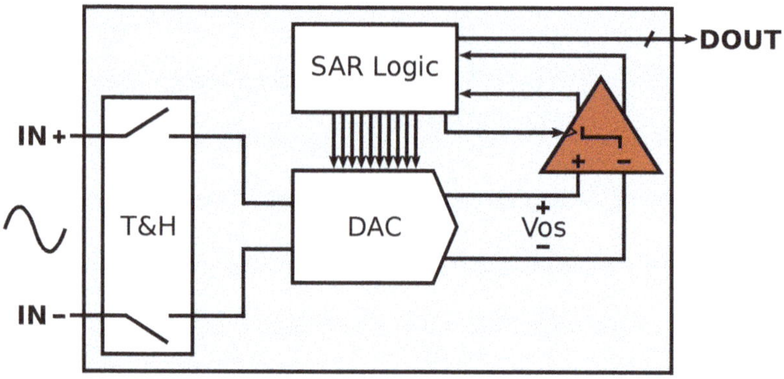

Figure 1. Top-level SAR ADC block diagram with its Track-and-Hold (T&H), DAC, comparator and SAR logic. A low-offset (V_{OS}), low-noise, and high-speed comparator determines the total performance and the accuracy on the output data (DOUT).

In this work, a dynamic self-calibration loop is proposed, able to reduce the comparator input-referred offset by a factor of ten. Its accuracy is limited only by the comparator noise, while its short critical path ensures no speed degradation. Finally, its compact size barely loads the comparator output. The loop's dual-sided implementation greatly enhances the calibration range, while its highly digital nature makes it readily scalable into deep-submicron nodes. Combined with a multi-stage, high-speed, low-noise dynamic comparator, >10 GHz operation is demonstrated with <1 mV input-referred noise (10 b accuracy), rendering it a perfect candidate for multi-GHz, high-resolution ADCs.

This paper is organized as follows. Section 2 describes the concept of the proposed dynamic comparator calibration. Section 3 discusses the circuit level implementation of the calibration and the comparator topology, supported by simulation data. Section 4 summarizes the measurement results along with a state-of-the-art comparison. Finally, Section 5 draws the conclusions of this work.

2. Dynamic Offset Calibration

The top-level architecture of the proposed offset calibration principle and its timing diagram are illustrated in Figure 2. The loop comprises a clocked comparator, two switched-capacitor calibration units (one for each side), and two offset compensating devices M_{SP}-M_{SN}. The calibration is performed simultaneously on both sides of the comparator (dual-sided), which maximizes the calibration range.

During calibration mode (CAL_EN is high), the common-mode voltage V_{CM} is applied to both comparator inputs. For a positive offset voltage V_{OS}, the differential comparator output will be positive. The output sign is sensed by the two calibration units, which start subtracting charge from C_{CALN} and adding charge to C_{CALP}, forcing nodes CALP/CALN to move in opposite directions to cancel this offset. When their difference reaches a certain value aV_{OS}, with a >1 depending on the size ratio of M_{SP}/M_{SN} and the input transistors (see Section 3.2), the comparator differential output changes sign alternately. This means that the offset has been compensated, and the comparator now sees an input difference dictated only by noise. During conversion mode (CAL_EN is low), C_{CALP}/C_{CALN} store the offset value, allowing the comparator to operate with canceled offset and decide correctly down to the noise level.

The comparator with the proposed calibration circuit can be easily incorporated in an ADC, where the already available periodic sampling clock can be used as CAL_EN, avoiding extra circuitry to generate that signal. Upon starting up the ADC, the calibration gradually corrects the comparator offset in multiple sampling cycles, by moving small packets of charge in each cycle until the required CALP/CALN difference is reached. This allows the offset calibration to run continuously in the background, tracking supply noise, which can affect the input-referred offset, offering a true dynamic cancellation, while not interfering with the ADC operation.

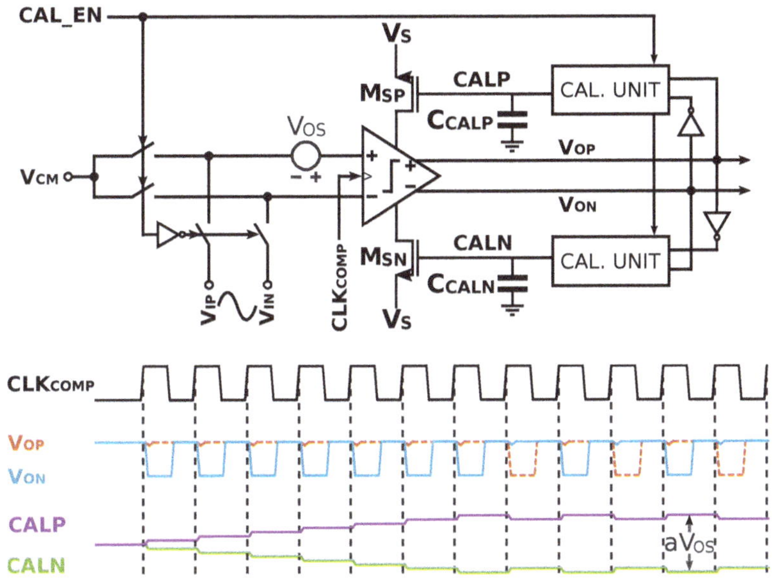

Figure 2. Top-level illustration of the proposed calibration (**top**) with its conceptual timing sequence (**bottom**).

3. Circuit Realization

3.1. Dynamic Self-Calibrating Loop

The single-ended version transistor-level implementation of the fully-differential self-calibration unit is shown in Figure 3. The unit consists of only eight switches plus one inverter. To eliminate the loading at the comparator output, all devices are minimum sized, while the fully-dynamic structure minimizes power overhead. Further, the delay between the comparator output and CALP/CALN is kept to a minimum of two transistors, such that the calibration loop does not impose a limitation on the total comparator speed. Unlike [21–23], there is no need for extra biasing circuitry to set the common-mode voltage on CALP/CALN. Here, it is gradually approaching the value set by proper sizing of the switches' on-resistance.

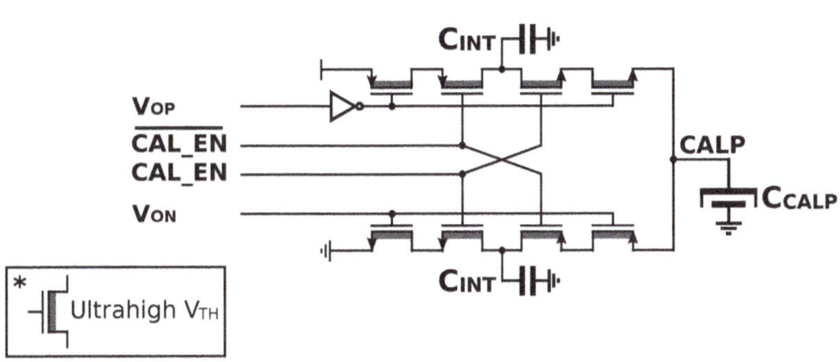

Figure 3. Transistor-level implementation of the self-calibration unit (single-ended shown for simplicity).

Charge is re-distributed between one of the internal capacitors C_{INT} and C_{CALP} and the amount of charge moved (calibration step) is controlled by the ratio of these capacitors and the time the calibration loop has available in each cycle, resulting in a wide compensation range. To reduce leakage on C_{CALP}, this capacitor has been constructed strictly as a Metal-Oxide-Metal (MOM) capacitor. Furthermore, ultrahigh V_{TH} transistors are employed at the expense of more cycles required to compensate a certain offset value. In this way, the calibration step in each cycle is traded-off with the number of cycles. This is never a problem when testing an ADC, since there is always an allocated start-up time, after which useful data are collected and processed. The ultimate accuracy limitation of the proposed calibration technique is the comparator sensitivity to various conditions (V_{DD} and/or V_{CM}). Therefore, a high-speed, low-noise, and low-input sensitivity comparator is needed to yield optimal results.

3.2. Comparator Core

The schematic of the comparator circuit where the proposed calibration is employed is shown in Figure 4. The comparator core incorporates a first amplification stage followed by a second amplifier/half latch and the final latch, in a fully-dynamic structure for low power operation [24]. The multi-stage configuration allows for a more orthogonal optimization of each stage for various trade-offs, which allows the comparator to simultaneously achieve both high speed and low noise.

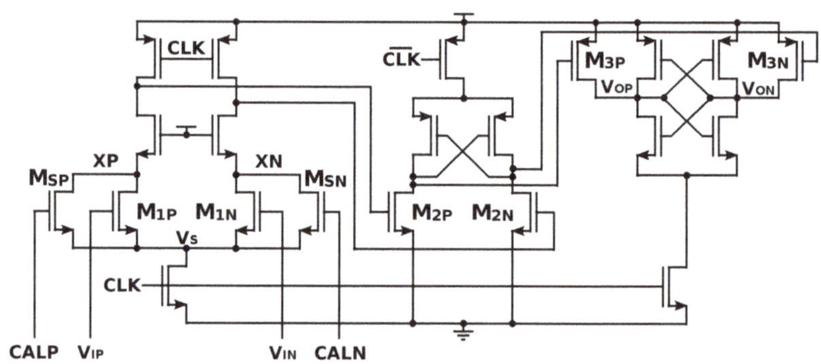

Figure 4. Comparator core with the extra offset compensating pair.

The required offset calibration pair M_{SP}/M_{SN} is connected in parallel to the main input pair M_{1P}-M_{1N}. The differential gate voltage of the extra pair is varied in the opposite direction to that of the main pair, in order to reverse the offset. This additional input pair leaks the charge from XP/XN without integration, which deteriorates the comparator noise performance. Therefore, the dimensioning of the offset canceling pair is an important trade-off in terms of noise and calibration range. Larger transistors result in a larger calibration range, but also larger charge leakage/noise. In this design, the sizes of M_{SP}/M_{SN} are chosen to be eight-times smaller than the main input pair M_{1P}/M_{1N}, to minimize the charge leakage, thus the noise degradation. This translates to a maximum of 125 mV offset compensation range (aV_{OS} = 1 V in Section 2) for a common-mode voltage of 0.5 V, which is large enough to allow a low-power and high-speed comparator design.

The implemented self-calibrating comparator performance in terms of delay and noise has been characterized with extracted simulations and compared to the comparators from [10,11,25,26], scaled to 28 nm (Figure 5). For the comparator delay, the Overdrive Recovery Test (ORT) [27,28] has been used, while the noise has been characterized with both pss + pnoise and transient simulations. The operating conditions were V_{DD} = 1.0 V and V_{CM} = 0.5 V. The proposed design achieved more than 20% faster regeneration time for small inputs due to the increased gain in the signal path and showed a lower input dependency for a wide range of voltages compared to [10,11,25,26] (Figure 5a). To achieve similar regeneration times, the tail, as well as the latching transistors of the works in [10,11,25,26] had to be upscaled, whose combined contribution increased the total input-referred noise by more than 15% with respect to this design, as shown in Figure 5b.

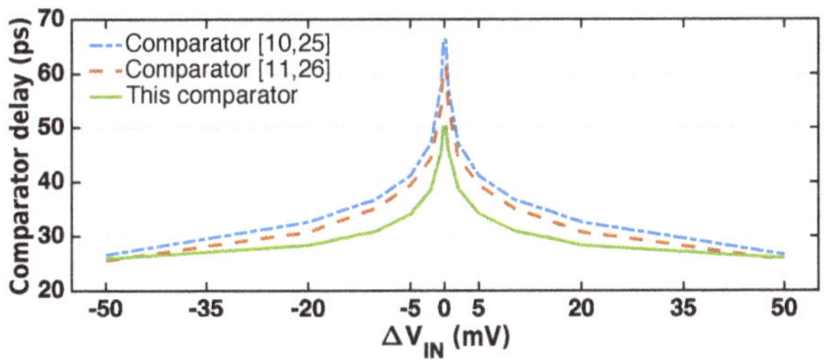

(**a**)

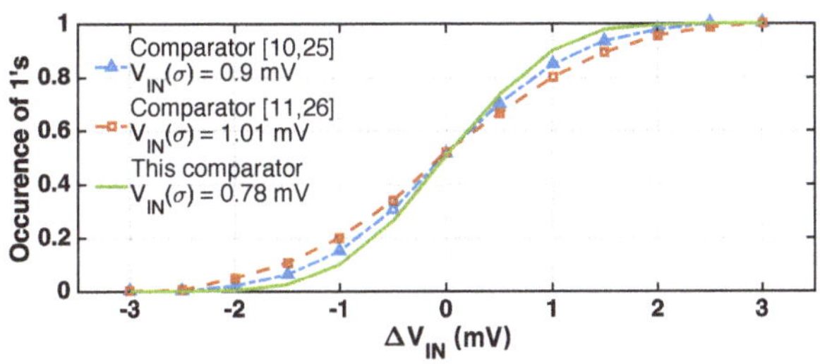

(**b**)

Figure 5. Simulated delay versus ΔV_{IN} for the same offset/noise (**a**) and cumulative noise distribution for similar delay (**b**) for [10,11,25,26] and the proposed design.

The offset for the three circuits with similar regeneration times has also been characterized through Monte Carlo simulations on 100 samples (Figure 6). A servo-loop has been used, which senses the comparator output and feeds back to the input the opposite offset value until the comparator goes into a metastable state. For the designed V_{CM} of 0.5 V, the 1-σ raw value for both [10,11,25,26] was larger than 11 mV, while it was 9.8 mV for the proposed design (Figure 6a). After enabling the proposed calibration, the offset was improved to 0.69 mV, set by the designed C_{INT}/C_{CAL} ratio and noise, without compromising the rest of the specifications.

To show the effectiveness of the calibration for different common mode conditions, the comparator V_{CM} was varied between 0.4 V (Figure 6b) and 0.6 V (Figure 6c). It is seen that the accuracy of the calibration loop remained functional for a wide range of common mode voltages, dictated only by the comparator noise.

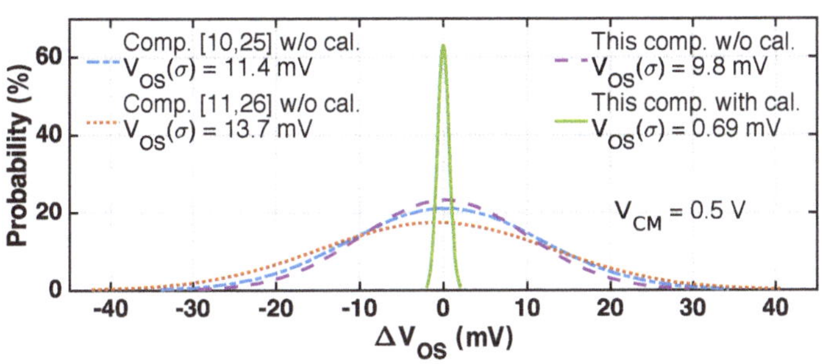

(**a**)

Figure 6. *Cont.*

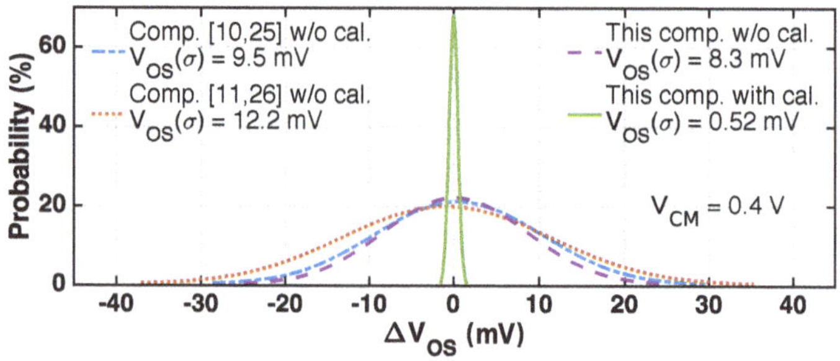

(b)

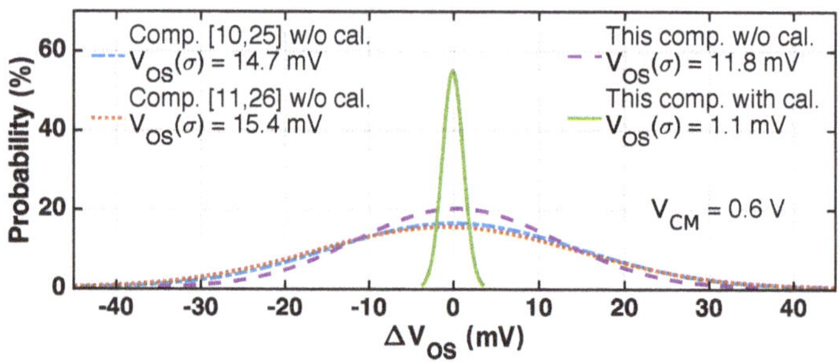

(c)

Figure 6. Simulated offset distribution for a V_{CM} of 0.5 V (**a**), 0.4 V (**b**), and 0.6 V (**c**); for [10,11,25,26] and the proposed design.

4. Experimental Results

4.1. Measurement Setup

The measurement setup used to evaluate the comparator performance is shown in Figure 7. A low phase noise signal source (Agilent E8257D) was used to generate the up to 11 GHz sinusoidal clock signal. This signal was converted into to a square pulse through on-chip CML + CMOS circuitry. An identical signal source was employed to generate the comparator input signal. Both input and clock signals were converted into differential signals by two identical wideband hybrids and AC-coupled to the chip through custom-designed bias-tees and phase-matched cables. A dual-channel source-meter was used to bias the differential comparator input, for easier noise and offset extraction.

The signal generators were locked together and with a 63 GHz bandwidth scope (DSOZ634A), serving as a data analyzer, which captured the differential output at full speed. The captured data were then processed on a PC in MATLAB. First, the comparator noise was characterized by observing the data, while the raw offset was subtracted from the comparator by applying different DC voltages from the source-meter. After noise characterization, the calibration loop was enabled and the calibrated comparator offset, as well as speed were evaluated.

The required supply and bias voltages for the different chip domains were generated with dedicated low-noise Low-Dropout Regulators (LDOs) on a custom bias board and provided to the chip after sufficient low-pass filtering.

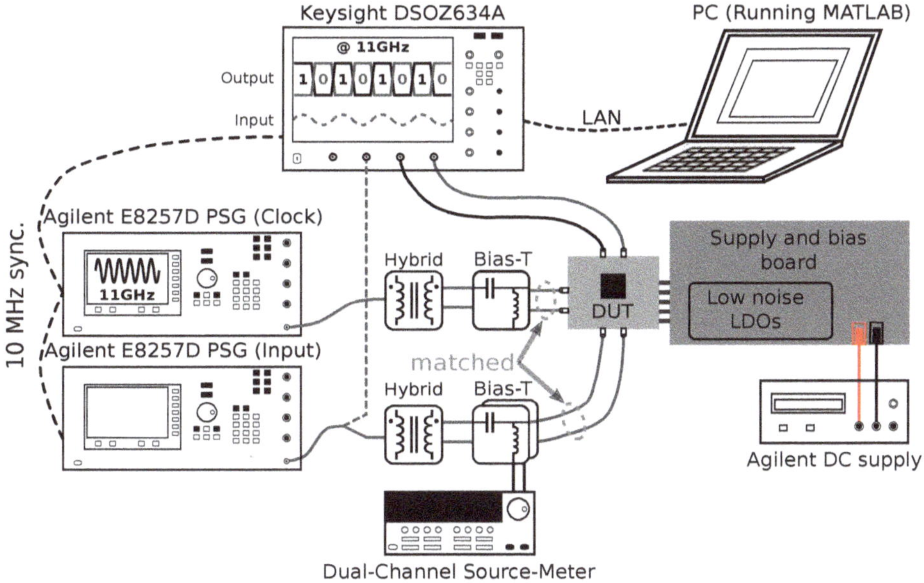

Figure 7. Measurement setup of the proposed self-calibrating comparator. LDO, Low-Dropout Regulator.

4.2. Measurement Results

The prototype self-calibrating comparator was realized in a single-poly ten-metal (1P10M) 28 nm bulk CMOS process and occupied an area of $35.5 \times 15.2\,\mu m^2$ (Figure 8). Most of the area was taken up by C_{CALP}/C_{CALN}, which is insignificant when used in an ADC. The measured power consumption of 0.89 mW at 1 V and 11 GHz clock frequency (F_{CLK}) partitions into 0.87 mW for the comparator core and only 0.02 mW for the calibration logic, less than 5% overhead.

Figure 9 illustrates the measured input noise versus ΔV_{IN} when varying V_{DD} (top) and V_{CM} (bottom), respectively, at 11 GHz. The noise was extracted by counting the percentage of positive decisions with increasing the differential input voltage, having first subtracted the comparator offset. The calibration loop was disabled for this measurement. The comparator measured a 1-σ noise voltage of 0.82 mV_{rms} for V_{DD} = 1 V and V_{CM} = 0.5 V, which varied by only 0.13/−0.14 mV when V_{DD} changed from 0.9 V to 1.1 V and −0.19/0.18 mV when V_{CM} changed from 0.4 V to 0.6 V.

The offset voltage of the comparator was measured across 15 chips operating at 11 GHz with V_{CM} = 0.5 V for the maximum compensation range (see Section 3.2), as shown in Figure 10. The offset was extracted by sweeping the input voltage of the comparator until the ratio of zeroes and ones was ∼50%. For the raw offset value, the calibration loop was disabled, while for the compensated value, the calibration was activated prior to collecting the data. Thanks to the proposed dual-sided calibration technique, the offset voltage was drastically reduced to 0.99 mV from the uncalibrated 10.3 mV (>10× improvement), without compromising the comparator speed, verifying its smooth integration in single- or multi-comparator ADCs. As expected, the calibration accuracy was ultimately limited by the comparator noise, which also matched nicely with the simulated results.

The maximum speed of the comparator for small inputs, close to the noise level, was characterized by observing the frequency above which increased differential input was required to preserve correct digital outputs. An eye diagram is shown in Figure 11 for an F_{CLK} of 11 GHz and a coherent Nyquist sinusoidal input frequency of 5.46 GHz (= (8133/16,384) × 11 GHz), such that the comparator can capture any input voltage value over the full-scale range. With this setup, no metastability errors were detected for voltages outside the comparator noise levels, measured over one million time samples.

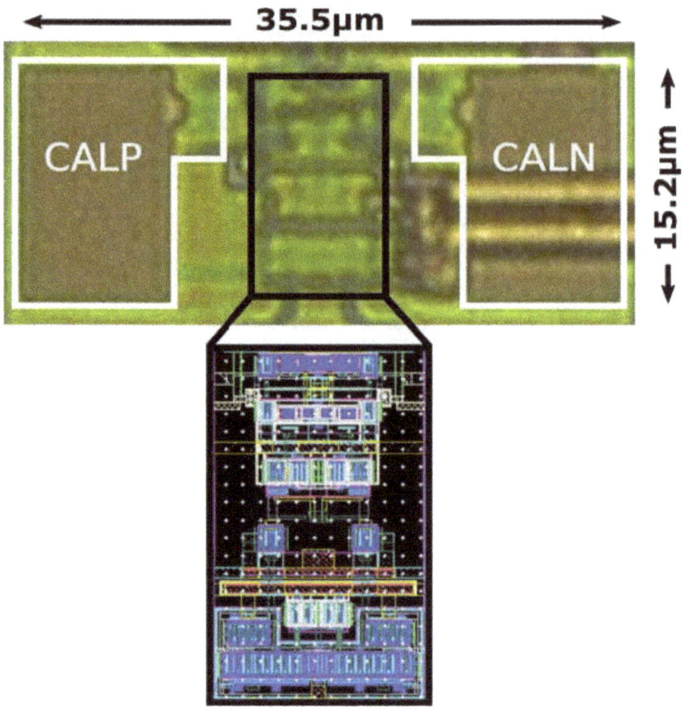

Figure 8. Die photo of the 28 nm self-calibrating comparator with a layout view of the comparator core.

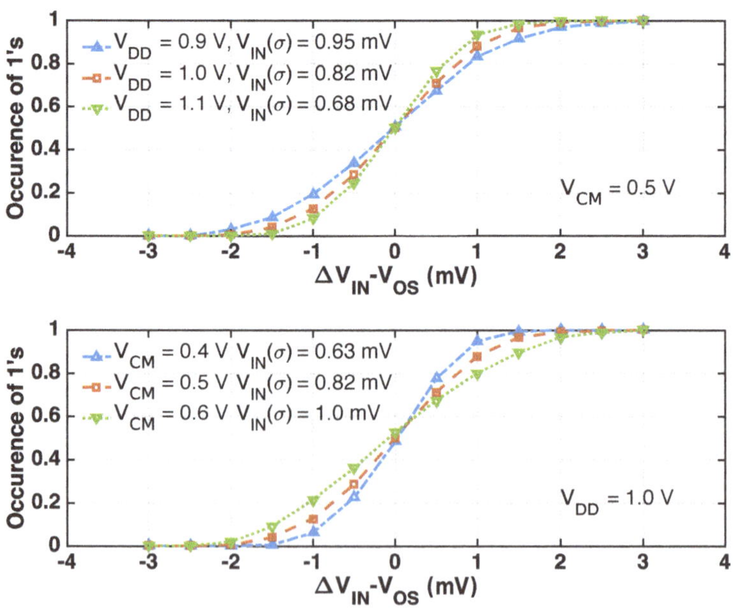

Figure 9. Measured cumulative noise distribution versus differential input at 11 GHz for varying V_{DD} (**top**) and varying V_{CM} (**bottom**).

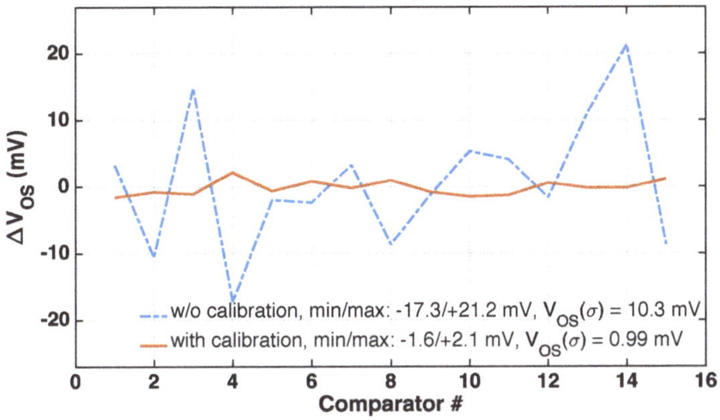

Figure 10. Measured raw and calibrated comparator offset voltage with the proposed calibration at 11 GHz.

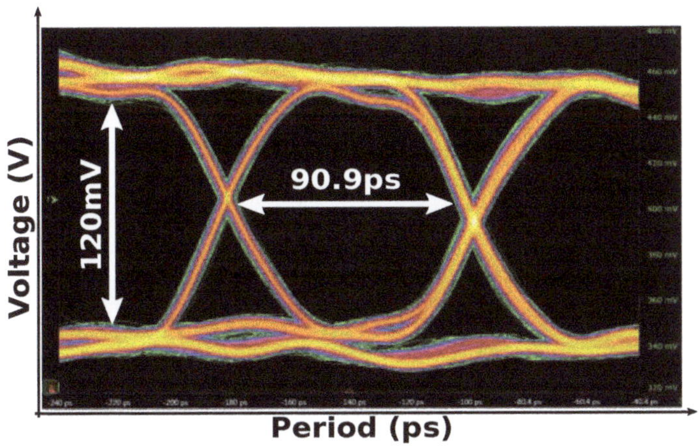

Figure 11. Eye diagram of the comparator output at 11 GHz for a coherent Nyquist input frequency.

This work compares favorably with state-of-the-art comparators, summarized in Table 1. This design achieved the highest reported operating frequency and the lowest delay slope, as well as the smallest sensitivity to V_{DD} and V_{CM} variations, compared to previously-measured published works. It also exhibited a very low input-referred noise and calibrated offset with state-of-the-art energy/comparison, demonstrating the effectiveness of the proposed calibration technique and its nearly zero overhead.

Table 1. Performance summary and comparison with state-of-the-art comparators.

	This Work	[11]	[21]	[29]	[30]	[31]	[22]
Technology (nm)	**28 nm**	90 nm	90 nm	65 nm	65 nm	65 nm	65 nm
Supply (V)	**1.0**	1.2	1.2	1.2	1.0	1.2	1.2
Delay/log(ΔV_{IN})	**12 ps/dec**	44 ps/dec	24 ps/dec	20 ps/dec	N.A.	N.A.	16 ps/dec
Maximum F_{CLK} (GHz)	**11.0**	2.0	1.0	7.0	7.2	4.0	1.5
Input-referred noise (mV)	**0.82**	1.5	1.0	15.0	200.0	50.0	0.32
Sensitivity to V_{DD} (mV)	**+0.13/−0.14**	N.A.	N.A.	N.A.	N.A.	N.A.	N.A.
Sensitivity to V_{CM} (mV)	**−0.19/+0.18**	N.A.	−0.2/+0.2	N.A.	N.A.	N.A.	N.A.
Uncalibrated offset (mV)	**10.3**	13.0	13.7	22.0	N.A.	N.A.	11.6
Calibrated offset (mV)	**0.99**	13.0	1.69	22.0	N.A.	3.0	0.53
Energy/comparison (fJ)	**81**	113	40	185	63	114	61

5. Conclusions

A high-speed, low-noise dynamic comparator with a dual-sided self-calibrating loop has been presented. The proposed dynamic calibration tremendously reduces the input offset by 10×, limited only by the comparator noise, without compromising its speed or significantly increasing its power. Combined with the implemented multi-stage comparator to enable better optimization between various trade-offs, the highest reported maximum frequency of 11 GHz is realized with only 0.82 mV and 0.99 mV input noise and offset, respectively, consuming only 0.89 mW from a 1 V supply. The prototype occupies a total area of only 0.00054 mm^2. In summary, the proposed circuit is an ideal candidate for any high speed, low noise/offset, power-/area-efficient mixed-signal system and can be adapted to any comparator structure. Moreover, its fully-dynamic implementation ensures 100% drawback-free scalability to lower technology nodes.

Future research will involve realizing a faster and lower noise comparator circuit to incorporate the proposed calibration loop. Finally, more transistor stacking will be employed in the circuit of Figure 3 to realize a finer calibration step and reduce the leakage on CALP/CALN without increasing C_{CALP}/C_{CALN}.

Author Contributions: A.R. designed, simulated, taped-out, and measured the circuits described in this manuscript under the supervision of M.S. (Michiel Steyaert), and F.T. and M.S. (Maarten Strackx) contributed to the technical discussions during the design time.

Funding: This research was partially funded by Nokia Bell Labs, Antwerp, Belgium.

Acknowledgments: The authors would like to thank Nokia Bell Labs, Antwerp Belgium, for partially supporting this work.

Conflicts of Interest: The authors declare no conflicts of interest.

References

1. Wang, Z.; Su, F.; Wang, Y.; Li, Z.; Li, X.; Yoshimura, R.; Naiki, T.; Tsuwa, T.; Saito, T.; Wang, Z.; et al. A 130 nm FeRAM-based parallel recovery nonvolatile SOC for normally-OFF operations with 3.9× faster running speed and 11× higher energy efficiency using fast power-on detection and nonvolatile radio controller. In Proceedings of the 2017 Symposium on VLSI Circuits, Kyoto, Japan, 5–8 June 2017; pp. C336–C337.
2. Choi, S.; Huh, Y.; Park, S.; Yoon, K.; Bang, J.; Shin, S.; Ju, Y.; Yang, Y.; Yoon, J.; Ahn, C.; et al. A Quasi-Digital Ultra-Fast Capacitor-Less Low-Dropout Regulator Based on Comparator Control for x8 Current Spike of PCRAM Systems. In Proceedings of the 2018 IEEE Symposium on VLSI Circuits, Honolulu, HI, USA, 18–22 June 2018; pp. 107–108.
3. Lee, C.; Lee, J.; Kim, K.; Heo, J.; Baek, J.; Cha, G.; Moon, D.; Lee, D.; Park, J.; Lee, S.; et al. Dual-Loop Two-Step ZQ Calibration for Dynamic Voltage–Frequency Scaling in LPDDR4 SDRAM. *IEEE J. Solid-State Circuits* **2018**, *53*, 2906–2916. [CrossRef]
4. Norimatsu, T.; Kawamoto, T.; Kogo, K.; Kohmu, N.; Yuki, F.; Nakajima, N.; Muto, T.; Nasu, J.; Komori, T.; Koba, H.; et al. A 25 Gb/s multistandard serial link transceiver for 50 dB-loss copper cable in 28 nm CMOS. In Proceedings of the 2016 IEEE International Solid-State Circuits Conference (ISSCC), San Francisco, CA, USA, 31 January–4 February 2016; pp. 60–61.
5. Aurangozeb; Hossain, A.K.M.D.; Hossain, M. Channel adaptive ADC and TDC for 28 Gb/s PAM-4 digital receiver. In Proceedings of the 2017 IEEE Custom Integrated Circuits Conference (CICC), Austin, TX, USA, 30 April–3 May 2017; pp. 1–4.
6. Yi, I.; Chae, M.; Hyun, S.; Bae, S.; Choi, J.; Jang, S.; Kim, B.; Sim, J.; Park, H. A Time-Based Receiver With 2-Tap Decision Feedback Equalizer for Single-Ended Mobile DRAM Interface. *IEEE J. Solid-State Circuits* **2018**, *53*, 144–154. [CrossRef]
7. Gandara, M.; Guo, W.; Tang, X.; Chen, L.; Yoon, Y.; Sun, N. A pipelined SAR ADC reusing the comparator as residue amplifier. In Proceedings of the 2017 IEEE Custom Integrated Circuits Conference (CICC), Austin, TX, USA, 30 April–3 May 2017; pp. 1–4.

8. Shim, M.; Jeong, S.; Myers, P.D.; Bang, S.; Shen, J.; Kim, C.; Sylvester, D.; Blaauw, D.; Jung, W. Edge-Pursuit Comparator: An Energy-Scalable Oscillator Collapse-Based Comparator With Application in a 74.1 dB SNDR and 20 kS/s 15 b SAR ADC. *IEEE J. Solid-State Circuits* **2017**, *52*, 1077–1090. [CrossRef]
9. Ding, Z.; Zhou, X.; Li, Q. A 0.5–1.1 V 10 B Adaptive Bypassing SAR ADC Utilizing Oscillation Cycle Information of VCO-Based Comparator. In Proceedings of the 2018 IEEE Symposium on VLSI Circuits, Honolulu, HI, USA, 18–22 June 2018; pp. 93–94.
10. Wicht, B.; Nirschl, T.; Schmitt-Landsiedel, D. Yield and speed optimization of a latch-type voltage sense amplifier. *IEEE J. Solid-State Circuits* **2004**, *39*, 1148–1158. [CrossRef]
11. Schinkel, D.; Mensink, E.; Klumperink, E.; van Tuijl, E.; Nauta, B. A Double-Tail Latch-Type Voltage Sense Amplifier with 18 ps Setup+Hold Time. In Proceedings of the 2007 IEEE International Solid-State Circuits Conference. Digest of Technical Papers, San Francisco, CA, USA, 11–15 February 2007; pp. 314–605.
12. Rabbi, F.; Das, S.; Hossain, Q.D.; Pathan, N.S. Design of a Low-Power Ultra High Speed Dynamic Latched Comparator in 90-nm CMOS Technology. In Proceedings of the 2018 International Conference on Computer, Communication, Chemical, Material and Electronic Engineering (IC4ME2), Rajshahi, Bangladesh, 8–9 February 2018; pp. 1–4.
13. Rezapour, A.; Shamsi, H.; Abbasizadeh, H.; Lee, K. Low Power High Speed Dynamic Comparator. In Proceedings of the 2018 IEEE International Symposium on Circuits and Systems (ISCAS), Florence, Italy, 27–30 May 2018; pp. 1–5.
14. Aiello, O.; Crovetti, P.; Alioto, M. Fully Synthesizable, Rail-to-Rail Dynamic Voltage Comparator for Operation down to 0.3 V. In Proceedings of the 2018 IEEE International Symposium on Circuits and Systems (ISCAS), Florence, Italy, 27–30 May 2018; pp. 1–5.
15. Pelgrom, M.J.M.; Duinmaijer, A.C.J.; Welbers, A.P.G. Matching properties of MOS transistors. *IEEE J. Solid-State Circuits* **1989**, *24*, 1433–1439. [CrossRef]
16. Pelgrom, M. *Analog-to-Digital Conversion*, 3rd ed.; Springer: Berlin, Germany, 2017.
17. Krämer, M.; Janssen, E.; Doris, K.; Murmann, B. A 14-Bit 30-MS/s 38-mW SAR ADC Using Noise Filter Gear Shifting. *IEEE Trans. Circuits Syst. II* **2017**, *64*, 116–120. [CrossRef]
18. Verbruggen, B.; Iriguchi, M.; Craninckx, J. A 1.7 mW 11 b 250 MS/s 2x interleaved fully dynamic pipelined SAR ADC in 40 nm digital CMOS. In Proceedings of the 2012 IEEE International Solid-State Circuits Conference, San Francisco, CA, USA, 19–23 February 2012; pp. 466–468.
19. Vaz, B.; Lynam, A.; Verbruggen, B.; Laraba, A.; Mesadri, C.; Boumaalif, A.; Mcgrath, J.; Kamath, U.; Torre, R.D.L.; Manlapat, A.; et al. 16.1 A 13 b 4 GS/s digitally assisted dynamic 3-stage asynchronous pipelined-SAR ADC. In Proceedings of the 2017 IEEE International Solid-State Circuits Conference (ISSCC), San Francisco, CA, USA, 5–9 February 2017; pp. 276–277.
20. Vaz, B.; Verbruggen, B.; Erdmann, C.; Collins, D.; Mcgrath, J.; Boumaalif, A.; Cullen, E.; Walsh, D.; Morgado, A.; Mesadri, C.; et al. A 13 Bit 5 GS/S ADC with Time-Interleaved Chopping Calibration in 16 NM FinFET. In Proceedings of the 2018 IEEE Symposium on VLSI Circuits, Honolulu, HI, USA, 18–22 June 2018; pp. 99–100.
21. Miyahara, M.; Asada, Y.; Paik, D.; Matsuzawa, A. A low-noise self-calibrating dynamic comparator for high-speed ADCs. In Proceedings of the 2008 IEEE Asian Solid-State Circuits Conference, Fukuoka, Japan, 3–5 November 2008; pp. 269–272.
22. Chan, C.H.; Zhu, Y.; Chio, U.F.; Sin, S.W.; Seng-Pan, U.; Martins, R.P. A reconfigurable low-noise dynamic comparator with offset calibration in 90 nm CMOS. In Proceedings of the IEEE Asian Solid-State Circuits Conference, Jeju, Korea, 14–16 November 2011; pp. 233–236.
23. Chan, C.; Zhu, Y.; Ho, I.; Zhang, W.; Seng-Pan, U.; Martins, R.P. A 5 mW 7 b 2.4 GS/s 1-then-2b/cycle SAR ADC with background offset calibration. In Proceedings of the 2017 IEEE International Solid-State Circuits Conference (ISSCC), San Francisco, CA, USA, 5–9 February 2017; pp. 282–283.
24. Ramkaj, A.; Strackx, M.; Steyaert, M.; Tavernier, F. A 36.4 dB SNDR @ 5 GHz 1.25 GS/s 7 b 3.56 mW single-channel SAR ADC in 28 nm bulk CMOS. In Proceedings of the ESSCIRC 2017 43rd IEEE European Solid State Circuits Conference, Leuven, Belgium, 11–14 September 2017; pp. 167–170.
25. Luu, D.; Kull, L.; Toifl, T.; Menolfi, C.; Brändli, M.; Francese, P.A.; Morf, T.; Kossel, M.; Yueksel, H.; Cevrero, A.; et al. A 12-bit 300-MS/s SAR ADC With Inverter-Based Preamplifier and Common-Mode-Regulation DAC in 14-nm CMOS FinFET. *IEEE J. Solid-State Circuits* **2018**, *53*, 3268–3279. [CrossRef]

26. Bindra, H.S.; Lokin, C.E.; Annema, A.; Nauta, B. A 30fJ/comparison dynamic bias comparator. In Proceedings of the ESSCIRC 2017 43rd IEEE European Solid State Circuits Conference, Leuven, Belgium, 11–14 September 2017; pp. 71–74.
27. Razavi, B. *Principles of Data Conversion System Design*; Wiley-IEEE Press: New York, NY, USA, 1995.
28. Nasrollahpour, M.; Yen, C.; Hamedi-hagh, S. A high-speed, low-offset and low-power differential comparator for analog to digital converters. In Proceedings of the 2017 International SoC Design Conference (ISOCC), Seoul, Korea, 5–8 November 2017; pp. 220–221.
29. Goll, B.; Zimmermann, H. A 65 nm CMOS comparator with modified latch to achieve 7 GHz/1.3 mW at 1.2 V and 700 MHz/47 μW at 0.6 V. In Proceedings of the 2009 IEEE International Solid-State Circuits Conference - Digest of Technical Papers, San Francisco, CA, USA, 8–12 February 2009; pp. 328–329.
30. Abbas, M.; Furukawa, Y.; Komatsu, S.; Takahiro, J.Y.; Asada, K. Clocked comparator for high-speed applications in 65 nm technology. In Proceedings of the 2010 IEEE Asian Solid-State Circuits Conference, Beijing, China, 8–10 November 2010; pp. 1–4.
31. Kong, L.; Lu, Y.; Alon, E. A multi-GHz area-efficient comparator with dynamic offset cancellation. In Proceedings of the 2011 IEEE Custom Integrated Circuits Conference (CICC), San Jose, CA. USA, 17–20 September 2011; pp. 1–4.

electronics

MDPI

Article

High-Linearity Self-Biased CMOS Current Buffer

Javier Alejandro Martínez-Nieto [1,*], María Teresa Sanz-Pascual [1], Nicolás Medrano-Marqués [2], Belén Calvo-López [2] and Arturo Sarmiento-Reyes [1]

[1] Electronics Department, National Institute of Astrophysics, Optics and Electronics (INAOE), 72840 Puebla, Mexico; materesa@inaoep.mx (M.T.S.-P.); jarocho@inaoep.mx (A.S.-R.)

[2] Group of Electronic Design (GDE), University of Zaragoza, 50009 Zaragoza, Spain; nmedrano@unizar.es (N.M.-M.); becalvo@unizar.es (B.C.-L.)

* Correspondence: almartinez@inaoep.mx; Tel.: +34-632-191-658

Received: 7 November 2018; Accepted: 6 December 2018; Published: 11 December 2018

Abstract: A highly linear fully self-biased class AB current buffer designed in a standard 0.18 μm CMOS process with 1.8 V power supply is presented in this paper. It is a simple structure that, with a static power consumption of 48 μW, features an input resistance as low as 89 Ω, high accuracy in the input–output current ratio and total harmonic distortion (THD) figures lower than −60 dB at 30 μA amplitude signal and 1 kHz frequency. Robustness was proved through Monte Carlo and corner simulations, and finally validated through experimental measurements, showing that the proposed configuration is a suitable choice for high performance low voltage low power applications.

Keywords: class AB operation; CMOS; current mirror; current buffer; quasi floating gate; low power

1. Introduction

Current mirrors are required not only to generate and replicate bias currents, but also as core cells in many analog and mixed signal applications: current conveyors, current feedback operational amplifiers or current-mode filters, among others, are based on this basic current processing block [1–9]. Unfortunately, the power consumption of current-mode circuits proportionally increases as the number of active branches where the current is replicated increases. This limitation, critical for the current low-voltage low-power IC design scenario set by the driving portable market, can be circumvented through class AB operation, which makes it possible to dynamically handle current levels higher than the quiescent bias current [10–14]. Furthermore, self-biasing may be used to establish the DC current in the circuit without any additional bias circuitry in order to optimize the power consumption [15].

The goal of this work is to accomplish a reliable fully self-biased class AB current buffer design. It relies on an active input to attain very low input impedance and high linearity, which is further increased by the coupling of the input and output branches through a single transistor. Preliminary results from a not fully self-biased implementation, i.e., requiring extra bias generation for the cascode transistors and the input amplifier, are presented in [16]. This paper presents the complete fully self-biased design, providing more insight into the operation principle and the actual implementation of the required amplifier and the corresponding compensation network, considering both a single-stage and a two-stage differential amplifier. Simulations including process variations and mismatch effects, as well as experimental results, validate the reliability of the proposed approach.

The circuit was characterized and compared with two other widely used class AB buffers designed with the same technology, same power supply and for the same input current range. The first is a quasi-floating gate current buffer (QFG-CB) and the second is a current-conveyor based current buffer (CC-CB). These topologies were chosen for their class AB operation as well as for their ability to keep the input node at a constant DC input voltage V_{dc} (virtual ground), as the proposed circuit does. This is a desirable characteristic in many cases, and becomes essential in some particular configurations based on MOS current dividers [17–19]. A particular case where this feature is exploited is the sign circuit required within the neuron of an analog neural system used to calibrate sensors [20,21] (see Figure 1). This sign circuit is required to determine the direction of the current flowing through a multiplier, thus allowing both positive and negative synaptic weights [22]. This particular application motivated the design of the proposed self-biased buffer configuration, with the key requirements of providing the highest possible accuracy and linearity response with a reduced power consumption and a compact size.

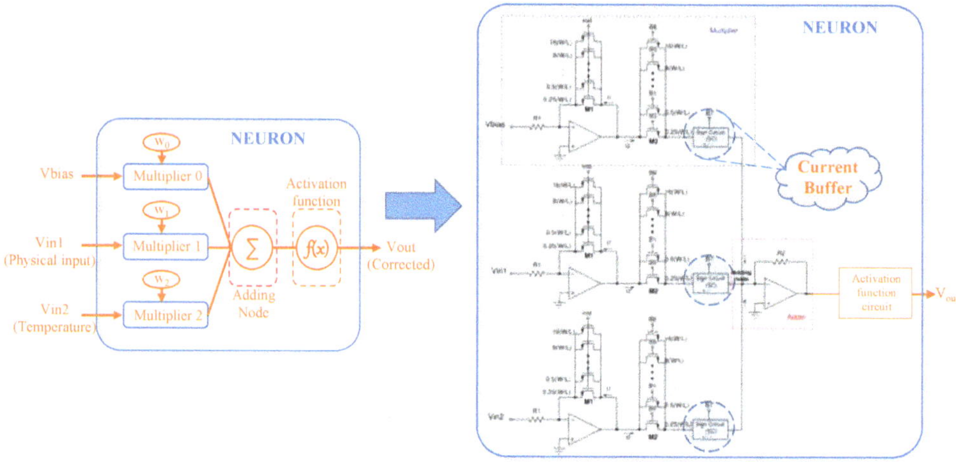

Figure 1. Neuron configuration implemented within the neural network-based microelectronic circuit for sensor calibration.

The paper is organized as follows: Section 2 presents the operation principle of the proposed buffer. The differential amplifiers and the compensation techniques used to ensure the buffer stability are also presented in this section. In Section 3, the current buffer is thoroughly characterized for both a single-stage and a two-stage amplifier as active input components to show the corresponding trade-offs. A comparison with two other widely used class AB current buffers with a well defined input voltage is also made. Measurement results of the integrated current buffer prototype and a comparison with other integrated circuits are presented in Section 4 and, finally, conclusions are drawn in Section 5.

2. Proposed Self-Biased Current-Buffer

The proposed self-biased current buffer (SB-CB) is shown in Figure 2. A Differential Amplifier (DA) sets the input voltage at V_{dc} and establishes a virtual ground at this node. The quasi-floating gate (QFG) approach is used to achieve class AB operation [23–30], since this technique requires no additional current and adds minimum hardware penalty, leading to a power efficient and compact solution. In static conditions, the bias current I_{Bias} is determined by the dimensions of the PMOS (P-type metal-oxide-semiconductor) transistors M_{p1} and M_{p2}, which are diode-connected and equally sized. Therefore, the same current flows through each NMOS (N-type metal-oxide-semiconductor) transistor M_{n1} and M_{n2}, whereas M_1 sinks twice the bias current.

Under dynamic conditions, the PMOS transistors act as dynamic current sources. If the input current flows out of the buffer, the current flowing through M_{n1} and M_{n2} decreases and so does the tail current in transistor M_1. Due to the RC coupling formed by capacitance C and resistances R_{large}, the gate voltage of M_{p1} and M_{p2} drops and their current driving capability increases. Hence, the bias current of the buffer is lower than the input current that can be handled. Neglecting channel-length modulation, the current transfer function is given by:

$$\frac{I_{out}}{I_{in}} = \frac{(g_{m_{n1}} + g_{m_{n2}})g_{m_{p2}} + A_d g_{m_1} g_{m_{n2}}}{(g_{m_{n1}} + g_{m_{n2}})g_{m_{p1}} + A_d g_{m_1} g_{m_{n1}}} \tag{1}$$

where A_d is the gain of the differential amplifier and g_{m_i} is the transconductance of transistor M_i. If a unity current gain, i.e., a current buffer, is required, the transconductance ratios $g_{m_{p2}}/g_{m_{p1}}$ and $g_{m_{n2}}/g_{m_{n1}}$ must both be equal to 1.

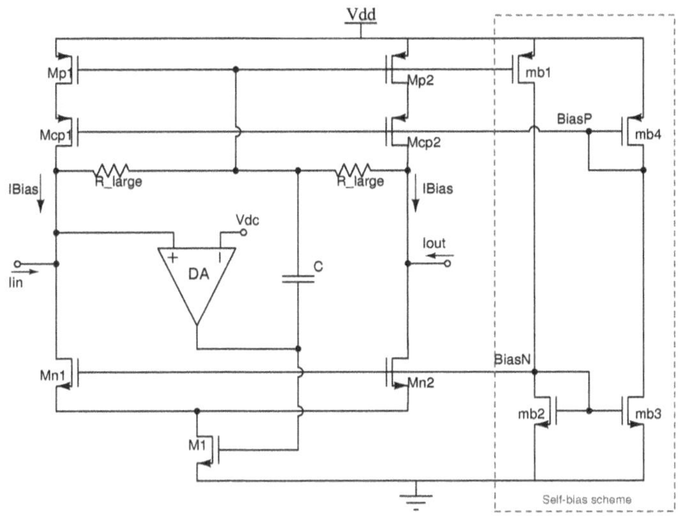

Figure 2. Proposed Self-Biased Current Buffer (SB-CB).

The input resistance R_{in} is the parallel of the equivalent resistance R_{inP} seen from the input to V_{DD}, and the equivalent R_{inN} seen from the input to ground:

$$R_{inP} = \frac{g_{m_{cp1}} r_{o_{cp1}} r_{o_{p1}}}{1 + g_{m_{p1}} r_{o_{cp1}} (g_{m_{cp1}} r_{o_{p1}} - 1)} \tag{2}$$

$$R_{inN} = \frac{2r_{o_{n1}}}{1 + A_d g_{m_1} r_{o_{n1}}} \tag{3}$$

$$R_{in} = R_{inP} || R_{inN} \approx \frac{2}{2g_{m_{p1}} + A_d g_{m_1}} \tag{4}$$

As expected, R_{in} can be reduced by increasing the differential amplifier gain A_d. The output resistance R_{out} is given by:

$$R_{out} = \frac{2}{g_{m_{p1}}} || \left(2r_{o_{n1}} + \frac{1}{g_{m_1}}\right) \approx \frac{2}{g_{m_{p1}}} \tag{5}$$

R_{out} is dominated by the equivalent resistance of the diode connection of transistor M_{p1}, so it may be lower than in other current buffer implementations. However, as shown below, a 2.4 MΩ output resistance was achieved in our design, which is still suitable for many applications.

The proposed SB-CB was designed in a standard 0.18 µm CMOS process with 1.8 V supply voltage. The transistor sizes are shown in Table 1. The channel length is $L \geq 1$ µm in all cases in order to reduce mismatch effects. The sizes were chosen so the buffer would be able to handle input currents up to 15 µA amplitude with a nominal bias current $I_{Bias} = 8$ µA. The coupling Metal-Insulator-Metal

(MIM) capacitor has a value $C = 1$ pF. The resistances R_{large} were implemented with minimum-size diode-connected MOS transistors in the cutoff region [31], as they do not need to have a precise value as long as the cutoff frequency $f_c = 1/[2\pi R_{large}C]$ is lower than the signal frequency. Cascode transistors improve the accuracy in the current copy, and the self-bias scheme shown in Figure 2 was used to establish the required $BiasP$ and $BiasN$ voltages [32]. Finally, an NMOS transistor not shown in the figure was connected to the input node as start-up circuit.

Table 1. Transistors aspect ratios for the proposed buffer.

Transistor	W/L (μm/μm)
M_{p1}, M_{p2}	2/1
M_1, M_{cp1}, M_{cp2}	20/1
M_{n1} M_{n2}	15/1
m_{b1}	0.72/3
m_{b2}, m_{b3}	2/2
m_{b4}	0.54/2

To analyze the stability of the SB-CB, it must be noted that the open-loop gain is given by:

$$A_{ol} = A_d \cdot A_{cs} \tag{6}$$

where A_d is the gain of the differential amplifier DA and A_{cs} is the gain of the common-source stage, i.e., transistor M_1:

$$A_{cs} \approx g_{m_1} \cdot \frac{1 + 2g_{m_{p1}} r_{o_{n1}}}{2g_{m_{p1}}(1 + g_{m_{p1}} r_{o_{n1}})} \tag{7}$$

First, a current buffer SB-CB1 where the DA is a single-stage PMOS differential pair with active load will be considered. When opening the feedback loop, a two-stage configuration results, as shown in Figure 3. To ensure stability, Miller compensation is applied. The bias current is set to 500 nA and derived from the current buffer itself. The amplifier shows 40 dB gain and the buffer is compensated with a Miller capacitance $C_{comp} = 300$ fF, attaining 82° phase margin for BW = 4.1 MHz.

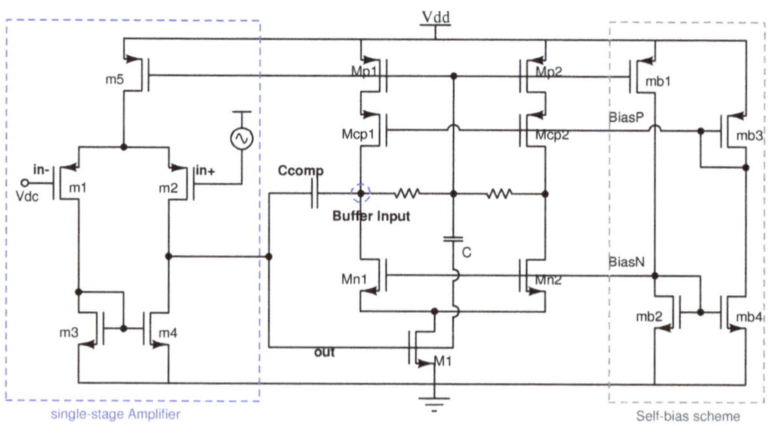

Figure 3. Open-loop configuration with a single-stage amplifier.

As shown in Equation (4), a higher gain differential amplifier will decrease the input impedance. Furthermore, the linearity is expected to increase by the virtual ground set at the input node. Therefore, a two-stage amplifier was also designed to explore the impact of the amplifier on the overall performance of the buffer. If the DA is a two-stage amplifier, the open-loop configuration turns into a three-stage amplifier, as shown in Figure 4. To achieve stability, nested-Miller compensation can be applied. This technique requires the second stage in the differential amplifier not to invert the signal, so the amplifier has to be accordingly designed [33–35].

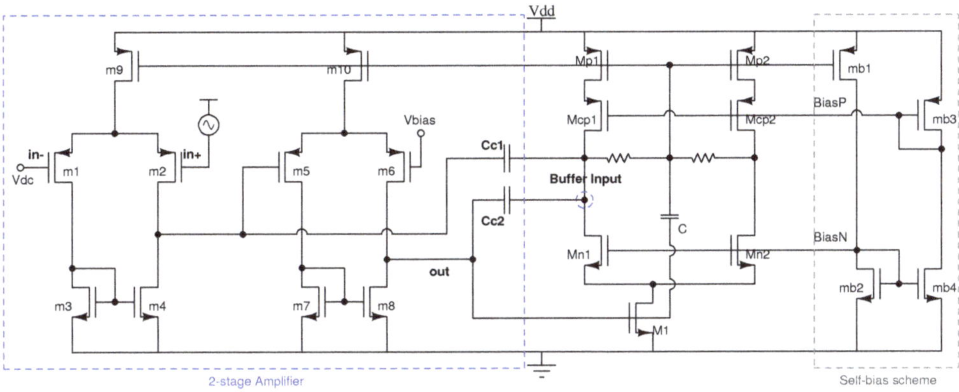

Figure 4. Open-loop configuration with a two-stage amplifier.

As shown in Figure 4, the two-stage DA was implemented with two cascaded PMOS differential pairs. An additional differential pair, not shown in the figure, was used to set the required bias voltage V_{bias} at the negative input of the second stage so the current distribution through its branches is symmetrical. Again, the bias currents were derived from the current buffer itself. Each differential pair is biased with 500 nA and the two-stage DA gain is 78 dB. The compensation capacitors values are $C_{C1} = 400$ fF and $C_{C2} = 100$ fF. The phase margin with the nested-Miller compensation is $PM = 62°$ for a bandwidth (BW) of 3 MHz.

3. Performance Characterization

For the sake of comparison, simulations were carried out for the self-biased buffer both with a single-stage amplifier (SB-CB1) and a two-stage amplifier (SB-CB2) as DA.

Figure 5 shows the output current and the relative error in the copy of current as a function of the input current. The SB-CB2 shows lower relative error in the transfer current. Considering a minimum input current $I_{in} = 100$ nA, the maximum relative error is 0.09% for the SB-CB2 and 0.24% for the SB-CB1. If the minimum input current is reduced to $I_{in} = 10$ nA, the maximum relative error increases to 0.75% for the SB-CB2 and 2.08% for the SB-CB1.

As for linearity, both current buffers show very low harmonic distortion. The THD for a 15 μA amplitude input current remains below −60 dB up to 100 kHz for the SB-CB2 and up to 30 kHz for the SB-CB1. Figure 6 shows THD versus frequency for both configurations.

Figure 7 shows the time response to a 30 μA_{pp} input current step for both SB-CBs. For the SB-CB1, the rise time is 1.23 μs and the fall time is 898 ns, both considering the response within 0.1% of the output signal. As for the SB-CB2 the rise time is 947 ns and the fall time is 1.13 μs under the same conditions.

Table 2 summarizes the main electrical characteristics of the proposed buffers SB-CB1 and SB-CB2. As expected, SB-CB2 shows higher linearity and lower input resistance than SB-CB1 with a slight increment in power consumption. Table 2 also shows the characteristics of two other widely used class AB current buffers with a virtual ground at the input node. For a fair comparison, these buffers were redesigned in the same 0.18 μm CMOS process with 1.8 V supply and for the same input current range $I_{in} = \pm 15$ μA.

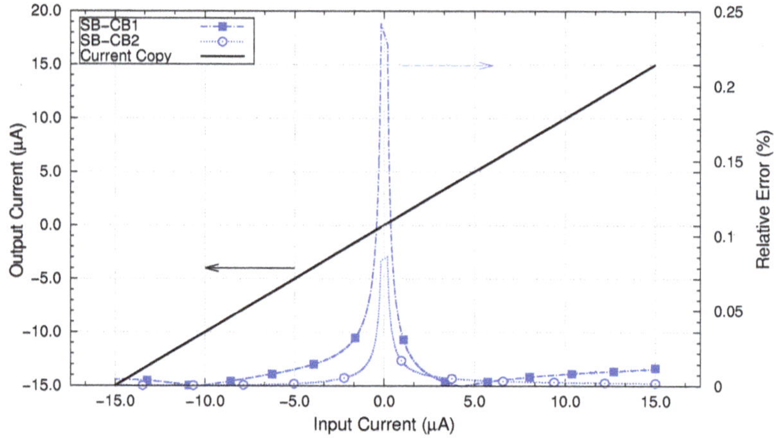

Figure 5. Output current and current transfer e_r as a function of I_{in}.

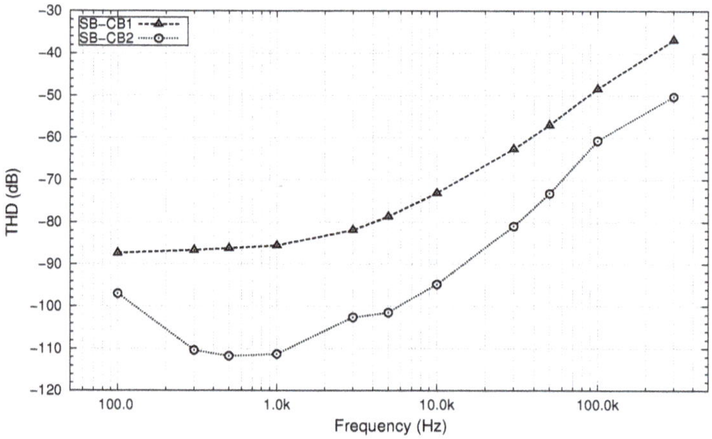

Figure 6. THD for a 30 μA_{pp} input current versus frequency.

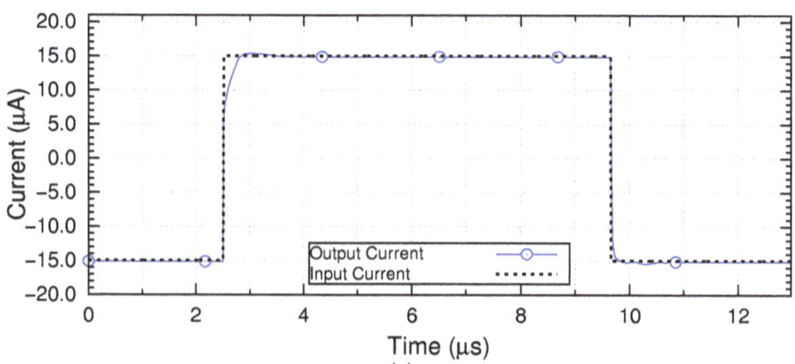

(a)

Figure 7. *Cont.*

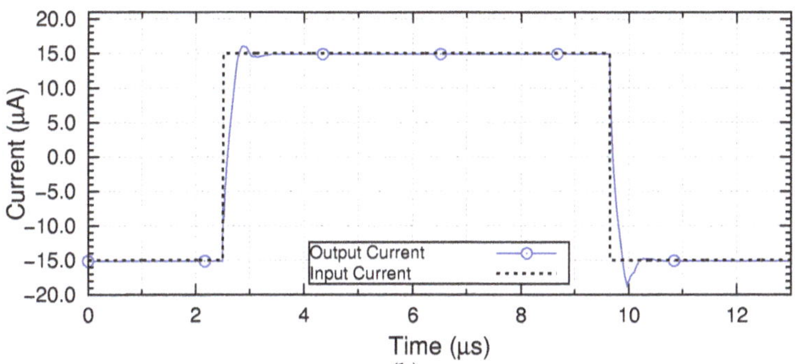

(b)

Figure 7. Response of the proposed circuit to an input current step: (**a**) SB-CB1; and (**b**) SB-CB2.

Table 2. Class AB current buffer characteristics.

Circuit	I_{Bias} (μA)	THD (dB) $I_{in}=30\mu A_{pp}$@1 kHz	Max. Power (μW) Static	Max. Power (μW) Dynamic	BW (MHz)	Rin (Ω)	Rout (MΩ)	$e_{r_{max}}$(%) $I_{in}=0.2\mu A_{pp}$	Settling Time at 0.1% (μs)	Active * Area (μm^2)
Proposed SB-CB1	8	−85.6	30.9	49.1	3.8	483	2.4	0.24	1.23	(MOS) 118 (MIM) 1404
Proposed SB-CB2	8	−111.3	32.4	51.6	2.6	8.3	2.4	0.09	1.13	(MOS) 118 (MIM) 1404
QFG-CB	3	−103.8	14.6 †	33.4 †	2.2	26.9	29.8	0.08	1.67	(MOS) 176 † (MIM) 1404
CC-CB	5	−50.1	24.7 †	59.3 †	1.0	448.5	63.0	1.74	1.16	(MOS) 630 †

* Estimated area by considering the number of MOS transistors and their sizes, and MIM-capacitors. † Bias circuit not considered.

The Quasi-Floating Gate Current-Buffer (QFG-CB) is presented in [26] and, as in the proposed SB-CB, the bias transistors act as dynamic current sources. The two-stage differential amplifier shown in Figure 4 was used in the design of the QFG-CB, and, again, nested-Miller compensation was used to ensure stability. The second configuration considered for comparison is the Current Conveyor based Current Buffer (CC-CB) [36–41]. Figure 8 shows both the schematic circuits and the transistor sizes of the aforementioned class AB current buffers. According to Table 2, the QFG-CB and the proposed circuit have higher estimated active area than the CC-CB due the MIM capacitors used for the QFG technique. However, if only the number of transistors is considered, the proposed circuit has the smallest area.

The bias current I_{Bias} is lowest for the QFG-CB, which results in the lowest power consumption, both static and dynamic. However, it should be mentioned that both the QFG-CB and CC-CB require additional biasing schemes which are not considered in the comparison.

The QFG-CB and the proposed SB-CB2 show the lowest relative error in the copy of current. At an input current I_{in} = 100 nA, the relative error remains below 0.1% for both circuits, and, even considering an input current I_{in} = 10 nA, the relative error remains below 0.8% in both cases, whereas the error of the CC-CB rises to 45%, which is unbearable in practical cases. As for the THD@30 μA$_{pp}$, it remains below −60 dB up to 100 kHz both for the SB-CB2 and for the QFG-CB. The CC-CB, in contrast, shows a THD higher than −55 dB even at low frequencies.

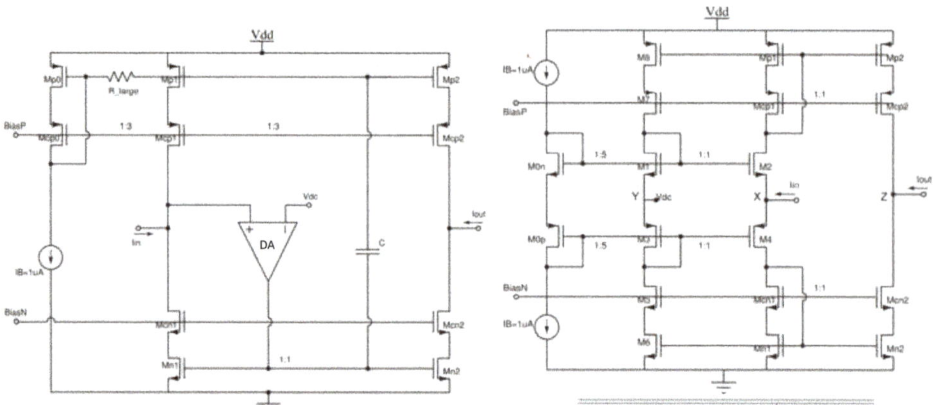

Transistor	W/L [$\mu m/\mu m$]
M_{p0}, M_{p1}, M_{p2}	12/1
M_{cp0}, M_{cp1}, M_{cp2}	10/1
M_{n1}, M_{n2}, M_{cn1}, M_{cn2}	5/1

(a)

Transistor	W/L [$\mu m/\mu m$]
M_{p1}, M_{p2}, M_{cp1}, M_{cp2}, M_7, M_8	20/1
M_{p0}, M_3, M_4	30/1
M_{n0}, M_{n1}, M_{n2}, M_3, M_4, M_6	10/1
M_{cn1}, M_{cn2}, M_5	15/1

(b)

Figure 8. Class AB current buffers with a virtual ground at the input node and their transistor sizes: (**a**) QFG-CB; and (**b**) CC-CB.

The proposed buffer shows the lowest input resistance, thanks both to the negative feedback established by the amplifier and to the diode-connection of the PMOS transistors. However, as expected, it also shows the lowest output resistance. A transistor working in saturation could be added in series with the diode-connected transistor to increase R_{out}. To keep the circuit symmetry, it would be necessary to also add another transistor to the input branch, but, from Equations (2)–(4), it can be seen that the input resistance may still be very low as long as the amplifier gain is sufficiently high. Finally, the proposed buffer shows the highest bandwidth.

To prove the robustness of the proposed self-biased buffers, corner process simulations were carried out and Table 3 shows the results. To ensure proper operation under all conditions, even when the bias current is reduced because of process variations, the transistors M_{p1} and M_{p2} were oversized in the design stage. The bias current I_{Bias} decreases down to 6.5 µA in the slow-slow corner but performance is not affected and the THD for a 30 μA_{pp} input current at 1 kHz remains below −80 dB for all cases. In the fast-fast corner, I_{Bias} increases up to 10 µA, therefore increasing the total power consumption to 38.7 µW for the SB-CB1 and 41.6 µW for the SB-CB2. As for the QFG-CB and CC-CB topologies, their robustness to process variations depends on the robustness of the external biasing circuit.

Finally, Monte Carlo simulations were carried out to verify the circuit operation under mismatch. The mean value and the standard deviation of main electrical parameters considering 500 samples are summarized in Table 4. In the proposed buffers, SB-CB1 and SB-CB2, the mean value for the gain distribution is practically 1 with the same 0.7% standard deviation. The SB-CB1 shows a higher mean offset value than SB-CB2, but the latter presents a higher standard deviation. As for THD, the mean value is lower than −66 dB for both circuits considering a 15 µA amplitude and 1 kHz frequency input signal. Linearity is therefore primarily degraded by mismatch and, according to these results, the actual THD is almost the same for the single-stage and the two-stage implementations. The SB-CB2 implementation may still be preferred if a very low input resistance is required, as is the case for example in configurations based on MOS current dividers [17,18]. Table 4 also shows that I_{Bias} is very robust to mismatch variations, and therefore so is the overall power consumption.

Table 3. THD and static power considering process variations.

Process Corner	I_{Bias} [μA]	SB-CB1		SB-CB2	
		Power [μW]	THD [dB] *	Power [μW]	THD [dB] *
typical	8.0	30.9	−85.6	32.4	−111.3
slow NMOS-slow PMOS	6.5	24.8	−85.0	26.7	−108.3
fast NMOS-fast PMOS	10.0	38.7	−87.1	41.6	−115.5
slow NMOS-fast PMOS	9.0	35.3	−86.5	38.2	−113.7
fast NMOS-slow PMOS	7.3	28.4	−84.7	30.4	−109.3

* THD@30 μA_{pp}@1 kHz.

Table 4. Monte Carlo analysis results.

Monte Carlo Analysis	SB-CB1		SB-CB2		QFG-CB		CC-CB	
	Mean	σ	Mean	σ	Mean	σ	Mean	σ
I_{Bias} (μA)	8.0	0.1	8.0	0.1	—	—	—	—
Gain	1.000	0.007	1.000	0.007	1.000	0.004	1.007	0.002
Offset (nA)	−1.1	124.3	−0.5	132.8	−0.45	73.07	−4.81	29.96
THD (dB)	−66.4	6.1	−67.0	6.7	−56.1	5.4	−49.1	1.2

By comparing the proposed SB-CB2 with the two other buffers, results show that the three implementations have a mean value in gain of nearly 1, showing the CC-CB the lowest standard deviation and the SB-CB2 the highest. The proposed SB-CB2 and the QFG-CB show similar offset mean value, but the SB-CB2 shows again the highest standard deviation. In Figure 9 the THD distribution is represented for all three implementations. The CC-CB shows the worst mean value of THD but the lowest standard deviation. The proposed self-biased buffer, in turn, is the most sensitive to mismatching, but still shows the highest linearity.

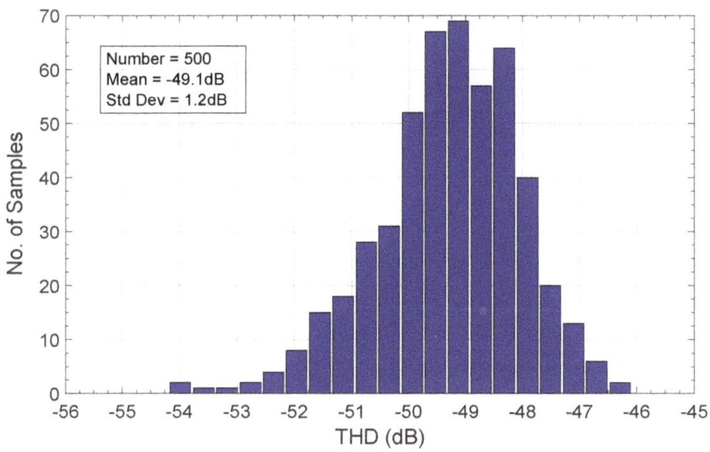

(a)

Figure 9. *Cont.*

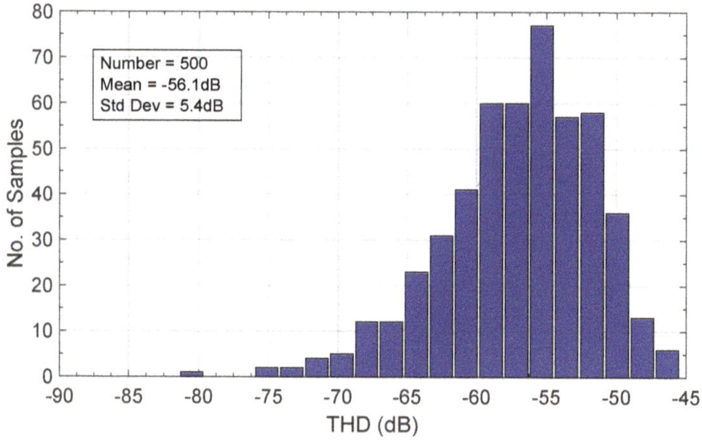

(**b**)

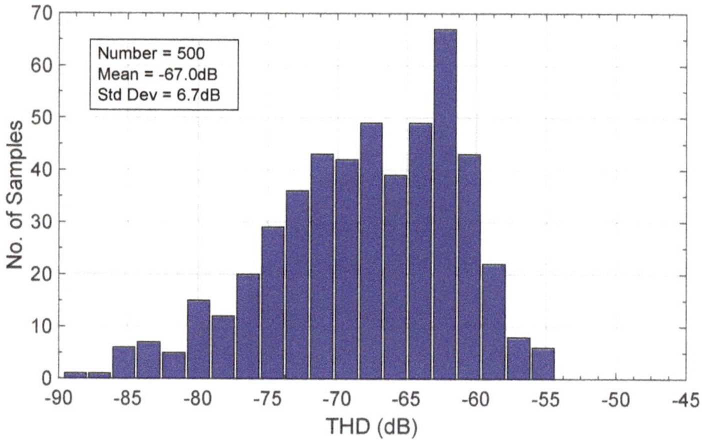

(**c**)

Figure 9. THD@30 μA_{pp}@1 kHz considering mismatch for (**a**) the CC-CB, (**b**) the QFG-CB and (**c**) the proposed SB-CB2.

4. Experimental Results

The self-biased current buffer SB-CB2 was integrated in the UMC (United Microelectronics Corporation) 0.18 µm CMOS technology with 1.8 V power supply. Figure 10 shows the microphotograph of the circuit and the layout. The circuit implementation occupies an area of 143 µm × 43 µm and exhibits a power consumption of 48 µW. Accordingly, the bias current is estimated to be 12 µA, which is a bit higher than expected from the results in Table 3. This increase in the bias current in turn results in an increase in the current capability of the buffer.

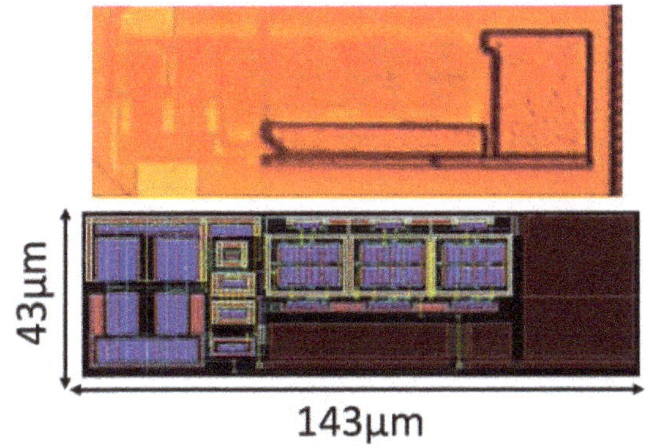

Figure 10. Integrated self-biased current buffer SB-CB2.

A PCB (Printed Circuit Board) was designed to carry out the characterization process. Figure 11a shows this PCB, and Figure 11b shows the photograph of the test setup. As the circuit processes the signal in the current domain, current conversion is necessary at both the input and the output. By means of a 10 kΩ resistance connected at the input node, the input current was generated, whereas the output current was measured through an external transimpedance amplifier configured with a TL081 integrated circuit [42]. This is detailed in Figure 12, which shows a block diagram of the interconnections within the PCB, as well as the methodology followed to carry out the experimental measurements after the circuit has been fabricated.

(**a**) (**b**)

Figure 11. Photograph of the setup used for the characterization: (**a**) PCB; and (**b**) test setup.

First, the current buffer was characterized under static conditions to obtain the DC characteristics and verify that the prototype is properly biased. Then, the time response was observed in the oscilloscope to test the current capability and accuracy of the buffer, as well as the settling time and input resistance. Finally, the frequency response and the harmonic distortion were characterized.

The circuit response to a 60 μA_{pp} sine input current at 1 kHz frequency is shown in Figure 13. This is the maximum output current that the buffer can handle before the signal starts getting distorted. The input–output characteristic is shown in Figure 14 for a -30 µA to $+30$ µA current range. A maximum relative error $e_r = 1.35\%$ is obtained, as also shown in Figure 14.

The input resistance was estimated from the response in the time domain, by measuring the input node voltage and calculating the derivative with respect to the input current. A 89 Ω input resistance was obtained.

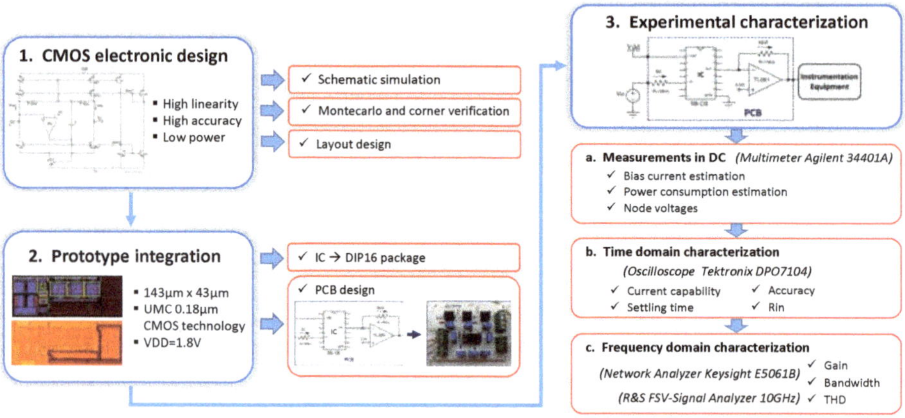

Figure 12. Experimental characterization block diagram.

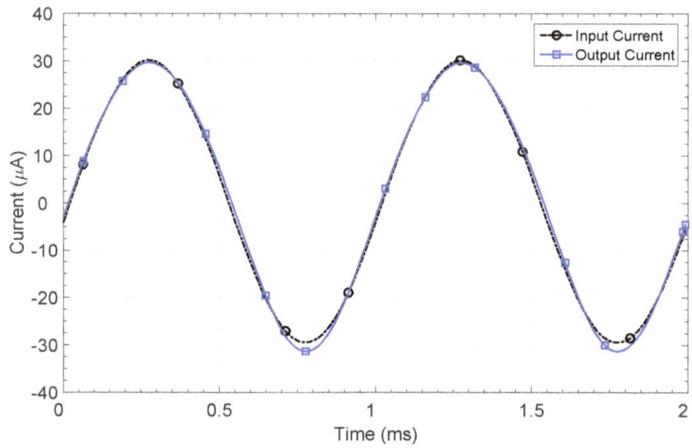

Figure 13. Integrated current buffer SB-CB2 measurement in the time domain.

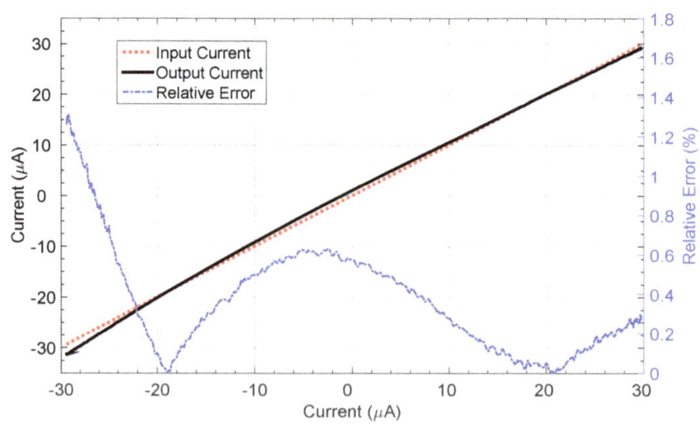

Figure 14. Integrated current buffer SB-CB2 response considering a −30 µA to 30 µA input range: output current and relative error e_r in the current copy.

If a 60 μA$_{pp}$ input current step is considered, the circuit shows a rise time of 8.6 μs and a fall time of 8.4 μs, both considering the response within 0.1% of the output signal. Figure 15 shows the oscilloscope screenshots of the buffer response to both the rising and falling edges of the input step for this dynamic characterization.

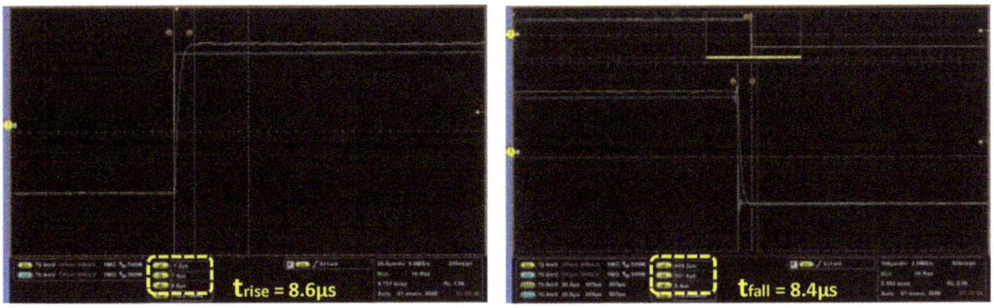

(a) (b)

Figure 15. Response of the integrated prototype to an input current step: (**a**) rising edge; and (**b**) falling edge.

The THD characterization was done using the signal analyzer ROHDE & SCHWARZ FSV-Signal Analyzer (10 Hz–6 GHz) [43]. Figure 16 shows the spectrum analyzer screenshots when considering a 60 μA$_{pp}$ sine input signal at 1 kHz (Figure 16a) and 10 kHz (Figure 16b). Both the frequency spectrum and the THD calculation are shown, considering ten harmonic components. The integrated prototype shows a −61 dB THD for the 60 μA$_{pp}$ input current at 1 kHz, and −53 dB at 10 kHz. These values correspond to the distortion specifications of the signal generator, so lower distortion values are actually expected.

Finally, the transfer function in the frequency domain was determined using the network analyzer E5061B ENA [44], as shown in Figure 17. Note that the bandwidth was reduced because of the parasitic capacitances of the chip package and the interconnection setup used for the characterization.

The self-biased current buffer electrical characteristics are summarized in Table 5, where a comparison with other topologies found in the literature is also presented. All the buffers presented in the table are based on the quasi-floating gate technique.

Note that, although the proposed circuit requires the highest bias current, it does not have a significant impact on the final consumption. Furthermore, the bias circuit of the other topologies has not been considered when estimating their power consumption.

The buffer in [26] and the proposed circuit show the lowest input resistance of 25 Ω and 89 Ω, respectively, so that a virtual ground is set at the input node, and therefore a higher linearity is observed when the maximum input current is considered in each case. The best experimental distortion figure is obtained in [26], at the cost of increased power consumption, which is almost three times the proposed SB-CB consumption. The buffers in [28,29] both present competitive power consumption, but with a rather high R_{in} (934 Ω and 4.8 kΩ, respectively). A higher distortion of −40 dB is observed in [28] for the maximum input current; even if a lower input current of 30 μA amplitude is considered, the THD is not higher than −53 dB. Similarly, the buffer presented in [29] shows a THD of −41 dB for a current I_{in} = 100 μA$_{pp}$.

Finally, the proposed SB-CB shows the lowest integration area, whereas the circuit presented in [29] has the highest dimensions because it uses three capacitors to achieve the class-AB operation.

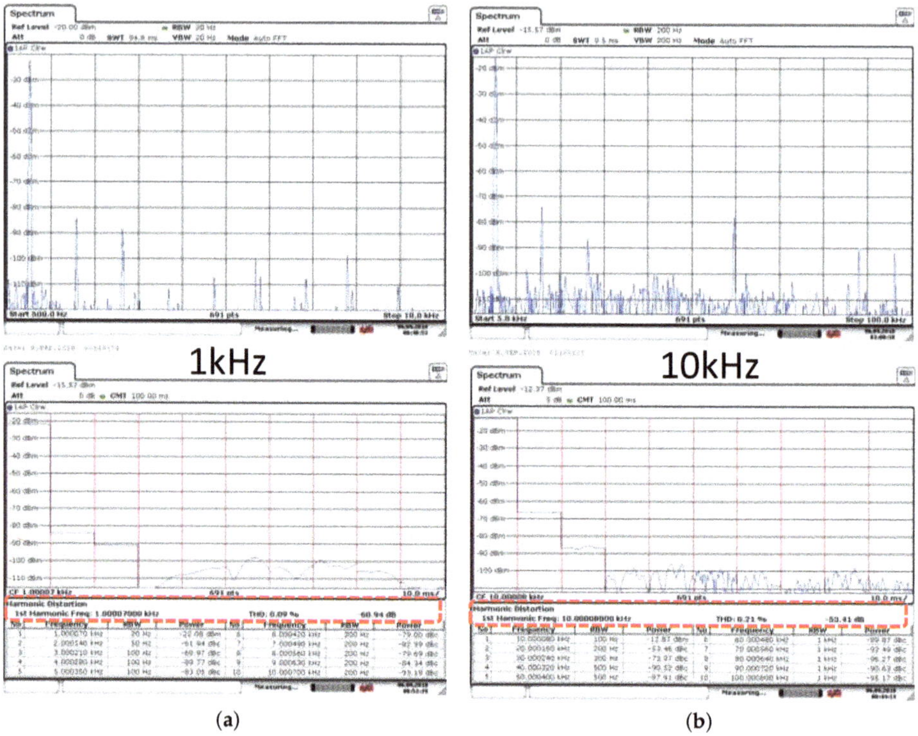

Figure 16. Integrated current buffer THD characterization for a 60 μA_{pp} input current at: (**a**) 1 kHz; and (**b**) 10 kHz considering ten harmonics.

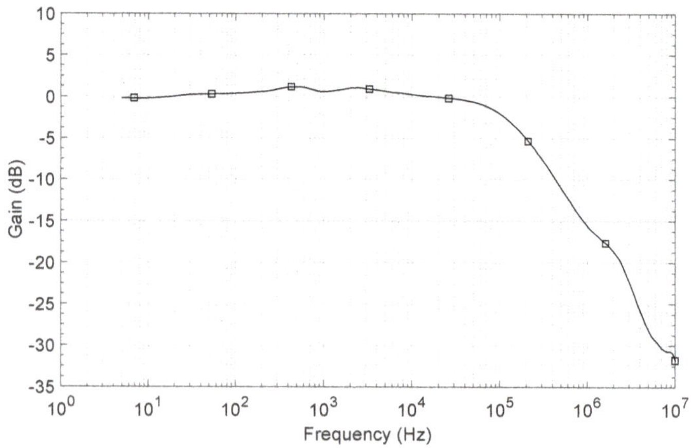

Figure 17. Integrated current buffer SB-CB2 frequency response.

Table 5. Electrical characteristics of the integrated SB-CB prototype and comparison with other circuits.

Parameter	This Work	Lopez-Martin'08 [26]	Suadet'13 [28] †	Esparza'14 [29]
CMOS Technology	0.18 μm	0.5 μm	0.18 μm	0.5 μm
Power Supply (V)	1.8	3.3	0.5	1.2
I_{Bias} (μA)	12	10	6	10
THD (dB)	<-61@60 μA_{pp}@1 kHz <-53@60 μA_{pp}@10 kHz	-59@200 μA_{pp}@120 kHz	-40@96 μA_{pp}@1 MHz	-41@100 μA_{pp} **
Power Consumption (μW)	48	165	8.2	36
BW (MHz)	2.6 †	120 †	230	72.4 †
Rin (Ω)	89	25	934	4.8k †
Rout (MΩ)	2.4 †	—	1.13	7.2 †
$e_{r_{max}}$ (%)	1.35%@$I_{in} = 60$ μA_{pp}	—	—	—
Settling Time (μs)	8.6	—	—	—
Area (μm^2)	6149	18,200	—	25,020

† Simulation results. ** Operation frequency not mentioned.

5. Conclusions

A self-biased class AB 1.8 V–0.18 μm CMOS current buffer based on the QFG approach is proposed in this paper. It shows the lowest input resistance and highest linearity when compared to other class AB current buffers with a virtual ground at the input node, at a cost of higher power consumption. However, as the proposed topology is self-biased, it does not require any additional circuitry, whereas other buffers require a biasing scheme. Monte Carlo and process corner simulations show that, even though the proposed buffer is more sensitive to process variations, it still shows the best performance in terms of linearity.

The integrated prototype was able to copy an input current ranging from −30 μA to +30 μA with a maximum relative error of 1.35% and 48 μW static power consumption. The prototype has a reduced area of 143 × 43 μm^2, making it a viable solution for battery-operated systems where minimum dimensions and low power operation are mandatory. The THD for the same amplitude input current remains below −53 dB up to 10 kHz, showing a high linearity characteristic even when the maximum input current is considered. The circuit also has a very low input resistance $R_{in} = 89\ \Omega$, thus setting a virtual ground at the input node, a relatively high output impedance and a circuit response time of 8.6 μs.

Author Contributions: Conceptualization, J.A.M.-N., M.T.S.-P. and B.C.-L.; Methodology, J.A.M.-N., M.T.S.-P. and A.S.-R.; Investigation, J.A.M.-N. and M.T.S.-P.; Validation, J.A.M.-N. and N.M.-M.; Writing—Original draft preparation, J.A.M.-N.; Writing—review and editing, J.A.M.-N., M.T.S.-P., N.M.-M., B.C.-L. and A.S.-R.; and Supervision, M.T.S.-P. and N.M.-M.

Funding: This work was supported by CONACYT 362674 Ph.D. Grant, CONACYT project No. CB-2015-257985 and MINECO-FEDER, UE (TEC2015-65750-R) Research Project.

Conflicts of Interest: The authors declare no conflict of interest. The founding sponsors had no role in the design of the study; in the collection, analyses, or interpretation of data; in the writing of the manuscript, and in the decision to publish the results.

References

1. Kumgern, M.; Wareechol, E.; Phasukkit, P. Quadrature oscillator and universal filter based on translinear current conveyors. *AEU Int. J. Electron. Commun.* **2018**, *94*, 69–78. [CrossRef]
2. Cini, U. A low-offset high CMRR current-mode instrumentation amplifier using differential difference current conveyor. In Proceedings of the 2014 21st IEEE International Conference on Electronics, Circuits and Systems (ICECS), Marseille, France, 7–10 December 2014; pp. 64–67. [CrossRef]

3. Esparza-Alfaro, F.; Pennisi, S.; Palumbo, G.; Lopez-Martin, A. Low-power class-AB CMOS voltage feedback current operational amplifier with tunable gain and bandwidth. *IEEE Trans. Circuits Syst. II Express Briefs* **2014**, *61*, 574–578. [CrossRef]
4. Swamy, M.N.S. Modified CFOA, its transpose, and applications. *Int. J. Circuit Theory Appl.* **2016**, *44*, 514–526. [CrossRef]
5. Reshma, P.G.; Gopi, V.P.; Babu, V.S.; Wahid, K.A. Analog CMOS implementation of FFT using cascode current mirror. *Microelectron. J.* **2017**, *60*, 30–37. [CrossRef]
6. Tsirimokou, G.; Psychalinos, C. Ultra-low voltage fractional-order circuits using current mirrors. *Int. J. Circuit Theory Appl.* **2016**, *44*, 109–126. [CrossRef]
7. Sotner, R.; Jerabek, J.; Langhammer, L.; Dvorak, J. Design and Analysis of CCII-Based Oscillator with Amplitude Stabilization Employing Optocouplers for Linear Voltage Control of the Output Frequency. *Electronics* **2018**, *7*, 157. [CrossRef]
8. Mowlavi, S.; Baharmast, A.; Sobhi, J.; Koozehkanani, Z. A novel current-mode low-power adjustable wide input range four-quadrant analog multiplier. *Integration* **2018**, *63*, 130–137. [CrossRef]
9. Lopez-Martin, A.; Garde, M.P.; Carvajal, R.G.; Ramírez-Angulo, J. On the Optimal Current Followers for Wide-Swing Current-Efficient Amplifiers. In Proceedings of the IEEE International Symposium on Circuits and Systems (ISCAS), Florence, Italy, 27–30 May 2018; pp. 1–5. [CrossRef]
10. Lopez-Martin, A.J.; Acosta, L.; Garcia-Alberdi, C.; Carvajal, R.G.; Ramirez-Angulo, J. Power-efficient analog design based on the class AB super source follower. *Int. J. Circuit Theory Appl.* **2012**, *40*, 1143–1163. [CrossRef]
11. Pourashraf, S.; Ramírez-Angulo, J.; Lopez-Martin, A.J.; González-Carvajal, R. A super class-AB OTA with high output current and no open loop gain degradation. In Proceedings of the IEEE 60th International Midwest Symposium on Circuits and Systems (MWSCAS), Boston, MA, USA, 6–9 August 2017; pp. 815–818.
12. Bruschi, P.; Navarrini, D.; Piotto, M. A class-AB CMOS operational amplifier for application as rail-to-rail high current drive output buffer. In Proceedings of the 28th European Solid-State Circuits Conference (ESSCIRC), Florence, Italy, 24–26 September 2002; pp. 731–734.
13. Kawahito, S.; Tadokoro, Y. CMOS class-AB current mirrors for precision current-mode analog-signal-processing elements. *IEEE Trans. Circuits Syst. II Analog Digit. Signal Process.* **1996**, *43*, 843–845. [CrossRef]
14. Zhao, X.; Wang, Y.; Jia, D.; Dong, L. Ultra-high current efficiency single-stage class-AB OTA with completely symmetric slew rate. *AEU-Int. J. Electron. Commun.* **2018**, *87*, 65–69. [CrossRef]
15. Grasso, A.D.; Marano, D.; Esparza-Alfaro, F.; Lopez-Martin, A.J.; Palumbo, G.; Pennisi, S. Self-biased dual-path push-pull output buffer amplifier for LCD column drivers. *IEEE Trans. Circuits Syst. I Regul. Pap.* **2014**, *61*, 663–670. [CrossRef]
16. Martinez-Nieto, J.A.; Sanz-Pascual, M.T.; Medrano-Marques, N.J.; Calvo-Lopez, B. Self-biased class-AB CMOS current buffer. In Proceedings of the IEEE 7th Latin American Symposium on Circuits Systems (LASCAS), Florianopolis, Brazil, 28 February–2 March 2016; pp. 255–258.
17. Bult, K.; Geelen, G. An inherently linear and compact most-only current-division technique. In Proceedings of the Solid-State Circuits Conference, Digest of Technical Papers, San Francisco, CA, USA, 19–21 February 1992; pp. 198–199. [CrossRef]
18. Pun, K.P.; Choy, C.S.; Chan, C.F.; da Franca, J. Digital frequency tuning technique based on current division for integrated active RC filters. *Electron. Lett.* **2003**, *39*, 1366–1367. [CrossRef]
19. Wang, W.; Jia, S.; Pan, T.; Wang, Y. Design of low-power high-speed dual-modulus frequency divider with improved MOS current mode logic. In Proceedings of the IEEE International Conference on Electron Devices and Solid State Circuits (EDSSC), Shenzhen, China, 6–8 June 2018; pp. 1–2.
20. Martínez-Nieto, J.A.; Sanz-Pascual, M.T.; Medrano-Marqués, N.J. Integrated mixed mode neural network implementation. In Proceedings of the European Conference on Circuit Theory and Design (ECCTD), Catania, Italy, 4–6 September 2017; pp. 1–4.
21. Martínez-Nieto, A.; Medrano, N.; Sanz-Pascual, M.T.; Calvo, B. An accurate analysis method for complex IC analog neural network-based systems using high-level software tools. In Proceedings of the IEEE 9th Latin American Symposium on Circuits & Systems (LASCAS), Puerto Vallarta, Mexico, 25–28 February 2018; pp. 1–4.

22. Martinez-Nieto, A.; Sanz-Pascual, M.T.; Marquez, A.; Perez-Bailon, J.; Calvo, B.; Medrano, N. A CMOS Mixed Mode Non-Linear Processing Unit for Adaptive Sensor Conditioning in Portable Smart Systems. *Procedia Eng.* **2016**, *168*, 1689–1692. [CrossRef]
23. Raj, N.; Singh, A.K.; Gupta, A.K. High performance current mirrors using quasi-floating bulk. *Microelectron. J.* **2016**, *52*, 11–22. [CrossRef]
24. Ramirez-Angulo, J.; Lopez-Martin, A.; Carvajal, R.; Chavero, F. Very low-voltage analog signal processing based on quasi-floating gate transistors. *IEEE J. Solid-State Circuits* **2004**, *39*, 434–442. [CrossRef]
25. Ramirez-Angulo, J.; Lopez-Martin, A.J.; Carvajal, R.G.; Calvo, B. Class-AB Fully Differential Voltage Followers. *IEEE Trans. Circuits Syst. II Express Briefs* **2008**, *55*, 131–135. [CrossRef]
26. Lopez-Martin, A.; Ramirez-Angulo, J.; Carvajal, R.; Algueta, J. Compact class AB CMOS current mirror. *Electron. Lett.* **2008**, *44*, 1335–1336. [CrossRef]
27. Garde, M.P.; Lopez-Martin, A.J.; Carvajal, R.G.; Ramirez-Angulo, J. Super class AB RFC OTA with adaptive local common-mode feedback. *Electron. Lett.* **2018**, *54*, 1272–1274. [CrossRef]
28. Suadet, A.; Kasemsuwan, V. A compact class-AB bulk-driven quasi-floating gate current mirror for low voltage applications. In Proceedings of the 13th International Symposium on Communications and Information Technologies (ISCIT), Surat Thani, Thailand, 4–6 September 2013; pp. 298–302.
29. Esparza-Alfaro, F.; Lopez-Martin, A.; Carvajal, R.G.; Ramirez-Angulo, J. Highly linear micropower class AB current mirrors using Quasi-Floating Gate transistors. *Microelectron. J.* **2014**, *45*, 1261–1267. [CrossRef]
30. Rana, C.; Afzal, N.; Prasad, D. A High Performance Bulk Driven Quasi Floating Gate MOSEFT Based Current Mirror. *Procedia Comput. Sci.* **2016**, *79*, 747–754. [CrossRef]
31. Ramirez-Angulo, J.; Lopez-Martin, A.; Carvajal, R.; Torralba, A.; Jimenez, M. Simple class-AB voltage follower with slew rate and bandwidth enhancement and no extra static power or supply requirements. *Electron. Lett.* **2006**, *42*, 784–785. [CrossRef]
32. Sanchez-Gonzalez, L.; Ducoudray-Acevedo, G. High accuracy self-biasing cascode current mirror. In Proceedings of the 49th IEEE International Midwest Symposium on Circuits and Systems, San Juan, Puerto Rico, 6–9 August 2006; Volume 1, pp. 465–468.
33. Cannizzaro, S.O.; Grasso, A.D.; Mita, R.; Palumbo, G.; Pennisi, S. Design procedures for three-stage CMOS OTAs with nested-miller compensation. *IEEE Trans. Circuits Syst. I Regul. Pap.* **2007**, *54*, 933–940. [CrossRef]
34. Grasso, A.D.; Palumbo, G.; Pennisi, S. Three-stage CMOS OTA for large capacitive loads with efficient frequency compensation scheme. *IEEE Trans. Circuits Syst. II Express Briefs* **2006**, *53*, 1044–1048. [CrossRef]
35. Fan, X.; Mishra, C.; Sanchez-Sinencio, E. Single miller capacitor frequency compensation technique for low-power multistage amplifiers. *IEEE J. Solid State Circuits* **2005**, *40*, 584–592. [CrossRef]
36. Parnklang, J.; Nanthanawanitch, W.; Titiroongruang, W. CMOS current follower circuits. In Proceedings of the IEEE Region 10 Conference TENCON 99, Cheju Island, Korea, 15–17 September 1999; Volume 2, pp. 1030–1033.
37. Kurashina, T.; Ogawa, S.; Watanabe, K. A high performance class-AB current conveyor. In Proceedings of the 1998 IEEE International Conference on Electronics, Circuits and Systems, Lisboa, Portugal, 7–10 September 1998; Volume 3, pp. 143–146. [CrossRef]
38. Zatorre, G.; Medrano, N.; Sanz, M.T.; Calvo, B.; Martinez, P.; Celma, S. Designing adaptive conditioning electronics for smart sensing. *IEEE Sens. J.* **2010**, *10*, 831–838. [CrossRef]
39. Molinar-Solís, J.E.; García-Lozano, R.Z.; Hidalgo-Cortes, C.; Rocha-Perez, J.M.; Díaz-Sánchez, A. A very compact CMOS class AB current mirror for low voltage applications. In Proceedings of the IEEE 4th Colombian Workshop on Circuits and Systems (CWCAS), Barranquilla, Colombia, 1–2 November 2012; pp. 1–4.
40. Palmisano, G.; Pennisi, S. Dynamic biasing for true low-voltage CMOS class AB current-mode circuits. *IEEE Trans. Circuits Syst. II Analog Digit. Signal Process.* **2000**, *47*, 1569–1575. [CrossRef]
41. Julien, M.; Bernard, S.; Soulier, F.; Kerzerho, V.; Cathebras, G. Breaking the speed-power-accuracy trade-off in current mirror with non-linear CCII feedback. *Microelectron. J.* **2018**. [CrossRef]
42. Texas Instruments. *TL08xx JFET-Input Operational Amplifiers*; Rev.3. Dallas, TX, USA, May 2015. Available online: http://www.ti.com/lit/ds/slos081i/slos081i.pdf (accessed on 7 November 2018).

43. ROHDE & SCHWARZ. *R&S FSV Signal and Spectrum Analyzer Operating Manual*; Test and Measurements: Munich, Germany, 2011. Available online: http://www.eava.ee/$\sim$laborid/side/Spektri_A/FSV13/FSV_Operating.pdf (accessed on 7 November 2018).
44. Keysight Technologies. *Keysight E5061B ENA Vector Network Analyzer, 100 kHz to 1.5/3 GHz, 5 Hz to 500 M/1.5 G/3 GHz*; Keysight Technologies: Santa Rosa, CA, USA, 2018. Available online: https://literature.cdn.keysight.com/litweb/pdf/5990-4392EN.pdf (accessed on 7 November 2018).

electronics

MDPI

Article

Demodulation of Angular Position and Velocity from Resolver Signals via Chebyshev Filter-Based Type III Phase Locked Loop

Huan Liu and **Zhong Wu** *

School of Instrumentation and Optoelectronic Engineering, Beihang University, Beijing 100191, China; liuhuan@buaa.edu.cn

* Correspondence: wuzhong@buaa.edu.cn; Tel.: +86-10-8233-9703

Received: 26 October 2018; Accepted: 21 November 2018; Published: 25 November 2018

Abstract: A high-accuracy demodulation algorithm is required to estimate angular position and angular velocity from resolver signals. In order to improve the estimation accuracy of conventional phase-locked loop (PLL) based demodulation method, a Chebyshev filter-based type III PLL method is proposed in this paper. The proposed method makes PLL become a system of type III tracking loop, which could greatly reduce the theoretical constant deviation in the estimation results of conventional type II PLL in case of variable speed. Meanwhile, the eigenvalues of type III PLL are placed to be the same position as those of a Chebyshev low-pass filter. In this way, demodulation parameters with stronger filter properties can be obtained to effectively suppress the high-frequency measurement noise in resolver signals. Thus, the proposed method can achieve higher demodulation precision compared with the conventional ones. Simulations and experiments are performed to validate the proposed demodulation method.

Keywords: demodulation; phase-locked loop; Chebyshev filter; measurement noise suppression

1. Introduction

Modern control algorithms for servomotors require accurate feedback information of both angular position and angular velocity. Usually, the angular position and velocity can be measured by shaft sensors, such as optical encoders, magnetic encoders, and resolvers. Compared with other sensors, resolvers have gained more attention owing to their various advantages such as simple and rugged structure, low cost, high accuracy, resistance to harsh environments [1–3]. Normally, a resolver generates two amplitude-modulated analog signals with rotor position information. Therefore, high-accuracy resolver-to-digital conversion (RDC) is required in order to extract rotor position and velocity from the resolver signals [4].

The special integrated circuit (IC)—such as AD2S80 series, AD2S83, AD2S1210 and PGA411-Q1—can conduct the RDC process to obtain angular position and velocity, and they are quite convenient to use. These commercial RDC ICs are mainly based upon type-II tracking loop technique, can continuously track the inputs and convert the input sine and cosine signals into a digital representation of rotor position and velocity. However, the cost of the RDC ICs is very high, even higher than the resolver itself; meanwhile, the typical RDC ICs have a limited bandwidth of 300–600 Hz [5], and it is difficult to adjust the parameters of RDC ICs to meet users' specific requirements.

In order to reduce costs and facilitate bandwidth adjustment of RDC, software-based RDC can be adopted [6,7]. To reduce the sampling pressure of CPU, generally, it is necessary to pre-detect the sine and cosine envelope of resolver outputs, and then demodulate the envelope signals to derive rotor position and velocity.

The commonly used demodulation methods may be divided into two categories, which are: (a) trigonometric method and (b) phase-locked loop (PLL) method [8]. The trigonometric method, also known as arctangent method, is simple and easy to implement. However, the trigonometric method only yields the unfiltered rotor position [9–12]. Usually, rotor velocity is obtained by a differential operation to rotor position. Nevertheless, the differential operation could amplify the noise in the resolver signals and cause large estimation error. A low-pass filter can be added to suppress the noise but at the expense of adding phase lag, which makes the trigonometric method lose its advantage of quick response.

The PLL method is a closed-loop strategy, can accurately track the rotor position and rotor velocity simultaneously, which makes it widely used in RDC ICs and software-based RDC. The estimates of rotor position and velocity are continuously corrected by the tracking loop. Most of RDC ICs adopt type II tracking loop as PLL structure, and this technique has been extended to software-based RDCs [13–16]. A PLL tracking algorithm on the basis of the type II tracking loop was proposed in [13] to achieve resolver-to-digital conversion with high precision. In [14], an RDC design using autotuning filters was presented to mitigate the resolver signal error on the speed output of the tracking loop. In [15], a novel and high-performance PLL-based resolver converter was proposed to measure angles in the full 360° range. A PLL converter using PI controller was designed in [16] for resolvers and sine/cosine encoders. Besides, the angle tracking observer (ATO) methods presented in [17–19] are also PLL systems.

Both the type II tracking loop and ATO can track the rotor position and rotor velocity smoothly and accurately. Also, they have better performance in disturbance attenuation compared with the trigonometric method. However, the demodulation accuracy is still restricted by the system structure of conventional PLL methods. On the one hand, the estimation accuracies of type II PLL would be much lower when the rotor velocity varies quickly. This is because the type II system has inherent theoretical estimation errors. On the other hand, all the PLL-based RDC designs reviewed so far, suffer from the fact that there is a trade-off between dynamic performance and noise-suppression capability when designing the parameters of PLL. In other words, to quickly track the input and achieve better dynamic performance with little phase delay, the gain of PLL are usually set large to have wide closed-loop bandwidth, but high gain will make the PLL system extremely sensitive to the noise in resolver signals. Moreover, it may cause overshoot and instability if the parameters are not properly selected.

In this paper, a demodulation algorithm via Chebyshev filter-based Type III PLL is presented to estimate angular position and angular velocity from resolver signals. The proposed design is of a type III system, and it can obtain better filter characteristics without compromising dynamic performance, so as to achieve higher estimation precision in demodulation. Simulation and experiments are performed to verify the proposed method.

The rest of the paper is organized as follows. In Section 2, the principle of resolver and software-based RDC are introduced, and the problem of the paper is formulated. In Section 3, a Chebyshev filter-based type III PLL design is proposed for the demodulation of angular position and angular velocity, and the performance of the proposed method is analyzed by using Bode diagram. In Section 4, simulation and experiment are carried out to verify the effectiveness of the proposed demodulation method. Finally, conclusions are given in Section 5.

2. Resolver Principles and Problem Formulation

2.1. Principle of Resolver and Software-Based RDC

Resolvers are commonly used as shaft position sensing apparatus in servomotor control systems, and the schematic structure of a sine/cosine resolver is shown in Figure 1. It consists of one rotating winding (fixed on the rotor) and two stator winding. The rotor is directly installed on the motor shaft. If the rotating winding is supplied with a sinusoidal excitation signal V_{ex}, ideally, the two stator windings generate two orthogonal amplitude-modulated signals, which can be described as

$$\begin{cases} y_{\sin} = kE \sin \omega_e t \sin \theta \\ y_{\cos} = kE \sin \omega_e t \cos \theta \end{cases} \tag{1}$$

where k is the transformation ratio of the resolver; E and ω_e denote the amplitude and frequency of the excitation signal, respectively; θ is the rotor position of the resolver.

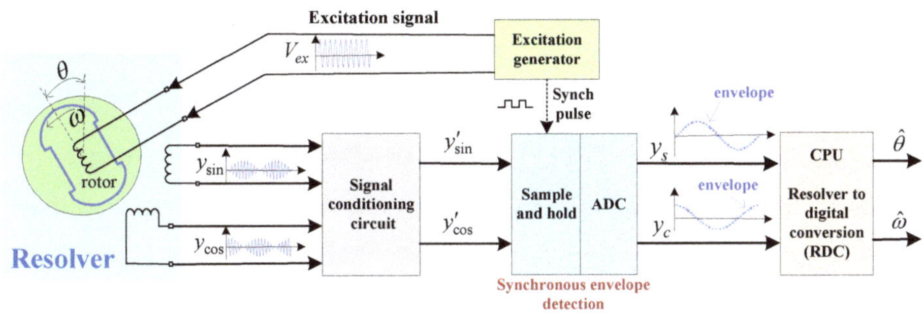

Figure 1. Schematic diagram of resolver and software-based RDC.

As shown in (1), the outputs of the resolver are two signals proportional to $\sin\theta$ and $\cos\theta$. Hence, to extract rotor position and velocity from the resolver output signals, further detection and demodulation are required. Detection refers to detecting the amplitude envelops from the resolver outputs; demodulation refers to the process of obtaining angular position and velocity from the envelop signals. Figure 1 illustrates the principle of software-based RDC. In order to relieve the pressure of microprocessor on sampling, the detection part is usually implemented by peripheral hardware circuits. After signal conditioning circuit, as in [20], the resolver signals are then synchronously demodulated by sampling at the peak of the sinusoid excitation signal in the sample and hold circuit. Next, the signals are sampled by A/D converters (ADC), which produces the digital value of sine and cosine envelope signals given by

$$\begin{cases} y_s = A \sin \theta \\ y_c = A \cos \theta \end{cases} \tag{2}$$

where A is the amplitude of the detected envelope signals. Then, the envelope signals are demodulated by the software-based RDC algorithm to derive rotor position and velocity.

It should be noted that in practical applications, the resolver envelope signals are usually not ideal sine and cosine signals, where measurement errors and noise disturbances inevitably exist [21]. The resolver measurement errors (including amplitude deviation, DC offsets and phase shift) can be calibrated effectively [22]. Whereas, it is difficult to suppress the noise disturbance (including excitation signal interference and white noise in the circuits). Consequently, the actual resolver envelope signals can be written as

$$\begin{cases} y_s = A \sin \theta + n_s \\ y_c = A \cos \theta + n_c \end{cases} \tag{3}$$

where n_s and n_c are measurement noise. The influence of noise disturbance should be considered in the study on demodulation algorithm.

2.2. Review of Conventional PLL-Based Demodulation Method

The principle of PLL-based demodulation method is shown in Figure 2a. The PLL consists of three parts: a phase detector (PD), a loop filter (LPF), and a voltage-controlled oscillator (VCO).

In software-based RDC, the VCO part can be modeled as an integrator. Define $\hat{\theta}$ and $\hat{\omega}$ as the estimates of angular position and velocity, and $\widetilde{\theta} = \theta - \hat{\theta}$ as the estimation error of angular position, then the output signal of PD can be given by

$$\varepsilon = \sin\theta\cos\hat{\theta} - \cos\theta\sin\hat{\theta} = \sin(\theta - \hat{\theta}) = \sin\widetilde{\theta} \tag{4}$$

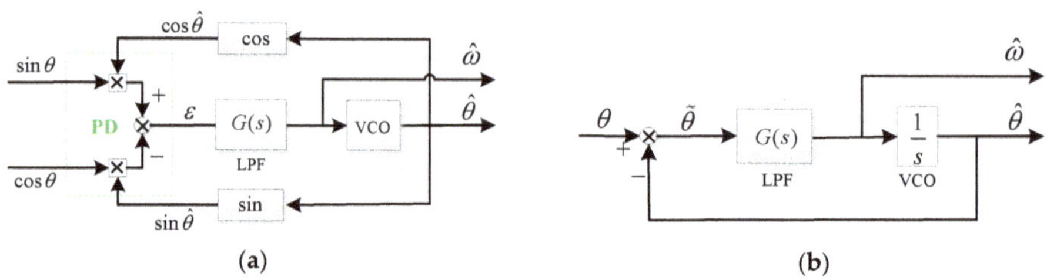

Figure 2. Block diagram of conventional PLL-based demodulation method. (**a**) Actual form; (**b**) simplified form.

When $\widetilde{\theta}$ is small enough, $\varepsilon = \sin\widetilde{\theta} \approx \widetilde{\theta}$. Based on this assumption, Figure 2a can be simplified as Figure 2b. $G(s)$ plays an important role in the PLL structure of Figure 2, and it affects the performance of PLL in tracking angular position and angular velocity.

Most of RDC ICs adopt type II tracking loop as PLL structure, and it can be extended to software-based RDC. Taking the PLL structure of AD2S1210 [23] as an example, the simplified form of system response block diagram of AD2S1210 is shown in Figure 3.

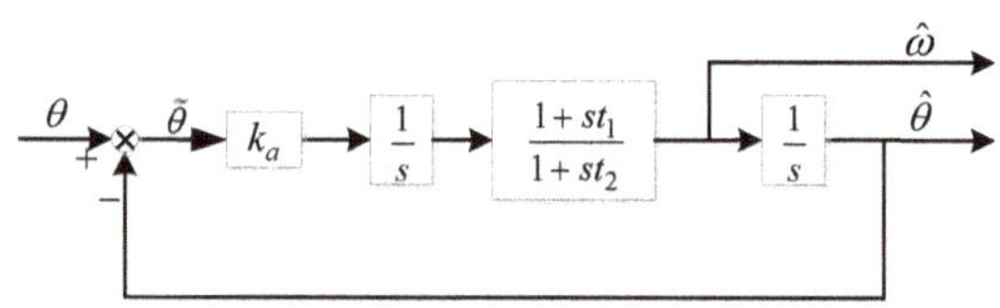

Figure 3. Simplified form of system response block diagram of AD2S1210.

According to Figure 3, the open loop transfer function of AD2S1210 can be described as

$$\frac{G(s)}{s} = \frac{k_a}{s^2} \times \frac{1 + st_1}{1 + st_2} \tag{5}$$

where $k_a > 0$ is the open-loop gain, $1 + st_1/1 + st_2$ is the compensation filter, and $t_1 > t_2$ are the compensation time constants. Equation (5) shows that AD2S1210 is a Type II system. The transfer functions from $\theta(s)$ to $\hat{\theta}(s)$, $\omega(s)$ to $\hat{\omega}(s)$ can be expressed as

$$G_{\theta 1}(s) = \frac{\hat{\theta}(s)}{\theta(s)} = \frac{k_a(1 + st_1)}{t_2 s^3 + s^2 + k_a t_1 s + k_a}, \; G_{\omega 1}(s) = \frac{\hat{\omega}(s)}{\omega(s)} = \frac{k_a(1 + st_1)}{t_2 s^3 + s^2 + k_a t_1 s + k_a} \tag{6}$$

The error transfer functions are

$$E_{\theta 1}(s) = \frac{\widetilde{\theta}(s)}{\theta(s)} = \frac{t_2 s^3 + s^2}{t_2 s^3 + s^2 + k_a t_1 s + k_a}, \; E_{\omega 1}(s) = \frac{\widetilde{\omega}(s)}{\omega(s)} = \frac{t_2 s^3 + s^2}{t_2 s^3 + s^2 + k_a t_1 s + k_a} \tag{7}$$

According to the final-value theorem, when the rotor runs at a constant velocity, AD2S1210 can achieve no-difference estimation of the rotor position and rotor velocity. When the rotor rotates at a constant acceleration of A rad/s^2, the steady state error of rotor velocity estimation is also equal to 0, but for position estimate, there exists a certain tracking error which is equal to A/k_a.

Equation (6) also indicates that AD2S1210 behaves like a third-order low-pass filter. At low frequencies, s^3 along with s^2 is very small and overwhelmed by the other terms, so (6) reduces to 1, which means that the AD2S1210 generates so little of an effect that the rotor position and velocity can be accurately estimated. Whereas, at high frequencies, s^n is very large, inducing attenuation and phase delay. Consequently, the gains of AD2S1210 are normally designed large enough to raise the effective bandwidth of RDC, which minimizes the phase delay, but this will make the system more sensitive to noise.

3. Design of Chebyshev Filter-Based Type III PLL for Demodulation

To improve the performance of PLL in demodulation algorithm, and meanwhile enhance its noise suppression ability, in this section, a Chebyshev filter-based type III PLL design is proposed.

3.1. Design of Type III PLL

Figure 4a shows the block diagram of Type III PLL design for the demodulation of angular position and angular velocity. q_1, q_2, q_3 are coefficients of the LPF in the PLL structure. The design of these coefficients will be explained in next part. As analyzed in Section 2.2, Figure 4a can be simplified as Figure 4b. In this demodulation method, the PLL is designed to be a type III system. Thus, the theoretical constant deviation problem of type II tracking loop in case of variable speed can be improved. Concrete analysis is as follows.

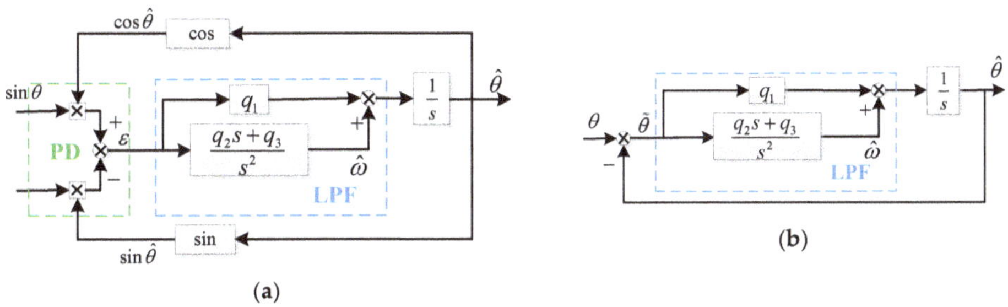

Figure 4. Block diagram of the Type III PLL design for demodulation. (**a**) Actual form; (**b**) Simplified form.

From Figure 4b, the transfer functions from $\theta(s)$ to $\hat{\theta}(s)$, $\omega(s)$ to $\hat{\omega}(s)$ can be expressed as

$$G_{\theta 2}(s) = \frac{\hat{\theta}(s)}{\theta(s)} = \frac{q_1 s^2 + q_2 s + q_3}{s^3 + q_1 s^2 + q_2 s + q_3},\ G_{\omega 2}(s) = \frac{\hat{\omega}(s)}{\omega(s)} = \frac{q_2 s + q_3}{s^3 + q_1 s^2 + q_2 s + q_3} \tag{8}$$

Then, the error transfer function can be written as

$$E_{\theta 2}(s) = \frac{\tilde{\theta}(s)}{\theta(s)} = \frac{s^3}{s^3 + q_1 s^2 + q_2 s + q_3},\ E_{\omega 2}(s) = \frac{\tilde{\omega}(s)}{\omega(s)} = \frac{s^3 + q_1 s^2}{s^3 + q_1 s^2 + q_2 s + q_3} \tag{9}$$

According to the error transfer function in (9), when the rotor runs at a constant velocity, the proposed Chebyshev filter-based method can achieve no-difference estimation of the rotor position and rotor velocity. When the rotor rotates at a constant acceleration of A rad/s^2, the steady state error of rotor position estimate and velocity estimate are also equal to 0.

3.2. Parameter Design of Type III PLL via Chebyshev Filter

After designing the type III PLL structure, the demodulation performance depends mainly on its parameters. As stated in precious section, in practical applications, there inevitably exist measurement noises in resolver signals. Seeing that, we need to carefully design the parameters of type III PLL to achieve the best estimation results.

Equation (8) indicates that type III PLL is essentially equivalent to a low-pass filter, which filters out the high-frequency noise that may exist in the envelop signals. Therefore, in this paper, the problem of parameter design for type III PLL is transformed into a filter design problem. The design of low-pass filter should meet the following principles: (1) in the low frequency range, the frequency characteristics of filter should be as close as possible to 1; (2) while in the high frequency range, it should be as close to 0 as possible. That is, as close as possible to the characteristics of the ideal low-pass filter, thereby, not only can it obtain good estimation accuracy, but also effectively suppress high-frequency measurement noise.

3.2.1. Introduction to Chebyshev Filter

Here, we design the parameters of type III PLL on the basis of Chebyshev low-pass filter, whose characteristic is closer to the ideal low-pass filter [24]. The amplitude-frequency characteristic of the nth-order Chebyshev low-pass filter is as follows [25]

$$|H(\omega)| = \sqrt{\frac{1}{1+\varepsilon^2 T_n^2(\omega/\omega_0)}} \tag{10}$$

where ω_0 is the passband cut-off frequency, $T_n(\omega)$ is the Chebyshev polynomial, and ε is the passband ripple factor. If the ripple factor is expressed in dB, define

$$\xi = -20\lg\frac{1}{\sqrt{1+\varepsilon^2}} = 10\lg(1+\varepsilon^2) \tag{11}$$

Then, the filter can be called a ξ(dB) Chebyshev filter. For instance, 1 dB Chebyshev filter means that $\xi = 1(\varepsilon = 0.50885)$.

The amplitude-frequency characteristic of Chebyshev filter is determined by ω_0 and ξ. ω_0 can be selected properly according to the dynamic requirements of the system and the frequency range of the practical noise. As for ξ, its value affects the gain fluctuation and high-frequency amplitude characteristics in the passband of the filter. If the ω_0 are set as 100 rad/s, then, the bode diagram of the third-order Chebyshev filter with different ξ is depicted in Figure 5. It shows that in the frequency range after 100 rad/s, the larger the ξ, the smaller the high-frequency amplitude is, the better the noise suppression ability, but the fluctuation in the passband is intensified. Decreasing ξ can reduce the amplitude fluctuation and phase angle lag in the passband, so as to get better dynamic property.

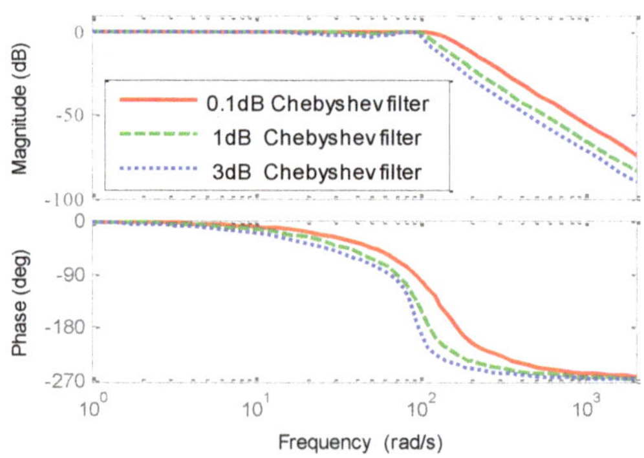

Figure 5. Bode diagram of the third-order Chebyshev filter with different ζ.

After designing ξ, the transfer function of nth-order Chebyshev filter can be expressed as

$$H(s) = \frac{a_n\omega_0^n}{s^n + a_1\omega_0 s^{n-1} + \cdots + a_{n-1}\omega_0^{n-1}s + a_n\omega_0^n} \tag{12}$$

Take the third-order Chebyshev filter for instance, its transfer function is

$$H(s) = \frac{a_3\omega_0^3}{s^3 + a_1\omega_0 s^2 + a_2\omega_0^2 s + a_3\omega_0^3} \tag{13}$$

3.2.2. Parameter Design of Type III PLL

According to the coefficients of the third-order Chebyshev filter in (13), corresponding to (8), we can get the following parameters for type III PLL

$$\begin{cases} q_1 = a_1\omega_0 \\ q_2 = a_2\omega_0^2 \\ q_3 = a_3\omega_0^3 \end{cases} \tag{14}$$

In this way, the eigenvalues of type III PLL are placed to be the same position as those of a Chebyshev low-pass filter. Thus, demodulation parameters with better filter characteristics can be obtained. As analyzed in Section 3.2.1, the parameters can be easily tuned according to practical requirements and noise characteristics.

The coefficients of the third-order Chebyshev filter under different ξ are listed in Table 1. Take 1 dB third-order Chebyshev filter ($\xi = 1$) for instance, the parameters of type III PLL can be set as

$$\begin{cases} q_1 = 0.98834\omega_0 \\ q_2 = 1.23841\omega_0^2 \\ q_3 = 0.49131\omega_0^3 \end{cases} \tag{15}$$

Table 1. Coefficients of the third-order Chebyshev filter.

ξ(dB)	a_1	a_2	a_3
0.1	1.93881	2.62949	1.63805
0.5	1.25291	1.53490	0.71569
1	0.98834	1.23841	0.49131
2	0.73782	1.02219	0.32689
3	0.59724	0.92835	0.25059

Therefore, after choosing ξ there exists only one parameter ω_0 that needs to be adjusted.

3.3. Performance Analysis of the Proposed Method

As for the speed servo system, to compare the performance of conventional type II tracking loop (AD2S1210) and the proposed Chebyshev filter-based type III PLL method, their parameters are set to make the closed-loop 3dB bandwidth of $G_\omega(s)$ identical when using the two methods. According to the parameters given in [23], the parameters of type II tracking loop are chosen as $k_a = 46.3 \times 10^3$, $t_1 = 8 \times 10^{-3}$s, $t_2 = 728 \times 10^{-6}$s. Take 1dB Chebyshev filter-based type III PLL for instance, to have the same 3dB bandwidth, ω_0 is set as $\omega_0 = 378$rad/s. According to (15), the coefficients of the proposed method are $q_1 = 0.98834 \times 378$, $q_2 = 1.23841 \times 378^2$, $q_3 = 0.49131 \times 378^3$. Then, according to the transfer function in (6) and (8), the amplitude–frequency characteristics of angular position and velocity are illustrated in Figure 6 when using the two demodulation methods.

The transfer functions from $\theta(s)$ to $\hat{\theta}(s)$, $\omega(s)$, to $\hat{\omega}(s)$ in (6) and (8) reflect the dynamic performance and noise suppression effect of the two demodulation methods. As depicted in Figure 6b, the closed-loop 3dB bandwidth of $G_\omega(s)$ is identical ($\omega_{bw} = 601$rad/s). It can be seen that compared with type II tracking loop, the curve of amplitude–frequency characteristic goes down in the high-frequency range by the proposed method, which indicates that the proposed method can suppress high-frequency noise more effectively in velocity estimates $\hat{\omega}$. While in Figure 6a, for angular position estimates $\hat{\theta}$, the noise-suppression ability of the proposed method is worse but with larger bandwidth. Normally, the angular velocity estimates contain more noises than angular position estimates. Thus, the estimation accuracy of velocity would be improved greatly, and the angular position estimation results would be a little worse.

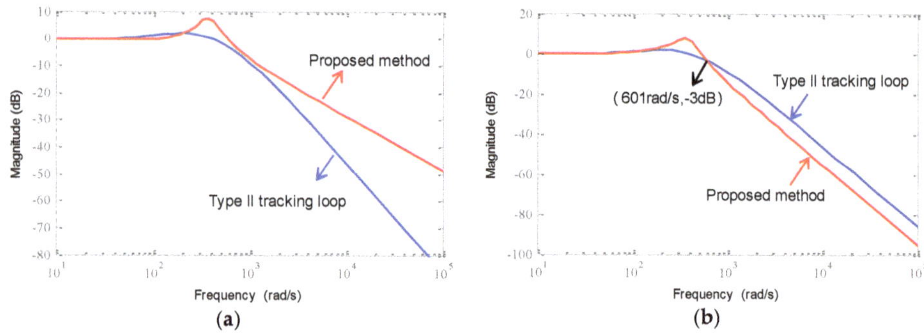

Figure 6. Amplitude-frequency characteristics of the two demodulation methods. (**a**) Angular position transfer function; (**b**) angular velocity transfer function.

Similarly, the amplitude–frequency characteristics of position and velocity error transfer function are depicted in Figure 7. As analyzed in [26], the error transfer functions from $\theta(s)$ to $\widetilde{\theta}(s)$, $\omega(s)$ to $\widetilde{\omega}(s)$ in (7) and (9) show the ability in suppressing un-modeled dynamics. According to the results shown in Figure 7a,b, the proposed method has better performance in suppressing un-modeled dynamics.

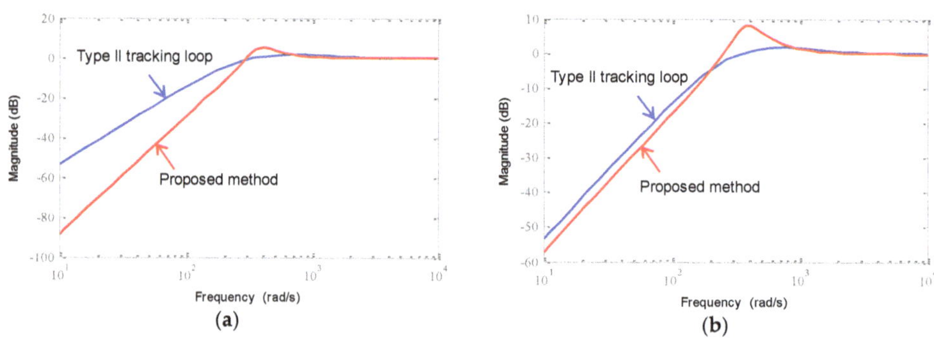

Figure 7. Amplitude-frequency characteristics of the two demodulation methods. (**a**) Angular position error transfer function; (**b**) angular velocity error transfer function.

From the above analysis, it can be seen that on the whole, the demodulation accuracy and noise-suppression ability can be improved by the proposed Chebyshev filter-based type III PLL method compared with the typical type II tracking loop method.

4. Simulation and Experimental Results

To verify the proposed demodulation algorithm for resolver signals, simulations and experiments are carried out.

4.1. Simulation

Figure 8 shows the semi-physical simulation platform. The resolver simulator takes a digital signal processor (DSP) TMS320F28335 (Texas Instruments Company, Dallas, TX, USA) as the core, and it can produce two envelope signals according to the preset value of angular position and velocity.

Then, the envelope signals are converted into two analog signals (y_s and y_c) by Digital to Analog Converter (DAC). After that, the resolver envelopes are sampled by Analogl to Digital Converter (ADC) in the signal acquisition circuit, and then uploaded to the upper computer for the demodulation algorithm through the USB interface. The difference between the semi-physical simulation and the real motor experiment is that the actual angular position and angular velocity of the resolver are available, so that the performance of the two demodulation methods (type II tracking loop and the proposed Chebyshev filter-based type III PLL method) can be compared.

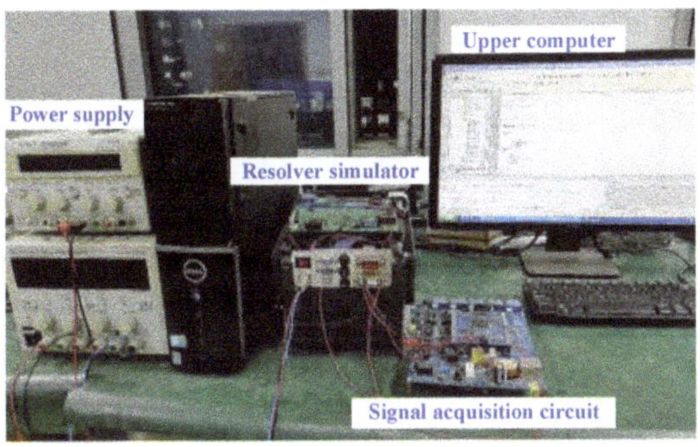

Figure 8. Semi-physical simulation platform.

As stated in previous section, demodulation parameters are set to make the closed-loop 3dB bandwidth of $G_\omega(s)$—the transfer function of estimated velocity—identical when using the two methods. Take 1dB Chebyshev filter-based demodulation method for instance, the parameters are set as Section 3.3. The semi-physical simulation is carried out on the condition of two cases: constant speed (2π rad/s) and constant acceleration ($10\pi\cdot$t rad/s).

- Case 1: Constant Speed (2π rad/s)

In this case, the rotor position and velocity estimation errors of the two demodulation methods are presented in Figure 9. Meanwhile, the average (AVG) and standard deviation (STD) of estimation errors are given in Table 2 to evaluate the performance of the two demodulation methods.

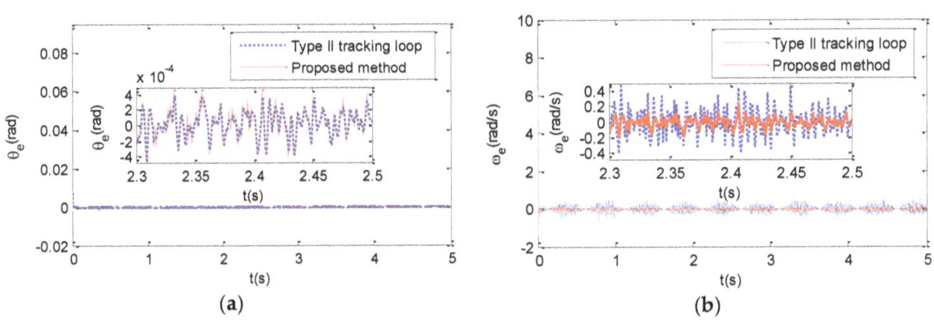

Figure 9. Estimation errors of the two demodulation methods in Case 1. (**a**) Rotor position estimation error; (**b**) rotor velocity estimation error.

As shown in Figure 9b, due to the closed-loop 3 dB bandwidths of $G_\omega(s)$ are identical, the convergence rates of velocity estimation error ω_e for the two demodulation methods are equal. In addition, from the enlarged views, we can see that the steady-state error curve of the proposed method is smoother compared with the type II tracking loop method. It means that the proposed method can attenuate the effects of measurement noise in resolver signals to some extent, thus, improve the steady-state accuracy of demodulation algorithm. While, for the estimation results of rotor position in Figure 9a, as analyzed in Section 3.3, the noise suppression capability is a little worse when using the proposed method, but its dynamic performance is improved.

The above analysis can also be proved by Table 2, from which we can see that the AVG of position and velocity estimation error is about the same, but the STD of velocity estimation error is reduced by 63% by the proposed method.

Table 2. Estimation error in simulation.

Cases	Estimation Error	Type II Tracking Loop		Proposed Method	
		AVG	STD	AVG	STD
Case 1	Position estimation error (rad)	1.951×10^{-7}	9.302×10^{-5}	1.877×10^{-7}	1.073×10^{-4}
	Velocity estimation error (rad/s)	-4.073×10^{-5}	0.0927	-2.112×10^{-5}	0.0340
Case 2	Position estimation error (rad)	2.714×10^{-4}	9.416×10^{-5}	3.424×10^{-7}	1.085×10^{-4}
	Velocity estimation error (rad/s)	5.738×10^{-5}	0.0936	3.207×10^{-5}	0.0348

- Case 2: Constant Acceleration ($10\pi \cdot t$ rad/s)

In the case of constant acceleration, through the two demodulation methods, the rotor position and velocity estimation results are presented in Figure 10 and Table 2, which indicate that compared with the type II tracking loop method, the rotor position and velocity estimation errors are reduced by more than 60% when using the proposed method. As shown from the enlarged views in Figure 10b, the error curve of rotor velocity is smoother, which means that the proposed method can effectively suppress the high-frequency noise in the resolver signals. Therefore, the proposed method can achieve higher estimation precision than the type II tracking loop method. Moreover, it can be seen from Figure 10a and the AVG of estimation errors in Table 2 that, when motor runs at a constant acceleration, the proposed method can greatly reduce the theoretical constant deviation in the estimation results of type II tracking loop method (from 2.714×10^{-4} to 3.424×10^{-7}). Therefore, in the case of variable speed, the proposed method can be adopted to accurately estimate rotor position and velocity.

From the above results, we may conclude that the proposed Chebyshev filter-based type III PLL method in this paper can achieve better performance no matter in the case of constant speed or the case of variable speed.

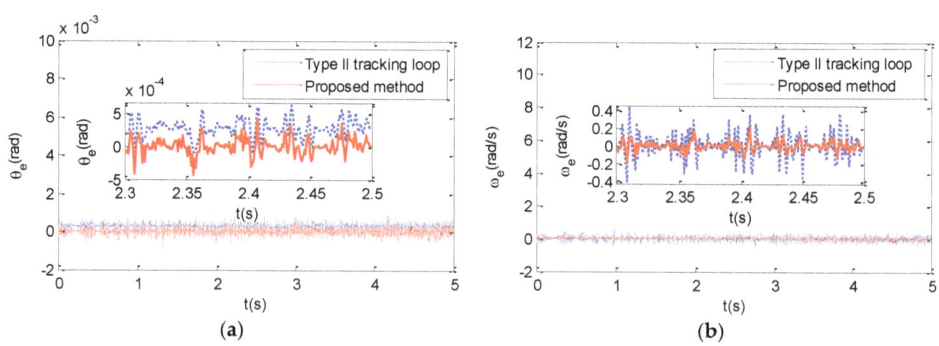

Figure 10. Estimation errors of the two demodulation methods in Case 2. (**a**) Rotor position estimation error; (**b**) rotor velocity estimation error.

4.2. Experiment

The experimental platform is illustrated in Figure 11, in which the PMSM (Infranor, Zurich, Switzerland) is equipped with a resolver. The parameters of PMSM and resolver are listed in Table 3.

In the experiment, the drive and control board, which take DSP TMS320F28335 (Texas Instruments Company, Dallas, TX, USA) as the core, is used to drive the PMSM and implement signals processing unit for resolver outputs.

The PMSM is controlled to rotate at $\omega = 2\pi\text{rad/s}$ in the experiment. After signal conditioning and synchronous envelope detection circuits, the detected resolver envelopes are sampled by ADC, and then uploaded to the upper computer for demodulation algorithm through USB. The demodulation parameters are set as the simulation part. The derived rotor position and velocity are depicted in Figure 12a,b, respectively. Also, the AVG and STD of estimated velocity are computed to evaluate the performance of the proposed method as listed in Table 4.

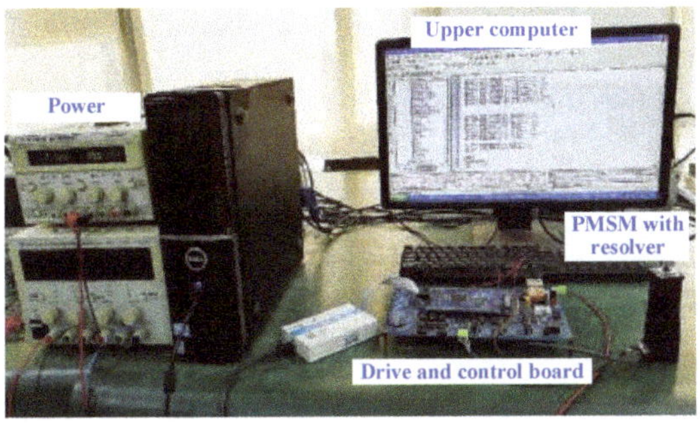

Figure 11. Experimental platform.

Table 3. PMSM and resolver parameters.

PMSM		Resolver	
Pole pairs	2	Pole pairs	1
Rated voltage	110 V(AC)	Input voltage	5 V ± 0.2 V (AC)
Rated speed	3000 r/min	Input freguency	10 kHz
Torque constant	0.15 Nm/A	Ouput voltage	>2 V
Phase resistance	8 Ω	Transformer raio	0.5 ± 5%
Phase inductance	10 mH	Electrical error	$\leq 10'$

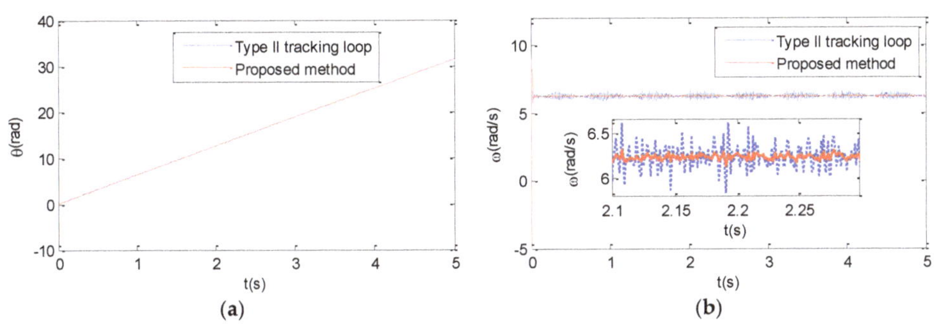

Figure 12. Estimated rotor position and velocity of the two demodulation methods. (**a**) Rotor position estimates; (**b**) rotor velocity estimates.

Table 4. Velocity estimates in experiment.

Demodulation Methods	AVG	STD
Type II tracking loop	6.23	0.105
Proposed method	6.23	0.0319

Note that the actual values of rotor position and velocity are unavailable in this physical experiment, so that the rotor position and velocity estimation errors cannot be obtained directly. The advantages of the proposed method can also be presented from the rotor velocity estimation results in Figure 12b. From the enlarged views in Figure 12b, it is obvious that the velocity estimation curve is smoother when using the proposed Chebyshev filter-based type III PLL method. It can also be seen from Table 4 that the STD of velocity estimates by the proposed method is smaller than that by the type II tracking loop (reduced by about 70%). The proposed method is equally effective when the PMSM is working at other rates. Therefore, it can be concluded that compared with type II tracking loop, the proposed Chebyshev filter-based type III PLL method can improve the demodulation performance to a certain extent due to its stronger noise suppression capability.

5. Conclusions

In order to improve the angular position and velocity estimation accuracy of PLL-based demodulation method, this paper designs a Chebyshev filter-based type III PLL method for demodulation, which makes PLL become a system of type III, and meanwhile, the proposed method has stronger filter property to effectively suppress the high-frequency measurement noise in the resolver signals. Thus, the proposed method has higher demodulation accuracy compared with the conventional ones. Furthermore, the proposed Chebyshev filter-based parameter design method can also provide a theoretical guidance and reference for parameter selecting in other applications.

Author Contributions: H.L. and Z.W. designed the method and wrote the paper; H.L. performed the experiments and analyzed the data.

Funding: This work was partially supported by The National Natural Science Foundation of China (10772011).

Conflicts of Interest: The authors declare that there is no conflict of interests regarding the publication of this paper.

References

1. Han, S.; Han, S. Resolver angle estimation using parameter and state estimation. *Measurement* **2016**, *93*, 460–464. [CrossRef]
2. Benammar, M.; Ben-Brahim, L.; Alhamadi, M.A. A novel resolver-to-360° linearized converter. *IEEE Sens. J.* **2004**, *4*, 96–101. [CrossRef]
3. Caruso, M.; Di Tommaso, A.O.; Genduso, F.; Miceli, R.; Galluzzo, G.R. A DSP-Based Resolver-To-Digital Converter for High-Performance Electrical Drive Applications. *IEEE Trans. Ind. Electron.* **2016**, *63*, 4042–4051. [CrossRef]
4. Bergas-Jané, J.; Ferrater-Simón, C.; Gross, G.; Ramírez-Pisco, R.; Galceran-Arellano, S.; Rull-Duran, J. High-Accuracy All-Digital Resolver-to-Digital Conversion. *IEEE Trans. Ind. Electron.* **2012**, *59*, 326–333. [CrossRef]
5. Ellis, G. Encoders and Resolvers. In *Control System Design Guide*, 4th ed.; Butterworth-Heinemann: Boston, MA, USA, 2012; Volume 14, pp. 285–311.
6. Khaburi, D.A. Software-Based Resolver-to-Digital Converter for DSP-Based Drives Using an Improved Angle-Tracking Observer. *IEEE Trans. Instrum. Meas.* **2012**, *61*, 922–929. [CrossRef]
7. Guo, C.; Wu, C.; Ni, F.; Liu, H. Software-based resolver-to-digital conversion and online fault compensation. In Proceedings of the 2016 IEEE International Conference on Mechatronics and Automation, Harbin, China, 7–10 August 2016; pp. 344–349.
8. Sivappagari, C.M.R.; Konduru, N.R. Review of RDC soft computing techniques for accurate measurement of resolver rotor angle. *Sens. Transducers* **2013**, *150*, 1–11.
9. Sarma, S.; Agrawal, V.K.; Udupa, S. Software-based resolver-to-digital conversion using a dsp. *IEEE Trans. Ind. Electron.* **2008**, *55*, 371–379. [CrossRef]
10. Benammar, M.; Khattab, A.; Saleh, S.; Bensaali, F.; Touati, F. A Sinusoidal Encoder-to-Digital Converter Based on an Improved Tangent Method. *IEEE Sens. J.* **2017**, *17*, 5169–5179. [CrossRef]

11. Pecly, L.; Schindeler, R.; Cleveland, D.; Hashtrudizaad, K. High-precision resolver-to-velocity converter. *IEEE Trans. Instrum. Meas.* **2017**, *66*, 2917–2928. [CrossRef]
12. Karabeyli, F.A.; Alkar, A.Z. Enhancing the accuracy for the open-loop resolver to digital converters. *J. Electr. Eng. Technol.* **2018**, *13*, 192–200.
13. Sun, J.D.; Cao, G.Z.; Huang, S.D.; Qiu, H. Software-based resolver-to-digital converter using the PLL tracking algorithm. In Proceedings of the International Conference on Ubiquitous Robots and Ambient Intelligence, Xi'an, China, 19–22 August 2016; pp. 719–723.
14. Qamar, N.A.; Hatziadoniu, C.J.; Wang, H. Speed Error Mitigation for a DSP-Based Resolver-to-Digital Converter Using Autotuning Filters. *IEEE Trans. Ind. Electron.* **2015**, *62*, 1134–1139. [CrossRef]
15. Benammar, M.; Gonzales, A.S.P. A Novel PLL Resolver Angle Position Indicator. *IEEE Trans. Instrum. Meas.* **2016**, *65*, 123–131. [CrossRef]
16. Alemadi, N.; Benbrahim, L.; Benammar, M. A new tracking technique for mechanical angle measurement. *Measurement* **2014**, *54*, 58–64. [CrossRef]
17. Harnefors, L.; Nee, H.P. A general algorithm for speed and position estimation of AC motors. *IEEE Trans. Ind. Electron.* **2000**, *47*, 77–83. [CrossRef]
18. Zhang, J.; Wu, Z. Composite state observer for resolver-to-digital conversion. *Meas. Sci. Technol.* **2017**, *28*, 065103. [CrossRef]
19. Raymundo, C.G.; João, O.P.P.; Suemitsu, W.I.; Soares, J.O. Improved demultiplexing algorithm for hardware simplification of sensored vector control through frequency-domain multiplexing. *IEEE Trans. Ind. Electron.* **2017**, *64*, 6538–6548.
20. Tiapkin, M.G.; Balkovoi, A.P. High resolution processing of position sensor with amplitude modulated signals of servo drive. In Proceedings of the 2017 IEEE Conference of Russian Young Researchers in Electrical and Electronic Engineering (EIConRus), St. Petersburg, Russia, 1–3 February 2017; pp. 1042–1047.
21. Alipour-Sarabi, R.; Nasiri-Gheidari, Z.; Tootoonchian, F.; Oraee, H. Performance Analysis of Concentrated Wound-Rotor Resolver for Its Applications in High Pole Number Permanent Magnet Motors. *IEEE Sens. J.* **2017**, *17*, 7877–7885. [CrossRef]
22. Wu, Z.; Li, Y. High-Accuracy Automatic Calibration of Resolver Signals via Two-Step Gradient Estimators. *IEEE Sens. J.* **2018**, *18*, 2883–2891. [CrossRef]
23. *AD2S1210 Data Sheet: Variable Resolution, 10-Bit to 16-Bit R/D Converter with Reference Oscillator*; Analog Devices, Inc.: Norwood, MA, USA, 2008; Available online: www.analog.com (accessed on 25 October 2018).
24. Kawakami, M. Nomographs for Butterworth and Chebyshev Filters. *IEEE Trans. Circuit Theory* **1963**, *10*, 288–289. [CrossRef]
25. Yahagi, T.; Wang, Y. *Digital Filters and Signal Processing*; The Science Publishing Company: Beijing, China, 2003; pp. 23–36.
26. Liu, H.; Wu, Z. On estimation algorithm of angular velocity for servo motors with resolvers. In Proceedings of the Chinese Control and Decision Conference, Shenyang, China, 9–11 June 2018; pp. 4019–4024.

electronics

MDPI

Article

An Accurate DDS Method Using Compound Frequency Tuning Word and Its FPGA Implementation

Yuqing Hou, Changlong Li and Sheng Tang *

School of Information Science and Technology, Northwest University, Xi'an 710000, China; houyuqin@nwu.edu.cn (Y.H.); licongg@stumail.nwu.edu.cn (C.L.)

* Correspondence: tangsheng@nwu.edu.cn; Tel.: +86-132-0169-3218

Received: 2 October 2018; Accepted: 14 November 2018; Published: 16 November 2018

Abstract: Because of its high resolution, low cost, small volume, low power dissipation and less conversion time consumption, the direct digital synthesizer (DDS) method has been applied more and more in the fields of frequency synthesis and signal generation. However, only a limited number of precise frequency signals can be synthesized by the traditional DDS, for the reason that its accumulator modulus is fixed, and its frequency tuning word must be integer. In this paper, a precise DDS method using compound frequency tuning word is proposed, which improves the accuracy of synthesized signals at any frequency points on the premise of guaranteeing the stability of synthesized signals. In order to verify the effectiveness of the new method, a DDS frequency synthesizer based on FPGA is designed and implemented. Taking the rubidium atomic clock PRS10 as standard frequency source, the experiments shows that the frequency stability of the synthesized signal is better than 8.0×10^{-12}/s, the relative frequency error is less than 4.8×10^{-12}, and that the frequency accuracy is improved by three orders of magnitude compared with the traditional DDS method.

Keywords: direct digital synthesizer (DDS); frequency tuning word (FTW); stability; accuracy

1. Introduction

With the development of society, the functionality and complexity of electronic devices is increasing. Frequency synthesizer, as a common signal generator, has been widely applied in many fields. Related researches have more and more requirements for the accuracy and stability of the frequency synthesizer, especially in the fields of satellite positioning, aerospace, surveying and mapping, guidance and high-speed communication [1,2]. The signal generator is a kind of instrument with a long history. With the birth of electronic technology, signal generation circuits have appeared in the 1920s. By the 1940s, there were standard signal generators that were mainly used to measure various receivers, and pulse signal generators were invented. Since the 1960s, signal generators have developed rapidly. In this period, analog electronics technology was generally used. The circuit of signal generators was composed of discrete components. RC and LC signal generation circuits play an important role in the development process. The RC circuit composed of resistor R and capacitor C can produce sine waves with continuous amplitude and adjustable frequency. The LC signal generating circuit composed of inductor L and capacitor C can produce less high-order harmonics and better output waveform. With the development of science and technology, digital circuits have entered the field of signal synthesis, and the way of signal synthesis has made rapid progress, many signal synthesis methods have been designed. Modern synthetic signal methods include direct analog frequency synthesis, phase-locked frequency synthesis and direct digital frequency synthesis. There are three kinds of synthesized frequency methods: direct analog frequency synthesis, phase-locked frequency synthesis and direct digital frequency synthesis [3]. Direct analog frequency synthesis uses one or more different transistors to design RC oscillator or LC oscillator as reference signal sources, and the output signals are directly generated by frequency doubling, frequency division and mixing, and the signals obtained by this method have the characteristics of high frequency stability and fast frequency conversion. But both of the hardware debugging and the spurious suppression are not easy in the implementation of this method. Phase-locked frequency synthesis, also known as indirect synthesis method, uses one or more standard frequency sources to generate a large number of harmonics or combined frequencies by mixing and dividing harmonic generators. Then, the phase-locked loop (PLL) is used to lock the frequency of the voltage controlled oscillator (VCO) to a certain harmonic or combination frequency. The required frequency output is indirectly generated by a voltage-controlled oscillator. The advantage of this method is that the phase-locked loop is equivalent to a narrow band tracking filter. Therefore, it can select the desired frequency signal well, suppress the spurious components, and avoid the use of a large number of filters, which are conducive to integration and miniaturization. The disadvantage of this method is slow response [4,5]. Direct digital frequency synthesis is based on the concept of phase to synthesize frequencies and adopts the technology of digital sampling and storage. Because the direct digital synthesizer (DDS) is an open-loop system without any feedback link, the frequency conversion time is very short. Besides, the method can be realized digitally and conveniently, and it is small and light in weight. The DDS method in the design of frequency synthesizer has gradually become a mainstream method in the current field of electronic measurement and testing [6].

Normally there are two main ways to design and implement a real frequency synthesizer. One is to use a dedicated DDS integrated chip to synthesize frequency. The other is to use FPGA to achieve DDS frequency synthesis. The first method is usually realized by using a microprocessor to drive the DDS integrated chip. In the given working mode of the DDS integrated chip, the internal circuit calculates the operating parameters of the kernel and synthesizes the frequency. Then, the analog signal is obtained by further processing by the later stage [7,8]. Due to the technical limitation of the DDS integrated chip, there are unavoidable performance defects in this kind of frequency synthesizer. The dedicated DDS integrated chip has a fixed number of phase accumulator bits and lacks flexibility. The number of its phase accumulators is relatively small, and the frequency resolution is relatively low. For example, we often use chip AD9913 to generate frequency signals. The bit width of its internal phase accumulator is 32 bits. When the reference frequency is 100 MHz, the resolution is 0.023 Hz. Even though 0.023 Hz is not a bad resolution for most applications, the typical accumulator-based DDS is not capable of generating some useful frequencies (like precisely 10 MHz) and cannot meet the high precision requirements of some special equipment or engineering [9]. When expecting to get precisely 10 MHz, the AD9913 can only produce an approximate frequency of 9.999999986030161380 MHz.

FPGA is a new type of digital circuit. Its circuit function is programmable and customizable, which is different from the traditional integrated circuit with the structure and function of a fixed circuit. FPGA technology has overturned the traditional design, tape-out and packaging process of digital circuits. New digital circuits are developed directly on the finished FPGA chip. It overcomes the shortcomings of the internal structure of the DDS chip and improves the flexibility of the chip, which enlarges the user range and application fields of special digital circuits. Each logic gate in the FPGA chip performs a logical operation at every clock cycle. Therefore, FPGA is essentially a very large-scale parallel computing device, which is very suitable for developing DDS devices with high speed, high accuracy and high flexibility [10]. The DDS device based on FPGA is usually composed of a phase accumulator, a waveform memory and a digital multiplier [11,12]. The phase accumulator accumulates the frequency tuning word loop to get the phase address; the size of the frequency tuning word determines the output frequency value; waveform memory stores a sampling point for a periodic output waveform; the phase address of the phase accumulator acts as the reading address of the waveform memory, from which the waveform sample points are taken out to form the digital waveform, and then the analog signal is obtained by further processing [13].

As mentioned above, the frequency synthesis principles of the dedicated DDS integrated chip and the FPGA method are basically the same. Both of them use the phase accumulator to recursively sum the clock rate and frequency tuning word, and synthesize the frequency by means of the look-up table. The phase accumulator modulus of traditional DDS is usually a fixed value, and the frequency tuning words must also be positive integers, so this method can only synthesize a limited number of precise frequency signals. However, some practical projects and systems (as shown in Figure 1) usually require precise frequency synthesizers to generate their reference frequency signals. Sometimes, these reference frequency signals must have a special frequency value. For example, in the communication system, in order to ensure the accuracy and effectiveness of information transmission, each base station generally needs to configure a communication clock subsystem. The communication clock subsystem is the basic guarantee for efficient and orderly operation of the whole system. A 5 MHz or 10 MHz frequency synthesizer is usually the frequency reference of these clock subsystems. Another example, the reference frequencies onboard satellite of some navigation satellite systems is 10.23 MHz. Any frequency error or deviation in the reference frequency will directly affect the navigation satellite system's performance and such errors accumulated over time will result in a significantly large user ranging error varying up to several meters [14]. These special frequency signals such as 5 MHz, 10 MHz and 10.23 MHz are hard to generate accurately by traditional DDS methods and devices. To solve this problem, a DDS method using compound frequency tuning word is designed and implemented in this paper, which is expected to further improve the accuracy of synthetic frequency under the condition of guaranteeing frequency stability.

Figure 1. Applications of high precision DDS.

2. Defects of Traditional DDS

The traditional DDS relies on the accumulator to recursively sum the frequency tuning word at the clock rate, obtaining the instantaneous value of the signal by means of a look-up table [15]. As shown in Figure 2, the method produces a time series of digital words at the output of the accumulator that increases linearly until the accumulator rolls over at its maximum value of 2^C. Hence, the accumulator output has a fixed modulus 2^C. Usually the accumulator output is truncated to P-bits (using only the MSBs) to reduce the size and complexity of the angle-to-amplitude conversion block that immediately follows the accumulator. This causes the time series of digital words produced by the accumulator to appear at the input to the angle-to-amplitude converter as P-bits words ranging in value from 0 to $(2^P - 1)$ [16]. The accumulator output sequence range 0 to $(2^P - 1)$ maps to one revolution on the unit circle, that is, it linearly maps binary values from 0 to $(2^P - 1)$ to radian angles from 0 to 2π. This mapping arrangement allows the angle-to-amplitude converter to translate the P-bits words to Y-bits amplitude values (Y) in a very efficient manner, and finally a low-pass filter is used to obtain a desired sinusoidal signal [17].

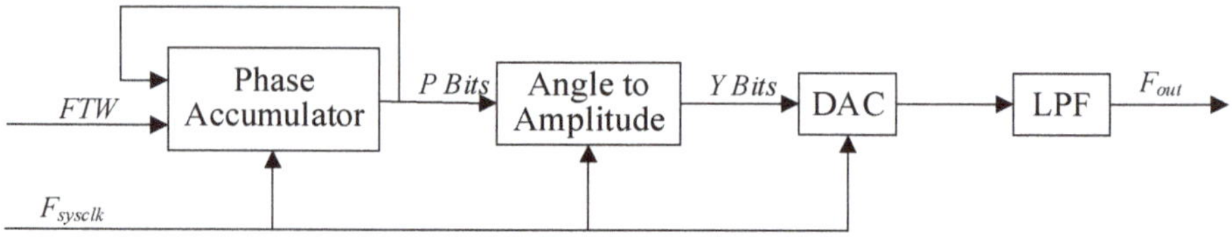

Figure 2. Functional block diagram of traditional DDS method.

The Y-bits digital amplitude sequence signal output by the converter is converted into an analog signal by a Y-bit DAC (Digital to Analog Converter, DAC) chip, and finally a low-pass filter is used to obtain a desired sinusoidal signal. We see the translation process relied on Equation (1):

$$x = A\sin(\frac{2k\pi}{2^P}) \tag{1}$$

where: A is the signal amplitude; P is the number of bits taken from the accumulator; k is the binary value of those bits at any given instant; x is the amplitude value corresponding to the address at a given time.

The following Equation expresses the frequency of the sinusoid that appears at the DAC output for a traditional accumulator-based DDS [18]:

$$F_{out} = \frac{FTW}{2^C}F_{sysclk} \tag{2}$$

where: F_{out} is the synthesized frequency; F_{sysclk} is the sampling frequency; FTW is the frequency turning words; $FTW < 2^{C-1}$.

The integer, frequency turning words, is a determining condition for controlling the output frequency. Since FTW, by definition, is an integer, then F_{out} is constrained to the following set of frequencies:

$$F_{out} \in \left\{0, \frac{F_{sysclk}}{2^C}, \frac{2F_{sysclk}}{2^C}, \frac{3F_{sysclk}}{2^C}, \ldots\ldots \frac{(2^C-1)F_{sysclk}}{2^C}\right\} \tag{3}$$

Inspection of the Equation (3) indicates that the modulus of the DDS accumulator, determines both the frequency resolution of the DDS and the number of possible output frequencies. It can be seen that the ratio of the output frequency to the sampling frequency must satisfy:

$$\frac{F_{out}}{F_{sysclk}} = \frac{FTW}{2^C} \tag{4}$$

Because the accumulator bit width is fixed, the frequency tuning word cannot be decimal, and the output frequency cannot be arbitrary value. Now for output frequencies that are integer submultiples of the sample rate (for example, $F_{sysclk}/10$), F_{out} can be expressed as $F_{out} = F_{sysclk}/Q$ (Q is an integer). Substituting F_{sysclk}/Q for F_{out} in Equation (4) leads to:

$$\frac{1}{Q} = \frac{FTW}{2^C} \tag{5}$$

Solving for FTW yields $FTW = 2^C/Q$. Because FTW and Q must both be integers, the only values of Q that satisfy Equation (5) can be expressed as 2^K, where K is an integer. In practical applications, the frequency tuning word that controls the output signal can be expressed as:

$$FTW = \frac{2^C}{2^K} = 2^{C-K} \quad (6)$$

In the specific DDS implementation, it can be divided into two categories:

I. When the Equation (6) is satisfied, the DDS can output an accurate frequency. To demonstrate, assume that C is 32, $Q = 16 = 2^C$, FTW = 268,435,456, the available frequency turning word is an integer, and the address covers range from 0 to 2^C. When accumulating a loop, the next cycle returns to the initial value and the exact frequency value can be output.

II. When the Equation (6) is not satisfied, the frequency tuning word is a decimal number, and the actual accumulation takes an integer. After one cycle, the first sampling point cannot return to the initial point, so the sampling points of each period are different within a certain range, resulting in different waveform amplitudes and unstable waveforms. At the same time, there is phase loss after accumulating one cycle, the cycle increases, the frequency decreases, and an error occurs. For example, F_{sysclk} = 100 MH, F_{out} = 10 MHz. In this case, $F_{out}/F_{sysclk} = F_{sysclk}/N = 1/10$, FTW = 429,496,729.6. A traditional accumulator-based DDS, regardless of the capacity of its accumulator, is not capable of synthesizing exactly 10 MHz. The closest frequency that a 32-bit accumulator-based DDS can get to 10 MHz with F_{sysclk} = 100 MHz is 9.99999998603016138 MHz, which is smaller than the expect value and has an absolute error of 0.01396983862 Hz. Such a frequency error is intolerable in some special precision equipment or engineering.

3. Accurate DDS Method Using Compound Frequency Tuning Word

Because the phase accumulator modulus of the traditional DDS is a fixed value, the frequency tuning word must also be a positive integer, which causes the traditional DDS method to only synthesize a finite number of precise frequency signals. In response to this problem, this paper proposes a DDS method using compound frequency tuning word. As visible in Figure 3, the phase accumulator recursively sums the frequency tuning word component X at a clock rate, and the obtained value is combined with the frequency tuning word components A and B of the auxiliary accumulator in the address generator. In the DDS method using the compound frequency tuning word, the compound frequency tuning word has three parts: X, A and B. The main working process of the address generator is as follows: the frequency tuning word X is added with a value A after B times of addition, and the subsequent operation is carried out on the basis of the new phase address. The m-bits addresses are obtained by truncating the obtained address to the low bit, and it is sent to the phase-to-amplitude converter to output the Y-bits amplitude of the address mapping. After the DA chip, the digital signal is converted into an analog signal, and finally the signal is filtered and amplified for output.

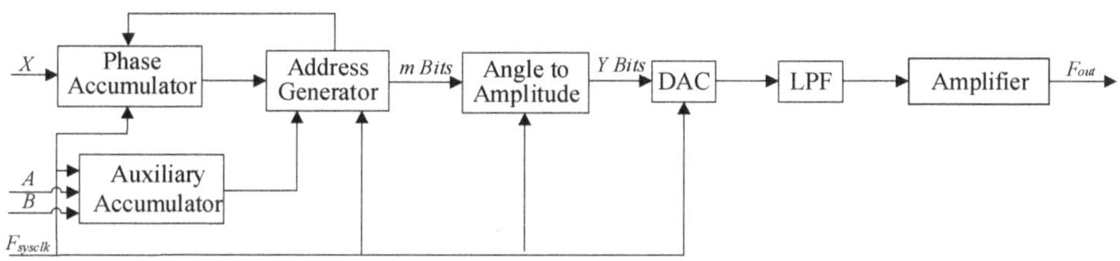

Figure 3. Functional block diagram of new DDS method.

The Equation for calculating the exact DDS output frequency of the compound frequency tuning word:

$$\frac{F_{out}}{F_{sysclk}} = \frac{M}{N} \quad (7)$$

where M, N are integers, $M < N/2$. The frequency ratio of the DDS method using compound frequency tuning word is very similar to that of the traditional DDS, but N in the DDS of the compound frequency tuning word is not required to be a power of 2, it can be any integer.

The relationship between the output frequency of the DDS method using compound frequency tuning word and the sampling frequency is as follows:

$$\frac{F_{out}}{F_{sysclk}} = \frac{M}{N} = \frac{X + \frac{Y}{N}}{2^C}, \ 0 < N < 2^C \tag{8}$$

In the Equation (8), the right end is a compound frequency tuning word, X represents an integer part. After N sampling, the difference between the actual phase and the maximum phase is Y. In the DDS method using compound frequency tuning word, the processing of the fractional part is mainly added, so that the output frequency eliminates the error or minimizes the error. When the output frequency and the system clock are determined, M and N are unique in Equation (7) (M and N are mutual primes). The X value (integer part) can be obtained by the Equation of the conventional DDS. Then find the remainder:

$$\begin{cases} X = \left[\frac{2^C M}{N}\right] \\ Y = M2^C - XN \end{cases} \tag{9}$$

In Equation (9), X is the frequency tuning word after rounding, the elements of X to the nearest integers towards zero. After N sampling, the difference between the actual phase and the maximum phase is the remainder, which will be further simplified:

$$\frac{Y}{N} = \frac{A}{B}, \ A < B, B > 0 \tag{10}$$

In Equation (10), A and B are prime numbers, and both are integers. A and B are processed in the auxiliary accumulator in the DDS method using compound frequency tuning word.

When M/N approximating $M = 1$, the corresponding N is a value near the actual number of sampling or the actual number of sampling when the address is overflowed. Assuming that N' is the actual number of sampling times for an address overflow, then N' is equal to the value near N or N, that is, the number of sampling times N' in a cycle completes a summation from 0 to the maximum address value, and when N' samples overflow for the next cycle, all sampling points in each cycle are coincident, then A is 0, B can be any positive integer, and the compound frequency control word $FTW' = \{X, 0, B\}$ is obtained. The compound frequency tuning word DDS can accurately output frequency. At this point, the phase difference between the two adjacent addresses is the phase represented by the low 16 bits of the accumulator.

If the sampling points of each cycle cannot overlap, DDS cannot accurately output the desired frequency. At this time:

I. If the simplification Equation (10) satisfies $B < N'$ and the B value is a number less than 10, the compound frequency tuning word is $FTW' = \{X, A, B\}$, after a period of sampling, the remainder is divided into uniform B equal parts, it is evenly inserted into the whole sampling process, each time A value is inserted, so that a period of sampling coverage range of 0 to 2^C. All the sampling points in the next cycle coincide with the first sampling point to ensure that the sampling points in each cycle are exactly the same, and the DDS can output an accurate frequency. The method of inserting A and B values is shown in Figure 4. For example, when the system clock = 100 MHz and the output frequency = 10 MHz, $C = 32$, in this paper, the frequency tuning word = 429,496,729, $Y = 6$, $M = 1$, $N = 10$, $A = 3$ and $B = 5$, they can be obtained, that is, the compound frequency tuning word FTW = {429,496,729, 3, 5}. During the same period of sampling, the fourth sampling address is $4X$. The new sampling method is adopted, the fifth sampling address is $(5X + 3)$, and $(5X + 3)$ is the new addressing basis for the accumulation; The ninth sampling address is $(9X + 3)$, the tenth sampling address is $(10X + 6)$, that is the maximum value of the phase address, so that the sampling points in a cycle cover 0 to 2^C, can ensure that the next cycle to take the same sampling points.

II. If the simplification Equation (10) does not satisfy $B < N'$, then the Equation (9) is approximately reduced to B less than 10 and a value of A is obtained. At this time, the compound frequency tuning word is $FTW' = \{X, A, B\}$, which ensures that the DDS synthesis frequency is closer to the expected value. For example, when the system clock F_{sysclk} = 100 MHz, the output frequency is required to be F_{out} = 1.024 MHz, the frequency tuning word = 43,980,465 can be found. We can get $Y/N = 347/3125$, and the constraint can be used to meet the requirements of $A/B = 1/9$. Finding the compound frequency tuning word is FTW' = {43,980,465, 1, 9}, we can get a relatively high precision. In the two cases mentioned above, When the address is only X-accumulating, the phase difference between the two adjacent sampling points is the phase represented by the lower 16-bits in the accumulator. When the sample address is $(X + A)$, the phase difference between the two adjacent sampling points is 16 bits plus A.

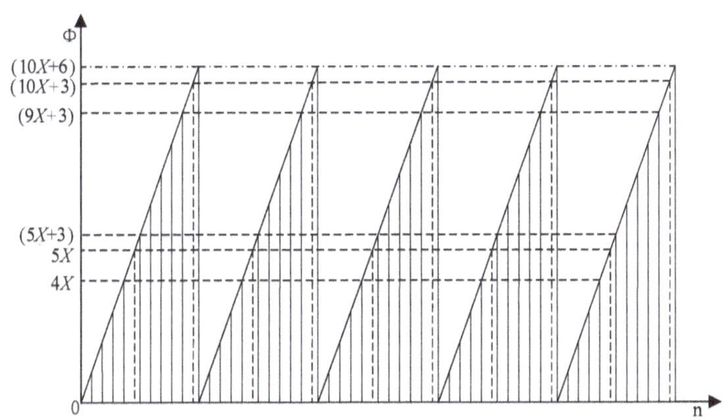

Figure 4. Address sampling method of the new DDS method. Address sampling model of the new DDS method for outputting 10 MHz signal in 100 MHz reference clock.

The main difference between the two methods is the different ways to generate addresses. The address of traditional DDS relies on the accumulator to recursively sum the frequency tuning word at the clock rate; The DDS method using compound frequency tuning word is under the control of the reference clock. When the frequency control word X is accumulated, each time B is accumulated, an A value is added to the address to get an adjustable address.

4. Development of DDS Frequency Synthesizer Based on FPGA

In order to verify the validity of the DDS method using compound frequency tuning word, this paper designs and implements an FPGA based frequency synthesizer. Thanks to the flexible programmability of FPGA, the traditional DDS method and the DDS method proposed by this paper using compound frequency tuning word can be repeatedly erased on the platform to facilitate comparison experiments. The frame structure of frequency synthesizer based on FPGA is shown in the Figure 5, which specifically consists of the reference clock module, the FPGA module, the DA module, the filter module and the amplification driver module. The Modules of DDS frequency synthesizer based on FPGA is shown as Figure 6 (The size of the circuit board is 142 × 62 mm). The reference clock module uses SRS's rubidium atomic clock PRS10, whose sine wave output has a frequency of 10 MHz, the amplitude is 0.7 V, and a stability of 1.52×10^{-12}/s (as shown in Figure 7). The 10 MHz output signal of rubidium atomic clock is processed by frequency division module in FPGA to obtain 100 MHz clock signal as sampling clock of frequency synthesizer. The FPGA module adopts Altera's

EP1C12Q240I7. All of the digital logic circuits such as phase accumulator, auxiliary accumulator, address generator and phase-to-amplitude converter are designed and implemented in EP1C12Q240I7. The DA module uses Maxim's 16-bit parallel input DAC chip (MAX5885) to convert the digital waveform output from the EP1C12Q240I7 into an analog waveform output. The filter module is designed to filter out clutter and other interfering signals in the output waveform of the DAC module. It uses a seventh-order elliptic filter to improve the quality of the synthesized waveform. The amplifier driver module is designed and implemented by TI's ultra-low noise integrated operational amplifier OPA847, which makes the device has a large driving capability, and the output synthesized frequency signal has an amplitude of not less than 3 Vpp under a load of 50 Ω.

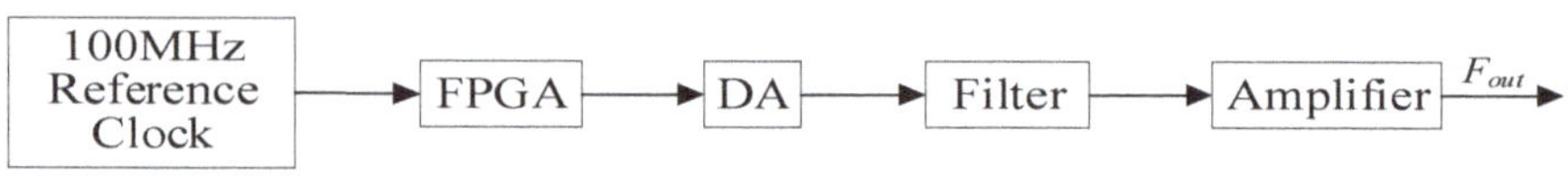

Figure 5. Block diagram of DDS frequency synthesizer based on FPGA.

(a)

(b)

Figure 6. Modules of DDS frequency synthesizer based on FPGA. (**a**) Standard frequency source rubidium atomic clock PRS10; (**b**) FPGA implementation platform of DDS.

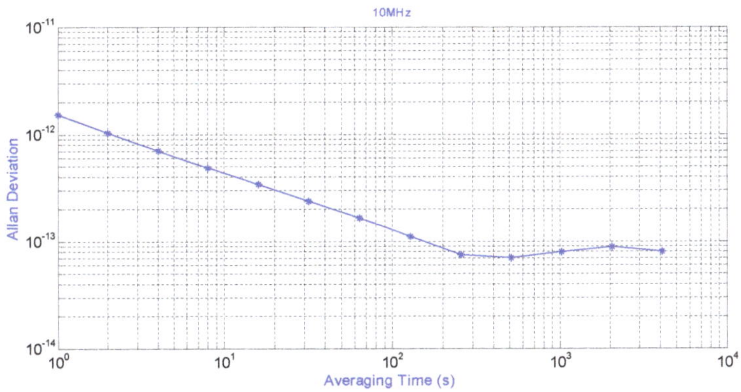

Figure 7. Stability of rubidium atomic clock PRS10.

5. Experimental Results and Analysis

In the fields of power electronics and frequency standards, there are special requirements for the frequency synthesizer output frequency. For example, in order to meet the needs of battery monitoring and management of electric vehicles, Kadirvel K. et al. proposed an IC chip, whose maximum recommended clock drive value is just 2.048 MHz [19]. In order to reduce the influence of the Dick effect, Wang et al. selected a 5 MHz signal with ultra-low phase noise as a reference of their microwave generator in the study of the influence of Dick effect in cold atomic clock in an integrating sphere [20].

But in practical engineering applications, a traditional DDS cannot accurately synthesize a standard frequency signal. Suppose that the phase accumulator of the traditional DDS is 32 bits, the closest output frequency and its error when synthesizing special frequency points such as 1.024 MHz, 2.048 MHz 5 MHz, and 10 MHz are shown in Table 1.

Table 1. Output capability of traditional DDS in special frequency points.

Special Frequency Points Example (MHz)	Frequency Tuning Word	Theoretical Value of Output Frequency (MHz)	Frequency Error (Hz)
1.024	43,980,465	1.023999997414648523	0.00258535146
2.048	87,960,930	2.047999994829297065	0.00517070293
5	214,748,364	4.999999981373548548	0.01864645149
10	429,496,729	9.999999986030161380	0.01396983861

As shown in the third column of Table 1, the frequency values represent the theoretical outputs of traditional DDS. In order to test the real performance of traditional DDS and the output stability and accuracy of the new DDS method proposed in this paper, we designed an experimental platform just as Figure 8 shown.

The platform mainly includes reference clock PRS10, distribution amplifier HP5087A, the FPGA implementation platform of DDS, Keysight 53220A counter, TimeLab test software and Stable32 test software. The photograph of the real experimental platform is shown as Figure 8b. After the system is powered on for about 9 min, the frequency-locked circuit indicates that the rubidium clock outputs stable reference signal and can start the follow-up operation. The standard frequency source of the rubidium atomic clock is sent through a distribution amplifier to get two signals, one of which is input to the FPGA implementation platform of DDS as its standard frequency source. The other is input to the counter as its standard frequency source. The device under test and measuring equipment with common standard frequency source can ensure the credibility of the measurement results.

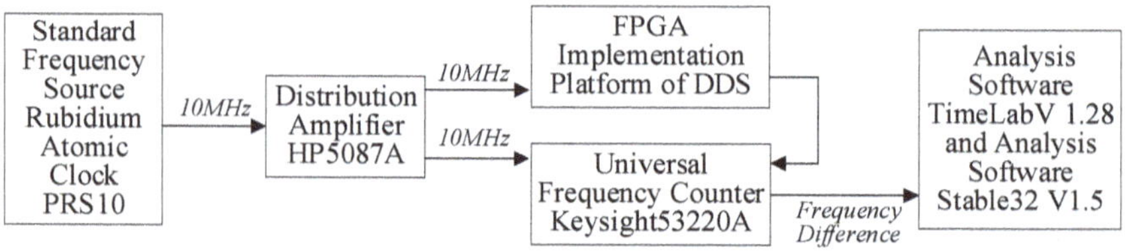

(a)

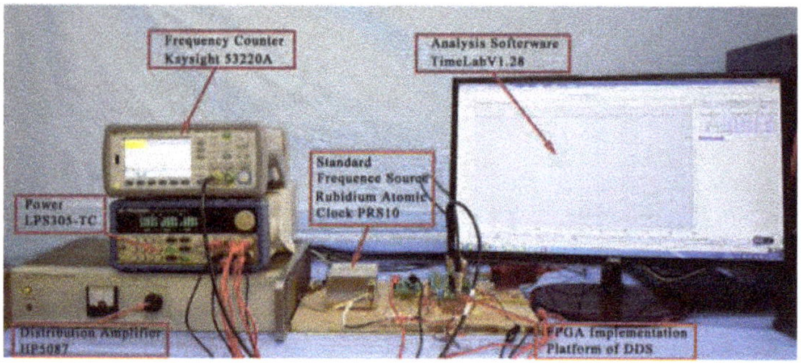

(b)

Figure 8. Experimental platform for testing the characteristics of the DDS. (**a**) Block diagram of the experimental platform; (**b**) Photograph of the experimental platform.

Then, the DDS method using the compound frequency tuning word is used to synthesize the frequency points of 1.024 MHz, 2.048 MHz, 5 MHz and 10 MHz. After nearly 11 h of measurement, its characteristics are analyzed. According to the test data, Stable32 is used to analyze the stability index of our new method. At the same time, experiments in Figure 9 show that the DDS method using compound frequency tuning word has better stability. The stability test results for different frequency points show that the new DDS method proposed in this paper has a good stability index (short term stability as an example, better than 8.0×10^{-12}/s).

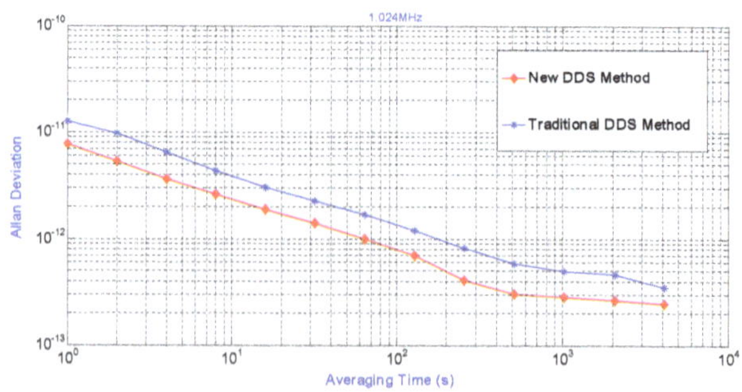

(a)

Figure 9. *Cont.*

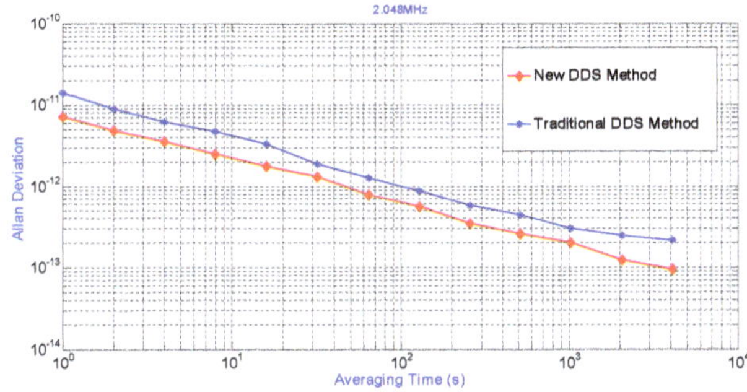

(**b**)

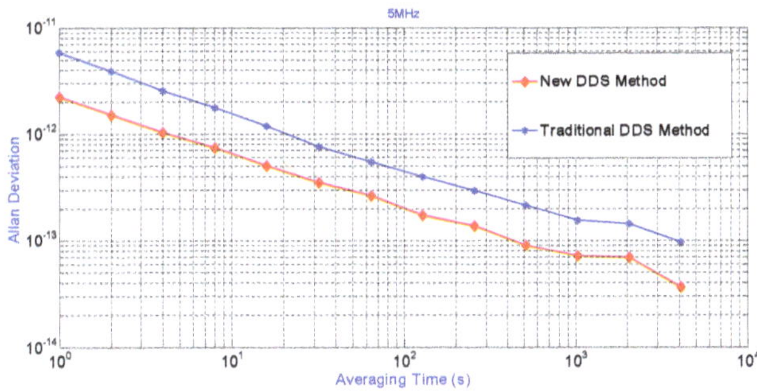

(**c**)

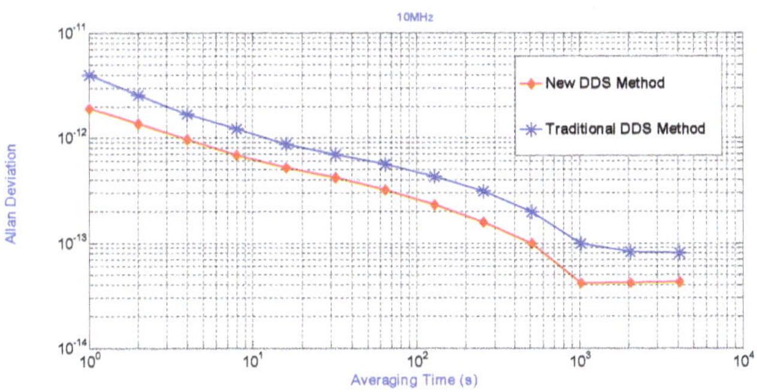

(**d**)

Figure 9. Stability of different frequency points. (**a**) Stability of 1.024 MHz; (**b**) Stability of 2.048 MHz; (**c**) Stability of 5 MHz; (**d**) Stability of 10 MHz.

By using the same experiment platform shown in Figure 7, the accuracy indexes of the DDS method using compound frequency tuning word and the traditional DDS frequency synthesis method are obtained and compared. The experiment results are shown in Figure 10, the abscissa represents the average time of measurement and the ordinate represents the relative frequency difference. The experimental results show that the relative frequency deviation of the synthesized frequency is about 4.80×10^{-12}, which is three orders of magnitude lower than the traditional DDS frequency synthesis method (the relative frequency deviation is about 2.00×10^{-9}). It can be proved that the DDS using compound frequency tuning word has higher frequency output accuracy than the traditional DDS method.

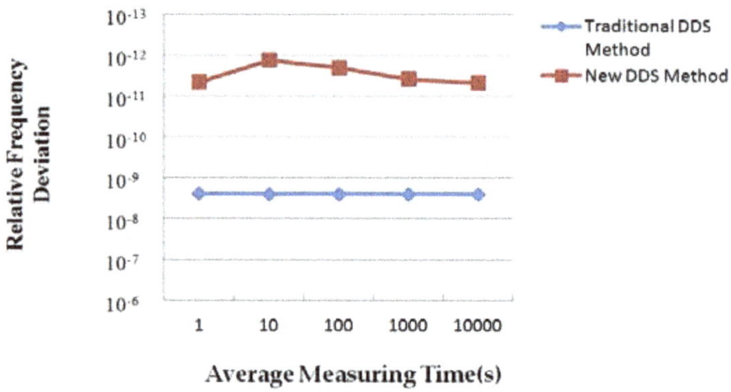

(a)

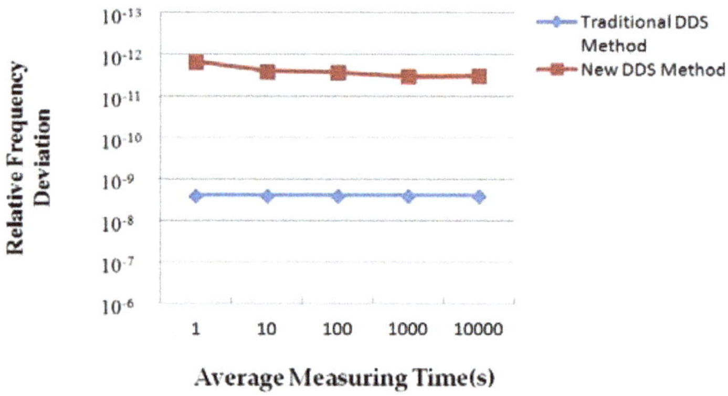

(b)

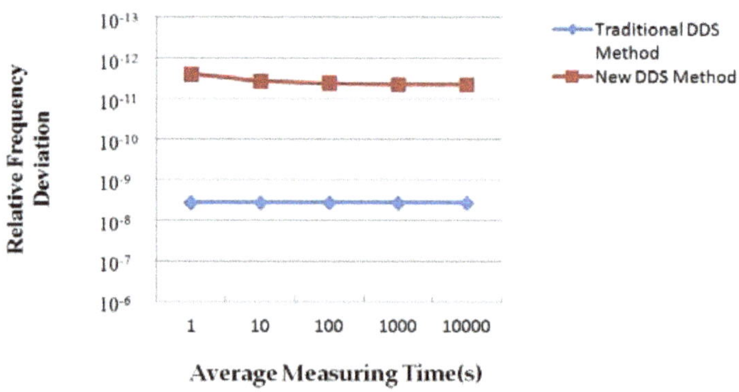

(c)

Figure 10. *Cont.*

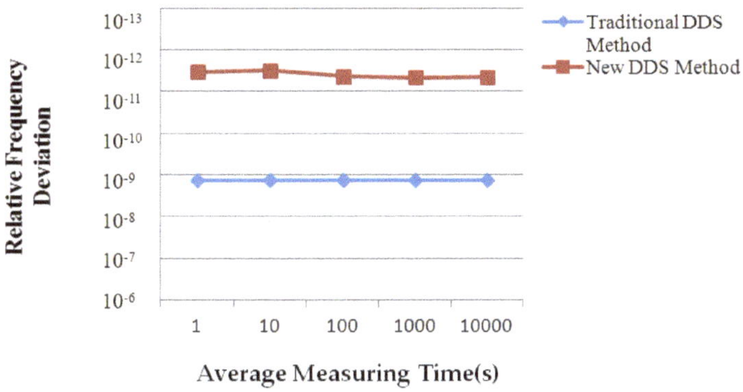

(**d**)

Figure 10. Measurement of accuracy. (**a**) 1.024 MHz signal accuracy; (**b**) 2.048 MHz signal accuracy; (**c**) 5 MHz signal accuracy; (**d**) 10 MHz signal accuracy.

6. Conclusions

This paper proposed a precise DDS method using the compound frequency tuning word and the implementation scheme of FPGA. In this method, the compound frequency tuning word is flexible and changeful, which overcomes several defects of typical DDS, such as the phase accumulator modulus is fixed value, the frequency tuning word must be positive integer, and synthesizing only a limited number of accurate signals. Then the proposed DDS method of compound frequency tuning word is realized by using FPGA. By adding an auxiliary accumulator and an address generator to process the relevant data, the method of accumulating phase address in synthetic frequency is adjusted to improve the quality of synthetic frequency signal. Taking advantage of the flexibility of FPGA in digital circuit design and the advantages of parallel computing architecture design, the new method and the traditional method can be implemented alternately on the FPGA platform to facilitate comparison experiments. The experimental measurement and analysis of the experimental data show that the proposed DDS method using compound frequency tuning word has higher accuracy of the synthesis frequency under the premise of ensuring the stability index. The related methods and technical schemes proposed in this paper are expected to provide references for high-precision frequency synthesizer or signal generator engineering.

The accuracy of the frequency signal is always an important target in the field of frequency standard comparison. Researchers have been looking for various methods to improve the quality of the synthesized signal. The most direct way to improve the signal quality is to increase the bit width and improve the resolution of the accumulator, but this method requires the chip to have a large memory space. With the same memory space of ROM, a new algorithm can be designed to improve the amplitude quantization to improve the accuracy of synthetic signals. At the same time, the influence of noise on signal synthesis is also great. We should synthesize the previous frequency synthesis methods. A new method of frequency synthesis and some noise reduction algorithms are designed to improve the quality of frequency synthesis. These are also the focus of our work in the future. We hope to make breakthroughs as soon as possible.

Author Contributions: All of the authors were working together very tightly. The theoretical works on the design of the DDS using compound frequency tuning word were mainly driven by S.T. and Y.H. The transformation of the DDS method into FPGA-synthesizer was done by C.L. C.L. was also responsible for performing the experiments. S.T. and C.L. analyzed the experimental results and wrote the paper together.

Funding: The Project supported by Natural Science Basic Research Plan in Shaanxi Province of China (Program NO. 2018JM1025), the service local special plan project of Shaanxi Education Department (Program NO. 17JF027) and the National Natural Science Foundation of China (Program NO. 11403018).

Conflicts of Interest: The authors declare no conflict of interest.

References

1. Abdelfattah, O.; Gal, G.; Roberts, G.W.; Shih, I.; Shih, Y.C. A top–down design methodology encompassing components variations due to wide-range operation in frequency synthesizer PLLS. *IEEE Trans. Very Large Scale Integr. Syst.* **2016**, *24*, 2050–2061. [CrossRef]
2. Qiu, Y.; Zhao, L.; Zhang, F. Design of 0.35-ps RMS Jitter 4.4–5.6-GHz Frequency Synthesizer with Adaptive Frequency Calibration Using 55-nm CMOS Technology. *Circuits Syst. Signal Process.* **2018**, *37*, 1479–1504. [CrossRef]
3. Zhang, Y.; Wang, H. Design of a System to Generate a Four Quadrant Signal at High-Frequency. *Intell. Autom. Soft Comput.* **2017**, *24*. [CrossRef]
4. Taheri, H.E.; Ehsanian, M. A new adaptive bandwidth, adaptive jitter frequency synthesizer using programmable charge pump circuit. *Anal. Integr. Circuits Signal Process.* **2018**, *96*, 373–384. [CrossRef]
5. Guo, S.; Gui, P.; Liu, T.; Zhang, T.; Xi, T.; Wu, G.; Fan, Y.; Morgan, M. A Low-Voltage Low-Phase-Noise 25-GHz Two-Tank Transformer-Feedback VCO. *IEEE Trans. Circuits Syst. I Regul. Pap.* **2018**, *99*. [CrossRef]
6. Rust, J.; Bärthel, M.; Paul, S. On high-accuracy direct digital frequency synthesis using linear function approximation. In Proceedings of the 2016 24th European Signal Processing Conference (EUSIPCO), Budapest, Hungary, 29 August–2 September 2016; pp. 672–676.
7. Du, Y.; Li, W.; Ge, Y.; Li, H.; Deng, K. Note: A high-frequency signal generator based on direct digital synthesizer and field-programmable gate array. *Rev. Sci. Instrum.* **2017**, *88*, 096103. [CrossRef] [PubMed]
8. Delorme, N.; Blanc, C.L.; Dezzani, A.; Bély, M.; Ferret, A.; Laminette, S. A NEMS-Array Control IC for Subattogram Mass Sensing Applications in 28 nm CMOS Technology. *IEEE J. Solid-State Circuits* **2015**, *1*, 249–258.
9. Leitner, S.; Wang, H.; Tragoudas, S. Design Techniques for Direct Digital Synthesis Circuits with Improved Frequency Accuracy Over Wide Frequency Ranges. *J. Circuits Syst. Comput.* **2017**, *26*, 1750035. [CrossRef]
10. He, J.; Jiang, J.; Li, N. Design of DDS Signal Generator Based on FPGA. *Comput. Meas. Control* **2017**, *2*, 063.
11. Sotiriadis, P.P. Single-Bit All-Digital Frequency Synthesis Using Homodyne Sigma-Delta Modulation. *IEEE Trans. Ultrason. Ferroelectr. Freq. Control* **2017**, *64*, 463–474. [CrossRef] [PubMed]
12. Kwiatkowski, P.; Różyc, K.; Sawicki, M.; Jachna, Z.; Szplet, R. 5 ps jitter programmable time interval/frequency generator. *Metrol. Meas. Syst.* **2017**, *1*, 57–68. [CrossRef]
13. Hu, P.F.; Shen, L.; Han, F.; Yang, F.; Song, M.J.; Zhang, L. Development of the data acquisition system for terahertz spectrometer. *Trans. Inst. Meas. Control* **2018**, *3*, 805–811. [CrossRef]
14. Khare, A.; Arora, R.; Banik, A.; Banik, A.; Mehta, S.D. Autonomous Rubidium Clock Weak Frequency Jump Detector for Onboard Navigation Satellite System. *IEEE Trans. Ultrason. Ferroelectr. Freq. Control* **2016**, *63*, 326–335. [CrossRef] [PubMed]
15. Madheswaran, M. An Improved Direct Digital Synthesizer Using Hybrid Wave Pipelining and CORDIC algorithm for Software Defined Radio. *Circuits Syst. Signal Process.* **2013**, *3*, 1219–1238. [CrossRef]
16. Huang, J.M.; Chen, Z.; Guo, H.; Han, K. FPGA Implementation of a Novel Type DDS Based on CORDIC Algorithm. *Adv. Intell. Soft Comput.* **2011**, *105*, 183–188.
17. Ryabov, I.V.; Tolmachev, S.V.; Chernov, D.A. A direct digital synthesizer of compound wideband signals. *Instrum. Exp. Tech.* **2014**, *57*, 420–425. [CrossRef]
18. Guo, X.; Wu, D.; Zhou, L.; Wu, J. A 2-GHz 32-bit ROM-based direct-digital frequency synthesizer in 0.13 μm CMOS. *Analog. Integr. Circuits Signal Process.* **2018**, *94*, 127–138. [CrossRef]
19. Kadirvel, K.; Carpenter, J.; Huynh, P.; Ross, J.M. A stackable, 6-cell, Li-ion, battery management IC for electric vehicles with 13, 12-bit ΣΔ ADCs, cell balancing, and direct-connect current-mode communications. *IEEE J. Solid-State Circuits* **2014**, *49*, 928–934. [CrossRef]
20. Wang, X.M.; Meng, Y.L.; Wang, Y.N.; Wan, J.Y.; Yu, M.Y. Dick Effect in the Integrating Sphere Cold Atom Clock. *Chin. Phys. Lett.* **2017**, *34*, 063702. [CrossRef]

electronics

MDPI

Article

Hardware Considerations for Tensor Implementation and Analysis Using the Field Programmable Gate Array

Ian Grout [1,*] and Lenore Mullin [2]

1 Department of Electronic and Computer Engineering, University of Limerick, V94 T9PX Limerick, Ireland
2 Department of Computer Science, College of Engineering and Applied Sciences, University at Albany, State University of New York, Albany, NY 12222, USA; lenore@albany.edu
* Correspondence: Ian.Grout@ul.ie; Tel.: +353-61-202-298

Received: 18 October 2018; Accepted: 8 November 2018; Published: 13 November 2018

Abstract: In today's complex embedded systems targeting internet of things (IoT) applications, there is a greater need for embedded digital signal processing algorithms that can effectively and efficiently process complex data sets. A typical application considered is for use in supervised and unsupervised machine learning systems. With the move towards lower power, portable, and embedded hardware-software platforms that meet the current and future needs for such applications, there is a requirement on the design and development communities to consider different approaches to design realization and implementation. Typical approaches are based on software programmed processors that run the required algorithms on a software operating system. Whilst such approaches are well supported, they can lead to solutions that are not necessarily optimized for a particular problem. A consideration of different approaches to realize a working system is therefore required, and hardware based designs rather than software based designs can provide performance benefits in terms of power consumption and processing speed. In this paper, consideration is given to utilizing the field programmable gate array (FPGA) to implement a combined inner and outer product algorithm in hardware that utilizes the available hardware resources within the FPGA. These products form the basis of tensor analysis operations that underlie the data processing algorithms in many machine learning systems.

Keywords: Inner and outer product; tensor; FPGA; hardware

1. Introduction

Embedded system applications are today demanding greater levels of digital signal processing (DSP) capabilities whilst providing low-power operation and with reduced processing times for complex signal processing operations found typically in machine learning [1] systems. For example, facial recognition [2] for safety and security conscious applications is a noticeable every day example, and many smartphones today incorporate facial recognition software applications for phone and software app access. Embedded environmental sensors, as an alternative application, can input multiple sensor data values over a period of time and, using DSP algorithms, can analyze the data and autonomously provide specific outcomes. Although these applications may differ, within the system hardware and software, these are simply algorithms accessing data values that need to be processed. The system does not need to know the context of the data it is obtaining. Data processing is rather concerned with how effectively and efficiently it can obtain, store, and process the data before transmitting a result to an external system. This requires not only an understanding of regular access patterns in important internet of things (IoT) algorithms, but also an ability to identify similarities amongst such algorithms. Research presented herein shows how scalar operations, such as plus and times, extended to all scalar operations, can be defined in a single circuit that implements all scalar operations extended to: (i) n-dimensional tensors (arrays); (ii) the inner product, (matrix multiply is a 2-d instance) and the outer product, both on n-dimensional arrays (the Kronecker Product is a 2-d instance); and (iii) compressions, or reductions, over arbitrary dimensions. However, even more relationships exist. One of the most compute intensive operations in IoT is the Khatri-Rao, or parallel Kronecker Product, which, from the perspective of this research, is an outer product projected to a matrix, enabling contiguous reads and writes of data values at machine speeds.

In terms of the data, when this data is obtained, it must be stored in the available memory. This will be a mixture of cache memory within a suitably selected software programmed processor (microcontroller (μC), microprocessor (μP), or digital signal processor (DSP)), locally connected external volatile or non-volatile memory connected to the processor, memory connected to the processor via local area network (LAN), or via some form of Cloud based memory (Cloud storage). Identifying what to use and when is the challenge. Ideally, the data would be stored in specific memory locations so that the processor can optimally access the stored input data, process the data, and store the result (the output data) again in memory in suitable new locations, or overwriting existing data in already utilized memory. Knowing and anticipating cache memory misses, for example, enable a design that minimizes overhead(s), such as signal delays, energy, heat, and power.

In many embedded systems implemented today, the software programmed processor is the commonly used programmable device to perform complex tasks and interface to input and output systems. The software approach has been developed over the last number of years and is supported through tools (usually available via an integrated development environment (IDE)) and programming language constructs, providing the necessary syntax and semantics to perform the required complex tasks. However, increasingly, the programmable logic device (PLD) [3] that allows for a hardware configuration to be downloaded into the PLD in terms of digital logic operations is utilized. Figure 1 shows the target device choices available to the designer today. Alternatively, an application specific integrated circuit (ASIC) solution whereby a custom integrated circuit is designed and fabricated could be considered. Design goals include not only semantic, denotational, and functional descriptions of a circuit, but also an operational description (how to build the circuit and associated memory relative to access patterns of important algorithms).

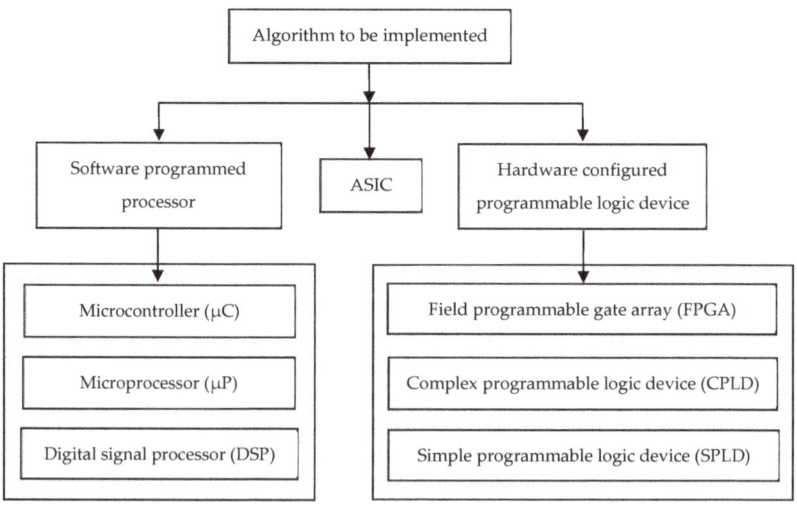

Figure 1. Programmable/configurable device choices for implementing digital signal processing operations in hardware and software.

In this paper, consideration is given to a general algorithm, and the resultant circuit, for an n-dimensional inner and outer product. This algorithm (circuit) builds upon scalar operations, thus creating a single IP (intellectual property) core that utilizes an efficient memory access algorithm. The field programmable gate array (FPGA) is used as the target hardware and the Xilinx® [4] Artix-7 [5] device is utilized in this case study. The two algorithms, the matrix multiplication, and Tensor Product (Kronecker Product), are foundational to essential algorithms in AI and IoT. The paper is presented in a way to discuss the necessary links between the computer science (algorithm design and development) and the engineering (circuit design, implementation, test, and verification) actions that need to be undertaken as a single, combined approach to system realization.

The paper is structured as follows. Section 2 will introduce and discuss algorithms for complex data analysis with a focus on tensor [6] analysis. An approach using tensor based computations with dimension data arrays that are to be developed and processed is introduced. Section 3 will discuss memory considerations for tensor analysis operations, and Section 4 will introduce the use of the FPGA in implementing hardware and hardware/software co-design realizations of tensor computations. Section 5 will provide a case study design created using the VHDL (Very High Speed Integrated Circuit (VHSIC) Hardware Description Language (HDL)) [7] for synthesis and implementation within the FPGA. The design architecture, simulation results, and physical prototype test results are presented, along with a discussion into implementation possibilities. Section 6 will conclude the paper.

2. Algorithms for Tensor Analysis

2.1. Introduction

In this section, data structures using tensor notation are introduced and discussed with the need to consider and implement high performance computing (HPC) applications [8], such as required in artificial intelligence (AI), machine learning (ML), and deep learning (DL) systems [9]. The section commences with an introduction to tensors and then followed by a discussion into the use of tensors in HPC applications. The algorithms foundational to IoT (Matrix Multiply, Kronecker Product, and Compressions (Reductions)) are targeted with the need for a unified n-dimensional inner and outer product circuit that can optimally identify and access suitable memories to store input and processed data.

2.2. Tensors as Algebraic Objects

As the need for IoT [10] and AI solutions grows, so does the need for High Performance Tensor (HPT) operations [11]. Tensors often provide a natural and compact representation for multidimensional data. For example, a function with five parameters can be thought of as a five-dimensional array. This is a particularly useful approach to structuring complex data sets for analysis.

With the complexity of tensor analysis requirements in real-world scenarios, there is a need for suitable hardware and software platforms to effectively and efficiently perform tensor analysis operations. Although there is a plethora of tensor platforms available for use, all the platforms are built upon tensors using various software programming languages, approaches, and performances. Selecting and obtaining the right programming language and hardware platform to run tensor computation programs on is not a trivial task. Fortunately, numerous efforts are underway to identify hot spots and build firmware and hardware. These efforts are built upon over 10 years of national and international workshops (e.g., [12,13]) uniting scientists to address these issues.

Tensors are algebraic objects that describe linear and multi-linear relationships. Tensors can be represented as multidimensional arrays. A tensor is denoted by its rank from 0 upwards. Each rank represents an array of a particular dimension. This idea is shown in Table 1 that identifies the tensor rank, its mathematical entity, and an example realization using the Python language [14], using Python lists to hold the data (in the examples, using integer numbers). A scalar value representing a magnitude (e.g., the speed of a moving object) is a tensor of rank 0. A rank 1 tensor is a vector representing a magnitude and direction (e.g., the velocity of a moving object: Speed and direction of motion). Matrices ($n \times m$ arrays) have two dimensions and are rank 2 tensors. A three-dimensional ($n \times m \times p$) array can be visualized as a cube and is a rank 3 tensor. Tensors with ranks greater than 3 can readily be created and analysis performed on the data they hold would be performed by accessing the appropriate element within the tensor and performing a suitable mathematical operation before storing the result in another tensor.

In a physical realization of this process, the tensor data would be stored in a suitable size memory, the data would be accessed (typically using a software programmed processor), and the computation would be undertaken using fixed- or floating-point arithmetic. This entire process should, ideally, stream data contiguously, and ideally anticipate where cache memory misses might occur, thus minimizing overhead up and down the memory hierarchy. For example, in an implementation using cache memory, L1 cache memory miss could also miss in L2, L3, and Page memory.

Table 1. Tensor rank (0 to n) with an example code using Python lists.

Rank.	Mathematical Entity	Example Realization in Python Code
0	Scalar (magnitude only)	A = 1
1	Vector (magnitude and direction)	B = [0,1,2]
2	Matrix (two dimensions)	C = [[0,1,2], [3,4,5], [6,7,8]]
3	Cube (three dimensions)	D = [[[0,1,2], [3,4,5]], [[6,7,8], [9,10,11]]]
n	n dimensions	...

A tensor rank, or a tensor's dimensionality, can be thought of in at least two ways. The more traditional way being, as the number of rows and columns change in a matrix, so does the dimension. Even with that perspective, computation methods often decompose such matrices into blocks. Conceptually, this can be thought of as "lifting" the dimension of an array. Further blocking "lifts" the dimension even more. Another way of viewing a tensor's dimensionality is by the number of arguments in a function input over time. The most general way to view dimensionality is to combine these attributes. The idealized methods for formulating architectural components are chosen to match the arithmetic and memory access patterns of the algorithms under investigation. In this paper, the n-dimensional inner and outer products are considered. Thus, in this case, what might be thought of as a two-dimensional problem can be lifted to perhaps eight or more dimensions to reflect a physical implementation, considering the memory as registers, the levels of cache memory, RAM (random access memory), and HDD (hard disk drive). With that formulation, it is possible to create deterministic cost functions validated by experimentation, and, ideally, an idealized component construction can be realized that meets desired goals, such as heat dissipation, time, power, and hardware cost.

When an algorithm is run, the hardware implementing the algorithm will access available memory. In Figure 2, a prototypical graph of how an algorithm that does not have cache memory misses or page memory faults is presented. The shape of the graph changes as it moves through the memory hierarchy. This identifies the time requirements associated with the different memories from L1 cache memory through to disk (HDD). Note the change in the slope with memory type. The slope reflects how attributes, such as speed, cost, and power, would affect performance. Algorithm execution (memory access) time is, however, relative to the L1 cache memory chosen. For example, it could be nanoseconds, microseconds, milliseconds, seconds, minutes, or hours as the data moves further up the memory hierarchy. Often, performance is related to a decrease in arithmetic operations, i.e., a reduction of arithmetic complexity. In an ideal computing environment, where memory and computation would have the same costs, this would be the case. Unfortunately, it is also necessary to be concerned with the cost of data input/output (I/O). In parallel to this, it is a necessity to consider memory access patterns and how these relate to the levels of memory. Pre-fetching is one way to alleviate delays. However, often the algorithm developer must rely on the attributes of a compiler and hope the compiler is pre-fetching data in an optimum manner. The developer must trust that this action is performed correctly. This is becoming harder to achieve given that machines are becoming ever more complex and compiler writers are getting scarcer. Empirical methods of experimentation reveal graphs, such as the one shown in Figure 2. Such diagnostic methods allow the algorithm developer to observe the performance of a particular algorithm running on a machine. It is then possible to look at memory speed, size, and other cost factors to put together a model of how we might improve performance through "dimension lifting". That said, the goal is always to try to keep the slope linear, i.e., the linear part of a polynomial curve such that the slope is minimized.

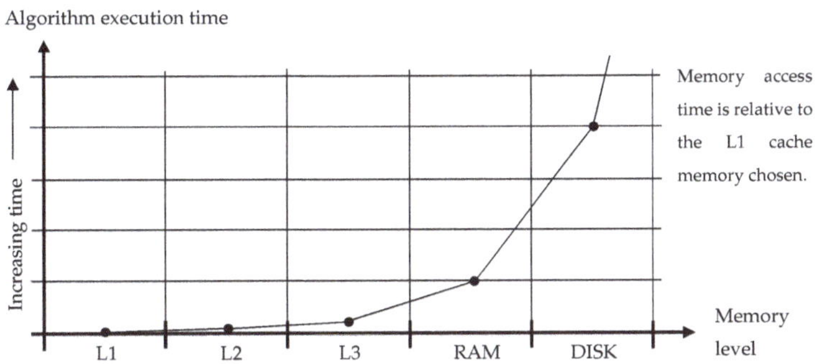

Figure 2. Algorithm execution (memory access) of time vs. memory hierarchy.

Presently, the goal is to achieve a situation where the graph is polynomial, avoiding exponential behavior, such as the one in Figure 2, using HDDs. A co-design approach, complemented with dimension lifting and analysis, as discussed above, can be used to calculate upper and lower bounds of algorithms relative to their data size, memory access patterns, and arithmetic. The goal is to ensure performance stays as linear as possible. This type of information enables the algorithm developers insight into what memories to select for use, i.e., what type and size of memory should be used to keep the slope constant. This, of course, would include pre-fetching, buffering, and timings to feed the prior levels at memory speed. If this is not possible, given the available memory choices, the slope change can be minimized.

2.3. Machine Learning, Deep Learning, and Tensors

Tensor and machine learning communities have provided a solid research infrastructure, reaching from the efficient routines for tensor calculus to methods of multi-way data analysis, i.e., from tensor decompositions to methods for consistent and efficient estimation of parameters of probabilistic models. Some tensor-based models have the characteristic that if there is a good match between the model and the underlying structure in the data, the models are much more interpretable than alternative techniques. Their interpretability is an essential feature for the machine learning techniques to gain acceptance in the rather engineering intensive fields of automation and control of cyber-physical systems. Many of these systems show intrinsically multi-linear behavior, which is appropriately modeled by tensor methods, and tools for controller design can use these models. The calibration of sensors delivering data and the higher resolution of measured data will have an additional impact on the interpretability of models.

Deep learning is a subfield of machine learning that supports a set of algorithms inspired by the structure and function of the human brain. Tensorflow™ [15], PyTorch [16], Keras [17], MXNet [18], The Microsoft Cognitive Toolkit (CNTK) [19], Caffe [20], Deeplearning4j [21], and Chainer [22] are machine learning frameworks that are used to design, build, and train deep learning models. Such frameworks continue to emerge. These frameworks support numerical computations on multidimensional data arrays, or tensors, e.g., point-wise operations, such as add, sub, mul, pow, exp, sqrt, div, and mod. They also support numerous linear algebra operations, such as Matrix-Multiply, Kronecker Product, Cholesky Factorization, LU (Lower-Upper) Decomposition, singular-value decomposition (SVD), and Transpose. The programs would be written in various languages, such as Python, C, C++, and Java. These languages also include libraries/packages/modules that have been developed to support high-level tensor operations, in many cases under the umbrellas of machine learning and deep learning.

2.4. Tensor Hardware

Google's introduction of a Tensor Processing Unit (TPU) [23] that works in conjunction with TensorFlow emphasizes that there is a need for fast tensor computation. That need will only grow exponentially as the use of AI increases. Consequently, what would an idealized processor for tensors look like? What would idealized software defined hardware look like? What are important pervasive algorithms? Two workshops, one at the NSF (National Science Foundation) [12] in America, and another at Dagstuhl [13], validated and promoted how tensors are used in numerous domains, considering AI and IoT in general. Charles Van Loan, a co-organizer of the NSF Workshop, emphasized the importance of The Kronecker Product. He called it the *Product of the Times*. The algorithm (circuit) presented herein is foundational to this very important algorithm. The goal is to develop designs that could be used to build a Universal Algebraic Unit© (UAU©) that could support all the mathematics in numerical libraries, such as NumPy, which most, if not all, applications mentioned above, use and rely on for performance. There are two challenges in the design and development of applications that require tensor support: Optimal software and hardware, necessitating a co-design approach. Due to the ubiquitous nature of tensors, a co-design approach is used to achieve the goals of the work.

2.5. Contribution of this Paper

This paper demonstrates the Matrix Multiplication and Kronecker Product that are both built from a common algorithm, the outer product. This design is unique in that is provides:

- A general approach to inner and outer product, n dimensional, $0 \leq$ n;
- a general approach relative to scalar operations other than + and $\times$; and
- a demonstration of how the design enables speed-up for Kronecker Products

The design presented in this paper is for an n-dimensional inner and outer product, e.g., for *2-d* matrix multiply, which builds upon the scalar operations of + and $\times$ [24]. Some operations may be realized in hardware, firmware, or software. This generalized inner product is defined using reductions and outer products [24], and reduces to three loops independent of conformable argument dimensions and shapes. This is due to Psi Reduction, where it is possible to, through linear and multilinear transformations, reduce an array expression to a normal form. Then, through "dimension lifting" of a normal form, idealized hardware can be realized where the size of buffers relative to speed and size of connecting memories, DMA (Direct Memory Access) hardware (contiguous and strided), and other memory forms, when a problem size is known, or conjectured, and details of hardware are available and known.

2.6. The Kronecker Family of Algorithms

With an ability to build an idealized Kronecker Product, it is possible to address multiple Kronecker Products, parallel Kronecker Products, and outer products of Kronecker Products. These algorithms are used throughout the models built by mathematicians. Moreover, they are often used many times in sequence, necessitating an optimization study. If strides are required, as in the classical approach, performance will suffer. The Kronecker Product is viewed as an outer product, no matter how many there are in an expression. Consequently, it is not necessary to be concerned with strided access until the end when the outer product result is transposed and reshaped, thus saving energy and time. It is then possible to capitalize on contiguous access streaming from component to component. The analysis may consider time, space, speed, and other parameters, such as energy and heat, to determine cost.

3. Memory Considerations for Tensor Analysis

3.1. Introduction

In order to understand memory considerations, it is important to understand the algorithms that dominate tensor analysis: Inner Products (Matrix Multiply), and Outer Products (Kronecker or Tensor Product). Others include transformations and selections of components. Models in AI and IoT [13] are dominated by multiple Kronecker Products, parallel Kronecker Products (Khatri-Rao), and outer products of Kronecker Products (Tracey Singh), in conjunction with compressions over dimensions. Memory access patterns are well known. Moving on from an algorithmic specification to an optimized software or hardware instantiation of that algorithm requires maximizing the data structures that represent the algorithm in conjunction with the memory(ies) of a computer.

3.2. Computer Memory Access

From the onset of computing, computer scientists and mathematicians have discussed the complexity of an algorithm that translates to finding the least amount of arithmetic to perform. In an ideal world, where memories had the same speed no matter where they were, the computation effort would be based on the complexity of the algorithm. In fact, in the early days of computing, that was the case where memory was only one clock cycle away from the CPU (central processing unit). This is not true now. Now, what matters is the least amount of arithmetic and an optimal use of memory. From an engineering point of view, this means an understanding of the algorithm operation from a memory access pattern perspective. Moreover, through that understanding, it is possible to create an optimal, predictive, and reproducible performance.

3.3. Cache Memory: Memory Types and Caches in a Typical Processor System

Over the years, memory has become faster in conjunction with memory types becoming more diverse. Architectures now support multiple, non-uniform memories, multiple processors, and multiple networks, and those architectures are combined to form complex, multiple networks. In an IoT application, there may be a case that one application requires the use of a substantial portion of the available resources and those resources must have a reproducible capacity. Figure 3 presents a view of the different memories that may be available in an IoT application, from processor to the Cloud. This view is based on a software programmed processor approach. Different memory types (principle of operation, storage capacity, speed, cost, ability to retain the data when the device power supply is removed (volatile vs. non-volatile memory), and physical location in relation to the processor core) would be considered based on the system requirements. The fastest memories with the shortest read and write times would be closest to the processor core, and are referred to as the cache memory. Figure 3 considers the cache memory as three levels (L1, L2, and L3), where the registers are closest to the core and on the same integrated circuit (IC) die as the processor itself before the cache memory would be accessed. L1 cache memory would be SRAM (static RAM) fabricated onto the same IC die as the processor, and would be limited in the amount of data it could hold. The registers and L1 cache memory would be used to retain the data of immediate use by the processor. External to the processor would be external cache memory (L2 and L3), where this memory may be fast SRAM with limited data storage potential or slower dynamic RAM (DRAM) that would have a greater data storage potential. RAM is volatile memory, so for data retention when the power supply is removed, non-volatile memory types would be required: EEPROM (electrically erasable programmable read only memory), Flash memory based on EEPROM, and HDD would be local memory followed by external memory connected to a local area network (a "network drive") and Cloud memory. However, there are costs associated with each memory type that would need to factored into a cost function for the memory.

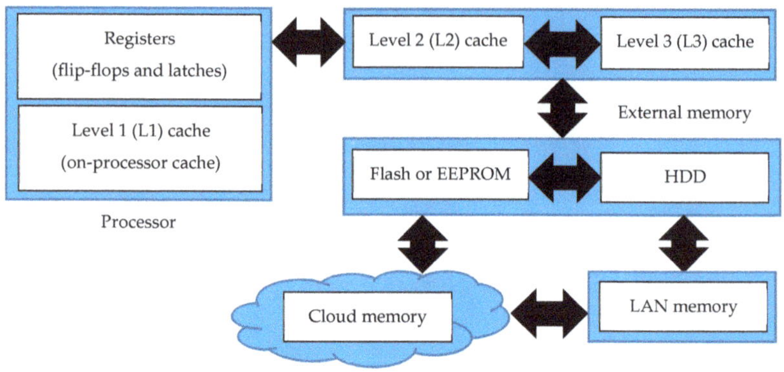

Figure 3. Availability of memory types in IoT applications.

3.4. Cache Misses and Implications

To help understand why cache memory misses, page faults, and other memory faults cause delays in computation, reference to the *1-d* fast Fourier transform (FFT) can be made. Theory states that a length *n* FFT has an *n log n* complexity and so an idealized computation time could be determined from this assumption. However, the computation could take significantly longer to complete, depending on a number of hardware related issues that include the availability of cache memory, the associativity of the cache, how many levels of memory there are, and the size of the problem. For example, if a four-way associative cache is used, a radix 4 FFT might be selected based on the size of the available associative cache. However, suppose a radix 2 is used. If the input vector for the FFT was 0 1 2 3 4 5 6 7 8 9 10 11 12 13 14 15, and the cache could fit only the first eight values, that means that on the 4th cycle, there is a cache miss. Knowing that the data could be reshaped and transposed to obtain data locality, i.e., 0 8 1 9 2 10 3 11 4 12 5 13 6 14 7 15, the operation could be completed with the data locally stored. However, as the input data set size increases, this not only results in cache misses, it also results in page faults, significantly slowing down the performance of an algorithm, which can be viewed graphically as an exponential rise [25]. For signal processing applications, the implication is that there will be a computation time increase.

3.5. Cost Functions for Memory Access

Usually cost functions are based on statistical methods. However, the analysis used in this work create a Normal Form that depicts the levels of memory desired relative to the access patterns of algorithms. With this, it is possible, a priori, to define the implementation requirements, such as maximum heat dissipation, power, cost, and time. With this information, as an FPGA designer, it is possible to utilize the available hardware resources and add the right types and levels of memory, the number of FPGAs linked together, and use FPGAs with other forms of processing unit. Such considerations would come from knowledge of the hardware and knowledge of algorithm requirements. Through experimental methods, developed by one of the co-authors, it can be seen that each level of memory as a Normal Form moves through the memory relative to its access patterns and arithmetic. What can be seen for any algorithm is that the curves, referring to Figure 2, start out constant, but then move to become a linear curve(s) while computation is still in real memory. Then, it is noticed that for each small piece of linearity, the slope gets steeper, indicating a change in memory speed. Thus, an evolution of a polynomial curve is seen that finally goes exponential when the access is to HDD. In parallel, if the available sizes and speeds of the various architectural components available, such as registers, buffers, and memories, are known, then it is possible to "dimension lift" the Normal Form to include all these attributes. Thus, performance can be predicted and verified via suitably designed experiments.

4. The Field Programmable Gate Array (FPGA)

4.1. Introduction

In this section, the FPGA is introduced as a configurable hardware device that has an internal circuit structure that can be configured to different digital circuit or system architectures. It can, for example, be configured to implement a range of digital circuits from a simple combinational logic circuit through to a complex processor architecture. With the available hardware resources and ability to describe the circuit/system design using a hardware description language (HDL), such as VHDL or Verilog [26], the designer can implement custom design architectures that are optimized to a set of requirements. For example, it is possible to describe a processor architecture using VHDL or Verilog, and to synthesize the design description using a set of design synthesis constraints into a logic description that can then be targeted to a specific FPGA device (design implementation, "place and route"). This processor, which is hardware, would then be connected to a memory containing a program for the processor to run, the memory may be registers (flip-flops), available memory macros within the FPGA or external memory devices connected to the pins of the FPGA. Therefore, it would be possible to implement a hardware only design or a hardware/software co-design. In addition, if adequate hardware resources were available, more than one processor could be configured into the FPGA and a multi-processor device therefore developed.

4.2. Programmable Logic Devices (PLDs)

The basic concept of the PLD is to provide a programmable (configurable) IC that enables the designer to configure logic cells and interconnect within the device itself to form a digital electronic circuit that is housed within a single packaged IC. In this, the hardware resources (the available hardware for use by the designer) will be configured to implement a required functionality. By changing the hardware configuration, the PLD will operate a different function. Hardware configured PLDs are becoming increasingly popular due to the potential benefits in terms of logic potential (obsolescence), rapid prototyping capabilities in digital ASIC design (early stage prototyping, design debugging, and performance evaluation), and design speed benefits, where PLD based hardware can implement the same functions as a software programmed processor, but in a reduced time. Concurrent (parallel) operations can be built into the PLD circuit configuration that would otherwise be implemented sequentially within a processor. This is particularly important for computationally expensive mathematical operations, such as the FFT, digital filtering, and other mathematical operations that require complex data sets to be analyzed in a short time. Table 2 summarizes available devices and their vendors. It is not, however, a trivial task to select the right device for a specific application or range of applications.

Table 2. PLD vendors and devices [27].

Vendor	FPGA	SPLD/CPLD	Company Homepage
Xilinx®	Virtex, Kintex, Artix and Spartan	CoolRunner-II, XA CoolRunner-II and XC9500XL	https://wwwxilinxcom/
Intel®	Stratix, Arria, MAX, Cyclone and Enpirion	—	https://wwwintelcom/content/www/us/en/fpga/deviceshtml
Atmel Corporation (Microchip)	AT40Kxx family FPGA	ATF15xx ATF25xx, ATF75xx CPLD families and ATF16xx, ATF22xx SPLD families	https://wwwmicrochipcom/
Lattice Semiconductor	ECP, MachX and iCE FPGA families	ispMACH CPLD family	http://wwwlatticesemicom/
Microsemi	PolarFire, IGLOO, IGLOO2, ProASIC3, Fusion and Rad-Tolerant FPGA families	—	https://wwwmicrosemicom/product-directory/fpga-soc/1638-fpgas

4.3. Hardware Functionality within the FPGA

Each FPGA provides a set hardware resources available to the designer where the use of specific resources would be considered to obtain a required performance in a specific application. However, this does rely on the use of the correct FPGA for the application and the knowledge of the designer in using these available hardware resources.

There are specific advantages in selecting an FPGA for use rather than an off-the-shelf processor. By selecting the appropriate hardware architecture, high speed DSP operation, such as digital filtering and FFT operations, can be achieved, which might not be possible in software. This is partly due to the ability to create a custom design architecture and partly due to concurrent operation, which means that operations in hardware can be run in parallel as well as sequentially. A typical FPGA also has a high number of digital input and output pins for connecting to peripheral devices with programmable I/O standards. This allows for flexibility in the types of peripheral devices, such as memory and communications ICs, that could be connected to the FPGA. Within the device, as well as programmable logic circuits, built-in memories for data storage are available, which have an immediate and temporary use, i.e., for cache memory scenarios. The DSP operations are supported using built-in hardware multipliers, and fast fixed-point and floating-point calculations can be implemented. In some FPGAs, built-in analog-to-digital converters (ADCs) are available for analog input sampling as well as IP blocks, such as FFT and digital filter blocks. These resources give the ability to develop a custom design architecture suited to the specific application. The FPGA is configured by downloading a design configuration as a sequence of binary logic values (sequence of 0's and 1's). The configuration would be initially created as a file using the FPGA design tools that is then downloaded into the device. The configuration values are stored in memory within the device, where the memory may be volatile or non-volatile:

- **Volatile memory:** When data are stored within the memory, the data are retained in the memory whilst the memory is connected to a power supply. Once the power supply has been removed, then the contents of the memory (the data) are lost. The early FPGAs utilized volatile SRAM based memory.
- **Non-volatile memory:** When data are stored within the memory, the data are retained in the memory even when the power supply has been removed. Specific FPGAs available today utilize Flash memory for holding the configuration.

5. Inner and Outer Product Implementation in Hardware Using the FPGA Case Study

5.1. Introduction

In this section, the design, simulation, and physical prototype testing of a single IP core that implements the inner and outer products are presented. The idea here is to have a hardware macro cell, or core, that can be accessed from an external digital system (e.g., a software programmed processor that can pass the computation tasks to this cell whilst it performs other operations in parallel). The input array data are stored as constants within arrays in the *ipOpCore* module, as shown in Figure 4, and are therefore, in this case, read-only. However, in another application, then it would be necessary to allow the arrays to be read-write for entering new data to be analyzed and then the design would be modified to allow array A and B data to be loaded into the core, either as serial or parallel data. Hence, the discussion provided in this section relates to the specific case study. In addition, a single result output could be considered and the need for test data output might not be a requirement. The motivation behind this work is to model tensors as multi-dimensional arrays and to analyze these using tensor analysis in hardware. This requires a suitable array access algorithm to be developed, the use of suitable memory for storing data in a specific application, and a suitable implementation strategy. In this paper, the inner and outer products are only considered using the FPGA as the target device, an efficient algorithm to implement the inner and outer products in a single circuit implemented in hardware is used, and appropriate embedded FPGA memory resources to enable fast memory access are used.

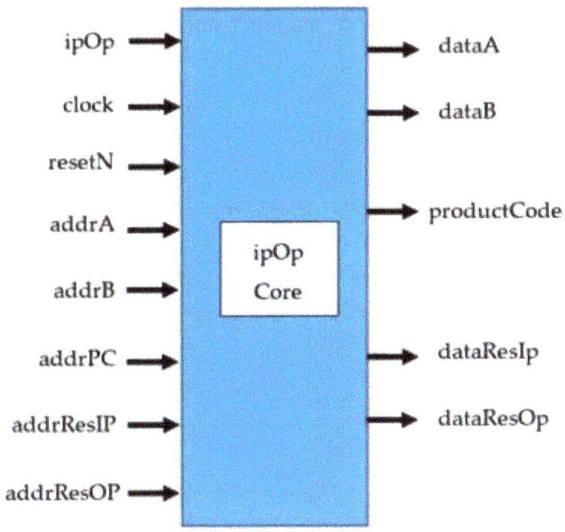


```
ENTITY ipOpCore IS
PORT( ipOp        : IN  STD_LOGIC;
      clock       : IN  STD_LOGIC;
      resetN      : IN  STD_LOGIC;
      addrA       : IN  STD_LOGIC_VECTOR(7 DOWNTO 0);
      addrB       : IN  STD_LOGIC_VECTOR(7 DOWNTO 0);
      addrPC      : IN  STD_LOGIC_VECTOR(7 DOWNTO 0);
      addrResIp   : IN  STD_LOGIC_VECTOR(7 DOWNTO 0);
      addrResOp   : IN  STD_LOGIC_VECTOR(7 DOWNTO 0);
      dataA       : OUT STD_LOGIC_VECTOR(7 DOWNTO 0);
      dataB       : OUT STD_LOGIC_VECTOR(7 DOWNTO 0);
      productCode : OUT STD_LOGIC_VECTOR(7 DOWNTO 0);
      dataResIp   : OUT STD_LOGIC_VECTOR(7 DOWNTO 0);
      dataResOp   : OUT STD_LOGIC_VECTOR(7 DOWNTO 0));
END ENTITY ipOpCore;
```

Figure 4. ipOpCore case study design.

The design shown in Figure 4 was created to allow for both product results to be independently accessed during normal runtime operation and for specific internal data to be accessed for test and debug purposes. The design description was written in VHDL as a combination of behavioral, RTL, and structural code targeting the Xilinx® Artix-7 (XC7A35TICSG324-1L) FPGA. This specific device was chosen for practical reasons as it contains hardware resources suited for this application. The design, however, is portable and is readily transferred to other FPGAs, or to be part of a larger digital ASIC design, if required. For any design implementation, the choice of hardware, and potentially software, to use would be based on a number of considerations. The FPGA was mounted on the Digilent® Arty A7-35T Development Board and was chosen for the following reasons:

The FPGA considered is used in other project work and as such, the work described in this paper could readily be incorporated into these projects. Specifically, sensor data acquisition using the FPGA and data analysis within the FPGA projects would benefit from this work where the algorithm and memory access operations used in this paper would provide additional value to the work undertaken.

1. The development board used provided hardware resources that were useful for project work, such as the 100 MHz clock, external memory, switches, push buttons, light emitting diodes (LEDs), expansion connectors, LAN connection, and a universal serial bus (USB) interface for FPGA configuration and runtime serial I/O.
2. The development board was physically compact and could be readily integrated into an enclosure for mobility purposes and operated from a battery rather than powered through the USB +5 V power.
3. The Artix-7 FPGA provided adequate internal resources and I/O for the work undertaken and external resources could be readily added via the expansion connectors if required.
4. For memory implementation, the FPGA can use the internal look-up tables (LUTs) as distributed memory for small memories, can use internal BRAM (Block RAM) for larger memories, and external volatile/non-volatile memories connected to the I/O.
5. For computation requirements, the FPGA allows for both fixed-point and floating-point arithmetic operations to be implemented.

6. For an embedded processor based approach, the MicroBlaze CPU can be instantiated within the FPGA for software based implementations.

The I/O for this module are as follows:

ipOp	User to select whether the inner or outer product is to be performed.
clock	Master clock (100 MHz).
resetN	Master, asynchronous active low reset.
addrA	Array A address for reading array contents (input tensor A).
addrB	Array B address for reading array contents (input tensor B).
addrPC	Array PC address for reading array contents (product code array).
addrResIp	Address of inner product for reading array contents (output tensor IP).
addrResOp	Address of outer product for reading array contents (output tensor OP).
dataA	Array A data element being accessed (for test purposes only).
dataB	Array B data element being accessed (for test purposes only).
productCode	Input array size and shape information for algorithm operation.
dataResIp	Inner product result array (Serial read-out).
dataResOp	Outer product result array contents (serial readout).

These I/O signals can be categorized as input *control*, input *address*, and output *data*.

5.2. Design Approach and Target FPGA

The operation of the combined inner and outer product is demonstrated by reference to a case study design that implements the necessary memory and algorithms functions within a single IP core. Given that these functions are to be mapped to a custom design architecture and configured within the FPGA, a range of possible solutions can be created. The starting point for the design is the computation to perform. Consider the tensor product of two arrays (A and B), where A is a 3×3 array and B is a 3×2 array. For demonstration purposes, the numbers are limited to being 8-bit signed integers rather than real numbers. The principle of evaluation is the same for both number types, but the HDL coding style to be adopted would be different. Therefore, the possible numbers considered would be integer values in the range of -128_{10} to $+127_{10}$. Internally within the VHDL code, these values were modelled as INTEGER data types that were suitable for simulation and synthesis. For synthesis, the integer numbers were translated to an 8-bit wide STD_LOGIC_VECTOR data type. This meant that the physical digital circuit utilized an 8-bit data bus and this size bus was selected as a standard width for all array input addresses and output data. Fixed-point, 2's complement arithmetic was also implemented. Whilst the data range was limited in size, this approach was chosen as the purpose of the work was to implement and demonstrate the algorithm and memory utilization. The VHDL code was written such that the data range and array sizes were readily adjusted within the array definitions and no modification to the algorithm code was required. Floating-point arithmetic rather than fixed-point arithmetic could be used by coding a floating-point multiplier for matrix multiplication operations (e.g., [28,29]), and modelling the data as floating point numbers rather than simple fixed-point scalar numbers as used here. Considering arrays A and B, these two arrays can be operated on to form the tensor product as both the inner product and the outer product:

$$A = \begin{bmatrix} 0 & 1 & 2 \\ 3 & 4 & 5 \\ 6 & 7 & 8 \end{bmatrix} \qquad B = \begin{bmatrix} 0 & 1 \\ 2 & 3 \\ 4 & 5 \end{bmatrix}$$

The tensor product for A and B is noted as:

$$C = A \otimes B$$

The result of the inner product, C_{ip}, is:

$$C_{ip} = A \otimes B = \begin{bmatrix} 10 & 13 \\ 28 & 40 \\ 46 & 67 \end{bmatrix}$$

The result of the outer product, C_{op}, is:

$$C_{op} = A \otimes B = \begin{bmatrix} 0 & 0 & 0 & 1 & 0 & 2 \\ 0 & 0 & 2 & 3 & 4 & 6 \\ 0 & 0 & 4 & 5 & 8 & 10 \\ 0 & 3 & 0 & 4 & 0 & 5 \\ 6 & 9 & 8 & 12 & 10 & 15 \\ 12 & 15 & 16 & 20 & 20 & 25 \\ 0 & 6 & 0 & 7 & 0 & 8 \\ 12 & 18 & 14 & 21 & 16 & 24 \\ 24 & 30 & 28 & 35 & 32 & 40 \end{bmatrix}.$$

The above products were initially developed using C and Python coding where the data in C were stored in arrays and in Python were stored in lists. The combined inner/outer product algorithm was verified through running the algorithm with different data sets and verifying the software simulation model results with manual hand calculation results. Once the software version of the design was verified, the Python code functionality was manually translated to a VHDL equivalent. The two key design decisions to make were:

1. How to model the arrays for early-stage evaluation work and how to map the arrays to hardware in the FPGA.
2. How to design the algorithm to meet timing constraints, such as maximum processing time, number of clock cycles required, hardware size considerations, and the potential clock frequency, with the hardware once it is configured within the FPGA.

In this design, the data set was small and so VHDL *arrays* were used for both the early-stage evaluation work and for synthesis purposes. In VHDL, the input and results arrays were defined and initialized as follows:

```
TYPE array_1by4  IS ARRAY (0 TO 3)  OF INTEGER;
TYPE array_1by6  IS ARRAY (0 TO 5)  OF INTEGER;
TYPE array_1by9  IS ARRAY (0 TO 8)  OF INTEGER;
TYPE array_1by36 IS ARRAY (0 TO 35) OF INTEGER;
TYPE array_1by54 IS ARRAY (0 TO 53) OF INTEGER;

CONSTANT arrayA          : array_1by9 := (0, 1, 2, 3, 4, 5, 6, 7, 8);
CONSTANT arrayB          : array_1by6 := (0, 1, 2, 3, 4, 5);

SIGNAL  arrayResultIp     : array_1by6  := (0, 0, 0, 0, 0, 0);
SIGNAL  arrayResultOp     : array_1by54 := (0, 0, 0, 0, 0, 0, 0, 0, 0,
                                            0, 0, 0, 0, 0, 0, 0, 0, 0,
                                            0, 0, 0, 0, 0, 0, 0, 0, 0,
                                            0, 0, 0, 0, 0, 0, 0, 0, 0,
                                            0, 0, 0, 0, 0, 0, 0, 0, 0,
                                            0, 0, 0, 0, 0, 0, 0, 0, 0);
```

These are one-dimensional arrays suited for ease of memory addressing, appropriate for the algorithm operation, synthesizable into logic, and have a direct equivalence in the C and Python software models. The input arrays (arrayA and arrayB) contain the input data. The results arrays (arrayResultIp (inner product) and arrayResultOp (outer product)) were initialized with 0′s. It was not necessary, in this case, to map to any embedded BRAM or external memory as the data set size was small and easily mapped by the synthesis tool to distributed RAM within the FPGA. The PC (product code) array is not shown above, but this is an array that contains the shape and size of arrays, A and B. For the algorithm, with direct mapping to VHDL from the Python code, the inner product and outer product each required a set number of clock cycles. Figure 5 shows a simplified timing diagram identifying the signals required to implement the inner/outer product computation. Once the computation has been completed, the array contents could then be read out one element at a time. For evaluation purposes, all array values were made accessible concurrently, but could readily be made available serially via a multiplexor arrangement to reduce the number of output signals required in the design.

A computation run would commence with the run control signal being pulsed 0-1-0 with the product selection input ipOp set to either logic 0 (inner product) or logic 1 (outer product). In this implementation, the inner product required 18 clock cycles and the outer product required 54 clock cycles to complete. The array data read-out operations are not, however, shown in Figure 5. The data values were defined using the INTEGER data type for modelling and simulation purposes, and these values were mapped to an 8-bit STD_LOGIC_VECTOR data type for synthesis into hardware. The 8-bit width data bus was sufficient to account for all data values in this study.

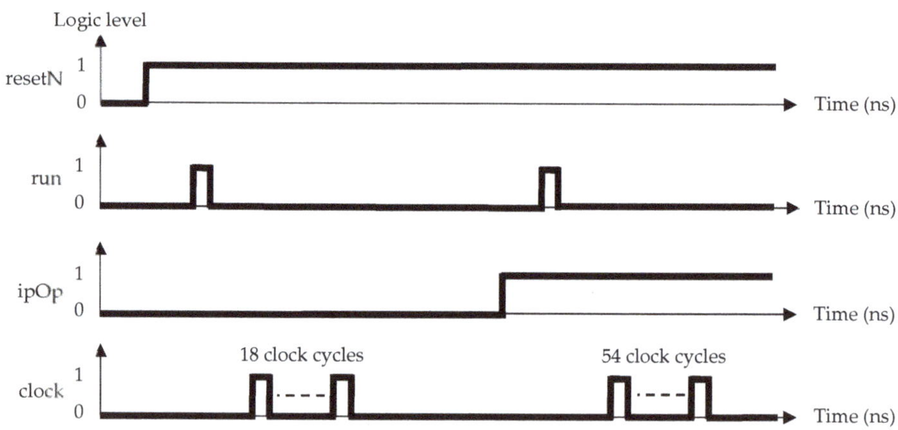

Figure 5. Computation control signal timing diagram.

5.3. System Architecture

How the memory and algorithm would generally be mapped to a hardware-only or a hardware/software co-design would be dependent on the design requirements, specification resulting from the requirements identification, available hardware, and the designer. Therefore, a range of possible solutions would be possible, but in this design, a hardware-only solution was a design requirement. The memory was modelled as VHDL arrays, and the algorithm was implemented using a counter and state machine arrangement. Both the inner and outer products were to be selectable for computation that required a design decision as to whether a single memory space for both products or separate memory spaces for each product would be suitable. Given the relatively small size of the data set and to support design evaluation, separate memory spaces for the inner and outer products were developed. However, an alternative implementation could utilize a single memory space. The system architecture is shown in Figure 6. Here, the *ipOpCore* module implements the memory computation (2′s complement number multiplication) whilst the *control unit* module implements the system control and algorithm. The *control unit* module input *control* signals are:

ipOp	User to select whether the inner or outer product is to be performed;
clock	Master clock (100 MHz);
resetN	Master, asynchronous active low reset; and
run	Initiate a computation run (0-1-0 pulse)

Figure 7 shows a simplified view of the elaborated VHDL code schematic that was generated by the Xilinx® Vivado v2015.3 (HL WebPACK Edition) software. This schematic shows the two modules (ipOpCore (I0) and controlUnit (I1)) that connect together to form the top-level design with 44 inputs and 40 outputs. The target FPGA was the Xilinx® Artix-7 mounted on the Digilent® Arty A7-35T Development Board. This board is shown in Figure 8 that identifies the key features of the board used and provided a convenient hardware platform to undertake the required design development and experiments. The FPGA was provided with an on-board 100 MHz clock module for the clock and the resetN signal was provided by one of the available on-board push buttons. On the board, the array address and data signals would be available internally within the FPGA (to connect to a system that would be integrated within the FPGA alongside this design) or to the external header pins on the development board (for connecting to another system external to the FPGA).

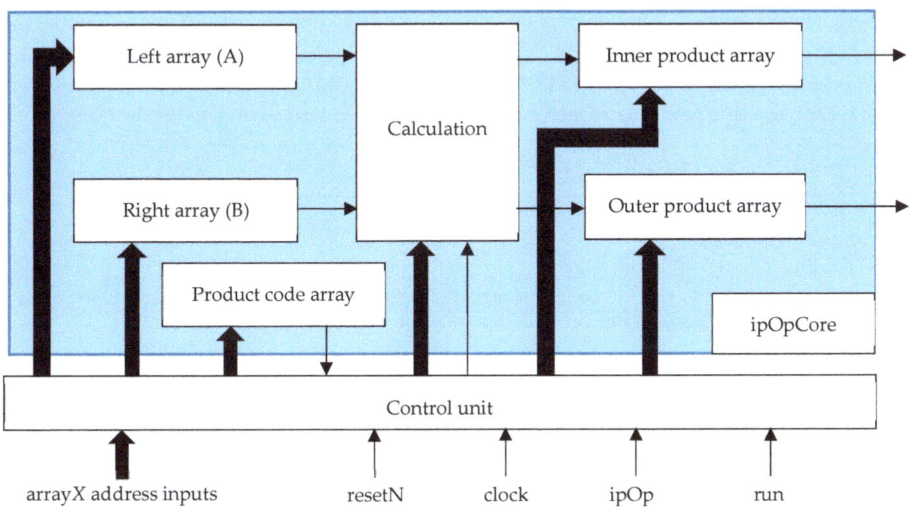

Figure 6. Simplified system block diagram.

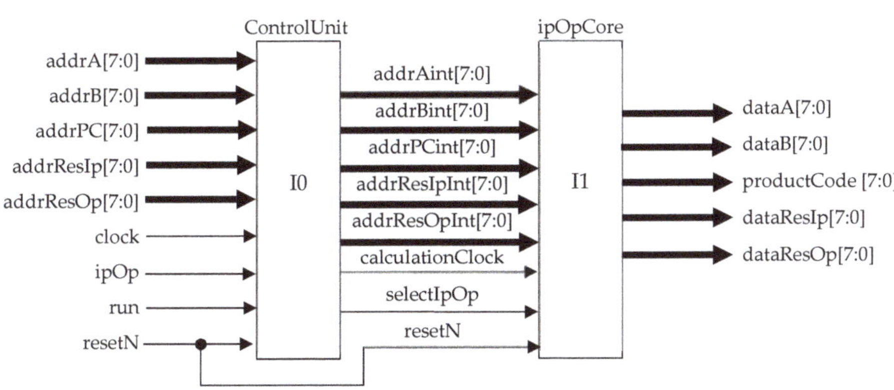

Figure 7. Simplified schematic view of the elaborated VHDL code.

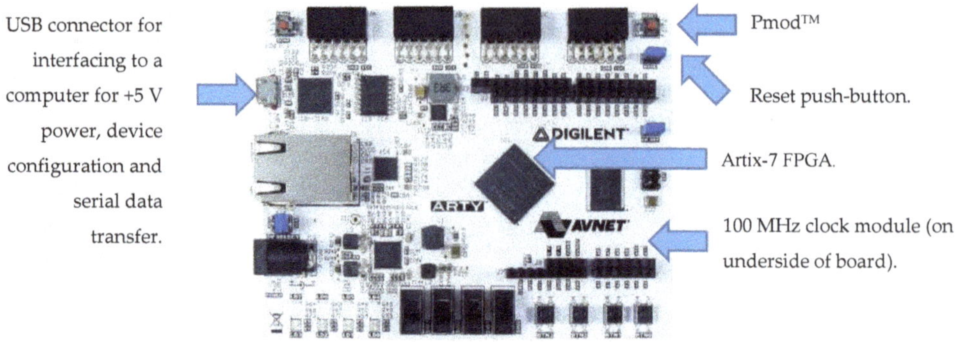

Figure 8. Xilinx® Artix-7 FPGA on the Digilent® Arty board identifying key components used in the experimentation.

The design must eventually be implemented within the FPGA and this is a two-step process. Firstly, the VHDL code is synthesized and then the synthesized design is implemented in the target FPGA. The synthesis and implementation operations can be run using the default settings, or the user can set constraints to direct the tools. In this case study, the default tool settings were used and Table 3 identifies the hardware resources required after synthesis and implementation for the design.

Table 3. Artix-7 FPGA resource utilization in the case study design.

Item	Use	Number Used
Package pin	Input	44
	Output	40
	Design synthesis results	
Post-synthesis I/O	Inputs	23 *
	Outputs	40
Slice LUTs	Total used	454
	LUT as logic	442
	LUT as memory (distributed RAM)	12
Slice registers	Total used	217
	Slice register as flip-flop	217
Other logic	Clock buffer	2
	Design implementation results	
Post-implementation I/O	Inputs	23
	Outputs	40
Slice LUTs	Total used	391
	LUT as logic	379
	LUT as memory (distributed RAM)	12
Slice registers	Total used	217
	Slice register as flip-flop	217
Other logic	Clock buffer	2

* Note that the number of inputs required in the design after synthesis do not include the address input bits that were always a constant logic 0 in this case study. This was due to the standard 8-bit address bus used for all input addresses and the sizes of the arrays meant that most significant bits (MSBs) of the array addresses were not required. Note also that post-implementation, the number of slice LUTs required was less than that post-synthesis.

5.4. Design Simulation

Design simulation was undertaken to ensure that the correct values were stored, calculated, and accessed. The Xilinx® Vivado software tool was used for design entry and simulation was performed using the built-in Vivado simulator. A VHDL test bench was used to perform the computation and array data read-out operations. Figure 9 shows the complete simulation run where the clock frequency in simulation was set to 50 MHz (the master clock frequency divided by two).

This simulation clock frequency was selected to allow for external control signals to be provided from an external system operating at 100 MHz to be provided on the falling edge of the 50 MHz clock.

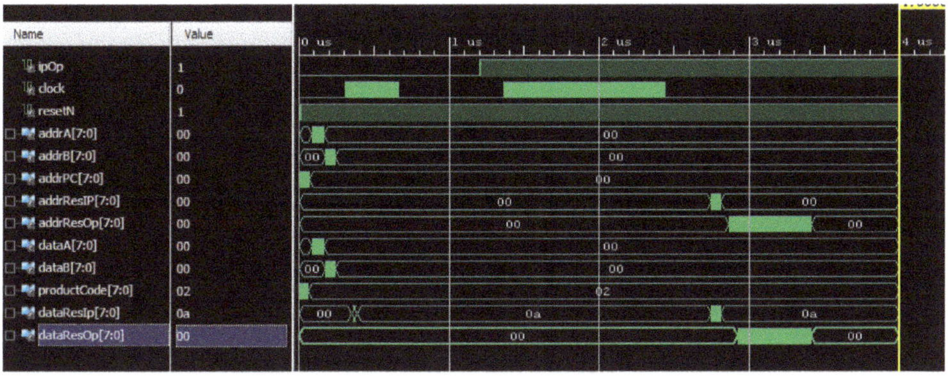

Figure 9. Simulation study results: Computation and results read-out.

For the inner product data read-out, Figure 10 shows the simulation results for all nine product array element values (dataResIp) being read out of the arrayResultIp array. The iPOp control signal is not used (set to logic 1 in the simulation test bench) as it is only used for the computation, the clock is held at logic 0 as it is also only used for the computation, and the reset signal is not asserted (resetN = 1). The inner product array address (addrResIp) is provided to access each element in the array sequentially.

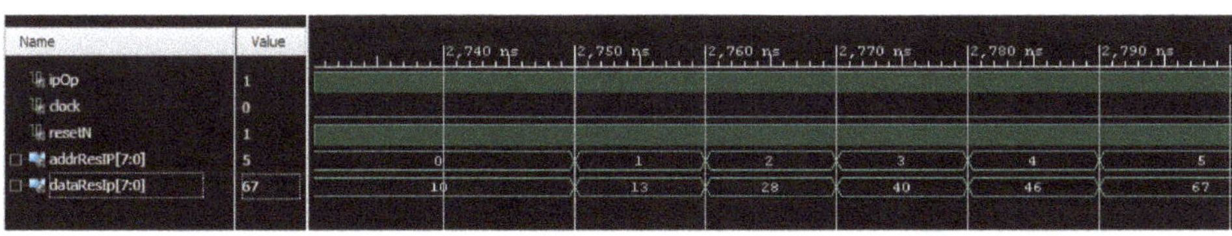

Figure 10. Simulation study results: Inner product.

This shows the specific results for the complete inner product as follows:

$$C_{ip} = A \otimes B = \begin{bmatrix} 10 & 13 \\ 28 & 40 \\ 46 & 67 \end{bmatrix}$$

For the outer product data read-out, Figure 11 shows the simulation results for the last 13 values (dataResOp) being read out of the arrayResultOp array. The iPOp control signal is not used (set to logic 1 in the simulation test bench) as it is only used for the computation, the clock is held at logic 0 as it is also only used for the computation, and the reset signal is not asserted (resetN = 1). The outer product array address (addrResOp) is provided to access each element in the array sequentially.

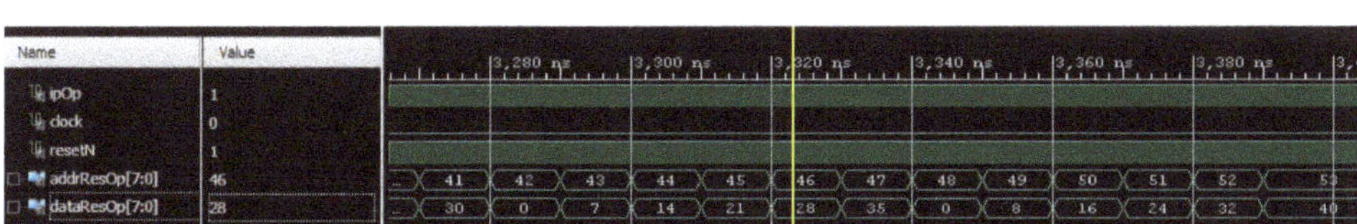

Figure 11. Simulation study results: Outer product (final set of results read-out only).

This shows the specific results for the last 13 values in the results array as follows:

$$C_{op} = A \otimes B = \begin{bmatrix} . & . & . & . & . & . \\ . & . & . & . & . & . \\ . & . & . & . & . & . \\ . & . & . & . & . & . \\ . & . & . & . & . & . \\ . & . & . & . & . & . \\ . & . & 0 & 7 & 0 & 8 \\ . & . & 14 & 21 & 16 & 24 \\ . & 30 & 28 & 35 & 32 & 40 \end{bmatrix}$$

5.5. Hardware Test Set-Up

Design analysis was in the main performed using simulation to determine correct functionality and signal timing considering the initial design description prior to synthesis (behavioral simulation), the synthesized design (post-synthesis simulation), and the implemented design (post-implementation simulation). This is a typical simulation approach that is supported by the FPGA simulator for verifying the design operation at different steps in the design process. Given that the design is intended to be used as a block within a larger digital system, the simulation results would give an appropriate level of estimating the signal timing and the circuit power consumption.

In addition to the simulation study, the design was also implemented within the FPGA and signal monitored using the development board connectors (the Pmod^{TM} (peripheral module) connectors) using a logic analyzer and oscilloscope. This test arrangement is shown in Figure 12.

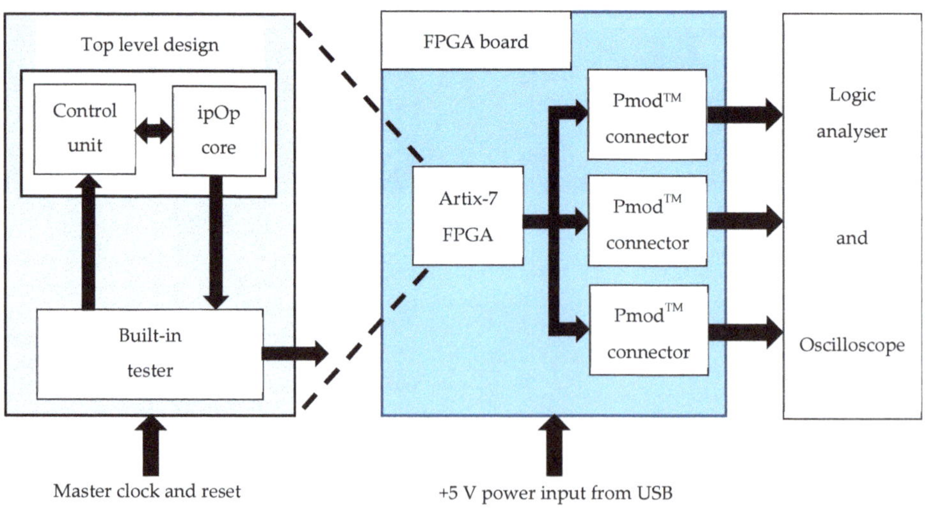

Figure 12. Embedded hardware tester.

To generate the top-level design module input signals, a built-in tester circuit was developed and incorporated into the FPGA. This was a form of a *built-in self-test* (BIST) [30] circuit that generated the control signals identified in Figure 5 and allowed the internal array address and data signals to be accessed. The tester circuit was set-up to continuously repeat the sequence in Figure 5 rather than run just once and so did not require any user set input control signals to operate. With the number of address and data bits required (40 address bits and 40 data bits) for the five arrays that exceeded the number of Pmod^{TM} connections available, these signals were multiplexed to eight address and eight data bits within the built-in tester and the multiplexor control signals were output for identifying the array being accessed. The control signals were also accessible on the Pmod™ connectors for test purposes.

Figure 13 shows a simplified schematic view of the elaborated VHDL code, where I0 is the top-level design module and I1 is the built-in tester module.

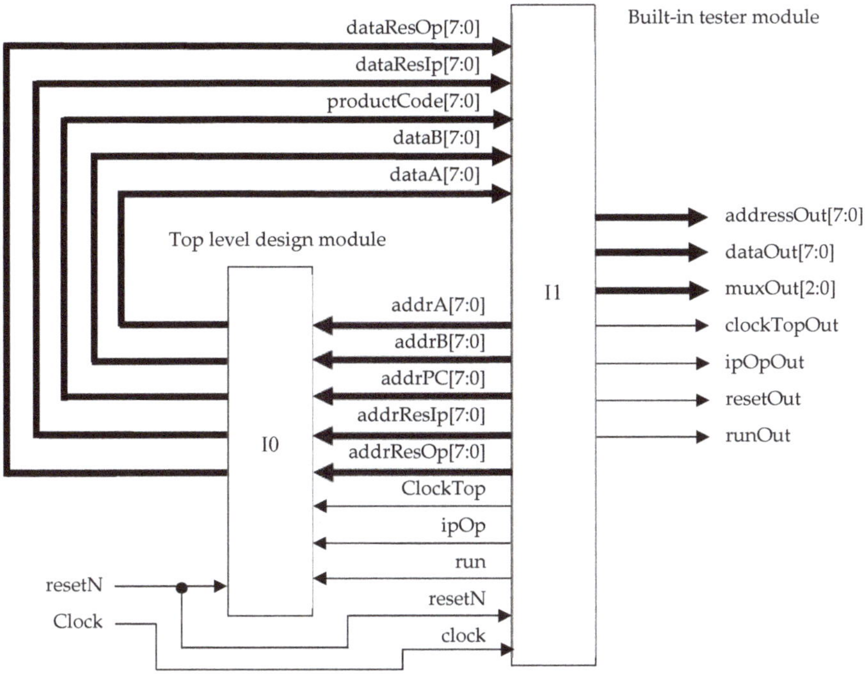

Figure 13. Embedded hardware tester: Simplified schematic view of the elaborated VHDL code.

The hardware test arrangement was useful to verify that the signals were generated correctly and matched the logic levels expected during normal design operation. However, it was necessary to reduce the speed of operation to account for non-ideal electrical parasitic effects that caused ringing of the signal. In this specific set-up, speed of operation of the circuit when monitoring the signals using the logic analyzer and oscilloscope was not deemed important, so the 100 MHz clock was internally divided within the built-in tester circuit to 2 MHz in the study. However, further analysis could determine how fast the signals could change if the Pmod™ connector was required to connect external memory for larger data sets.

Figure 14 shows the logic analyzer test set-up with the Artix-7 FPGA Development Board (bottom left) and the Digilent® Analog Discovery "USB Oscilloscope and Logic Analyzer" (top right) [31]. The 16 digital inputs for the logic analyzer function were available for use and a GND (ground, 0) connection was required to monitor the test circuit outputs. Internal control signals (ClockTop, ipOp, run, and resetN) were also available for monitoring in this arrangement.

The Digilent® Waveforms software [32] was utilized to control the test hardware and view the results. Figure 15 shows the logic analyzer output in Waveforms. Here, one complete cycle of the test process is shown, where the calculations are initially performed and the array outputs then read. The logic level values obtained in this physical prototype test agreed with the results from the simulation study. The data values are shown as a combined integer number value (data, top) and the values of the individual bits (7 down to 0).

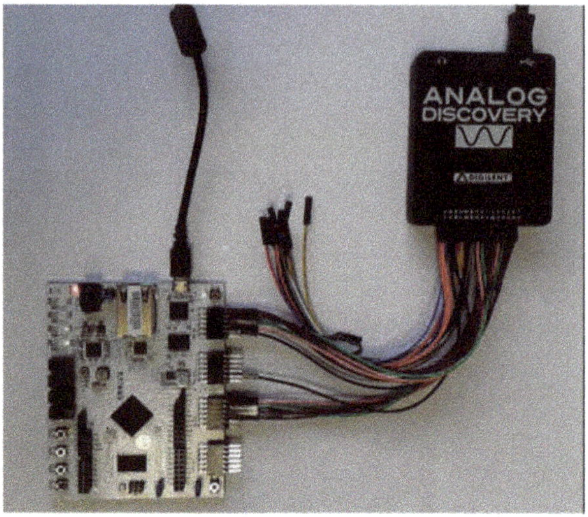

Figure 14. Logic analyzer test set-up using the Digilent® Analog Discovery.

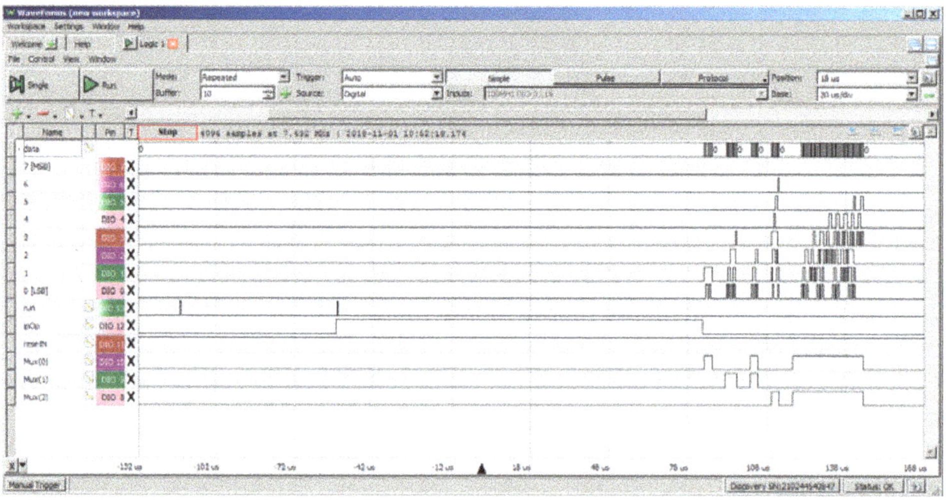

Figure 15. Logic analyzer test results using the Digilent® Analog Discovery: Complete cycle.

The run signal (a 0-1-0 pulse) initiates the computation that is selected by the ipOp signal at the start of the cycle. The data readout on the 8-bit data bus can be seen towards the end of the cycle as both a bus value and individual bit values.

Figure 16 shows the data readout operation towards the end of the cycle. The data output identifies the values for array A (nine values), array B (six values), and the inner product result array (six values) as identified in Section 5.2.

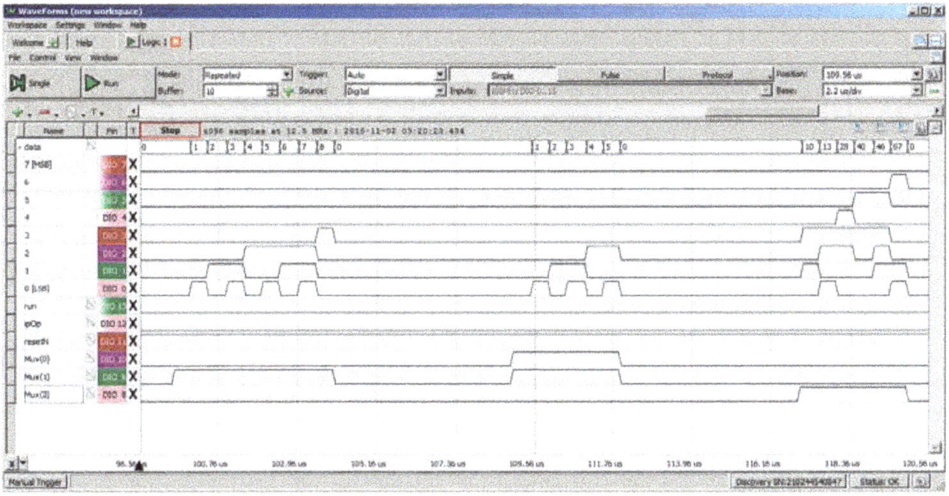

Figure 16. Logic analyzer test results using the Digilent® Analog Discovery: Data readout.

5.6. Design Implementation Considerations

This case study design has presented one example implementation of the combined inner and outer product algorithm. The study focused on creating a custom hardware only design implementation rather than developing the algorithm in software to run on a suitable processor architecture. The approach taken to create the hardware design was to map the algorithm operations in software to a hardware equivalence. The hardware design was created using two main modules:

1. The *computation* module.
2. The *control* module. The *control* module was required to receive control signals from an external system and transform these to internal control signals for the *computation* module.

The *computation* module itself was modelled as two separate sub-modules as this was based on the underlying structure of the problem that was to efficiently access data from memory for running a computation on data held in specific memory locations:

1. The *memory* module.
2. The *algorithm* module.

For a specific application, the *memory* module would be used for storing input data, intermediate results data, and final (output) data. For this design, the physical memory used was internal to the FPGA using distributed memory within the LUTs given the size of the data set, the availability of hardware resources within the FPGA, and the synthesis tool that automatically determined what hardware resources were to be used. The memory was modelled using VHDL arrays where the input data arrays held constant values and the intermediate and output data arrays held variables. In a different scenario, the memory modelling in VHDL might be different. For example, explicitly targeting internal BRAM cells and external memory attached to the FPGA pins. Such an approach would resemble a standard processor architecture with different levels of memory. The internal latches, flip-flops, distributed RAM, and BRAM cells within the FPGA would map to cache memory internal to the processor and external memory to attached memory devices as depicted in Figure 3.

The designer would have design choices when considering the *algorithm* module. One approach would be to use a standard processor architecture that would be software programmed and mapped to hardware resources within the FPGA. Depending on the device, the processor may be an embedded core (a so-called *hard core*) or may be an IP block that can be instantiated in a design and synthesized into the available FPGA logic (a so-called *soft core*). For example, in Xilinx® FPGAs, then the MicroBlaze 32-bit RISC (reduced instruction set computer) CPU can be instantiated into a custom design. It is also possible to have, if the hardware resources are sufficient, instantiated multiple *soft cores* within the FPGA. This would allow for a multi-processor solution and on-chip processor-to-processor communications with parallel processing. A second approach would be to develop a custom architecture solution that maps the algorithm and memory modules to the user requirements, giving a choice to implement sequential or parallel (concurrent) operations. This provides a high level of flexibility for the designer, but requires a different design approach, thinking in terms of hardware rather than software operations. A third approach would be to create a hardware-software co-design incorporating custom architecture hardware and standard processor architectures working concurrently.

A final consideration in implementation would to be identify example processor architectures and target hardware used in machine and deep learning applications, where their benefits and limitations for specific applications could be assessed. For example, in software processor applications, then the CPU is used for tensor computations where a GPU (graphics processing unit) is not available. GPUs have architectures and software programming capabilities that are better than a CPU for applications, such as gaming, where high-speed data processing and parallel computing operations are required. An example GPU is the Nvidia® Tensor Core [33].

6. Conclusions

In this paper, the design and simulation of a hardware block to implement a combined inner and outer product was introduced and elaborated. This work was considered in the context of developing embedded digital signal processing algorithms that can effectively and efficiently process complex data sets. The FPGA was used as the target hardware and the product algorithm developed as VHDL modules. The algorithm was initially evaluated using C and Python code before translation to a hardware description in VHDL. The paper commenced with a discussion into tensors and the need for effective and efficient memory access to control memory access times and the cost associated with such memory access operations. To develop the ideas, the FPGA hardware implementation developed was an example design that paralleled an initial software algorithm (C and Python coding) used for algorithm development. The design was evaluated in simulation and hardware implementation issues were discussed.

Author Contributions: The authors contributed equally to the work undertaken as a co-design effort. Co-design is not just giving tasks to scholars in separate yet complementary disciplines. It is an ongoing attempt to communicate ideas across disciplines and a desire to achieve a common goal. This type of research not only helps both disciplines but it gives definitive proof of how that collaboration achieves greater results, and mentors students and other colleagues to do the same. The case study design was a hardware implementation by Ian Grout using the FPGA of the algorithm developed by Lenore Mullin. Both authors contributed equally to the background discussions and the writing of the paper.

Funding: This research received no external funding.

Conflicts of Interest: The authors declare no conflict of interest.

References

1. Earley, S. Analytics, Machine Learning, and the Internet of Things. *IT Prof.* **2015**, *17*, 10–13. [CrossRef]
2. Corneanu, C.A.; Simón, M.O.; Cohn, J.F.; Guerrero, S.E. Survey on RGB, 3D, Thermal, and Multimodal Approaches for Facial Expression Recognition: History, Trends, and Affect-Related Applications. *IEEE Trans. Pattern Anal. Mach. Intell.* **2016**, *38*, 1548–1568. [CrossRef] [PubMed]
3. Xilinx®. FPGA Leadership across Multiple Process Nodes. Available online: https://www.xilinx.com/products/silicon-devices/fpga.html (accessed on 9 October 2018).
4. Xilinx®. Homepage. Available online: https://www.xilinx.com/ (accessed on 9 October 2018).
5. Xilinx®. Artix-7 FPGA Family. Available online: https://www.xilinx.com/products/silicon-devices/fpga/artix-7.html (accessed on 9 October 2018).

6. Daniel Fleisch. *A Student's Guide to Vectors and Tensors*; Cambridge University Press: Cambridge, UK, 2011; ISBN-10 0521171903, ISBN-13 978-0521171908.
7. Institute of Electrical and Electronics Engineers. *IEEE Std 1076-2008—IEEE Standard VHDL Language Reference Manual*; IEEE: New York, NY, USA, 2009; ISBN 978-0-7381-6853-1, ISBN 978-0-7381-6854-8.
8. Kindratenko, V.; Trancoso, P. Trends in High Performance Computing. *Comput. Sci. Eng.* **2011**, *13*, 92–95. [CrossRef]
9. Lane, N.D.; Bhattacharya, S.; Mathur, A.; Georgiev, P.; Forlivesi, C.; Kawsar, F. Squeezing Deep Learning into Mobile and Embedded Devices. *IEEE Pervasive Comput.* **2017**, *16*, 82–88. [CrossRef]
10. Mullin, L.; Raynolds, J. Scalable, Portable, Verifiable Kronecker Products on Multi-scale Computers. In *Constraint Programming and Decision Making. Studies in Computational Intelligence*; Ceberio, M., Kreinovich, V., Eds.; Springer: Cham, Switzerland, 2014; Volume 539.
11. Gustafson, J.; Mullin, L. Tensors Come of Age: Why the AI Revolution Will Help HPC. 2017. Available online: https://www.hpcwire.com/2017/11/13/tensors-come-age-ai-revolution-will-help-hpc/ (accessed on 9 October 2018).
12. Workshop Report: Future Directions in Tensor Based Computation and Modeling. 2009. Available online: https://www.researchgate.net/publication/270566449_Workshop_Report_Future_Directions_in_Tensor-Based_Computation_and_Modeling (accessed on 9 October 2018).
13. Tensor Computing for Internet of Things (IoT). 2016. Available online: http://drops.dagstuhl.de/opus/volltexte/2016/6691/ (accessed on 9 October 2018).
14. Python.org, Python. Available online: https://www.python.org/ (accessed on 9 October 2018).
15. Tensorflow™. Available online: https://www.tensorflow.org/ (accessed on 9 October 2018).
16. Pytorch. Available online: https://pytorch.org/ (accessed on 9 October 2018).
17. Keras. Available online: https://en.wikipedia.org/wiki/Keras (accessed on 9 October 2018).
18. Apache MxNet. Available online: https://mxnet.apache.org/ (accessed on 9 October 2018).
19. Microsoft Cognitive Toolkit (MTK). Available online: https://www.microsoft.com/en-us/cognitive-toolkit/ (accessed on 9 October 2018).
20. CAFFE: Deep Learning Framework. Available online: http://caffe.berkeleyvision.org/ (accessed on 9 October 2018).
21. DeepLearning4J. Available online: https://deeplearning4j.org/ (accessed on 9 October 2018).
22. Chainer. Available online: https://chainer.org/ (accessed on 9 October 2018).
23. Google, Cloud TPU. Available online: https://cloud.google.com/tpu/ (accessed on 9 October 2018).
24. Lenore, M.; Mullin, R. A Mathematics of Arrays. Ph.D. Thesis, Syracuse University, Syracuse, NY, USA, 1988.
25. Mullin, L.; Raynolds, J. Conformal Computing: Algebraically connecting the hardware/software boundary using a uniform approach to high-performance computation for software and hardware. *arXiv* **2018**. Available online: https://arxiv.org/pdf/0803.2386.pdf (accessed on 1 November 2018).
26. Institute of Electrical and Electronics Engineers. *IEEE Std 1364™-2005 (Revision of IEEE Std 1364-2001), IEEE Standard for Verilog® Hardware Description Language*; IEEE: New York, NY, USA, 2006; ISBN 0-7381-4850-4, ISBN 0-7381-4851-2.
27. Ong, Y.S.; Grout, I.; Lewis, E.; Mohammed, W. Plastic optical fibre sensor system design using the field programmable gate array. In *Selected Topics on Optical Fiber Technologies and Applications*; IntechOpen: Rijeka, Croatia, 2018; pp. 125–151, ISBN 978-953-51-3813-6.
28. Dou, Y.; Vassiliadis, S.; Kuzmanov, G.K.; Gaydadjiev, G.N. 64 bit Floating-point FPGA Matrix Multiplication. In Proceedings of the 2005 ACM/SIGDA 13th International Symposium on Field-Programmable Gate Arrays, Monterey, CA, USA, 20–22 February 2005. [CrossRef]
29. Amira, A.; Bouridane, A.; Milligan, P. Accelerating Matrix Product on Reconfigurable Hardware for Signal Processing. In *Field-Programmable Logic and Applications, Proceedings of the 11th International Conference, FPL 2001, Belfast, UK, 27–29 August 2001*; Lecture Notes in Computer Science; Brebner, G., Woods, R., Eds.; Springer: Berlin/Heidelberg, Germany, 2001; pp. 101–111, ISBN 978-3-540-42499-4 (print), ISBN 978-3-540-44687-3 (online).
30. Hurst, S.L. *VLSI Testing—Digital and Mixed Analogue/Digital Techniques*; The Institution of Engineering and Technology: London, UK, 1998; pp. 241–242, ISBN 0-85296-901-5.
31. Digilent®. Analog Discovery. Available online: https://reference.digilentinc.com/reference/instrumentation/analog-discovery/start?redirect=1 (accessed on 1 November 2018).

32. Digilent®. Waveforms. Available online: https://reference.digilentinc.com/reference/software/waveforms/waveforms-3/start (accessed on 1 November 2018).
33. Nvidia. Nvidia Tensor Cores. Available online: https://www.nvidia.com/en-us/data-center/tensorcore/ (accessed on 1 November 2018).

electronics

MDPI

Article

Acoustic-Based Fault Diagnosis of Commutator Motor

Adam Glowacz

Department of Automatic Control and Robotics, Faculty of Electrical Engineering, Automatics, Computer Science and Biomedical Engineering, AGH University of Science and Technology, al. A. Mickiewicza 30, 30-059 Kraków, Poland; adglow@agh.edu.pl

Received: 4 October 2018; Accepted: 2 November 2018; Published: 5 November 2018

Abstract: In the paper, the author presents acoustic-based fault diagnosis of a commutator motor (CM). Five states of the commutator motor were considered: healthy commutator motor, commutator motor with broken rotor coil, commutator motor with shorted stator coils, commutator motor with broken tooth on sprocket, commutator motor with damaged gear train. A method of feature extraction MSAF-15-MULTIEXPANDED-8-GROUPS (Method of Selection of Amplitudes of Frequency Multiexpanded 8 Groups) was described and implemented. Classification methods, such as nearest neighbour (NN), nearest mean (NM), self-organizing map (SOM), backpropagation neural network (BNN) were used for acoustic analysis of the commutator motor. The paper provides results of acoustic analysis of the commutator motor. The results had a good recognition rate. The results of acoustic analysis were in the range of 88.4–94.6%. The NM classifier and the MSAF-15-MULTIEXPANDED-8-GROUPS provided TE_{RCM} = 94.6%.

Keywords: commutator motor; fault diagnosis; method; technique; signal processing; acoustic

1. Introduction

Fault diagnosis of electrical rotating motors has been extensively investigated since the 20th century, and can increase the reliability and safety of electrical rotating motors. Condition monitoring of electrical motors are very important for industry, reducing loss due to unforeseen faults and damage. Unforeseen faults and damage of electrical rotating motors lead to the loss of production and income. Unfortunately, stator and rotor are the most important components in electrical rotating motors. Stator and rotor faults appear very often. Stators and rotors of electrical rotating motors must be monitored. Condition monitoring guarantees safe operation of machines and prevents unforeseen breakdowns. Acoustic signals contain a lot of diagnostic information, and can be used for detection of faults. Therefore, acoustic signals and signal processing methods should be deeply studied for proper recognition. Scientists developed many diagnostic methods of fault diagnosis. They are used for various types of machines and faults. Faults of electrical rotating motors (stator faults, rotor faults, broken rotor bar, ring cracking, bearing failures, rotor shaft failure, air-gap irregularities, broken teeth on sprocket) can be diagnosed by vibration [1–12] and acoustic signals [13–22]. Electric current analysis [23–31] and thermal analysis [32–34] are mostly used for limited faults, such as stator faults, rotor faults, and bearing failures. Acoustic signals are difficult to process, because they are very noisy (for example, several operating motors generate many acoustic signals). The advantage of acoustic-based fault diagnosis is non-invasive measurement (for example, we can measure an acoustic signal two meters from the machine). Vibration-based fault diagnosis is similar, but we have to put our measuring device close to the machine. Vibration signals are less noisy than acoustic signals.

The paper presents acoustic-based fault diagnosis of the commutator motor (CM). Five states of the commutator motor (CM) were considered: CM with shorted stator coils (Figures 1a and 2a), CM with broken rotor coil (Figures 1b and 2b), healthy CM (Figure 1c), CM with broken tooth on sprocket (Figure 3), CM with damaged gear train (Figure 4).

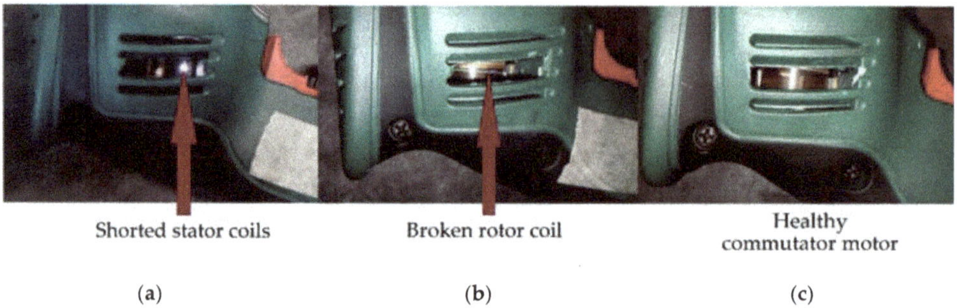

Figure 1. (**a**) Commutator motor (CM) with shorted stator coils; (**b**) CM with broken rotor coil; (**c**) healthy CM.

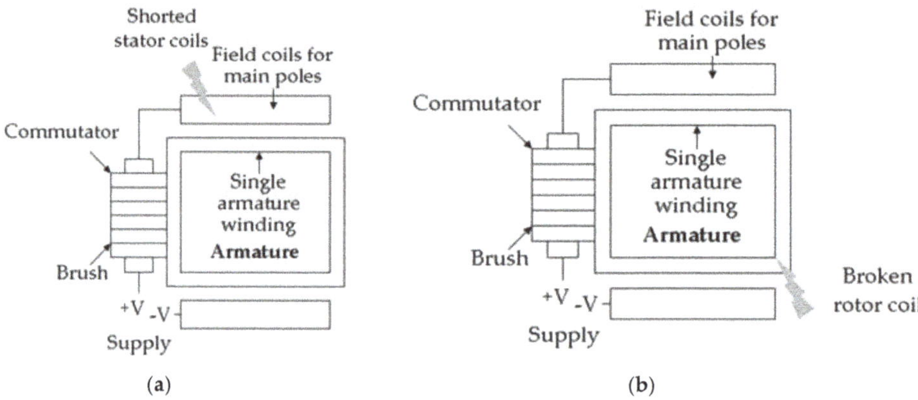

Figure 2. (**a**) CM with shorted stator coils; (**b**) CM with broken rotor coil.

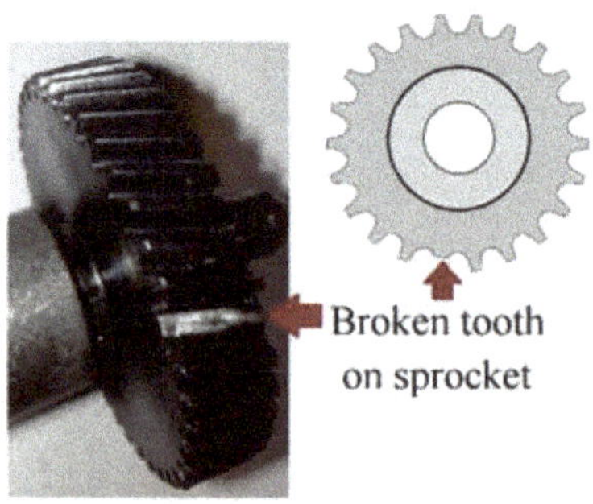

Figure 3. Broken tooth on sprocket of the electric impact drill.

Figure 4. Damaged gear train of the electric impact drill.

The described approach consists of methods of signal processing, such as amplitude normalization, FFT, the MSAF-15-MULTIEXPANDED-8-GROUPS, nearest neighbour (NN), or nearest mean (NM) or SOM (self-organizing map) or BNN (backpropagation neural network). The paper provided the results of acoustic analysis of the CM.

2. Acoustic-Based Fault Diagnosis Technique of the Commutator Motor

Acoustic-based fault diagnosis technique was based on pattern recognition. It used a pre-processing step, feature extraction step, and classification step. A block diagram of acoustic-based fault diagnosis technique was shown in Figure 5.

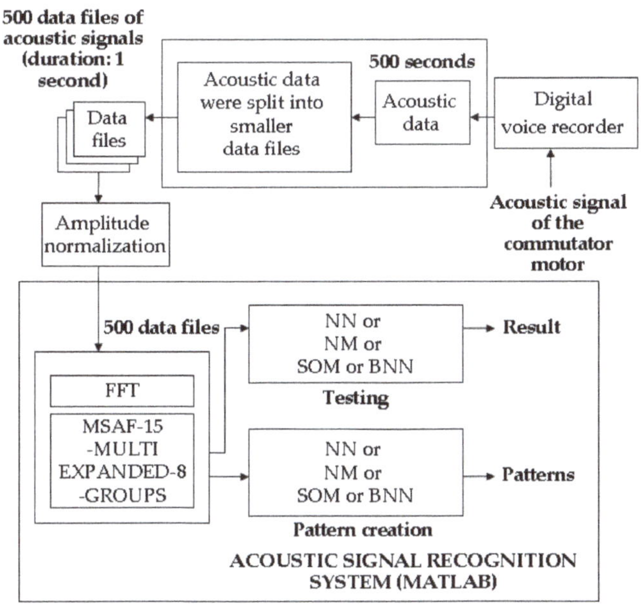

Figure 5. Flowchart of acoustic-based fault diagnosis techniques of the CM using the MSAF-15-MULTIEXPANDED-8-GROUPS.

Acoustic signals were measured using ZALMAN ZM-MIC1 microphone. The following steps of signal processing were used: recording of acoustic signal of the CM, split of soundtrack into smaller data files, amplitude normalization, FFT, the MSAF-15-MULTIEXPANDED-8-GROUPS, NN classifier or NM classifier or SOM or BNN. Recording of the acoustic signal of the CM was carried out using digital voice recorder (format: WAVE, number of channel-1, sampling frequency-44,100 Hz). There was also possibility to record acoustic signals using a capacity microphone with a computer (Figures 6 and 7). Next, splitting the soundtrack into smaller data files was carried out. The obtained data files (1 second samples) were processed by amplitude normalization, windowing (window size of 32,768), FFT, and the MSAF-15-MULTIEXPANDED-8-GROUPS. The MSAF-15-MULTIEXPANDED-8-GROUPS selected 1–15 frequency components, forming feature vectors consisting of 1–15 frequency components. Computed feature vectors were used for pattern creation and testing. Methods such as NN, NM, SOM, and BNN were used for data classification.

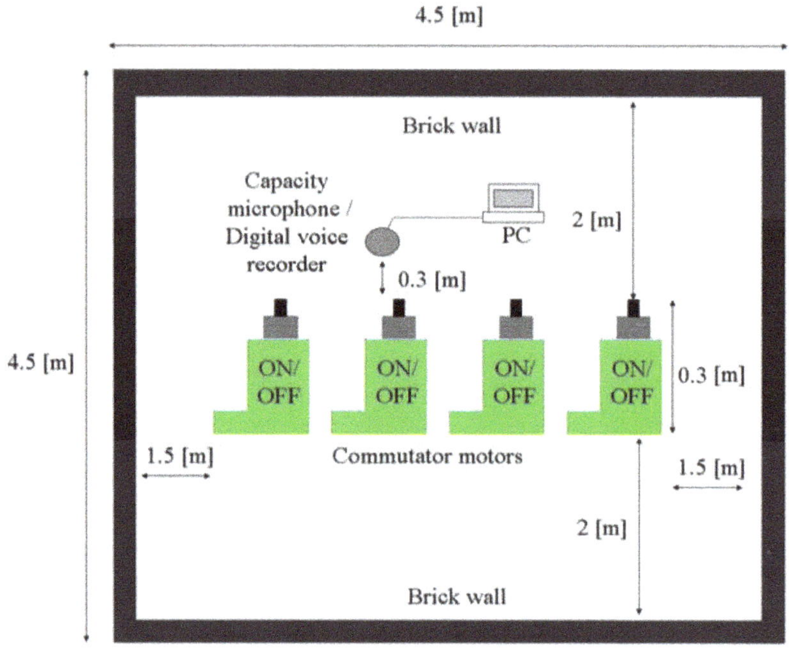

Figure 6. Experimental setup of analysis of acoustic signals of commutator motors.

Figure 7. Capacity microphone (ZALMAN ZM-MIC1 microphone) and the commutator motor (electric impact drill).

2.1. Method of Selection of Amplitudes of Frequency Multiexpanded 8 Groups

The Method of Selection of Amplitudes of Frequency Multiexpanded 8 Groups (MSAF-15-MULTIEXPANDED-8-GROUPS) depended on differences of spectra of acoustic signals. Differences of spectra of acoustic signals depended on generated acoustic signals of the CM. Generated acoustic signals depended on type of the motor, motor size, rotor speed, and analysed faults of the motor. The author analysed 5 states of the CM (healthy CM, CM with broken rotor coil, CM with shorted stator coils, CM with broken tooth on sprocket, and CM with damaged gear train). Steps of the MSAF-15-MULTIEXPANDED-8-GROUPS were listed below:

1. Compute frequency spectra of acoustic signals of commutator motors (the author used 6 one-second samples for state A, 6 one-second samples for state B, 6 one-second samples for state C, 6 one-second samples for state D). Computed frequency spectrum of state A (healthy CM) was described as vector of 16,384 elements **hcm** = [hcm_1, hcm_2, ..., hcm_{16384}]. Computed frequency spectrum of state B (CM with broken rotor coil) was denoted as vector of 16,384 elements **cmbrc** = [$cmbrc_1$, $cmbrc_2$, ..., $cmbrc_{16384}$]. Computed frequency spectrum of state C (CM with shorted stator coils) was expressed as vector of 16,384 elements **cmssc** = [$cmssc_1$, $cmssc_2$, ..., $cmssc_{16384}$]. Computed frequency spectrum of state D (CM with broken tooth on sprocket) was expressed as vector of 16384 elements **cmbts** = [$cmbts_1$, $cmbts_2$, ..., $cmbts_{16384}$]. Computed frequency spectrum of state E (CM with damaged gear train) was expressed as vector of 16,384 elements **cmdgt** = [$cmdgt_1$, $cmdgt_2$, ..., $cmdgt_{16384}$].
2. Compute differences of computed frequency spectra of states A, B, C, D, E: **hcm** − **cmbrc**, **hcm** − **cmssc**, **cmbrc** − **cmssc**, **cmbts** − **hcm**, **cmbts** − **cmbrc**, **cmbts** − **cmssc**, **cmdgt** − **hcm**, **cmdgt** − **cmbrc**, **cmdgt** − **cmssc**, **cmdgt** − **cmbts**.
3. Compute absolute values: |**hcm** − **cmbrc**|, |**hcm** − **cmssc**|, |**cmbrc** − **cmssc**|, |**cmbts** − **hcm**|, |**cmbts** − **cmbrc**|, |**cmbts** − **cmssc**|, |**cmdgt** − **hcm**|, |**cmdgt** − **cmbrc**|, |**cmdgt** − **cmssc**|, |**cmdgt** − **cmbts**|.
4. Select 15 maximum differences of computed frequency spectra of states A, B, C, D, E: $\max_1$(|**hcm** − **cmbrc**|), $\max_2$(|**hcm** − **cmbrc**|), ... , $\max_{15}$(|**hcm** − **cmbrc**|), $\max_1$(|**hcm** − **cmssc**|), $\max_2$(|**hcm** − **cmssc**|), ... , $\max_{15}$(|**hcm** − **cmssc**|), $\max_1$(|**cmbrc** − **cmssc**|), $\max_2$(|**cmbrc** − **cmssc**|), ... , $\max_{15}$(|**cmbrc** − **cmssc**|), $\max_1$(|**cmbts** − **hcm**|), $\max_2$(|**cmbts** − **hcm**|), ... , $\max_{15}$(|**cmbts** − **hcm**|), $\max_1$(|**cmbts** − **cmbrc**|), $\max_2$(|**cmbts** − **cmbrc**|), ... , $\max_{15}$(|**cmbts** − **cmbrc**|), $\max_1$(|**cmbts** − **cmssc**|), $\max_2$(|**cmbts** − **cmssc**|), ... , $\max_{15}$(|**cmbts** − **cmssc**|), $\max_1$(|**cmdgt** − **hcm**|), $\max_2$(|**cmdgt** − **hcm**|), ... , $\max_{15}$(|**cmdgt** − **hcm**|), $\max_1$(|**cmdgt** − **cmbrc**|), $\max_2$(|**cmdgt** − **cmbrc**|), ... , $\max_{15}$(|**cmdgt** − **cmbrc**|), $\max_1$(|**cmdgt** − **cmssc**|), $\max_2$(|**cmdgt** − **cmssc**|), ... , $\max_{15}$(|**cmdgt** − **cmssc**|), $\max_1$(|**cmdgt** − **cmbts**|), $\max_2$(|**cmdgt** − **cmbts**|), ... , $\max_{15}$(|**cmdgt** − **cmbts**|). The result of feature extraction method MSAF-15 is the vector consisted of 1–15 frequency components. Let us see following example using the MSAF-15. There are 5 states of the CM: A, B, C, D, E. Five frequency spectra of acoustic signals of the CM ((FS-A), (FS-B), (FS-C), (FS-D), (FS-E)-frequency spectra of states A, B, C, D, E) were computed. The MSAF-15 computed frequency components 100, 160, 200, 260, 300 Hz for difference |(FS-A) − (FS-B)|. The MSAF-15 computed frequency components 120, 160, 210, 260, 310 Hz for difference |(FS-A) − (FS-C)|. The MSAF-15 computed frequency components 120, 170, 220, 230, 310 Hz for difference |(FS-B) − (FS-C)|. The MSAF-15 computed frequency components 400, 410, 420, 430, 440 Hz for difference |(FS-D) − (FS-A)|. The MSAF-15 computed frequency components 405, 415, 425, 435, 445 Hz for difference |(FS-D) − (FS-B)|. The MSAF-15 computed frequency components 410, 415, 420, 425, 430 Hz for difference |(FS-D) − (FS-C)|. The MSAF-15 computed frequency components 500, 505, 510 Hz for difference |(FS-E) − (FS-A)|. The MSAF-15 computed frequency components 515, 520, 525 Hz for difference |(FS-E) − (FS-B)|. The MSAF-15 computed frequency components 530, 540, 550 Hz for difference |(FS-E) − (FS-C)|. The MSAF-15 computed frequency components 560, 570, 580 Hz for difference |(FS-E)

– (FS-D)|. None of common frequency components were selected for the presented example. The MSAF-15-MULTIEXPANDED-GROUPS extends the MSAF-15 method. A parameter called *TCoF-TS* (Threshold of common frequency components-training sets) was used. This parameter was defined as: *TCoF-TS* = (number of required common frequency components of analysed training sets)/(number of analysed differences).

5. Set the parameter *TCoF-TS*. This parameter affects the number of common frequency components. Let us consider following example using the MSAF-15-MULTIEXPANDED. Four training sets of acoustic training samples are given: (A1, B1, C1, D1, E1), (A2, B2, C2, D2, E2), (A3, B3, C3, D3, E3), (A4, B4, C4, D4, E4), where A1, A2, A3, A4—denoted 4 acoustic training samples of state A; B1, B2, B3, B4—denoted 4 acoustic training samples of state B; C1, C2, C3, C4—denoted 4 acoustic training samples of state C; D1, D2, D3, D4—denoted 4 acoustic training samples of state D; E1, E2, E3, E4—denoted 4 acoustic training samples of state E. The MSAF-15-MULTIEXPANDED computed frequency components (FS-A1, FS-B1, FS-C1, FS-D1, FS-E1), (FS-A2, FS-B2, FS-C2, FS-D2, FS-E2), (FS-A3, FS-B3, FS-C3, FS-D3, FS-E3), (FS-A4, FS-B4, FS-C4, FS-D4, FS-E4), where FS-A1, FS-A2, FS-A3, FS-A4—denoted 4 frequency spectra of state A, FS-B1, FS-B2, FS-B3, FS-B4—denoted 4 frequency spectra of state B, FS-C1, FS-C2, FS-C3, FS-C4—denoted 4 frequency spectra of state C, FS-D1, FS-D2, FS-D3, FS-D4—denoted 4 frequency spectra of state D, FS-E1, FS-E2, FS-E3, FS-E4—denoted 4 frequency spectra of state E. Next, 40 differences between frequency spectra are computed: |(FS-A1) – (FS-B1)|, |(FS-A1) – (FS-C1)|, |(FS-B1) – (FS-C1)|, |(FS-D1) – (FS-A1)|, |(FS-D1) – (FS-B1)|, |(FS-D1) – (FS-C1)|, |(FS-E1) – (FS-A1)|, |(FS-E1) – (FS-B1)|, |(FS-E1) – (FS-C1)|, |(FS-E1) – (FS-D1)|, ... , |(FS-A4) – (FS-B4)|, |(FS-A4) – (FS-C4)|, |(FS-B4) – (FS-C4)|, |(FS-D4) – (FS-A4)|, |(FS-D4) – (FS-B4)|, |(FS-D4) – (FS-C4)|, |(FS-E4) – (FS-A4)|, |(FS-E4) – (FS-B4)|, |(FS-E4) – (FS-C4)|, |(FS-E4) – (FS-D4)|. Let us consider following example. If we set *TCoF-TS* = 4/40 = 0.1, then the MSAF-15-MULTIEXPANDED selects frequency components found 4 times for 40 differences. If we set *TCoF-TS* = 6/40 = 0.15, then the MSAF-15-MULTIEXPANDED selects frequency components found 6 times for 40 differences. The MSAF-15-MULTIEXPANDED found frequency component 160 Hz-6 times, frequency component 210 Hz-4 times. The MSAF-15-MULTIEXPANDED selects 160, 210 Hz (if *TCoF-TS* = 4/40 = 0.1). The MSAF-15-MULTIEXPANDED selects 160 Hz (if *TCoF-TS* = 6/40 = 0.15). The MSAF-15-MULTIEXPANDED selects none of frequency components (if *TCoF-TS* = 8/40 = 0.2). The parameter *TCoF-TS* depends on analysed signal.
6. Select groups of common frequency components. The MSAF-15-MULTIEXPANDED-8-GROUPS used 8 groups. Each group of common frequency components consists of the best frequency components for recognition. Let us analyse following example. There are 4 states of the CM: A, B, C, D (for 5 states it will be similarly). The MSAF-15-MULTIEXPANDED-8-GROUPS found: frequency component 210 Hz (4 times), frequency component 160 Hz (6 times), frequency component 400 Hz (7 times). The frequency component 210 Hz is good for recognition of |(FS-A) – (FS-B)|, |(FS-A) – (FS-C)|, |(FS-B) – (FS-C)|. The frequency component 160 Hz is good for recognition of |(FS-D) – (FS-A)|, |(FS-D) – (FS-B)|, |(FS-D) – (FS-C)|. The frequency component 400 Hz is good for recognition of |(FS-A) – (FS-B)|. Essential frequency components are 210 Hz and 160 Hz. The frequency component 400 Hz is not good for analysis. The essential frequency components 160 Hz and 210 Hz form 1 group of essential frequency components.
7. Use 1–8 computed groups.
8. Form a feature vector.

The author presented the MSAF-15-MULTIEXPANDED-8-GROUPS in Figure 8.

The author used 30 one-second training samples for proper pattern creation process. The author used 6 training sets (30 one-second samples). Computed absolute values of differences: |**hcm** – **cmbrc**|, |**hcm** – **cmssc**|, |**cmbrc** – **cmssc**|, |**cmbts** – **hcm**|, |**cmbts** – **cmbrc**|, |**cmbts** – **cmssc**|, |**cmdgt** – **hcm**|, |**cmdgt** – **cmbrc**|, |**cmdgt** – **cmssc**|, |**cmdgt** – **cmbts**| were depicted in Figures 9–18 (rotor speed-3000 rpm, training set 6).

The MSAF-15-MULTIEXPANDED-8-GROUPS found 28 essential frequency components: 48, 50, 79, 81, 97, 101, 128, 157, 159, 1469, 1471, 1672, 1926, 1927, 1934, 1935, 1939, 1942, 1953, 1957, 1958, 1961, 1978, 2038, 2039, 2042, 2059, 2547 Hz for *TCoF-TS* = 0.125 (3/18 = 0.125). Computed essential frequency components were presented in Tables 1–5 (Figures 19–23).

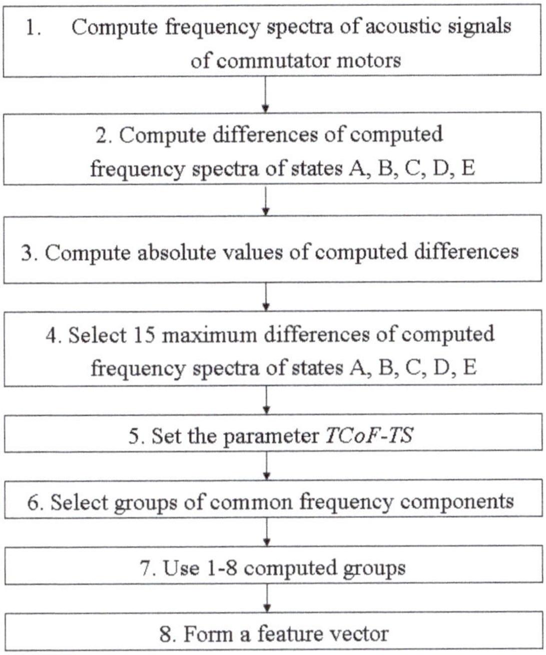

Figure 8. Flowchart of the MSAF-15-MULTIEXPANDED-8-GROUPS.

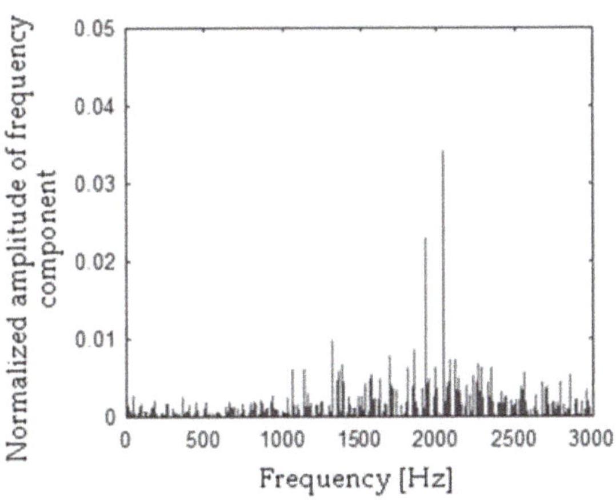

Figure 9. Absolute values of difference of frequency spectra (|**hcm** − **cmbrc**|) using the MSAF-15-MULTIEXPANDED-8-GROUPS.

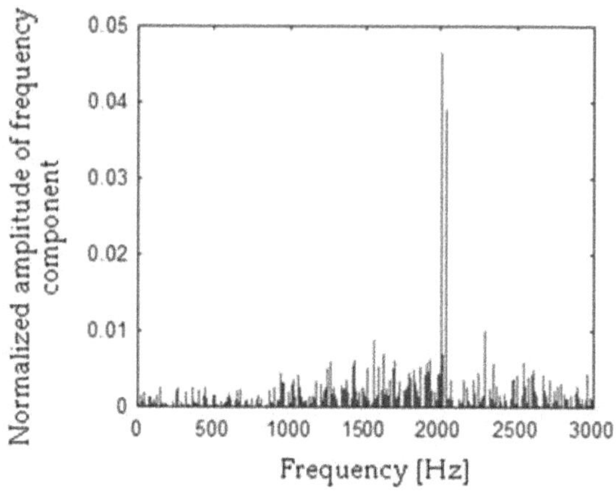

Figure 10. Absolute values of difference of frequency spectra (|**hcm** − **cmssc**|) using the MSAF-15-MULTIEXPANDED-8-GROUPS.

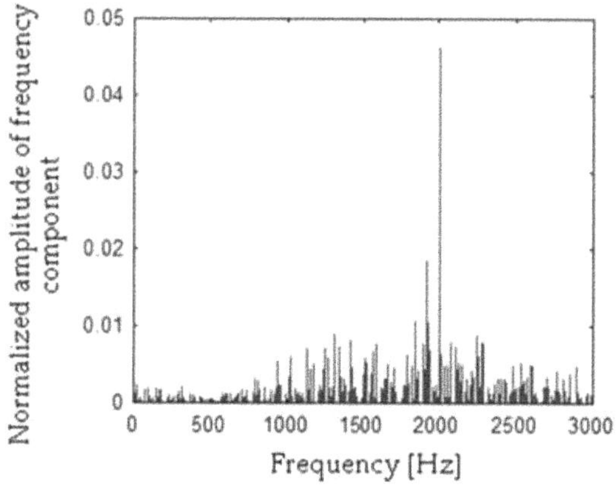

Figure 11. Absolute values of difference of frequency spectra (|**cmbrc** − **cmssc**|) using the MSAF-15-MULTIEXPANDED-8-GROUPS.

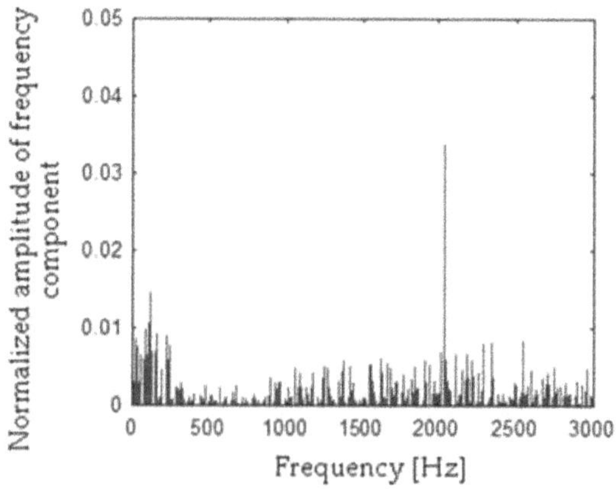

Figure 12. Absolute values of difference of frequency spectra (|**cmbts** − **hcm**|) using the MSAF-15-MULTIEXPANDED-8-GROUPS.

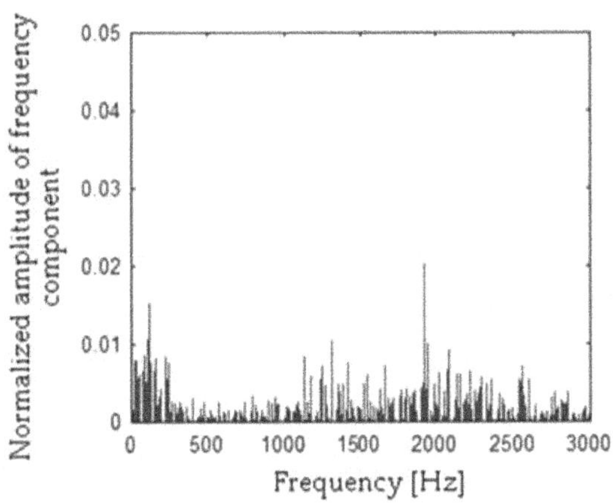

Figure 13. Absolute values of difference of frequency spectra (|**cmbts** − **cmbrc**|) using the MSAF-15-MULTIEXPANDED-8-GROUPS.

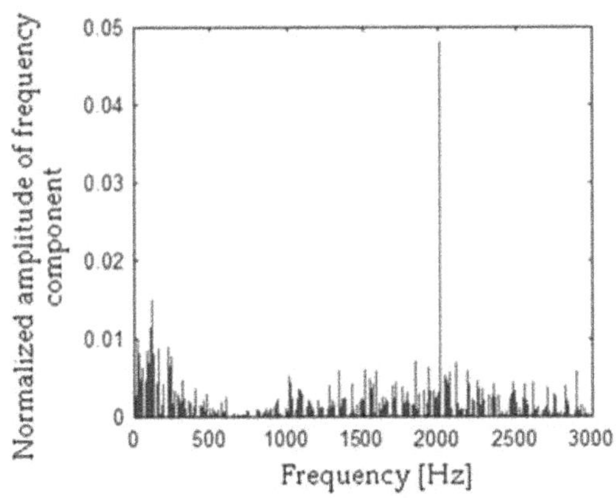

Figure 14. Absolute values of difference of frequency spectra (|**cmbts** − **cmssc**|) using the MSAF-15-MULTIEXPANDED-8-GROUPS.

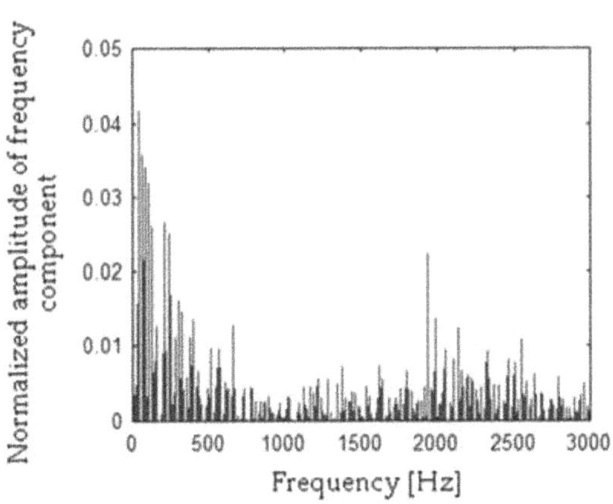

Figure 15. Absolute values of difference of frequency spectra (|**cmdgt** − **hcm**|) using the MSAF-15-MULTIEXPANDED-8-GROUPS.

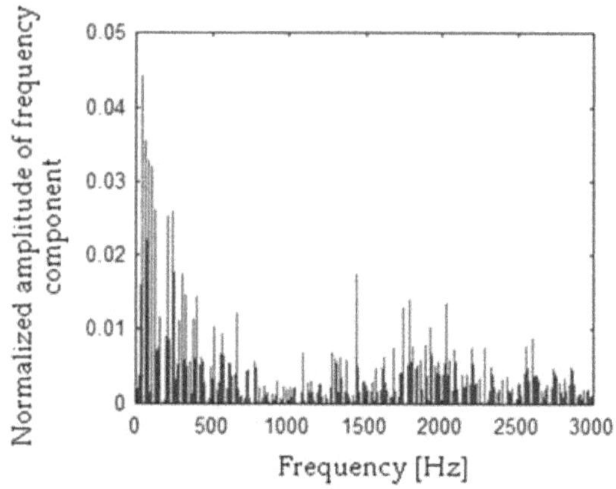

Figure 16. Absolute values of difference of frequency spectra (| **cmdgt** − **cmbrc** |) using the MSAF-15-MULTIEXPANDED-8-GROUPS.

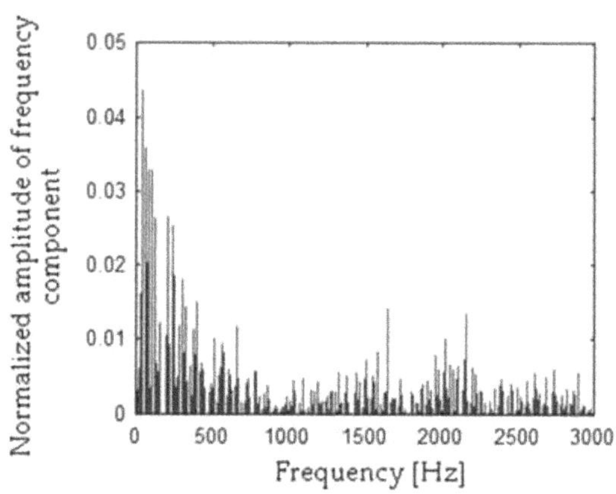

Figure 17. Absolute values of difference of frequency spectra (| **cmdgt** − **cmssc** |) using the MSAF-15-MULTIEXPANDED-8-GROUPS.

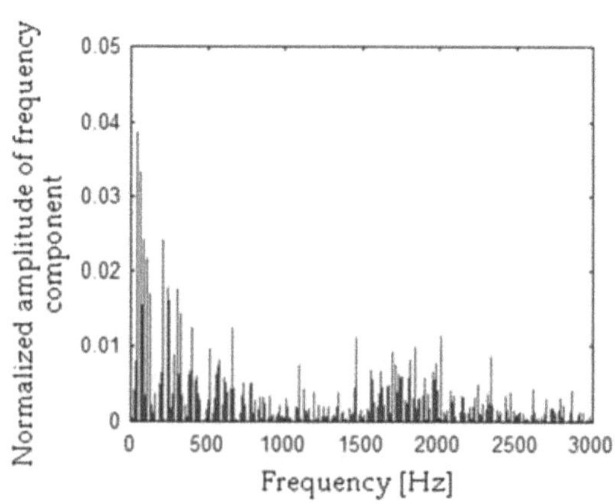

Figure 18. Absolute values of difference of frequency spectra (| **cmdgt** − **cmbts** |) using the MSAF-15-MULTIEXPANDED-8-GROUPS.

Table 1. Computed essential frequency components for vector **hcm** (healthy CM).

Value of Feature			
0.006776	0.011506	0.003938	0.006896
0.003093	0.007896	0.005020	0.006722
0.009599	0.006721	0.034127	0.041374
0.006398	0.001086	0.006096	0.002857
0.004421	0.002796	0.005071	0.005120
0.006776	0.011506	0.003938	0.006896
0.003093	0.007896	0.005020	0.006722

Table 2. Computed essential frequency components for vector **cmbrc** (CM with broken rotor coil).

Value of Feature			
0.005696	0.009528	0.026896	0.004757
0.005556	0.010468	0.026023	0.012042
0.003759	0.009350	0.006706	0.007029
0.005175	0.006020	0.013454	0.003632
0.008541	0.000666	0.005728	0.005724
0.005696	0.009528	0.026896	0.004757
0.005556	0.010468	0.026023	0.012042

Table 3. Computed essential frequency components for vector **cmssc** (CM with shorted stator coils).

Value of Feature			
0.002803	0.002988	0.008468	0.004678
0.004008	0.002770	0.001949	0.006048
0.021171	0.004456	0.005253	0.002312
0.005484	0.005549	0.003781	0.004390
0.005037	0.000740	0.008683	0.004213
0.002803	0.002988	0.008468	0.004678
0.004008	0.002770	0.001949	0.006048

Table 4. Computed essential frequency components for vector **cmbts** (CM with broken tooth on sprocket).

Value of Feature			
0.007080	0.007388	0.006834	0.001410
0.006244	0.007172	0.003946	0.005266
0.002335	0.002029	0.004501	0.007593
0.020768	0.016296	0.016799	0.018125
0.003584	0.011559	0.007451	0.003169
0.007080	0.007388	0.006834	0.001410
0.006244	0.007172	0.003946	0.005266

Table 5. Computed essential frequency components for vector **cmdgt** (CM with damaged gear train).

Value of Feature			
0.002066	0.002245	0.001124	0.000820
0.000845	0.000655	0.002080	0.001540
0.002178	0.001563	0.000810	0.001933
0.002259	0.001234	0.002635	0.003116
0.002729	0.006634	0.000739	0.004000
0.051576	0.012309	0.007059	0.015156
0.019388	0.008816	0.028623	0.016604

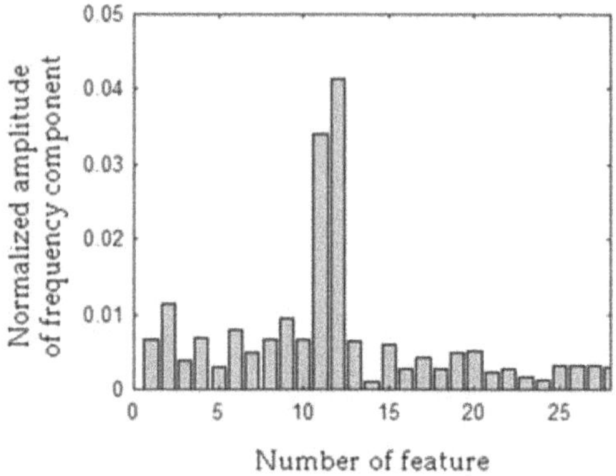

Figure 19. Computed essential frequency components for vector **hcm** (healthy CM).

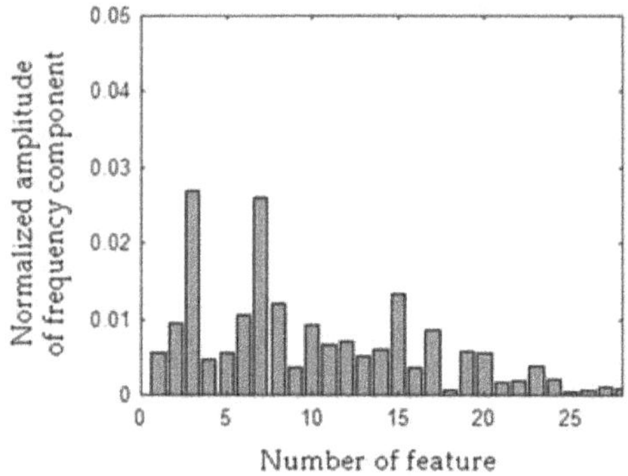

Figure 20. Computed essential frequency components for vector **cmbrc** (CM with broken rotor coil).

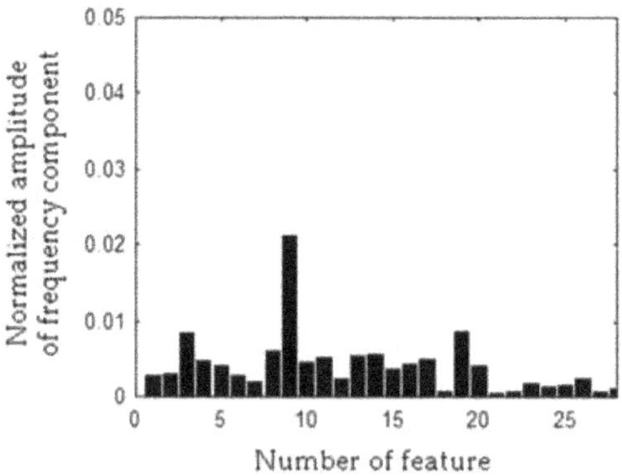

Figure 21. Computed essential frequency components for vector **cmssc** (CM with shorted stator coils).

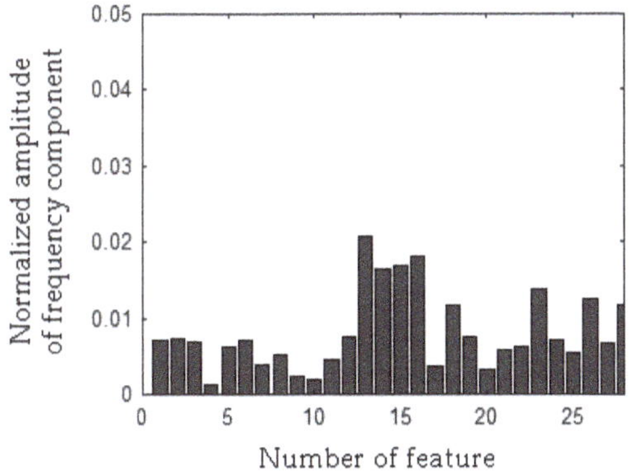

Figure 22. Computed essential frequency components for vector **cmbts** (CM with broken tooth on sprocket).

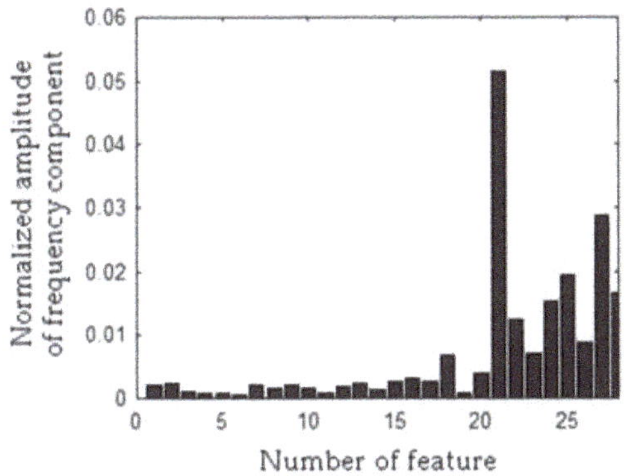

Figure 23. Computed essential frequency components for vector **cmdgt** (CM with damaged gear train).

Found essential frequency components were classified by the NN classifier [35,36], NM classifier, SOM [37], BNN [38–44]. There was possibility to use another classifier such as naive Bayes, support vector machine [45–47], linear discriminant analysis [48], fuzzy classifiers [49,50], and fuzzy c-means clustering [51]. The results of recognition depended on number of found essential frequency components and selected classification method.

2.2. Nearest Neighbour Classifier

A classification step can be achieved by the nearest neighbour (NN) classifier. This method of classification is well-known. It is used in economics, telecommunication, pattern recognition, fault diagnosis, and image recognition. The NN classifier can classify data (linearly separable and non-linearly separable) with high recognition rate. The NN is simple to implement, and it requires a few training feature vectors for proper classification. It uses labels for classification of test feature vectors. The classifier uses metric distance to compare two vectors (training and test feature vectors). There are many distance metrics to compare training and test vector. Distance metrics such as Euclidean, Manhattan, and Minkowski had similar results. In this paper, the classification step was carried out using Manhattan distance (1):

$$d(\mathbf{x}-\mathbf{cmbrc}) = \sum_{i=1}^{1} |(x_i - cmbrc_i)| \quad (1)$$

where d(**x-cmbrc**)—computed distance, unknown test feature vector $\mathbf{x} = [x_{36}, x_{37}, x_{59}, x_{60}, x_{72}, x_{75}, x_{95}, x_{117}, x_{118}, x_{1092}, x_{1094}, x_{1243}, x_{1432}, x_{1433}, x_{1438}, x_{1439}, x_{1442}, x_{1444}, x_{1452}, x_{1455}, x_{1456}, x_{1458}, x_{1471}, x_{1515}, x_{1516}, x_{1518}, x_{1531}, x_{1894}]$ and training feature vector $\mathbf{cmbrc} = [cmbrc_{36}, cmbrc_{37}, cmbrc_{59}, cmbrc_{60}, cmbrc_{72}, cmbrc_{75}, cmbrc_{95}, cmbrc_{117}, cmbrc_{118}, cmbrc_{1092}, cmbrc_{1094}, cmbrc_{1243}, cmbrc_{1432}, cmbrc_{1433}, cmbrc_{1438}, cmbrc_{1439}, cmbrc_{1442}, cmbrc_{1444}, cmbrc_{1452}, cmbrc_{1455}, cmbrc_{1456}, cmbrc_{1458}, cmbrc_{1471}, cmbrc_{1515}, cmbrc_{1516}, cmbrc_{1518}, cmbrc_{1531}, cmbrc_{1894}]$. The result of classification was depended on the nearest distance $d()$. Description of the NN classifier is available in following articles [35,36].

2.3. Nearest Mean Classifier

Similar to the NN classifier, the nearest mean (NM) classifier is also based on computed distance. It uses average feature vector instead of training feature vectors. Average feature vector **afv** is defined as (2)

$$\mathbf{afv} = \frac{1}{p}\sum_{i=1}^{p} y_i \tag{2}$$

where **afv**—average feature vector, p—number of essential frequency components, and y_i—value of essential frequency component with i index.

The nearest distance between test and average feature vector is computed. Next, the label occurring with specific average feature vector is the label for the test feature vector. The NM classifier can classify data with a high recognition rate. For the classification step, the author used Manhattan distance (1).

2.4. Self-Organizing Map

The self-organizing map was used for machine learning. The self-organizing map (SOM) is a clustering method. It does not use labels for classification of test feature vectors. The SOM is an unsupervised neural network. It is used for clustering data, meteorology and oceanography, project prioritization, selection, and failure analysis. For example, the SOM is used for meteorological applications such as climate change analysis, precipitation, snow, wind, air temperature, etc. The SOM analysis has been applied for computed feature vectors of acoustic signals. Similarity between feature vectors depends on the location of features on the two-dimensional map (nodes). The SOM uses training step and testing step. In the training step, weights of nodes are changed (at the beginning, values of weights are random). The author used the following self-organizing map (Figure 24). It was implemented in MATLAB. Description of the SOM is available in the following article [37].

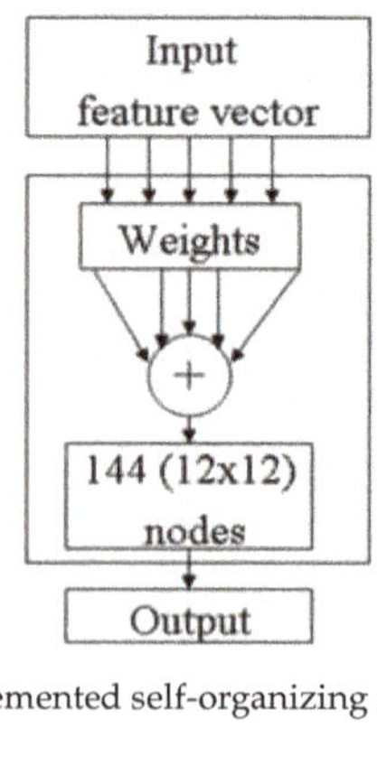

Figure 24. Implemented self-organizing map—144 nodes.

2.5. Backpropagation Neural Network

Backpropagation neural network (BNN) was also used for machine learning. It is a supervised learning method. It is also a well-known method of data classification. It has been used for many applications, such as speaker recognition, image recognition, signal recognition, control, prediction, computer games, robots, etc. The applied backpropagation algorithm has been described in the literature [38–44]. The author implemented a backpropagation neural network (Figure 25).

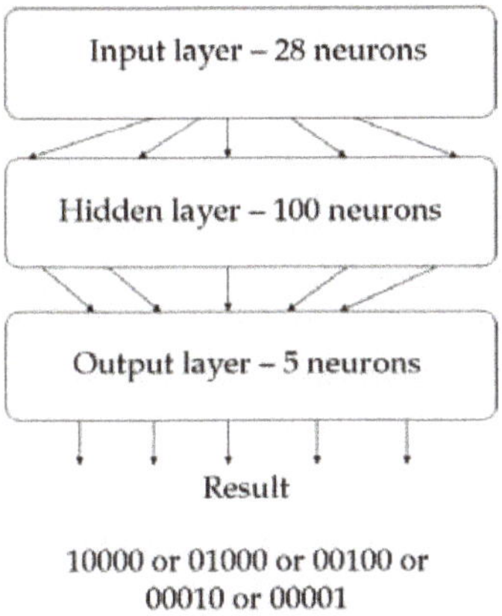

Figure 25. Implemented backpropagation neural network (BNN).

Layers of BNN had following number of neurons: input layer—28, hidden layer—100, output layer—5. The values of output neurons were 10000—healthy CM, 01000—CM with shorted stator coils, 00100—CM with broken rotor coil, 00010—CM with broken tooth on sprocket, and 00001—CM with damaged gear train. More information about BNN can be found in following papers [38–44].

3. Results of Acoustic-Based Fault Diagnosis Technique of the Commutator Motor

Measurements of acoustic signals of commutator motors were conducted in the workshop. The author measured and analysed 5 states of the CM: healthy CM, CM with shorted stator coils (Figure 26a), CM with broken rotor coil (Figure 26b), CM with broken tooth on sprocket (Figure 27), and CM with damaged gear train (Figure 28). Analysed commutator motors had the following parameters: W_{oM} = 1.84 kg, P_{oM} = 500 W, RS_{oM} = 3000 rpm, V_{oM} = 230 V, f_{oM} = 50 Hz, where W_{oM}—weight of the motor, P_{oM}—rated power of the motor, RS_{oM}—rotor speed, V_{oM}—supply voltage of the motor, and f_{oM}—current frequency of the motor.

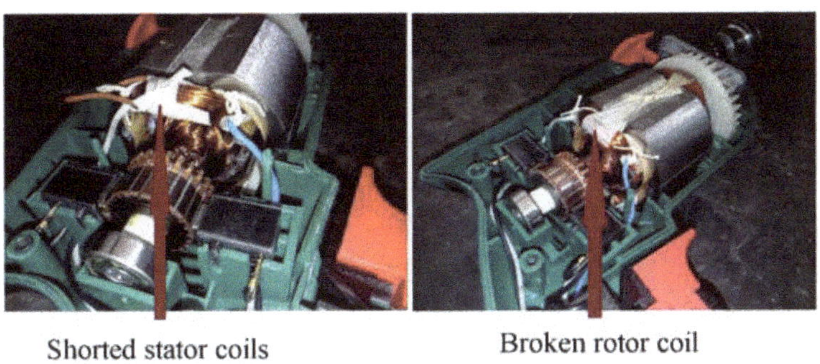

(**a**) (**b**)

Figure 26. (**a**) CM with shorted stator coils; (**b**) CM with broken rotor coil.

Figure 27. CM with broken tooth on sprocket.

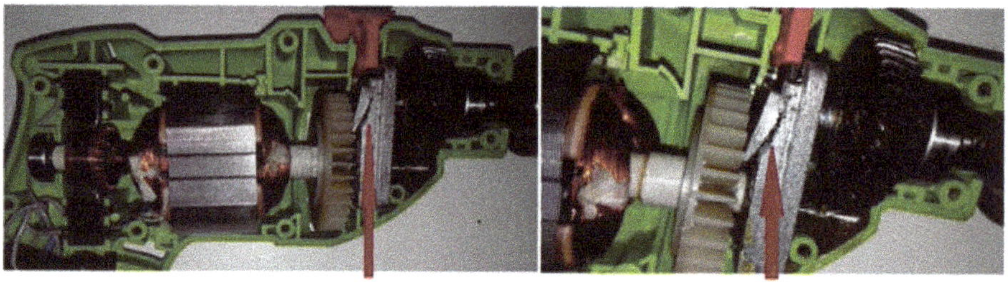

Figure 28. CM with damaged gear train.

The author used 30 one-second training samples for proper pattern creation process. The author used 500 one-second test samples for proper testing process. Training and test samples of the CM were processed and analysed. The author used technique presented in Section 2 for proper fault diagnosis. Acoustic data of the CM were analysed using efficiency of recognition (E_{RCM}). It was defined as (3):

$$E_{RCM} = (N_{TSCMTP})/(N_{ATSCM}) \cdot 100\% \tag{3}$$

where: E_{RCM}—efficiency of recognition of the CM for defined class, N_{TSCMTP}—number of test samples of the CM for defined class tested properly, N_{ATSCM}—number of all test samples of the CM for defined class.

Total efficiency of recognition of the CM (TE_{RCM}) was introduced to evaluate efficiency of recognition of all states of the CM. It was expressed as (4):

$$TE_{RCM} = (E_{RCM1} + E_{RCM2} + E_{RCM3} + E_{RCM4} + E_{RCM5})/5 \tag{4}$$

where TE_{RCM}—total efficiency of recognition of the CM, E_{RCM1}—efficiency of recognition of the healthy CM, E_{RCM2}—efficiency of recognition of the CM with broken rotor coil, E_{RCM3}—efficiency of recognition of the CM with shorted stator coils, E_{RCM4}—efficiency of recognition of the CM with broken tooth on sprocket, and E_{RCM5}—efficiency of recognition of the CM with damaged gear train.

Acoustic signal analysis of the CM is presented in Tables 6–9. The acoustic signal analysis of the CM using the MSAF-15-MULTIEXPANDED-8-GROUPS and the NN are presented in Table 6 (28 analysed frequency components).

Table 6. Acoustic signal analysis of the CM using the MSAF-15-MULTIEXPANDED-8-GROUPS and the NN.

Type of Acoustic Signal	E_{RCM} [%]
Healthy CM	90
CM with broken rotor coil	87
CM with shorted stator coils	94
CM with broken tooth on sprocket	100
CM with damaged gear train	100
	TE_{RCM} [%]
	94.2

Table 7. Acoustic signal analysis of the CM using the MSAF-15-MULTIEXPANDED-8-GROUPS and the NM.

Type of Acoustic Signal	E_{RCM} [%]
Healthy CM	89
CM with broken rotor coil	91
CM with shorted stator coils	93
CM with broken tooth on sprocket	100
CM with damaged gear train	100
	TE_{RCM} [%]
	94.6

Table 8. Acoustic signal analysis of the CM using the MSAF-15-MULTIEXPANDED-8-GROUPS and the SOM.

Type of Acoustic Signal	E_{RCM} [%]
Healthy CM	87
CM with broken rotor coil	86
CM with shorted stator coils	81
CM with broken tooth on sprocket	90
CM with damaged gear train	98
	TE_{RCM} [%]
	88.4

Table 9. Acoustic signal analysis of the CM using the MSAF-15-MULTIEXPANDED-8-GROUPS and the BNN.

Type of Acoustic Signal	E_{RCM} [%]
Healthy CM	85
CM with broken rotor coil	88
CM with shorted stator coils	82
CM with broken tooth on sprocket	91
CM with damaged gear train	99
	TE_{RCM} [%]
	89

The acoustic signal analysis of the CM using the MSAF-15-MULTIEXPANDED-8-GROUPS and the NM are shown in Table 7 (28 analysed frequency components).

The acoustic signal analysis of the CM using the MSAF-15-MULTIEXPANDED-8-GROUPS and the SOM were presented in Table 8 (28 analysed frequency components).

The acoustic signal analysis of the CM using the MSAF-15-MULTIEXPANDED-8-GROUPS and the BNN were presented in Table 9 (28 analysed frequency components).

The method of feature extraction the MSAF-15-MULTIEXPANDED-8-GROUPS and selected classifiers provided high recognition rates (TE_{RCM} in the range of 88.4–94.6%).

The MSAF-15-MULTIEXPANDED-8-GROUPS used 28 frequency components of the CM. The NN classifier and the MSAF-15-MULTIEXPANDED-8-GROUPS provided TE_{RCM} = 94.2%. The NM classifier and the MSAF-15-MULTIEXPANDED-8-GROUPS provided TE_{RCM} = 94.6%. The SOM (self-organizing map) and the MSAF-15-MULTIEXPANDED-8-GROUPS provided TE_{RCM} = 88.4%. The BNN and the MSAF-15-MULTIEXPANDED-8-GROUPS provided TE_{RCM} = 89%.

Self-organizing map and backpropagation neural network are trained. The training is different each time. The NN classifier and the NM classifier had the same results each time. Moreover, feature vectors had small differences between them. The NN classifier and the NM classifier were better for the recognition of close feature vectors.

4. Conclusions

The article presented acoustic-based fault diagnosis technique of the CM. Five states of the CM were considered: healthy CM, CM with broken rotor coil, CM with shorted stator coils, CM with broken tooth on sprocket, and CM with damaged gear train. The method of feature extraction MSAF-15-MULTIEXPANDED-8-GROUPS was described and implemented. Classifiers NN, NM, SOM, and BNN were used for acoustic analysis of the CM.

Analysed values of TE_{RCM} were in the range of 88.4–94.6%. The NM classifier and the MSAF-15-MULTIEXPANDED-8-GROUPS provided TE_{RCM} = 94.6%. The implementation of the proposed fault diagnosis technique had low cost. Laptop and microphones are available for $300. Other types of rotating electric motors (such as DC motors, synchronous motors, induction motors) can also be diagnosed using acoustic analysis.

The proposed acoustic-based fault diagnosis technique has its limitations. If the motor runs too quietly, it is difficult to use the mentioned technique and microphone. However, the presented acoustic-based fault diagnosis technique is appropriate for acoustic signals of rotating motor and other types of acoustic signals (for example acoustic signal of an engine). The proposed acoustic-based fault diagnosis technique can be extended to detect more complicated faults. Future research will focus on the analysis of new feature extraction methods, other types of faults, and other diagnostic signals, such as vibrations.

Funding: This research was funded by AGH University of Science and Technology, grant No. 11.11.120.714.

Acknowledgments: This work has been supported by AGH University of Science and Technology, grant No. 11.11.120.714. The author thanks unknown reviewers for their valuable suggestions.

Conflicts of Interest: The author declares no conflict of interest.

References

1. Li, Z.X.; Jiang, Y.; Hu, C.; Peng, Z. Recent progress on decoupling diagnosis of hybrid failures in gear transmission systems using vibration sensor signal: A review. *Measurement* **2016**, *90*, 4–19. [CrossRef]
2. Kim, K.C. Analysis on Electromagnetic Vibration for Interior Permanent Magnet Synchronous Motor Due to Temperature and Loads. *Adv. Sci. Lett.* **2017**, *23*, 9767–9772. [CrossRef]
3. Zhao, H.M.; Deng, W.; Yang, X.H.; Li, X.M. Research on a vibration signal analysis method for motor bearing. *Optik* **2016**, *127*, 10014–10023. [CrossRef]
4. Moosavian, A.; Najafi, G.; Ghobadian, B.; Mirsalim, M. The effect of piston scratching fault on the vibration behavior of an IC engine. *Appl. Acoust.* **2017**, *126*, 91–100. [CrossRef]
5. Cruz-Vega, I.; Rangel-Magdaleno, J.; Ramirez-Cortes, J.; Peregrina-Barreto, H. Automatic progressive damage detection of rotor bar in induction motor using vibration analysis and multiple classifiers. *J. Mech. Sci. Technol.* **2017**, *31*, 2651–2662. [CrossRef]
6. Armentani, E.; Sepe, R.; Parente, A.; Pirelli, M. Vibro-Acoustic Numerical Analysis for the Chain Cover of a Car Engine. *Appl. Sci.* **2017**, *7*, 610. [CrossRef]
7. Jafarian, K.; Mobin, M.; Jafari-Marandi, R.; Rabiei, E. Misfire and valve clearance faults detection in the combustion engines based on a multi-sensor vibration signal monitoring. *Measurement* **2018**, *128*, 527–536. [CrossRef]

8. Siljak, H.; Subasi, A. Berthil cepstrum: A novel vibration analysis method based on marginal Hilbert spectrum applied to artificial motor aging. *Electr. Eng.* **2018**, *100*, 1039–1046. [CrossRef]
9. Zurita-Millan, D.; Delgado-Prieto, M.; Saucedo-Dorantes, J.J.; Carino-Corrales, J.A.; Osornio-Rios, R.A.; Ortega-Redondo, J.A.; Romero-Troncoso, R.D. Vibration Signal Forecasting on Rotating Machinery by means of Signal Decomposition and Neurofuzzy Modeling. *Shock Vib.* **2016**, *2016*, 1–13. [CrossRef]
10. Martinez, J.; Belahcen, A.; Muetze, A. Analysis of the Vibration Magnitude of an Induction Motor with Different Numbers of Broken Bars. *IEEE Trans. Ind. Appl.* **2017**, *53*, 2711–2720. [CrossRef]
11. Saucedo-Dorantes, J.J.; Delgado-Prieto, M.; Ortega-Redondo, J.A.; Osornio-Rios, R.A.; Romero-Troncoso, R.D. Multiple-Fault Detection Methodology Based on Vibration and Current Analysis Applied to Bearings in Induction Motors and Gearboxes on the Kinematic Chain. *Shock Vib.* **2016**, *2016*, 1–13. [CrossRef]
12. Li, Y.; Chai, F.; Song, Z.X.; Li, Z.Y. Analysis of Vibrations in Interior Permanent Magnet Synchronous Motors Considering Air-Gap Deformation. *Energies* **2017**, *10*, 1259. [CrossRef]
13. Delgado-Arredondo, P.A.; Morinigo-Sotelo, D.; Osornio-Rios, R.A.; Avina-Cervantes, J.G.; Rostro-Gonzalez, H.; Romero-Troncoso, R.D. Methodology for fault detection in induction motors via sound and vibration signals. *Mech. Syst. Signal Process.* **2017**, *83*, 568–589. [CrossRef]
14. Dong, J.N.; Jiang, J.W.; Howey, B.; Li, H.D.; Bilgin, B.; Callegaro, A.D.; Emadi, A. Hybrid Acoustic Noise Analysis Approach of Conventional and Mutually Coupled Switched Reluctance Motors. *IEEE Trans. Energy Convers.* **2017**, *32*, 1042–1051. [CrossRef]
15. Stief, A.; Ottewill, J.R.; Orkisz, M.; Baranowski, J. Two Stage Data Fusion of Acoustic, Electric and Vibration Signals for Diagnosing Faults in Induction Motors. *Elektronika ir Elektrotechnika* **2017**, *23*, 19–24. [CrossRef]
16. Xia, K.; Li, Z.R.; Lu, J.; Dong, B.; Bi, C. Acoustic Noise of Brushless DC Motors Induced by Electromagnetic Torque Ripple. *J. Power Electron.* **2017**, *17*, 963–971. [CrossRef]
17. Sangeetha, P.; Hemamalini, S. Dyadic wavelet transform-based acoustic signal analysis for torque prediction of a three-phase induction motor. *IET Signal Process.* **2017**, *11*, 604–612. [CrossRef]
18. Prainetr, S.; Wangnippanto, S.; Tunyasirut, S. Detection Mechanical Fault of Induction Motor Using Harmonic Current and Sound Acoustic. In Proceedings of the 2017 International Electrical Engineering Congress (iEECON), Pattaya, Thailand, 8–10 March 2017.
19. Uekita, M.; Takaya, Y. Tool condition monitoring for form milling of large parts by combining spindle motor current and acoustic emission signals. *Int. J. Adv. Manuf. Technol.* **2017**, *89*, 65–75. [CrossRef]
20. Baghayipour, M.; Darabi, A.; Dastfan, A. An Experimental Model for Extraction of the Natural Frequencies influencing on the Acoustic Noise of Synchronous Motors. In Proceedings of the 8th Power Electronics, Drive Systems & Technologies Conference (PEDSTC), Mashhad, Iran, 14–16 February 2017; pp. 125–130.
21. Islam, M.R.; Uddin, J.; Kim, J.M. Acoustic Emission Sensor Network Based Fault Diagnosis of Induction Motors Using a Gabor Filter and Multiclass Support Vector Machines. *Ad Hoc Sens. Wirel. Netw.* **2016**, *34*, 273–287.
22. Binojkumar, A.C.; Saritha, B.; Narayanan, G. Experimental Comparison of Conventional and Bus-Clamping PWM Methods Based on Electrical and Acoustic Noise Spectra of Induction Motor Drives. *IEEE Trans. Ind. Appl.* **2016**, *52*, 4061–4073. [CrossRef]
23. Singh, G.; Naikan, V.N.A. Detection of half broken rotor bar fault in VFD driven induction motor drive using motor square current MUSIC analysis. *Mech. Syst. Signal Process.* **2018**, *110*, 333–348. [CrossRef]
24. Aimer, A.F.; Boudinar, A.H.; Benouzza, N.; Bendiabdellah, A. Use of the root-ar method in the diagnosis of induction motor's mechanical faults. *Revue Roumaine des Sciences Techniques, Serie Electrotechnique et Energetique* **2017**, *62*, 134–141.
25. Tian, L.S.; Wu, F.; Shi, Y.; Zhao, J. A Current Dynamic Analysis Based Open-Circuit Fault Diagnosis Method in Voltage-Source Inverter Fed Induction Motors. *J. Power Electron.* **2017**, *17*, 725–732. [CrossRef]
26. Gangsar, P.; Tiwari, R. Analysis of Time, Frequency and Wavelet Based Features of Vibration and Current Signals for Fault Diagnosis of Induction Motors Using SVM. In Proceedings of the ASME Gas Turbine India Conference, Bangalore, India, 7–8 December 2017.
27. Yu, L.; Zhang, Y.T.; Huang, W.Q.; Teffah, K. A Fast-Acting Diagnostic Algorithm of Insulated Gate Bipolar Transistor Open Circuit Faults for Power Inverters in Electric Vehicles. *Energies* **2017**, *10*, 552. [CrossRef]
28. Sun, H.; Yuan, S.Q.; Luo, Y.; Guo, Y.H.; Yin, J.N. Unsteady characteristics analysis of centrifugal pump operation based on motor stator current. *Proc. Inst. Mech. Eng. Part A J. Power Energy* **2017**, *231*, 689–705. [CrossRef]

29. Spanik, P.; Sedo, J.; Drgona, P.; Frivaldsky, M. Real Time Harmonic Analysis of Recuperative Current through Utilization of Digital Measuring Equipment. *Elektronika ir Elektrotechnika* **2013**, *19*, 33–38. [CrossRef]
30. Glowacz, A.; Glowacz, W.; Glowacz, Z. Recognition of armature current of DC generator depending on rotor speed using FFT, MSAF-1 and LDA. *Eksploatacja i Niezawodnosc* **2015**, *17*, 64–69. [CrossRef]
31. Gutten, M.; Korenciak, D.; Kucera, M.; Sebok, M.; Opielak, M.; Zukowski, P.; Koltunowicz, T.N. Maintenance diagnostics of transformers considering the influence of short-circuit currents during operation. *Eksploatacja i Niezawodnosc* **2017**, *19*, 459–466. [CrossRef]
32. Lopez-Perez, D.; Antonino-Daviu, J. Application of Infrared Thermography to Failure Detection in Industrial Induction Motors: Case Stories. *IEEE Trans. Ind. Appl.* **2017**, *53*, 1901–1908. [CrossRef]
33. Resendiz-Ochoa, E.; Osornio-Rios, R.A.; Benitez-Rangel, J.P.; Morales-Hernandez, L.A.; Romero-Troncoso, R.D. Segmentation in Thermography Images for Bearing Defect Analysis in Induction Motors. In Proceedings of the 2017 IEEE 11th International Symposium on Diagnostics for Electrical Machines, Power Electronics and Drives (SDEMPED), Tinos, Greece, 29 August–1 September 2017; pp. 572–577.
34. Munoz-Ornelas, O.; Elvira-Ortiz, D.A.; Osornio-Rios, R.A.; Romero-Troncoso, R.J.; Morales-Hernandez, L.A. Methodology for Thermal Analysis of Induction Motors with Infrared Thermography Considering Camera Location. In Proceedings of the IECON 2016—42nd Annual Conference of the IEEE Industrial Electronics Society, Florence, Italy, 23–26 October 2016; pp. 7113–7118.
35. Dong, S.J.; Luo, T.H.; Zhong, L.; Chen, L.L.; Xu, X.Y. Fault diagnosis of bearing based on the kernel principal component analysis and optimized k-nearest neighbour model. *J. Low Freq. Noise Vib. Act. Control* **2017**, *36*, 354–365. [CrossRef]
36. Godoy, W.F.; da Silva, I.N.; Goedtel, A.; Palacios, R.H.C.; Lopes, T.D. Application of intelligent tools to detect and classify broken rotor bars in three-phase induction motors fed by an inverter. *IET Electr. Power Appl.* **2016**, *10*, 430–439. [CrossRef]
37. Merainani, B.; Rahmoune, C.; Benazzouz, D.; Ould-Bouamama, B. A novel gearbox fault feature extraction and classification using Hilbert empirical wavelet transform, singular value decomposition, and SOM neural network. *J. Vib. Control* **2018**, *24*, 2512–2531. [CrossRef]
38. Lopes, T.D.; Goedtel, A.; Palacios, R.H.C.; Godoy, W.F.; de Souza, R.M. Bearing fault identification of three-phase induction motors bases on two current sensor strategy. *Soft Comput.* **2017**, *21*, 6673–6685. [CrossRef]
39. Negrov, D.; Karandashev, I.; Shakirov, V.; Matveyev, Y.; Dunin-Barkowski, W.; Zenkevich, A. An approximate backpropagation learning rule for memristor based neural networks using synaptic plasticity. *Neurocomputing* **2017**, *237*, 193–199. [CrossRef]
40. Caesarendra, W.; Wijayaa, T.; Tjahjowidodob, T.; Pappachana, B.K.; Weec, A.; Izzat Roslan, M. Adaptive neuro-fuzzy inference system for deburring stage classification and prediction for indirect quality monitoring. *Appl. Soft Comput.* **2018**, *72*, 565–578. [CrossRef]
41. Jia, F.; Lei, Y.G.; Guo, L.; Lin, J.; Xing, S.B. A neural network constructed by deep learning technique and its application to intelligent fault diagnosis of machines. *Neurocomputing* **2018**, *272*, 619–628. [CrossRef]
42. Zhang, W.; Li, C.H.; Peng, G.L.; Chen, Y.H.; Zhang, Z.J. A deep convolutional neural network with new training methods for bearing fault diagnosis under noisy environment and different working load. *Mech. Syst. Signal Process.* **2018**, *100*, 439–453. [CrossRef]
43. Lee, J.H.; Delbruck, T.; Pfeiffer, M. Training Deep Spiking Neural Networks Using Backpropagation. *Front. Neurosci.* **2016**, *10*, 508. [CrossRef] [PubMed]
44. Zajmi, L.; Ahmed, F.Y.H.; Jaharadak, A.A. Concepts, Methods, and Performances of Particle Swarm Optimization, Backpropagation, and Neural Networks. *Appl. Comput. Intell. Soft Comput.* **2018**, *2018*, 1–7. [CrossRef]
45. Johnson, J.M.; Yadav, A. Complete protection scheme for fault detection, classification and location estimation in HVDC transmission lines using support vector machines. *IET Sci. Meas. Technol.* **2017**, *11*, 279–287. [CrossRef]
46. Zhang, C.; Peng, Z.X.; Chen, S.; Li, Z.X.; Wang, J.G. A gearbox fault diagnosis method based on frequency-modulated empirical mode decomposition and support vector machine. *Proc. Inst. Mech. Eng. Part C J. Mech. Eng. Sci.* **2018**, *232*, 369–380. [CrossRef]

47. Widodo, A.; Yang, B.S. Support vector machine in machine condition monitoring and fault diagnosis. *Mech. Syst. Signal Process.* **2007**, *21*, 2560–2574. [CrossRef]
48. Glowacz, A. Diagnostics of Rotor Damages of Three-Phase Induction Motors Using Acoustic Signals and SMOFS-20-EXPANDED. *Arch. Acoust.* **2016**, *41*, 507–515. [CrossRef]
49. Valis, D.; Zak, L. Contribution to prediction of soft and hard failure occurrence in combustion engine using oil tribo data. *Eng. Fail. Anal.* **2017**, *82*, 583–598. [CrossRef]
50. Valis, D.; Zak, L.; Pokora, O.; Lansky, P. Perspective analysis outcomes of selected tribodiagnostic data used as input for condition based maintenance. *Reliab. Eng. Syst. Saf.* **2016**, *145*, 231–242. [CrossRef]
51. Yan, X.P.; Xu, X.J.; Sheng, C.X.; Yuan, C.Q.; Li, Z.X. Intelligent wear mode identification system for marine diesel engines based on multi-level belief rule base methodology. *Meas. Sci. Technol.* **2018**, *29*, 05110. [CrossRef]

electronics

MDPI

Article

Seismic Random Noise Attenuation Method Based on Variational Mode Decomposition and Correlation Coefficients

Yaping Huang [1,2,*], Hanyong Bao [3] and Xuemei Qi [1]

1 School of Resources and Geosciences, China University of Mining and Technology, Xuzhou 221116, China; qixuemei@cumt.edu.cn
2 Key Laboratory of Coal Methane and Fire Control, Ministry of Education, China University of Mining and Technology, Xuzhou 221116, China
3 Research Institute of Petroleum Exploration and Development, Jianghan Oilfield Company, SINOPEC, Wuhan 430223, China; geobhy@sina.com
* Correspondence: yphuang@cumt.edu.cn; Tel.: +86-158-5248-4072

Received: 6 October 2018; Accepted: 26 October 2018; Published: 28 October 2018

Abstract: Seismic data is easily affected by random noise during field data acquisition. Therefore, random noise attenuation plays an important role in seismic data processing and interpretation. According to decomposition characteristics of seismic signals by using variational mode decomposition (VMD) and the constraint conditions of correlation coefficients, this paper puts forward a method for random noise attenuation in seismic data, which is called variational mode decomposition correlation coefficients VMDC. Firstly, the original signals were decomposed into intrinsic mode functions (IMFs) with different characteristics by VMD. Then, the correlation coefficients between each IMF and the original signal were calculated. Next, based on the differences among correlation coefficients of effective signals and random noise as well as the original signals, the corresponding treatment was carried out, and the effective signals were reconstructed. Finally, the random noise attenuation was realized. After adding random noise to simple sine signals and the synthetic seismic record, the improved complementary ensemble empirical mode decomposition (ICEEMD) and VMDC were used for testing. The testing results indicate that the proposed VMDC has better random noise attenuation effects. It was also used in real-world seismic data noise attenuation. The results also show that it could effectively improve the signal-to-noise ratio (SNR) of seismic data and could provide high-quality basic data for further interpretation of seismic data.

Keywords: VMD; signal analysis; ICEEMD; IMF; random noise; attenuation

1. Introduction

The seismic signal is a typical nonlinear and nonstationary signal. The seismic exploration process is affected by various factors. There are effective waves and large amounts of random noise in seismic data. Therefore, effective treatment for random noise attenuation could not only improve the signal-to-noise ratio (SNR) and quality of seismic data, but also provide benefit for further interpretation of seismic data, lithology parameter inversion and seismic attributes analyses [1]. At present, there are the median filter method, f-x prediction filter method, polynomial fitting method, wavelet transform and empirical mode decomposition (EMD) method for random noise attenuation in seismic exploration. They have their own advantages and disadvantages. The median filter is a smoothing-based method. By this method, the basic frequency-domain signals tend to shift to low-frequency signals, and the high-frequency signals may be damaged [2,3]. In case of relatively low SNR in high frequency, the f-x prediction filter method may easily cause severe distortion of high-frequency signals and reduce the fidelity of signals and SNR of the seismic profile [4]. The polynomial fitting method requires the original seismic signals to have good continuity. The false seismic events may occur after data processing [5]. In the application of a wavelet transform, the selection of generating functions and de-noising thresholds has significant impacts on the effects of random noise attenuation [6]. EMD has the problems of end effect and mode mixing, which may lead to unsatisfactory effects of random noise attenuation [7].

Variational mode decomposition (VMD) is an adaptive signal processing method put forward by Dragomiretskiy. Compared with EMD, it has stronger noise-resistance ability. Moreover, it could successfully separate two harmonics with very similar frequencies, and the separating effects are not affected by the sampling frequency [8–11]. Li et al. introduced the principles of VMD and proposed a lateral consistency preserved VMD method [12]. Liu et al. studied the seismic time-frequency representation based on VMD [13]. Li et al. proposed a hybrid de-noising method based on thresholding variational mode decomposition [14]. Li et al. used VMD to analyze the depositional sequence characterization [15]. Jia et al. proposed a method to improve the resolution by using generalized S-transform based on VMD [16]. Zhao et al. extracted intrinsic mode functions (IMFs) based on VMD from seismic amplitudes to constrain self-organizing map facies analysis [17]. Lyu et al. analyzed the discontinuities with VMD-based coherence [18].

Combining VMD with correlation coefficients, this paper developed a new method for seismic data random noise attenuation. VMD was firstly used to decompose the original signals into IMFs with different characteristics. Then, the correlation coefficients between each IMF component and the original signal were calculated. The corresponding treatment was carried out based on differences among correlation coefficients of effective signals, random noise and original signals. The effective signals were reconstructed. Finally, the random noise attenuation was achieved. The results show that the VMDC method performs well in seismic random noise attenuation.

2. Methods

2.1. The Improved Complementary Ensemble Empirical Mode Decomposition ICEEMD

EMD was developed by Huang et al., and is a powerful analytical tool for nonlinear nonstationary signals [19]. However, it has the problems of end effect and mode mixing. ICEEMD was put forward by Tary et al., and this method can solve the above problems to some extent [20–23]. The calculation steps of this method are as follows:

(1) Use EMD to calculate the local mean of the i-th iteration $x^i = x + \varepsilon_0 w^i$, to get the first residual error.

$$r_1 = (1/I)\sum_{i=1}^{I} M[x + \varepsilon_0 E_1(w^i)] \tag{1}$$

(2) Calculate the first IMF.

$$IMF_1 = x - r_1 \tag{2}$$

(3) Calculate the second residuals and the second IMF.

$$r_2 = (1/I)\sum_{i=1}^{I} M(r_1 + \varepsilon_1 E_2(w^i)) \tag{3}$$

$$IMF_2 = r_1 - r_2 \tag{4}$$

(4) When $k = 3, \ldots, K$, Calculate the k-th residual error.

$$r_k = (1/I)\sum_{i=1}^{I} M(r_{k-1} + \varepsilon_{k-1} E_k(w^i)) \tag{5}$$

(5) Calculate the k-th IMF.

$$IMF_k = r_{k-1} - r_k \tag{6}$$

In the above formula, $E_k(\bullet)$ represents the operator generating the k-th IMF, $M(\bullet)$ represents the operator generating local mean of signal, x is the input signal, w^i is the decomposition of white noise with zero mean unit variance, ε_k is a constant greater than zero, r_i is the i-th residuals and I is the number of iterations.

2.2. VMD

In order to avoid the frequency mixture issue of the EMD [19], Dragomiretskiy et al. proposed a signal decomposition method with varying scales, which is the VMD method [8]. Compared with EMD, it has a solid mathematical basis and could be used to effectively solve the mode mixing problem. By VMD, the original signals could be decomposed into k band-limited signals u_k with the center frequency of ω_k, where k is the default decomposition scale. It is assumed that each mode function u_k is a limited bandwidth near its center frequency. The adaptive decomposition of the signal is realized by searching the optimal solution of the constrained variational model. The center frequency and bandwidth of each IMF are constantly updated in the iterative solution of the variational model. According to frequency-domain characteristics of actual signals, the adaptive decomposition of the signal band could be completed and some narrow-band IMFs could be obtained.

The steps of estimating the bandwidth of u_k are in the following [8]:

(1) To calculate the analytic function of each u_k and obtain the corresponding one-sided frequency spectrum by Hilbert transform.

(2) To adjust the estimated central spectrum by adding exponential terms and modulate the frequency spectrum of each mode into the corresponding basic frequency band.

(3) To estimate the bandwidth by Gaussian smoothness of the demodulated signal and gradient energy criterion.

By following the aforementioned steps, the obtained constrained variational problem is as follows:

$$\begin{array}{c} \min\limits_{\{u_k\},\{\omega_k\}} \left\{ \sum\limits_k \left\| \partial_t \left[(\delta(t) + \frac{j}{\pi t}) * u_k(t) \right] \exp(-j\omega_k t) \right\|_2^2 \right\}, \\ s.t. \quad \sum\limits_k u_k = x(t) \quad , \end{array} \tag{7}$$

where $\{u_k\} = \{u_1, u_2, \ldots, u_k\}$ is the function of each mode. $\{\omega_k\} = \{\omega_1, \omega_2, \ldots, \omega_k\}$ is the center frequency and $\sum\limits_k = \sum\limits_{k=1}^{K}$ is the sum of each mode.

(4) To transform the above constrained variational problem into an unconstrained variational problem by introducing the Lagrange multiplier $\lambda(t)$ and two-penalty factor, the formula of the augmented Lagrange multiplier could be obtained, as follows:

$$\begin{array}{l} L(\{u_k\}, \{\omega_k\}, \lambda) = \\ \alpha \sum\limits_k \left\| \partial_t \left[(\delta(t) + \frac{j}{\pi t}) u_k(t) \right] \exp(-j\omega_k t) \right\|_2^2 + \\ \| x(t) - \sum\limits_k u_k(t) \|_2^2 + \left\langle \lambda(t), x(t) - \sum\limits_k u_k(t) \right\rangle \end{array} . \tag{8}$$

The alternating direction method of multipliers is applied to solve the above variational problems. The iterative optimization of u^{k+1}, ω_k^{k+1} and λ^{k+1} could get the saddle points of the augmented Lagrange multiplier. The iteration steps are as follows:

(1) To initialize $u^1, \omega^1, \lambda^1, n = 0$.

(2) If $n = n + 1$, to perform the whole loop.

(3) To execute the first inner loop, update u_k according to $\omega_k^{k+1} = \underset{u^k}{\operatorname{argmin}} L(\{u_{i<k}^{n+1}\}, \{u_{i\geq k}^{n}\}, \{\omega_i^n\}, \lambda^n)$.

(4) $k = k+1$, and repeat step (3) until the completion of the first loop when $k = K$.

(5) To execute the second inner loop and update λ according to $\lambda^{n+1} = \lambda^n + \tau(x(t) - \sum_k u_k^{n+1})$.

(6) To repeat step 2 and step 5 until meeting $\sum_k (\|u_k^{n+1} - u_k^n\|_2^2 / \|u_k^n\|_2^2) < \varepsilon$. The whole looping will end and k IMFs could be obtained.

2.3. Correlation Coefficient Method

The Pearson correlation coefficient (PCC) is a statistical method to quantitatively measure correlations between two random variables. One of its important mathematical characteristics is that the variations of positions and scales will not cause the changes of correlation coefficients, so it is suitable for correlation evaluation of geophysical data [24–26]. The Pearson correlation coefficient could be expressed as follows [27]:

$$\begin{aligned}\rho &= \frac{\operatorname{cov}(X,Y)}{\sigma_X \sigma_Y} = \frac{E(XY) - E(X)E(Y)}{\sqrt{E(X^2) - E^2(X)}\sqrt{E(Y^2) - E^2(Y)}} \\ &= \frac{\sum (x_i - \overline{x})(y_i - \overline{y})}{\sqrt{\sum (x_i - \overline{x})^2}\sqrt{\sum (y_i - \overline{y})^2}}\end{aligned} \tag{9}$$

where $\operatorname{cov}(X, Y)$ refers to covariance of X and Y. σ_X and σ_Y are the standard deviations. $\overline{x} = E(X)$ and $\overline{y} = E(Y)$ are the expected values of X and Y, respectively. The bigger the absolute values of the correlation coefficients, the stronger the correlation. The closer the correlation coefficient is to 1 or -1, the stronger the correlation between X and Y. The closer the correlation coefficient is to 0, the weaker the correlation between X and Y. Generally, the correlation intensity among variables could be judged according to Table 1, as follows:

Table 1. Person correlation coefficients and correlation intensity.

Absolute Value of Pearson Correlation Coefficient (PCC)	Correlation Intensity
0.8–1.0	Extremely strong correlation
0.6–0.8	Strong correlation
0.4–0.6	Medium correlation
0.2–0.4	Weak correlation
0.0–0.2	Extremely weak correlation or no correlation

Table 1 shows that when the correlation coefficient is greater than 0.4, X and Y have good correlation. If the correlation coefficient is less than 0.2, the correlation coefficient of X and Y is poor. When the correlation coefficient is 0.2–0.4, the correlation between X and Y is general.

2.4. Random Noise Attenuation Method Based on VMD and Correlation Coefficients

It was assumed that a random noise signal was made of a noise-free signal and random noise. The VMD algorithm was firstly used to decompose the target signals into various IMFs which may include effective signals, effective signals with partial noise, and noise signals. Formula (9) was used to calculate the correlation coefficients between each IMF and the original signal. Then, on this basis, the effective signals were reconstructed. The reconstruction principles are as follows:

(1) The correlation coefficient of less than 0.2 represented that the IMF component had no correlation with the original signal, and had only the random noise, so the IMF component did not participate in signal reconstruction.

(2) The correlation coefficient of bigger than 0.4 represented that there was good correlation between the IMF component and the original signal, so the IMF component participated in signal reconstruction.

(3) The correlation coefficient between 0.2 and 0.4 represented that there was weak correlation between the IMF component and the original signal. The IMF component contained the effective signal and random noise. It was decomposed into $s_n = s_k + n_k$, where s_k denotes the effective signal and n_k denotes the residual random noise. Then, the VMD was used for IMF treatment to get s_k and n_k. s_k participated in the signal reconstruction. The process of random noise attenuation based on VMDC is shown in Figure 1.

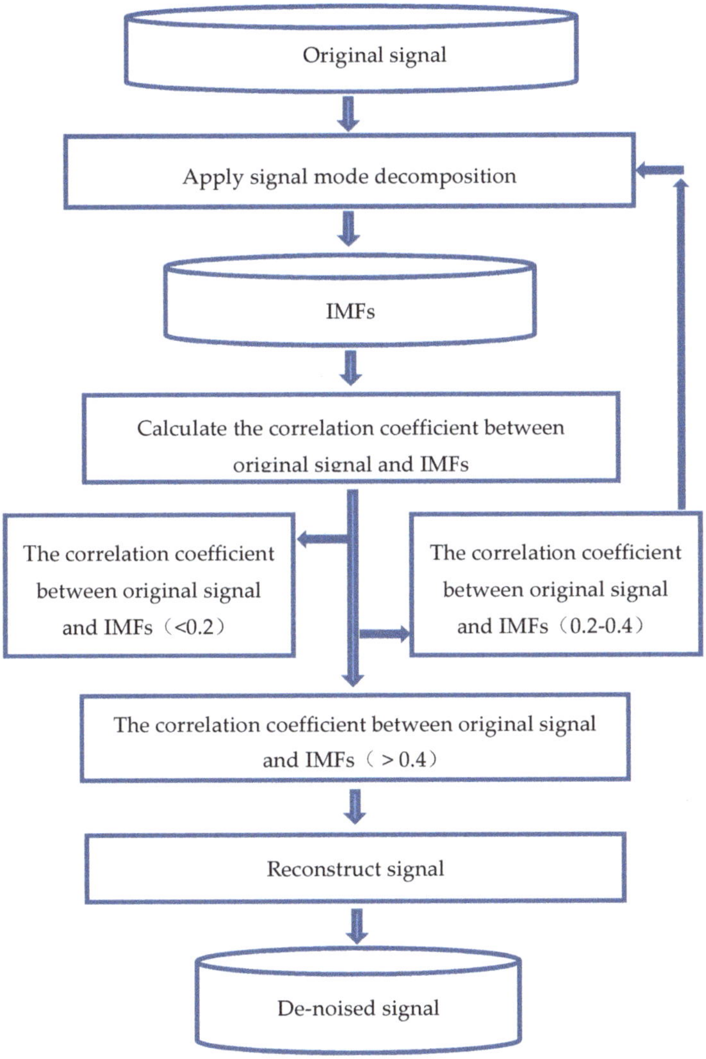

Figure 1. Flow chart of VMDC.

3. Theoretical Model Test

3.1. Simple Signal Model Test

The simple signal was expressed as $z(t) = x1(t) + x2(t)$, where $x1(t) = \sin(12 \times \pi \times t)$ and $x2(t) = \sin(32 \times \pi \times t)$. The number of sampling points was 1000. ICEEMD and VMD were used for signal decomposition. The results are shown in Figures 2 and 3. Developed by Colominas et al., ICEEMD has developed the EMD and could solve the problems of modal mixing in the application of EMD [20,28]. As shown in Figures 2 and 3, ICCEMD decomposes signals according to the frequency from high to low, while the VMD was the opposite. When the signal did not contain noise, the correlation coefficients between the decomposed signals by ICEEMD and VMD and the original signals could reach more than 0.97. In terms of calculation efficiency, ICEEMD took 21.96 s while VMD took only 2.15 s.

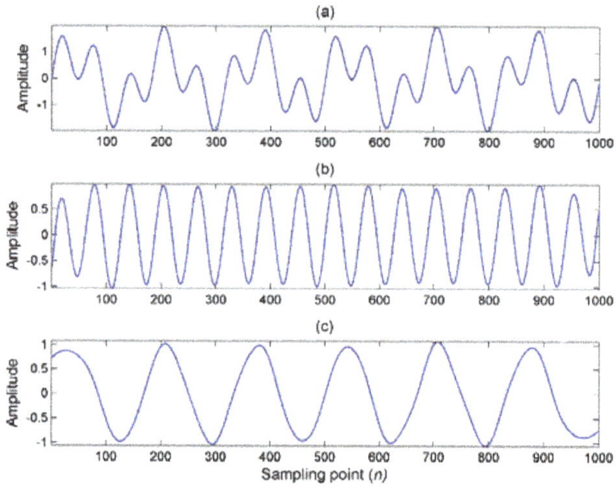

Figure 2. Decomposition of the artificially mixed signal by improved complementary ensemble empirical mode decomposition (ICEEMD): (**a**) the mixed signal; (**b**) IMF1 (intrinsic mode function 1) of the signal; (**c**) IMF2 of the signal.

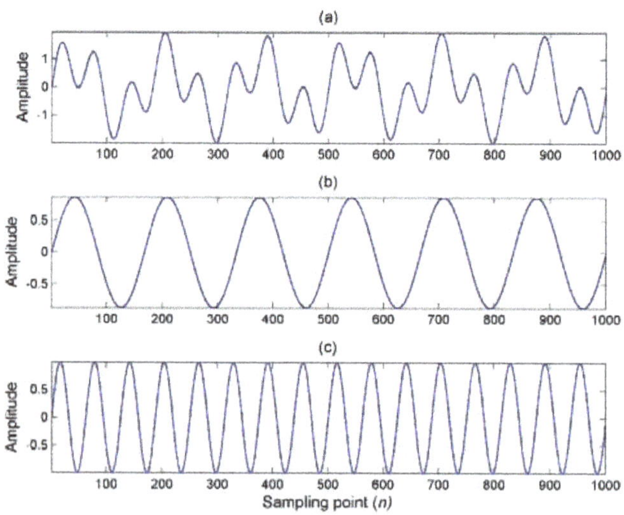

Figure 3. Decomposition of the artificially mixed signal by variational mode decomposition (VMD): (**a**) the mixed signal; (**b**) IMF1 of the signal; (**c**) IMF2 of the signal.

The above results show that both ICEEMD and VMD have good decomposition effects, and VMD has higher calculation efficiency.

20% random noise was added to the above simple signal using z(t) = x1(t)+x2(t)+0.2*rand(t). Figures 4 and 5 show the decomposition results by ICEEMD and VMD, respectively. In Figure 4, some IMF components contained modal mixing, which affected the reconstructed signals after superposition of the IMFs. By VMD, the correlation coefficients between each IMF and the original signal were calculated to be 0.73, 0.70 and 0.03. According to the principles of signal reconstruction in Section 2.3, IMF3 did not participate in signal reconstruction. Figure 6 shows the reconstruction results of IMF1 and IMF2. The blue line denotes the signals without noise, and the red line denotes the reconstruction results. The correlation coefficient between the decomposition signal and the original signal was 0.99 and the root mean square error (RMSE) was only 0.0985. In terms of calculation efficiency, ICEEMD took 22.90 s while VMD only took 0.51 s.

The above results indicate that VMD still has good decomposition effects with random noise in signals. Moreover, it is more beneficial to suppress random noise with higher operating efficiency.

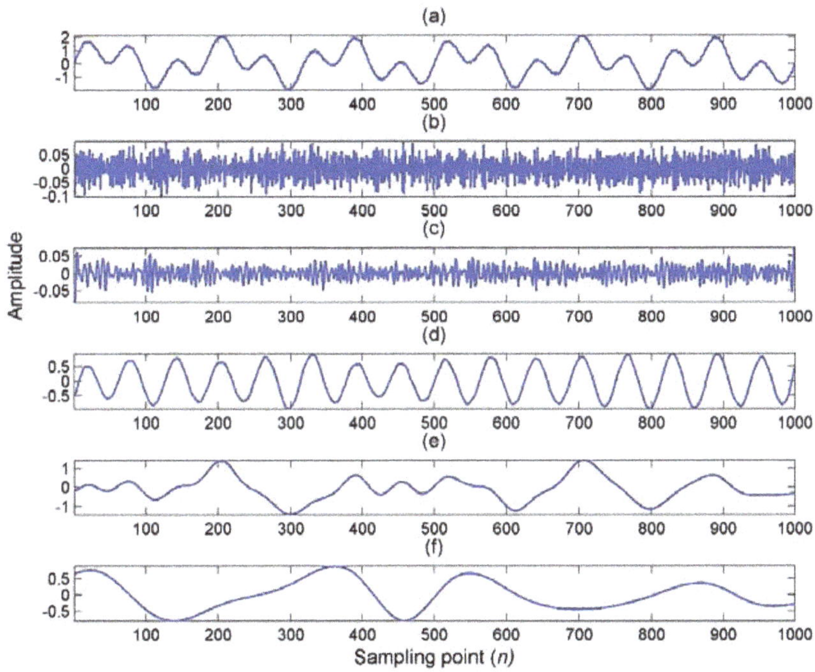

Figure 4. Decomposition of the artificially mixed signal with 20% additive noise by ICEEMD: (**a**) the signal; (**b**) IMF1 of the signal; (**c**) IMF2 of the signal; (**d**) IMF3 of the signal; (**e**) IMF4 of the signal; and (**f**) IMF5 of the signal.

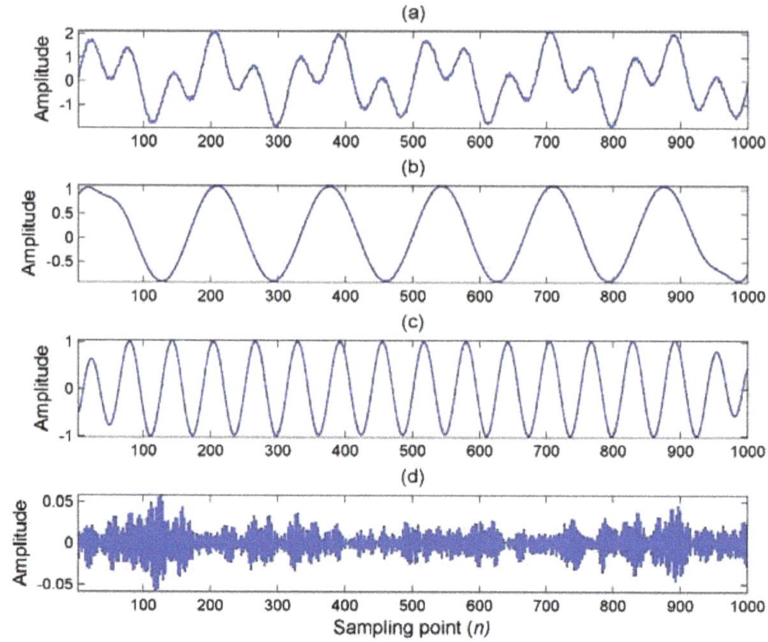

Figure 5. Decomposition of the artificially mixed signal with 20% additive noise by VMD: (**a**) the signal; (**b**) IMF1 of the signal; (**c**) IMF2 of the signal; (**d**) IMF3 of the signal.

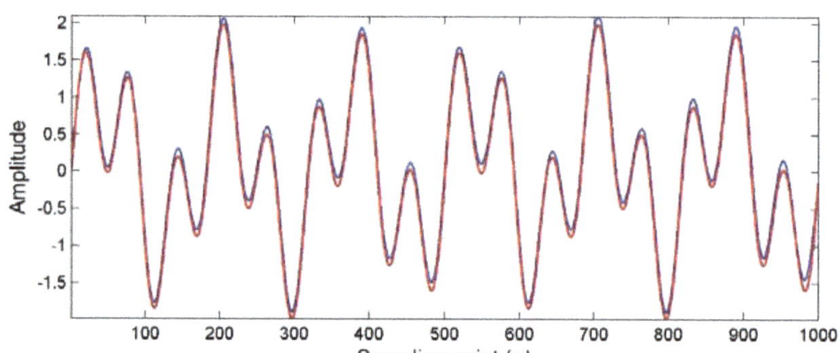

Figure 6. Reconstruction results by VMDC (red) and the original signals without noise (blue).

3.2. Synthetic Seismic Records Model Test

To further test the new method's application effects in seismic data processing, the synthetic seismic records model test was conducted after adding random noise. This paper established synthetic seismic records with the sampling interval of 0.1 ms and the dominant frequency of the wavelet of 45 HZ, and the synthetic seismic records with 20% random noise, as shown in Figure 7. The comparison in Figure 7 shows that after adding noise, the SNR of the synthetic seismic records was reduced, the events became blurred and some information of seismic horizons was masked by the random noise.

The decomposition of synthetic seismic records with 20% random noise by ICEEMD could obtain the seismic records and noise profiles, as shown in Figure 8. It could be seen that there was strong random noise in the profile after suppressing the random noise. Figure 9 shows the seismic records after suppressing the random noise by the VMDC. Compared with the noise attenuation results by

ICEEMD and VMDC, the de-noising effects by the VMDC are superior to the effects by ICEEMD. In Figure 9, the seismic events were better restored and the SNR was greatly improved. In addition, the noise in Figure 8 contained a small amount of effective waves, while noise was dominant in Figure 9.

The above results show that, under the synthetic seismic records with the random noise, the reconstructed events by the VMDC are more obvious and continuous, and the random noise reduction effects are better.

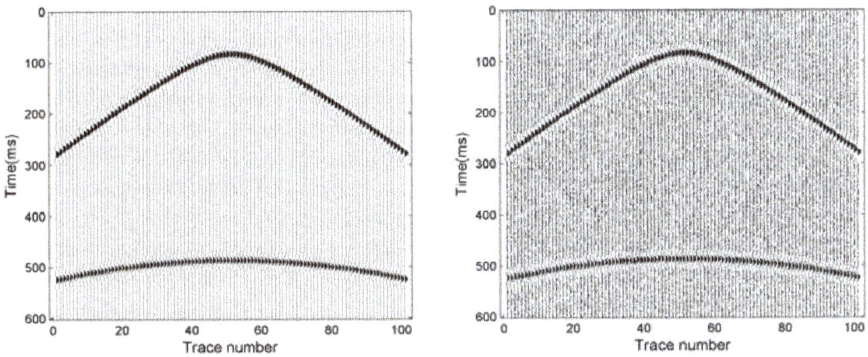

Figure 7. Synthetic seismic records without noise (**left**) and with 20% noise (**right**).

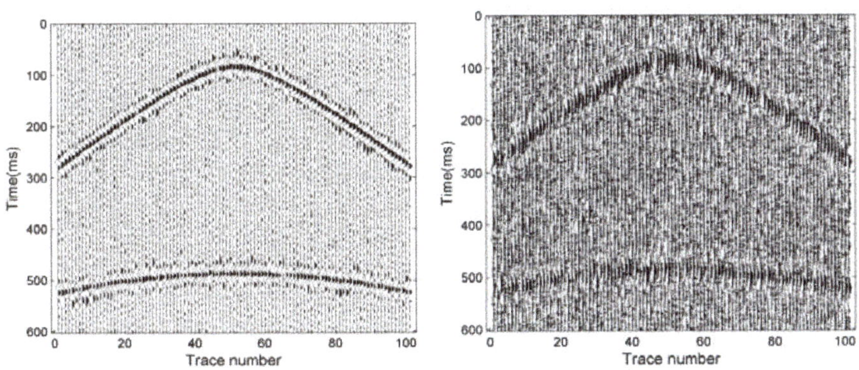

Figure 8. Random noise attenuation results (**left**) and the noise (**right**) by ICEEMD.

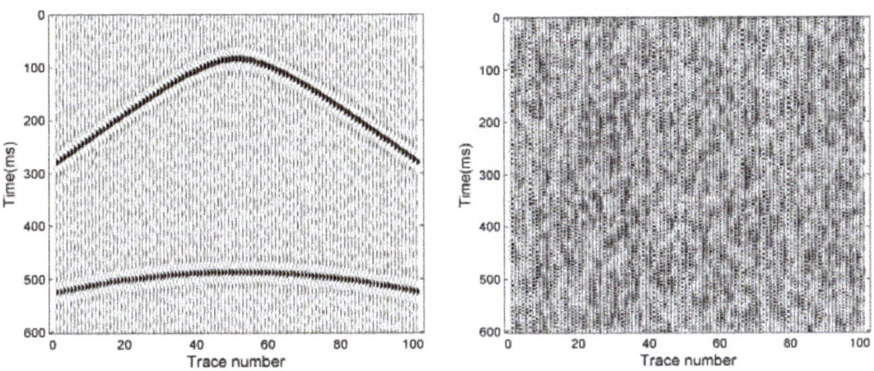

Figure 9. Random noise attenuation results (**left**) and the noise (**right**) by VMDC.

4. Case Study

To fully verify the application effects of the proposed VMDC in real-world seismic data, this paper selected the 3-D seismic data in Inner Mongolia in China to carry out the test. The acquisition of seismic data was in the winter, so the wind was blowing very hard. At the same time, the gangue field and air shafts were under construction. What is worse, the random noise interference was more serious in this area because there were many vehicles in adjacent industrial areas. Figure 10 shows the actual seismic profiles with random noise. It could be found that the existing random noise reduced the SNR of the seismic data and influenced the continuity of events in seismic records.

The ICEEMD and the VMDC were used for noise attenuation of post-stack seismic data. The signal reconstruction process was introduced by taking the 50th channel as an example. Firstly, ICEEMD and VMDC were used to decompose seismic signals. Then, by ICEEMD, the correlation coefficients between each IMF and the original signal were calculated as 0.1432, 0.5058 and 0.7329. By VMD, the correlation coefficients between each IMF and the original signal were calculated as 0.7221, 0.6611 and 0.1156. Therefore, the IMF2 and IMF3 decomposed by ICEEMD, and IMF1 and IMF2 decomposed by VMDC, participated in the signal constructions. The signals in other seismic traces were reconstructed by similar steps, as shown in Figures 11 and 12. Through the comparative analysis of Figures 10–12, it is found that both ICEEMD and the VMDC could suppress the random noise to a certain extent, and the latter could significantly enhance the continuity of events.

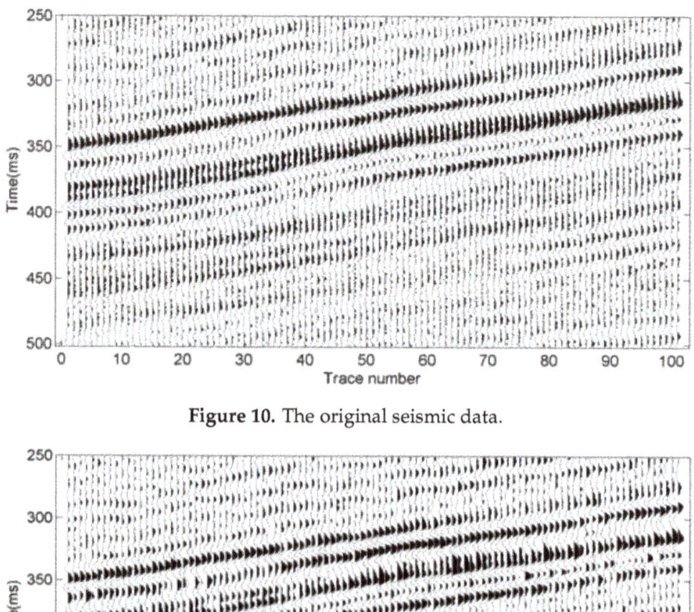

Figure 10. The original seismic data.

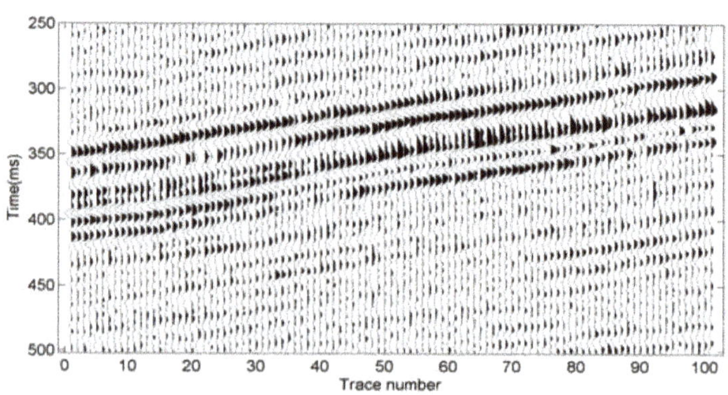

Figure 11. Random noise attenuation results by ICEEMD.

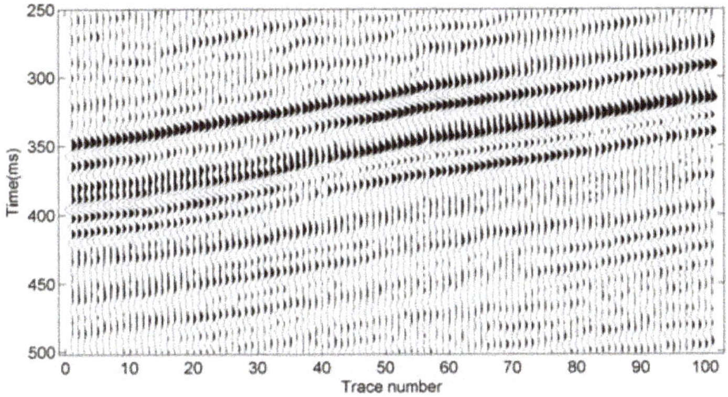

Figure 12. Random noise attenuation results by VMDC.

Above all, the VMDC method proposed in this paper has obvious noise attenuation effects and could make the events more clear and continuous. What is more, it could improve the SNR of seismic data and the smoothing of each channel of the seismic record. It also could better reflect shapes of strata. It is shown that the method could achieve good effects in random noise attenuation in real-world seismic data.

5. Conclusions

This paper proposed a new method for seismic random noise attenuation based on VMD and correlation coefficients, called VMDC. Under the situation of simple sine signals without noise, both ICEEMD and VMDC have better decomposition results, and VMDC has better calculation efficiency. After adding random noise to the simple sine signals and the synthetic seismogram, the testing results show that the VMDC has better noise attenuation effects. The application results in real-world seismic data indicate that the new method could significantly improve the SNR of seismic data, enhance the continuity of events as well as provide reliable basic data for further seismic data interpretation.

Author Contributions: Y.H. conceived and designed the experiments; Y.H. and H.B. performed the experiments; X.Q. and H.B. analyzed the data; Y.H. and X.Q. wrote the paper.

Funding: This research was supported by the Fundamental Research Funds for the Central Universities grant number 2018QNA45 and A Project Funded by the Priority Academic Program Development of Jiangsu Higher Education Institutions (PAPD).

Conflicts of Interest: The authors declare no conflicts of interest.

References

1. Yan, Z.H.; Luan, X.W.; Wang, Y.; Pan, J.; Fang, G.; Shi, J. Seismic random noise attenuation based on empirical mode decomposition of fractal dimension. *Chin. J. Geophys.* **2017**, *60*, 2845–2857. (In Chinese) [CrossRef]
2. Liu, C.; Liu, Y.; Yang, B.J.; Wang, D.; Sun, J.G. A 2D multistage median filter to reduce random seismic noise. *Geophysics* **2006**, *71*, V105–V110. [CrossRef]
3. Liu, Y.K. Noise reduction by vector median filtering. *Geophysics* **2013**, *78*, 79–86. [CrossRef]
4. Abbas, K.; Abdolrahim, J. Random noise reduction by F-X deconvolution. *J. Earth* **2010**, *5*, 61–68.
5. Liu, G.C.; Chen, X.H.; Li, J.Y.; Du, J.; Song, J.W. Seismic noise attenuation using nonstationary polynomial fitting. *Appl. Geophys.* **2011**, *8*, 18–26. [CrossRef]
6. Zhang, Z.H.; Sun, C.Y.; Tang, J.; Xiao, G.R.; Li, L.J. A denoising method based on combined Curvelet and Wavelet transform. In Proceedings of the Beijing 2014 International Geophysical Conference and Exposition, Beijing, China, 21–24 April 2014.

7. Huang, Y.P.; Di, H.B.; Malekian, R.; Qi, X.M.; Li, Z.X. Noncontact measurement and detection of instantaneous seismic attributes based on complementary ensemble empirical mode decomposition. *Energies* **2017**, *10*, 1655. [CrossRef]
8. Dragomiretskiy, K.; Zosso, D. Variational mode decomposition. *IEEE Trans. Signal Process.* **2014**, *62*, 531–544. [CrossRef]
9. Glowacz, A.; Glowacz, W.; Glowacz, Z.; Kozik, J. Early fault diagnosis of bearing and stator faults of the single-phase induction motor using acoustic signals. *Measurement* **2018**, *113*, 1–9. [CrossRef]
10. Glowacz, A. Fault diagnosis of single-phase induction motor based on acoustic signals. *Mech. Syst. Signal Process.* **2019**, *117*, 65–80. [CrossRef]
11. Li, Z.X.; Jiang, Y.; Guo, Q.; Hu, C.; Peng, Z.X. Multi-dimensional variational mode decomposition for bearing-crack detection in wind turbines with large driving-speed variations. *Renew. Energy* **2016**, *116*, 55–73. [CrossRef]
12. Li, F.Y.; Zhao, T.; Qi, X.; Marfurt, K.J.; Zhang, B. Lateral consistency preserved Variational Mode Decomposition. In *SEG Technical Program Expanded Abstracts 2016*; Society of Exploration Geophysicists: Dallas, TX, USA, 21 October 2016.
13. Liu, W.; Cao, S.Y.; Chen, Y.K. Applications of variational mode decomposition in seismic time-frequency analysis. *Geophysics* **2016**, *81*, V365–V378. [CrossRef]
14. Li, F.Y.; Zhang, B.; Verma, S.; Marfurt, K.J. Seismic signal denoising using thresholded variational mode decomposition. *Explor. Geophys.* **2017**, *49*, 450–461. [CrossRef]
15. Li, F.Y.; Zhang, B.; Zhai, R.; Zhou, H.L.; Marfurt, K.J. Depositional sequence characterization based on seismic variational mode decomposition. *Interpretation* **2017**, *5*, SE97–SE106. [CrossRef]
16. Jia, J.F.; Chen, X.H.; Jiang, S.H.; Jiang, W.; Zhang, J. Resolution enhancement in the generalized S-transform domain based on variational-mode decomposition of seismic data. In Proceedings of the International Geophysical Conference, Qingdao, China, 17–20 April 2017.
17. Zhao, T.; Li, F.Y.; Marfurt, K.J. Constraining self-organizing map facies analysis with stratigraphy: An approach to increase the credibility in automatic seismic facies classification. *Interpretation* **2017**, *5*, T163–T171. [CrossRef]
18. Lyu, B.; Li, F.Y.; Zhao, T.; Marfurt, K.J. Highlighting discontinuities with variational-mode decomposition-based coherence. In *SEG Technical Program Expanded Abstracts 2018*; Society of Exploration Geophysicists: Anaheim, CA, USA, 19 October 2018.
19. Huang, N.E.; Shen, Z.; Long, S.R.; Wu, M.C.; Shih, H.H.; Zheng, Q.; Liu, H.H. The empirical mode decomposition and the Hilbert spectrum for nonlinear and non-stationary time series analysis. *Proc. R. Soc. Lond. A Math. Phys. Eng. Sci.* **1998**, *454*, 903–995. [CrossRef]
20. Tary, J.B.; Herrera, R.H.; Han, J.; Baan, M.V.D. Spectral estimation—What is new? what is next? *Rev. Geophys.* **2014**, *52*, 723–749. [CrossRef]
21. Han, J.; Mirko, V.D.B. Microseismic and seismic denoising via ensemble empirical mode decomposition and adaptive thresholding. *Geophysics* **2015**, *80*, KS69–KS80. [CrossRef]
22. Wang, Z.G.; Gao, J.H.; Wang, P.; Jiang, X.D. The analytic wavelet transform with generalized Morse wavelets to detect fluvial channels in the Bohai Bay Basin, China. *Geophysics* **2016**, *81*, O1–O9. [CrossRef]
23. Chen, W.; Song, H. Automatic noise attenuation based on clustering and empirical wavelet transform. *J. Appl. Geophys.* **2018**, *159*, 649–665. [CrossRef]
24. Chen, Z.A.; Wu, X.Y. Accuracy of measuring velocity improved by correlative analysis method. *Prog. Geophys.* **2001**, *16*, 101–103. (In Chinese) [CrossRef]
25. Cui, Z.J.; Li, Z.X.; Chen, Z.L. A study on the new method for determining small earthquake sequence type-Correlation analysis of spectral amplitude. *Chin. J. Geophys.* **2012**, *55*, 1718–1724. (In Chinese) [CrossRef]
26. Yin, C.C.; Sun, S.Y.; Cao, X.H.; Liu, Y.H.; Chen, H. 3D joint inversion of magnetotelluric and gravity data based on local correlation constraints. *Chin. J. Geophys.* **2018**, *61*, 358–367. (In Chinese) [CrossRef]

27. Rodgers, J.L.; Nicewander, W.A. Thirteen ways to look at the correlation coefficient. *Am. Stat.* **1988**, *42*, 59–66. [CrossRef]
28. Colominas, M.A.; Schlotthauer, G.; Torres, M.E. Improved complete ensemble emd: A suitable tool for biomedical signal processing. *Biomed. Signal Process. Control* **2014**, *14*, 19–29. [CrossRef]

MDPI
St. Alban-Anlage 66
4052 Basel
Switzerland
Tel. +41 61 683 77 34
Fax +41 61 302 89 18
www.mdpi.com

Electronics Editorial Office
E-mail: electronics@mdpi.com
www.mdpi.com/journal/electronics

ent.com/pod-product-compliance
LLC
926
4B/151
3 9 2 8 2 9 4 4 *